Understanding **Platelet-Rich Fibrin**

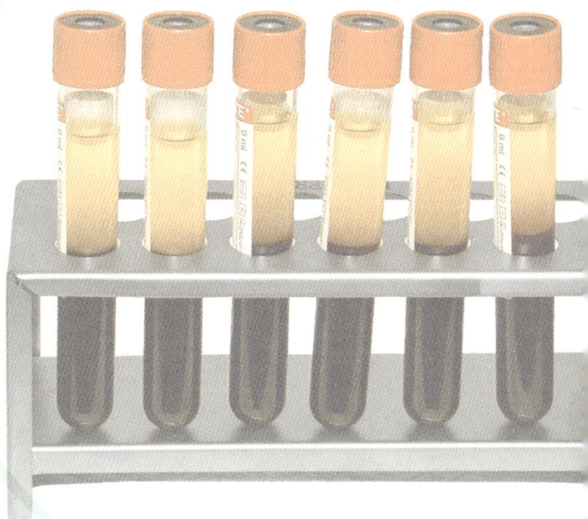

Edited by
Richard J. Miron, DDS, BMSc, MSc, PhD, Dr med dent
Group Leader, The Miron Research Lab
Lead Educator, Advanced PRF Education (www.prfedu.com)
Venice, Florida

Department of Periodontology
University of Illinois at Chicago
Chicago, Illinois

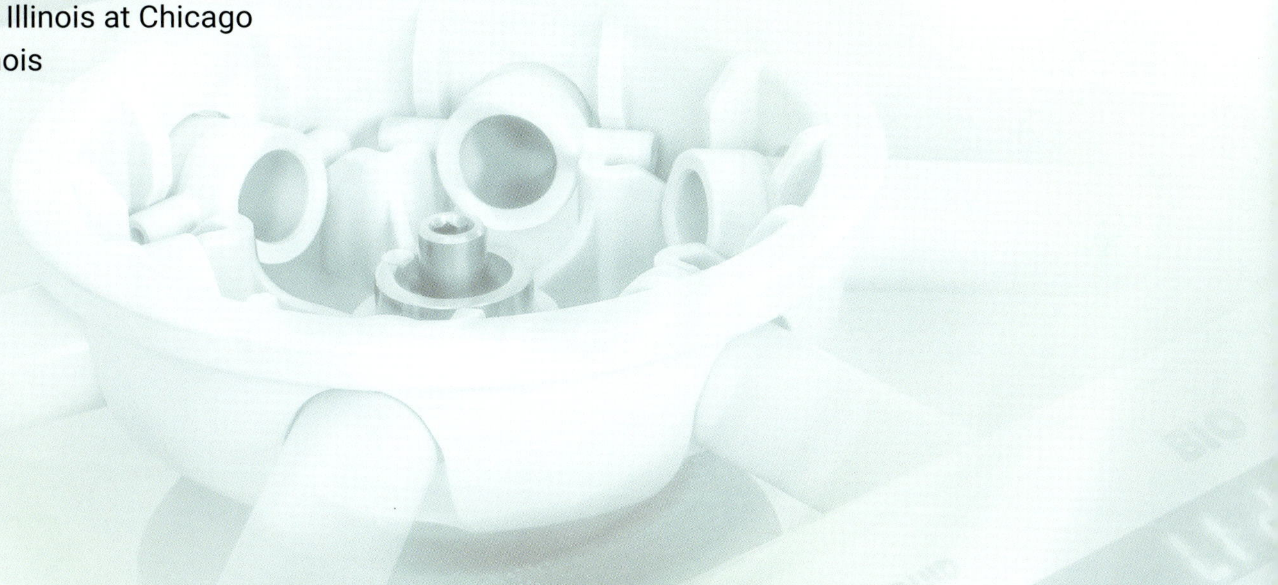

Biology editors: M. Fujioka-Kobayashi | R. Gruber | T. Kawase | Y. Zhang | **Periodontology editors:** V. Moraschini | A. Sculean | Y. Shirakata | H.-L. Wang | **Implant editors:** L. Canullo | L. Bessa | H. Gluckman | M.A. Pikos | **Oral surgery editors:** A. Garg | N. Saulacic | B. Schaller

QUINTESSENCE PUBLISHING

Berlin | Chicago | Tokyo
Barcelona | London | Milan | Mexico City | Moscow | Paris | Prague | Seoul | Warsaw
Beijing | Istanbul | Sao Paulo | Zagreb

Library of Congress Cataloging-in-Publication Data

Names: Miron, Richard J. (Richard John), 1983- editor.
Title: Understanding platelet-rich fibrin / edited by Richard J. Miron.
Description: Batavia, IL : Quintessence Publishing Co Inc, [2021] | Includes bibliographical references and index. | Summary: "This book outlines the science behind platelet-rich fibrin and then details how to use it in clinical practice to optimize healing outcomes and promote tissue regeneration. Applications include gingival recessions, intrabony and furcation defects, extraction site management, implant dentistry, guided bone regeneration, sinus grafting, oral and maxillofacial surgery, regenerative endodontics, facial esthetics, and medicine"-- Provided by publisher.
Identifiers: LCCN 2020035734 | ISBN 9781647240493 (hardcover)
Subjects: MESH: Platelet-Rich Fibrin | Fibrin--therapeutic use | Tissue Engineering | Bone Regeneration
Classification: LCC QP91 | NLM WH 400 | DDC 612.1/15--dc23
LC record available at https://lccn.loc.gov/2020035734

QUINTESSENCE PUBLISHING
USA

Quintessence Publishing Co, Inc
411 N Raddant Road
Batavia, IL 60510
www.quintpub.com

5 4 3 2 1

Editor: Leah Huffman
Design: Sue Zubek
Production: Angelina Schmelter

Printed in the USA

Contents

Dedication

Le prix Fatiha Chandad

Ce prix fut créé en l'honneur de la Dre Fatiha Chandad, qui a travaillé comme professeure à l'Université Laval pendant plus de trois décennies. Elle a également été vice-doyenne aux études supérieures et à la recherche et directrice du Groupe de recherche en écologie buccale (GREB) au cours des dernières années de sa carrière.

Dre Chandad était une personne lumineuse possédant une passion pour la vie remarquable par son désir d'aider les autres à cheminer. Sa passion pour les sciences combinée à sa volonté à contribuer au développement des élèves, ont été adoptées par des étudiants du monde entier qu'elle a su encadrés au cours de sa carrière. Elle était animée d'une modestie et d'une humilité frappantes; qualités que peu d'individus possèdent dans le monde d'aujourd'hui.

Chaque année, le prix Fatiha Chandad est décerné à un étudiant en quatrième année de médecine dentaire qui a montré des traits de personnalité similaires à ceux du Dr Chandad. Un étudiant qui a su aider d'autres camarades de classe sans motifs personnels et a offert son aide sans réserve dans des situations difficiles ou des périodes de stress lors de leur parcours à la Faculté de Médecine Dentaire.

I would like to dedicate this textbook specifically to Dr Fatiha Chandad and the Dental Faculty at the University of Laval in Quebec, Canada, who devoted countless hours to training students like me in dentistry.

Dr Chandad, Dean of Dental Research, was someone who motivated me and more importantly made it possible for students to work within her laboratory on research projects during their 4 years of dental studies. It was here that I first started my research activities on PRP/PRF and became fascinated with research as a whole. Dr Chandad is one of the only people I have ever met never wanting to be recognized for her achievements, instead insisting that her students be recognized and at the forefront of their own success. It was during these times that I was awarded the prestigious Hatton Award in Canada and was later named the IADR Young Investigator of the Year in Implant Dentistry.

In appreciation of her contributions to my career, 100% of the royalties from this textbook will be donated to create a scholarship named after her for new dental graduates at the University of Laval. *Milles fois merci pour tout tes efforts!*

Foreword

Many important medical/dental discoveries were stumbled upon by pure chance. An example of this is osseointegration, which launched the modern era of dental implants and orthopedic-embedded prostheses. It occurred when orthopedic researcher Per-Ingvar Brånemark found that his titanium research cages in sheep were so completely ingrown with bone that he couldn't remove them. Similarly, platelet-rich plasma (PRP)/platelet-rich fibrin (PRF) was discovered by the serendipity of observing accelerated and more complete healing in patients who developed hematomas. From that simple observation in the 1980s, the components of the blood clot responsible for the advanced healing have since been determined to be the complete and active growth factors in the alpha granules of viable platelets and several cell adhesion molecules and homing signals in the fibrin clot.

From the early pioneering work in the 1990s to the present, the benchmark of platelet numbers and the functional characteristics of the growth factors, cell adhesion molecules, and homing signals of PRP/PRF have progressed to a mature science and easy-to-use point-of-care devices.

Today, PRP/PRF devices are able to predictably concentrate platelets to known therapeutic levels by a variety of technologic means and to include or exclude leukocytes as per the needs of the wound. PRP/PRF has become a mainstay in bone regeneration in dental implantology and jaw reconstruction, for tendon and joint repair in orthopedics, for soft tissue healing in wound care centers, and in plastic surgery; it has thus benefited hundreds of thousands of patients worldwide.

As one of several individuals who were there at the beginning of the PRP/PRF discovery and who helped to move it along the way somewhat, I am delighted to see that the next generation of clinical researchers as published in *Understanding Platelet-Rich Fibrin* have brought it to the next level.

This text is written for the clinician to understand how and why PRF promotes healing of both bone and soft tissue as well as how to apply it to improve their own results. Written with a balanced blend of science and clinical applications by the most experienced and accomplished PRP/PRF scientists and clinicians of the day, and beautifully illustrated, *Understanding Platelet-Rich Fibrin* is a book for this decade that transcends all specialties of dentistry and many of medicine.

Robert E. Marx, DDS
Professor of Surgery and Chief of OMFS
University of Miami
Miller School of Medicine
Miami, Florida

Preface

Over 20 years ago, platelet concentrates entered into the medical field as a means to deliver autologous growth factors responsible for favoring wound healing. During this time span, it has gained widespread acceptance in many fields of medicine due to its more natural delivery system.

Most notably, the past 5 years have seen a tremendous increase in publications on PRF, with over 200 scientific peer-reviewed papers being published each and every year. During this span, a marked increase in our understanding of PRF therapy has been made with respect to selection of appropriate centrifugation devices, impact of tube chemistry on clotting, the optimization of protocols to better concentrate PRF, and even the ability to extend the working properties of PRF from 2–3 weeks toward 4–6 months using a simple heating process. Collectively, we continue to gather new knowledge, and as a result, PRF therapy has become one of the fastest-growing therapeutic options in dentistry. Thousands of users have now benefited from this technology, and this number is only expected to continue to increase.

This book is very different from others in its concept design. More than a dozen expert researchers and clinicians alike were gathered as editors across their different fields of expertise. As section editors, these true experts of their respective disciplines were able to produce a much higher overall quality of this textbook. I am grateful for their encouraging team spirit, their effort in bringing this book to an entirely new level, and their level of professionalism and mindset that ultimately led to this comprehensive textbook. I am also grateful to the numerous clinicians who have provided videos to better educate/demonstrate surgical techniques and concepts with PRF, which will greatly enhance the learning experience of the reader.

The book is divided into four primary sections, including *(1)* biology of PRF, *(2)* periodontology, *(3)* implant dentistry, and *(4)* additional dental and medical applications. The book aims to take the reader from a basic biologic understanding of PRF through explanations of the various protocols utilized followed by application of these concepts in numerous clinical scenarios.

I therefore am thrilled to introduce our textbook titled *Understanding Platelet-Rich Fibrin*. I hope you enjoy learning the many aspects centered around the use of PRF in regenerative dentistry.

Acknowledgments

To my parents, family, and friends who have all sacrificed far too often in my pursuit of a career in academic dentistry.

To my classmates, colleagues, and mentors who constantly raised the bar and strived for better.

To Quintessence Publishing for your thorough input in the editing, illustrations, and design of this textbook.

To my family at Lakewood Ranch Dental in Florida who have made clinical practice as enjoyable an experience as can be on a *daily* basis.

To all leaders and researchers alike who have contributed enormously to the field of PRP/PRF and laid the foundation for this textbook to be written.

To the faculty in the Department of Biomedical Sciences and Cell Biology at the University of Western Ontario (London, Canada; BMSc, MSc), the Dental School at the University of Laval (Quebec; DDS), the Department of Music at Berklee College (Boston; MMus), the Department of Periodontology at the University of Bern (PhD, Dr med dent), the Department of Oral Implantology at Wuhan University (China; postdoctoral research fellow), the Plastic Surgery Department at Queen Mary University (London; clinical masters in facial esthetics), and the Department of Periodontology at the University of Illinois at Chicago (clinical masters in periodontology). Your education and mentorship has provided endless opportunities.

And lastly, to the team at Advanced PRF Education (www.prfedu.com) for making excellence in teaching a top priority.

Contributors

Fabrice Baudot, DDS, MSc
Private Practice
Saint-Gély-du-Fesc, France

Luis Bessa, DDS
Director, North Clinic
Porto, Portugal

Mark Bishara, DDS
Private Practice
Bowmanville, Ontario, Canada

Thomas Boas, MSc (Econ)
CEO, Puremed
Roskilde, Denmark

Luigi Canullo, DDS, PhD
Independent Researcher
Rome, Italy

Marco Antonio Castro Pinto, DDS, MSc
Professor, Department of Reconstructive Dentistry
Montemorelos University School of Dentistry
Nuevo Léon, Mexico

Raluca Cosgarea, DDS, MSc, PhD
Professor, Department of Prosthetic Dentistry
Iuliu Haţieganu University
Cluj-Napoca, Romania

Catherine Davies, MBBCh, MBA
Private Practice Specializing in Facial Esthetics
Johannesburg, South Africa

Massimo Del Fabbro, MD, PhD
Professor, Department of Biomedical, Surgical, and Dental Sciences
University of Milan
Milan, Italy

Scott Delboccio, DMD
Private Practice
Naples, Florida

Anika Dham
Research Student, Nova Southeastern University
Fort Lauderdale, Florida

Jonathan Du Toit, DDS, MSc
Department of Periodontics and Oral Medicine
Faculty of Health Sciences
University of Pretoria
Pretoria, South Africa

Meizi Eliezer, DDS, MSc, PhD
Research Associate, Department of Periodontology
School of Dental Medicine
University of Bern
Bern, Switzerland

Masako Fujioka-Kobayashi, DDS, PhD
Professor, Department of Oral and Maxillofacial Surgery
School of Life Dentistry at Tokyo
The Nippon Dental University
Tokyo, Japan

Maria Elisa Galarraga-Vinueza, DDS, MSc, PhD
Professor, School of Dentistry
Universidad de las Américas (UDLA)
Quito, Ecuador

Arun K. Garg, DMD
Private Practice Limited to Implantology
Miami, Florida

Stefan Gerber, MD, DDS, MSc, PhD
Assistant Professor, Department of Cranio-Maxillofacial Surgery
University of Bern
Bern, Switzerland

Ezio Gheno, DDS, PhD
Post-Graduation Program in Dentistry
Fluminense Federal University
Niterói, Rio de Janeiro, Brazil

Alfonso Gil, DDS, MSc
Resident, Fixed and Removable Prosthodontics and Dental Material Science
University of Zurich
Zurich, Switzerland

Howard Gluckman, BDS, MChD (OMP)
Specialist in Periodontics, Implantology, and Oral Medicine
Director of Implant & Aesthetic Academy
Cape Town, South Africa

Reinhard Gruber, PhD
Professor, Department of Oral Biology
Medical University of Vienna
Vienna, Austria

Thomas Lau Hansen, PhD
Puremed
Roskilde, Denmark

Tommy Hardon, DVM
Head Veterinarian, Haslev Dyreklinik
Haslev, Denmark

David Lee Hill, DDS, MSc
Private Practice
Chapel Hill, North Carolina

Søren Jepsen, DDS, MSc, PhD
Director of the Department of Periodontology
University of Bonn
Bonn, Germany

Valerie Kanter, DDS, MSc
Professor, Department of Endodontics
University of California, Los Angeles
Los Angeles, California

Dwayne Karateew, DDS, MSc
Director of Program in Periodontics
University of Illinois at Chicago
Chicago, Illinois

Tomoyuki Kawase, DDS, PhD
Professor, Division of Oral Bioengineering
Institute of Medicine and Dentistry
Niigata University
Niigata, Japan

Johan Lenz, DVM
Veterinarian, Jonas Tornell Veterinär
Ängelholm, Sweden

Marius Leretter, DDS, PhD
University of Medicine and Pharmacy of Timişoara
Vice Dean of Dental School
Timişoara, Romania

Victoria Lima, DDS, MSc
Research Fellow, Division of Periodontics
Institute of Science and Technology
São Paulo State University (UNESP)
São Paulo, Brazil

Richard J. Martin, DDS
Private Practice Limited to Oral and Facial Surgery
Lewisville, Texas

Yuriy May, DMD
Private Practice
Farmington, Connecticut

Brian Mealey, DDS, MS
Professor and Graduate Program Director, Department of Periodontics
University of Texas Health Science Center at San Antonio
San Antonio, Texas

Jacob Coakley Meyer, DVM
Veterinarian, Charlottenlund Dyrehospital
Charlottenlund, Denmark

Richard J. Miron, DDS, BMSc, MSc, PhD, Dr med dent
Group Leader, The Miron Research Lab
Lead Educator, Advanced PRF Education
Venice, Florida

Department of Periodontology
University of Illinois at Chicago
Chicago, Illinois

Omid Moghaddas, DDS, MSc
Assistant Professor, Department of Periodontology
Dental Faculty, Tehran Medical Sciences
Islamic Azad University
Tehran, Iran

Naheed Mohamed, DMD, MSD
Private Practice
Oakville, Ontario, Canada

Vittorio Moraschini, DDS, MSc, PhD
Professor, Department of Oral Surgery
Dental School, Fluminense Federal University
Niterói, Rio de Janeiro, Brazil

Ana Paz, DDS, MS
Private Practice
Lisbon, Portugal

Michael A. Pikos, DDS, MSc
Director, Pikos Institute
Tampa, Florida

Nikola Saulacic, DDS, PhD
Assistant Professor, Department of Cranio-Maxillofacial Surgery
University of Bern
Bern, Switzerland

Benoît Schaller, Dr med, Dr med dent
Professor, Department of Cranio-Maxillofacial Surgery
University of Bern
Bern, Switzerland

Anton Sculean, DMD, Dr med dent, MS, PhD
Professor and Executive Director and Chairman
Department of Periodontology
University of Bern
Bern, Switzerland

Senthil Selvan, DDS
Director, Jeya Dental Clinic
Theni, India

Samer Shaltoni, DMD, MS
Private Practice Limited to Oral Surgery
Buffalo Grove, Illinois

Yoshinori Shirakata, DDS, PhD
Associate Professor, Department of Periodontology
Kagoshima University Graduate School of Medical and Dental Sciences
Kagoshima, Japan

Miguel Stanley, DDS
Private Practice
Lisbon, Portugal

Robert Talac, MD, PhD
Director, Renaxis Spine and Orthopedic Clinic
Houston, Texas

Mustafa Tunali, DDS, PhD
Professor, Department of Periodontology
Haydarpasa Training Hospital
Gulhane Military Medical Academy
Istanbul, Turkey

Delia Tuttle, DDS, MS
Private Practice
Lake Elsinore, California

Hom-Lay Wang, DDS, MSD, PhD
Professor and Director of Graduate Periodontics
Department of Periodontics and Oral Medicine
University of Michigan School of Dentistry
Ann Arbor, Michigan

Hudi Xu, DDS, PhD
Research Associate, Department of Dental Implantology
School of Stomatology
Wuhan University
Wuhan, China

Yufeng Zhang, MD, DDS, PhD
Professor, Department of Dental Implantology
School of Stomatology
Wuhan University
Wuhan, China

Abbreviations

The abbreviations listed here are used throughout the book and are NOT always spelled out in the chapters for ease of reading.

ALP	alkaline phosphatase	**L-PRF**	leukocyte PRF
AM	amniotic membrane	**LPS**	lipopolysaccharide
A-PRF	advanced PRF	**LSCC**	low-speed centrifugation concept
BoP	bleeding on probing	**mRNA**	messenger RNA
BMP	bone morphogenetic protein	**MRONJ**	medication-related osteonecrosis of the jaw
CAF	coronally advanced flap	**MSC**	mesenchymal stem cell
CAL	clinical attachment level	**OFD**	open flap debridement
CBC	complete blood count	**ONJ**	osteonecrosis of the jaw
CEJ	cementoenamel junction	**PD**	probing depth
C-PRF	concentrated-PRF	**PDGF**	platelet-derived growth factor
CTG	connective tissue graft	**PPE**	personal protective equipment
DBBM	deproteinized bovine bone mineral	**PPP**	platelet-poor plasma
DFDBA	demineralized freeze-dried bone allograft	**PRF**	platelet-rich fibrin
ECM	extracellular matrix	**PRGF**	plasma rich in growth factors
EDTA	ethylenediaminetetraacetic acid	**PRP**	platelet-rich plasma
EGF	epidermal growth factor	**PTFE**	polytetrafluoroethylene
EMD	enamel matrix derivative	**RBC**	red blood cell
e-PRF	extended-PRF	**RBH**	residual bone height
ePTFE	expanded polytetrafluoroethylene	**RCF**	relative centrifugal force
FDA	US Food and Drug Administration	**RCT**	randomized controlled trial
FDBA	freeze-dried bone allograft	**rpm**	revolutions per minute
GBR	guided bone regeneration	**RT-PCR**	real-time polymerase chain reaction
GF	growth factor	**SD**	standard deviation
H&E	hematoxylin-eosin stain	**SE**	standard error
hPDLC	human periodontal ligament cell	**SEM**	scanning electron microscopy
H-PRF	PRF obtained through horizontal centrifugation	**TGF-β**	transforming growth factor β
IGF	insulinlike growth factor	**TMJ**	temporomandibular joint
IL	interleukin	**TNF-α**	tumor necrosis factor α
i-PRF	injectable-PRF	**T-PRF**	titanium-prepared PRF
ISQ	implant stability quotient	**VEGF**	vascular endothelial growth factor
KTW	keratinized tissue width	**WBC**	white blood cell

1

Evolution of Platelet Concentrates

Contributors

Richard J. Miron

Chapter Highlights

- *Evolution of PRF and the reasons for its discovery*
- *Discussion of PRP vs PRGF vs PRF vs L-PRF, A-PRF, etc*
- *Biologic background of key steps involved during wound healing*

Video 1-1

Platelet concentrates were derived more than 20 years ago following the discovery that platelets themselves act as key regulators during the wound healing process. Initial attempts were first made to concentrate these cells using anticoagulants and a centrifugation device; the resulting biomaterial was called *platelet-rich plasma* (PRP). Shortly thereafter, protocols were developed with the aim of avoiding the use of anticoagulants altogether, because clotting is a pivotal step during the wound healing cascade; the resulting biomaterial was called *platelet-rich fibrin* (PRF). Today, platelet concentrates have become incredibly relevant worldwide, with their use spanning across nearly every field of regenerative medicine. Furthermore, one of the main growth factors (GFs) found in platelets—platelet-derived growth factor (PDGF)—has been commercialized as a ready-made laboratory recombinant protein under the trade name GEM 21S (Lynch Biologics). Thus, as medicine has continued to evolve and progress, an obvious and clear trend favoring GF use has been established. Furthermore, by modifying centrifugation devices and spin protocols of PRP/PRF, a greater ability to concentrate not only platelets but also leukocytes became possible, further favoring tissue regeneration. This chapter takes a deep look at the years of research leading to the significant advancement that has been made in this field. The evolution from PRP to PRF, including pioneering concepts such as the low-speed centrifugation concept and horizontal centrifugation, are discussed in terms of their ability to favor higher cell content, GF concentration, and ultimately better wound healing.

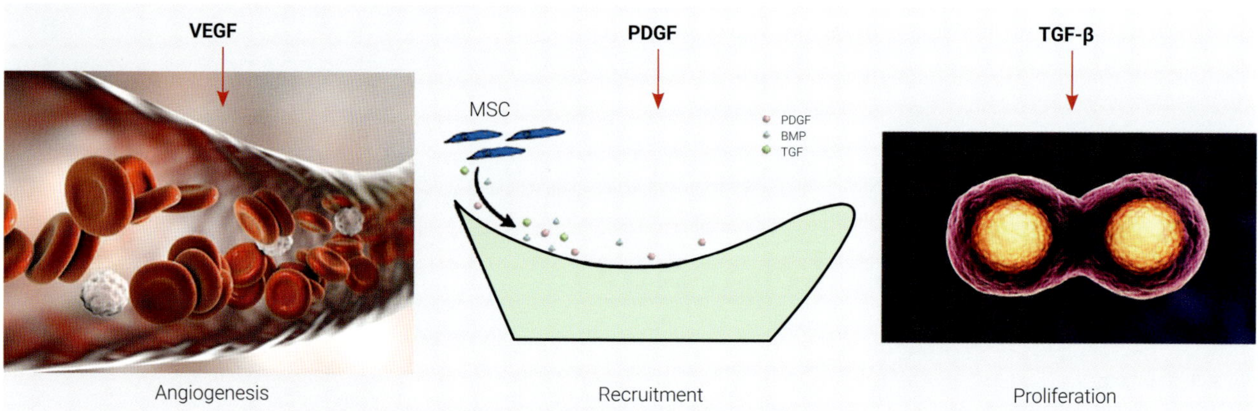

Fig 1-1 The three main GFs that are released from PRF include VEGF, a known inducer of angiogenesis; PDGF, a known inducer of cell recruitment; and TGF-β1, a known stimulator of cell proliferation. MSC, mesenchymal stem cell.

Platelet concentrates have been utilized in medicine for over two decades because of their ability to rapidly secrete autologous GFs and ultimately speed wound healing. They have gained tremendous momentum as a regenerative agent derived from autologous sources capable of stimulating tissue regeneration in a number of medical fields.[1,2] Many years ago, it was proposed that by concentrating platelets using a centrifugation device, GFs derived from blood could be collected from a platelet-rich plasma layer and later utilized in surgical sites to promote local wound healing.[1,2] Today, it has been well established that platelet concentrates act as a potent mitogen capable of the following (Fig 1-1):

- Speeding the revascularization of tissues (angiogenesis)
- Recruiting various cells including stem cells
- Inducing the prompt multiplication of various cell types found in the human body (proliferation)

Wound healing is a complex biologic process whereby many cell types interact with one another as well as their local extracellular matrix (ECM) in order to repair and regenerate damaged tissues.[3–6] While many regenerative agents currently exist on the market to help speed tissue regeneration, it is important to note that the majority are derived from other human sources (allografts) and animal byproducts. These naturally create a foreign body reaction when implanted into host tissues. While the majority of such biomaterials do certainly favor improved healing, it has generally been recognized and accepted that the gold standard for the majority of tissue-regenerative procedures in basically every field of medicine has been the use of autogenous tissues.

Specifically in dentistry, platelet concentrates were introduced over 20 years ago by Robert E. Marx and colleagues with the aim of concentrating blood proteins as a natural source of GFs that would stimulate vascularization (angiogenesis) and tissue ingrowth based on the fact that blood supply is pivotal for tissue regeneration of all tissues.[7] Wound healing has been described as a four-step process that includes *(1)* hemostasis, *(2)* inflammation, *(3)* proliferation, and *(4)* maturation[8–10] (Fig 1-2). Each phase overlaps one another and encompasses various microenvironments, including different cell types that assist in wound healing. Noteworthy are the implications of immune cells during biomaterial integration. In a study titled "OsteoMacs: Key players around bone biomaterials," osteal macrophages were discussed as being key and pivotal cells during the wound healing process.[11] Thus, as tissue biology has continued to evolve, platelet concentrates have also seen significant advancement with respect to their ability to favor healing by incorporating immune cells (leukocytes). Various systematic reviews from multiple fields of medicine have now demonstrated their ability to support tissue regeneration across many tissue types and cell types. This chapter reviews the evolution of platelet concentrates.

PRP (1990s)

The use of platelet concentrates has slowly and gradually gained popularity over time, with a dramatic increase

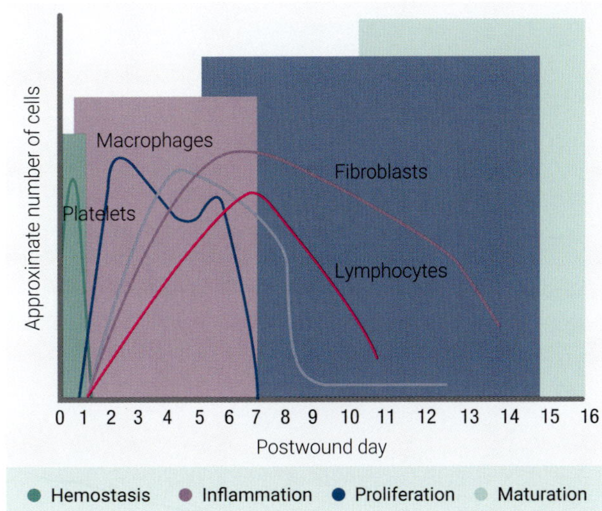

Fig 1-2 Four phases of wound healing: *(1)* hemostasis, *(2)* inflammation, *(3)* proliferation, and *(4)* maturation. Noteworthy are the overlaps between each of the phases and the population of cells found in each category. Whereas lymphocytes typically arise at 7 days, the ability of PRF to introduce a high number at day 0 acts to speed the regenerative phase during this process.

being observed in the past 5 to 10 years. This parallels precisely the massive increase in research articles being published on the topic. Despite this, it is important to review and highlight the pioneering work conducted by Marx and colleagues over 20 years ago, without which none of this textbook would exist.[12–14]

Platelet-rich plasma (PRP), as its name implies, was designed to accumulate platelets in supraphysiologic doses within the plasma layer following centrifugation. The main aim of PRP was to isolate and further concentrate the highest quantity of platelets and their associated GFs for regenerative purposes, thereafter reimplanting this specialized supraconcentrate at sites of local injury. This concept has been the basis of thousands of research articles, with their protocols being utilized to favor wound healing in millions of patients.

Initial protocols typically ranged in duration from 30 minutes to 1 hour based on the centrifugation/collection systems and protocols utilized. The original concept was pioneered by Harvest Technology, where it was shown that over 95% platelet concentration could be accumulated, having the potential to help favor the regenerative phase of many cell types including soft tissues, epithelial cells, periodontal ligament cells, and bone cells.[15,16] Because these initial protocols were lengthy, anticoagulants were added to the blood collection tubes. These typically were various forms of concentrated bovine thrombin or sodium citrate.

Despite its growing success and continued use after its discovery, several reported limitations existed with these initial formulations of PRP. The 30-minute or longer technique was generally considered lengthy for routine dental or medical practice, and more importantly, the use of anticoagulants was shown to limit wound healing from reaching its maximum potential. Simply put, when injury is created following an open wound, a blood clot is one of the first steps that occurs in order for healing to take place. Shortly thereafter, cells and GFs get trapped within this newly formed ECM, and the wound healing process/cascade begins. By limiting the body's ability to form a stable clot, wound healing is limited. Several studies have now demonstrated the superior outcomes of platelet-rich fibrin (PRF) when compared to PRP simply by removing anticoagulants from their formulations.[17–21] Even the pioneering research team behind the plasma rich in growth factors (PRGF) concept (Anitua et al) have since demonstrated more physiologic healing ability with anticoagulant removal.[17]

Another drawback of PRP was the fact that it remained liquid by nature (due to the use of anticoagulants), so when it was combined with biomaterials, a much faster delivery of GFs was observed (Fig 1-3). While an initial burst of GFs is typical of PRP therapy, a slower release of GFs over an extended period of time has been shown to better stimulate cell growth and tissue regeneration.[22,23]

Much advancement related to PRP therapy has been made over the past 20 years, and two excellent textbooks have been written by its pioneers—*Dental and Craniofacial Applications of Platelet-Rich Plasma* by Robert E. Marx and Arun K. Garg (Quintessence, 2005), and *Autologous Blood Concentrates* by Arun K. Garg (2018). Its breakthrough features include the novel ability to concentrate platelets

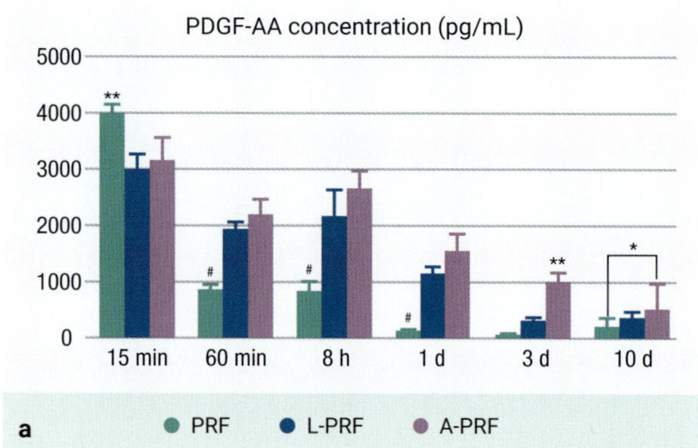

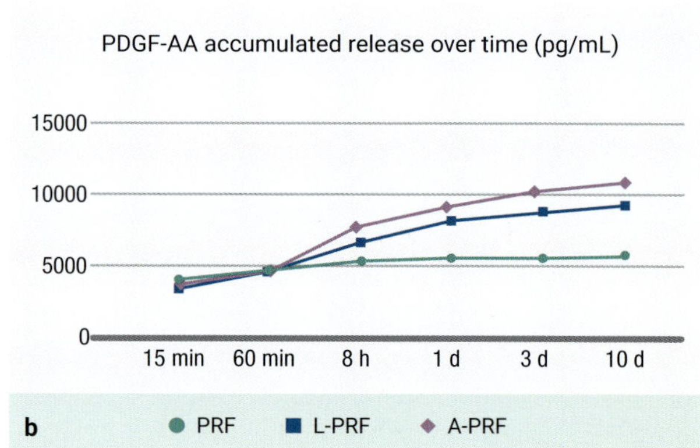

Fig 1-3 *(a and b)* GF release from PRP and PRF at each time point of PDGF-AA over a 10-day period. Notice that while PRP has significantly higher GF release at early time points, over a 10-day period, significantly higher levels are most commonly found with A-PRF due to the slow and gradual release of GFs utilizing slower centrifugation speeds. (Adapted from Kobayashi et al.[19])

to supraphysiologic doses and further stimulate tissue regeneration across virtually all tissue types. For these reasons, PRP has not surprisingly been utilized in practically every field of medicine.

Snapshot of PRP

- Marx was the first to show that a concentration of platelets could favor tissue regeneration in the oral cavity.
- A subsequent device was brought to market thanks to these breakthrough research projects conducted at the University of Miami (Harvest system).
- PRP is credited for having exponentially grown the entire field of platelet concentrates, including its subcategories such as PRF.

L-PRF (2000–2010)

Because the anticoagulants utilized in PRP prevented clotting, pioneering work performed by Dr Joseph Choukroun and Dr David Dohan Ehrenfest led to the development of PRF.[24] The aim was to develop a second-generation platelet concentrate focused on anticoagulant removal. Because anticoagulants were removed, a much quicker working time was needed, and centrifugation had to begin shortly after blood draw (otherwise, the blood would naturally clot). Furthermore, high g-force centrifugation protocols were initially utilized in an attempt to separate blood layers prior to clotting. The final spin cycle (initial studies ranged from 2500–3000 rpm for 10–12 minutes = ~700g) resulted in a plasma layer composed of a fibrin clot with entrapment of platelets and leukocytes. The main advantage of this fibrin matrix was its ability to release GFs over an extended period of time while the fibrin clot was being degraded.[25] Over the years, PRF has been termed *L-PRF* (for *leukocyte platelet-rich fibrin*) due to the discoveries that several leukocytes remained incorporated in PRF and that white blood cells play a central and key role in the tissue healing process. The most commonly utilized protocol today is a spin cycle at 3000 rpm for 10 minutes or 2700 rpm for 12 minutes (RCF-max = ~700g, RCF-clot = ~400g).

Several other advantages also existed during clinical use because it avoided the need for dual-spin protocols requiring pipetting or various specialized tube compartments, which made the overall procedure much more user-friendly, cheaper, and faster when compared to PRP. Original protocols were purposefully designed to spin at high centrifugation speeds with the main aim of phase separation to occur as quickly as possible in order to separate the red corpuscle base layer from the upper plasma layer prior to clotting. Following centrifugation, a platelet-rich fibrin mesh was formed, giving it the working name *PRF*[26–28] (Fig 1-4). PRF has since been highly researched, with over 1,000 publications dedicated to this topic alone.

Additionally, research teams from around the world have demonstrated the impact of leukocytes on tissue healing.[29–34] While it was once thought that the additional benefit of leukocyte incorporation into PRF was its main properties in improved host defense to foreign pathogens,[29–34] it has since been shown in well-conducted basic research studies

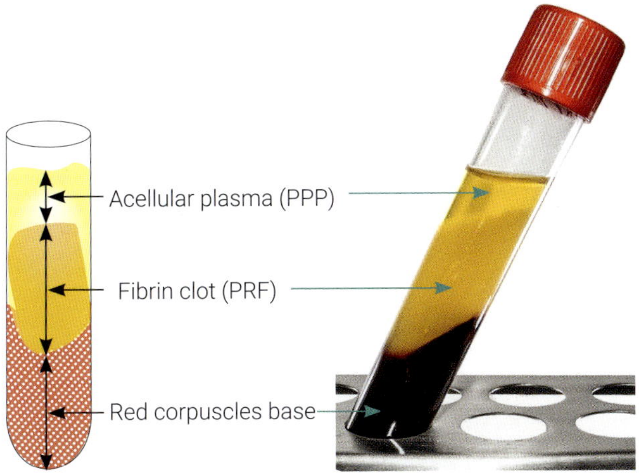

Fig 1-4 Layers produced after centrifugation of whole blood. A PRF clot forms in the upper portion of tubes after centrifugation.

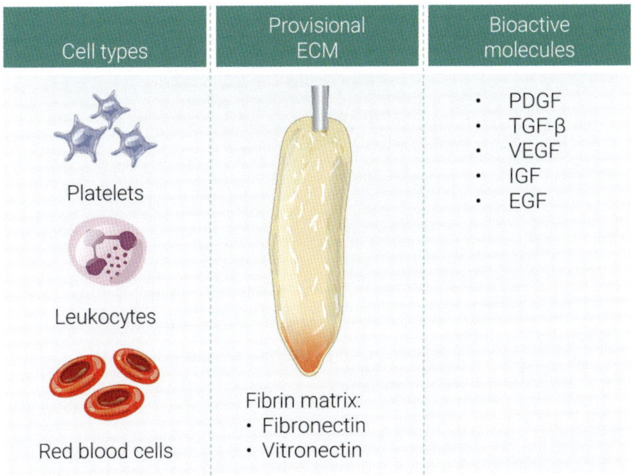

Fig 1-5 Three main components of PRF all derived naturally from the human body. These include *(1)* cell types (platelets, leukocytes, and red blood cells); *(2)* a provisional ECM 3D scaffold fabricated from autologous fibrin (including fibronectin and vitronectin); and *(3)* a wide array of over 100 bioactive molecules, including most notably PDGF, TGF-β, VEGF, IGF, and EGF.

that leukocytes are pivotal to tissue regeneration and favor faster wound healing also.[11,35–37] In dentistry, where the oral cavity is filled with bacteria and microbes, the inclusion of leukocytes was initially thought to play a pivotal role in wound healing by participating in the phagocytosis of debris, microbes, and necrotic tissues, as well as directing the future regeneration of these tissues through the release of several cytokines and GFs and orchestrating cell-to-cell communication between many cell types.

Tissue engineering with PRF

Tissue engineering has been an emerging discipline over the past decade, with major breakthroughs routinely being made every year. At its simplest foundation, tissue engineering requires three parameters: *(1)* a scaffold responsible to support tissue ingrowth, *(2)* cells that may act to promote tissue regeneration, and *(3)* GFs that stimulate the overall wound healing events. Unlike the majority of biomaterials currently available on the market, PRF actually contains each of these three properties (Fig 1-5). For comparative purposes, routine bone allografts contain a scaffold (mineralized cortical/cancellous bone) and GFs embedded in its bone matrix (such as bone morphogenetic protein 2 [BMP-2]) but have no cells. Recombinant human GFs typically have a GF (for instance, rhBMP-2) and a carrier (collagen sponge) but also lack cells. Certain stem technologies typically contain cells and also a delivery system (for instance a nanocarrier delivery system) but lack GFs. The ability to actually contain each of the three tissue engineering properties within a single biomaterial is quite rare and, more importantly, usually extremely expensive (think recombinant GFs and/or stem cell technology).

PRF, on the other hand, is a particularly simple and inexpensive way to utilize the three principles of tissue engineering by utilizing a 3D scaffold (fibrin) that incorporates both regenerative host cells (platelets and leukocytes) and various GFs. These include PDGF, TGF-β, and VEGF, each of which is crucial during the regeneration process. Furthermore, the concentrated leukocytes (as opposed to simply platelets) in PRF have been well implicated as key regulators of tissue healing and formation.[26–28,31,38]

Snapshot of PRF

- PRF is considered a second-generation platelet concentrate with a longer GF release profile.
- Centrifugation protocols are shorter and do not need any chemical additives such as anticoagulants.
- PRF falls more in line with tissue engineering principles in that it is not only an accumulation of cells and GFs but also a scaffold (fibrin matrix).
- PRF incorporates leukocytes, which are key cells in pathogen defense and biomaterial integration.

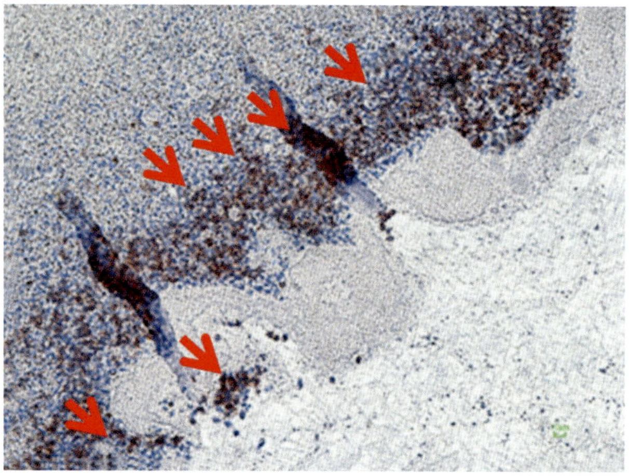

Fig 1-6 Histologic observation of leukocytes following centrifugation. Resulting white blood cells have been shown to be contained basically in the layers between the plasma PRF layer and the red blood cell clot. This finding demonstrated quite clearly that the g-force was excessive, necessitating the development of newer protocols aimed to improve the retention of leukocytes within the PRF matrix. (Reprinted with permission from Ghanaati et al.[39])

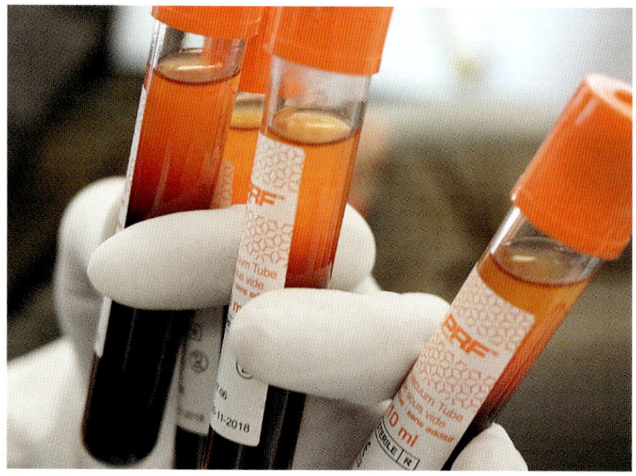

Fig 1-7 Newer centrifugation protocols allow production of a liquid formulation of PRF found in the top 1- to 2-mL layer of centrifugation tubes following a 3- to 5-minute protocol. This liquid can be collected in a syringe and reinjected into defect sites or mixed with biomaterials to improve their bioactive properties. (Reprinted with permission from Davies and Miron.[40])

A-PRF and i-PRF (2014–2018)

While much of the research performed in the late 2000s and early 2010s was dedicated to the clinical uses and indications of L-PRF discussed later in this textbook, major discoveries were made several years later from basic research laboratories. Following extensive clinical use and research with the original L-PRF protocol, it was discovered in 2014 by Dr Shahram Ghanaati that centrifugation carried out at relatively high centrifugation speeds (~700g) led to the great majority of leukocytes being located either at the buffy coat zone (between the red blood cell layer and the upper plasma layer) or more commonly at the bottom of centrifugation tubes (Fig 1-6).[39] It was expressed that the longer the centrifugation time is carried out, the more likely it is that cells get pushed further down the centrifugation tube. Similarly, the faster the spin centrifugation speed (higher g-force), the greater the proportion of cells found in the lower levels of centrifugation tubes.

Pioneering research within his laboratory led to the development of an advanced PRF (A-PRF) whereby lower centrifugation speeds (~200g) led to a higher accumulation of platelets and leukocytes more evenly distributed throughout the upper PRF layers. These newer protocols more favorably led to a higher release and concentration of GFs over a 10-day period when compared to PRP or L-PRF.[19] In 2015 to 2017, our research team further demonstrated that optimization of PRF could be achieved by reducing not only centrifugation speed but also the time involved. The A-PRF protocol was therefore modified from 14 minutes at 200g as originally described in 2014 down to an 8-minute protocol.[19]

Following an array of basic research studies on this topic, it was observed that by further reducing the g-force and also the time, it was possible to obtain a plasma layer that had not yet converted into fibrin (ie, scientifically liquid fibrinogen but often referred to as *liquid-PRF* for simplicity). In a study titled "Injectable platelet rich fibrin (i-PRF): Opportunities in regenerative dentistry?",[20] it was demonstrated that at lower centrifugation speeds and times (~60g for 3 minutes), a liquid-PRF (termed *injectable-PRF* or *i-PRF*) could be obtained. While these protocols typically produced minimal volumes (~1.0–1.5 mL), it was shown that both platelets and leukocytes were even more highly concentrated when compared to L-PRF or A-PRF (Fig 1-7).[40] This liquid-PRF layer could be utilized clinically for approximately 15 to 20 minutes, during which time fibrinogen and thrombin had not yet converted to a fibrin matrix (ie, remained liquid). This has since been utilized for injection into various joints/spaces similar to PRP, however with the reported advantages of a longer GF release time. Furthermore, the concept of "sticky" bone was also developed. Importantly, a different type of tube (plastic) was needed to minimize clotting, as will be discussed in detail in chapter 5.

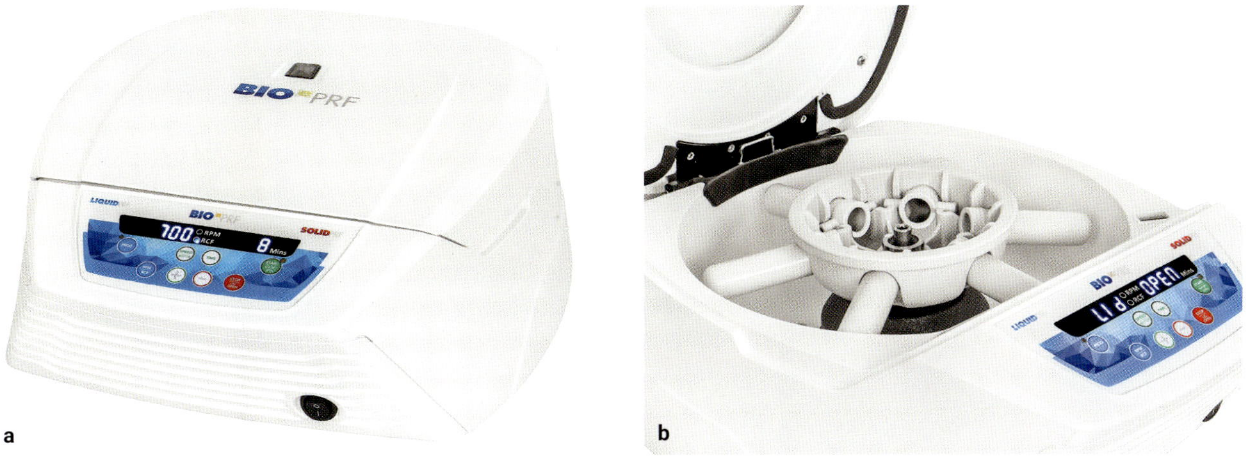

Fig 1-8 *(a)* Clinical photograph of a Bio-PRF centrifuge. *(b)* Photograph demonstrating the horizontal centrifugation concept. The tubes are inserted vertically (up and down), but once the device begins to rotate, the tubes swing out completely horizontally. This favors better blood cell layer separation with higher platelet and GF concentrations.

Snapshot of A-PRF and i-PRF

- Original L-PRF protocols were shown to be too fast, leading to all the cells being accumulated only at the buffy coat zone, with the majority of leukocytes found within the red blood cell layer.
- The low-speed centrifugation concept was shown in 2014 to favor a higher concentration of cells within PRF membranes.
- By further lowering speed and time, a liquid-PRF formulation became available, commonly known as *injectable-PRF* (or i-PRF).

H-PRF and C-PRF (2019–Present)

Very recently, our research group discovered through a series of basic laboratory experiments that horizontal centrifugation led to significantly greater concentrations of platelets and leukocytes when compared to currently available fixed-angle centrifugation devices most commonly utilized to produce L-PRF and A-PRF. Simply, horizontal centrifuges are routinely utilized in high-end research laboratories as well as in medical hospitals because of their greater ability to separate layers based on density (Fig 1-8; see also chapters 2 and 3). Unlike fixed-angle centrifugation systems whereby the tubes are actually inserted at a 45-degree angle, in horizontal centrifugation systems (often referred to as *swing-out bucket centrifugation*), the tubes have the ability to swing out to 90 degrees once they are in rotation (Video 1-2). Amazingly, the original PRP systems developed by Harvest and Marx utilized and still use this technology.

Video 1-2

In 2019, an article on the topic demonstrated clearly that horizontal centrifugation could lead to up to a four-times greater cell content when compared to fixed-angle centrifugation.[41] This represented a marked ability to greatly concentrate cells found within PRF, which were primarily being accumulated on the back distal surfaces of PRF tubes (Fig 1-9). The major disadvantage of fixed-angle centrifugation is that during the spin cycle, cells are typically driven along the back wall of the centrifugation tubes at high g-forces (Fig 1-10). This also exposes cells to higher compressive forces against the back wall, and

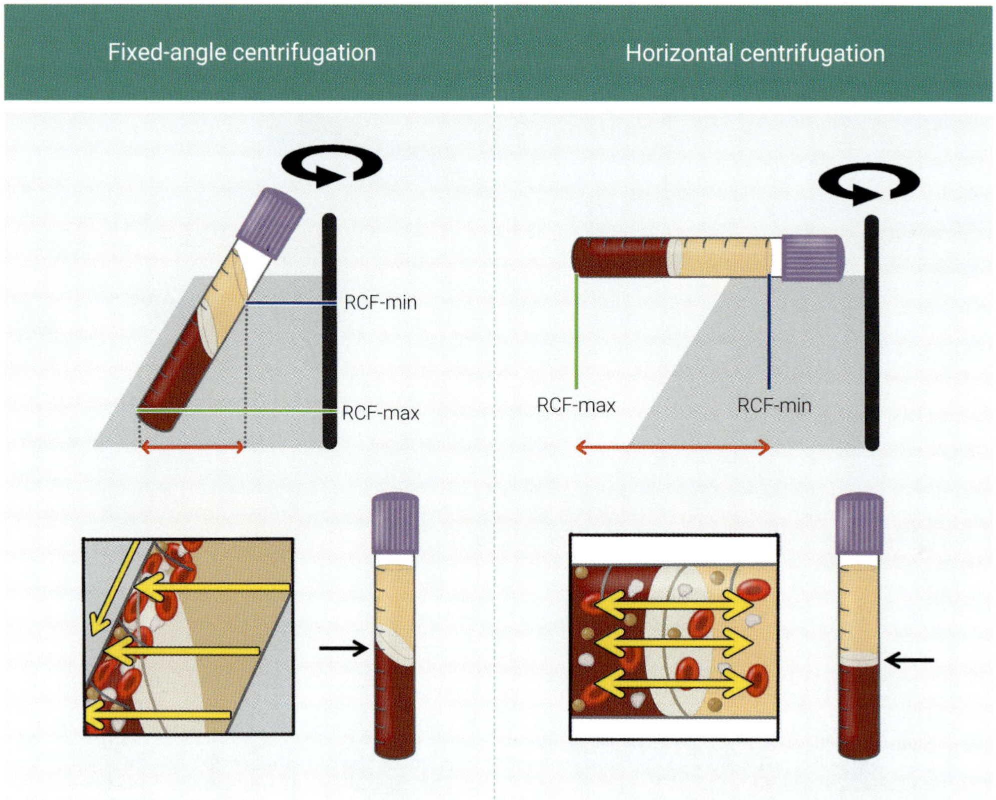

Fig 1-9 Illustrations comparing fixed-angle and horizontal centrifuges. With horizontal centrifugation, a greater separation of blood layers based on density is achieved owing to the greater difference in RCF-min and RCF-max. Following centrifugation on fixed-angle centrifuges, blood layers do not separate evenly, and as a result, an angled blood separation is observed. In contrast, horizontal centrifugation produces even separation. Owing to the large RCF values (~200g–700g), the cells are pushed toward the outside and downward. On a fixed-angle centrifuge, cells are pushed toward the back of centrifugation tubes and then downward/upward based on cell density. These g-forces produce additional shear stress on cells as they separate based on density along the back walls of centrifugation tubes. In contrast, horizontal centrifugation allows for the free movement of cells to separate into their appropriate layers based on density, allowing for better cell separation as well as less trauma/shear stress on cells. (Modified from Miron et al.[41])

cells must then separate by traveling either up or down the inclined centrifugation slope based on their respective cell density differences. Because red blood cells are larger and heavier than platelets and leukocytes, they travel downward, whereas lighter platelets travel toward the top of the tube where PRF is collected. This makes it relatively difficult for the small cell types such as platelets and leukocytes to reach the upper layer, especially granted that red blood cells outnumber in particular white blood cells typically by 1,000-fold (see chapter 2). Therefore, it is not possible to reach optimal accumulation of platelets or leukocytes using a fixed-angle centrifuge.

Furthermore, by utilizing a novel method to quantify cell types found in PRF, it was possible to substantially improve standard i-PRF protocols that favored only a 1.5- to 3-fold increase in platelets and leukocytes. Noteworthy is that several research groups began to show that the final concentration of platelets was only marginally improved in i-PRF when compared to standard baseline values of whole blood.[41,42] In addition, significant modifications to PRF centrifugation protocols have further been developed, demonstrating the ability to improve standard i-PRF protocols toward liquid formulations that are significantly more concentrated (C-PRF) with over 10- to 15-times greater concentrations of platelets and leukocytes when compared to i-PRF (see chapters 2 and 3). Today, C-PRF has been established as the most highly concentrated PRF protocol described in the literature.

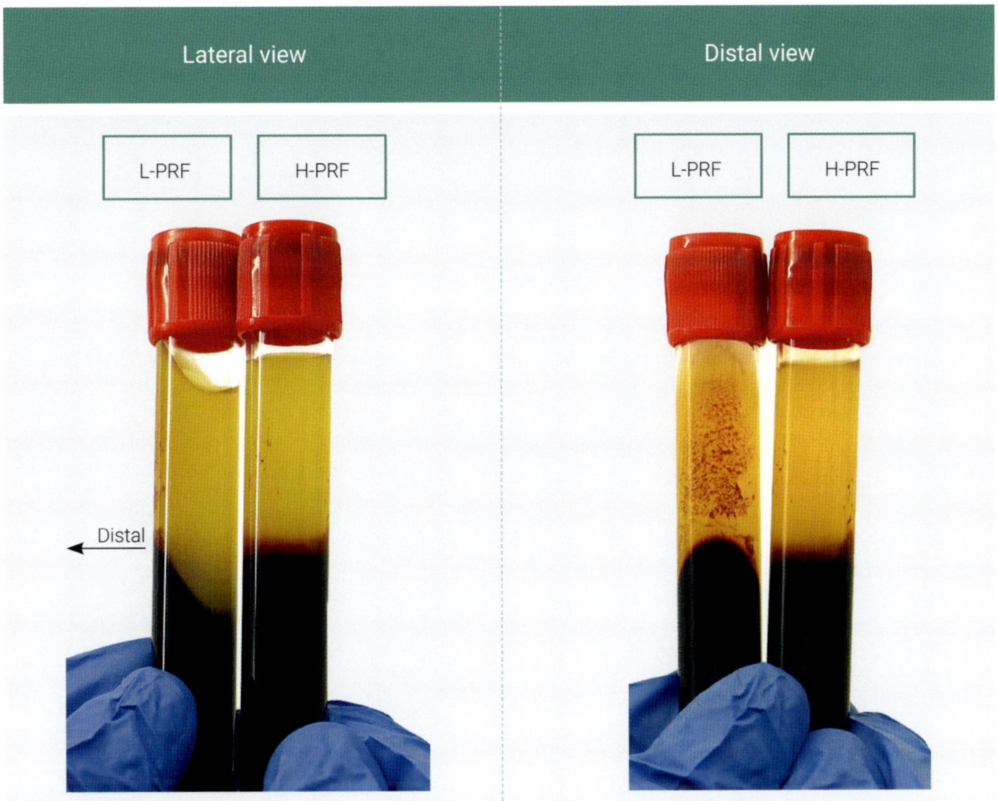

Fig 1-10 Visual representation of layer separation following either L-PRF or H-PRF protocols. L-PRF clots are prepared with a sloped shape, and multiple red dots are often observed on the distal surface of PRF tubes, while H-PRF results in horizontal layer separation between the upper plasma and lower red corpuscle layer.

Snapshot of H-PRF and C-PRF

- Horizontal centrifugation leads to up to a four-times greater accumulation of platelets and leukocytes when compared to fixed-angle centrifugation systems commonly utilized to produce L-PRF and A-PRF.
- Cells accumulate evenly when PRF is produced via horizontal centrifugation as opposed to along the back distal surface of PRF tubes on fixed-angle centrifuges.
- Standard i-PRF can be further improved with horizontal centrifugation.

Conclusion

Platelet concentrates have seen a wide and steady increase in popularity since they were launched more than two decades ago. While initial concepts launched in the 1990s led to the working name *platelet-rich plasma*, subsequent years and discoveries have focused more specifically on their anticoagulant removal (ie, PRF). Several recent improvements in centrifugation protocols, including the low-speed centrifugation concept and horizontal centrifugation, have led to increased concentrations of GFs and better healing potential. Both solid-PRF as well as liquid-based formulations now exist, with an array of clinical possibilities created based on the ability to accumulate supraphysiologic doses of platelets and blood-derived GFs. Future strategies to further improve PRF formulations and protocols are continuously being investigated to additionally improve clinical practice utilizing this technology.

References

1. Anfossi G, Trovati M, Mularoni E, Massucco P, Calcamuggi G, Emanuelli G. Influence of propranolol on platelet aggregation and thromboxane B_2 production from platelet-rich plasma and whole blood. Prostaglandins Leukot Essent Fatty Acids 1989;36:1–7.
2. Fijnheer R, Pietersz RN, de Korte D, et al. Platelet activation during preparation of platelet concentrates: A comparison of the platelet-rich plasma and the buffy coat methods. Transfusion 1990;30: 634–638.
3. Coury AJ. Expediting the transition from replacement medicine to tissue engineering. Regen Biomater 2016;3:111–113.
4. Dai R, Wang Z, Samanipour R, Koo KI, Kim K. Adipose-derived stem cells for tissue engineering and regenerative medicine applications. Stem Cells Int 2016;2016:6737345.
5. Rouwkema J, Khademhosseini A. Vascularization and angiogenesis in tissue engineering: Beyond creating static networks. Trends Biotechnol 2016;34:733–745.
6. Zhu W, Ma X, Gou M, Mei D, Zhang K, Chen S. 3D printing of functional biomaterials for tissue engineering. Curr Opin Biotechnol 2016;40:103–112.
7. Upputuri PK, Sivasubramanian K, Mark CS, Pramanik M. Recent developments in vascular imaging techniques in tissue engineering and regenerative medicine. Biomed Res Int 2015;2015:783983.
8. Gosain A, DiPietro LA. Aging and wound healing. World J Surg 2004;28:321–326.
9. Eming SA, Brachvogel B, Odorisio T, Koch M. Regulation of angiogenesis: Wound healing as a model. Prog Histochem Cytochem 2007;42:115–170.
10. Eming SA, Kaufmann J, Lohrer R, Krieg T. Chronic wounds: Novel approaches in research and therapy [in German]. Hautarzt 2007;58:939–944.
11. Miron RJ, Bosshardt DD. OsteoMacs: Key players around bone biomaterials. Biomaterials 2016;82:1–19.
12. de Vries RA, de Bruin M, Marx JJ, Hart HC, Van de Wiel A. Viability of platelets collected by apheresis versus the platelet-rich plasma technique: A direct comparison. Transfus Sci 1993;14:391–398.
13. Whitman DH, Berry RL, Green DM. Platelet gel: An autologous alternative to fibrin glue with applications in oral and maxillofacial surgery. J Oral Maxillofac Surgery 1997;55:1294–1299.
14. Marx RE, Carlson ER, Eichstaedt RM, Schimmele SR, Strauss JE, Georgeff KR. Platelet-rich plasma: Growth factor enhancement for bone grafts. Oral Surg Oral Med Oral Pathol Oral Radiol Endodontol 1998;85:638–646.
15. Jameson C. Autologous platelet concentrate for the production of platelet gel. Lab Med 2007;38:39–42.
16. Marx RE. Platelet-rich plasma: Evidence to support its use. J Oral Maxillofac Surg 2004;62:489–496.
17. Anitua E, Prado R, Troya M, et al. Implementation of a more physiological plasma rich in growth factor (PRGF) protocol: Anticoagulant removal and reduction in activator concentration. Platelets 2016;27:459–466.
18. Abd El Raouf M, Wang X, Miusi S, et al. Injectable-platelet rich fibrin using the low speed centrifugation concept improves cartilage regeneration when compared to platelet-rich plasma. Platelets 2019;30:213–221.
19. Kobayashi E, Fluckiger L, Fujioka-Kobayashi M, et al. Comparative release of growth factors from PRP, PRF, and advanced-PRF. Clin Oral Investig 2016;20:2353–2360.
20. Miron RJ, Fujioka-Kobayashi M, Hernandez M, et al. Injectable platelet rich fibrin (i-PRF): Opportunities in regenerative dentistry? Clin Oral Investig 2017;21:2619–2627.
21. Wang X, Zhang Y, Choukroun J, Ghanaati S, Miron RJ. Effects of an injectable platelet-rich fibrin on osteoblast behavior and bone tissue formation in comparison to platelet-rich plasma. Platelets 2018;29:48–55.
22. Lucarelli E, Beretta R, Dozza B, et al. A recently developed bifacial platelet-rich fibrin matrix. Eur Cell Mater 2010;20:13–23.
23. Saluja H, Dehane V, Mahindra U. Platelet-rich fibrin: A second generation platelet concentrate and a new friend of oral and maxillofacial surgeons. Ann Maxillofac Surg 2011;1:53–57.
24. Choukroun J, Adda F, Schoeffler C, Vervelle A. Une opportunité en paro-implantologie: Le PRF. Implantodontie 2001;42:e62.
25. Dohan Ehrenfest DM, Del Corso M, Diss A, Mouhyi J, Charrier JB. Three-dimensional architecture and cell composition of a Choukroun's platelet-rich fibrin clot and membrane. J Periodontol 2010;81:546–555.
26. Choukroun J, Diss A, Simonpieri A, et al. Platelet-rich fibrin (PRF): A second-generation platelet concentrate. Part IV: Clinical effects on tissue healing. Oral Surg Oral Med Oral Pathol Oral Radiol Endod 2006;101:e56–e60.
27. Dohan DM, Choukroun J, Diss A, et al. Platelet-rich fibrin (PRF): A second-generation platelet concentrate. Part I: Technological concepts and evolution. Oral Surg Oral Med Oral Pathol Oral Radiol Endod 2006;101:e37–e44.
28. Dohan DM, Choukroun J, Diss A, et al. Platelet-rich fibrin (PRF): A second-generation platelet concentrate. Part II: Platelet-related biologic features. Oral Surg Oral Med Oral Pathol Oral Radiol Endod 2006;101:e45–e50.
29. Martin P, Leibovich SJ. Inflammatory cells during wound repair: The good, the bad and the ugly. Trends Cell Biol 2005;15:599–607.
30. Tsirogianni AK, Moutsopoulos NM, Moutsopoulos HM. Wound healing: Immunological aspects. Injury 2006;37(suppl 1):S5–S12.
31. Adamson R. Role of macrophages in normal wound healing: An overview. J Wound Care 2009;18:349–351.
32. Davis VL, Abukabda AB, Radio NM, et al. Platelet-rich preparations to improve healing. Part I: Workable options for every size practice. J Oral Implantol 2014;40:500–510.
33. Davis VL, Abukabda AB, Radio NM, et al. Platelet-rich preparations to improve healing. Part II: Platelet activation and enrichment, leukocyte inclusion, and other selection criteria. J Oral Implantol 2014;40:511–521.
34. Ghasemzadeh M, Hosseini E. Intravascular leukocyte migration through platelet thrombi: Directing leukocytes to sites of vascular injury. Thromb Haemost 2015;113:1224–1235.
35. Batoon L, Millard SM, Raggatt LJ, Pettit AR. Osteomacs and bone regeneration. Curr Osteoporos Rep 2017;15:385–395.
36. Chang MK, Raggatt LJ, Alexander KA, et al. Osteal tissue macrophages are intercalated throughout human and mouse bone lining tissues and regulate osteoblast function in vitro and in vivo. J Immunol 2008;181:1232–1244.
37. Winkler IG, Sims NA, Pettit AR, et al. Bone marrow macrophages maintain hematopoietic stem cell (HSC) niches and their depletion mobilizes HSCs. Blood 2010;116:4815–4828.
38. Dohan DM, Choukroun J, Diss A, et al. Platelet-rich fibrin (PRF): A second-generation platelet concentrate. Part III: Leucocyte activation: A new feature for platelet concentrates? Oral Surg Oral Med Oral Pathol Oral Radiol Endod 2006;101:e51–e55.
39. Ghanaati S, Booms P, Orlowska A, et al. Advanced platelet-rich fibrin: A new concept for cell-based tissue engineering by means of inflammatory cells. J Oral Implantol 2014;40:679–689.
40. Davies C, Miron RJ. PRF in Facial Esthetics. Chicago: Quintessence, 2020.
41. Miron RJ, Chai J, Zheng S, Feng M, Sculean A, Zhang Y. A novel method for evaluating and quantifying cell types in platelet rich fibrin and an introduction to horizontal centrifugation. J Biomed Mater Res A 2019;107:2257–2271.
42. Varela HA, Souza JCM, Nascimento RM, et al. Injectable platelet rich fibrin: Cell content, morphological, and protein characterization. Clin Oral Investig 2019;23:1309–1318.

2

Biology of PRF: Fibrin Matrix, Growth Factor Release, and Cellular Activity

Contributors

Masako Fujioka-Kobayashi
Yufeng Zhang
Reinhard Gruber
Richard J. Miron

Chapter Highlights

- *What is PRF?*
- *How does PRF differ from PRP at the biologic and cellular level?*
- *What is the role of each cell type found in PRF?*
- *What is the role of each GF found in PRF?*
- *How does centrifugation speed and time affect PRF?*
- *What advantages exist for horizontal centrifugation versus fixed-angle centrifugation?*

Video 2-1

Much can be discussed with respect to the biology of PRF and its ability to impact tissue regeneration. During the natural wound healing process, vascularization of tissues plays a pivotal role, facilitating the invasion of incoming cells, growth factors (GFs), cytokines, and other regenerative factors. The main aim of platelet concentrates, discovered over two decades ago, is to favor new blood flow (angiogenesis) to damaged tissues, thereby improving their healing potential by delivering a supraphysiologic concentration of blood-derived cells (namely platelets) and regenerative GFs. This chapter takes a deep look into the actual separation of blood layers during the centrifugation process to provide the clinician a general overview of the cell types and GFs found in PRF, including their roles, and also discusses the effects of centrifugation speed and time on cell layer separation. Furthermore, the advantages of producing an autologous fibrin scaffold are presented as it being a key regulator of wound healing because of its autologous source and its ability to promote the slow and gradual release of GFs over time. The advantages of horizontal centrifugation versus fixed-angle centrifugation are also discussed based on recent data from various laboratories from around the world.

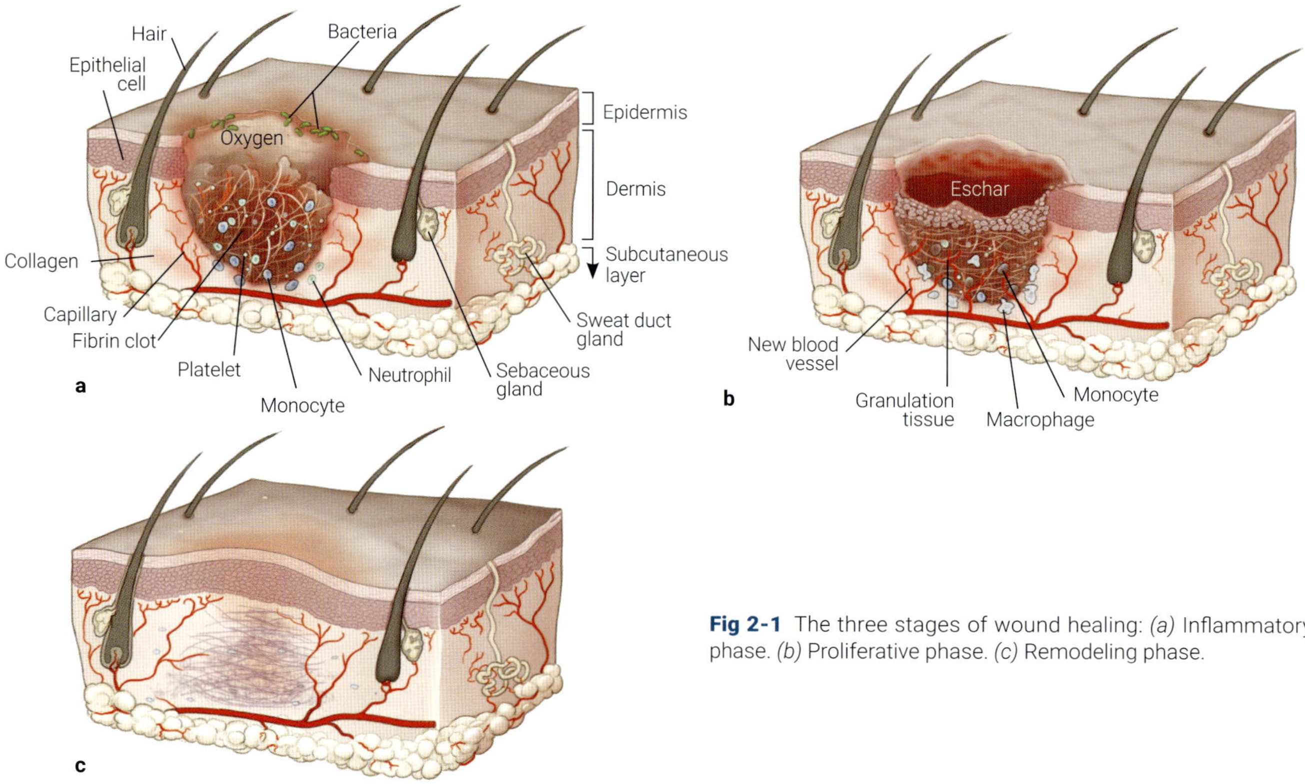

Fig 2-1 The three stages of wound healing: *(a)* Inflammatory phase. *(b)* Proliferative phase. *(c)* Remodeling phase.

The wound healing process is divided into three stages: the inflammatory phase, the proliferative phase, and the remodeling phase (Fig 2-1). The inflammatory phase starts at the time of injury and generally involves a wide array of cytokines and growth factors (GFs) that are released within the first 24 to 48 hours. Accordingly, a dynamic interaction occurs between endothelial cells, angiogenic cytokines, and the extracellular matrix (ECM) in an attempt to accelerate wound healing via an orchestrated delivery of multiple GFs in a well-controlled fashion.[1]

In general, blood provides essential components to the healing process that comprise both cellular and protein products that essentially are the base components of wound healing. During the healing process, blood will undergo clotting within a few minutes to prevent further blood loss. This is an important step that will be later discussed in the PRF tube section, because in order for clotting to occur and even be improved in both speed and quality (in particular in patients taking anticoagulants), a proper understanding of the clotting cascade is required. In its simplest of forms, oxygen helps improve blood clotting, and for this reason, the simple removal of centrifugation tube lids following the spin process will lead to faster clotting of PRF and a superior fibrin mesh. One of the major roles of platelets is to assist during hemostasis through a fibrin clot formation.[1,2] Not surprisingly, the additional use of PRF for wound healing (for example, following tooth removal and extraction socket healing) in patients taking anticoagulants can drastically improve the healing outcomes simply by improving clotting. Because PRF contains many platelets and a fibrin nucleus is already formed, bleeding has been shown to be significantly reduced postoperative when PRF is utilized in patients on anticoagulant therapy.[3]

Tips

- Oxygen helps improve blood clotting, so the simple removal of centrifugation tube lids following the spin process will lead to faster clotting of PRF.
- In patients undergoing anticoagulant therapy, the simple addition of PRF during surgery can help favor faster clotting, thereby reducing bleeding times postoperative.

Table 2-1 Properties of cells found in whole blood

	Platelets	WBCs	RBCs
Density (kg/m^3)	1040–1065	1055–1085	1095–1100
Frequency (1/µL)	200,000	5,000	5,000,000
Surface (μm^2)	28	330	140
Radius (µm)	11.5	5–7.5	4
Volume (μm^3)	14	200	92
Shape	Irregular disc	Spherical	Biconcave

Platelets also release various GFs and cytokines that further lead to tissue regeneration but also attract macrophages and neutrophils to the defect site. These cells are responsible for clearing debris, replacing necrotic tissue, and removing bacteria from the wound site.

The proliferative phase begins by day 3, where the blood clot within the wound is further supplied with a provisional matrix typically composed in part with fibrin, which facilitates cell migration, while the clot within the vessel lumen contributes to hemostasis.[2] Fibroblast cells are recruited to the wound site and begin producing new collagen in a random and somewhat disorganized order. Simultaneously, new blood vessel formation leads to new angiogenesis, and the wound gradually begins to gain initial stability.

During the third and final stage (the remodeling phase), disorganized collagen is replaced by newly organized collagen fibrils that provide enhanced stability and strength to the injured site, where tissue regeneration takes place[4] (see Fig 2-1).

Whole blood is comprised of four main components: blood plasma, red blood cells (RBCs), white blood cells (WBC), and platelets. Initially, platelets were reported as the major responsible component for the activation and release of crucial GFs for wound healing, including PDGF, coagulation factors, adhesion molecules, cytokines, and angiogenic factors. Their role has been extremely well described in the literature, so typically the entire field has been referred to as *platelet concentrates* or *platelet-rich plasma/fibrin*. Interestingly, however, over the years more attention has been placed on leukocytes, which are not only responsible for host defense but also highly implicated in the wound healing and regenerative phases.

Table 2-1 highlights the various cell types found in blood, including their density, frequency, and surface area. Note that while platelets are the lightest of the group, WBCs and RBCs are very similar in density. For these reasons they are also harder to separate in a centrifuge based on density. Noteworthy is the fact that per µL, there are 5,000,000 RBCs when compared to only 5,000 WBCs. Therefore, RBCs outnumber WBCs in a 1,000:1 ratio, which make them difficult to separate, especially on a fixed-angle centrifugation device as discussed later in this chapter (Video 2-2).

Leukocytes and RBCs are similar in density. This makes these two cell types extremely difficult to separate, especially because RBCs outnumber them 1,000 to 1.

Video 2-2

Cells in PRF

As shown in Table 2-1, the three main cell types found in PRP and PRF are platelets, leukocytes (WBCs), and RBCs. The entire initial goal of platelet concentrates was of course to concentrate platelets. Because they are the lightest of all cells found in blood, it was possible to utilize a centrifugation device to separate these layers based on their density. Lighter cells (platelets) could therefore be accumulated to the top, followed by leukocytes. Because RBCs are the densest of the group, they tend to migrate downward during the centrifugation process. In an ideal situation, the final PRF matrix should be composed of a high concentration of platelets, leukocytes, and fibrin. It has been shown that the initially developed PRF (also termed L-PRF for *leukocyte PRF*) concentrates contained greater than 90% platelets and more than 50% leukocytes within a high-density fibrin network when compared to whole blood.[5] By utilizing more advanced quantification devices and recently developed methods, our research team has been better able to harvest leukocytes specifically. The lower yield of leukocytes is typically a result of their more similar density to RBCs, making them harder to separate from and accumulate in the upper layers where PRF is collected. This is particularly difficult on fixed-angle centrifuges. Several other methods have been proposed to favor accumulation of cells, including shorter centrifugation times as well as lower centrifugation forces, as discussed later in this chapter (see section on the low-speed centrifugation concept).[6]

Leukocytes have been shown to be an integral component of PRF therapy and play a prominent role in wound healing. Studies from basic sciences and animal research have revealed how impactful a role leukocytes play during tissue regeneration by comparing PRP/PRF therapy with and without WBCs.[7–9] In these split design studies, the contralateral side receiving leukocytes performed significantly better, promoting researchers and clinicians to develop protocols to better incorporate or harvest leukocytes. Naturally, PRF contains a higher number of leukocytes when compared to the first-generation platelet concentrates PRP and PRGF.

While the role of leukocytes has been well described as host defense against incoming pathogens, they also play a central role in immune modulation of biomaterials and participate in the wound healing process due to their ability to secrete key immune cytokines such as IL-1β, IL-6, IL-4, and TNF-α.[2,10,11] They have been highly investigated in PRF therapy, with the impact of centrifugation speed and time affecting both their concentration and location, mainly owing to the fact that they are very similar in density and size to RBCs. Previously, it was demonstrated how faster protocols initially utilized to produce L-PRF were far too high in both g-force and time (2700 rpm for 12 minutes; ~700g).[6] This led to the histologic observation that the majority of cells were concentrated either at the buffy coat region or at the bottom of centrifugation tubes within the RBC layer component.[6] Based on these observations, it became clear that centrifugation speeds (g-forces) were evidently too high, pushing leukocytes especially down to the bottom of centrifugation tubes and away from the PRF clot. In order to redistribute leukocyte cell numbers across the entire PRF matrix, both a change in centrifugation speed and/or time (lower) as well as a change in the centrifugation device (horizontal centrifugation as opposed to fixed-angle) were deemed necessary to further improve platelet formulations, as reviewed later in this chapter.

Advantages of a 3D Fibrin Network

Fibrin is the activated form of a plasmatic molecule called *fibrinogen* that converts into fibrin with thrombin. Fibrin formation is one of the first key components to tissue wound healing. When an individual cuts himself, the first event taking place prior to any regeneration is fibrin clot formation. This is why patients on anticoagulant therapy typically do not heal quite as effectively because delayed clotting leads to delayed healing. The obvious advantage of PRF therapy is its ability to accumulate various cell types including platelets without anticoagulants, thereby improving clotting properties. Once a fibrin clot is formed during the centrifugation cycle, cells and GFs are able to be trapped within the 3D fibrin matrix, favoring the slower and gradual release of GFs from PRF over time.[12] Fibrin is a soluble fibrillary molecule that is present in high quantity both in plasma itself as well as in the α-granules of platelets. Fibrin therefore plays a determining role in platelet aggregation during hemostasis and is critical to healing. Not surprisingly, the use of fibrin alone (without GFs or living cells as a fibrin glue) has been shown to lead to matrix stabilization favoring tissue stability, cellular invasion, and ultimately tissue regeneration.[13–15] However, PRF has numerous advantages in that during the fibrin clot formation, a supraphysiologic concentration of platelets, leukocytes, and GFs are also present, forming a sort of "superclot" consisting of an intimate assembly of cytokines, glycanic chains, and structural glycoproteins enmeshed within a slowly polymerized fibrin network[16,17] (Fig 2-2).

> ***PRF forms a "superclot" consisting of an intimate assembly of cytokines, glycanic chains, and structural glycoproteins enmeshed within a slowly polymerized fibrin network.***

The fibrin scaffold produced following centrifugation has further been identified as a biologic 3D network with the

Fig 2-2 SEM examination of the fibrin clot revealing a dense and mature fibrin matrix with various cell types entrapped within its matrix.

ability for the fibrillar micropores to support cell migration, proliferation, differentiation, and delivery of GFs. Platelets have theoretically been described as being massively trapped within the fibrin network, and the release of GFs is largely dictated by the actual timespan in which the 3D PRF scaffold is broken down (typically within 2–3 weeks).[18] One of the major advantages of PRF when compared to PRP is the fact that by simply removing anticoagulants, a fibrin matrix is formed with a natural autologous delivery system capable of slowly and gradually releasing GFs over time.[7–9] This leads to GF delivery over a period of 2 to 3 weeks as opposed to only a few hours observed in PRP.[12]

Finally, stem cells exist naturally in whole blood, albeit at extremely low levels.[19,20] Stem cells have the ability and potential to differentiate into many cell types, including adipocytes, osteoblasts, and chondrocytes. Many commercial enterprises report that mesenchymal stem cells (MSCs) exist in extremely high numbers in PRF or that only certain protocols or machinery favor their accumulation, but these reports have not been validated in any high-quality peer-reviewed journal. While future research investigating the impact of MSCs in blood is necessary, it may represent a potential future strategy to isolate MSCs relatively easily at low cost.

> *The commercial claims that MSCs exist in extremely high numbers in PRF or that only certain protocols or machinery favors their accumulation have not been validated.*

Growth Factors in Blood

Naturally, GFs are critical to wound healing, and a variety of GFs have been commercialized as recombinant human sources once their roles were established. GFs are largely responsible for the migration of cells and also play a critical role in their adhesion, proliferation, and differentiation. While GFs exist in all tissues, it is important to note that blood serves as a main reservoir of numerous GFs and cytokines promoting angiogenesis and tissue regeneration for wound healing. It is also important to note that certain GFs may exist as inactive or partially active precursors that require proteolytic activation, or may further require binding to matrix molecules for activity or stabilization. For this reason, interfering with the natural clotting cascade such as when utilizing PRP may affect the bioactivity of certain GFs.[12] Typically, GFs also have extremely short biologic half-lives in the order of a few minutes.[21] The body has been trained to secrete various GFs in programmed orders to activate very complex cellular processes.[22] Unlike recombinant human GFs that typically only comprise a single GF, platelet concentrates create the opportunity to deliver many autologous GFs simultaneously. Furthermore, leukocytes are known immune cells capable of "sensing" their microenvironment during the regenerative phase. Together with platelets, leukocytes serve as a major cell type during the natural wound healing process. The GFs most accumulated and delivered in PRF include VEGF, PDGF, TGF-β1, EGF, and IGF.[23,24] Their individual roles are discussed below.

> *Unlike recombinant human GFs that typically only comprise a single GF, platelet concentrates create the opportunity to deliver many autologous GFs simultaneously.*

VEGF

VEGF is secreted by activated thrombocytes and macrophages to damaged sites to promote angiogenesis. The VEGF family is related to PDGF and includes VEGF-A, -B, -C, -D, and -E. VEGF has previously been isolated and utilized as a recombinant GF described as the most potent GF for angiogenesis of tissues, stimulating new blood vessel formation, and facilitating nutrients and increased blood flow to sites of injury.[25,26] It has potent effects on tissue remodeling, and the incorporation of recombinant human VEGF into various bone biomaterials alone has been demonstrated to increase new tissue regeneration.[27]

PDGF

PDGFs are essential regulators that mainly promote the migration, proliferation, and survival of mesenchymal cell lineages.[28–33] Platelets are of course the major source of PDGF, with various groups divided into homo- (PDGF-AA, PDGF-BB, PDGF-CC, and PDGF-DD) and heterodimeric (PDGF-AB) polypeptide dimers linked by disulfide bonds. They are mainly present in large quantities in platelet α-granules. Therefore, the ability to concentrate platelets in PRF leads to the massive accumulation and subsequent release of PDGF following centrifugation. It is important to note that because PDGF has an extremely short half-life (characterized by as little as 2 minutes), the ability for the PRF scaffold to act as a reservoir of GFs by protecting PDGF from matrix metalloproteinases actually drastically improves its release profile in living tissues when compared to PRP. Because it plays such a critical role in the mechanisms of physiologic healing, a recombinant source (rhPDGF-BB) was made commercially available with FDA clearance for the regeneration of various defects in medicine and dentistry.

Because PDGF has an extremely short half-life (characterized by as little as 2 minutes), the ability for the PRF scaffold to act as a reservoir of GFs by protecting PDGF from matrix metalloproteinases actually drastically improves its release profile in living tissues when compared to PRP.

TGF-β1

TGF-β is a large superfamily of more than 30 members that mediate tissue repair, immune modulation, and ECM synthesis.[34,35] TGF-β1 is the predominant isoform that supports cell proliferation of practically all cell types.[21] It further plays a role in angiogenesis, re-epithelialization, and connective tissue regeneration.[21] Platelets are known to be a major source of TGF-β production. Specific to bone healing and remodeling, TGF-β also exerts effects by being heavily released from autogenous bone.[21]

EGF

The EGF family is responsible for the chemotaxis and angiogenesis of endothelial cells and mitosis of mesenchymal cells. It further enhances epithelialization and markedly shortens the overall healing process time when administered as a recombinant source. EGF is typically secreted naturally by the body after acute injury and acts to significantly increase the tensile strength of wounds. EGF receptors exist on most human cell types, including those that play a critical role during wound repair such as fibroblasts, endothelial cells, and keratinocytes.[36]

IGF

IGFs are positive regulators of proliferation and differentiation of most cell types.[37] IGF is found in high levels in PRF because it is highly released from platelets during their activation and degranulation, leading to the differentiation of mesenchymal cells. IGF is also an extremely attractive GF because it constitutes the major axis of programmed cell apoptosis regulation by inducing survival signals protecting cells from many apoptotic stimuli.[37]

Comparative Analysis of PRP Versus PRF

Years ago, much research investigated the comparative analysis in bioactivity and tissue regenerative properties of PRP and PRF.[12] The aim was to determine if PRF was significantly better than previously utilized PRP with respect to the initial cell content, GF release, and clinical benefit. Naturally, the release profile of GFs has been an important and highly investigated research topic over the years and differs quite significantly between PRP and PRF. The typical advantage of PRF over PRP is that it enables the controlled release of GFs over an extended period of time because of its fibrin matrix. Therefore, when PRF is directly compared to PRP, a much longer GF delivery response has been observed[38] (Fig 2-3). In a first study on this topic performed by our research team, GF release from

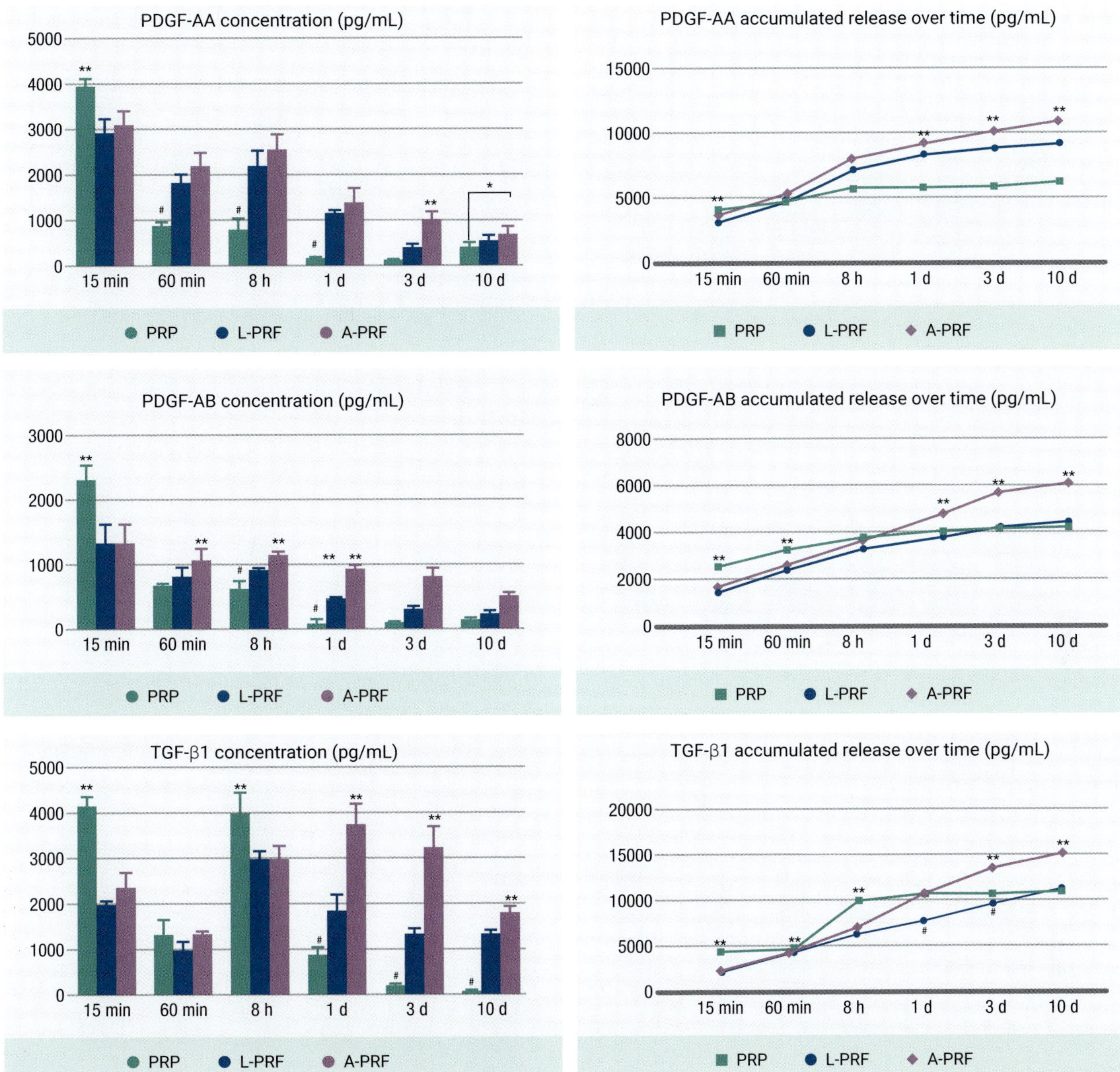

Fig 2-3 GF release from PRP and PRF at each time point for PDGF-AA, PDGF-AB, and TGF-β1 over a 10-day period. Notice that while PRP has significantly higher GF release at early time points, over a 10-day period, significantly higher levels are most commonly found with A-PRF due to the slow and gradual release of GFs following use of slower centrifugation speeds. (Adapted from Kobayashi et al.[12])

three different platelet concentrates including PRP, L-PRF, and A-PRF demonstrated a much higher total amount of GFs released from PRF when compared to PRP over a 10-day period. In order to precisely quantify each GF in a comparative fashion, PDGF, VEGF, TGF-β1, EGF, and IGF were compared at numerous time points including 15 minutes, 60 minutes, 8 hours, 24 hours, 3 days, and 10 days (see Fig 2-3). Interestingly, at an early time point (15 minutes), significantly higher numbers of GFs were released from PRP when compared to L-PRF or A-PRF, whereas both PRF groups had higher release at later time points and also total release of GFs. From 3 days and onward, the GF release of PRF far exceeded that of PRP.[12]

Overall, PRP can be recommended for fast delivery of GFs, whereas PRF is better suited for long-term delivery and favors better overall wound healing potential.

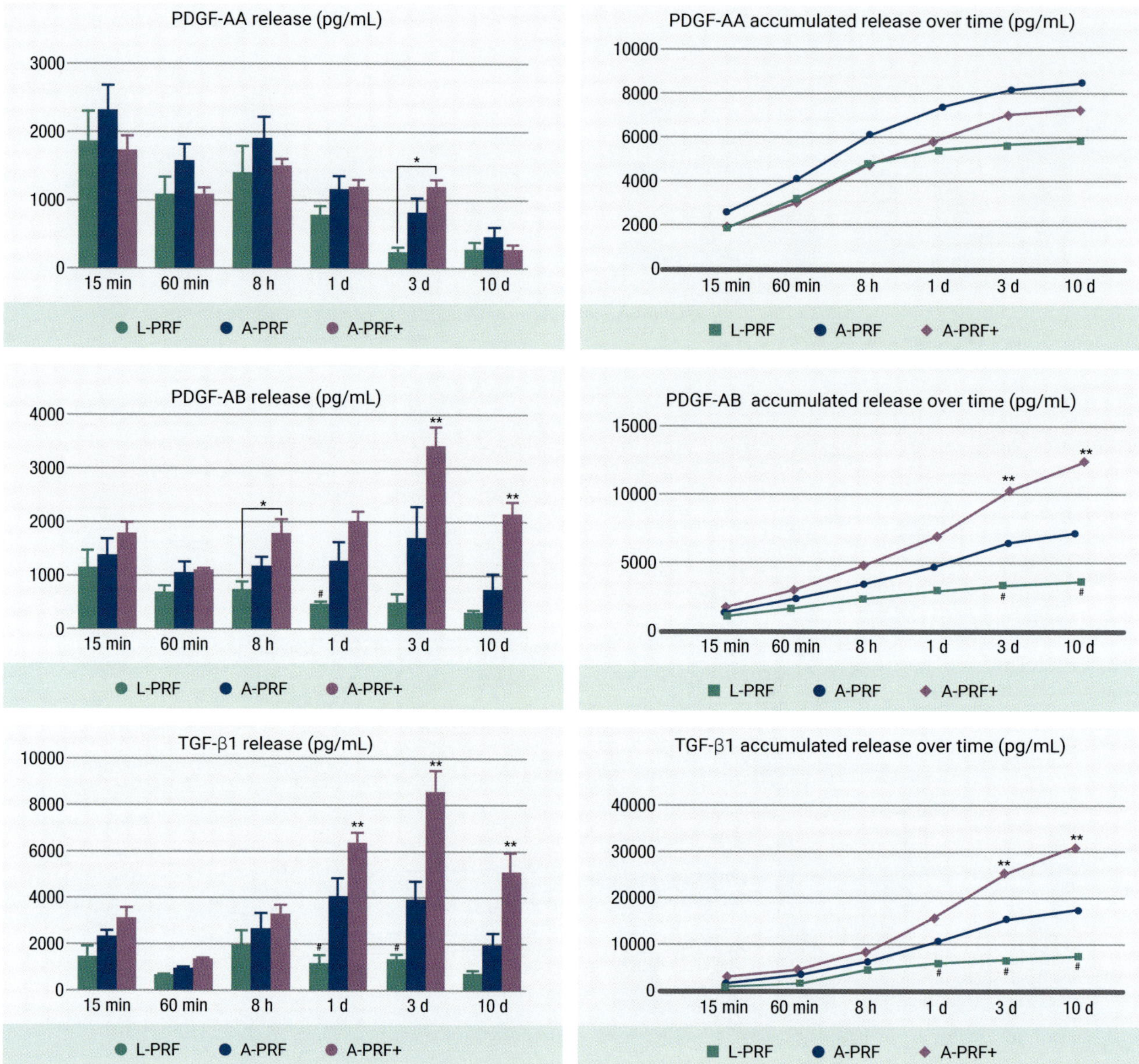

Fig 2-4 GF release resulting from the LSCC at each time point for PDGF-AA, PDGF-AB, and TGF-β1 over a 10-day period. In general, it was found that A-PRF+ demonstrated significantly the highest GF release when compared to all other modalities after a 10-day period. (Adapted from Fujioka-Kobayashi et al.[24])

The Low-Speed Centrifugation Concept

The evolution of PRF in the years 2014 to 2017 was highly focused on ways to improve centrifugation parameters in order to favor more GF release. Within the original L-PRF matrix, cells were surprisingly found gathered at the bottom of the PRF matrix or bottom of centrifugation tubes.[6] In simple terms, the faster and longer centrifugation is carried out, the more matter (ie, cells) is moved to the bottom of centrifugation tubes. With the development of A-PRF utilizing the low-speed centrifugation concept (LSCC), cell pull-down was reduced, which increased the total number of cells left contained within the top layer of PRF.

The faster and longer centrifugation is carried out, the more cells are pushed to the bottom of centrifugation tubes.

In a cell and GF analysis study on this topic by our group, the newer protocols for the production of A-PRF+,

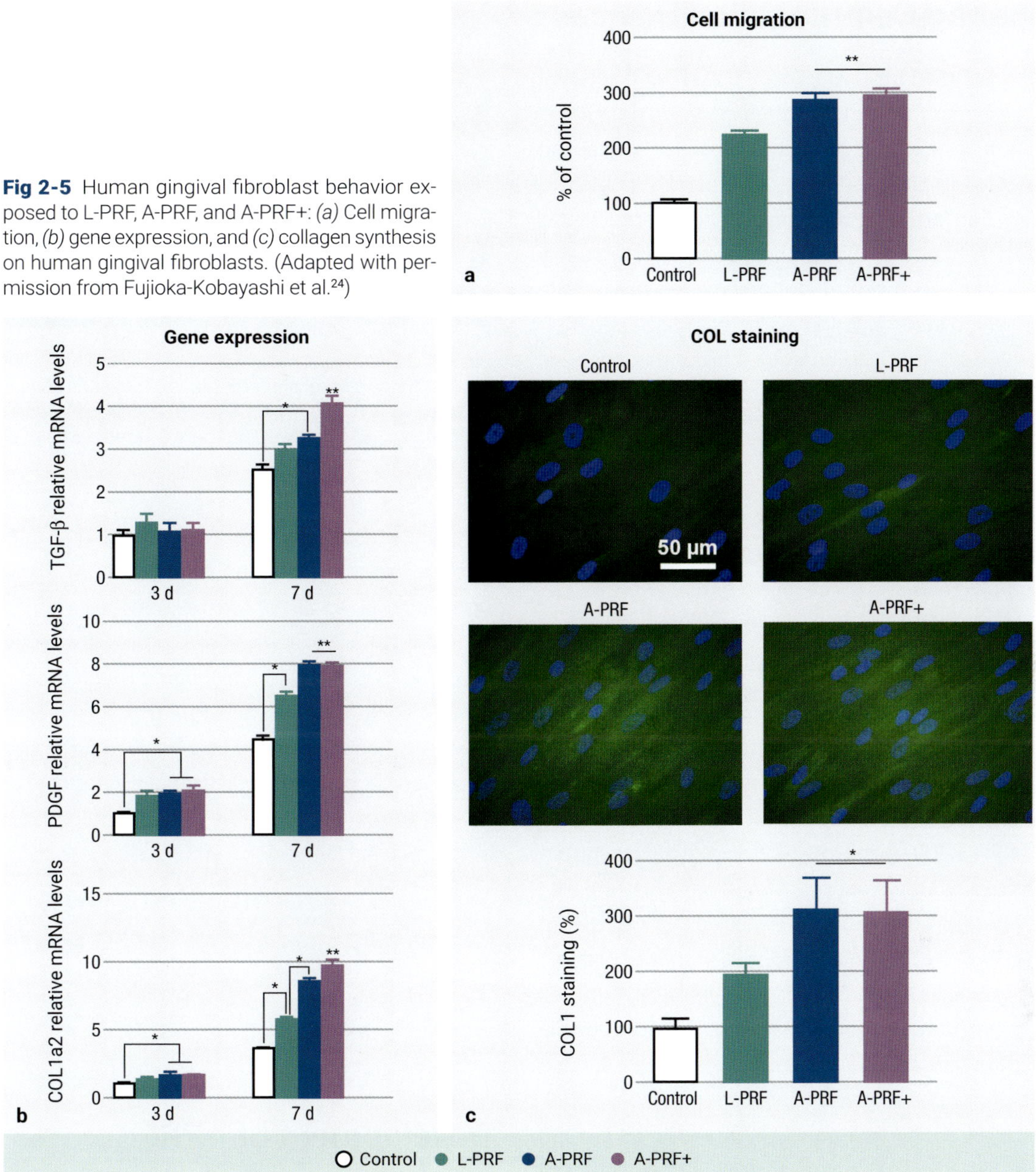

Fig 2-5 Human gingival fibroblast behavior exposed to L-PRF, A-PRF, and A-PRF+: *(a)* Cell migration, *(b)* gene expression, and *(c)* collagen synthesis on human gingival fibroblasts. (Adapted with permission from Fujioka-Kobayashi et al.[24])

which involves not only lower centrifugation speed but also less time (1300 rpm for 8 min), was shown to lead to even further increases in GF release of TGF-β1, PDGF-AA, PDGF-AB, PDGF-BB, VEGF, IGF, and EGF (Fig 2-4).[24] Therefore, an optimization of centrifugation parameters could be obtained simply by lowering the g-force and time, favoring better GF delivery.

Not surprisingly, when cells were cultured with the various formulations of PRF, it was also observed that centrifugation utilizing these lower speeds/time further led to greater cellular bioactivity. In a study titled "Optimized platelet-rich fibrin with the low-speed concept: Growth factor release, biocompatibility, and cellular response," our research team demonstrated that protocols at lower speeds and time not only led to significantly greater human fibroblast cell migration and proliferation compared to L-PRF but also demonstrated significantly higher mRNA levels of PDGF, TGF-β, and collagen1 at either 3 or 7 days (Fig 2-5). In addition, PRF membranes implanted subcutaneously that were fabricated using lower centrifugation speeds also generated greater and faster vascularization in vivo.[24]

Comparing Centrifugation Devices

One highly debated topic with much commercial interest has been the role of the centrifugation device on production of PRF. As such, there are general parameters that need to be respected to calculate relative centrifugal forces (RCF) effectively; this forms the basis of an entire chapter in this textbook (see chapter 4). While there are definitely quality differences between centrifugation devices, the debate between which centrifugation system to utilize is generally overhyped, leading to confusion among colleagues regarding how to accurately select, and more importantly, optimize a centrifugation system.

The debate surrounding which centrifugation system to utilize is generally overhyped, leading to confusion among colleagues regarding how to accurately select and, more importantly, optimize a centrifugation system.

As such, in 2018 our research team addressed this question in a study titled "Comparison of platelet-rich fibrin (PRF) produced using 3 commercially available centrifuges at both high (~ 700 g) and low (~ 200 g) relative centrifugation forces" with the specific aim to demonstrate that any centrifugation device could be utilized to produce PRF using the LSCC.[39] In that study, PRF was produced on three commonly utilized commercially available centrifuges including the IntraSpin (Intra-Lock), Duo Quattro (Process for PRF), and Salvin (Salvin Dental) devices (Fig 2-6). Two separate protocols were tested on each machine, including the original L-PRF protocol (~700g RCF-max [~400g RCF-clot] for 12 min) as well as the A-PRF+ protocol (~200g RCF-max [~130g RCF-clot] for 8 min). Each of the tested groups was compared for cell numbers, GF release, scanning electron microscopy (SEM) for morphologic differences, and clot size (both weight and length/width).

It was found that PRF clots produced utilizing the lower centrifugation speeds and time *(1)* contained a higher concentration of evenly distributed platelets, *(2)* secreted higher concentrations of GFs over a 10-day period, and *(3)* were smaller in size. This was irrespective of the centrifugation device utilized and consistently observed on all three devices. The greatest impact was found between the protocols utilized (up to a 200% difference). Furthermore,

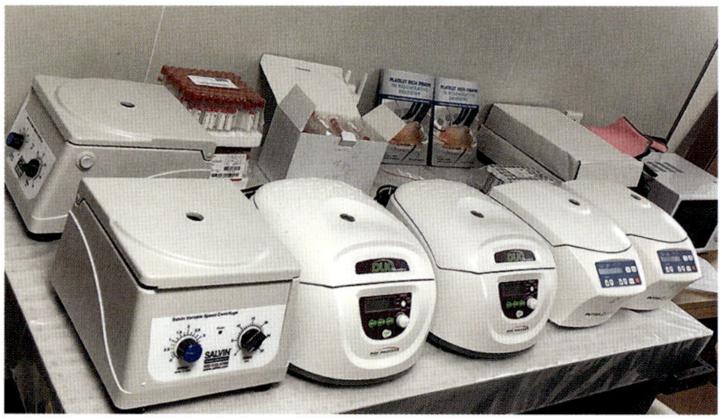

Fig 2-6 Experimental setup: Each of the centrifuges was utilized in duplicate for a total of six centrifuges. From each patient, a total of 18 tubes were drawn (2 per machine) and centrifuged accordingly at either high- or low-speed protocols. (Reprinted with permission from Miron et al.[39])

it was revealed that centrifugation carried out at lower speeds had a slightly less dense fibrin mesh with more open spaces to allow for cellular migration (Fig 2-7). Most importantly, it was revealed for the first time that *centrifugation tubes* actually had a much greater impact on the final size outcome of PRF clots when compared to centrifugation devices. This was the first time our team had observed such dramatic differences in the size outcomes of PRF tubes, and since then a plethora of research has since been performed on that topic alone (see chapter 5).

Research has clearly shown that PRF tubes matter much more than the centrifugation device used.

Development of i-PRF

One of the advantages of PRP is that it is liquid in nature, making it easy to utilize in combination with various bone biomaterials, most notably bone grafting materials. With PRF, as centrifugation speeds and times were being reduced further and further, a nonclotted liquid plasma layer was noticed prior to actual clot formation. This liquid-PRF layer is actually liquid fibrinogen that has not yet converted to fibrin. This liquid formulation of PRF was given the working name *liquid-PRF or injectable-PRF (i-PRF)* for simplicity (Fig 2-8a).[40] This layer can be quickly harvested (Fig 2-8b) and injected into a defect area. Interestingly, once liquid-PRF has converted to a solid state,

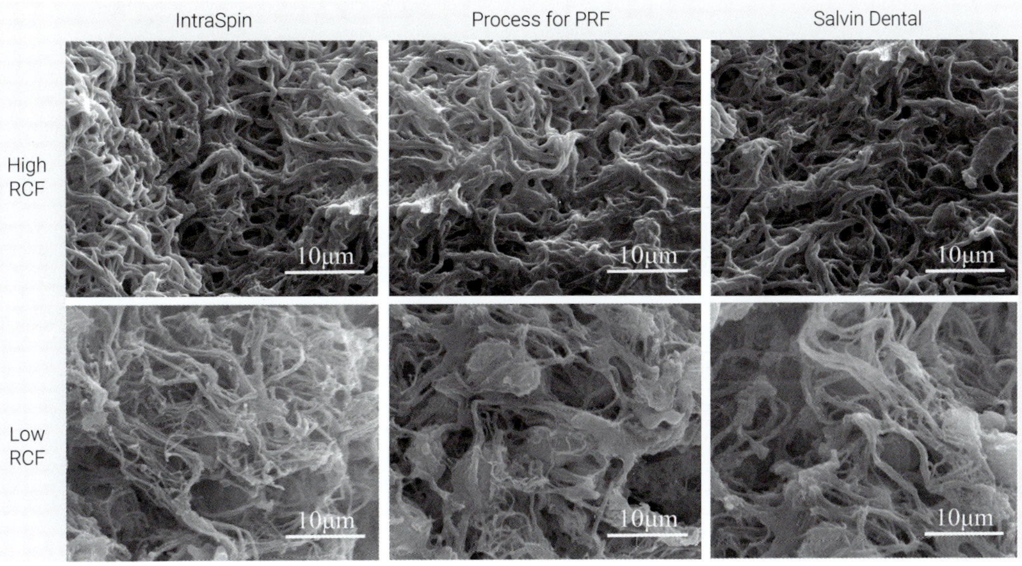

Fig 2-7 SEM of PRF clots produced on three different devices at either high-speed (~700g) or low-speed (~200g) protocols. Notice that the clots produced at high g-force typically were more densely packed with fibrin. (Reprinted with permission from Miron et al.[39])

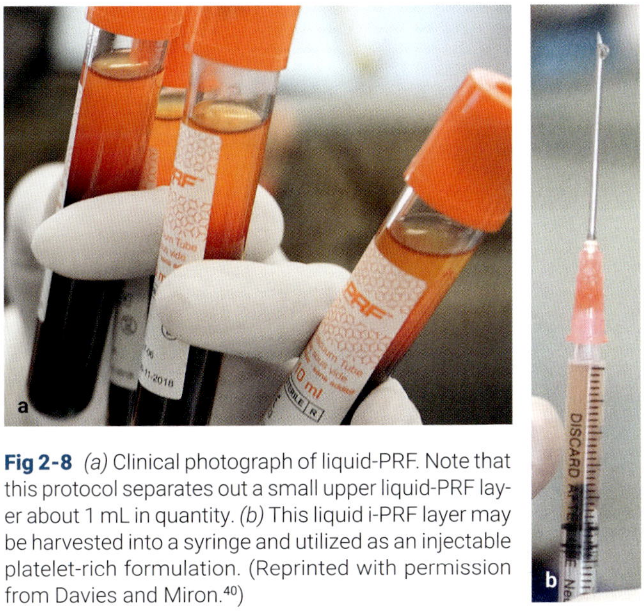

Fig 2-8 *(a)* Clinical photograph of liquid-PRF. Note that this protocol separates out a small upper liquid-PRF layer about 1 mL in quantity. *(b)* This liquid i-PRF layer may be harvested into a syringe and utilized as an injectable platelet-rich formulation. (Reprinted with permission from Davies and Miron.[40])

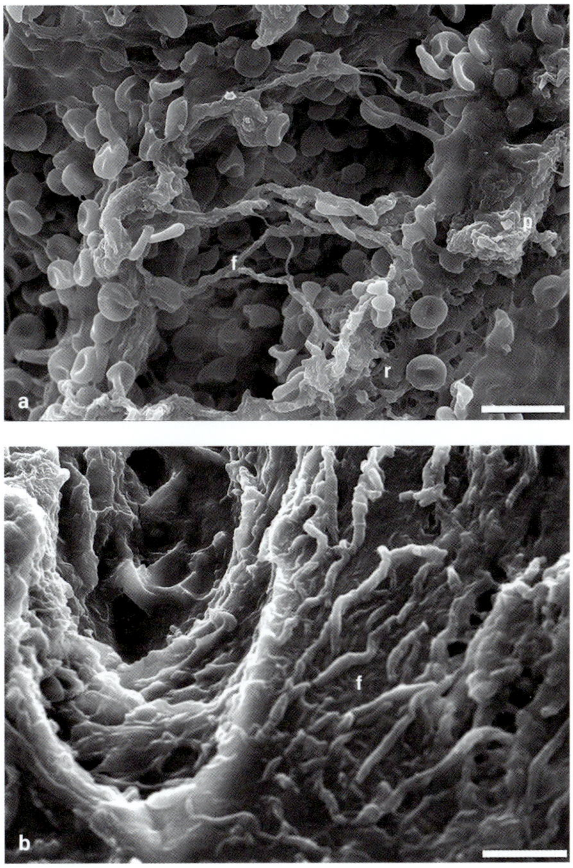

Fig 2-9 The surface *(a)* and cross-section *(b)* microstructures of the i-PRF. f, fibrin; p, platelet aggregates; r, RBC. Scale bar = 10 μm. (Reprinted with permission from Zhang et al.[41])

it forms the standard fibrin fibrillar PRF that most are familiar with, as depicted in Fig 2-9 by SEM.[41] Based on its potential for clinical applications, both preclinical and clinical studies have since been conducted to evaluate its regenerative potential.

Data from our laboratories first found that GF release from PRP was typically within the first hour, whereas i-PRF had a much more widespread release of GFs over time, similar to solid-PRF[42–44] (Fig 2-10). Unlike our previous studies comparing various solid-PRF membrane formulations,

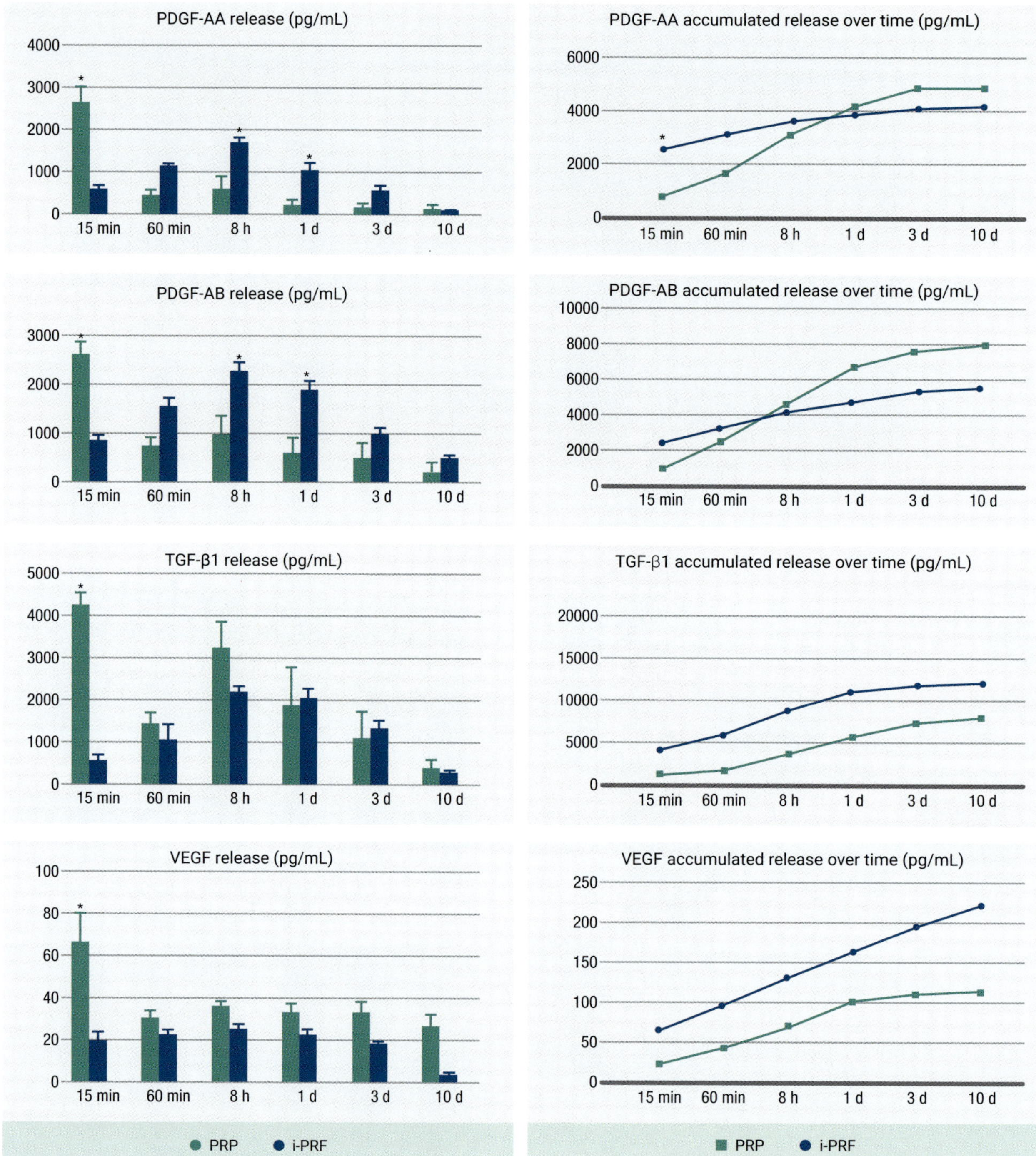

Fig 2-10 GF release from i-PRF compared with PRP at each time point for PDGF-AA, PDGF-AB, TGF-β1, and VEGF over a 10-day period. Note the varying GF release from PRP and i-PRF. Some GFs were in fact more highly released from PRP, which posed many questions several years ago. (Adapted from Miron et al.[44])

however, some GFs were in fact secreted in higher levels from PRP when compared to i-PRF, whereas others were more highly released from i-PRF. In 2015, i-PRF was developed and initially studied as a very short and slow centrifugation protocol of 700 rpm (60g) for 3 to 4 minutes. Following this protocol, i-PRF remained liquid for roughly 15 to 20 minutes. This new formulation was utilized for a variety of procedures including mixing with bone grafts to form a stable fibrin graft with improved handling and graft stability.

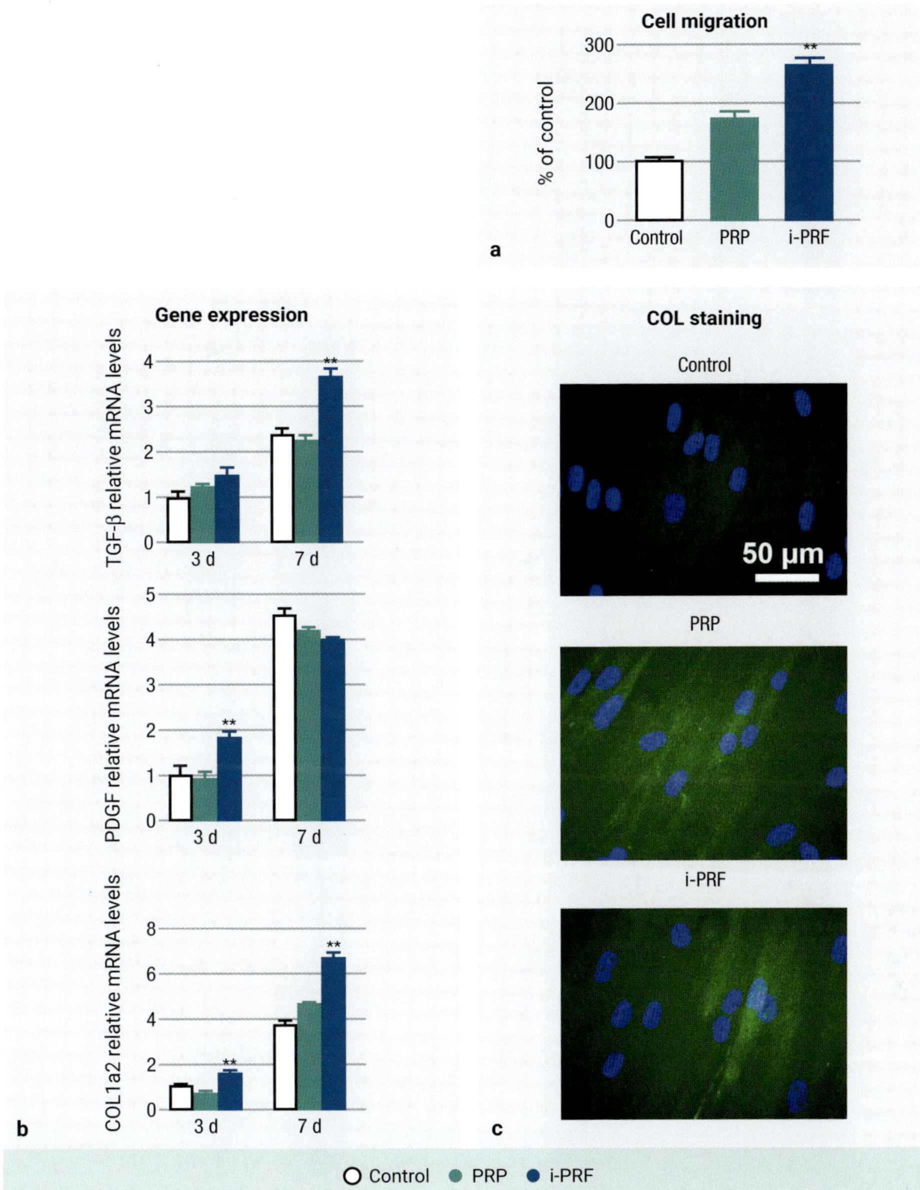

Fig 2-11 Human gingival fibroblast behavior exposed to i-PRF versus PRP: *(a)* Cell migration, *(b)* gene expression, and *(c)* collagen synthesis. (Adapted with permission from Miron et al.[44])

A variety of basic research studies have since demonstrated the regenerative potential of i-PRF when compared to PRP.[44–49] While both formulations exhibited high biocompatibility of human gingival fibroblasts as well as significantly induced higher cell migration when compared to control tissue-culture plastic in vitro (Fig 2-11),[44] it was found that i-PRF induced significantly greater cell migration, mRNA levels of TGF-β, and collagen1 expression.[45–49] It wasn't until years later that further means to improve i-PRF were developed, as highlighted later in this chapter.

Quantifying Cell Types in PRF

Pitfalls of current methods

Two pitfalls exist with quantifying cell types found within PRF research. Histologic studies have been infrequently performed and show variability and bias in results, because many times it becomes difficult to know the exact orientation of PRF relative to the histologic processing. It therefore becomes very difficult to assess where cells are actually

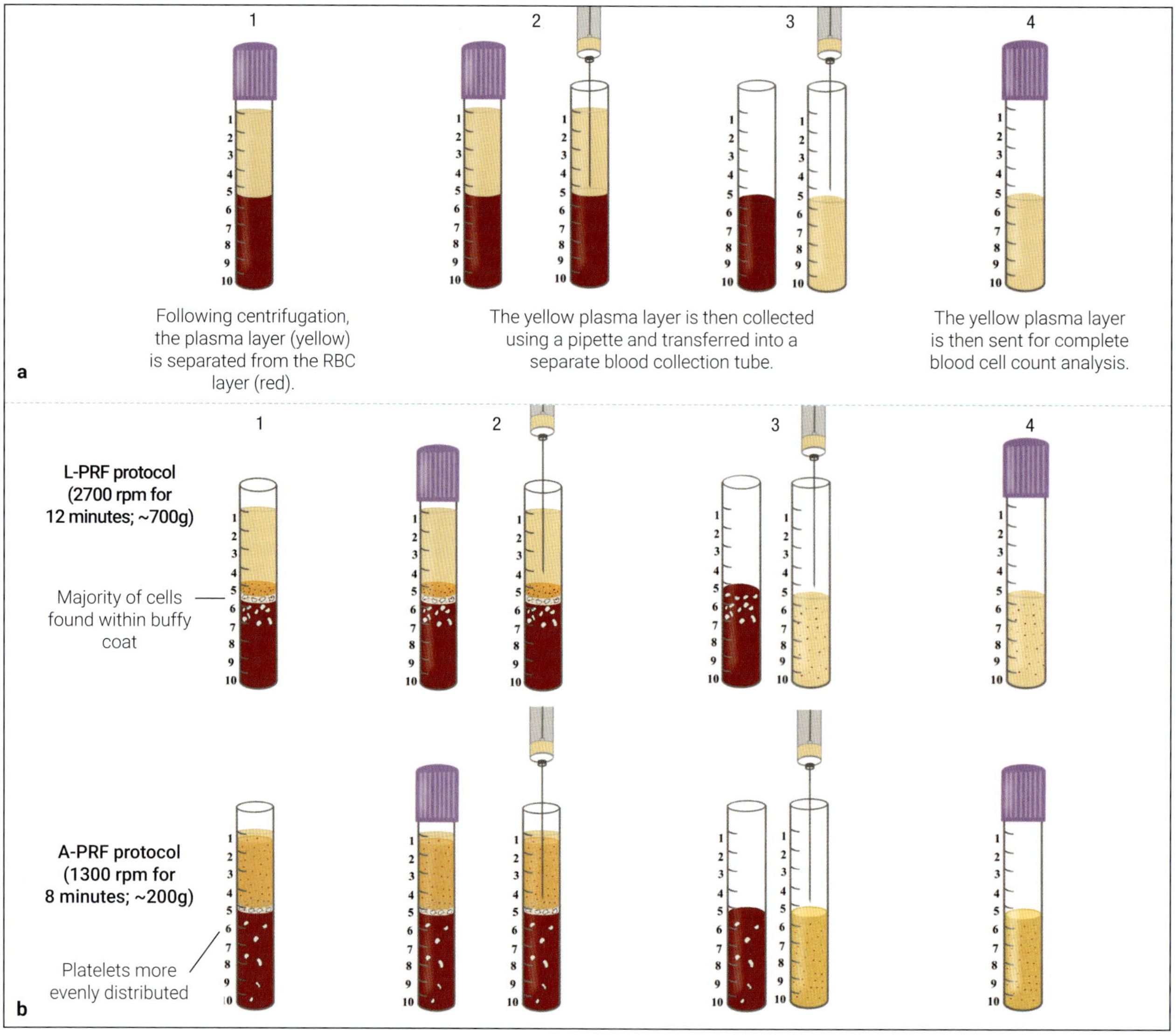

Fig 2-12 *(a)* One method to investigate cells found in PRP/PRF is to send the plasma layer following centrifugation for a full CBC. This CBC value can be compared to a control of whole blood to investigate its % increase. *(b)* One of the reported pitfalls of this technique is that the precise location of cells within the upper layer is not revealed. For instance, following the L-PRF protocol, the cells are gathered more precisely at the buffy coat zone, whereas the A-PRF protocol tends to more evenly distribute platelets. Therefore, this technique may not necessarily represent the most ideal scientific accuracy/data.

located following cell separation and histologic preparation. Another method commonly performed has been the use of complete blood counts (CBCs). Simply, following centrifugation, the plasma layer can be harvested and sent for a CBC and compared relative to whole blood (Fig 2-12a). While this accurately reports an increased ratio with respect to platelet and leukocyte numbers, one of the main drawbacks is that the method cannot determine if the cells are evenly distributed within the according PRF layers. As illustrated in Fig 2-12b, by utilizing this method, very similar final outcomes were found with only marginal improvements of roughly 20% visible between the two previously utilized protocols of L-PRF and A-PRF, yet histologically a much better distribution of platelets was observed by Ghanaati et al.[6] Therefore, a new method was proposed to better quantify cell types in platelet concentrates, as described next.

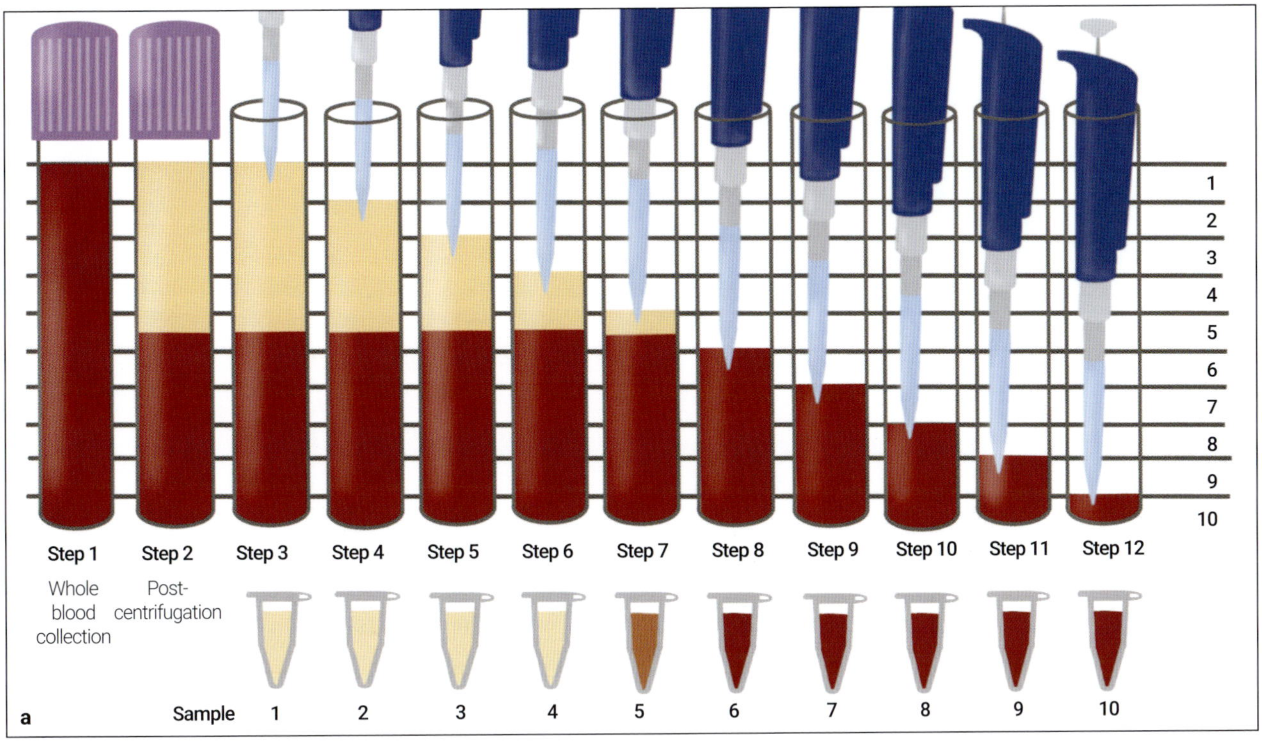

Fig 2-13 *(a)* Illustration demonstrating the proposed novel method to quantify cell types following centrifugation of PRF. Currently, one of the limitations is that whole blood is compared to the total plasma concentration following centrifugation. This, however, does not give a proper representation regarding the location of cells following centrifugation. By sequentially pipetting 1 mL of volume from the top layer downward, it is possible to send each of the 10 samples for CBC analysis and accurately determine the precise location of each cell type following centrifugation at various protocols. Notice that one layer (in this case layer 5) will contain some yellow plasma and RBCs. This is typically the location of the buffy coat, where a higher concentration of platelets is located.

A novel method to quantify PRF

Over the past few years, several commercially available centrifuges have further been brought to market. These vary in many factors including protocols, RCF values, tube-rotor angulation, rotor radius size, and tube composition. Each of these plays a role in the final obtained PRF membrane, yet little data is scientifically available displaying cell numbers and content in the various layers following centrifugation.

While much commercial debate exists on this topic, there has been no accurate method to quantify/determine the precise location of cells following centrifugation, with few histologic studies performed investigating cell numbers within the fibrin clots. In a pioneering research article published in 2019, our research team proposed a novel method to quantify cell numbers and concentration within the PRF scaffolds following centrifugation by utilizing a sequential pipetting methodology.[50] Standard protocols commonly utilized for the production of PRF were investigated according to their respective manufacturer's protocols. New to this research article, for the first time following centrifugation, 1-mL layers were sequentially pipetted from the upper layer of blood tubes toward the bottom of the tube until all 10 mL were harvested in sequential samples (Fig 2-13).[50] Each of these 10 samples from each centrifugation tube was then sent for CBC analysis to accurately quantify the cell numbers within each 1-mL blood layer and then compared according to cell numbers and concentrations. This study represented a novel experimental methodology to more accurately depict cell numbers in PRF following centrifugation using various protocols.

By utilizing this novel technique and investigating each 1 mL layer by layer, it was possible for the first time to investigate exactly each cell layer following centrifugation and determine the precise location of each blood cell type.

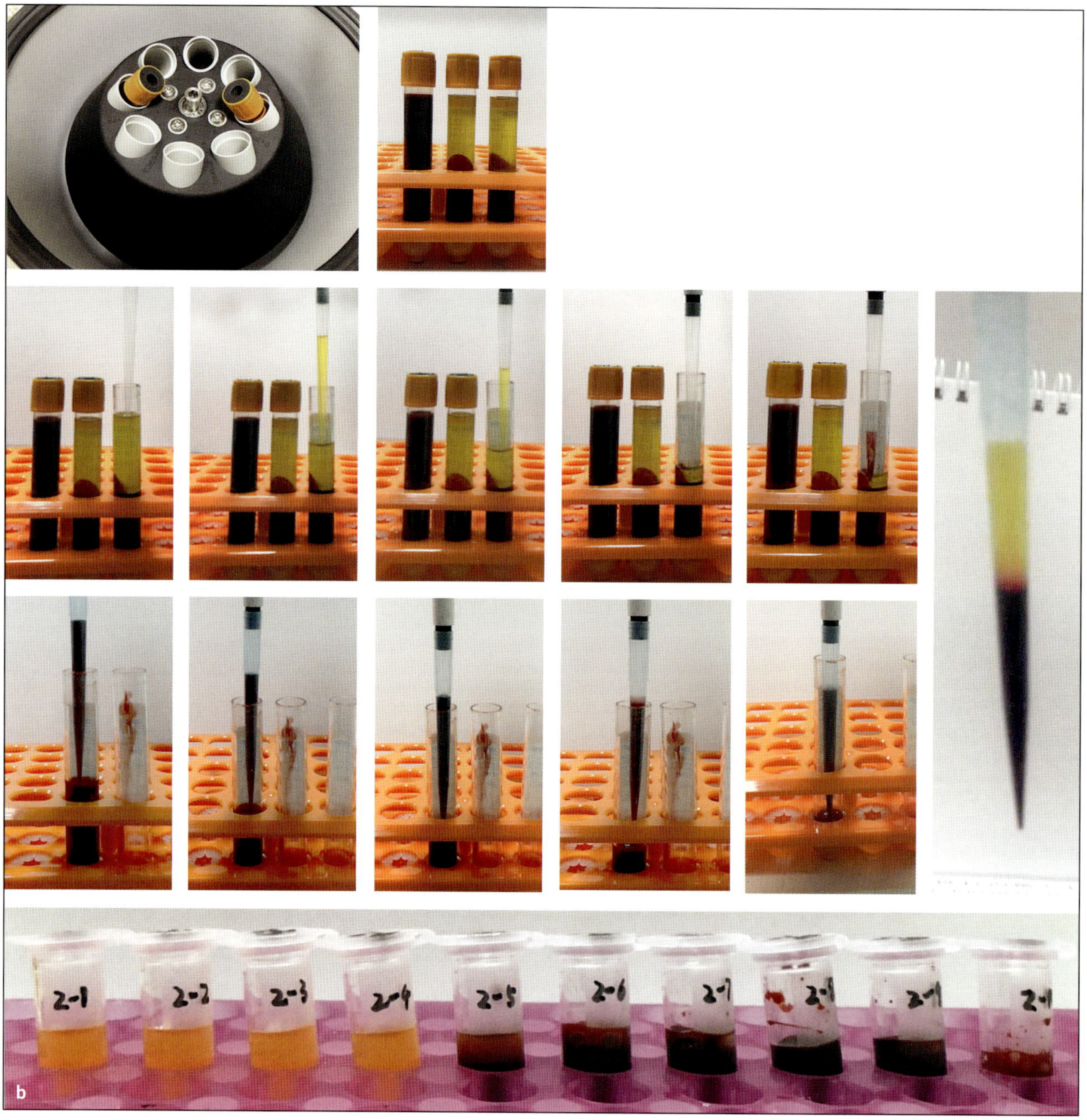

Fig 2-13 *(cont)* *(b)* Visual demonstration of the protocol. Following centrifugation with two 10-mL centrifugation tubes, blood layers are then separated. Thereafter, 1-mL samples are pipetted precisely from the upper layer downward. Notice that when layer 5 was drawn, it was possible to visualize the layer separation between the yellow plasma and RBC layers. This separation layer was noted for all samples. (Reprinted with permission from Miron et al.[50])

L-PRF protocol

Following centrifugation, 1-mL sequential layers were sent for CBC analysis according to Fig 2-14a. As Fig 2-14b shows, the original L-PRF protocol using the IntraSpin device (2700 rpm for 12 min; ~700g) with a 33-degree fixed-angle centrifuge revealed precisely that the number of leukocytes (control 6 × 10^9 cells/L) and platelets was significantly concentrated in layer 5 (~17 × 10^9 cells/L; *arrows* represent where the plasma and RBC layers separate). Interestingly, a threefold to fourfold increase in leukocyte number was observed specifically at this interface within the buffy coat. Notice, however, that no leukocytes

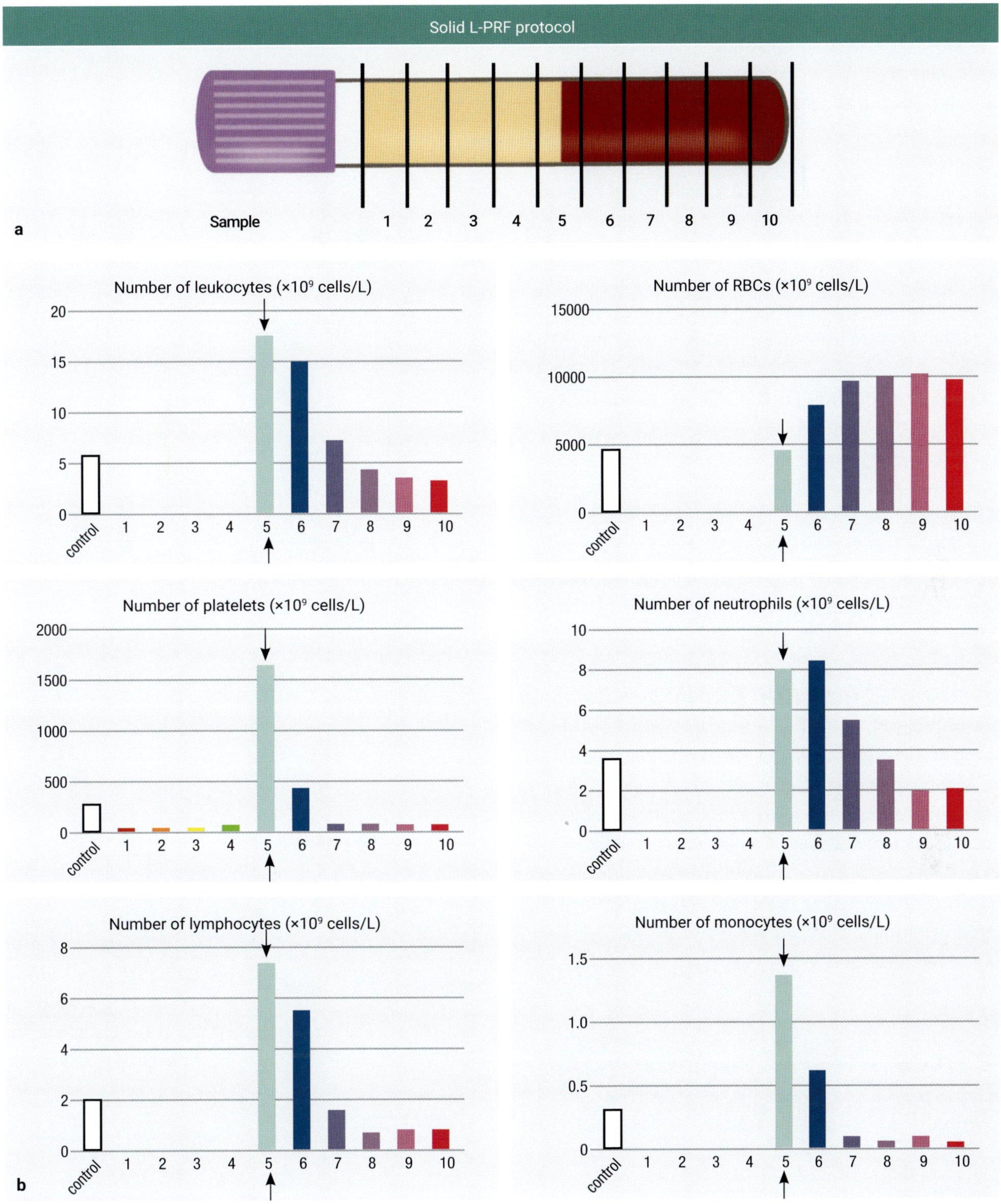

Fig 2-14 *(a)* Separation of the 10-mL tube into 10 1-ml layers for pipetting. *(b)* The concentration of cell types in each 1-mL layer utilizing the solid L-PRF protocol (2700 rpm for 12 minutes; ~700g). Notice that the majority of platelets accumulated directly within the 5th layer in the buffy coat. Furthermore, the highest concentration of leukocytes was also noted in this layer. The first 4 layers of this plasma layer were typically devoid of all cells. (Adapted with permission from Miron et al.[50])

were found in any of the upper 4 layers, displaying a very uneven PRF clot with respect to cell numbers. Almost all cells within the PRF clot were exclusively found within this 5th layer. Notice also that more leukocytes were found in the RBC layer below the PRF clot. A similar trend was also observed for lymphocytes, neutrophils, and monocytes (see Fig 2-14b). Naturally, all RBCs were found in layers 5 through 10 in the visually red layers. Platelets were accumulated once again precisely in layer 5 (six- to eightfold), within the buffy coat zone.

Because the majority of cells were found in layer 5, we were interested to determine if these cells were specifically found within the yellow plasma layer (within the PRF clot) or within the RBC layer. For this, a second blood tube was utilized; 500 μL of blood volume was collected just above the RBC layer within the buffy coat, and 500 μL was taken from the RBC layer. It was revealed that the majority of platelets were found within the yellow plasma layer (> 80%), whereas the majority of leukocytes and other WBCs were found within the red blood layer (Fig 2-15). This revealed that most leukocytes were in fact not found within the PRF layers utilizing the L-PRF protocol.

Interestingly, the final concentration of leukocytes found using the L-PRF protocol was 4.13 × 10^9 cells/L, whereas the control whole blood value from this patient was 6.125 × 10^9 cells/L, representing a 33% reduction in leukocyte concentration when compared to control blood. Platelet numbers were increased 1.61-fold. The total leukocyte and platelet content represent 33% and 80% of the total blood cells found, respectively, within this 10-mL blood sample. This meant that roughly 20% of platelets and 66% of leukocytes were actually located within the RBC layer (similar to the observed histologic results by Ghanaati et al in 2014[6]).

With the L-PRF protocol, the majority of leukocytes and platelets were not found within the plasma layer but rather in layer 5 within the buffy coat zone.

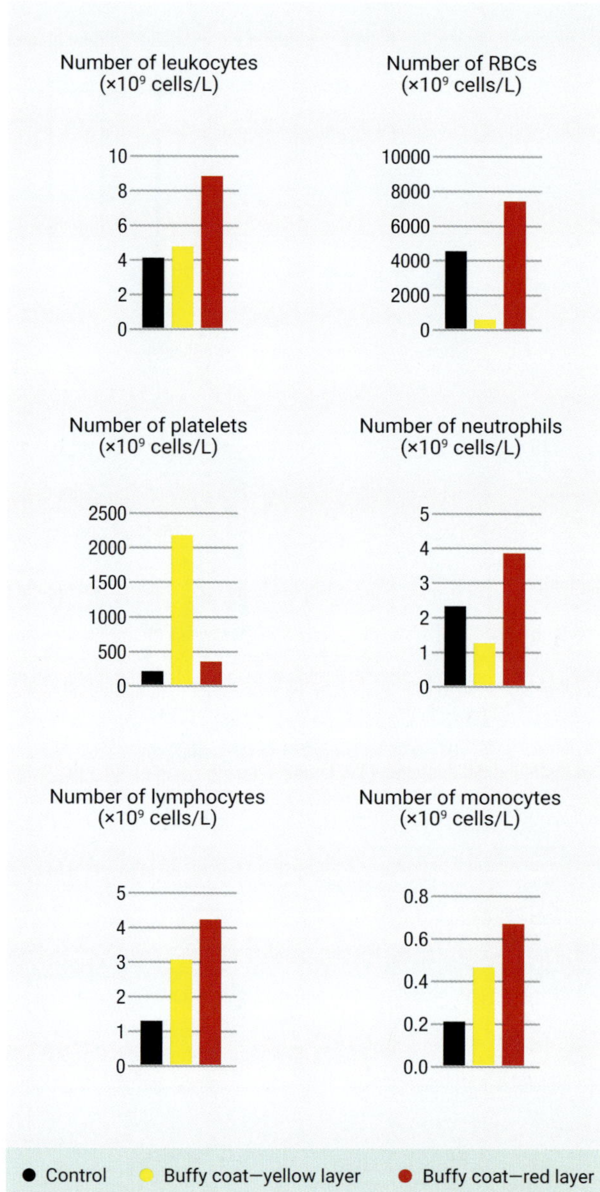

Fig 2-15 Layer 5 (the zone that incorporates the buffy coat containing a plasma and RBC component) demonstrated the cell-rich zone. Analysis of this zone revealed that many of the cells were in fact located in the red zone (especially leukocytes). The cell-rich zone contains a yellow buffy coat zone, but the red portion of this buffy coat also contains many cells. Following these findings, it is generally recommended to harvest a small portion of the red zone, specifically when drawing liquid-PRF (i-PRF), because many cells are located within this region. (Adapted from Miron et al.[50])

A-PRF protocol

Figure 2-16 depicts centrifugation following A-PRF protocols (1300 rpm for 8 minutes on a Duo Quattro centrifuge). Interestingly, the number of platelets were concentrated throughout the first four to five layers, unlike the L-PRF protocol. Here, a twofold increase in platelets was observed compared to a 1.6-fold increase utilizing the L-PRF protocol. More importantly, however, the platelets were found evenly distributed throughout the A-PRF plasma layers. When investigating leukocyte number, however, a significantly lower concentration (33% original values) as well as

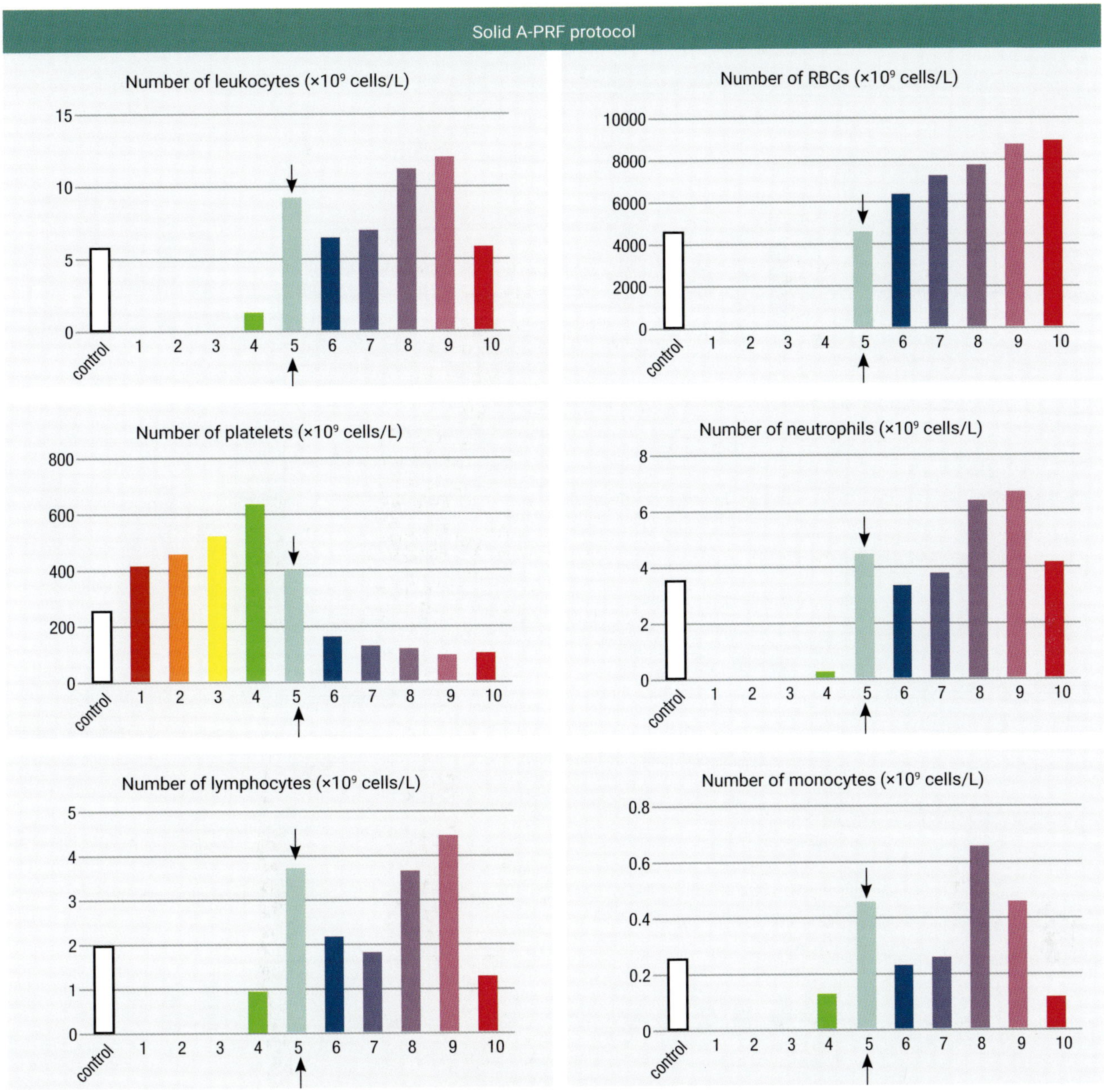

Fig 2-16 The concentration of cell types in each 1-mL layer utilizing the solid A-PRF protocol (1300 rpm for 8 minutes; ~200g). Notice that the platelets were more evenly distributed throughout the upper 5-mL plasma layer. Noteworthy, however, is that the majority of WBCs (leukocytes, neutrophils, lymphocytes, and monocytes) were not found in the upper plasma layer. (Adapted from Miron et al.[50])

total numbers (9.315 vs 20.65 × 10^9 cells/L) were found in the A-PRF group when compared to L-PRF. Therefore, it was initially suspected that either the g-force or the total time was not sufficient to adequately accumulate or separate the leukocytes utilizing the A-PRF protocol.[50] Once again, lower leukocytes in PRF were actually found when compared to whole blood.

While the A-PRF protocol with LSCC led to a higher concentration of platelets, it was not effectively capable of concentrating leukocytes.

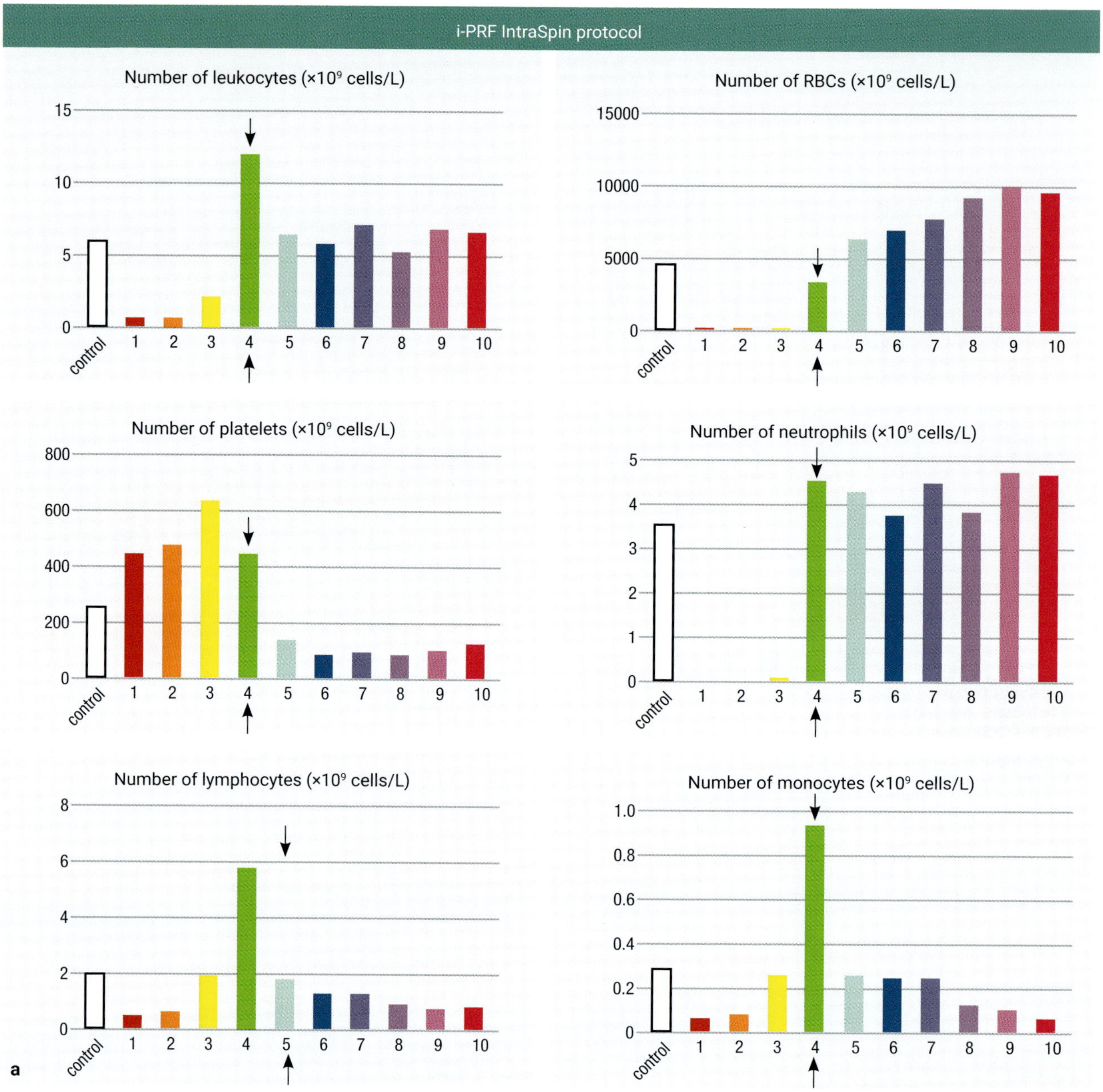

Fig 2-17 *(a)* The concentration of cell types in each 1-mL layer utilizing the i-PRF IntraSpin protocol (2700 rpm for 3 minutes; ~700g). Notice that most platelets are more evenly distributed utilizing this protocol when compared to the 12-minute solid-PRF IntraSpin protocol.

i-PRF protocol

Liquid-PRF protocols were then investigated and compared (Fig 2-17). The IntraSpin protocol (2700 rpm for 3 min; ~700g) is depicted in Fig 2-17a. Interestingly, this protocol accumulated platelets evenly throughout the PRF layer better than when utilizing the 12-minute protocol. Nevertheless, leukocytes were significantly lower once again when compared to whole blood, representing only 54% of the original control blood concentrations. This demonstrates that following centrifugation, *lower* numbers of leukocytes are found in L-PRF samples when compared to control blood in either L-PRF protocol. Platelet concentrates were increased 2.12-fold.

The i-PRF protocol recommended by Process for PRF (Duo Quattro centrifuge) produced a 1.23-fold increase in leukocyte concentration and a 2.07-fold increase in platelet concentration when compared to whole blood (see Fig 2-17b). The overall

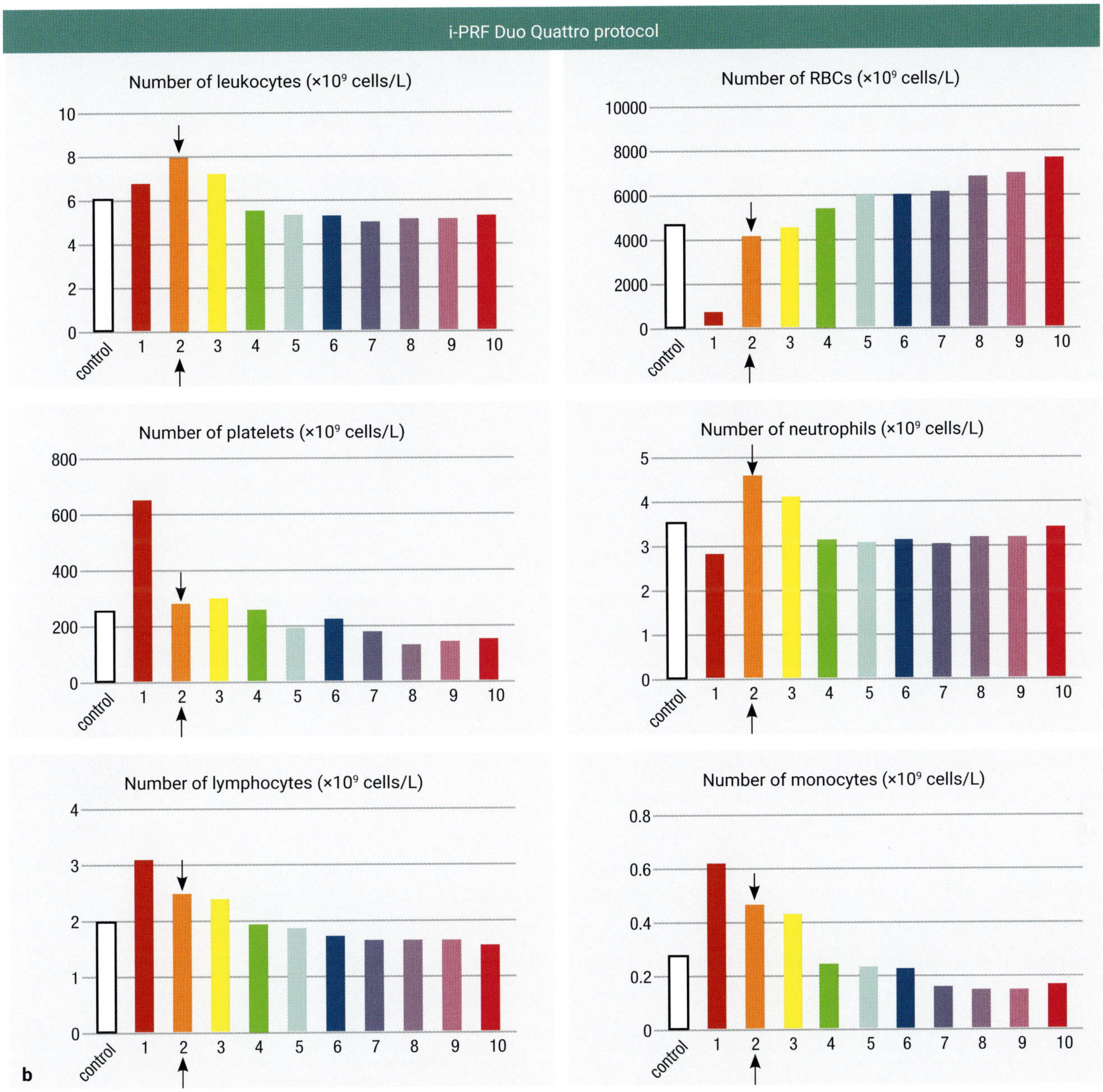

Fig 2-17 *(cont)* *(b)* The concentration of cell types in each 1-mL layer utilizing the i-PRF Duo Quattro protocol (800 rpm for 3 minutes; ~60g). Notice that very little change in platelet or leukocyte accumulation is observed utilizing this centrifugation cycle. A slight increase in platelets and leukocytes is, however, observed when compared to the control. (Adapted from Miron et al.[50])

accumulation demonstrated an 18% total leukocyte content and a 31% total platelet count when compared to whole blood. This represented an extremely low platelet yield, as all other protocols produced at least 80% total yield. (Keep in mind here that this means 70% of platelets are found within the red layer following the use of this LSCC and not in the upper, liquid-PRF layer.) Most notably, the change in cell density layer by layer, as depicted in Fig 2-17b, was almost unnoticeable. These findings revealed that the i-PRF protocol displayed an inability to concentrate cells effectively, and it was clear that improvements were needed.

The use of the original i-PRF protocol only accumulated on average 18% of leukocytes and 31% of platelets. It was clear improvements were needed.

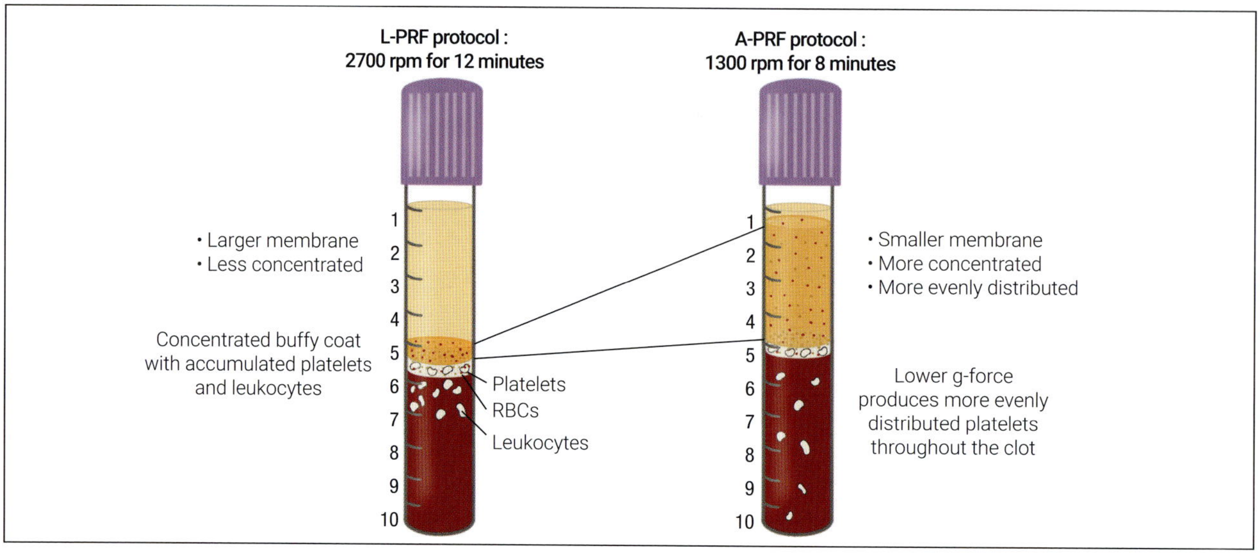

Fig 2-18 Summary of the findings comparing L-PRF and A-PRF protocols. While neither was typically able to collect leukocytes (reviewed later in the chapter), the lower centrifugation speeds using the A-PRF protocol allowed for the more even distribution of platelets in the upper layers. (Reprinted with permission from Miron et al.[52])

Discussion of findings

One of the most surprising findings was the observation that almost all platelets were accumulated in layer 5 using the conventional L-PRF protocols. Almost no platelets were observed in the first four layers following centrifugation, and the majority of leukocytes were found in the RBC layer, not in the PRF clot. This was a bit ironic, granted the working name *leukocyte platelet-rich fibrin* (as in, leukocyte-rich and platelet-rich fibrin). Previous studies have also shown that L-PRF protocols result in lower platelet and leukocyte numbers when compared to various other protocols produced at lower RCF values.[6,51]

> *It was certainly ironic that low levels of leukocytes were found in the PRF clot of L-PRF when compared to normal blood, considering the working name leukocyte PRF, which implies that the PRF clot would be leukocyte rich.*

While platelet concentrations were also lower in L-PRF vs A-PRF, it was revealed that the actual differences were not as drastic as previously reported.[51] Our study demonstrated more specifically that the platelets and leukocytes are in fact found *precisely* at the junction between the yellow plasma and RBC interface. Previous studies likely failed to collect *all* the liquid within the yellow plasma layer, and as a result, extremely low platelet and leukocyte values may have been reported. As clearly shown in the present study, failure to do so, especially at a g-force of ~700g (RCF-max) or greater, results in extremely low concentration values because the upper 4 mL of plasma is practically devoid of cells. Therefore, we demonstrated in this study that L-PRF protocols are in fact quite rich in platelets (~80–90% total), but these cells are found precisely within a 1-mL layer directly above the RBCs within the buffy coat. It also demonstrates the effectiveness of the present methodologic protocol for evaluating PRF protocols.

In this study, the A-PRF protocol resulted in a more evenly distributed upper platelet layer throughout the PRF plasma layer, further validating the LSCC (Fig 2-18). Tubes at lower g-forces centrifuged for less time consistently resulted in a better distribution of platelets throughout the PRF matrix in the upper 4 to 5 mL, whereas an uneven distribution of cells was found using the 700g by 12-minute protocol. It is therefore clinically recommended to avoid utilizing original L-PRF protocols for membrane fabrication because all the cells are entirely found in a thin layer at the base of the PRF clot. However, low leukocyte yields were observed utilizing these protocols.

This study also led to the observation that the manufacturer's recommended protocol for i-PRF (~60g for 3 minutes) was not adequately effective at separating cell types or producing high yields of platelets/leukocytes. Figure 2-17 demonstrates minimal change in cell layer changes following this short centrifugation cycle at low RCF values. Based on the data obtained within this study, a paradigm shift in our understanding of platelet concentrates with respect to the LSCC was noted. We now know that too-low RCF values/times will produce ineffective separation of blood layers, as demonstrated in these i-PRF protocols.

Based on these findings, we now know that too-low RCF values/times will produce ineffective separation of blood layers.

Pitfalls in i-PRF protocols

Several interesting findings were observed more recently with respect to the original i-PRF protocols. In 2016, when the first PRF textbook was written, Miron and Choukroun wrote the following: "Interestingly however, total growth factor release of PDGF-BB, VEGF, and TGF-β1 were significantly higher in PRP when compared to i-PRF. Methods to further understand these variations are continuously being investigated in our laboratory as well as others. It may be hypothesized that the differences in spin protocols are suggested to have collected slightly different cell populations and/or total growth factors responsible for the variations in release over time."

Back then, it remained puzzling to our research team why higher GF content was found in PRP despite the protocols utilized. Based on the results from the newly developed quantification method, it was revealed very clearly that the reason for these lower levels of GFs released from i-PRF was in fact owing to its inability to fully shift cells to their correct blood layers because centrifugation was carried out at too low a speed/time. Figure 2-18 demonstrates that only roughly 30% of total platelets are in fact accumulated in i-PRF, with 70% remaining in the lower RBC layer.[52] Furthermore, a separate study conducted by an independent group found only minimal improvements in platelet numbers following i-PRF protocols (less than 50% increase), with an actual reduction in leukocyte concentration as well as VEGF release.[53] Therefore, it must clearly be noted that with respect to the LSCC, data has now demonstrated that it is definitely possible to centrifuge too slowly and too little a time for effectiveness. It became clear that improvements could be made to these i-PRF protocols.

Optimization of i-PRF into C-PRF

Based on our findings that following L-PRF protocols, the majority of cells were massively accumulated at the buffy coat directly above the RBC layer (within 1 mL) with very few cells found throughout the upper four 1-mL layers,[50] we hypothesized that if we could specifically collect this 1-mL layer, we would create a much more liquid-PRF formulation rich in cells and GFs (Fig 2-19). In a study titled "A novel method for harvesting concentrated platelet-rich fibrin (C-PRF) with a 10-fold increase in platelet and leukocyte yield,"[52] we addressed two specific questions: *(1)* In what total volume were the majority of these cells located above the RBC layer within the buffy coat? and *(2)* What final concentration could be harvested by collecting only the cells found within this precise buffy coat region when compared to conventional i-PRF protocols?

Unlike the previous study, we decided to quantify PRF using 100-μL sequential layers (ie, 0.1 mL) to precisely investigate the exact location of cells (Fig 2-20). Because we had previously observed a massive cell accumulation within the buffy layer in a 1-mL sample range following the L-PRF protocol, we aimed to investigate precisely the volume in which these cells were located within this 1-mL layer. As such, we developed a novel methodologic approach whereby 100-μL sequential layers were pipetted starting from about 1.2–1.5 mL above the buffy coat down to the RBC layer (see Fig 2-20, depicted as +1 to +12 layers). Additionally, three layers were harvested within the RBC layer to determine the number of cells incorporated within this layer as well (see Fig 2-20, depicted as −1 to −3 layers). Each of these 100-μL layers was sent for CBC analysis.

The second tube from each group was utilized to determine the final concentration from the liquid version of the i-PRF yellow plasma layer. For L-PRF protocols, one tube was utilized to harvest 0.5 mL of a C-PRF (defined as the 0.5-mL buffy coat directly above the RBC layer). This layer was termed *concentrated-PRF* in reference to the harvesting of this concentrated buffy coat layer (see Fig 2-19). Similarly, 0.3 mL of C-PRF liquid was harvested from this layer as well.

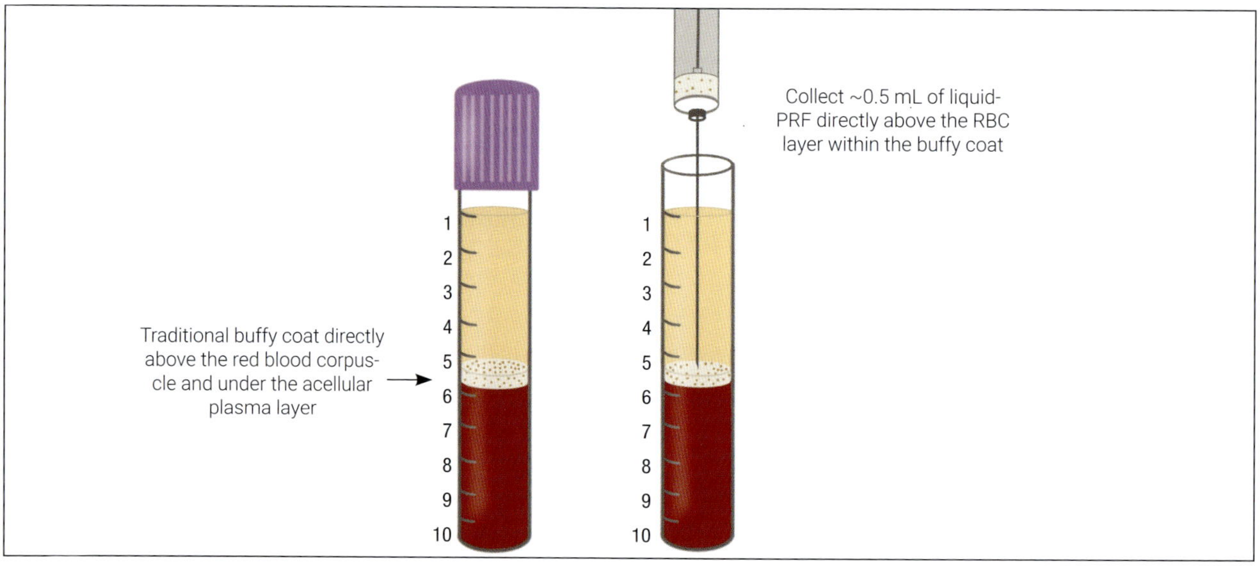

Fig 2-19 Proposed method to harvest C-PRF. Based on the finding that following L-PRF protocols all cells are accumulated within 300–500 μL above the RBC, this proposed method would collect this 0.3–0.5 mL of liquid C-PRF directly above the RBC junction for a highly concentrated liquid of platelets, leukocytes, and monocytes. (Reprinted with permission from Miron et al.[52])

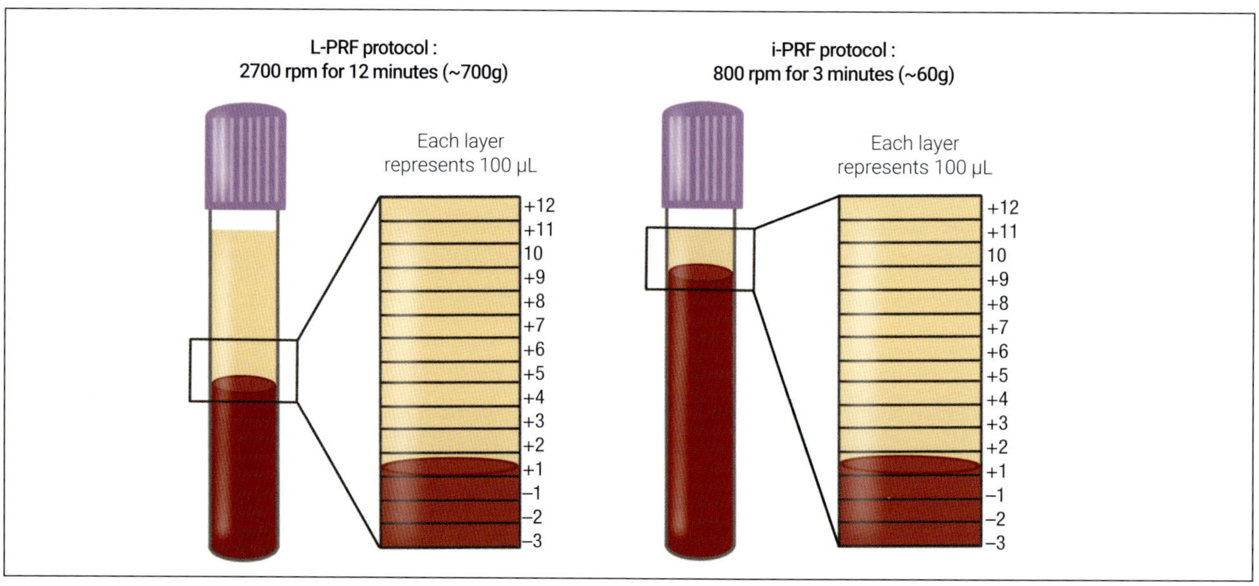

Fig 2-20 A second methodologic illustration depicting the sequential harvesting technique. Because the majority of cells accumulated within the 1-mL within the buffy coat following L-PRF protocols, we sought to investigate precisely the total volume of liquid (mL) above the buffy coat in which cells are concentrated. For the L-PRF protocols, 3.5 mL were removed followed by sequential 100-μL layers pipetted followed by CBC analysis. Three layers in the RBC layer were also harvested. In comparison, all plasma layers of the i-PRF protocol were also harvested in 100-μL sequential layers. Three RBC layers (100 μL each) were also collected. (Reprinted with permission from Miron et al.[52])

To address our first question regarding the precise volume in the buffy coat in which the increased cell numbers were observed, the first 3.5 mL of the upper plasma layer was removed (acellular layer) from the centrifugation tube (leaving 1.0–1.5 mL of remaining sample above the RBC layer). The sequential pipetting methodology was then utilized with 100-μL layers to accurately determine up to what layer above the RBC the cells were precisely located (see Fig 2-20). Furthermore, 300 μL within the RBC layer was also harvested and quantified in 100-μL sequential layers. In contrast, the entire i-PRF layer was collected starting from the upper 100-μL layer and

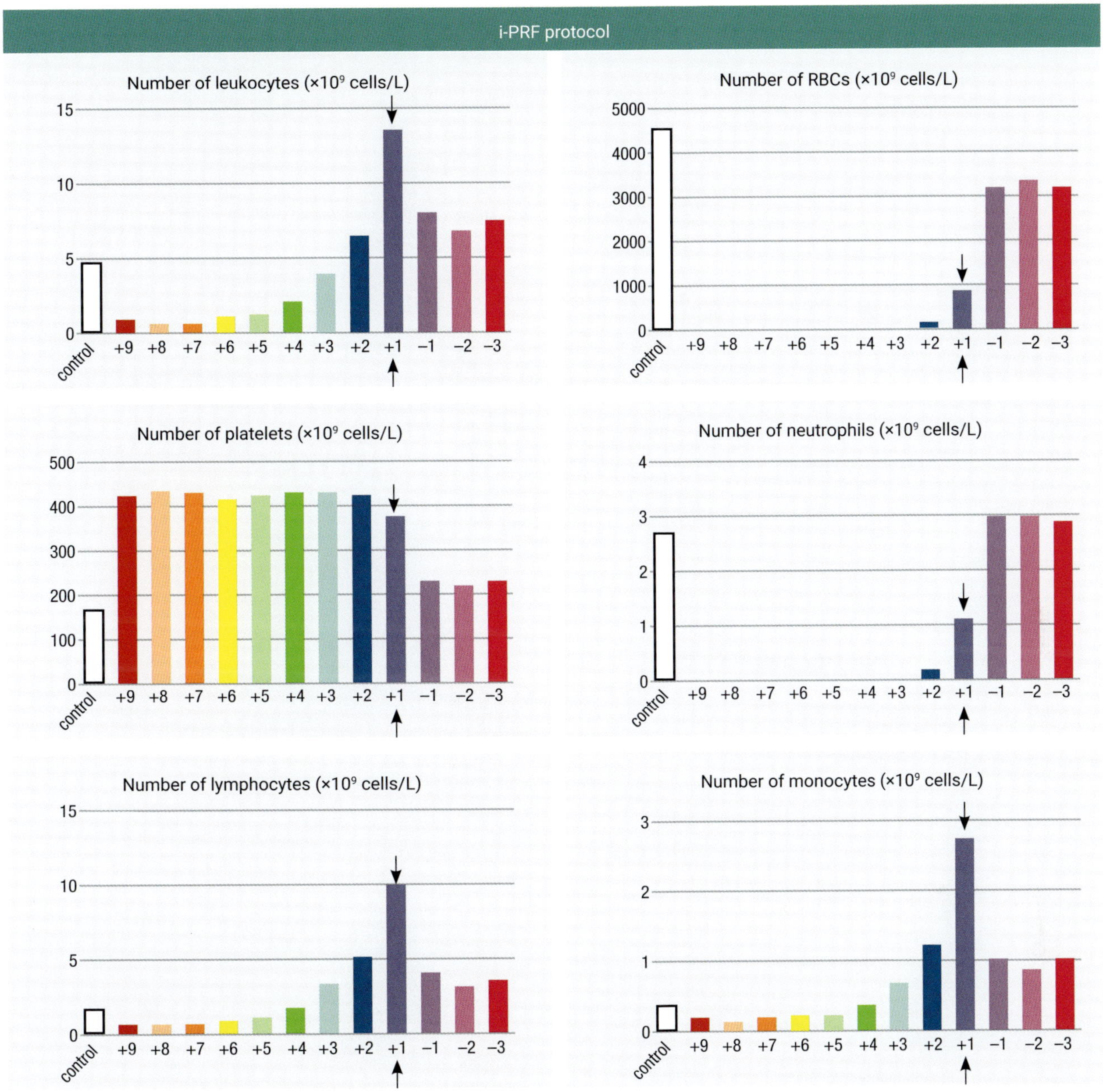

Fig 2-21 The concentration of cell types in each 100-µL layer utilizing the i-PRF protocol (800 rpm for 3 minutes; ~60g). Notice the significant increase in leukocytes and monocytes in the buffy coat layer (+1; *arrows*). (Adapted from Miron et al.[52])

sequentially pipetted until all plasma layers were collected. Once again, 300 µL was sequentially pipetted in 100-µL layers from the RBC layer.

Figure 2-21 demonstrates the results following sequential pipetting of 100-µL layers of the i-PRF protocol. Notice how specifically in layer +1, a threefold increase (from 5 to 15 × 10^9 cells/L) is found in leukocytes directly at the buffy coat layer (represented by *arrows*). Notice also the five- to sixfold increase in monocytes. The RBCs begin to accumulate at layer +1, and by layer −1 the sample is within the RBC layer. The remaining WBC and platelet levels drop within layer −1 after the yellow-red transition (see Fig 2-21). Following the i-PRF protocol, we observed a 2.5-fold increase (from baseline ~220 to ~550 × 10^9 platelets/L) in platelets in the top 13 layers (1.3 mL) following the i-PRF protocol and only a slight increase in leukocytes.

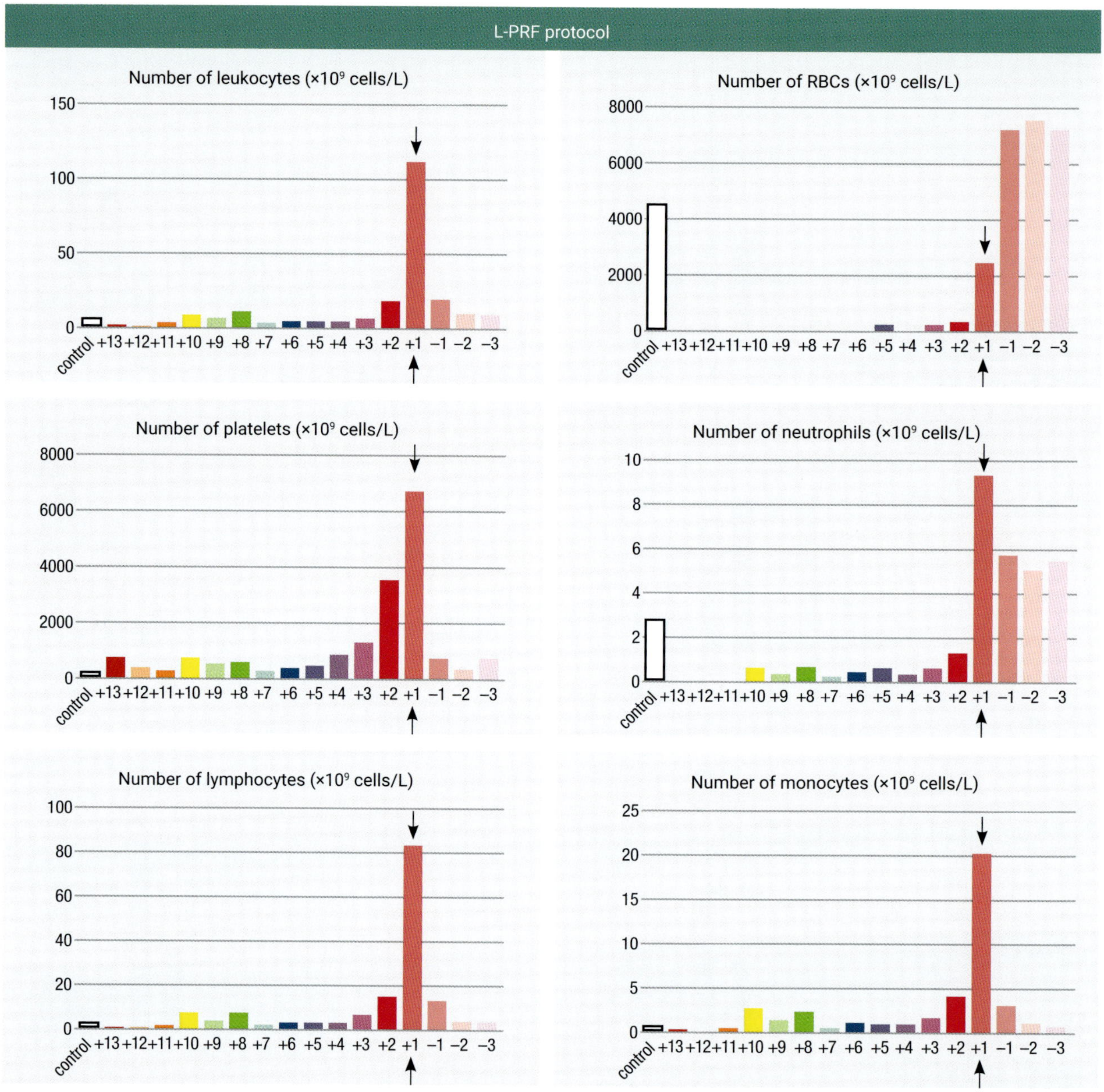

Fig 2-22 The concentration of cell types in each of the layers (100 μL each) above the RBC layer following the L-PRF protocol. Notice the massive increase in platelets (roughly a 20-fold increase) specifically at the buffy coat layer between the yellow plasma and RBC layers. Interestingly, all cells seemed to accumulate within the three to five layers (300–500 μL) above the RBC layer. (Adapted from Miron et al.[52])

Figure 2-22 demonstrates the results following sequential pipetting of 100-μL layers of the L-PRF protocol. Interestingly, almost all the cells accumulate within the three layers (ie, 300 μL) above the RBC layer. Most surprisingly, within this layer a massive increase in platelets, monocytes, leukocytes, and lymphocytes was observed. For instance, a roughly 225 to 6,000 × 10^9 platelets/L increase was observed, representing a > 25-fold increase in platelet concentration, specifically 100 μL above the RBC layer.

Based on these results, we assumed that a 0.3- to 0.5-mL layer of C-PRF could be preferentially collected within this buffy coat directly above the RBC layer. Figure 2-23a demonstrates that while the i-PRF protocol increases leukocyte numbers 1.23-fold, a marked and significant

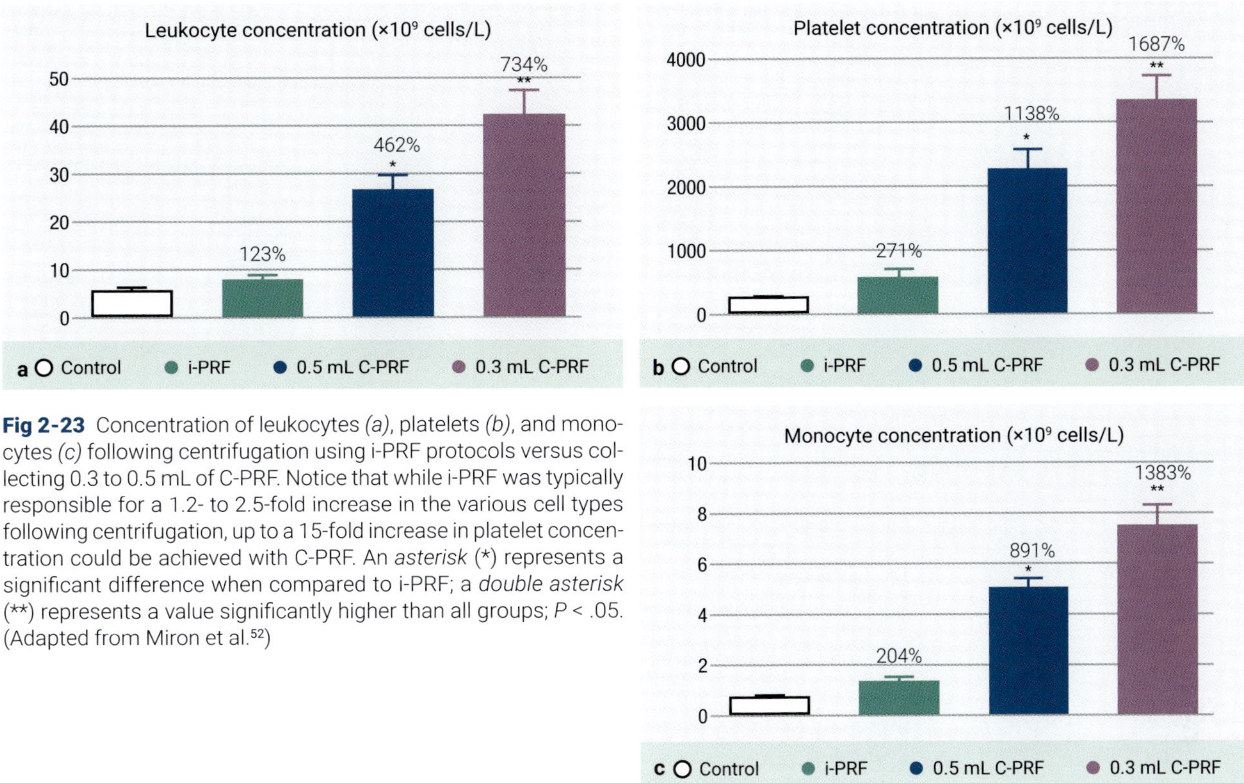

Fig 2-23 Concentration of leukocytes *(a)*, platelets *(b)*, and monocytes *(c)* following centrifugation using i-PRF protocols versus collecting 0.3 to 0.5 mL of C-PRF. Notice that while i-PRF was typically responsible for a 1.2- to 2.5-fold increase in the various cell types following centrifugation, up to a 15-fold increase in platelet concentration could be achieved with C-PRF. An *asterisk* (*) represents a significant difference when compared to i-PRF; a *double asterisk* (**) represents a value significantly higher than all groups; $P < .05$. (Adapted from Miron et al.[52])

Table 2-2 Leukocyte, platelet, and monocyte concentrations in whole blood compared to i-PRF and C-PRF

	Whole blood	i-PRF	0.5 mL C-PRF	0.3 mL C-PRF	i-PRF % increase	0.5 mL C-PRF % increase	0.3 mL C-PRF % increase
Leukocytes	6.0	7.8	26.7	42.4	123.8	461.7	733.7
Platelets	206.3	586.0	2327.9	3437.6	270.6	1138.2	1687.3
Monocytes	0.5	1.1	4.6	7.1	203.8	891.5	1383.4

The final three columns represent the percent increase in leukocyte, platelet, and monocyte values when compared to baseline controls. Notice that while i-PRF protocols are able to achieve a roughly 1.2- to 2.7-fold increase in cell types, C-PRF protocols, especially the 0.3-mL C-PRF group, are able to massively concentrate leukocytes (> 7-fold), platelets (> 16-fold), and monocytes (> 13-fold) when compared to controls. (Reprinted with permission from Miron et al.[52])

increase representing a 4.62- and 7.34-fold increase was observed with both 0.5 mL and 0.3 mL of C-PRF, respectively. Even more pronounced is that while i-PRF protocols have typically been shown to increase platelet yields between 200% and 300%, the C-PRF protocols massively increased platelet yields 1138% and 1687%, respectively (Fig 2-23b). A similar trend was also observed for monocytes (Fig 2-23c). The total values following averages from six patients are summarized in Table 2-2.

Chapter 3 discusses this new method of concentrating liquid-PRF directly from the buffy coat layer in more detail and reports on the marked improvement in cellular activity when compared to original i-PRF protocols.

Horizontal Centrifugation of PRF

Until 2019, the majority of centrifugation carried out for the production of PRF was performed on fixed-angle centrifuges. Horizontal centrifugation, on the other hand, is utilized frequently in research laboratories and in medical hospitals due to its superior ability to separate layers based on their density (Fig 2-24). In fact, the original PRP systems that date back 20 years were brought to market utilizing horizontal centrifugation for this very reason[54] (see chapters 3 and 4). In an attempt to investigate cell layer separation utilizing horizontal centrifugation, the same layer-by-layer sequential pipetting (1 mL each) was utilized as previously done with the L-PRF and A-PRF protocols.

Figure 2-25 depicts a 700g force performed on a horizontal centrifuge for 8 minutes (the selection of this protocol is presented in chapter 3). Interestingly, it was observed that more leukocytes, platelets, lymphocytes, and monocytes were much more evenly distributed throughout the PRF layers when compared to fixed-angle centrifugation. Unlike either the L-PRF or A-PRF protocols, a general increase in leukocyte numbers was observed (127% original values), and a 2.4-fold increase in platelet concentration was observed. This represents over a fourfold increase in leukocytes when compared to A-PRF protocols and a twofold increase when compared to L-PRF. Furthermore, this method concentrated 99.7% of all platelets and 53% of all leukocytes within the plasma layer.[50]

Within that study, horizontal centrifugation was proposed as a means to better separate cell layers based on density (Video 2-3). Two advantages were noted utilizing horizontal centrifugation. First, a completely horizontal tube produced from a swing-out bucket allows for the greatest differential between the minimum and maximum radius found within a centrifugation tube (see Fig 2-24a). This allows for a greater ability to separate cell layers based on disparities between the RCF-min and RCF-max produced within a tube. Second, a fixed-angle centrifuge results in more trauma to cells. Because centrifugation typically pushes cells outward and downward, cell layer separation is always observed in an angulated fashion toward the back distal surface of PRF tubes using fixed-angle centrifugtion (see Fig 2-24b). Furthermore, during the centrifugation process, cells are pushed toward the outer wall and then typically migrate either up or down the centrifugation tube based on density. Larger cells (such as RBCs) entrap smaller cells such as platelets during the centrifugation process and drag them downward along the back centrifugation wall into the RBC layer as a result of this cell accumulation against the back wall (see Fig 2-24c). In contrast, PRF produced via horizontal centrifugation and separation allows the direct flow-through of cells (see Fig 2-24c).

Video 2-3

Therefore, horizontal centrifugation allows cells to migrate freely throughout the blood layers. This allows for better separation of cell types (along with the greater differentiation in RCF values between RCF-min and RCF-max), resulting in higher final concentrations of desired cells (platelets and leukocytes) within the appropriate final blood layers. Furthermore, cells are less likely to suffer potential damage along the back wall of centrifugation tubes produced using high g-forces (~200–700g) following fixed-angle centrifugation. We therefore introduced this concept as "gentler centrifugation," whereby cells are more freely able to separate between layers without the necessary friction produced on the back wall of fixed-angle centrifuges such as those produced on the fixed-angle IntraSpin and Process for PRF devices.[50] This concept has been expanded with a series of research investigation launched to better optimize PRF. This is presented in greater detail in chapter 3 of this textbook.

During the centrifugation process on fixed-angle centrifuges, cells are pushed toward the outer distal wall and then typically migrate either up or down the centrifugation tube based on density. Larger cells (such as RBCs) entrap smaller cells such as platelets during the centrifugation process and drag them downward along the back centrifugation wall. These platelets and leukocytes don't make it to the upper PRF membrane.

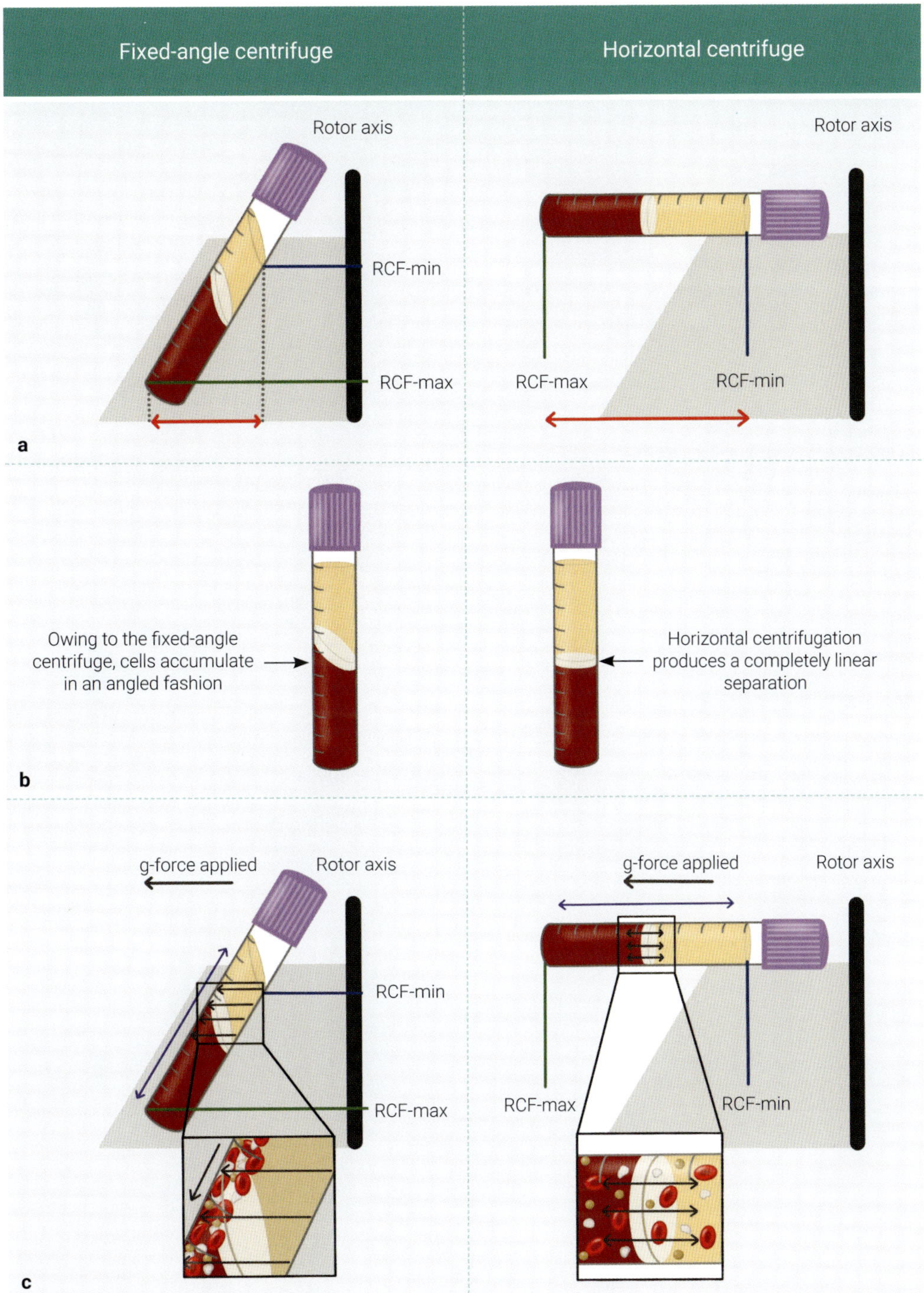

Fig 2-24 Illustrations comparing fixed-angle and horizontal centrifugation. *(a)* With fixed-angle centrifuges, a greater separation of blood layers based on density is achieved owing to the greater difference in RCF-min and RCF-max. *(b)* Following centrifugation on fixed-angle centrifuges, blood layers do not separate evenly, and as a result, an angled blood separation is observed. In contrast, horizontal centrifugation produces an even separation. *(c)* Because of the large RCF values (~200g–700g), cells are pushed toward the outside and downward. On a fixed-angle centrifuge, cells are pushed toward the back of centrifugation tubes and then downward/upward based on cell density. These g-forces produce additional shear stress on cells as they separate based on density along the back walls of centrifugation tubes. In contrast, horizontal centrifugation allows for the free mobility of cells to separate into their appropriate layers based on density, allowing for more optimal cell separation as well as less trauma/shear stress on cells. (Reprinted with permission from Miron et al.[50])

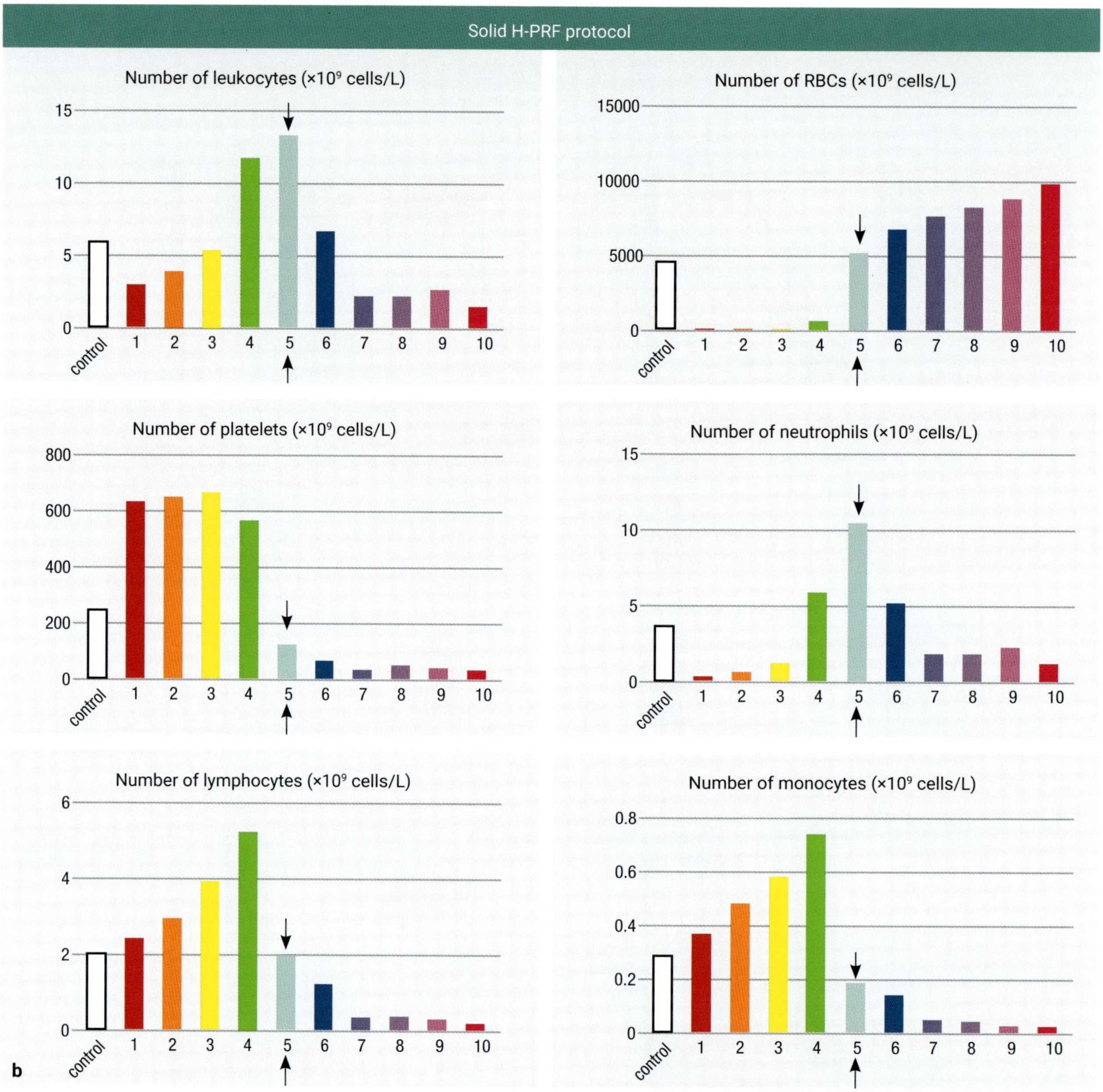

Fig 2-25 The concentration of cell types in each 1-mL layer utilizing the solid-PRF horizontal centrifugation protocol (700g for 8 minutes). Notice that most of the platelets as well as WBCs are now more evenly distributed throughout the upper plasma layer. (Adapted from Miron et al.[50])

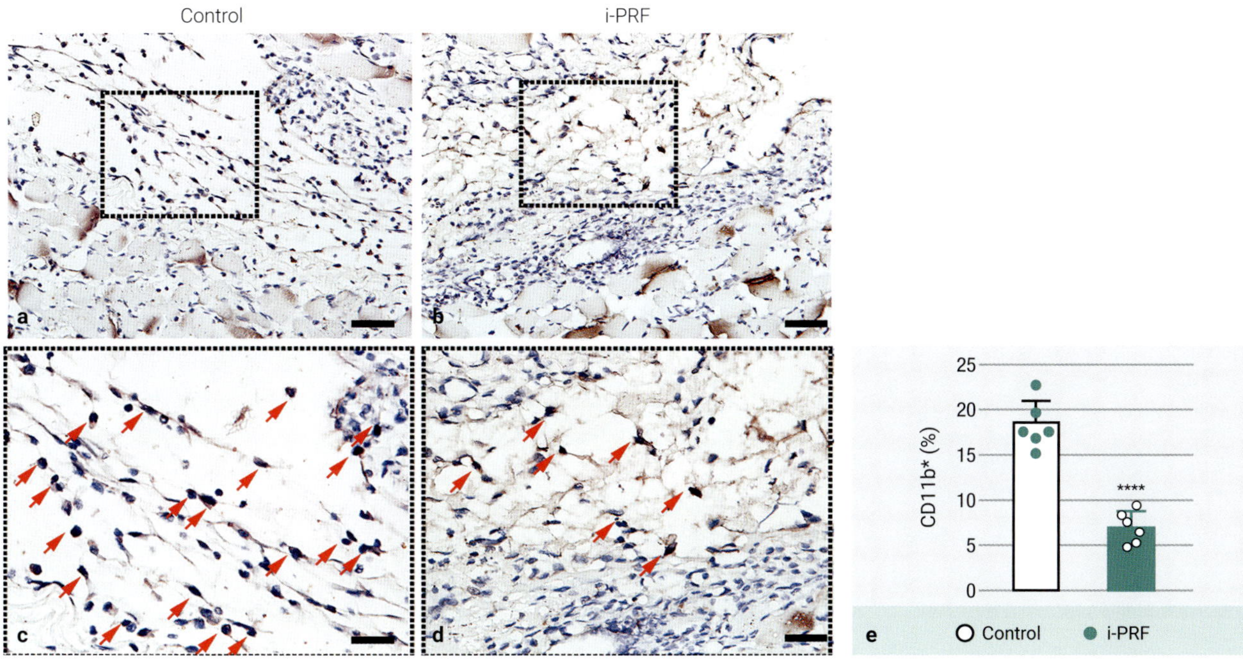

Fig 2-26 The expressions of CD11b in tissues from control *(a and c)* and i-PRF groups *(b and d)*. Scale bar in *a* and *b* = 20 µm; scale bar in *c* and *d* = 10 µm. *(e)* Quantification of the immunohistochemical staining was calculated by the percentage of CD11b-positive cells in all cells in the area of the same size. ****$P < .0001$. Error bars indicate SD. (Reprinted with permission from Zhang et al.[41])

Biologic Activity of PRF on Immune Cells

In an extensive systematic review by Strauss et al investigating the biologic properties of PRF, 1,746 studies were identified, of which 53 were included.[55] Because PRF is capable of improving angiogenesis in vivo, it was reported that PRF enhanced the proliferation, migration, adhesion, and osteogenic differentiation of a variety of different cell types along with cell signaling activation. Furthermore, it was concluded that PRF reduced inflammation, suppressed osteoclastogenesis, and increased the expression of various GFs in mesenchymal cells.[55]

Several interesting and very recent studies have investigated the effects of PRF and its involvement with macrophage polarization from proinflammatory M1 toward proresolving M2 phenotypes.[56] Macrophages are extremely important cells during the healing process and can be involved with either secretion of proinflammatory markers (M1) or proresolution markers (M2). These studies are extremely relevant to PRF.

In a study by Nasirzade et al, murine primary macrophages and a human macrophage cell line were exposed to saliva and lipopolysaccharides (LPS) with and without PRF lysates.[56] The expression of the proinflammatory M1 marker genes *IL-1β* and *IL-6* were greatly decreased, and PRF-conditioned medium enhanced the expression of tissue-resolution markers. It was therefore concluded that PRF holds an anti-inflammatory activity and shifts the macrophage polarization from an M1 toward an M2 phenotype.[56] In a second study on this topic, it was also reported that i-PRF reduced proinflammatory M1 phenotype of macrophages along with activated dendritic cells around muscle defects injected with bacterial suspension (Figs 2-26 and 2-27).[41]

While the study of PRF on various cell types such as immune cells has only begun, it appears that it acts to reduce the proinflammatory response and further decrease common oral cavity inflammatory responses to LPS. Thus, this work may in part explain the observed clinical decreases in postoperative pain reported in the clinical chapters discussed later in this textbook.

PRF appears to reduce the proinflammatory response and further decrease common oral cavity responses to LPS.

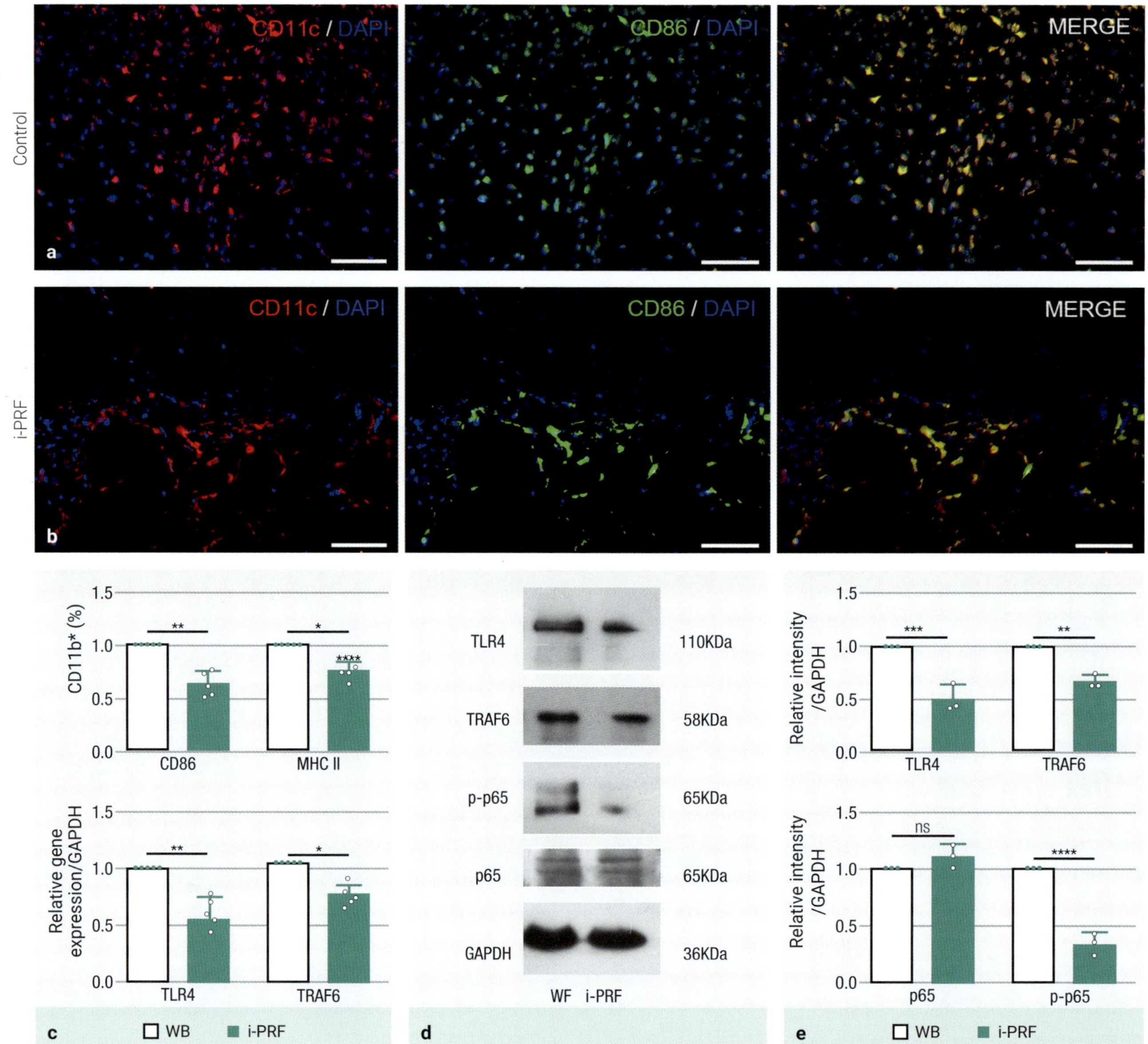

Fig 2-27 Effects of i-PRF on the maturation of dendritic cells. *(a and b)* Sections of immunofluorescence with antibodies against CD11c and CD86. Scale bar = 20 µm. *(c)* Relative expressions of maturation-related and inflammatory-related genes in DC2.4. *(d and e)* Immunoblotting and relative intensity of NF-κB signaling pathway and maturation-related markers of DC2.4 stimulated by the whole blood (WB) or i-PRF in the presence of LPS. *$P < .05$; **$P < .005$; ***$P < .0005$; ****$P < .0001$; ns, no significant difference. Error bars indicate SD. (Reprinted with permission from Zhang et al.[41])

Anti-inflammatory and Antibacterial Properties of PRF

Both the anti-inflammatory and antibacterial properties of PRF have been a topic of much interest in recent years following the clinical observation that PRF seems to reduce postoperative swelling and pain. In a study titled "Effects of liquid platelet-rich fibrin on the regenerative potential of hPDLCs cultured under inflammatory conditions," our group investigated the effects of PRF on human periodontal ligament cells (hPDLCs) under inflammatory conditions. As a model, hPDLCs were investigated using a migration and proliferation assay. To investigate hPDLC differentiation, alkaline phosphatase (ALP) assay, Alizarin Red Staining, and gene expression levels of *Runx2*, *Col1a1*,

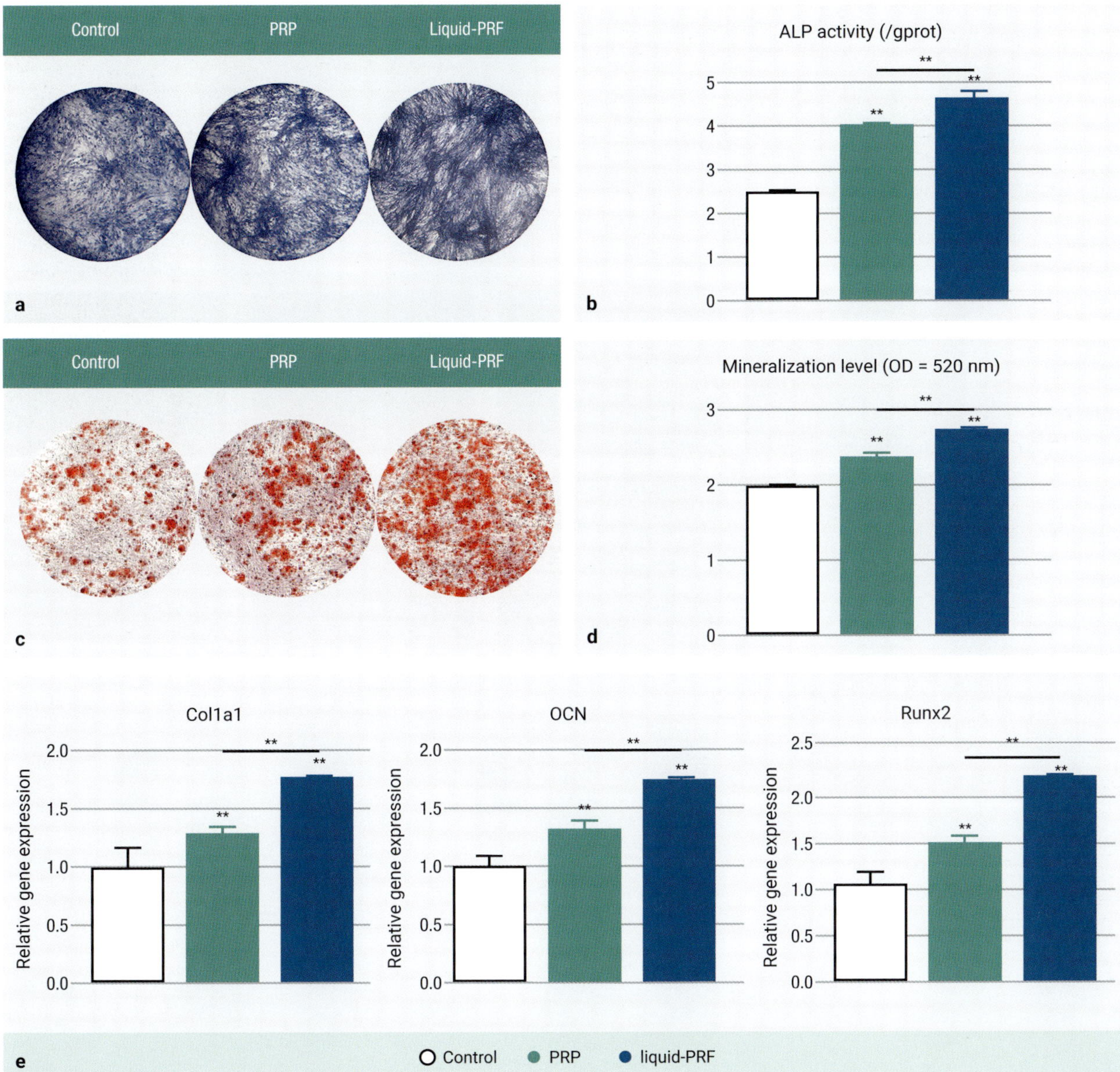

Fig 2-28 Differentiation of hPDLCs. *(a and b)* Effects of PRP and liquid-PRF on ALP activity were detected by ALP staining and ALP activity test, respectively. *(c)* Alizarin Red S staining showed the mineralized nodules in each group after induction for 14 days. *(d)* The semiquantification of mineralization level. *(e)* Relative gene expression levels of *Col1a1*, *OCN*, and *Runx2* after being treated with PRP or liquid-PRF for 14 days. Error bars correspond to the mean ± SD; significant differences are indicated: $*P < .05$; $**P < .01$.

and *OCN* were conducted. Furthermore, cells were cultured with LPS to induce an inflammatory condition to investigate the ability of PRF to impact inflammatory resolution. All assays were compared to PRP (lower in WBCs).

Osteogenic differentiation demonstrated that liquid-PRF significantly induced greater ALP activity and more mineralized nodules when compared to PRP and controls (Fig 2-28). According to the experimental timeline, cells were pretreated with or without LPS to induce an inflammatory condition for 7 days, and then liquid-PRF was added to the culture media for an additional incubation period of 7 days (Fig 2-29a). Immunofluorescence images demonstrated that LPS induced more p65 expression (a marker for inflammation; Fig 2-29b), while the addition of liquid-PRF decreased

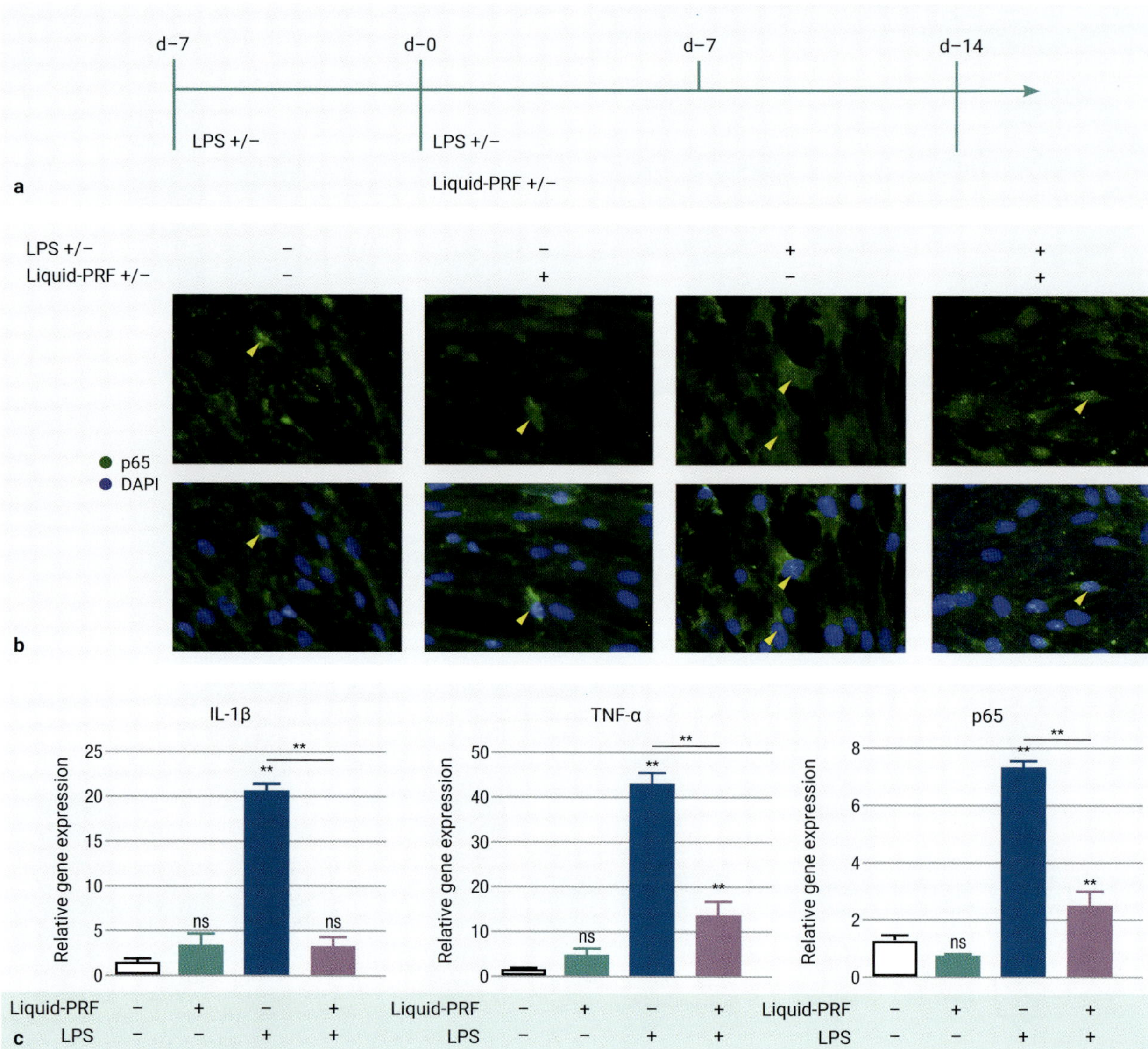

Fig 2-29 Liquid-PRF can decrease the inflammation induced by LPS. *(a)* The timeline of experimental inflammatory condition stimulation. *(b)* Immunofluorescence staining of p65 in hPDLCs after being cultured with or without LPS and/or liquid-PRF. *(c)* The relative gene expression levels of inflammatory markers including IL-1β, TNF-α, and p65. Error bars correspond to the mean ± SD; significant differences are indicated: $*P < .05$; $**P < .01$; ns, not statistically significant vs control group.

its expression level. Furthermore, other inflammation markers including *IL-1β* and *TNF-α* were also significantly downregulated, as confirmed by real-time polymerase chain reaction (RT-PCR; Fig 2-29c). In summary, it was concluded that liquid-PRF displayed an anti-inflammatory response when hPDLCs were cultured with LPS.

In a final experiment, it was observed that liquid-PRF promoted the osteogenic differentiation of hPDLCs even when cultured in an inflammatory environment. Briefly, cells pretreated and cultured with LPS resulted in an intense reduction in mineralization nodule formation (Figs 2-30a and 2-30b). As the previous experiment demonstrated the ability for liquid-PRF to decrease the inflammatory response, it was further found that liquid-PRF could actually reverse a decrease in mineralization observed by LPS and resulted in a significant upregulation

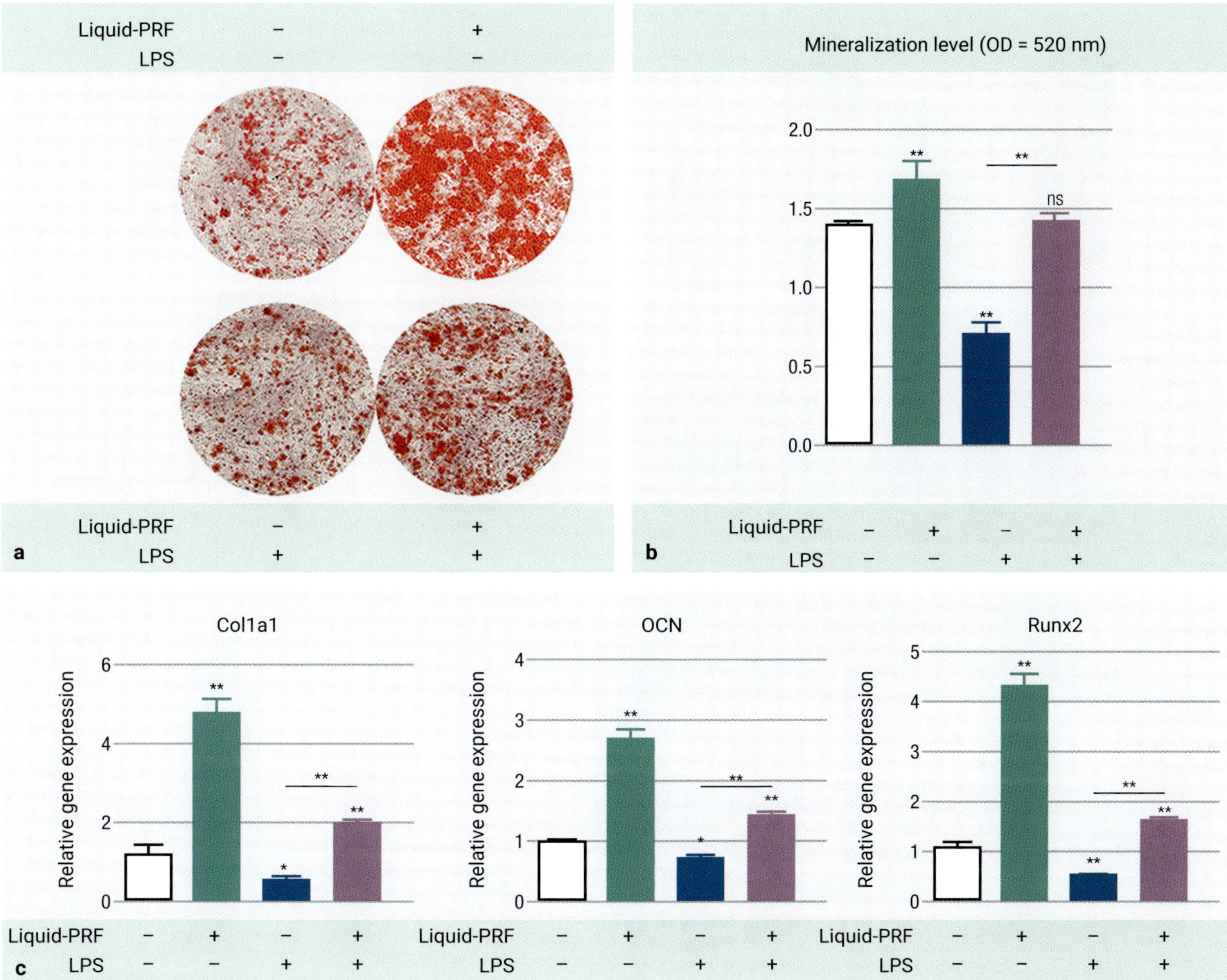

Fig 2-30 Liquid-PRF can promote the osteogenic potential of hPDLCs in an inflammatory environment induced by LPS. *(a)* Alizarin Red S staining indicated the odontoblastic differentiation of hPDLCs in the presence of LPS and/or liquid-PRF. *(b)* Mineralization level. *(c)* Gene expression levels of *Runx2*, *Col1a1*, and *OCN* in inflammatory condition. Error bars correspond to the mean ± SD; significant differences are indicated: **P* < .05; ***P* < 0.01; ns, not statistically significant vs control group.

of expression markers *Runx2*, *Col1a1*, and *OCN* (Fig 2-30c). These findings indicate that the anti-inflammatory effect and regenerative potential of liquid-PRF can counterbalance the negative inflammatory effect induced by LPS. Later chapters address these findings more specific to the periodontal field, because PRF has been shown to improve the regeneration of intrabony and furcation defects not only by improving GF release but also by counterbalancing the inflammatory response induced by LPS (see chapter 10).

Furthermore, it has also been shown that PRF exerts potent antibacterial properties. In another study by our group, PRF was separated into solid and leachate components, and the PRF clots were divided into five equal layers to explore the specific antibacterial aspects of PRF. Both antimicrobial tests and flow cytometric analysis revealed

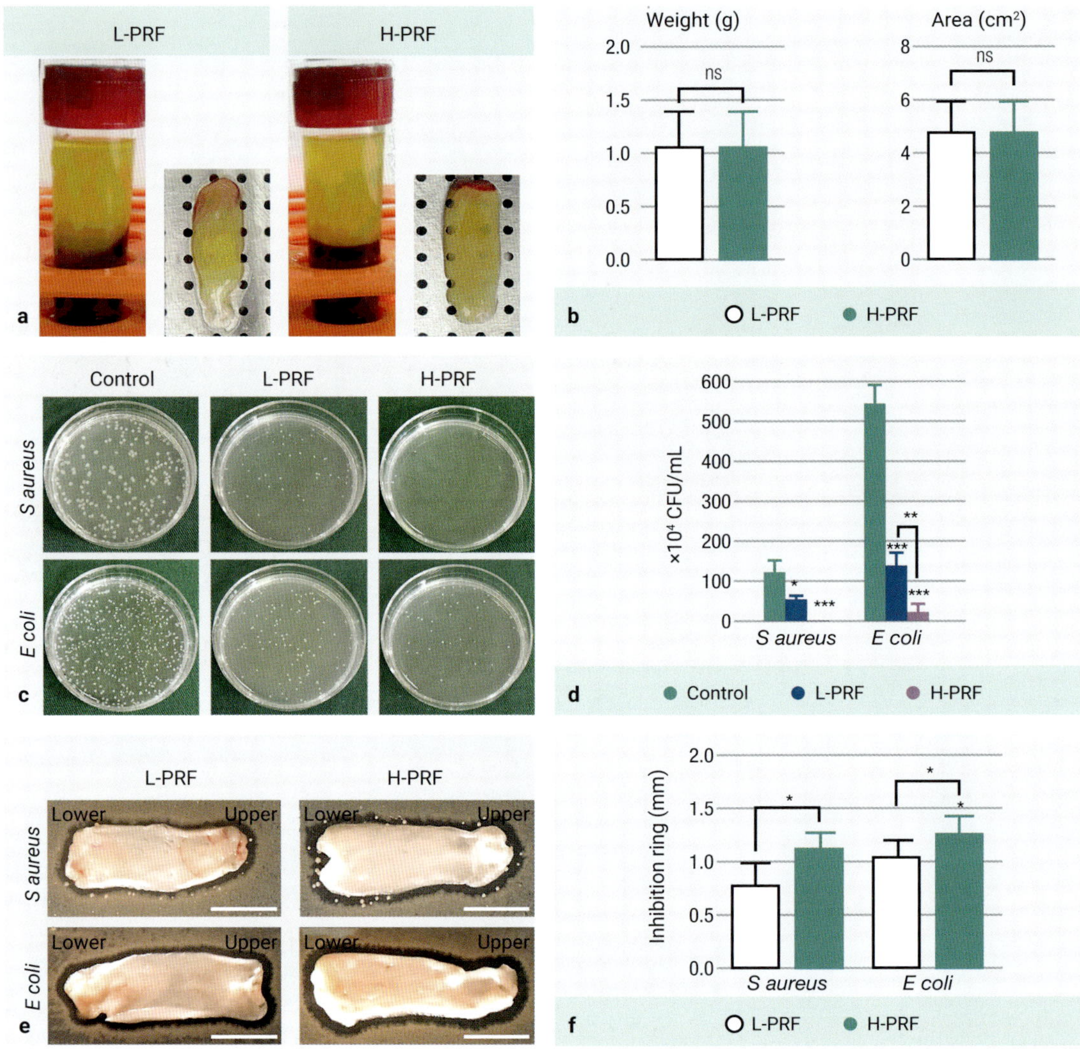

Fig 2-31 Antibacterial properties of L-PRF and H-PRF. *(a)* Photographs of PRF obtained after centrifugation by both protocols. Note the horizontal layer centrifugation in H-PRF versus the angled layering in L-PRF. *(b)* Weight and size measurements of PRF matrices. *(c and d)* Photographs and quantitative analysis of bacterial colony of *Staphylococcus aureus* and *Escherichia coli* incubated with L-PRF or H-PRF clots for 4 hours. *(e and f)* Photographs and quantification of the inhibition zone of L-PRF and H-PRF membranes incubated with *S aureus* or *E coli* after 24 hours. $^*P < .05$; $^{**}P < .01$; $^{***}P < .001$; ns, not statistically significant.

that PRF produced using horizontal centrifugation demonstrated a significantly better antibacterial effect than L-PRF and was strongly correlative with immune cell numbers and types (Fig 2-31). In addition, our results demonstrated that the antimicrobial ability of PRF clots were less efficient than the wet PRF containing leachate, which suggest a promising application guidance to retain the liquid components of PRF for better anti-infection properties during clinical use. This study is presented in greater detail in chapter 3.

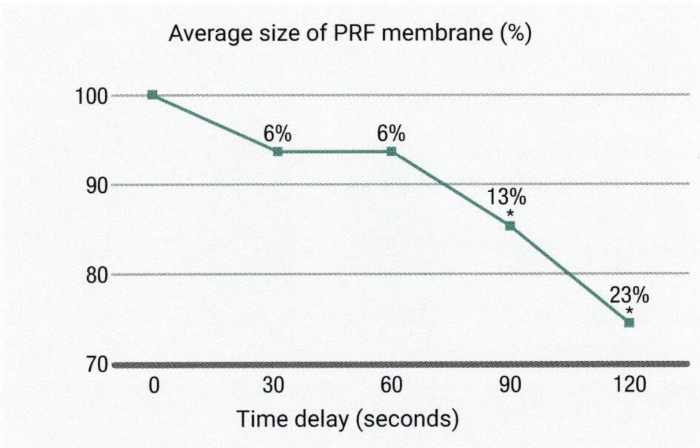

Fig 2-32 Average size of PRF membranes from 60 patients following blood draw after an initial wait period of 0, 30, 60, 90, and 120 seconds prior to centrifugation. Notice that after 90 seconds, the PRF membranes were significantly reduced in size (by 13%). Following a 120-second wait period, these membranes were further significantly reduced in size (by 23%) compared to the control (0-second wait period). $^*P < .05$ indicates a significant difference between 0 seconds from centrifugation and the investigated time period of 90 and 120 seconds. (Adapted from Miron et al.[57])

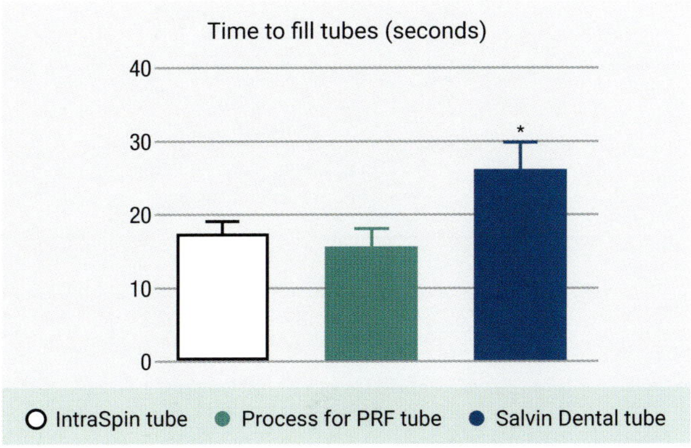

Fig 2-33 Required time interval to fill each manufacturer's PRF tube. Note that in general it takes roughly 15 seconds, but some manufacturers have slower-filling tubes. (Adapted from Miron et al.[39])

The Effect of Age, Sex, and Time on the Size Outcomes of PRF Membranes

Two topics that were heavily questioned for many years were *(1)* How long does the clinician have from the start of blood draw? and *(2)* Why do colleagues observe so much variability in clot size even from blood draws coming from the same patient? In 2019, we addressed this topic in a publication where the final PRF size outcomes were compared following centrifugation that took place after 0, 30, 60, 90, and 120 seconds in both male and female patients of different age categories. Each participant donated six vials of blood, and centrifugation was begun precisely after 0, 30, 60, 90, and 120 seconds.[57]

As depicted in Fig 2-32, by 90 seconds already a drop in membrane size of 13% was observed, and by 120 seconds this dropped even further. The entire goal of centrifugation is to separate layers based on density, so when blood remains sitting in a centrifugation tube for 120 seconds, it is certain that some of the fibrinogen and thrombin are beginning to convert into fibrin. Thereafter, when centrifugation begins, it becomes harder and harder to separate layers (and most importantly cell types as a result). Therefore, centrifugation should be carried out between 60 and 90 seconds after blood draw. It generally takes 15 seconds to fill the tube with blood to begin with, sometimes even longer (Fig 2-33).

> ***Always remember that the entire goal of centrifugation is to separate layers based on density. When blood sits in a centrifugation tube for 120 seconds, it is certain that some of the fibrinogen and thrombin are beginning to convert into fibrin.***

One noticeable trend was that the size of the membranes produced between male and female patients was different. On average, the size of PRF membranes produced by females was 17% larger than those produced in males (Fig 2-34). As the role of centrifugation is to separate blood layers transitionally over time, these differences were due

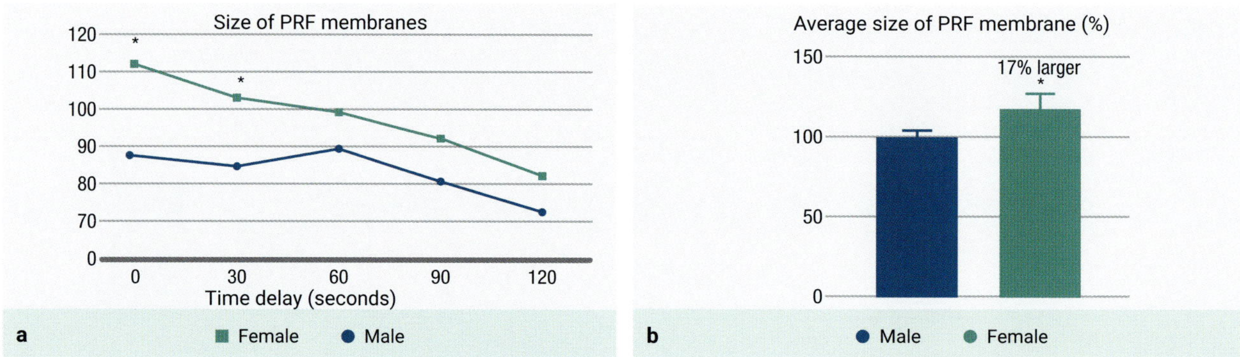

Fig 2-34 *(a)* Comparison of the average size of PRF membranes between males and females in 60 patients. Notice that at earlier time points, the female PRF membranes were significantly larger compared to male PRF membranes. *$P < .05$ indicates a significant difference between male and female PRF membrane sizes (%). *(b)* Bar graph representing the average size of PRF membranes from males and females in 60 patients. On average, the female PRF membranes were 17% larger. *$P < .05$ indicates a significant difference between male and female PRF membrane sizes (%). (Adapted from Miron et al.[57])

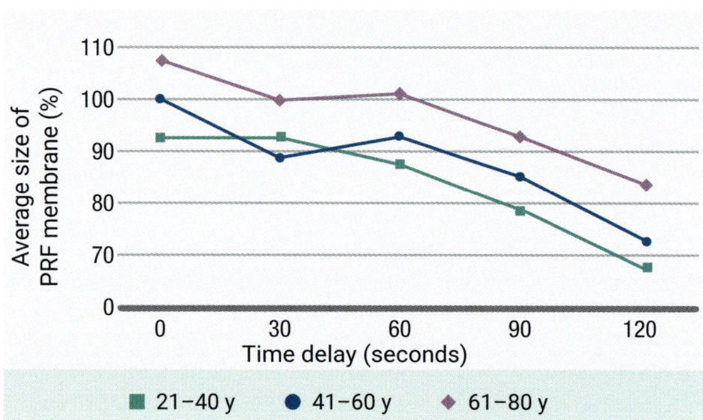

Fig 2-35 Comparison of the average size of PRF membranes between various age groups: 21–40 years, 41–60 years, and 61–80 years. While no significant differences were noticed between the groups, in general older patients produced larger membranes. (Adapted from Miron et al.[57])

to females generally reporting lower hematocrit levels within their peripheral blood compared to males.[58,59] The same trend was also observed in older patients (Fig 2-35). This is also due to the fact that as one ages, a lower hematocrit count is usually noted, and therefore the "density" of blood is lowered (fewer RBCs). Seemingly, it is easier for blood layers to separate accordingly.

Because centrifugation separates blood based on density, it is important to note that variability will exist. Females and older patients have less hematocrit when compared to men and younger individuals. As a result, PRF clots produced in older females will be significantly larger than young males (especially young athletes or those living at high altitude). Generally, in the younger male population or patients routinely living at high altitude, a 20% increase in the RCF values of each protocol is recommended.

Conclusion

The use of GFs in dentistry has gained tremendous momentum and popularity in recent years, especially because of the easily obtainable and low-cost group of platelet concentrates. Autologous PRF is a 100% natural blood-derived tissue engineering scaffold that is totally physiologic and safe and may be utilized for the purpose of wound healing. This chapter outlined the main GFs and cell types found in PRF and further demonstrated the massive effect of centrifugation parameters on the final cell layer separation in PRF. Major advancements with respect to first utilizing the LSCC and thereafter horizontal centrifugation have more recently optimized the final production of PRF. Future research remains ongoing to further highlight all the biologic properties and advantages of PRF, such as its ability to regulate immune cells as well as participate in antimicrobial defense. In summary, PRF serves as an excellent tissue engineering scaffold by fulfilling its three main criteria: scaffold (fibrin), cells (platelets and leukocytes), and GFs (PDGF, VEGF, TGF-β).

References

1. Guo S, Dipietro LA. Factors affecting wound healing. J Dent Res 2010;89:219–229.
2. Tonnesen MG, Feng X, Clark RA. Angiogenesis in wound healing. J Investig Dermatol Symp Proc 2000;5:40–46.
3. de Almeida Barros Mourao CF, Miron RJ, de Mello Machado RC, Ghanaati S, Alves GG, Calasans-Maia MD. Usefulness of platelet-rich fibrin as a hemostatic agent after dental extractions in patients receiving anticoagulant therapy with factor Xa inhibitors: A case series. J Oral Maxillofac Surg 2019;23:381–386.
4. Gurtner GC, Werner S, Barrandon Y, Longaker MT. Wound repair and regeneration. Nature 2008;453:314–321.
5. Dohan Ehrenfest DM, Del Corso M, Diss A, Mouhyi J, Charrier JB. Three-dimensional architecture and cell composition of a Choukroun's platelet-rich fibrin clot and membrane. J Periodontol 2010;81:546–555.
6. Ghanaati S, Booms P, Orlowska A, et al. Advanced platelet-rich fibrin: A new concept for cell-based tissue engineering by means of inflammatory cells. J Oral Implantol 2014;40:679–689.
7. Kawazoe T, Kim HH. Tissue augmentation by white blood cell-containing platelet-rich plasma. Cell Transplant 2012;21:601–607.
8. Perut F, Filardo G, Mariani E, et al. Preparation method and growth factor content of platelet concentrate influence the osteogenic differentiation of bone marrow stromal cells. Cytotherapy 2013;15:830–839.
9. Pirraco RP, Reis RL, Marques AP. Effect of monocytes/macrophages on the early osteogenic differentiation of hBMSCs. J Tissue Eng Regen Med 2013;7:392–400.
10. Gosain A, DiPietro LA. Aging and wound healing. World J Surg 2004;28:321–326.
11. Eming SA, Brachvogel B, Odorisio T, Koch M. Regulation of angiogenesis: Wound healing as a model. Prog Histochem Cytochem 2007;42:115–170.
12. Kobayashi E, Fluckiger L, Fujioka-Kobayashi M, et al. Comparative release of growth factors from PRP, PRF, and advanced-PRF. Clin Oral Investig 2016;20:2353–2360.
13. Chase AJ, Newby AC. Regulation of matrix metalloproteinase (matrixin) genes in blood vessels: A multi-step recruitment model for pathological remodelling. J Vasc Res 2003;40:329–343.
14. Mazzucco L, Borzini P, Gope R. Platelet-derived factors involved in tissue repair-from signal to function. Transfus Med Rev 2010;24:218–234.
15. Nguyen LH, Annabi N, Nikkhah M, et al. Vascularized bone tissue engineering: Approaches for potential improvement. Tissue Eng Part B Rev 2012;18:363–382.
16. Burnouf T, Goubran HA, Chen TM, Ou KL, El-Ekiaby M, Radosevic M. Blood-derived biomaterials and platelet growth factors in regenerative medicine. Blood Rev 2013;27:77–89.
17. Reed GL. Platelet secretory mechanisms. Semin Thromb Hemost 2004;30:441–450.
18. Dohan DM, Choukroun J, Diss A, et al. Platelet-rich fibrin (PRF): A second-generation platelet concentrate. Part II: Platelet-related biologic features. Oral Surg Oral Med Oral Pathol Oral Radiol Endod 2006;101:e45–e50.
19. Gruber R, Kandler B, Holzmann P, et al. Bone marrow stromal cells can provide a local environment that favors migration and formation of tubular structures of endothelial cells. Tissue Eng 2005;11:896–903.
20. Au P, Tam J, Fukumura D, Jain RK. Bone marrow-derived mesenchymal stem cells facilitate engineering of long-lasting functional vasculature. Blood 2008;111:4551–4558.
21. Clark RA. Fibrin and wound healing. Ann N Y Acad Sci 2001;936:355–367.
22. Shaikh FM, Callanan A, Kavanagh EG, Burke PE, Grace PA, McGloughlin TM. Fibrin: A natural biodegradable scaffold in vascular tissue engineering. Cells Tissues Organs 2008;188:333–346.
23. Nurden AT. Platelets, inflammation and tissue regeneration. Thromb Haemost 2011;105(suppl 1):S13–S33.
24. Fujioka-Kobayashi M, Miron RJ, Hernandez M, Kandalam U, Zhang Y, Choukroun J. Optimized platelet-rich fibrin with the low-speed concept: Growth factor release, biocompatibility, and cellular response. J Periodontol 2017;88:112–121.
25. Lozito TP, Taboas JM, Kuo CK, Tuan RS. Mesenchymal stem cell modification of endothelial matrix regulates their vascular differentiation. J Cell Biochem 2009;107:706–713.
26. Kato J, Tsuruda T, Kita T, Kitamura K, Eto T. Adrenomedullin: A protective factor for blood vessels. Arterioscler Thromb Vasc Biol 2005;25:2480–2487.
27. Shamloo A, Xu H, Heilshorn S. Mechanisms of vascular endothelial growth factor-induced pathfinding by endothelial sprouts in biomaterials. Tissue Eng Part A 2012;18:320–330.
28. Martin P, Leibovich SJ. Inflammatory cells during wound repair: The good, the bad and the ugly. Trends Cell Biol 2005;15:599–607.
29. Tsirogianni AK, Moutsopoulos NM, Moutsopoulos HM. Wound healing: Immunological aspects. Injury 2006;37(suppl 1):S5–S12.
30. Adamson R. Role of macrophages in normal wound healing: An overview. J Wound Care 2009;18:349–351.
31. Davis VL, Abukabda AB, Radio NM, et al. Platelet-rich preparations to improve healing. Part I: Workable options for every size practice. J Oral Implantol 2014;40:500–510.
32. Davis VL, Abukabda AB, Radio NM, et al. Platelet-rich preparations to improve healing. Part II: Platelet activation and enrichment, leukocyte inclusion, and other selection criteria. J Oral Implantol 2014;40:511–521.
33. Ghasemzadeh M, Hosseini E. Intravascular leukocyte migration through platelet thrombi: Directing leukocytes to sites of vascular injury. Thromb Haemost 2015;113:1224–1235.
34. Border WA, Noble NA. Transforming growth factor beta in tissue fibrosis. N Engl J Med 1994;331:1286–1292.
35. Bowen T, Jenkins RH, Fraser DJ. MicroRNAs, transforming growth factor beta-1, and tissue fibrosis. J Pathol 2013;229:274–285.
36. Babensee JE, McIntire LV, Mikos AG. Growth factor delivery for tissue engineering. Pharma Res 2000;17:497–504.
37. Giannobile WV, Hernandez RA, Finkelman RD, et al. Comparative effects of platelet-derived growth factor-BB and insulin-like growth factor-I, individually and in combination, on periodontal regeneration in *Macaca fascicularis*. J Periodontal Res 1996;31:301–312.
38. Eren G, Gurkan A, Atmaca H, Donmez A, Atilla G. Effect of centrifugation time on growth factor and MMP release of an experimental platelet-rich fibrin-type product. Platelets 2016;27:427–432.
39. Miron RJ, Xu H, Chai J, et al. Comparison of platelet-rich fibrin (PRF) produced using 3 commercially available centrifuges at both high (~ 700 g) and low (~ 200 g) relative centrifugation forces. Clin Oral Investig 2020;24:1171–1182.
40. Davies C, Miron RJ. PRF in Facial Esthetics. Chicago: Quintessence, 2020.
41. Zhang J, Yin C, Zhao Q, et al. Anti-inflammation effects of injectable platelet-rich fibrin via macrophages and dendritic cells. J Biomed Mater Res A 2020;108:61–68.
42. Lucarelli E, Beretta R, Dozza B, et al. A recently developed bifacial platelet-rich fibrin matrix. Eur Cell Mater 2010;20:13–23.

43. Saluja H, Dehane V, Mahindra U. Platelet-rich fibrin: A second generation platelet concentrate and a new friend of oral and maxillofacial surgeons. Ann Maxillofac Surg 2011;1:53–57.
44. Miron RJ, Fujioka-Kobayashi M, Hernandez M, et al. Injectable platelet rich fibrin (i-PRF): Opportunities in regenerative dentistry? Clin Oral Investig 2017;21:2619–2627.
45. Abd El Raouf M, Wang X, Miusi S, et al. Injectable-platelet rich fibrin using the low speed centrifugation concept improves cartilage regeneration when compared to platelet-rich plasma. Platelets 2019;30:213–221.
46. Chai J, Jin R, Yuan G, Kanter V, Miron RJ, Zhang Y. Effect of liquid platelet-rich fibrin and platelet-rich plasma on the regenerative potential of dental pulp cells cultured under inflammatory conditions: A comparative analysis. J Endod 2019;45:1000–1008.
47. Wang X, Yang Y, Zhang Y, Miron RJ. Fluid platelet-rich fibrin stimulates greater dermal skin fibroblast cell migration, proliferation, and collagen synthesis when compared to platelet-rich plasma. J Cosmet Dermatol 2019;18:2004–2010.
48. Wang X, Zhang Y, Choukroun J, Ghanaati S, Miron RJ. Behavior of gingival fibroblasts on titanium implant surfaces in combination with either injectable-PRF or PRP. Int J Mol Sci 2017;18:331.
49. Wang X, Zhang Y, Choukroun J, Ghanaati S, Miron RJ. Effects of an injectable platelet-rich fibrin on osteoblast behavior and bone tissue formation in comparison to platelet-rich plasma. Platelets 2018;29:48–55.
50. Miron RJ, Chai J, Zheng S, Feng M, Sculean A, Zhang Y. A novel method for evaluating and quantifying cell types in platelet rich fibrin and an introduction to horizontal centrifugation. J Biomed Mater Res A 2019;107:2257–2271.
51. Choukroun J, Ghanaati S. Reduction of relative centrifugation force within injectable platelet-rich-fibrin (PRF) concentrates advances patients' own inflammatory cells, platelets and growth factors: The first introduction to the low speed centrifugation concept. Eur J Trauma Emerg Surg 2018;44:87–95.
52. Miron RJ, Chai J, Zhang P, et al. A novel method for harvesting concentrated platelet-rich fibrin (C-PRF) with a 10-fold increase in platelet and leukocyte yields. Clin Oral Investig 2020;24:2819–2828.
53. Varela HA, Souza JCM, Nascimento RM, et al. Injectable platelet rich fibrin: Cell content, morphological, and protein characterization. Clin Oral Investig 2019;23:1309–1318.
54. Lourenço ES, Mourão CFAB, Leite PEC, Granjeiro JM, Calasans-Maia MD, Alves GG. The in vitro release of cytokines and growth factors from fibrin membranes produced through horizontal centrifugation. J Biomed Mater Res A 2018;106:1373–1380.
55. Strauss FJ, Nasirzade J, Kargarpoor Z, Stahli A, Gruber R. Effect of platelet-rich fibrin on cell proliferation, migration, differentiation, inflammation, and osteoclastogenesis: A systematic review of in vitro studies. Clin Oral Investig 2020;24:569–584.
56. Nasirzade J, Kargarpour Z, Hasannia S, Strauss FJ, Gruber R. Platelet-rich fibrin elicits an anti-inflammatory response in macrophages in vitro. J Periodontol 2020;91:244–252.
57. Miron RJ, Dham A, Dham U, Zhang Y, Pikos MA, Sculean A. The effect of age, gender, and time between blood draw and start of centrifugation on the size outcomes of platelet-rich fibrin (PRF) membranes. Clin Oral Investig 2019;23:2179–2185.
58. Zeng SM, Yankowitz J, Widness JA, Strauss RG. Etiology of differences in hematocrit between males and females: Sequence-based polymorphisms in erythropoietin and its receptor. J Gend Specif Med 2001;4:35–40.
59. Billett HH. Hemoglobin and hematocrit. In: Walker HK, Hall WD, Hurst JW (eds). Clinical Methods: The History, Physical, and Laboratory Examinations, ed 3. Boston: Butterworths, 1990.

3

Horizontal Versus Fixed-Angle Centrifugation of PRF: Optimization of C-PRF

Contributors

Richard J. Miron
Masako Fujioka-Kobayashi
Tomoyuki Kawase
Yufeng Zhang

Chapter Highlights

- *Advantages of horizontal centrifugation versus fixed-angle centrifugation*
- *Systematic evaluation of horizontal centrifugation using 24 different protocols*
- *Optimization of H-PRF in both liquid and solid formulations*
- *Histologic evaluation of L-PRF versus H-PRF*
- *Optimization of C-PRF using horizontal centrifugation*

The first two chapters of this textbook discussed the history of platelet concentrates and their commercial and biologic evolution. The aim of this chapter is to discuss centrifugation parameters that may best optimize the production of PRF. A great deal of work has been conducted at both the preclinical and clinical level to maximize the regenerative potential of PRF, and this chapter provides a more detailed overview of these methods and takes a closer look at the cell types found in PRF. It also discusses the work conducted by our laboratories to improve PRF using horizontal centrifugation by evaluating a range of 24 different protocols in a systematic and standardized way. Thereafter, comparative histologic evaluation compares PRF membranes produced following fixed-angle versus horizontal centrifugation. Lastly, data from the previous chapter is expanded upon to explain the concept of C-PRF.

Video 3-1

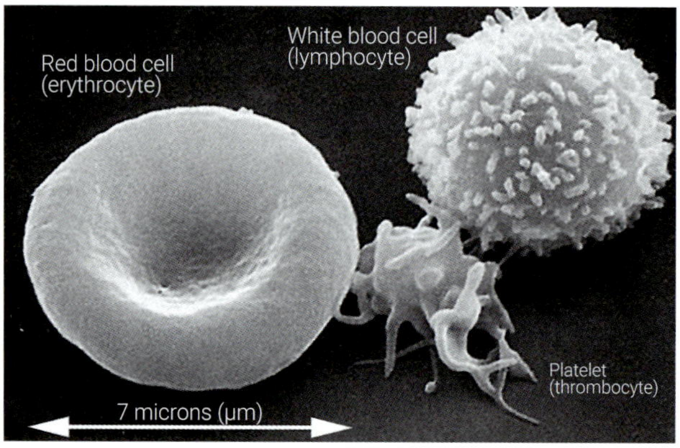

Fig 3-1 SEM showing the relative size of RBCs, platelets, and WBCs (leukocytes). Note the much larger size of RBCs and WBCs when compared to platelets as well as the similarity in size between WBCs and RBCs.

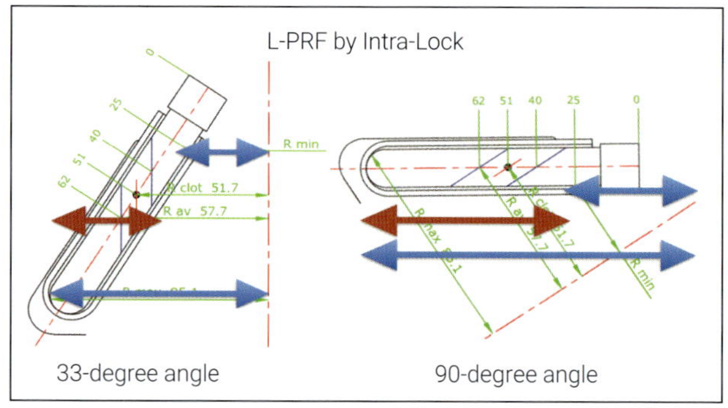

Fig 3-2 Graphic demonstrating the difference between the RCF-min and RCF-max on both fixed-angle and horizontal centrifugation systems. Note that because the tubes are completely horizontal on a horizontal centrifugation system, the gradient difference between the RCF-min and RCF-max is much greater, resulting in much better layer separation.

The previous chapter provided a brief overview of the advantages of horizontal centrifugation of PRF. Previously, publications by our group found that horizontal centrifugation was superior at accumulating platelets and leukocytes when compared to standard fixed-angle centrifugation utilized to produce PRF.[1] Both solid-based and liquid-based PRF matrices were obtained with up to a fourfold increase in platelet/leukocyte numbers and/or concentrations.[1]

This chapter discusses in greater detail the ability of horizontal-PRF (H-PRF) to better separate cell layers in blood based on their density. As previously shown (see Table 2-1), platelets are the lightest of the blood cells, with white blood cells (WBCs) and red blood cells (RBCs) similar in size and density (Fig 3-1). Therefore, the optimization of centrifugation parameters becomes extremely relevant should the clinician desire to maximize the harvest of cell types following centrifugation.

When cells are separated in a blood collection tube based on their density, it is important to understand that their ability to separate is based on differences in g-force produced at the RCF-min versus RCF-max of the tube (as reviewed in greater detail in chapter 4). Note that the greater the angulation of the tube, the greater the difference in g-force differential that exists between the RCF-min of the tube versus the RCF-max (Fig 3-2).

One avenue that is extremely relevant to the production of PRF has been the effect of the larger radius from the rotor found when centrifugation is carried out on a horizontal versus fixed-angle centrifuge. To illustrate this point further, an overlapping image of a centrifugation tube is provided in Fig 3-3. If a certain g-force can be expected at the RCF-clot, a much smaller RCF-max will be produced on a fixed-angle centrifuge (Fig 3-3a) when compared to a horizontal centrifuge (Fig 3-3b). While certain commercial entities such as the L-PRF system will report the g-force values specifically at the RCF-clot, note how in Fig 3-3b, the same RCF-clot value on both fixed-angle and horizontal systems displays a much greater RCF-max value if the tube is horizontal. This allows for higher g-force productions at the RCF-max on a horizontal centrifuge, which means faster and better ability for the heavier cells (ie, RBCs) to be pulled down to the bottom of the tube more effectively.

More recently, an introduction of longer tubes (13-mL tubes versus 10-mL) has been proposed as a way to further concentrate cells (Fig 3-3c). While this concept is true, note that should these tubes be utilized on a fixed-angle centrifugation device, the difference between the RCF-min and RCF-max will still equate to less than that obtained with a 10-mL tube in a horizontal centrifuge. These longer tubes still pose the main issue related to cells accumulating on the back distal surface of PRF tubes without proper cell layer separation, as reviewed in the previous chapter.

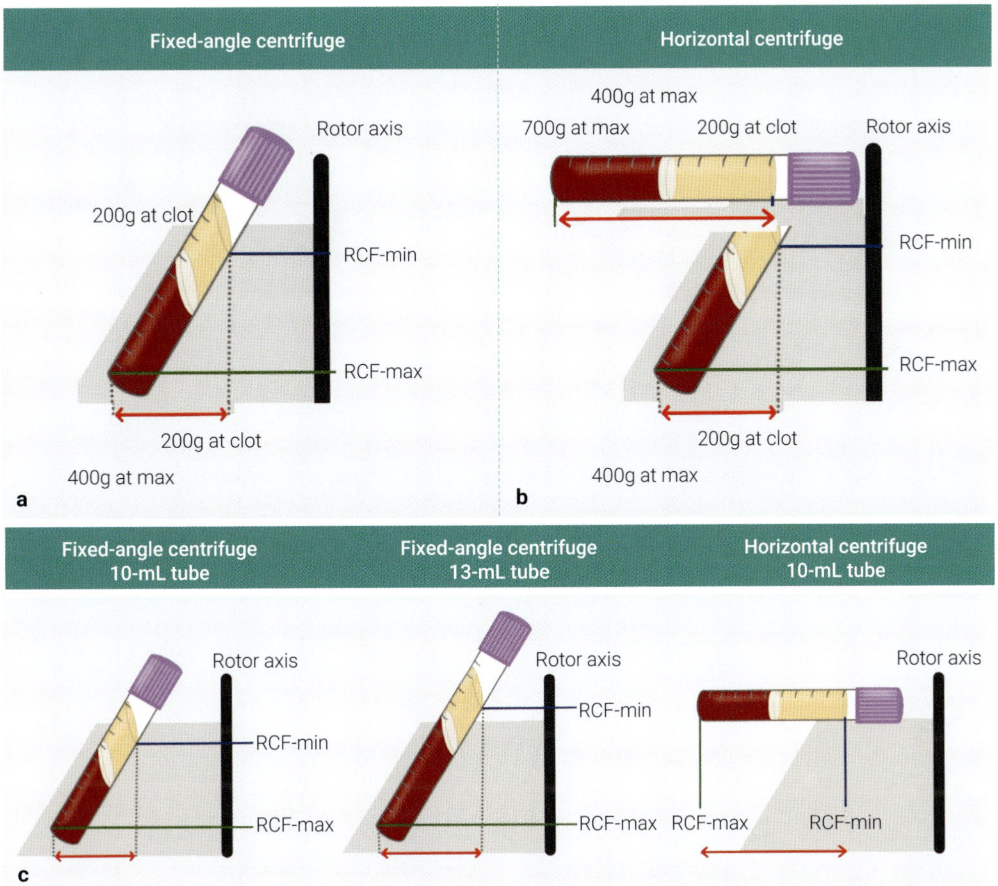

Fig 3-3 *(a and b)* Demonstration that even at the exact same RCF-clot value on either fixed-angle or horizontal centrifugation, the final RCF-max value on a fixed-angle centrifugation system is much smaller than that observed on a horizontal system due to the greater distance from the centrifuge rotor. *(c)* The introduction of 13-mL tubes has led to claims that they are better able to concentrate cells. While this is true compared to 10-mL tubes, if centrifugation remains carried out on a fixed-angle device, the difference between the RCF-min and RCF-max remains inferior to that on a horizontal centrifuge, even if a 10-mL tube is used.

Larger Radius, Higher RCF, Shorter Spin Time

While not peer-reviewed in the same way as our published studies, the Internet is rife with support for horizontal centrifugation. Drucker Diagnostics, for example, clearly demonstrate how tubes centrifuged in a horizontal manner experience a much larger radius, resulting in a higher RCF and more efficient pull-down forces (https://druckerdiagnostics.com/horizontal-vs-fixed-angle/). Furthermore, they have shown that the time required for complete centrifugation on a horizontal rotor is only two-thirds of the time required on a fixed-angle centrifuge (Fig 3-4). This is particularly important because PRF is subject to clotting over time, so saving a few critical minutes by switching to horizontal centrifuge can translate to better layer separation and greater accumulation of cells following the protocol.[2]

The time required for complete centrifugation on a horizontal rotor is only two-thirds of the time required on a fixed-angle centrifuge.

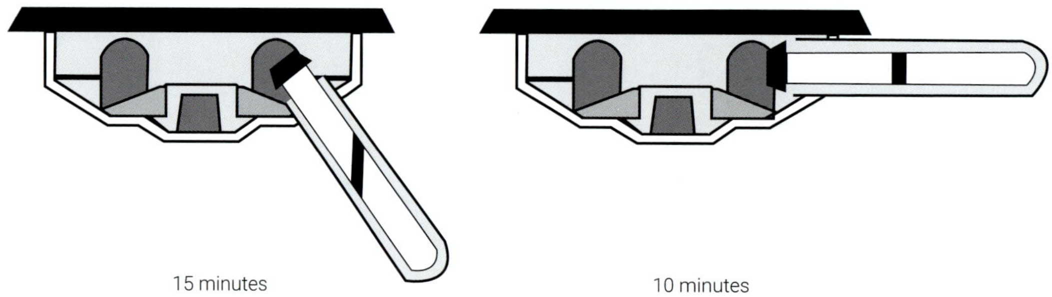

Fig 3-4 Graphic from Drucker Diagnostics demonstrating that centrifugation carried out on a horizontal centrifuge requires only two-thirds the time required on a fixed-angle centrifuge. Therefore, a 12-minute protocol on a fixed-angle rotor would take only 8 minutes with a horizontal system. (Adapted from https://druckerdiagnostics.com/horizontal-vs-fixed-angle/.)

Accumulation of Cells in PRF Tubes

One thing commonly found following centrifugation is the angle produced after completion of the spin cycle (Fig 3-5).[2] In a study by Takahashi et al, the distribution of cells and growth factors (GFs) in PRF following centrifugation using two different centrifugation devices was investigated.[3] Blood samples were obtained in tubes and immediately centrifuged to prepare PRF using two protocols. Both matrices were compressed, embedded in paraffin, and subjected to immunohistochemical examination.[3]

Following histologic assessment, it was observed that leukocytes and plasma proteins were localized on the back walls of PRF tubes (referred to as *distal surface*), including the interface corresponding to the buffy coat (Figs 3-6 and 3-7).[3] These cells were being accumulated on the back distal surfaces only when utilizing fixed-angle centrifugation due to the gravitational pull (see Fig 3-8; see also Video 3-2).

More intriguing was the fact that the type of tube (plain glass versus silica-coated plastic) had a significant impact on the final distribution of cells found within PRF clots when the clots produced at different centrifugation speeds were investigated histologically.[4] The specific role of tubes for the production of PRF has been such a hot topic as of late that an entire book chapter is dedicated solely to this topic (see chapter 5).

Cells were being accumulated on the back distal surfaces only when utilizing fixed-angle centrifugation.

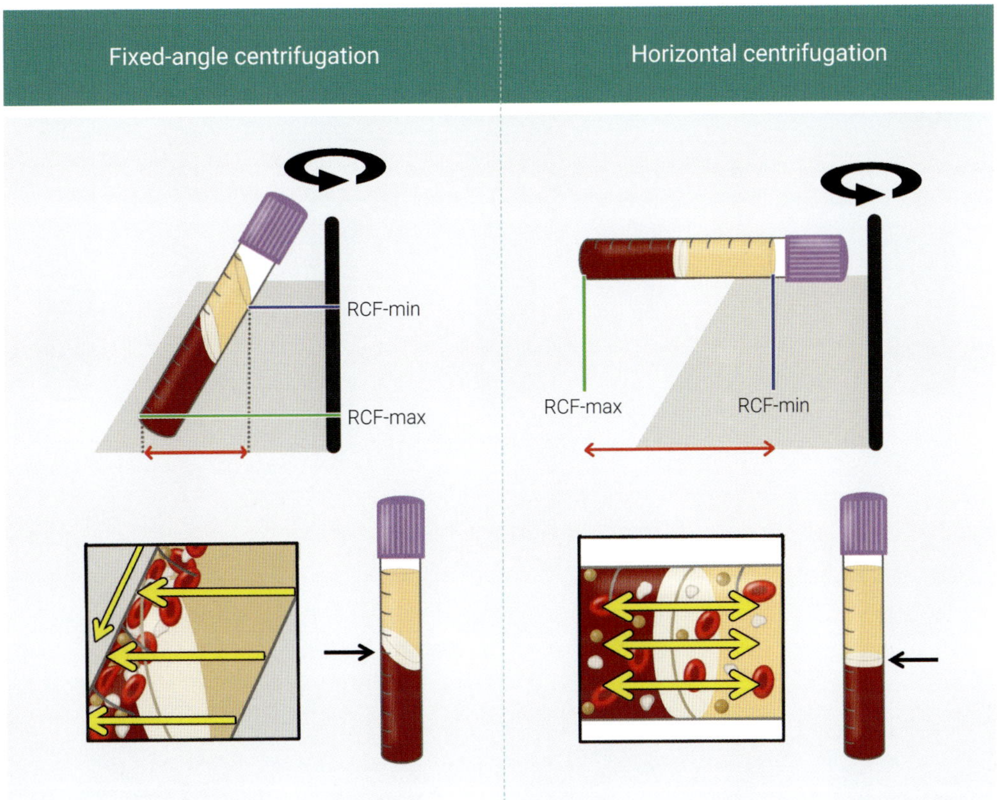

Fig 3-8 Illustrations comparing fixed-angle and horizontal centrifuges. With horizontal centrifugation, increased separation of blood layers based on density is achieved due to the increased difference in RCF-min and RCF-max. Following centrifugation on fixed-angle centrifuges, blood layers do not separate evenly, and as a result, an angled blood separation is observed. In contrast, horizontal centrifugation produces even separation. Owing to the large RCF values (~200g–700g), the cells are pushed toward the outside and downward. On a fixed-angle centrifuge, cells are pushed toward the back of centrifugation tubes and then downward/upward based on cell density. These g-forces produce additional shear stress on cells as they separate based on density along the back walls of centrifugation tubes. In contrast, horizontal centrifugation allows for the free movement of cells to separate into their appropriate layers based on density, allowing for better cell separation and less trauma/shear stress on cells. (Adapted from Miron et al.[1])

Video 3-2

with H-PRF, observing the morphology of cells and their localization on the surface of PRF clots by SEM and histologically by transaxial frozen sections by means of a film method.[5] It was consistently observed that L-PRF clots demonstrated a sloped separation between the upper plasma and the bottom RBC layers according to the angle of the rotor. Interestingly, red dots were often observed on the distal walls of the tubes in the upper layers, consisting of aggregations of RBCs, leukocytes, and platelets by SEM and histology[5] (Fig 3-9). Clots produced on the horizontal centrifuge showed much smoother cell layer distribution and separation along the tube surfaces when compared to L-PRF.

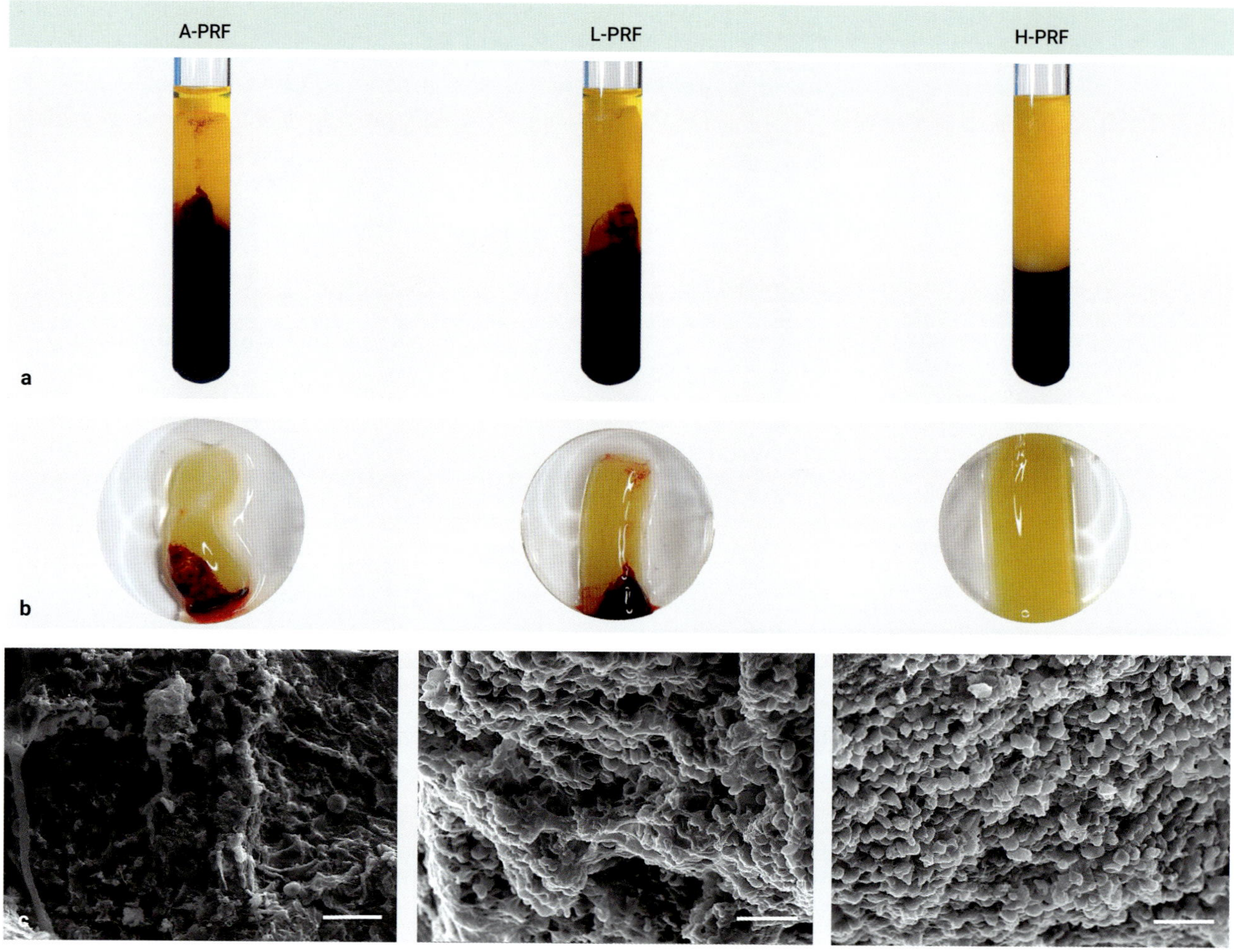

Fig 3-9 Characterization of A-PRF, L-PRF, and H-PRF. *(a and b)* The morphology and size of A-PRF, L-PRF, and H-PRF after centrifugation using the manufacturer's recommended centrifugation protocols and tubes. *(c)* SEM of A-PRF, L-PRF, and H-PRF clots. Scale bar = 20 μm. (Reprinted with permission from Zhang et al.[6])

Microscopic and histologic observation of L-PRF and H-PRF

PRF clots were further investigated by SEM and histologic assessment for cell distribution and surface configurations (Figs 3-10 to 3-13). Three distinct patterns were observed on the distal walls of L-PRF clots, including RBC clusters on the smooth or wavy fibrin clot surfaces as well as clusters of leukocytes, platelets, and crushed RBCs (see Fig 3-10). Histologic observation further confirmed these pattern types (see Fig 3-11). The border between the plasma and RBC layers included more dense fibrin networks covered with many blood cells (see Fig 3-10e). Many leukocytes were found at this layer (see Fig 3-11d).

Within H-PRF clots, two typical patterns were observed on the surface, including fewer blood cells on the smooth clot surfaces, with more found located on the rough surfaces (see Fig 3-12). Abundant platelets were found within the clots, with a few clusters located on the actual clot surface (see Fig 3-13). The border between the plasma and RBC layers included a dense fibrin network with many leukocytes.

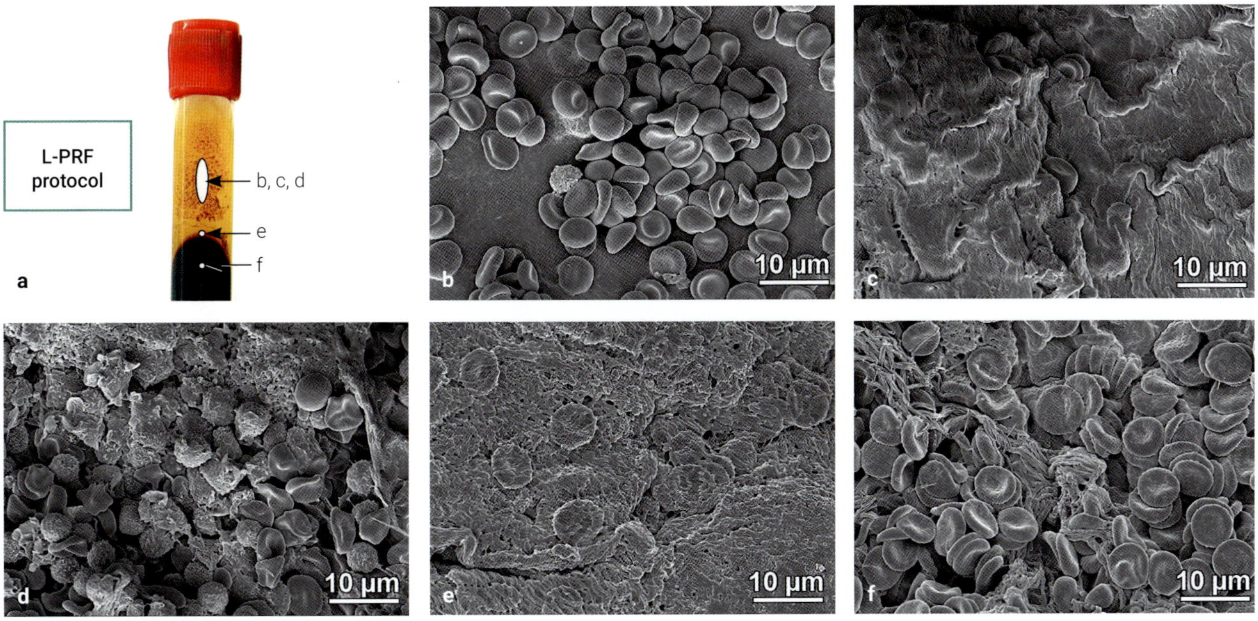

Fig 3-10 SEM images of the distal surface of PRF clots prepared utilizing the L-PRF protocol. *(a)* The described areas observed by SEM. The L-PRF clot surfaces showed typically three types, as shown in *b* to *d*. *(b)* Clusters of RBCs were observed overlaying a smooth clot surface. *(c)* A wavy surface was observed, including RBCs. *(d)* The rough surface included leukocytes, platelets, and crushed RBCs. *(e)* The dense fibrin networks were observed, including RBCs at the border between the yellow plasma and RBC layers. *(f)* Many RBCs were observed within a fibrin network in the RBC layer. (Reprinted with permission from Fujioka-Kobayashi et al.[5])

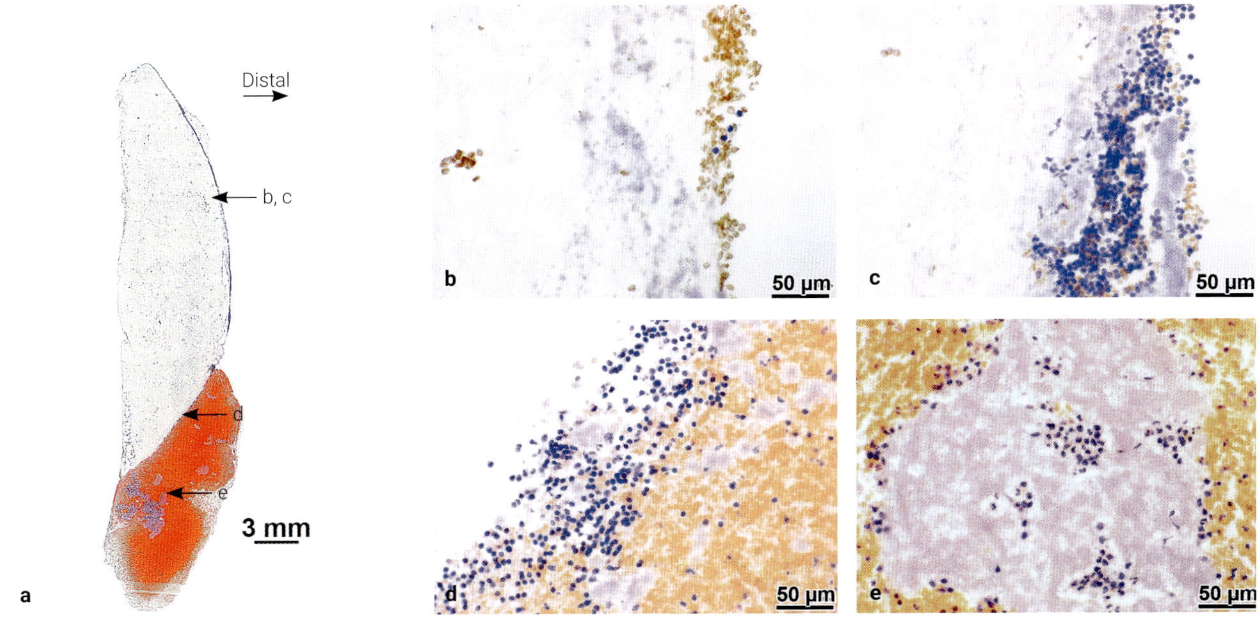

Fig 3-11 Histologic observation of the frozen section of L-PRF sectioned transaxially. *(a)* The panoramic view of the sections from the whole PRF clot including the RBC layer stained with hematoxylin. The L-PRF clots and RBC layer were separated by a fixed-angle centrifuge. The distal wall showed two typical patterns, shown in *b* and *c*. *(b)* Clusters of RBCs with a few leukocytes were located on fibrin networks on the distal surface. *(c)* The aggregated cluster consisting of platelets, leukocytes, and RBCs occasionally observed. *(d)* Many leukocytes were located at the border between the PRF clot and the RBC layer. *(e)* The aggregated clusters of cells containing leukocytes were occasionally observed within the RBC layer within the red buffy coat zone. (Reprinted with permission from Fujioka-Kobayashi et al.[5])

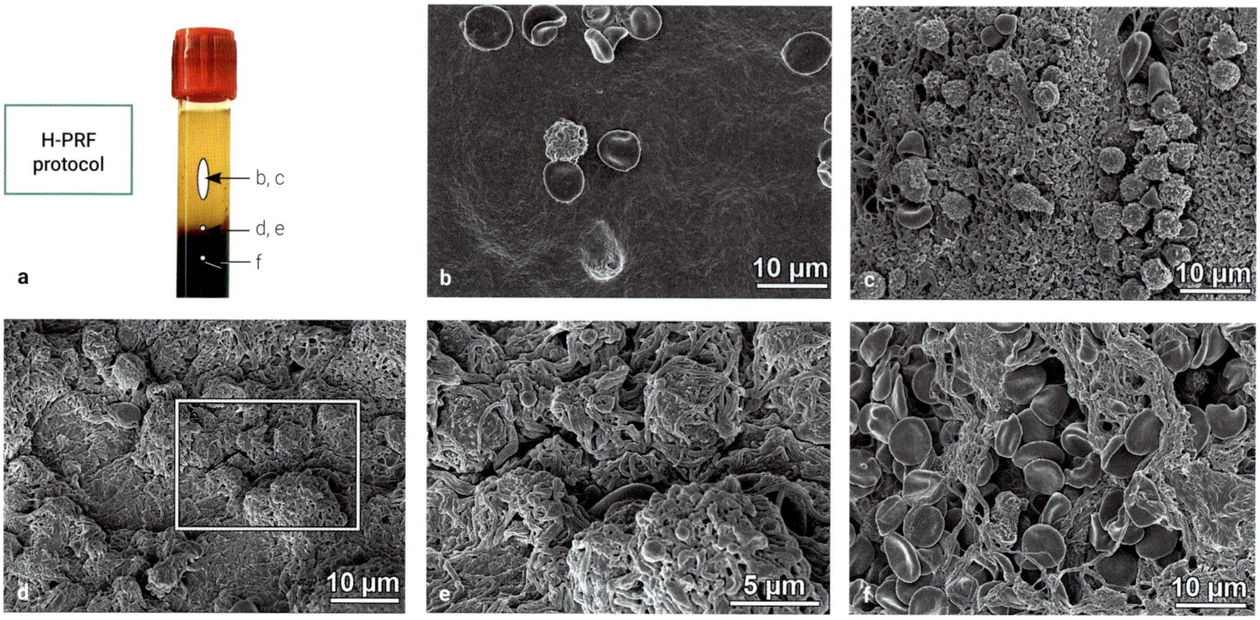

Fig 3-12 SEM images of the distal surface of PRF clots prepared utilizing the H-PRF protocol. *(a)* The described areas observed by SEM. The H-PRF clot surfaces showed typically two types, as shown in *b* and *c*. *(b)* Few leukocytes and RBCs were observed on the smooth surfaces. *(c)* The rough surfaces included leukocytes, platelets, and RBCs. *(d and e)* The fibrin networks twisted around the leukocytes with platelets at the border between the yellow plasma and RBC layers. *(f)* Many RBCs were observed within a fibrin network in the RBC layer. (Reprinted with permission from Fujioka-Kobayashi et al.[5])

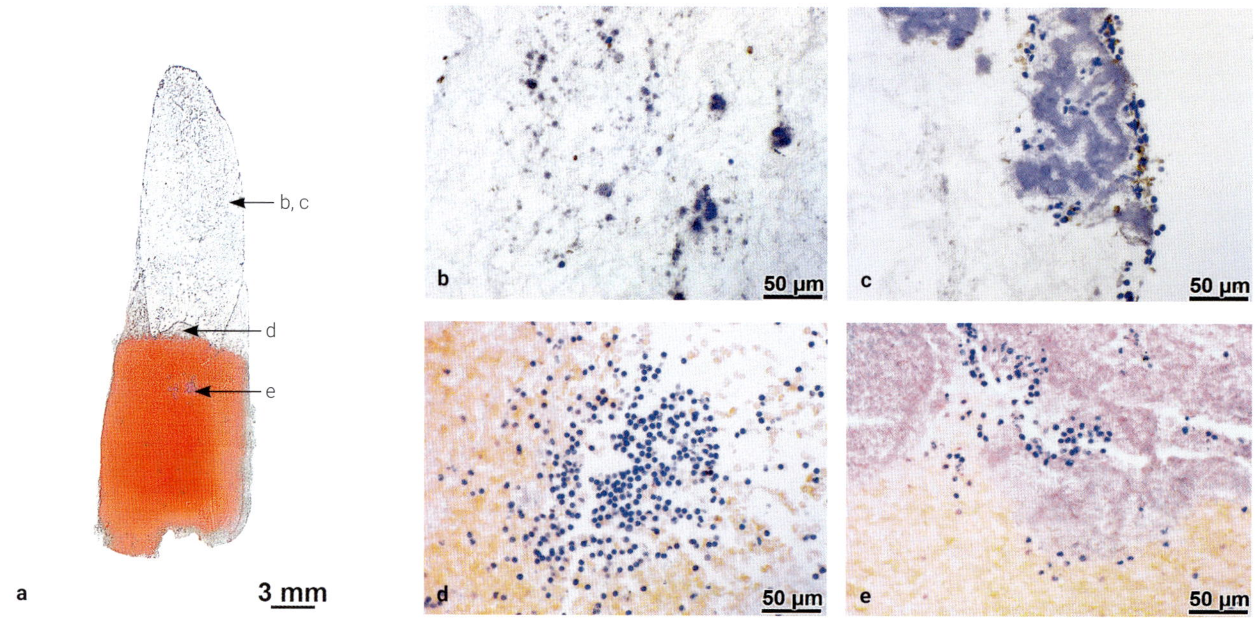

Fig 3-13 Histologic observation of the frozen section of H-PRF sectioned transaxially. *(a)* The panoramic view of the sections from the whole PRF clot including the RBC layer stained with hematoxylin. The H-PRF clots and RBC layers were separated evenly and horizontally with no obvious accumulation of cells on the distal surface. The clots showed two typical patterns shown in *b* and *c*. *(b)* The fibrin networks were observed in the clots with many platelets and few leukocytes/RBCs. *(c)* Aggregated clusters consisting of leukocytes and a few RBCs were occasionally observed. *(d)* Many leukocytes were located at the border between the clots and RBC layer. *(e)* Aggregated clusters of cells containing leukocytes were occasionally observed in the RBC layer within the red buffy coat zone. (Reprinted with permission from Fujioka-Kobayashi et al.[5])

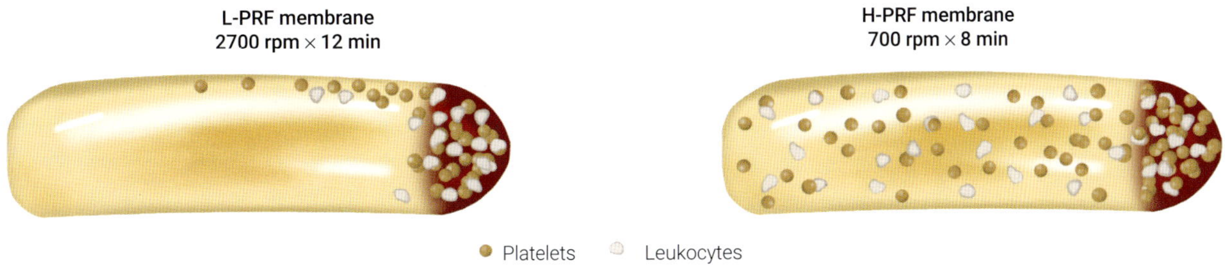

Fig 3-14 Graphic demonstrating cell distribution within PRF when centrifugation was carried out either by fixed-angle or horizontal centrifugation. Note that the majority of cells following the L-PRF protocol are found along the back distal surface of PRF tubes as well as primarily contained within the buffy coat layer. A more even distribution of cells was observed when horizontal centrifugation was utilized. (Reprinted with permission from Fujioka-Kobayashi et al.[5])

Interestingly, aggregated clusters of platelets with leukocytes were found in both L-PRF and H-PRF within the RBC layer, approximately 5 mm below the precise separation typically referred to as the *buffy coat zone*. A representative summary figure is provided in Fig 3-14.

Optimization of PRF Protocols

To this day, PRF has not been most efficiently optimized. Additionally, in the early 2000s, a variety of publications on the topic of PRF were utilized at different RCF/rpm parameters. As highlighted in chapter 4, many were actually using the same rpm values for devices with different rotor sizes (which completely changes the g-force) without understanding its pronounced impact on cell layer separation.

It is important to note that larger-radius centrifuges produce much greater g-force even at identical rpms.

One of the most common limitations to PRF is the fact that various protocols have never been investigated in studies. In 2014, Ghanaati et al discovered a way to further optimize the production of PRF using three different protocols and by gradually reducing RCF. By doing so, he discovered that more cells could be obtained in the upper PRF layers; this method has since been named the *low-speed centrifugation concept* (LSCC).[7]

In 2019, drastically better results were obtained utilizing horizontal centrifugation. While 20% to 30% better results were obtained with the LSCC, the ability to simply shift from fixed-angle to horizontal centrifugation led to as much as a fourfold increase in cells. Our research team then investigated 24 different protocols (instead of the original 3) to better optimize PRF.

Evaluation of 24 protocols for the production of PRF

While in previous studies few protocols were compared (generally at most three),[7] the desire and emphasis of our research group was to better investigate for the first time the effect of numerous centrifugation parameters on the final production of PRF. As such, 24 different protocols were evaluated (Fig 3-15).[8] All protocols were compared utilizing a recent method to quantify cells in PRF in 1-mL sequential layers pipetted from the upper layer downward until all 10 mL were harvested (see chapter 2 for methodology). In total, 960 complete blood counts (CBCs) were investigated. Both solid- and liquid-based PRF protocols were investigated following 24 protocols involving six RCF values (100g, 200g, 400g, 700g, 1000g, and 1200g) at four centrifugation times (3, 5, 8, and 12 minutes).

Figure 3-16 demonstrates the overall final volume of plasma of each of the protocols, along with the total yield and concentrations of platelets and leukocytes above baseline values. Note that of each of the protocols, 200g for 5 minutes led to the highest concentration of platelets/leukocytes. The best yield of leukocytes was achieved after centrifugation for 8 minutes utilizing the 700g, 1000g, and 1200g protocols. Of those, the highest concentration was achieved at 700g for 8 minutes (owing to the reduced total plasma volume). Another noteworthy trend that was apparent was that as centrifugation time was increased, a general increase in percent yield was observed; however, generally speaking, an overall decrease in concentration was also observed (see Fig 3-16). Furthermore, it was apparent

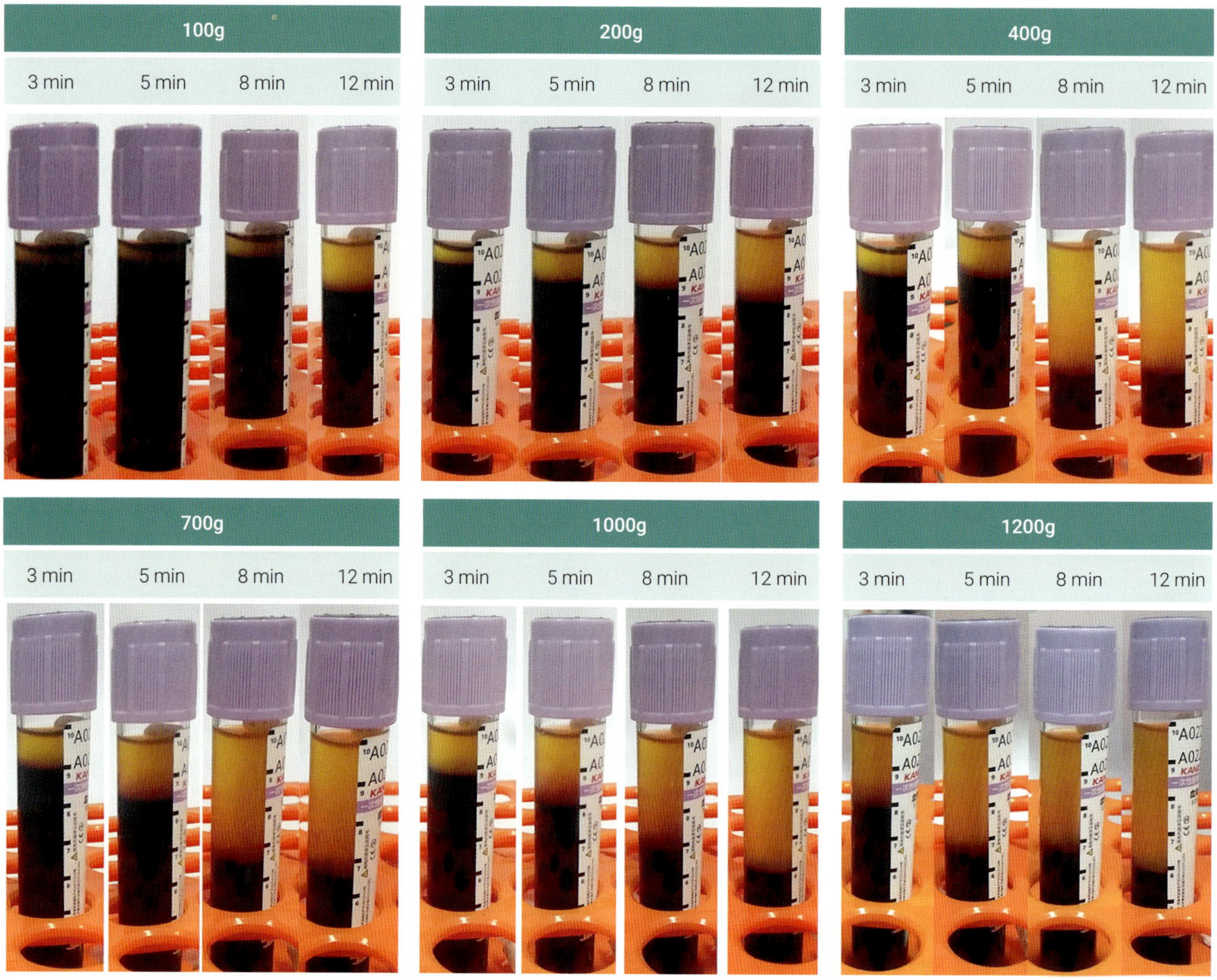

Fig 3-15 Clinical image demonstrating the plasma layer separation for the 24 protocols investigated in this study. Note that while some protocols reveal roughly identical plasma layer separation, the underlying cellular content in the various protocols may be drastically different. (Reprinted with permission from Miron et al.[8])

that certain centrifugation protocols that were too reduced in RCF (such as 100g) typically did not lead to adequate yield of cells. This relates with our group's previous work on i-PRF demonstrating that these low centrifugation speeds and time (~800 rpm for 3–4 minutes) led to substandard concentrations of platelets and leukocytes (see chapter 2). Protocols that were too fast or lengthy (1000g or more) led to a reduction in yield and/or concentrations (as more cells then got pushed into the bottom layers or the volume of total plasma led to a reduction in concentration).

To simplify Fig 3-16, certain time points were removed from the graph to facilitate its understanding for the reader. Figure 3-17 demonstrates only two protocols (700g and 1000g) over time. Note that in Fig 3-17a, a general increase of centrifugation time is associated with an increase in platelet yield. Note, however, in Fig 3-17b that an increase in centrifugation time actually decreases the concentration of platelets (because the plasma volume is increased, so even if the total yield of platelets remains the same or even slightly higher, the actual concentration decreases).

In Fig 3-18, the 100g protocol is included as well. Notice here how the yield is extremely low in platelets (Fig 3-18a) as well as in concentration (Fig 3-18b). This is a result of the speed cycle simply being so slow that it is unable to

Fig 3-16 *(a to e)* Evaluation of 24 protocols utilized for the production of PRF. Data includes final volume (mL), total leukocyte and platelet yields (% of the total from 10 mL), as well as concentration of leukocytes and platelets above baseline values (% increase). (Reprinted with permission from Miron et al.[8])

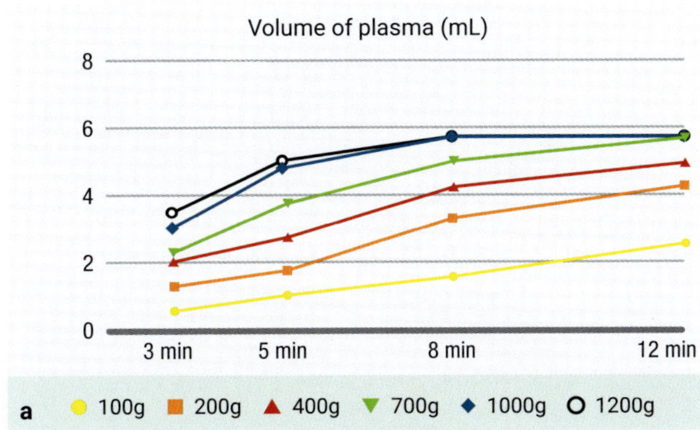

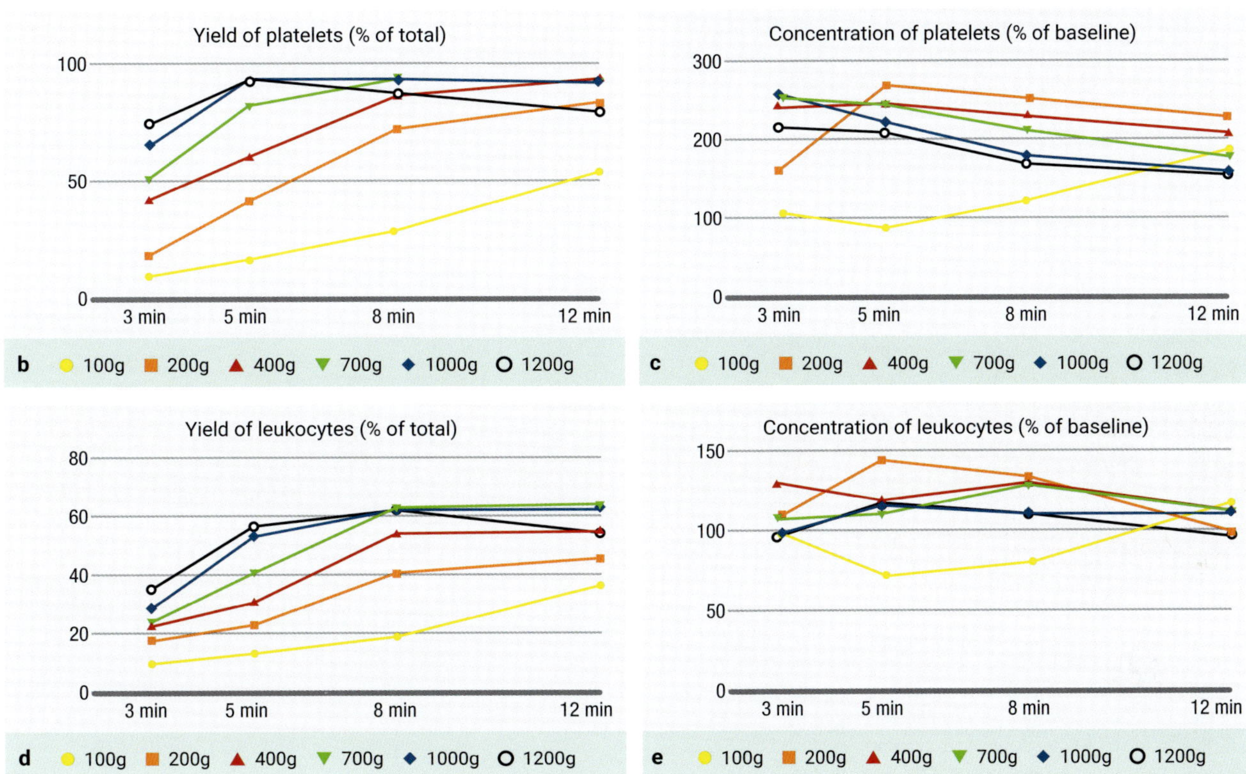

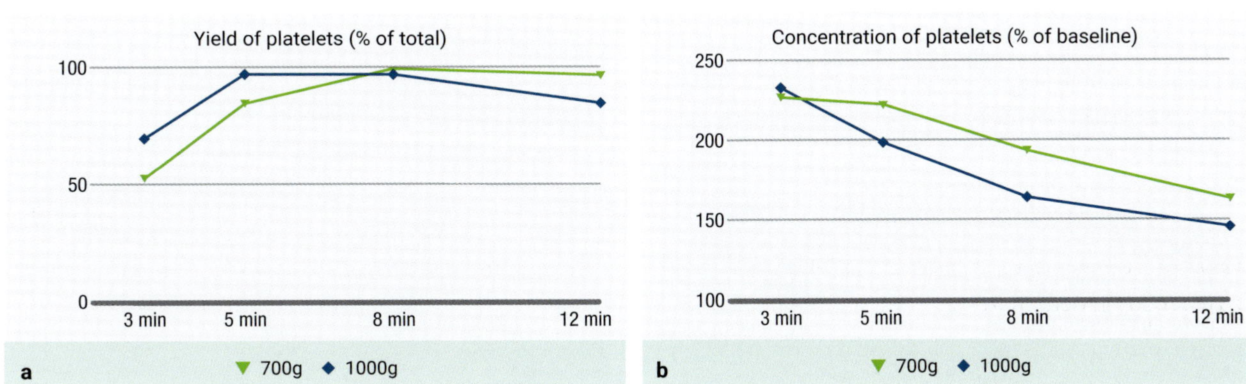

Fig 3-17 Evaluation of protocols utilizing both 700g and 1000g RCF for the production of PRF. Both total yield of platelets *(a)* as well as concentration of platelets *(b)* are depicted. (Reprinted with permission from Miron et al.[8])

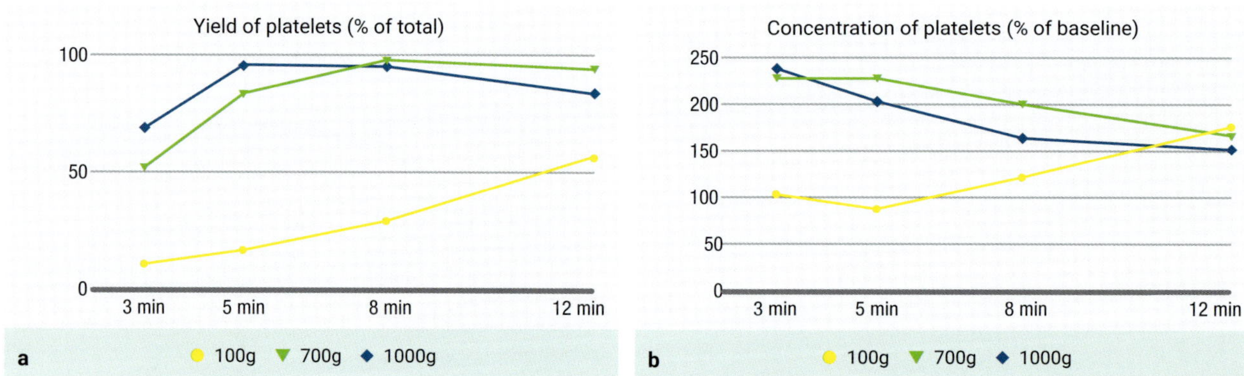

Fig 3-18 Previous graphs demonstrating 700g and 1000g protocols but with the addition of 100g results as well. It is very clear that the 100g protocol is not able to accumulate high yields of platelets *(a)*, and the concentration remains low *(b)*.

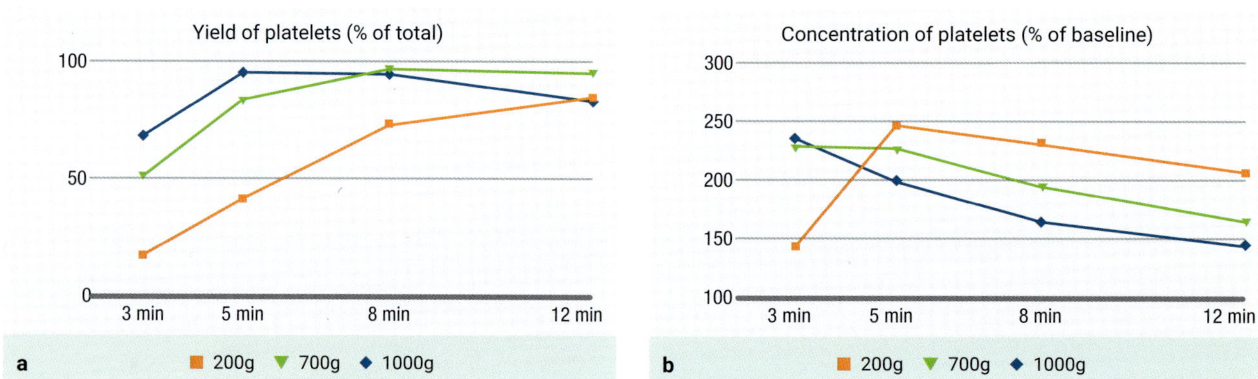

Fig 3-19 Evaluation of protocols utilizing both 700g and 1000g RCF for the production of PRF but now with the addition of the 200g protocols. Both total yield of platelets as well as concentration of platelets are depicted. Note that when a 200g protocol is utilized, while the yield is still low *(a)*, the concentration peaks at 5 minutes at a level higher than any other group *(b)* (Reprinted with permission from Miron et al.[8])

accumulate or concentrate platelets in the upper layer. As alluded to in chapter 2, it is possible to centrifuge *too* slowly to the point where platelets and leukocytes do not actually accumulate effectively in the upper layers. This is a common misconception that many clinicians maintain due to inaccurate information provided by various manufacturers.

Centrifugation carried out too slowly is a common misconception that many clinicians maintain due to inaccurate information provided by various manufacturers.

In Fig 3-19, observe the 200g protocol. Notice how at the higher g-force, the cells are actually able to accumulate more efficiently (Fig 3-19a). More importantly, observe the concentration of platelets in Fig 3-19b following a 200g protocol for 5 minutes. Here the platelets are actually most concentrated after 5 minutes, and thereafter, even though their yield continues to rise (see Fig 3-19a), the concentration actually begins to decrease because of the increase in liquid-PRF volume. Therefore, a 200g to 300g centrifugation cycle is most effectively able to concentrate platelets and leukocytes (300g is the one chosen after further optimization).

Discussion

In general, platelets were evenly distributed throughout a variety of protocols within the upper three to six plasma-rich layers (see Fig 3-16), but it was obvious that WBCs required more pristine fine-tuning to reach adequate harmony within the upper plasma layers. Protocols within the 400g to 700g (5- to 8-minute) range were better able to accumulate and distribute platelets/leukocytes more evenly throughout the upper layers.

Currently, one standard in the field of PRF is the novel use of injectable i-PRF.[9] Previously, our research group found that only slight increases in platelets and leukocytes were noted, with failure to adequately accumulate cells in the upper plasma layer owing to extremely low RCF values (60g) and centrifugation times (3–4 minutes) on fixed-angle centrifugation devices.[1] In a study titled "Injectable platelet-rich fibrin: Cell content, morphological, and protein characterization,"[10] Varela et al observed only a slight increase in platelets (less than 33%) and leukocytes following i-PRF protocols, with decreases in VEGF reported when compared to that in whole blood. Altogether, these studies confirm that previously utilized i-PRF protocols (~60g for 3–4 minutes on a fixed-angle centrifuge) are inadequately effective at separating blood cell layers due to their considerable reduction in centrifugation speed and time. Furthermore, protocols at 100g or lower are inefficient at accumulating platelets and leukocytes in the upper plasma layer, highlighting the limits of the LSCC.

It was also observed within this study that the use of PRF produced using a protocol of 200g for 5 minutes resulted in the highest concentration of platelets and leukocytes. Within these studies, up to a fourfold increase in platelet/leukocyte concentration and/or yield was observed when compared to the results of previously utilized i-PRF protocols produced on a fixed-angle centrifuge.[1] Further unpublished data has found that a 300g protocol for 5 minutes resulted in the highest concentration of liquid-PRF, as discussed further in chapter 5. Nevertheless, based on the C-PRF protocols established in chapter 2, it remained obvious that better concentrations could be achieved by further modifying centrifugation parameters.

Establishing C-PRF on a Horizontal Centrifuge

In a recent study titled "Improved growth factor delivery and cellular activity using concentrated platelet-rich fibrin (C-PRF) when compared to traditional injectable (i-PRF) protocols," Fujioka-Kobayashi et al aimed to investigate and optimize PRF in its most concentrated formulation.[11] As reviewed in chapter 2, the ability to centrifuge at higher centrifugation speeds and times leads to an accumulation directly at the buffy coat layer.[12] There was an approximately tenfold increase in baseline concentrations specifically in this 0.3- to 0.5-mL buffy coat layer directly above the RBC layer produced using higher centrifugation protocols.[12] The PRF obtained from this harvesting technique was given the working name *concentrated-PRF* (C-PRF). Figure 3-20 demonstrates a clinical photograph of standard liquid-PRF protocols versus those of C-PRF. It was hypothesized that based on the extensive increase in the yield of platelets and leukocytes, C-PRF would exhibit higher GF release as well as superior cellular activity. Therefore, the aim of this study was twofold. First, a new centrifugation protocol was developed on a horizontal centrifugation system with the aim of accumulating the greatest concentrations of platelets and leukocytes within the buffy coat. The second aim was to compare the total GF release of PRF obtained through this newly developed C-PRF protocol to that of PRF obtained through the clinically utilized liquid i-PRF protocol over a 10-day period and to investigate the regenerative properties of human gingival fibroblasts in vitro.

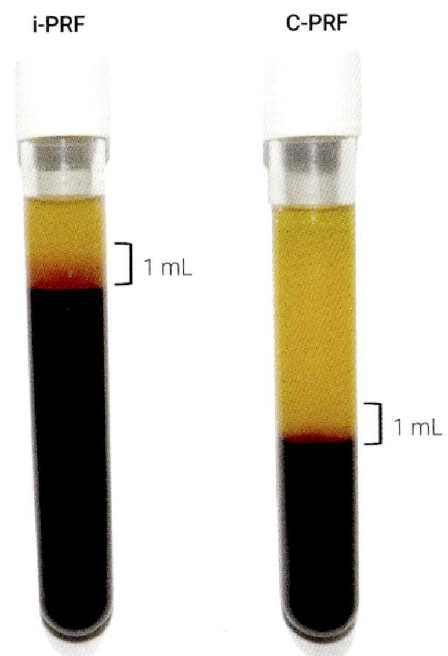

Fig 3-20 Visual representation of layer separation following either the i-PRF (300g for 5 minutes) or C-PRF protocol (3000g for 8 minutes). Plasma was collected from the buffy coat region within the 1-mL layer above the RBC layer. (Reprinted with permission from Fujioka-Kobayashi et al.[11])

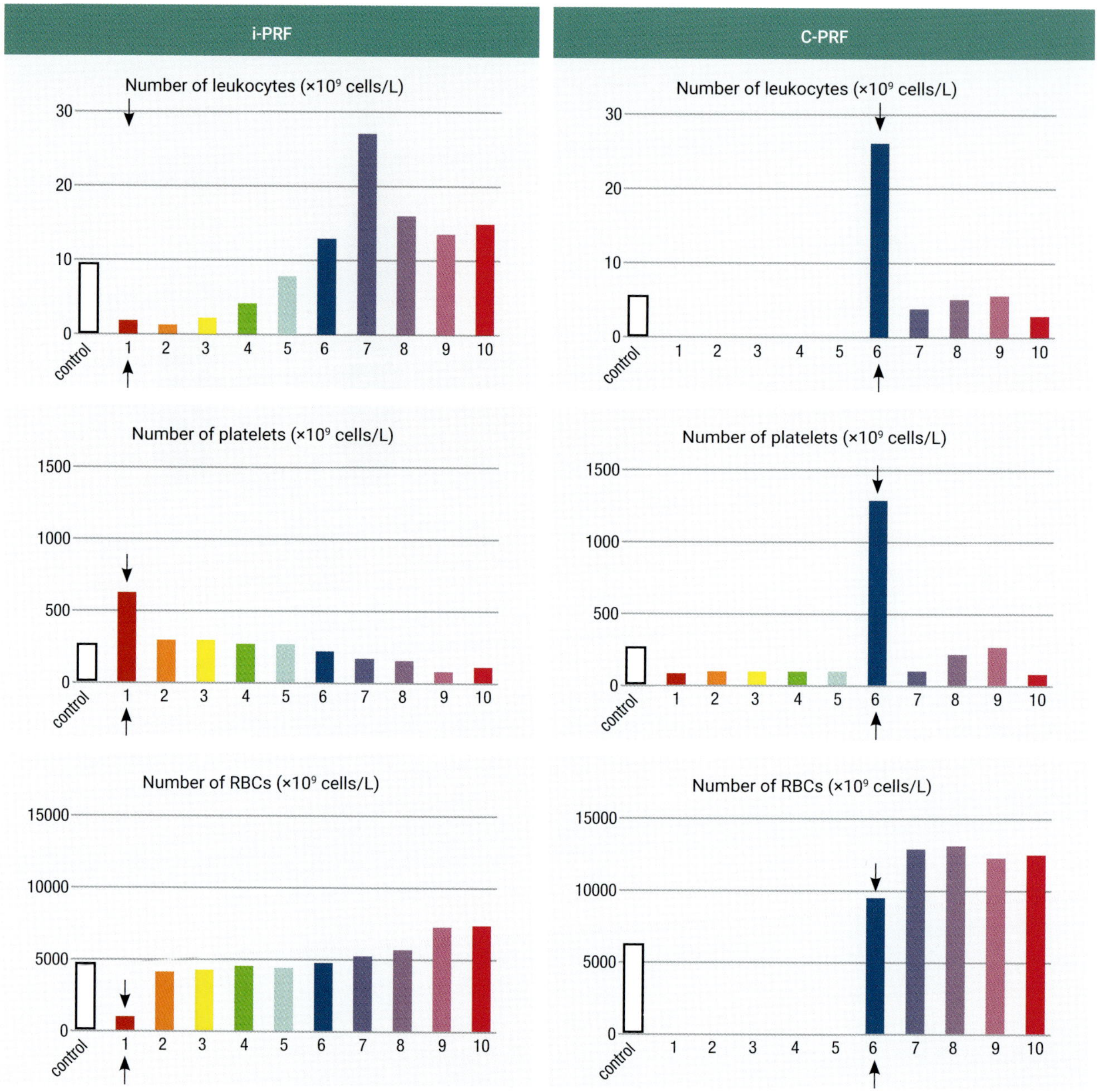

Fig 3-21 The concentrations of different cell types found in each 1-mL layer of the 10-mL tube obtained through the i-PRF protocol (300g for 5 minutes) and the C-PRF protocol (3000g for 8 minutes). Note that in the PRF obtained through the i-PRF protocol, the majority of platelets and leukocytes were located in the 1-mL buffy coat layer. In the PRF obtained through the C-PRF protocol, although higher concentrations of platelets and leukocytes were found in the upper 1-mL layer, the majority of the platelets and leukocytes were actually located in the RBC layers. (Reprinted with permission from Fujioka-Kobayashi et al.[11])

Optimization of C-PRF protocols

Prior to initiating the C-PRF experiments, protocols of 3000g for 5 minutes, 8 minutes, and 12 minutes were compared to optimize the accumulation of cells within the buffy coat layer. It was found that the 5-minute protocol was unable to concentrate all cells within the buffy coat layer, with the platelets remaining in the upper 4 mL. Following the 8-minute protocol, the sequential pipetting method results revealed that the majority of platelets and leukocytes were located within the buffy coat layer 6 region (Fig 3-21). No further advantage was observed following the 12-minute protocol (data not shown). For comparison purposes, a standard i-PRF protocol resulted in a slight concentration

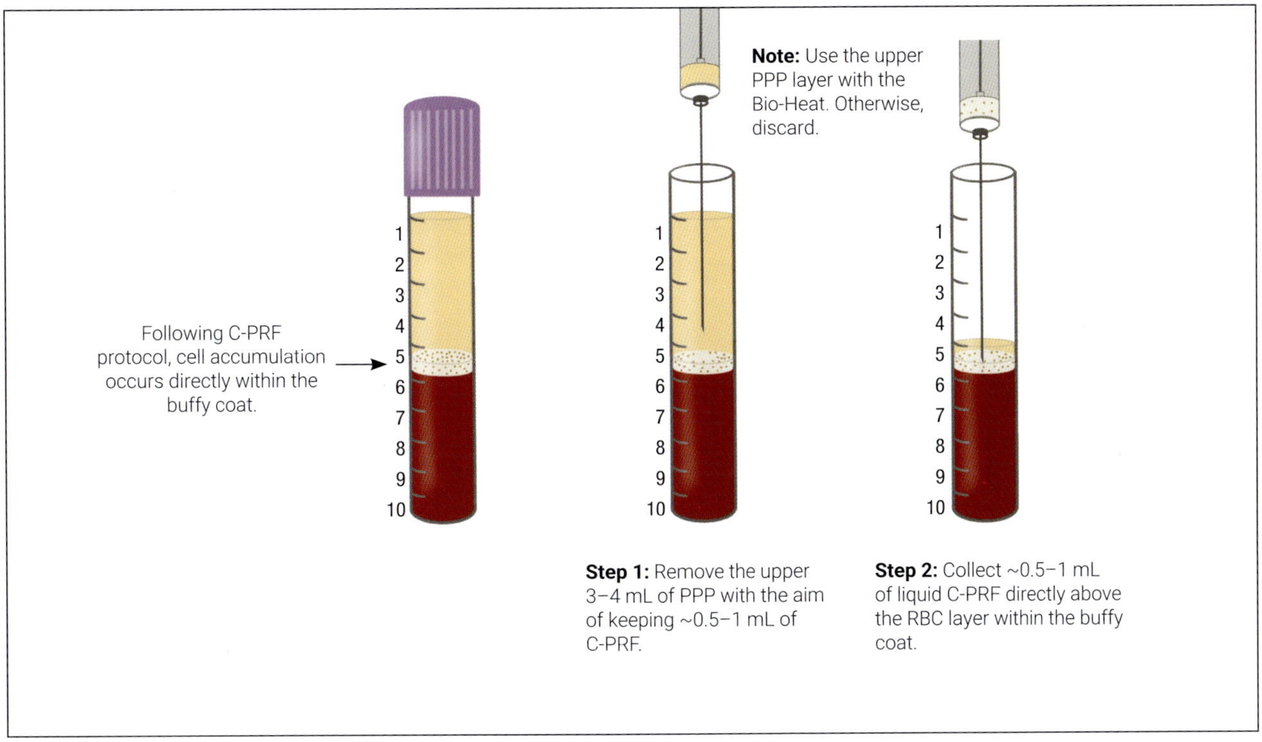

Fig 3-22 Method to collect and concentrate C-PRF. Following centrifugation at higher speeds (2000g for 8 minutes), the majority of cells are located directly at the buffy coat layer. Instead of attempting to remove this layer with a long needle into the deep layers, it is highly advised to first remove the upper 4 mL of platelet-poor plasma (PPP), followed by collection of the C-PRF buffy coat layer.

of platelets and leukocytes in the upper 1-mL layer from which i-PRF was harvested (see Fig 3-21). Note that many platelets/leukocytes, however, remained in the RBC layers. To harvest C-PRF, a 1-mL layer within this buffy coat layer was collected and processed for further analysis (Fig 3-22).

Tips

- In clinical practice, it is best to harvest C-PRF by first removing the upper 3 to 4 mL of platelet-poor plasma (discard it). The remaining C-PRF layer can then be taken much more easily. It is much more difficult to attempt to retrieve this buffy coat zone with 5 mL over top of it; it is harder to concentrate it, and too much volume is often collected.
- Many centrifuges may not reach the 3000g speed. A 2000g protocol for 8 minutes will also achieve a C-PRF layer.

Comparative GF release between i-PRF and C-PRF

In the first set of experiments, the release of GFs including PDGF-AA, PDGF-AB, PDGF-BB, TGF-β1, VEGF, EGF, and IGF-1 from i-PRF and C-PRF was investigated by ELISA (Fig 3-23). The release of all GFs over the entire 10-day (240-hour) period was significant for both protocols, with the C-PRF protocol resulting in up to two- to threefold higher quantities. This demonstrated clearly that this newly developed protocol had much greater regenerative potential when compared to previously utilized i-PRF protocols.

Biocompatibility and cellular activity of i-PRF and C-PRF

It was first observed that while i-PRF induced a twofold increase in cell migration when compared to that observed in the control, a significantly higher fourfold increase was observed when cells were cultured with C-PRF (Fig 3-24a).

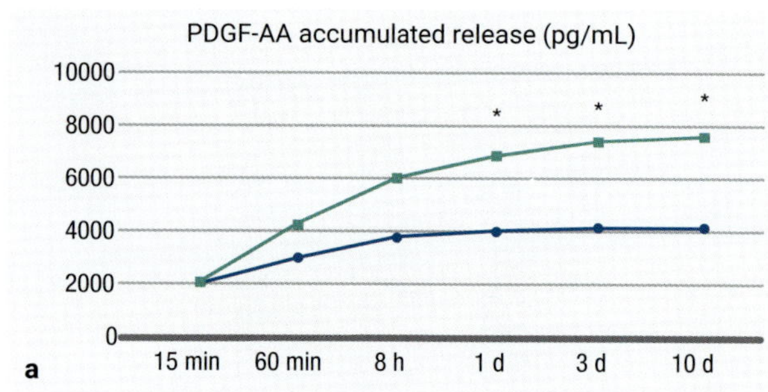

Fig 3-23 Protein quantification of PDGF-AA *(a)*, PDGF-AB *(b)*, PDGF-BB *(c)*, TGF-β1 *(d)*, VEGF *(e)*, EGF *(f)*, and IGF-1 *(g)* at each time point over a 10-day (240-hour) period for PRF obtained through the i-PRF protocol and the C-PRF protocol, as determined by ELISA. Data represents the mean ± SE; an *asterisk* (*) indicates a value significantly higher than the other group ($P < .05$). (Reprinted with permission from Fujioka-Kobayashi et al.[11])

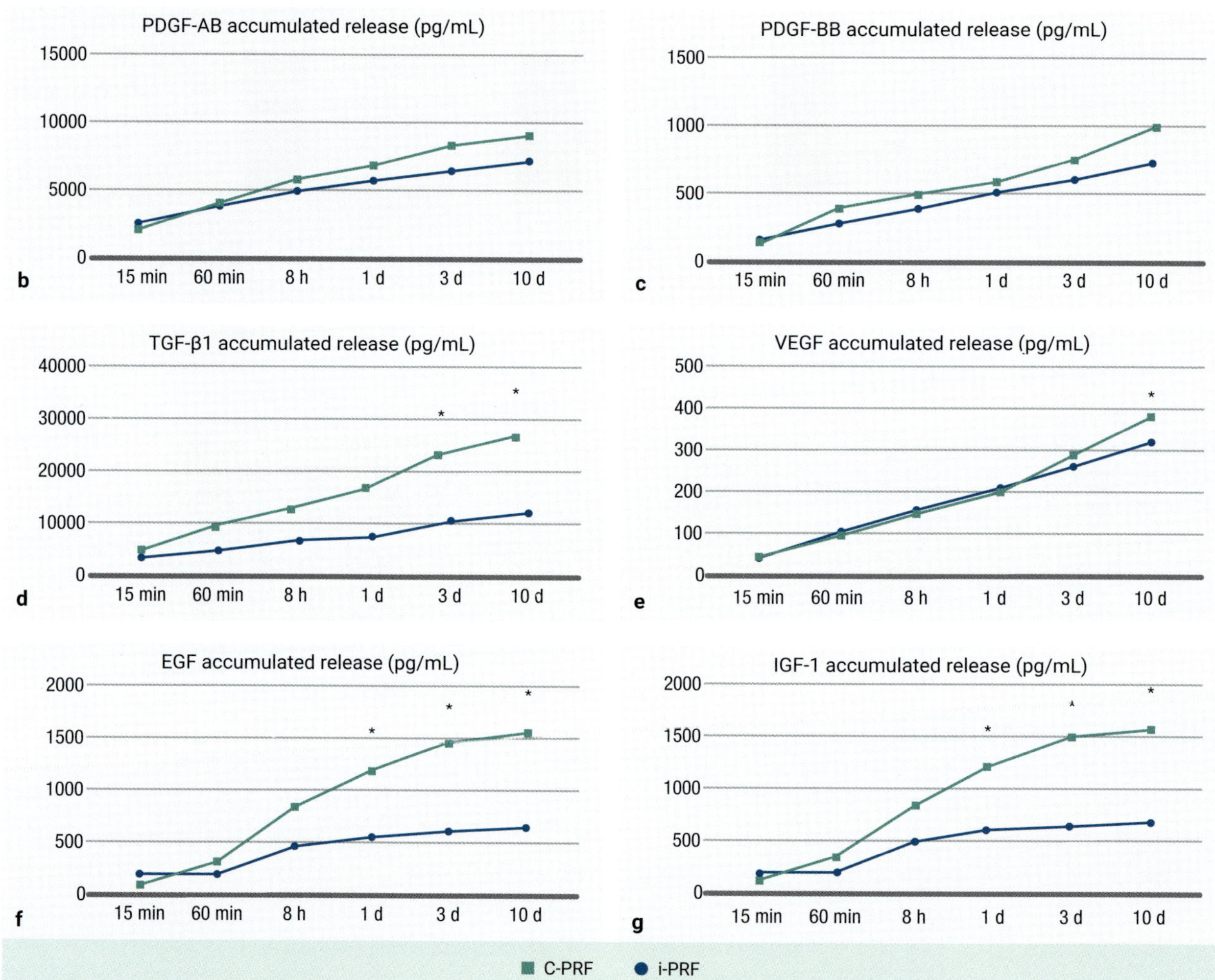

Furthermore, C-PRF also induced significantly higher cell proliferation at 3 and 5 days postseeding when compared to i-PRF (Fig 3-24b). Both i-PRF and C-PRF were able to significantly upregulate TGF-β 3 days postseeding (Fig 3-24c), and a significant 250% and 400% increase in the PDGF-AA was observed for i-PRF and C-PRF, respectively (Fig 3-24d). The analysis of COL1 immunostaining also revealed significantly higher COL1A staining for C-PRF when compared to i-PRF and control tissue culture plastic groups (Figs 3-24e and 3-24f).

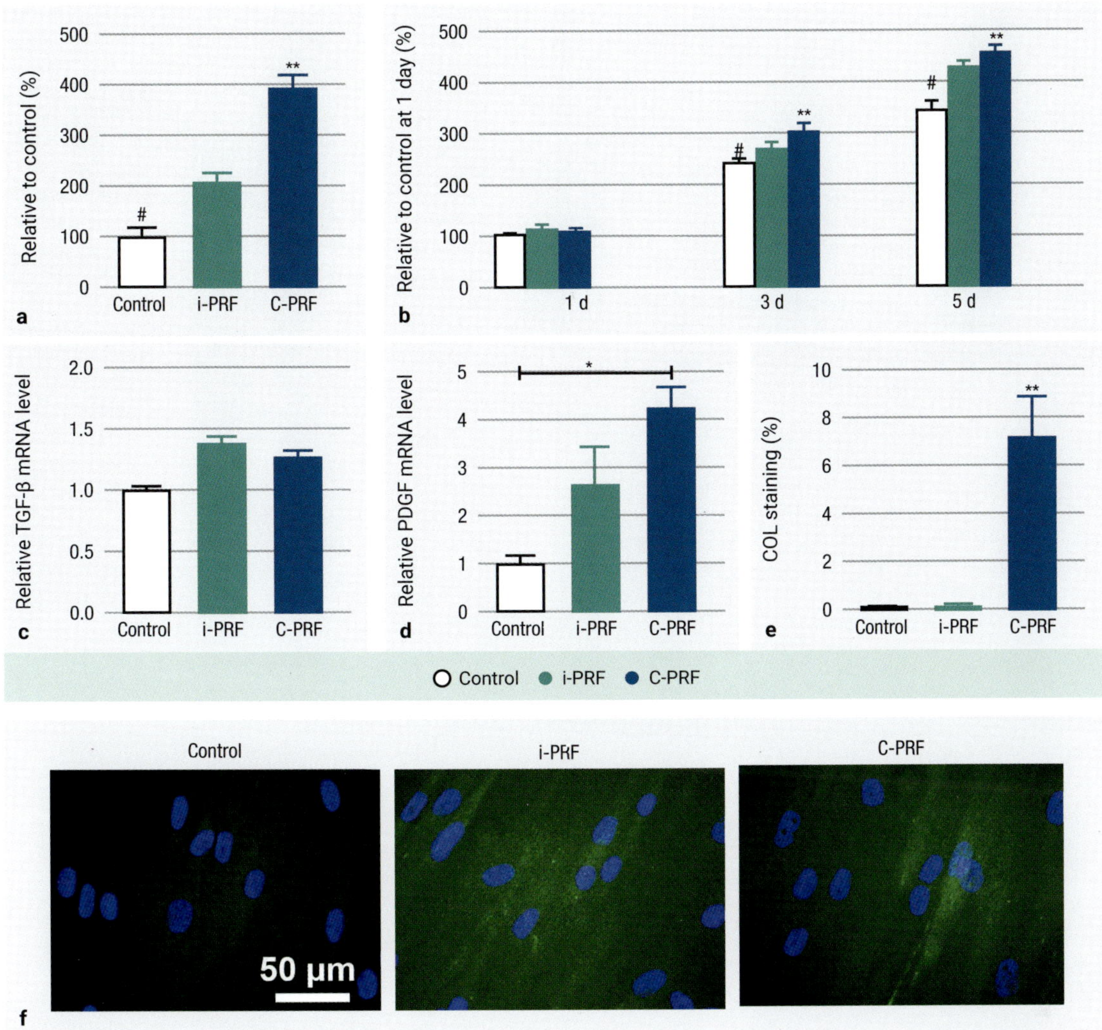

Fig 3-24 *(a and b)* Cell migration and proliferation at 1, 3, and 5 days in human gingival fibroblast (HGF-1) cells. *(c and d)* RT-PCR analysis of mRNA levels of TGF-β and PDGF in human gingival fibroblasts treated with i-PRF and C-PRF at 3 days. *(e and f)* Quantitative and representative staining of collagen1 at 14 days. Data represents mean ± SE; an *asterisk* (*) indicates a value significantly higher than the control group ($P < .05$); a *double asterisk* (**) indicates a value significantly higher than all other groups ($P < .05$); a *number sign* (#) indicates a value significantly lower than all groups ($P < .05$). (Reprinted with permission from Fujioka-Kobayashi et al.[11])

Conclusion

In summary, the results from the 24-protocol investigation demonstrated clearly that certain protocols were better at increasing the amount of platelet/leukocyte yield (400g to 700g for 8 minutes), whereas others were more effective at concentrating PRF (200g to 300g for 5 minutes). Therefore, new protocols were designed accordingly for solid-PRF and liquid-PRF. While histologically it was observed that horizontal centrifugation led to better cell layer separation of blood cells as well as more even distribution of cells throughout the PRF clot, it was also histologically observed that the majority of cells accumulated on the back distal surfaces of PRF tubes when fixed-angle centrifugation was carried out. Lastly, a new C-PRF formulation was evaluated demonstrating up to a threefold increase in GF release during a 10-day period and further elicited fourfold increases in gingival fibroblast migration, PDGF gene expression, and collagen1 synthesis when compared to standard i-PRF protocols.

References

1. Miron RJ, Chai J, Zheng S, Feng M, Sculean A, Zhang Y. A novel method for evaluating and quantifying cell types in platelet rich fibrin and an introduction to horizontal centrifugation. J Biomed Mater Res A 2019;107:2257–2271.
2. Miron RJ, Dham A, Dham U, Zhang Y, Pikos MA, Sculean A. The effect of age, gender, and time between blood draw and start of centrifugation on the size outcomes of platelet-rich fibrin (PRF) membranes. Clin Oral Investig 2019;23:2179–2185.
3. Takahashi A, Tsujino T, Yamaguchi S, et al. Distribution of platelets, transforming growth factor-beta1, platelet-derived growth factor-BB, vascular endothelial growth factor and matrix metalloprotease-9 in advanced platelet-rich fibrin and concentrated growth factor matrices. J Investig Clin Dent 2019;10:e12458.
4. Tsujino T, Masuki H, Nakamura M, et al. Striking differences in platelet distribution between advanced-platelet-rich fibrin and concentrated growth factors: Effects of silica-containing plastic tubes. J Funct Biomater 2019;10:43.
5. Fujioka-Kobayashi M, Kono M, Katagiri H, et al. Histological comparison of platelet rich fibrin clots prepared by fixed-angle versus horizontal centrifugation [epub ahead of print 18 April 2020]. Platelets doi: 10.1080/09537104.2020.1754382.
6. Zhang J, Yin C, Zhao Q, et al. Anti-inflammation effects of injectable platelet-rich fibrin via macrophages and dendritic cells. J Biomed Mater Res A 2020;108:61–68.
7. Ghanaati S, Booms P, Orlowska A, et al. Advanced platelet-rich fibrin: A new concept for cell-based tissue engineering by means of inflammatory cells. J Oral Implantol 2014;40:679–689.
8. Miron RJ, Chai J, Fujioka-Kobayashi M, Sculean A, Zhang Y. Evaluation of 24 protocols for the production of platelet-rich fibrin. BMC Oral Health (in press).
9. Miron RJ, Fujioka-Kobayashi M, Hernandez M, et al. Injectable platelet rich fibrin (i-PRF): Opportunities in regenerative dentistry? Clin Oral Investig 2017;21:2619–2627.
10. Varela HA, Souza JCM, Nascimento RM, et al. Injectable platelet rich fibrin: Cell content, morphological, and protein characterization. Clin Oral Investig 2019;23:1309–1318.
11. Fujioka-Kobayashi M, Katagiri H, Kono M, et al. Improved growth factor delivery and cellular activity using concentrated platelet-rich fibrin (C-PRF) when compared with traditional injectable (i-PRF) protocols [epub ahead of print 7 May 2020]. Clin Oral Investig doi: 10.1007/s00784-020-03303-7.
12. Miron RJ, Chai J, Zhang P, et al. A novel method for harvesting concentrated platelet-rich fibrin (C-PRF) with a 10-fold increase in platelet and leukocyte yields. Clin Oral Investig 2020;24:2819–2828.

4

Understanding Relative Centrifugal Force (G-Force)

Contributors

Richard J. Miron

Chapter Highlights

- *Understanding what RCF actually means*
- *Understanding how to calculate proper RCF and rpm values for any centrifugation system*
- *Current literature drawbacks based on misreporting of RCF values over the years*
- *Example calculations of RCF/rpm to help the practitioner optimize any centrifugation system*

Video 4-1

This chapter is designed with the sole purpose of providing clinicians a better means to understand centrifugation parameters that will permit the optimization of their centrifugation device irrespective of the centrifugation system and trade names associated. A very clear understanding of the calculations needed to convert revolutions per minute (rpm) into relative centrifugal force (RCF, or g-force) is provided utilizing simple and straightforward formulas.

Unfortunately, the quest for profit in company portfolios has led to much confusion among clinicians regarding centrifugation devices and protocols. This chapter aims to fill that void and provides the necessary steps to improve PRF protocols irrespective of the centrifugation device utilized. Furthermore, over the years, numerous scientific publications that are well cited in the literature have actually failed to accurately report g-force values in their articles, which has further led to the considerable confusion in the field. This chapter aims to clarify this confusion and provides clinicians with the necessary information to optimize any centrifugation system.

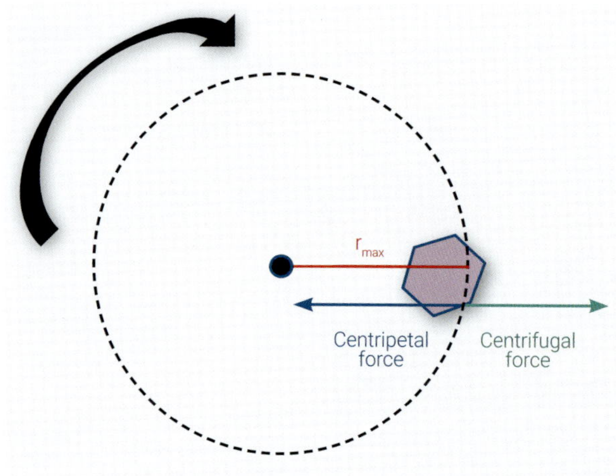

Fig 4-1 Schematic illustration of the centrifugation process during the rotation of an object *(purple symbol)* and the resulting centrifugal force *(green arrow)* and centripetal force *(blue arrow)* in relation to the maximum radius *(red line)*.

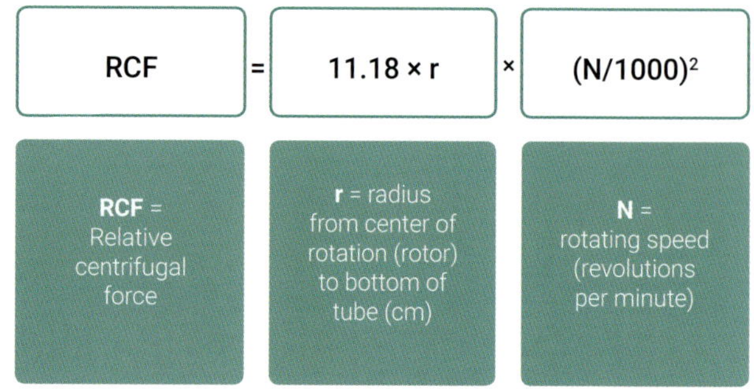

Fig 4-2 Easy formula for calculating RCF (g-force).

Definition and Calculation of RCF

The goal of centrifugation as a whole is simply to apply relative force in a circular fashion in order to separate layers based on their density. Centrifugation systems have been utilized for decades successfully and have become standard equipment in the majority of medical hospitals and research laboratories. As a basic researcher, you are taught to always report RCF values in scientific literature for obvious reasons. The reporting of g-forces during centrifugation allows for the transfer of centrifugation conditions irrespective of laboratory, country, or rotor size of the centrifugation device utilized. This is an internationally recognized way to transmit working conditions from one colleague to another for reproducibility purposes.

Simply put, during the spin cycle, the further away the object is from its spinning axis, the more g-force it receives during rotation (Fig 4-1). I personally give the following example in my courses to illustrate this point. Imagine holding a child tight in your arms and spinning them around you at 1 rotation per second. That child would receive a certain g-force. Now imagine that child holding onto a jump rope or cord and twirling/lassoing that child around a large room with the same speed of 1 rotation per second. Naturally that child would be traveling at a much greater velocity and would be subject to much higher g-forces. Therefore, the reporting of revolutions per minute (rpm) commonly utilized in PRF centrifugation in various research articles is entirely useless. The entire purpose of reporting actual RCF values is to be able to relate the g-force received on that object to another operator in another part of the world that may have a centrifugation device with a larger or smaller radius. With adequate reporting of RCF values, any colleague could adapt appropriate centrifugation conditions in order to reproduce the experiment.

> *The reporting of revolutions per minute (rpm) commonly utilized in PRF centrifugation parameters is entirely useless.*

The formula for RCF is a relatively simple one (Fig 4-2):

$$RCF = 11.18 \times r \times (N/1000)^2$$

where *N* is revolutions per minute and *r* is the radius in mm.[1] Therefore, the radius plays a multiplying role on the relative centrifugal g-force. As the radius is increased (like in the child example), g-force values are also greatly increased. Internationally, the g-force is reported at the RCF-max, where *r* represents the radius from the center of the rotor to the outermost region of the bottom of the tube.

Having worked in the PRF field for a decade, I cannot believe the great confusion and debates that have arisen because of this simple formula. I have also previously published on a number of errors in the reporting of g-force over the years (see Table 2-2),[2] which has led to considerable confusion and difficulty for clinicians trying to optimize their centrifugation device. Furthermore, RCF/rpm errors have further been incorrectly retranscribed in subsequent years. Entire companies have attempted to mask these parameters, which has created further confusion, leading

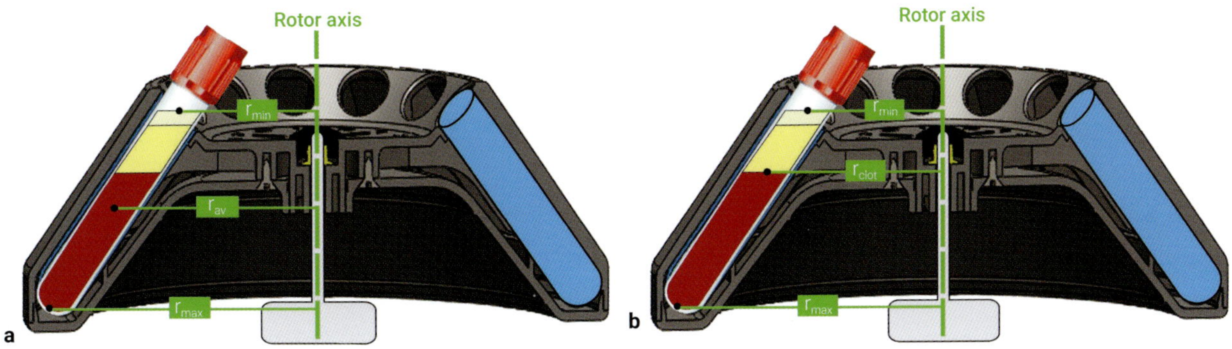

Fig 4-3 *(a)* Figure demonstrating the impact of the radius on RCF values depending on the location at which RCF values are calculated. Because radius from the rotor is a variable in the standard calculation for RCF, significant differences are therefore reported between RCF-min and RCF-max. *(b)* RCF-clot values were introduced in the late 2000s, further complicating the field, because the majority of studies have not reported the location at which RCF values were derived from. (Reprinted with permission from Miron et al.[2])

Table 4-1 **Representative table demonstrating inaccuracies in g-force reporting in previously published studies**

Device		RCF reported in study	RCF-clot	RCF-min	RCF-av	RCF-max
IntraSpin: Rotor angulation: 33 degrees Distance to rotor ▸			50 mm	40 mm	60 mm	80 mm
Previous studies by Pinto and Quirynen	2700 rpm	408g	408g	326g	489g	653g
Choukroun et al[3] (2006)	2500 rpm	280g	349g	280g	419g	559g
Dohan et al[4] (2006)	3000 rpm	400g	503g	402g	604g	805g

While the authors have made an argument that these original articles reported g-forces at 400g, the RCF values in these original articles were calculated at RCF-min and not RCF-clot, as they were reported. Furthermore, the article referenced by Choukroun et al did not even report a g-force of ~400g and instead reported a g-value of 280g. These studies have RCF-max values ranging from 559g to 805g. The L-PRF g-force utilized in our study performed by Fujioka-Kobayashi et al utilized an RCF-max value of 708g, well within this reference range. (Adapted from Miron et al.[2])

many clinicians to be unsure as to how to optimize their own centrifugation devices. I cannot imagine the number of patients that have received substandard treatment with PRF as a result of this. This chapter was written to fill this void.

Confusion Regarding RCF Values

One of the areas that has led to great confusion over the years is that RCF values have been calculated at various regions along a centrifugation tube. For these reasons, it is important to have a basic understanding of RCF values, including calculations to obtain RCF-min, RCF-max, and RCF-av. As depicted in Fig 4-3a representing a centrifuge (which typically range in angulation from 30 to 45 degrees), RCF values can easily be doubled between the RCF-min and RCF-max based on this increased radius distance (Table 4-1).[2–4] RCF-av represents the average g-force throughout an entire centrifugation tube. Internationally, the reporting of RCF values in scientific literature has always been at the RCF-max at the bottom of centrifugation tubes.

One of the problems with RCF reporting is that many articles related to PRF have often utilized RCF-min or RCF-clot values without reporting precisely where the RCF values were actually calculated.[5] Simply, if one states that an RCF value of 400g was utilized, internationally one assumes this value is reported at the RCF-max (international standard and most common in over 90% of the literature). Unfortunately, throughout PRF history, this has not always been the case, which has caused significant difficulty for researchers to further reproduce data. To further complicate these matters, when the general understanding of these mistakes were made in previous publications, several colleagues randomly introduced an "RCF-clot" value to the field in what I feel was an attempt to mask obvious mistakes previously made (Fig 4-3b). This is best exemplified by the fact that an IntraSpin

Table 4-2 Scientific literature reporting L-PRF at different spin cycles utilizing various centrifugation protocols with the according reported g-forces[3–21]*

Authors (year)	Protocol	Reported g-force
Choukroun et al[3] (2006)	**2500 rpm for 10 minutes**	**280g**
One of the first articles on PRF describing a centrifugation protocol of 2500 rpm for 10 minutes on a PC-O2 machine (same rotor size utilized on the current Intra-Lock IntraSpin system). Today it is known that the g-force reported in this study was in fact inaccurate because it was calculated at RCF-min. Based on the current rotor size of 80 mm, it is now known that this should have been represented at 559g at the RCF-max or 349g at the RCF-clot at 50 mm (both different from the reported values of 400g cited by Pinto and Quirynen).		
Dohan et al[4] (2006)	**3000 rpm for 10 minutes**	**~400g**
Though this article was published in a group of five articles with Dohan and Choukroun as first authors in 2006, it is interesting to note that the reported centrifugation protocols and g-forces vary between the publications. Had this original protocol utilized the RCF-clot at 50 mm, this report should have utilized an RCF of 503g at the clot (reported in the study at ~400g). Had this report calculated the g-force based on the international guidelines at the bottom of the centrifugation tube, the reported g-force would be ~805g.		
Simonpieri et al[6] (2009)	**rpm not reported for 12 minutes**	**400g**
One of the difficulties with this study is the lack of reported rpm values, with only the g-force reported. While the reported g-force is a much more important reported value when compared to rpm (owing to the potential differences in rotor sizes), it is impossible to know based on the information provided by these authors if the g-force was calculated at RCF-max or RCF-clot. This is precisely why reported g-forces need to be standardized to avoid future confusion.		
Dohan Ehrenfest et al[7] (2009)	**rpm not reported for 12 minutes**	**400g**
Three years following his first publication, Dohan et al changed his protocol to 12 minutes and 400g in standard L-PRF.		
Su et al[8] (2009)	**2700 rpm for 12 minutes**	**700g**
One of the first studies where the g-force was in fact accurately calculated at the bottom of centrifugation tubes on an Intra-Lock centrifuge machine spinning at the most reported L-PRF protocols of 2700 rpm for 12 minutes. This study utilized RCF-max at the bottom of centrifugation tubes.		

*Comments are made following certain studies to highlight their relevance.

machine (Intra-Lock) can be placed beside a PC-O2 machine from Hettich (its base model), and both will display different RCF-values at the same rpm (Video 4-2). As such, the field progressed rather slowly, with subsequent errors introduced regarding the reporting of RCF values (see Table 4-2).

RCF-clot values have inherent limitations with inaccuracies because the exact location of the clot can never be standardized at the same location, especially when liquid-PRF versus PRF membranes are created.[2] While internationally the standard to report g-forces is calculated at the RCF-max, surprisingly many colleagues have deviated from this standard, leaving countless clinicians unsure as to how to set their own centrifugation devices. This chapter aims to clarify these issues and enable clinicians to do just that.

Video 4-2

Initial studies investigating L-PRF

Table 4-2 lists a number of studies reporting the g-forces and centrifugation times over the years investigating L-PRF.[3–21] These studies were aimed at removing anticoagulants from blood concentrates. Unfortunately, a misrepresentation of the centrifugal g-forces is common in a number of these studies. Reports from the literature have shown that the majority of these initial studies that utilize the PC-O2 centrifuge have commonly reported either inaccurate g-forces or rpm values over the years. Table 4-2 further demonstrates that numerous reports utilizing the same centrifugation system with the same settings (time and rpm) have commonly reported different centrifugation g-forces (without mentioning where the g-force values were actually being calculated—at the RCF-min, RCF-clot, or RCF-max), creating much confusion in the field. Many studies have since retranscribed these inaccurate values (with little knowledge on how to accurately report g-force), which has not only led to confusion but also

Table 4-2 *(cont)*

<table>
<tr><th>Authors (year)</th><th>rpm and time</th><th>Reported g-force</th></tr>
<tr><td>Mazor et al[9] (2009)</td><td>rpm not reported for 12 minutes</td><td>400g</td></tr>
<tr><td>Dohan Ehrenfest et al[10] (2010)</td><td>3000 rpm for 10 minutes</td><td>Not reported</td></tr>
<tr><td colspan="3">Once again, though many years had passed since the L-PRF was introduced, studies by leaders in the field including Dohan Ehrenfest et al continued to report PRF at various centrifugation speeds and times as highlighted in the above study. No RCF value was reported. Even 10 years following the discovery of L-PRF by Choukroun and colleagues, reports in the literature still vary regarding the "standard" L-PRF protocol. Here the protocols were reported at 3000 rpm for 10 minutes, similar to the reports published by Dohan et al in 2006.</td></tr>
<tr><td>Simonpieri et al[11] (2011)</td><td>rpm not reported for 12 minutes</td><td>400g</td></tr>
<tr><td colspan="3">Once again the 400g for 12 minutes protocol was utilized, but it is not possible to know the actual spin time utilized without knowing if the RCF value was calculated at the RCF-min, RCF-clot, or RCF-max. This missing information is pivotal to reproduce the data.</td></tr>
<tr><td>Lekovic et al[12] (2012)</td><td>rpm not reported for 10 minutes</td><td>1000g</td></tr>
<tr><td colspan="3">In this study, Lekovic et al utilized a Labofuge 300 (Heraeus) centrifuge and did not report rpm but reported an RCF value of 1000g. It remains unknown if g-force was actually spun that high and if it was calculated at the bottom of centrifugation tubes or at the RCF-clot; however, a lack of description further complicates analysis and advancement of the field. Why was such a high RCF value chosen for the first time?</td></tr>
<tr><td>Kazemi and Fakhrjou[13] (2015)</td><td>3000 rpm for 10 minutes</td><td>Not reported</td></tr>
<tr><td colspan="3">Once again, reports in 2015 still utilize L-PRF at various centrifugation speeds and times.</td></tr>
<tr><td>Pinto et al[14] (2018)</td><td>2700 rpm for 12 minutes</td><td>~400g</td></tr>
<tr><td colspan="3">This study utilized an approximate RCF value of 400g in reference to the work by Dohan et al in 2006; however, calculations of RCF values were performed at the RCF-clot as opposed to the RCF-min (as was done in the Dohan et al study from 2006) or by utilizing RCF-max, which is the international standard.</td></tr>
<tr><td>Fujioka-Kobayashi et al[15] (2017)</td><td>2700 rpm for 12 minutes</td><td>708g</td></tr>
<tr><td colspan="3">While the RCF-max value of 708g falls within the range of original L-PRF values in studies by Choukroun et al and Dohan et al in 2006, the authors incorrectly transcribed an rpm value of 2700, most commonly utilized on the IntraSpin system.</td></tr>
<tr><td>Afat et al[16] (2018)</td><td>3000 rpm for 10 minutes</td><td>Not reported</td></tr>
<tr><td colspan="3">Even in 2018, several still utilized the initial published values by Dohan et al at 3000 rpm for 10 minutes. No report on which centrifuge was utilized and no report on the g-forces are provided. Both the device rotor size and RCF are critical parameters to be able to reproduce this work.</td></tr>
<tr><td>Nizam et al[17] (2018)</td><td>Nüve Laboratory Equipment for 12 minutes (rpm not reported)</td><td>400g</td></tr>
<tr><td colspan="3">Because another PRF centrifuge was utilized with no reported rpm, it is unknown where g-force was calculated, whether at the RCF-clot or RCF-max.</td></tr>
<tr><td>Tabrizi et al[18] (2018)</td><td>Intra-Lock system centrifuged at 28000 rpm for 12 minutes</td><td>Not reported</td></tr>
<tr><td colspan="3">Though g-force was not reported, the authors state using an Intra-Lock system spun at 28000 rpm. While the present authors are sure one extra 0 was accidentally added, this highlights the lack of attention paid during the peer-review process, warranting better standardization regarding the reported rpm/g-force for all studies.</td></tr>
<tr><td>Dohan Ehrenfest et al[19] (2018)</td><td>2700 rpm for 12 minutes</td><td>400g</td></tr>
<tr><td colspan="3">While Dohan et al reported in his original studies 3000 rpm for 10 minutes, here a common value of 2700 rpm is reported for 12 minutes with a PRF-clot value of 400g. Note that this g-force is equivalent to ~650g at the bottom of this same centrifugation tube, yet no mention of actually utilizing RCF-clot was provided.</td></tr>
<tr><td>Meschi et al[20] (2018)</td><td>Centrifuged at 702g in an Intra-Lock centrifuge</td><td>702g</td></tr>
<tr><td colspan="3">Work by Meschi et al utilized the Intra-Lock L-PRF system and accurately reported RCF values at ~700g at the base of the centrifugation tube (RCF-max). Therefore, until 2018, very little accurate reporting was actually provided.</td></tr>
<tr><td>Cortellini et al[21] (2018)</td><td>2700 rpm for 3 minutes</td><td>408g</td></tr>
<tr><td colspan="3">In this study by Cortellini et al, the authors introduce a very short centrifugation time but fail to recalculate the new RCF-clot owing to the new location of the PRF-layer separation. Inaccuracies such as these are subject to major confusion in the field. These miscalculations were the basis for much debate in the field.</td></tr>
</table>

Adapted from Miron et al.[2]

difficulty in effectively advancing the field. Our understanding today regarding the effects of centrifugation g-force on PRF-based matrices has greatly been enhanced with improvements made utilizing the low-speed centrifugation concept (LSCC). Thankfully, a consensus was reached in 2019 with the aim to more effectively report g-force values in future studies in an accurate and scientific manner and further advance the field.

Conclusions from Table 4-2

- Original g-forces for L-PRF were not reported correctly, either at the RCF-clot or at the RCF-max.
- Much variation in reported g-forces is found in the literature, even for studies using the exact same centrifuge with the same centrifugation settings (speed and time).
- Original protocols have modified their centrifugation protocols over the years to improve their formulations. Both have changed from reporting RCF-min values to either RCF-clot or RCF-max values.
- L-PRF protocols have varied in the literature from 2500–3000 rpm at a spin cycle ranging from 10 to 12 minutes.

Calculating g-force at the bottom of the tubes

Several authors have correctly reported g-force values based on the rotor radius and effectively reported either RCF-min, RCF-av, RCF-max, or RCF-clot. However, many authors have deviated from reporting g-forces (without proper notification) based on international centrifugation standards that effectively report g-force at the bottom of centrifugation tubes. While certainly the g-force applied to the PRF clot is different from that applied at the RCF-max, next is an overview of how this change not only deviates from standard reports in the field but is subject to inaccuracies leading to future confusion in the field.

Reporting RCF at the clot not only deviates from international standards but is also subject to inaccuracies leading to future confusion in the field.

Reporting RCF values at the clot

Anyone who has ever performed considerable work with PRF has certainly experienced great variability in PRF clot sizes and layer separation. As highlighted in chapter 2, simply the effect of patient sex and age can have a marked and significant impact on the size and layer separation of PRF.[22]

To illustrate this concept, Fig 4-4 demonstrates the effect of time on the position of PRF clots formed following spin cycles of 3, 6, 9, and 12 minutes at 2400 rpm (708g at the RCF-max on a Duo Quattro centrifuge [Process for PRF]).[2] Notice that while centrifugation was carried out at exactly the same speed, the position at which the clot is located is pushed further downward with each longer spin cycle. Because RCF-clot utilizes the radius where the clot is located, g-force must then be reported differently. When g-force is calculated based on international guidelines at the bottom of the centrifugation tubes (110 mm), 708g is utilized in all cases irrespective of the centrifugation time. This is most representative granted that all clots were created at the same centrifugation speed using the same machine and tubes.

Figure 4-5 further illustrates this concept. Based on the method of calculating RCF at the clot (RCF-clot),[23] even though the rpm remains unchanged, because of an increase in centrifugation time, the PRF clot is displayed downward within the centrifugation tube, affecting the radius at which the centrifugation g-force is calculated. This highlights the fact that though centrifugation is carried out at the exact same speed (2400 rpm), the RCF-clot in fact varies based on centrifugation time (from 425g to 515g). Because of this discrepancy, international guidelines to report g-forces on a centrifugation system have been set at the bottom of the centrifugation tubes in order to best report g-forces accurately.

The issue with reporting g-force miscalculations has been created primarily with respect to L-PRF protocols. Table 4-2 highlights how an otherwise excellent study conducted by Cortellini et al on liquid fibrinogen nevertheless led to significant additional confusion utilizing RCF-clot as a means to represent g-force.[21,23] Noteworthy, on a standard L-PRF clot produced with an Intra-Spin device, the authors report that the clot is located at approximately a 50-mm radius (subject to variation). Nevertheless, based on their calculations, this reported g-force equates to 408g at the clot (reported as RCF-clot).

Fig 4-4 Figure demonstrating the effects of centrifugation time on displacement of the PRF clot. Though all tubes were centrifuged at 2400 rpm on a Duo Quattro centrifuge (RCF-max = 708g), notice how following 3, 6, 9, and 12 minutes, the clot is displaced downward. In such a situation, RCF-clot cannot be standardized due to the different location of the PRF clot, affecting the radius at which RCF-clot is calculated. (Reprinted with permission from Miron et al.[2])

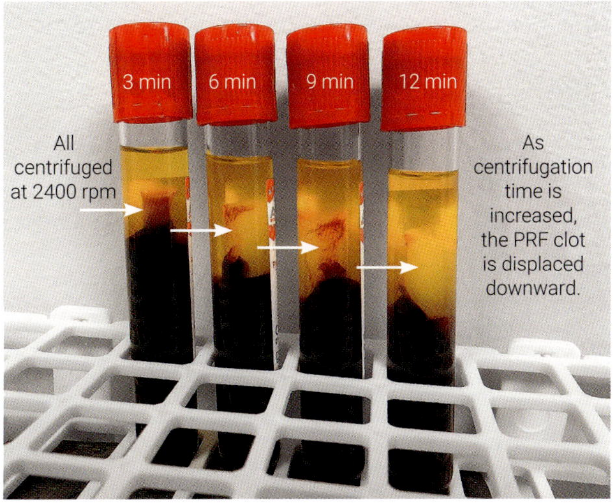

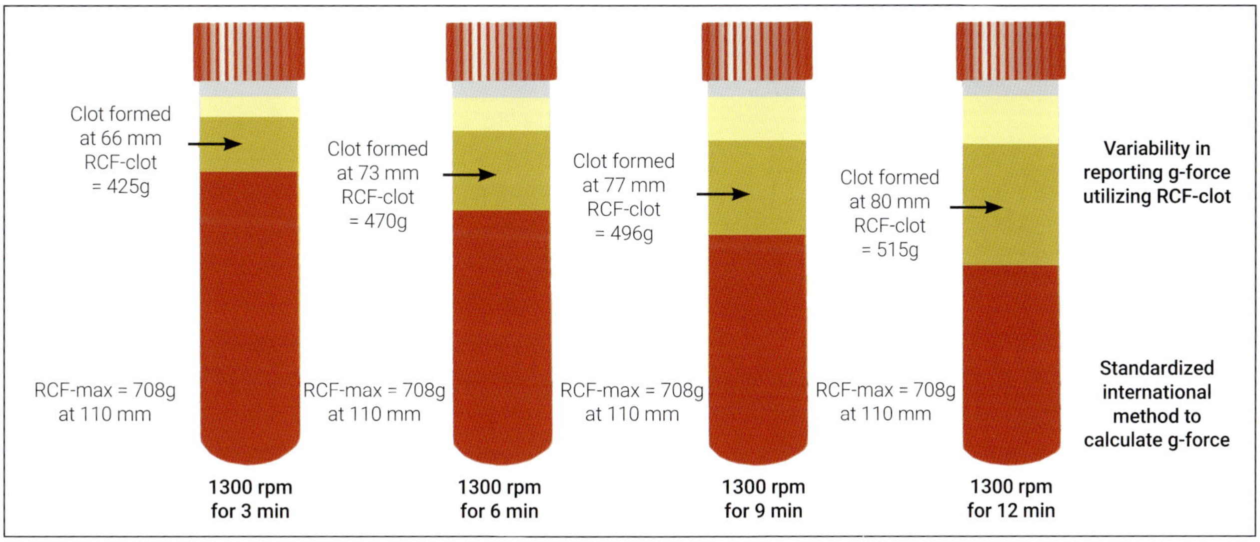

Fig 4-5 Illustration demonstrating once again the inability to accurately report g-forces at RCF-clot. While each of these tubes is centrifuged at 2400 rpm on a Duo Quattro centrifuge for 3, 6, 9, and 12 minutes, notice how the RCF-clot values change due to the increased radius of the PRF clot. Despite being centrifuged at the same speed, the RCF-clot values range from 425g to 515g. International guidelines recommend displaying g-force values at RCF-max (708g). Utilizing the RCF-max values accurately reports g-forces in each of these scenarios. (Reprinted with permission from Miron et al.[2])

Figure 4-6 illustrates the separation of layers at 2700 rpm when a 3-minute protocol was utilized in their study versus a 12-minute protocol. Notice the drastic difference in locations of the clots/separation layers based on this reported shorter centrifugation period.

The study was heavily criticized by many colleagues for having inaccurately reported a g-force of 408g at the RCF-clot with the assumption that after 3 minutes, the location of the clot would be precisely the same (50 mm from the radius) as the location of the clot produced after a 12-minute centrifugation period, as the authors had reported in previous studies.[21] As can clearly be observed in Fig 4-6, these clots are nowhere near the same location within the tubes. Had the authors used the RCF-max to report their data based on international centrifugation standards/guidelines, whether they centrifuged for 3 or 12 minutes at 2700 rpm, RCF-max would always be precisely reported at 653g. Instead, the authors assumed/utilized the same g-force equivalent to the clot location after a 12-minute spin cycle. This is not simply a miscalculation in their study but more importantly a limitation with reporting g-force at the RCF-clot.

Furthermore, it should also be noted that the location of the PRF clot may vary rather significantly from patient to patient depending on individual hematocrit levels of that patient.[22]

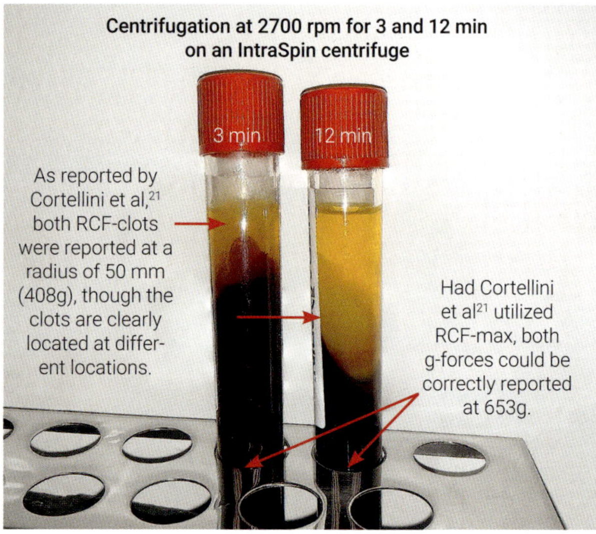

Fig 4-6 Figure demonstrating the effects of centrifugation time on location of PRF clots as reported by Cortellini et al[21] utilizing an IntraSpin centrifuge spun at 2700 rpm for 3 and 12 minutes. The authors report the exact same g-force at the RCF-clot, which assumed a 50-mm radius. Notice, however, that because of the different clot locations, error was created in their report due to having calculated the clot to report RCF values. Instead, had the authors reported g-force at the bottom of the centrifugation tubes based on international guidelines, both values would accurately be reported at 653g. (Reprinted with permission from Miron et al.[2])

For these reasons, it is logical that the standard method to report g-force in a consistent manner and to eliminate any variation is to report g-force at the bottom of the centrifugation tubes.

Biologic differences of PRF clots based on centrifugation parameters

It is also important to note that each centrifuge—due to differences in rotor size, angulation of tubes, composition of tubes, and vibration—will result in a different biologic PRF clot. Several authors are entirely correct in stating that even a centrifugation at the same rpm will exert different centrifugal forces if the centrifuge rotors have different radius sizes, bucket types, or bucket sizes. Figure 4-7 demonstrates that different g-forces will inevitably be reported at either the clot (RCF-clot) or end of the tube (RCF-max) if the tubes are angled differently. Therefore, a concept of "biological signature" of each PRF clot fabricated on various machines is certainly subject to differences. These differences, however, are generally overhyped and overmarketed by various commercial entities and associated speakers for financial gain. Nonetheless, for any scientific investigations into centrifugation speeds or time, it is best to use the same device to minimize additional factors like these.

The concept of a "biological signature" of each PRF clot fabricated on various machines is generally overhyped and overmarketed by various commercial entities and associated speakers for financial gain.

Standardizing RCF Value Reporting

A consensus report from 2019 determined that in the future, all articles on PRF should report the following parameters[5]:

1. Dimensions of the rotor (radius at the clot and end of the tube)
2. Rotor angulation for the tube holder
3. Revolutions per minute (rpm) and time
4. RCF value calculated at either the RCF-min, RCF-clot, or RCF-max
5. Composition and size of tubes utilized to produce PRF
6. Centrifugation model utilized

By simply stating each of these parameters, further research in the field may be standardized and reproduced. Effectively, should proper parameters not be followed accordingly, the production process of PRF will be different and the quality of the fibrin mesh as well as concentration of cells and growth factors will not be accurate or comparable. Consequently, much gained scientific knowledge will be lost.

Calculating RCF Values

The aim of this exercise is to provide the reader the opportunity to accurately calculate RCF values irrespective of the centrifugation device utilized. By using this formula, it is possible to calculate RCF values on any device (see Fig 4-2):

$$RCF = 11.18 \times r \times (N/1000)^2$$

Furthermore, entire websites are designed to calculate these values in a user-friendly format (Fig 4-8).

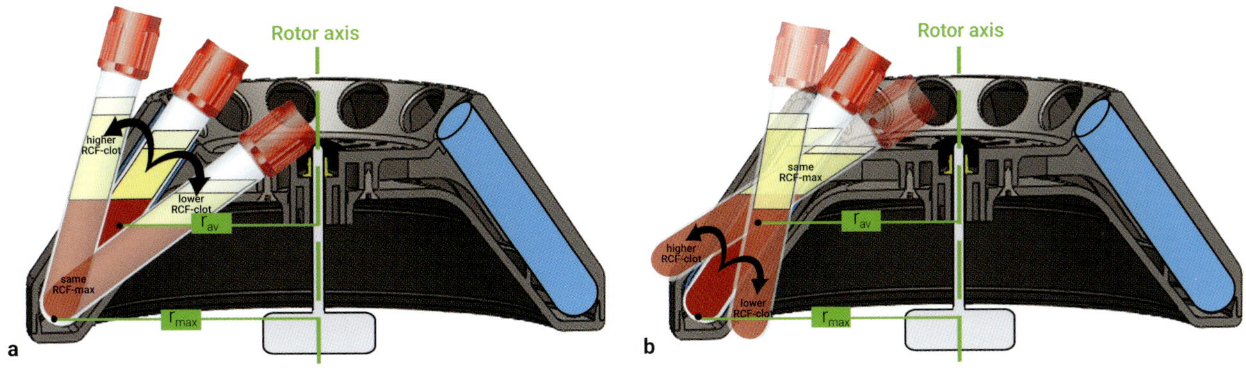

Fig 4-7 Illustration further demonstrating how no two centrifuges can report the same RCF-clot/RCF-max values if the angulation of tubes is different. *(a)* Notice how even if the RCF-max is identical, a change of tube angulation will result in a different RCF-clot. *(b)* Even if RCF-clot is the same, RCF-max is different, and the applied g-forces throughout that tube will also be subject to variation. Therefore, when performing scientific studies investigating the effects of centrifugation speed and time, it is most scientifically logical to utilize the same centrifugation device in order to minimize additional factors that may affect the results. (Reprinted with permission from Miron et al.[2])

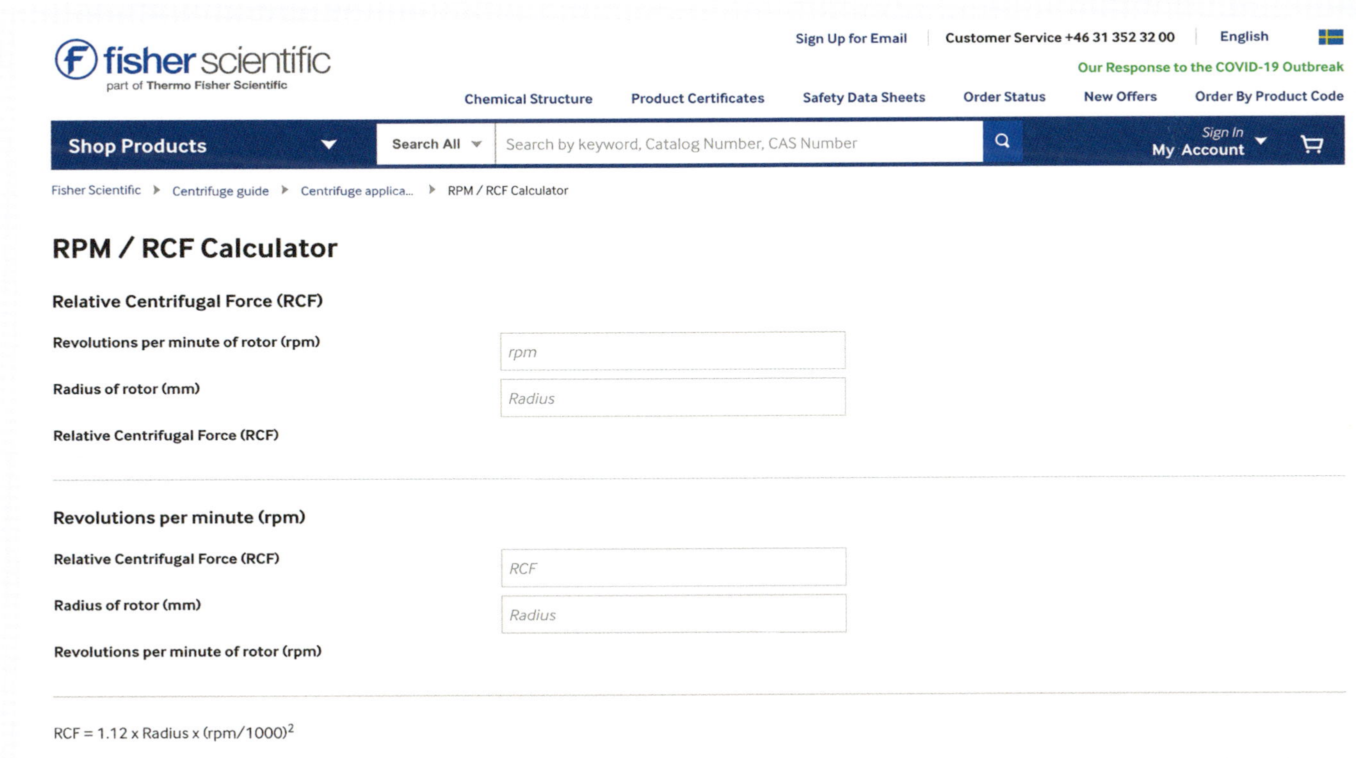

Fig 4-8 Screen capture of the Fisher Scientific webpage that demonstrates the ability to calculate RCF and rpm values extremely easily: https://www.fishersci.se/se/en/scientific-products/centrifuge-guide/centrifuge-applications-tools/rpm-rcf-calculator.html.

Table 4-3 lists the RCF/rpm values of both the Intra-Spin and Duo Quattro devices at various protocols.[5] The reader is recommended to use these values to practice calculating on their own using the Fisher Scientific website: https://www.fishersci.se/se/en/scientific-products/centrifuge-guide/centrifuge-applications-tools/rpm-rcf-calculator.html.

Table 4-3 RCF values for the two most frequently used centrifuges for the preparation of PRF (IntraSpin L-PRF centrifuge and Duo Quattro centrifuge)

Device	RCF-min	RCF-clot	RCF-av	RCF-max
IntraSpin: Rotor angulation: 33 degrees Distance to rotor ▸	40 mm	50 mm	60 mm	80 mm
2700 rpm	326g	408g	489g	653g
Duo Quattro: Rotor angulation: 40 degrees Distance to rotor ▸	63 mm	77 mm	87 mm	110 mm
1300 rpm	119g	145g	164g	208g
2700 rpm	513g	628g	708g	897g

If the rpm and rotor radius are known, a nomogram chart[24] can be used to determine the RCF. (Reprinted with permission from Miron et al.[5])

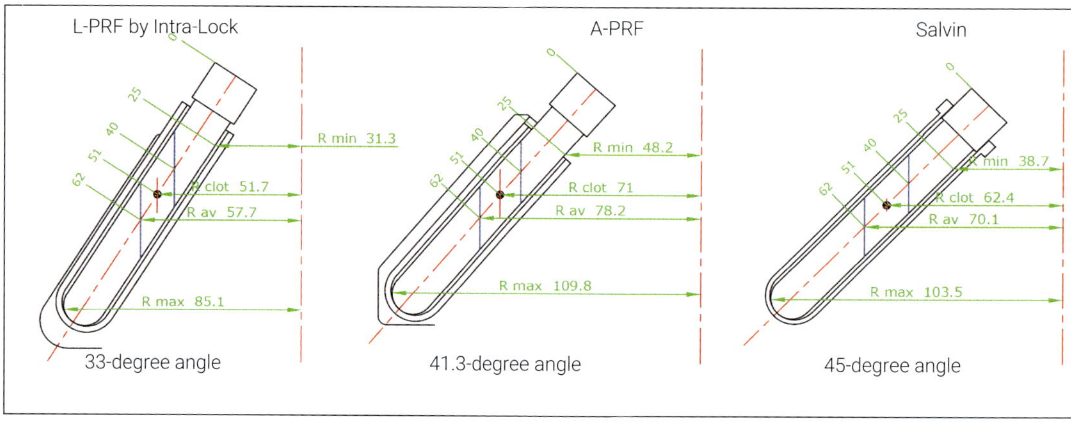

Fig 4-9 Quantification of the r-min, r-clot, and r-max for three different centrifugation devices including their angulations. From such drawings, it is possible to calculate the RCF-min, RCF-clot, and RCF-max for any centrifugation speed (RCF or rpm). (Reprinted with permission from Miron et al.[25])

Video 4-1

Based on the LSCC and work conducted by our group, a relative RCF value of 200g produced using a fixed-angle centrifugation device led to greater cell concentrations and an even distribution of platelets (see chapter 2). Based on this knowledge, it is now possible to take a Salvin centrifuge as presented in chapter 2 with a maximum radius of 103.5 mm (Fig 4-9) and easily calculate the optimized rpm for this fixed-angle centrifugation system (ie, 1314 rpm). This also clearly illustrates how a much smaller rotor (eg, one that has an r-max of 65 mm) would naturally require a significantly higher rpm to achieve the same RCF value (ie, 1657 rpm). Video 4-1 represents a demo calculation being performed.

On a horizontal system, because the tube lays completely horizontally, the difference between the RCF-min and RCF-max is significantly extended (see chapter 3). For this reason, the optimized RCF value on a horizontal centrifuge is naturally much greater at the RCF-max (eg, 700g). On the Bio-PRF horizontal system with a radius of 125 mm, this would equate to ~2220 rpm (700 RCF-max).

By doing this relatively simple exercise (as is done in the majority of basic research master/PhD programs), one can quickly gain the ability to rapidly convert rpm/RCF and optimize any centrifugation device.

Conclusion

This chapter provided an overview of RCF and the ability to rapidly calculate rpm/RCF values based on the rotor radius value. In the future, a better understanding of RCF/rpm calculations will serve the community better by providing an effective method to quickly optimize PRF across the world, irrespective of the centrifugation device utilized. It is further important that scientific colleagues working in the field and publishing on this topic accurately report RCF values in future studies, and it is strongly advised that at a minimum, the guidelines from the consensus report in 2019 be respected accordingly.

By simply stating each of these parameters, future research in the field will be better standardized and reproducible. The production process of PRF will be effectively improved worldwide, and the quality of the fibrin mesh as well as concentration of cells and growth factors will be enhanced. Consequently, should this knowledge fail to be accurately reported, much gained scientific knowledge will be lost and patient care compromised.

References

1. Beck DJ, Bibby BG. A centrifugal technique of measuring food retention. J Dent Res 1961;40:148–160.
2. Miron R, Choukroun J, Ghanaati S. Controversies related to scientific report describing g-forces from studies on platelet-rich fibrin: Necessity for standardization of relative centrifugal force values. Int J Growth Factors Stem Cells Dent 2018;1:80.
3. Choukroun J, Diss A, Simonpieri A, et al. Platelet-rich fibrin (PRF): A second-generation platelet concentrate. Part V: Histologic evaluations of PRF effects on bone allograft maturation in sinus lift. Oral Surg Oral Med Oral Pathol Oral Radiol Endod 2006;101:299–303.
4. Dohan DM, Choukroun J, Diss A, et al. Platelet-rich fibrin (PRF): A second-generation platelet concentrate. Part III: Leucocyte activation: A new feature for platelet concentrates? Oral Surg Oral Med Oral Pathol Oral Radiol Endod 2006;101:e51–e55.
5. Miron RJ, Pinto NR, Quirynen M, Ghanaati S. Standardization of relative centrifugal forces in studies related to platelet-rich fibrin. J Periodontol 2019;90:817–820.
6. Simonpieri A, Del Corso M, Sammartino G, Dohan Ehrenfest DM. The relevance of Choukroun's platelet-rich fibrin and metronidazole during complex maxillary rehabilitations using bone allograft. Part II: Implant surgery, prosthodontics, and survival. Implant Dent 2009;18:220–229.
7. Dohan Ehrenfest DM, Diss A, Odin G, Doglioli P, Hippolyte MP, Charrier JB. In vitro effects of Choukroun's PRF (platelet-rich fibrin) on human gingival fibroblasts, dermal prekeratinocytes, preadipocytes, and maxillofacial osteoblasts in primary cultures. Oral Surg Oral Med Oral Pathol Oral Radiol Endod 2009;108:341–352.
8. Su CY, Kuo YP, Tseng YH, Su CH, Burnouf T. In vitro release of growth factors from platelet-rich fibrin (PRF): A proposal to optimize the clinical applications of PRF. Oral Surg Oral Med Oral Pathol Oral Radiol Endod 2009;108:56–61.
9. Mazor Z, Horowitz RA, Del Corso M, Prasad HS, Rohrer MD, Dohan Ehrenfest DM. Sinus floor augmentation with simultaneous implant placement using Choukroun's platelet-rich fibrin as the sole grafting material: A radiologic and histologic study at 6 months. J Periodontol 2009;80:2056–2064.
10. Dohan Ehrenfest DM, Del Corso M, Diss A, Mouhyi J, Charrier JB. Three-dimensional architecture and cell composition of a Choukroun's platelet-rich fibrin clot and membrane. J Periodontol 2010;81:546–555.
11. Simonpieri A, Choukroun J, Del Corso M, Sammartino G, Dohan Ehrenfest DM. Simultaneous sinus-lift and implantation using microthreaded implants and leukocyte- and platelet-rich fibrin as sole grafting material: A six-year experience. Implant Dent 2011;20:2–12.
12. Lekovic V, Milinkovic I, Aleksic Z, et al. Platelet-rich fibrin and bovine porous bone mineral vs. platelet-rich fibrin in the treatment of intrabony periodontal defects. J Periodontal Res 2012;47:409–417.
13. Kazemi D, Fakhrjou A. Leukocyte and platelet rich plasma (L-PRP) versus leukocyte and platelet rich fibrin (L-PRF) for articular cartilage repair of the knee: A comparative evaluation in an animal model. Iran Red Crescent Med J 2015;17:e19594.
14. Pinto NR, Ubilla M, Zamora Y, Del Rio V, Dohan Ehrenfest DM, Quirynen M. Leucocyte- and platelet-rich fibrin (L-PRF) as a regenerative medicine strategy for the treatment of refractory leg ulcers: A prospective cohort study. Platelets 2018:29:468–475.
15. Fujioka-Kobayashi M, Miron RJ, Hernandez M, Kandalam U, Zhang Y, Choukroun J. Optimized platelet-rich fibrin with the low-speed concept: Growth factor release, biocompatibility, and cellular response. J Periodontol 2017;88:112–121.
16. Afat IM, Akdogan ET, Gonul O. Effects of leukocyte- and platelet-rich fibrin alone and combined with hyaluronic acid on pain, edema, and trismus after surgical extraction of impacted mandibular third molars. J Oral Maxillofac Surg 2018;76:926–932.
17. Nizam N, Eren G, Akcali A, Donos N. Maxillary sinus augmentation with leukocyte and platelet-rich fibrin and deproteinized bovine bone mineral: A split-mouth histological and histomorphometric study. Clin Oral Implants Res 2018;29:67–75.
18. Tabrizi R, Arabion H, Karagah T. Does platelet-rich fibrin increase the stability of implants in the posterior of the maxilla? A split-mouth randomized clinical trial. Int J Oral Maxillofac Surg 2018;47:672–675.
19. Dohan Ehrenfest DM, Pinto NR, Pereda A, et al. The impact of the centrifuge characteristics and centrifugation protocols on the cells, growth factors, and fibrin architecture of a leukocyte- and platelet-rich fibrin (L-PRF) clot and membrane. Platelets 2018;29:171–184.

20. Meschi N, Fieuws S, Vanhoenacker A, et al. Root-end surgery with leucocyte- and platelet-rich fibrin and an occlusive membrane: A randomized controlled clinical trial on patients' quality of life. Clin Oral Investig 2018;22:2401–2411.
21. Cortellini S, Castro AB, Temmerman A, et al. Leucocyte- and platelet-rich fibrin block for bone augmentation procedure: A proof-of-concept study. J Clin Periodontol 2018;45:624-634.
22. Miron RJ, Dham A, Dham U, Zhang Y, Pikos MA, Sculean A. The effect of age, gender, and time between blood draw and start of centrifugation on the size outcomes of platelet-rich fibrin (PRF) membranes. Clin Oral Investig 2018;22:2401–2411.
23. Pinto N, Quirynen M. Letter to the Editor: RE: Optimized platelet-rich fibrin with the low-speed concept: Growth factor release, biocompatibility, and cellular response. J Periodontal 2019;90:119–121.
24. Rickwood D. Centrifugation: A Practical Approach. Oxford: IRL, 1984.
25. Miron RJ, Xu H, Chai J, et al. Comparison of platelet-rich fibrin (PRF) produced using 3 commercially available centrifuges at both high (~700 g) and low (~200 g) relative centrifugation forces. Clin Oral Investig 2020;24:1171–1182.

5

Protocols for PRF

Contributors

Richard J. Miron
Masako Fujioka-Kobayashi
Yufeng Zhang

Chapter Highlights

- *Protocol and indications for the production of solid-PRF membranes*
- *Protocol and indications for the production of liquid-PRF*
- *Protocol and indications for the production of C-PRF*

Video 5-1

This chapter is a relatively short one designed to simplify protocols for the production of PRF. In general, only three protocols are necessary for PRF therapy. The first is the standard PRF membrane in which ideally a higher platelet and leukocyte yield is harvested with an even distribution of cells within the upper 4- to 5-mL PRF layers. This is best achieved using a horizontal centrifugation system (700g for 8 minutes). The second protocol is a liquid-PRF formulation capable of concentrating platelets and leukocytes within the upper 1-mL layer (previously known as i-PRF). By utilizing a horizontal centrifugation system, higher concentrations are ensured (smaller volume though a higher concentration of cells). This protocol is achieved utilizing a 300g protocol for 5 minutes. The final and third protocol is that of C-PRF, where cells are purposefully accumulated specifically toward the buffy coat layer using faster spin protocols. This is best achieved using a 2000g protocol for 8 minutes; a resulting 0.3- to 0.5-mL cell-rich zone may be collected exactly within the buffy coat. With these three simple protocols, the practitioner is able to perform practically all clinical procedures as highlighted within this chapter.

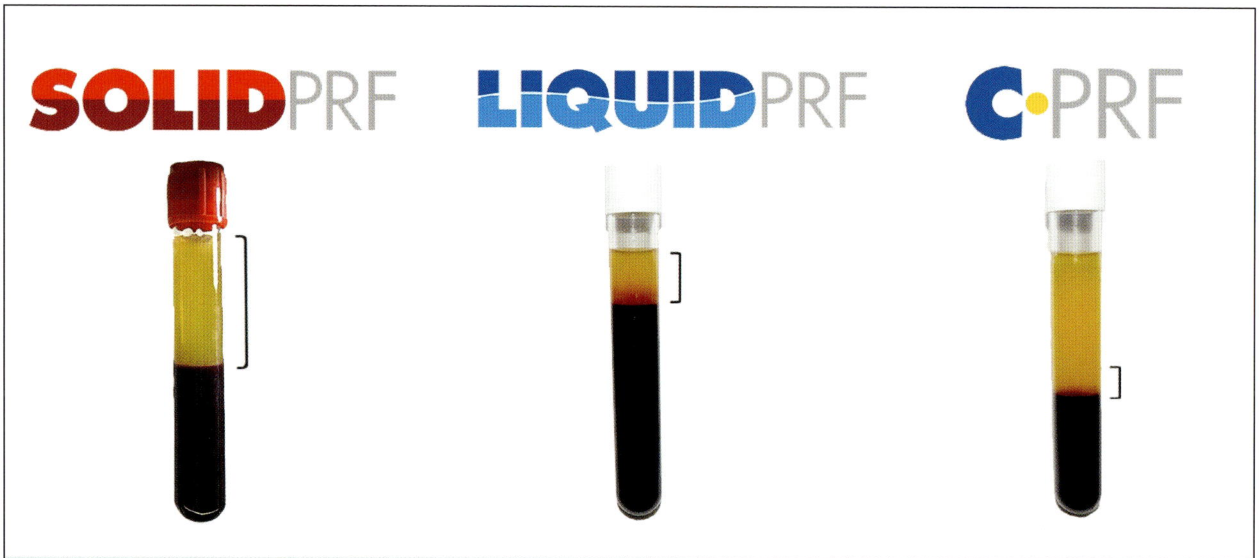

Fig 5-1 Three centrifugation protocols needed for the production of PRF.

This is an extremely short chapter, but it contains important clinical information concerning PRF protocols, their associated cell layer separation, and well as intended clinical applications. The aim is simply to better address precisely which PRF protocol should be performed for various clinical procedures. A clear layout and summary is presented with rationale for each protocol. In total, only three protocols are needed for PRF therapy (Fig 5-1); each of these and its various applications are presented below.

Solid-PRF Clots (700g for 8 Minutes)

The first and most relevant protocol for the production of solid-PRF in regenerative dentistry is the fabrication of PRF membranes (Fig 5-2). The aim of this protocol is to concentrate cells with a high yield in the upper four to five 1-mL layers of PRF. By utilizing a 700g protocol for 8 minutes, as highlighted in chapters 2 and 3, a maximum yield of platelets and leukocytes is achieved with a relatively even distribution of cells within the upper PRF layers (see Fig 5-2). For the production of PRF membranes, red tubes are utilized (hydrophilic plain glass tube), therefore inducing clot formation.

The main indications for producing PRF include but are not limited to: *(1)* the majority of dental procedures; *(2)* soft tissue wound healing around implants; *(3)* production of PRF fragments to mix with bone grafts; *(4)* sinus grafting procedures either utilized alone or in combination with bone grafts; *(5)* GBR procedures either as an outer barrier membrane on top of collagen or to mix with bone grafting particles; and *(6)* healing of complex wounds in medicine including diabetic wounds (Box 5-1). Therefore, as a dental clinician, this is the protocol most often utilized.

Furthermore, when the practitioner desires to make "sticky bone," this same protocol is utilized, and both liquid-PRF (blue or white) and solid-PRF (red) tubes are drawn simultaneously and centrifuged at the same time (700g for 8 minutes; see chapter 9). Always remember to draw liquid-PRF tubes first.

Always remember to draw liquid-PRF tubes first.

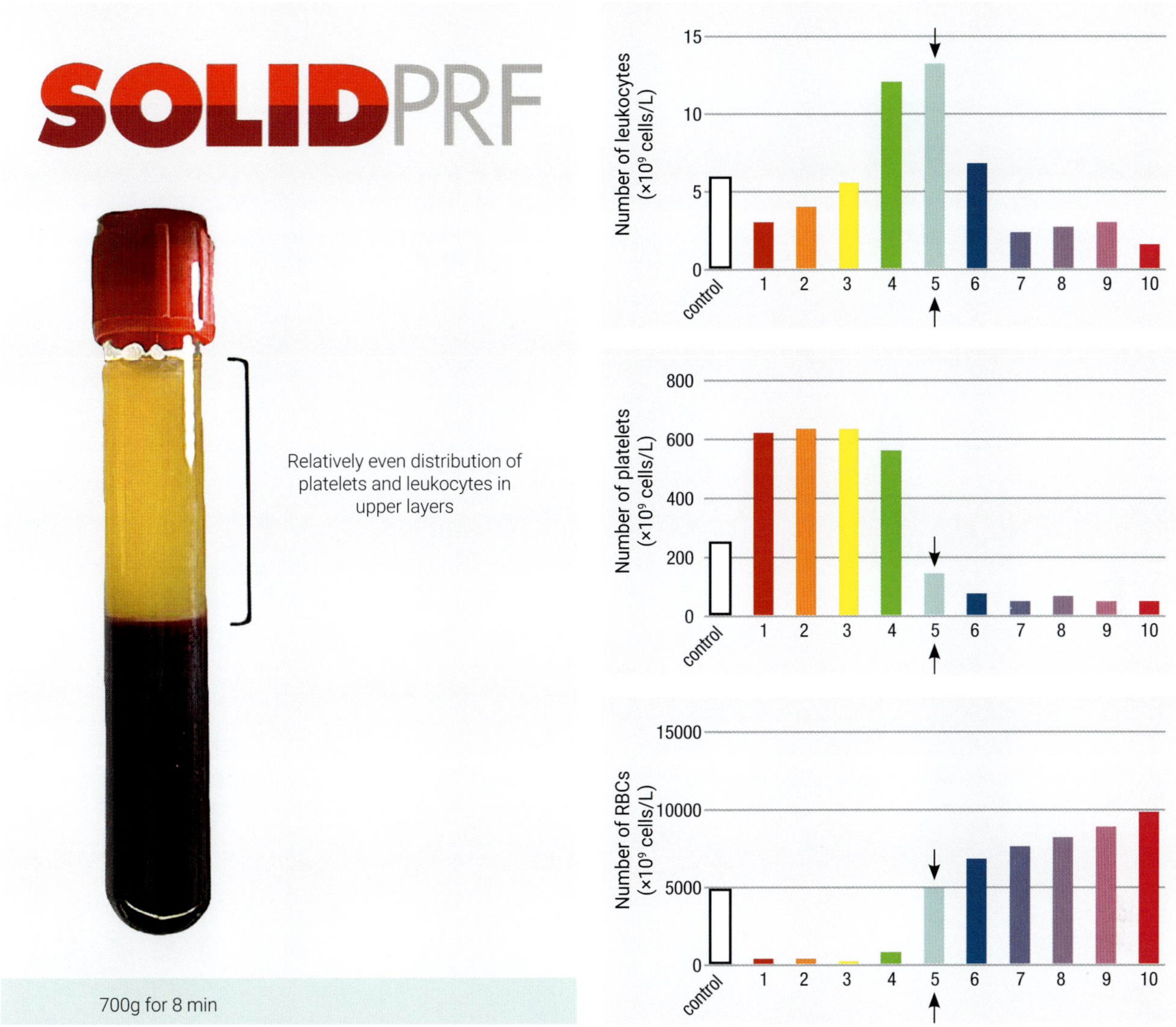

Fig 5-2 Solid-PRF protocols are produced at 700g for 8 minutes. As depicted in the bar graph, the aim is to evenly distribute platelets and leukocytes in the upper four layers of PRF membranes.

Box 5-1 Clinical indications for solid-PRF protocols utilizing red tubes

Relatively even distribution of platelets and leukocytes in upper layers

- Majority of dental procedures
- Soft tissue healing around implants
- Cutting of PRF to mix with bone grafts
- Sinus grafting
- GBR procedures
- Healing of complex diabetic wounds

Liquid-PRF (300g for 5 Minutes)

The goal and purpose of this protocol is to harvest a small volume of liquid with a higher concentration of cells. As observed in Fig 5-3, the plasma layer generally separates within the first 1 to 2 mL following this slower and shorter spin cycle where platelets and leukocytes are generally more concentrated within the upper layer. This is best achieved using a 300g protocol for 5 minutes using liquid-PRF tubes (white/blue tubes, hydrophobic PET). While many cells remain in the bottom red blood cell layers (including platelets and leukocytes), this generally is recommended as a slow spin cycle favoring higher concentrations when compared to solid-PRF protocols (700g for 8 minutes).

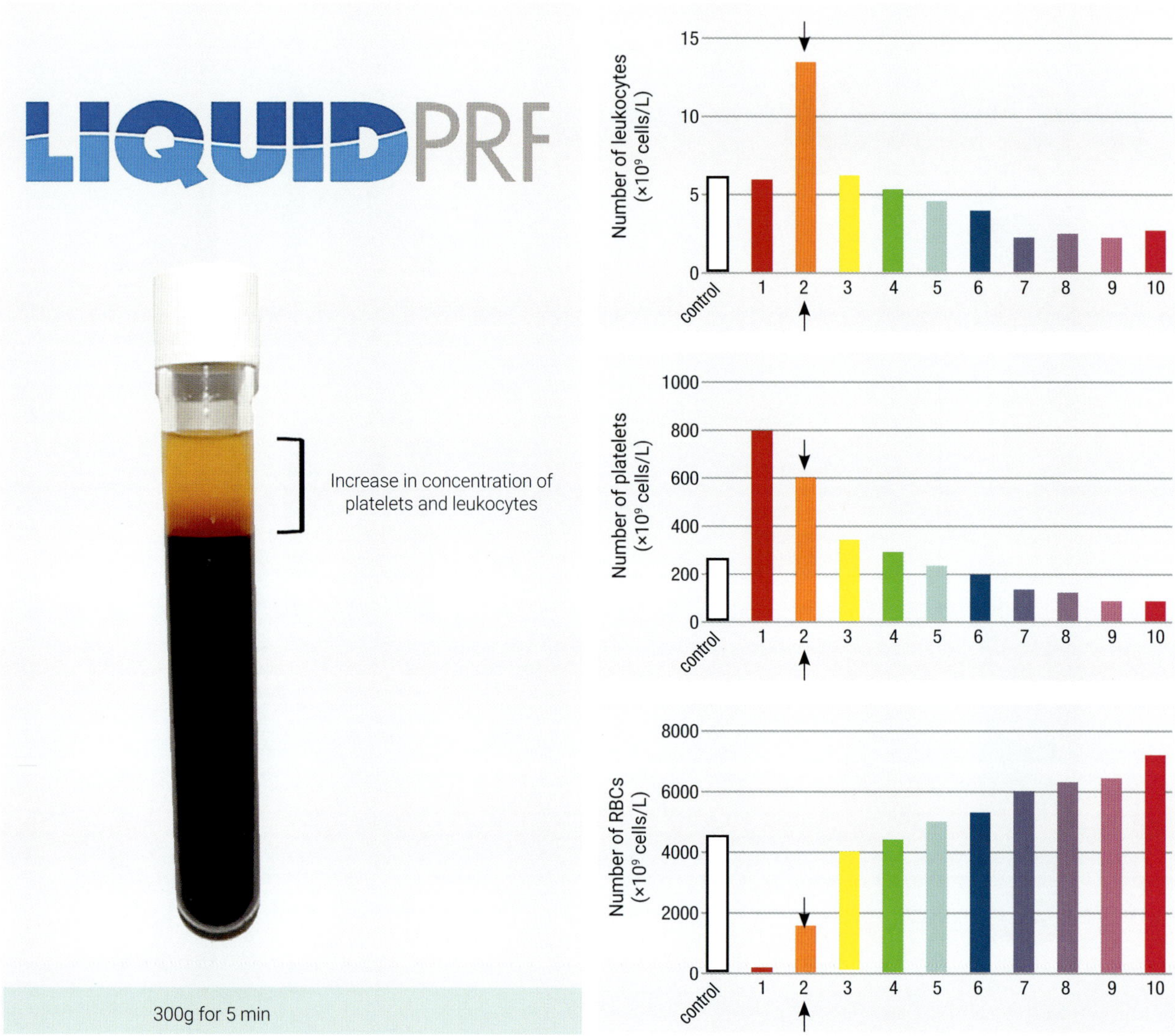

Fig 5-3 Liquid-PRF protocols are produced at 300g for 5 minutes. As depicted in the bar graph, the aim is to concentrate platelets and leukocytes in the upper 1.0- to 1.5-mL layer of liquid-PRF.

The main indications include: *(1)* mixing with bone grafts; *(2)* injecting into tissues such as knee joints; *(3)* microneedling with PRF or facial injections where red blood cells may cause hyperpigmentation issues; *(4)* nonsurgical periodontal therapy by injection of liquid-PRF into periodontal pockets; and *(5)* endodontic procedures (Box 5-2). If two centrifuges are available in the same clinic and the practitioner desires to make sticky bone, the horizontal centrifuge can be utilized to make solid-PRF membranes while the second centrifuge can be utilized to make a more concentrated final liquid-PRF utilizing this protocol (300g for 5 minutes).

Box 5-2 Clinical indications for liquid-PRF protocols utilizing white tubes

Increased concentration in platelets and leukocytes

- Mixing with bone grafts
- Injecting into tissues
- Microneedling with PRF
- Nonsurgical periodontal therapy
- Endodontic procedures

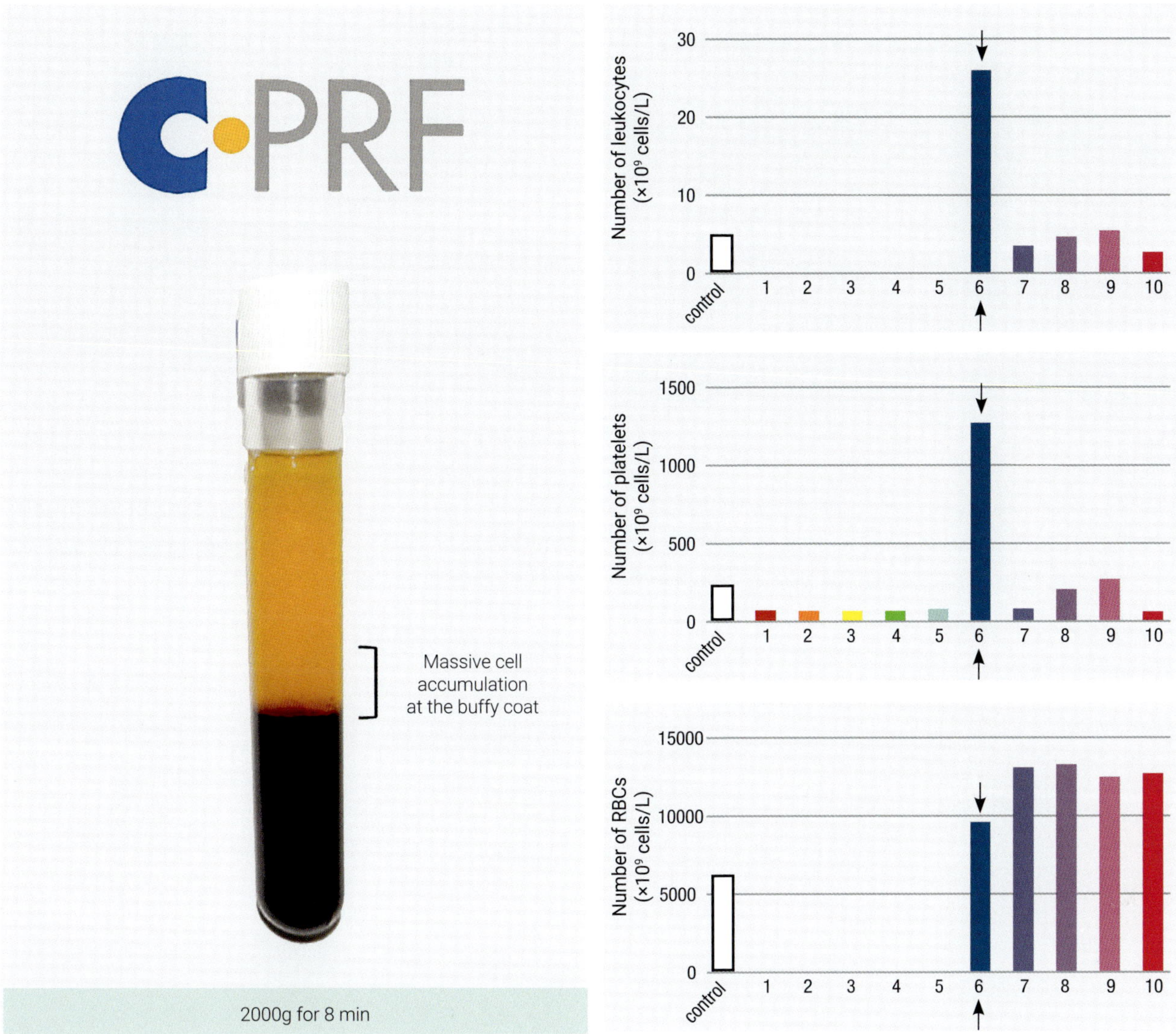

Fig 5-4 C-PRF protocols are produced at 2000g for 8 minutes. As depicted in the bar graph, the aim is to concentrate all platelets and leukocytes at the buffy coat.

C-PRF (2000g for 8 Minutes)

The goal and purpose of this protocol is to harvest the highest possible concentration of platelets and leukocytes (hence the term *concentrated-PRF*). As such, faster protocols are utilized to purposefully send the cells to a thin buffy coat layer rich in platelets and leukocytes (Fig 5-4). A standard protocol utilized is 2000g for 8 minute using liquid-PRF tubes (white/blue tubes, hydrophobic PET). Thereafter, the upper platelet-poor plasma layer composed primarily of albumin can be sent to the Bio-Heat device as reviewed in chapter 7 to extend the resorption properties of PRF (e-PRF). If not needed and the practitioner desires simply to collect C-PRF, the upper 3- to 4-mL layer can first be discarded, followed by collection of a 0.5- to 1-mL rich layer of platelets and leukocytes.

The main indications include: *(1)* filling facial tissues (esthetics) using Bio-Fillers and *(2)* injecting into joint spaces such as knees when the most concentrated liquid-PRF can be obtained (Box 5-3). If two centrifuges are available in the same clinic and the practitioner desires to make the Bio-Bone protocol (reviewed in chapter 7), the

Box 5-3 Clinical indications for C-PRF protocols utilizing white tubes

Massive cell accumulation at the buffy coat

- Filling facial tissues (esthetics)
- Used with GBR membranes to extend working properties of PRF from 2–3 weeks to 4–6 months using Bio-Heat technology
- Injecting into joint spaces

first horizontal centrifuge can be utilized to make C-PRF while the second centrifuge can be utilized to make a liquid-PRF for bone graft membranes (700g for 8 minutes; see chapter 7).

Conclusion

This chapter reviewed the three main protocols needed for PRF therapy. The first protocol (solid-PRF; 700g for 8 minutes) produces a PRF clot with an even distribution of cells. This protocol is utilized more frequently in dentistry for the production of PRF membranes and PRF plugs. The second protocol (liquid-PRF; 300g for 5 minutes) is utilized to generate a more concentrated injectable-PRF. This is most often utilized for injection into joint spaces for facial esthetic procedures. The final protocol (C-PRF; 2000g for 8 minutes) produces a concentrated-PRF with a massive accumulation of platelets and leukocytes directly at the buffy coat layer. This can be utilized similar to liquid-PRF where a higher dose of growth factors and cells are needed to generate tissues. With these three simple protocols, the practitioner is able to perform practically all clinical procedures.

6

Importance of Centrifugation Tubes for the Production of PRF

Contributors

Richard J. Miron*
Tomoyuki Kawase*
Hudi Xu
Mustafa Tunali
Anika Dham
Yufeng Zhang
Masako Fujioka-Kobayashi
Anton Sculean

*Authors contributed equally to this work.

Chapter Highlights

- *Importance of centrifugation tubes for the production of PRF*
- *Why certain tubes are hydrophilic versus hydrophobic*
- *Recent data on the incorporation of chemicals into PRF tubes*
- *The negative impact of silica and silicone on cell survival and apoptosis*
- *Future trends in the field and leading-edge research on the topic*

Video 6-1

It is hard to imagine that the effect of centrifugation tubes may in fact be more relevant for the production of PRF than the actual centrifugation device. Tubes, which are generally purchased and range in price from $0.85 to $1.55 each, have a massive implication in the final outcomes of both solid-PRF and liquid-PRF. Simply, more hydrophilic tubes (typically plain glass or silica-coated plastic) have been shown to be more effective for the production of solid-PRF (red caps), whereas hydrophobic tubes (typically PET plastic tubes) are utilized for the production of liquid-PRF (white or blue caps). The more hydrophilic the surface, the better-quality clot will occur (and the better the fibrin mesh), whereas the more hydrophobic, the more PRF will remain liquid in nature.

Unfortunately, the majority of tubes utilized for the production of PRF are not purposefully designed for PRF; they are generally laboratory test tubes utilized for lab testing, which have since been brought to market and utilized for human clinical applications (many without FDA/CE clearance). In a recent technical note on this topic, it was discovered that many PRF tubes are filled with chemical additives such as silica/silicone with unpredictable clinical outcomes. Many clinicians have noted an increased variability in PRF clot sizes, a decreased rate of clot formation (PRF remains liquid even after an adequate protocol is followed), or even an increased rate in the clinical signs of inflammation following the use of PRF. This chapter addresses these more recent issues as well as the research under investigation to develop hydrophilic/hydrophobic chemical-free tubes for the production of PRF.

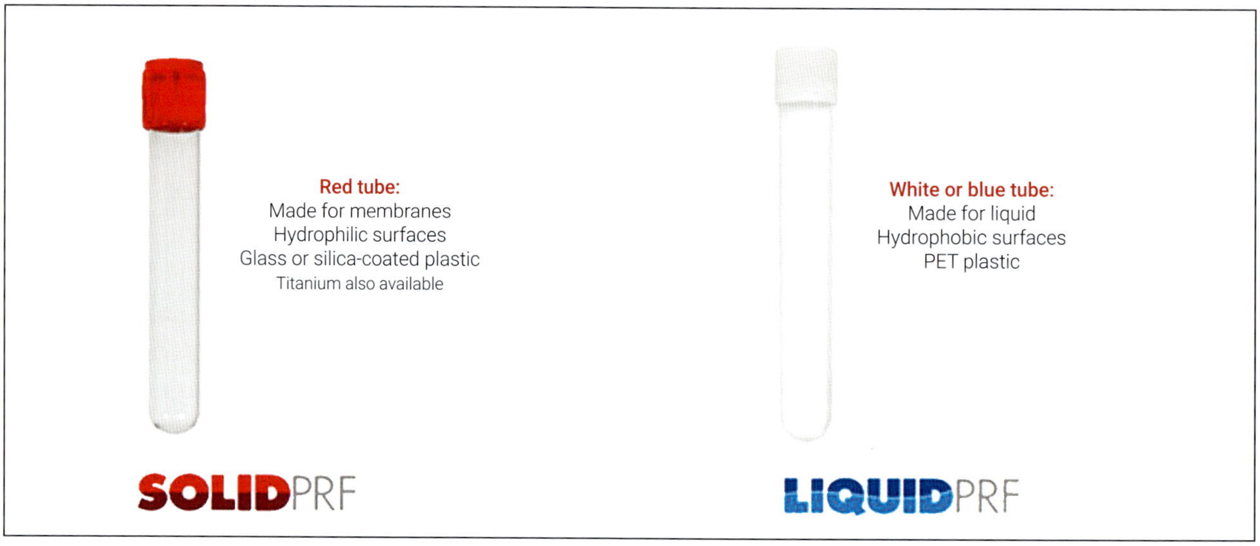

Fig 6-1 Centrifugation tubes with red caps are hydrophilic and promote faster clotting because the platelets accumulate on the tube walls. Tubes with white or blue caps are hydrophobic and repel platelets from the tube walls, making them better suited for liquid-PRF formulations.

The aim of this chapter is to present the marked impact of PRF tubes on the production of PRF as well as all the research that has been conducted on this topic to date. Typically, the production of PRF is best achieved with plain, chemical-free glass tubes. Unfortunately, a variety of centrifugation tubes commonly used for laboratory testing/diagnostics and not necessarily manufactured for human clinical use have been utilized in clinical practice for the production of PRF. Of great concern has been the number of companies that have utilized laboratory tubes that contain various chemical additives into their formulations.

A basic understanding of the hydrophilic and hydrophobic properties of these tubes is needed to understand why they impact PRF production. Think about this in the context of implants. Research has shown that when surface hydrophilicity was improved on implant surfaces, a reported faster improvement in osseointegration occurred.[1–8] This was mainly due to the fact that as hydrophilicity of implants was improved, platelets were able to better attach to these "water-loving" hydrophilic surfaces, facilitating clot formation and, as a result, a quicker regenerative process.

These same concepts apply to PRF tubes; the more hydrophilic the tube, the faster platelets can aggregate, favoring faster clotting (Fig 6-1). A wave of research has been dedicated to this topic to better understand the full impact of PRF tubes on the production of PRF (Video 6-1). On the other end of the spectrum, when PRF tubes are more hydrophobic (the typical material used is PET plastic tubes), these non-water-loving tube walls keep platelets off the walls and in suspended solution (see Fig 6-1). As a result, the cells bind much more slowly and clotting is delayed. These tubes are utilized more frequently for liquid injectable purposes. Therefore, the more hydrophilic the PRF tube, the faster clotting will occur. The more hydrophobic the tube, the longer PRF will remain liquid.

The more hydrophilic the PRF tube, the faster clotting will occur. The more hydrophobic the tube, the longer PRF will remain liquid.

Video 6-1

Importance of Clot Activation from PRF Tube Walls

Under natural wound healing conditions, once an injury occurs, the blood begins to condense, and endogenous and exogenous coagulation pathways are activated to repair damaged blood vessels.[9–11] During PRF clot formation in tubes, the main pathway to activate blood is via endogenous coagulation pathways associated with "surface contact" once blood interacts with the surface of the centrifuge tube walls.[12] The endogenous coagulation pathway relies on different enzyme groups that result in the activation of factor XII (FXII) to factor XIIa,[13] leading to the formation of a solidified clot entrapped with a higher concentration of platelets, immune cells, and growth factors.[11,14–17] Thus, platelets in particular have great implications on the formation of blood clots (or PRF clots) through adhesion to various modalities (damaged endothelium, tube walls, aggregating with other platelets) and by providing coagulation reaction steps that are all triggered by platelet adherence and activation.[9,18,19] Activated platelets in blood clots and PRF membranes release coagulation factors such as FXII, von Willebrand factor, factor V, growth factors, and cytokines that play important roles in complex physiologic processes, including recruitment and activation of fibroblasts, white blood cells, and macrophages as well as mesenchymal stem cells.[18,20] Therefore, if the strategy to develop platelet concentrate therapy is to increase the regenerative potential of platelets in blood, it is obvious that a higher number of activated platelets results in larger PRF clots and better biologic properties.[9]

During PRF formation, platelets in blood first adhere to the walls of the tube, which leads to their activation. Because the hydrophilicity/hydrophobicity of the tube surface is important for protein and cell adhesion, modifying the hydrophilicity/hydrophobicity of the tube surface as well as its chemical charge can be important for regulating platelet activation in PRF membrane preparation. On hydrophobic procoagulant surfaces, not only is the adherence of platelet compromised, but the coagulation factor (ie, FXII) adsorption is also significantly reduced by competing proteins, leading to a drastic reduction in efficiency of FXII contacts within the tube wall.[14,21] Hence, enhancing the hydrophilicity/chemical charge of the centrifugation tubes can be a promising avenue to increase the formation of a superior 3D PRF clot through promotion of platelet adhesion and activation via tube walls.[14,22,23] This can be accomplished either by modifying the surface characteristic of glass through surface treatments (see Study 5) or by modifying the tube wall material from glass to a more hydrophilic material such as titanium (see Study 6).

Of great concern, however, has been the recent trend whereby clinicians have been utilizing laboratory tubes for the production of PRF. These tubes often have chemical additives such as silicone and silica, which can have a negative impact on the production of PRF. Following, recommendations are provided to clinicians with the aim of further optimizing PRF clots/membranes by appropriately selecting quality PRF tubes based on a number of studies on this particular topic.

Study 1: Miron et al (2018)

The size of PRF membranes is dependent on centrifugation tubes and not centrifugation devices

The aim of this study was to compare and contrast the effect of three different centrifugation devices (IntraSpin, Process for PRF, and Salvin) on the final outcomes of PRF clots.[24] In this study, two centrifugation speeds were utilized for each device to compare and contrast the differences between centrifugation devices. One of the most surprising observations, however, was the fact that PRF membranes produced using glass tubes (BD-Salvin and Process for PRF) were larger than those produced using IntraSpin silica-coated plastic tubes. Clot size varied by as much as 200%, whereas the differences between the actual centrifugation devices were minimal. This clearly indicated that PRF tubes seemed to have a greater effect on the final size outcomes of PRF clots than PRF centrifugation devices.[24]

Clot size varied by as much as 200%, whereas the differences between the actual centrifugation devices were minimal.

To directly investigate the effects of PRF tubes on the final outcomes of PRF clot sizes, two tubes from each of the manufacturers were collected from a single participant in a random order (six tubes total).[24] Thereafter, each of the six

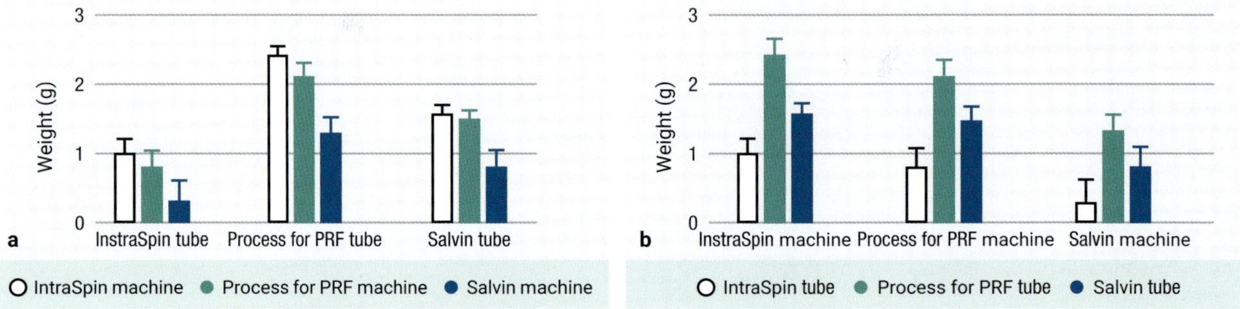

Fig 6-2 Final sizes of PRF clots produced utilizing three different centrifugation tubes in each of three different centrifugation devices (a total of nine tested groups). *(a)* In general, the IntraSpin centrifugation device produced slightly larger clots than the other devices (roughly 15%). *(b)* The glass tubes (Process for PRF, Salvin) produced the largest PRF clots—roughly 200% to 250% larger than those produced with plastic silica-coated IntraSpin tubes. (Reprinted with permission from Miron et al.[24])

tubes was placed into one centrifugation device. By utilizing such an approach, we could address the role of PRF tubes by specifically introducing only one variable by utilizing exactly the same centrifugation device at the same speed with the same patient blood. It therefore became possible to truly assess the effect of PRF tube on the final PRF outcome. For each single participant, this experiment was repeated a total of three times (on each of the three centrifugation devices) for a total of 18 blood tubes harvested from each patient (180 mL total). Each series of six tubes was placed into each centrifugation device.

Interestingly, the PRF centrifugation device had little effect on the final size outcomes of PRF membranes (~15% difference between various fixed-angle centrifuges; Fig 6-2). However, the differences in the PRF clots produced in the different tubes had a marked and pronounced effect on the final size outcomes of PRF tubes. Most surprisingly, it was observed that the IntraSpin tubes produced an approximately 200% to 250% smaller PRF clot than the other glass tubes. Furthermore, it was revealed for the first time that the centrifugation tubes are central to the quality production of PRF. Future research investigating tube characteristics thus became critically important for the future optimization of PRF.[24]

Clinical significance

In general, PRF clots produced in glass tubes are larger than those produced in silica-coated plastic tubes, and the selection of proper tubes has a much greater impact than the centrifugation device.

Study 2: Tsujino et al (2019)

Evidence for contamination of silica microparticles in A-PRF matrix prepared using silica-coated plastic tubes

A group of authors based in Japan aimed to better understand the effects of silica-coated plastic tubes and, more specifically, the effects of silica release and incorporation into the PRF matrix.[25] In this study, blood samples were collected into three different brands of silica-containing plastic tubes and were immediately centrifuged as per the protocol for A-PRF. A-PRF clots were fixed and examined using scanning electron microscopy (SEM) or enzymatically degraded to spectrophotometrically determine the amount of silica microparticles. Regardless of tube brand and individual donor, both SEM examination and spectrophotometric determination demonstrated that significant levels of silica microparticles (between 5% and 30%) were incorporated into the A-PRF matrix, which would be consequently incorporated into human tissues within implantation sites (Fig 6-3). Following enzymatic digestion of PRF clots, it was extremely apparent that the silica from the tube walls of plastic tubes, such as IntraSpin, was shed from the walls and incorporated within the PRF clots, leaving significant byproducts within these clots, as observed via SEM in Fig 6-3. It was reported by Tsujino et al that clinicians should not exclude the possibility that silica microparticles negatively influence tissue regeneration, and the authors further

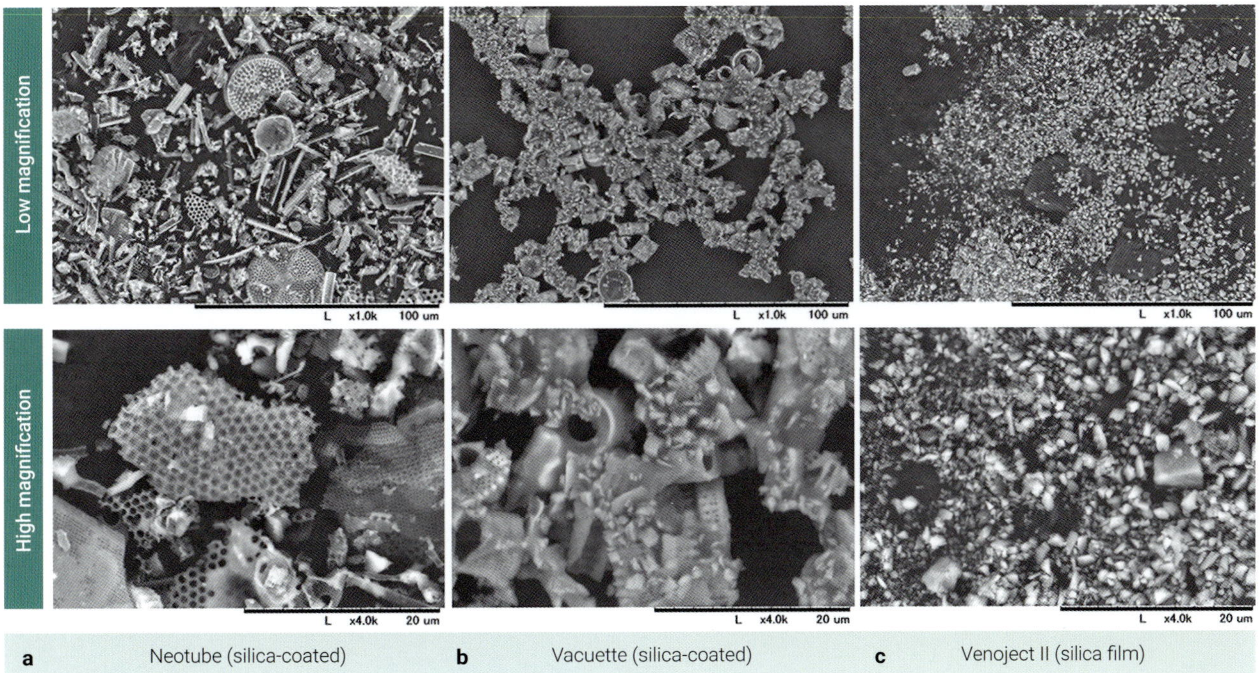

Fig 6-3 In this experiment, PRF clots were produced in three different commercially available tubes containing silica. Following centrifugation, the PRF clots were removed and enzymatically digested, and "leftover" remaining silica particles were visually assessed by SEM. These images show the SEM observations of silica microparticles contained in Neotubes *(a)*, Vacuette tubes *(b)*, and Venoject II tubes *(c)* at low *(top)* and high *(bottom)* magnification. Note the high incorporation of silica microparticles detached from PRF tube walls into PRF clots. (Reprinted with permission from Tsujino et al.[25])

recommended not using silica-containing tubes until their safety is assured.[25]

Clinical significance

In general, tubes that contain a silica coating may shed their coatings into PRF membranes, thus embedding silica particles at quite significant levels within PRF clots.

Study 3: Masuki et al (2020)

Acute cytotoxic effects of silica microparticles used for coating plastic blood-collection tubes on human periosteal cells

In a second study by the same group, the effect of silica microparticles released from PRF tube walls was then investigated on human periosteal cells for potential cytotoxicity.[26] To further assess the biosafety of the silica microparticles, the authors examined their effect on primary human periosteal cells derived from alveolar bone. Silica microparticles were obtained from silica-coated tubes and added to cell cultures. Cellular responses were monitored using a tetrazolium assay, phase-contract inverted microscopy, an immunofluorescence method, and SEM.[26]

It was found that the silica microparticles adsorbed onto the cell surface with seemingly high affinity and induced apoptosis of cells, resulting in a significant reduction in cell proliferation and viability (Figs 6-4 and 6-5).[26] These combined findings suggest that silica microparticles contained in plastic tubes for the purpose of blood coagulation may be hazardous where silica-contaminated PRF matrices are implanted. Future research is needed to better understand their use in clinical practice.

Clinical significance

In general, silica coatings actually have a high affinity for cell membrane walls, causing apoptosis and a reduction in cell proliferation.

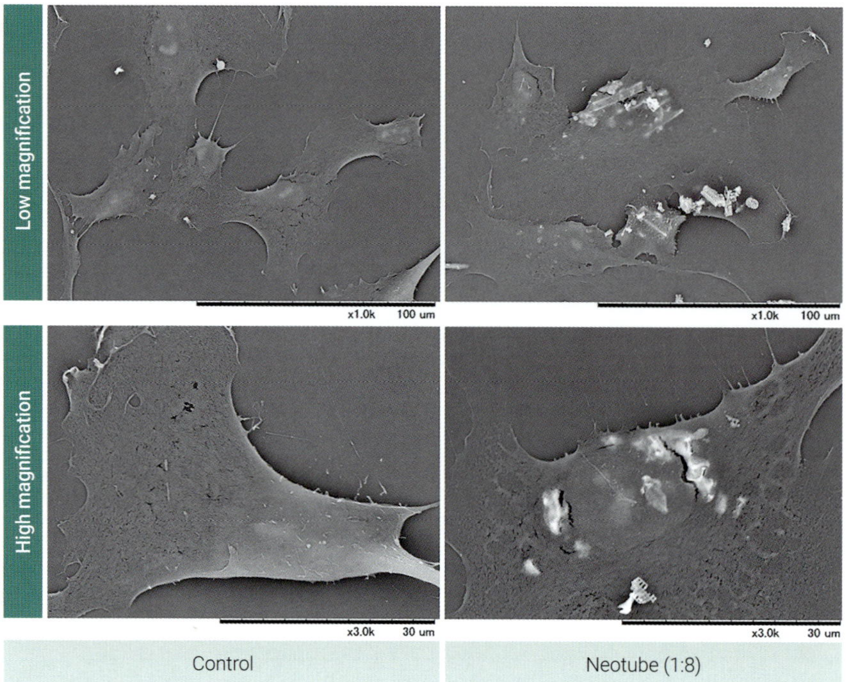

Fig 6-4 Microstructural images of human periosteal cells treated with silica microparticles. The cells treated with silica microparticles derived from Neotubes (1:8 dilution) for 24 hours were fixed and examined using SEM at low and high magnification. Note that the cells rapidly incorporated silica with high affinity. Similar observations were obtained from four other independent experiments, including those involving Vacuette's silica. (Reprinted with permission from Masuki et al.[26])

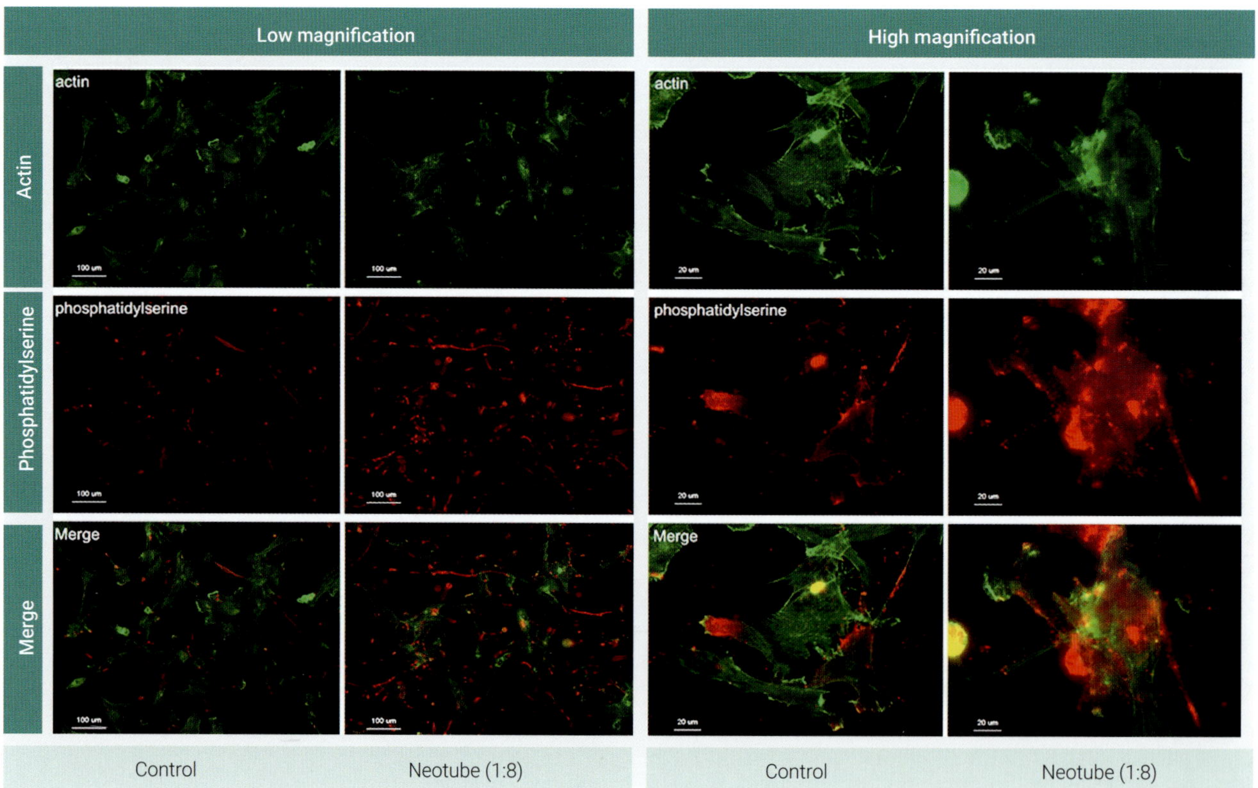

Fig 6-5 Fluorescence visualization of apoptosis in human periosteal cells treated with silica microparticles. The cells were treated with silica microparticles derived from Neotubes for 24 hours. The fixed cells were probed with PE-conjugated annexin V for detection of phosphatidylserine on the cell surface, which is accepted as a marker of apoptosis, and are shown at low and high magnification. The cells were counterstained with FITC-conjugated phalloidin to visualize cytoskeletal polymerized actin. Similar observations were obtained from four other independent experiments, including those involving Vacuette's silica. (Reprinted with permission from Masuki et al.[26])

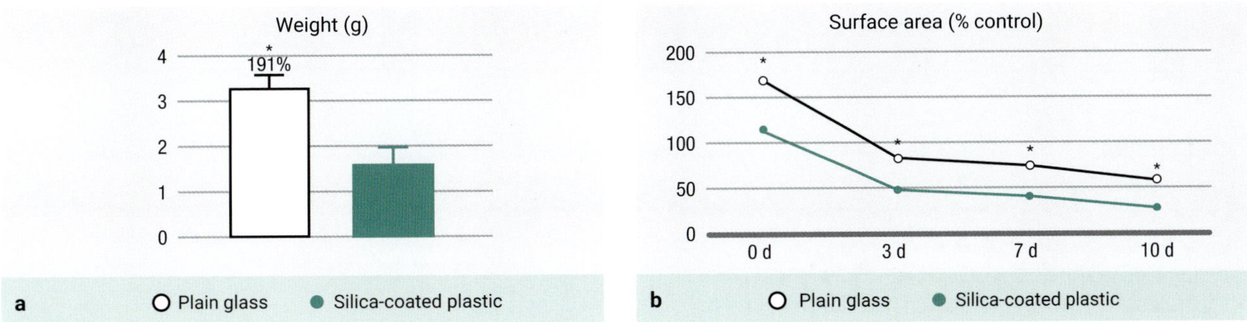

Fig 6-6 Comparative analysis of PRF tube clot weights and sizes from six individuals after centrifugation utilizing the same centrifugation speed and time and either plain glass tubes or silica-coated plastic tubes. *(a)* In general, the silica-coated plastic tubes on average decrease the final weight of PRF-based matrices nearly twofold. *(b)* Over a 10-day period, while the PRF clots slowly and gradually degraded over time in both groups, significantly increased amounts of membrane remained at 0, 3, and 7 days in the glass tubes. Both glass tubes and silica-coated plastic tubes were filled to 9 mL.

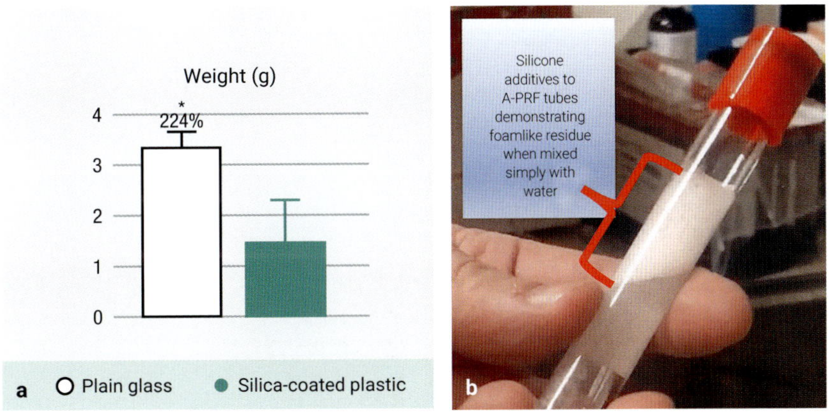

Fig 6-7 Effect of the addition of silicone to glass tubes. *(a)* Comparative weight analysis of PRF clots produced in plain glass tubes versus those produced in A-PRF glass tubes coated with silicone. Clot membranes increase more than 200% in plain glass tubes. Silicone may be more detrimental to standard clot size than silica. *(b)* Furthermore, when liquid is introduced into silicone-coated glass tubes, a foamlike residue is generally created above the liquid layer.

Study 4: Miron et al (2019)

Chemical additives to PRF tubes including silica and silicone negatively impact final PRF clot size

Following our group's previous publication highlighted in Study 1, a second study was performed investigating the effect of various tubes on the final size outcomes of PRF clots. Analysis over a 10-day period comparing the differences between PRF clots demonstrated that PRF clots produced in plastic silica-coated tubes were significantly smaller (roughly half the size) than those produced from glass tubes (Fig 6-6a). Furthermore, trends continued to favor plain chemical-free glass tubes over the entire 10-day study period (Fig 6-6b). This study highlighted that much future research on tube chemistry is necessary to better understand clot formation, even formation in response to plain chemical-free glass tubes fabricated via different methods and starting materials.

More recently, the addition of silicone has specifically been introduced at high levels into A-PRF glass tubes. Comparative analysis revealed that PRF clots produced in A-PRF glass tubes containing silicone-coated walls had drastically reduced clot size formation (by over twofold) compared to the size of those produced with standard plain glass tubes (Fig 6-7). Notably, some samples did not undergo any clotting after a standard 8-minute centrifugation cycle. Silicone addition results in the production of a residue when either blood or water is introduced into PRF tubes, as demonstrated in Fig 6-7b.

Clinical significance

Both silica and silicone addition to the interior of PRF walls led to a nearly twofold reduction in PRF clot sizes. Furthermore, A-PRF silicone tubes specifically left a pronounced foamlike residue following interactions with solutions.

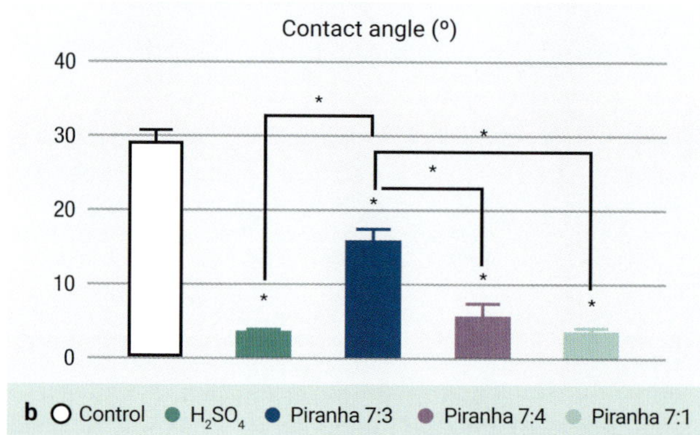

Fig 6-8 *(a)* Contact angle of glass surfaces for five groups including *(1)* control group, *(2)* H_2SO_4 group, *(3)* Piranha 7:3 group, *(4)* Piranha 4:1 group, and *(5)* Piranha 7:1 group. *(b)* Quantitative data of contact angle of glass surfaces in all groups. *$P < .05$.

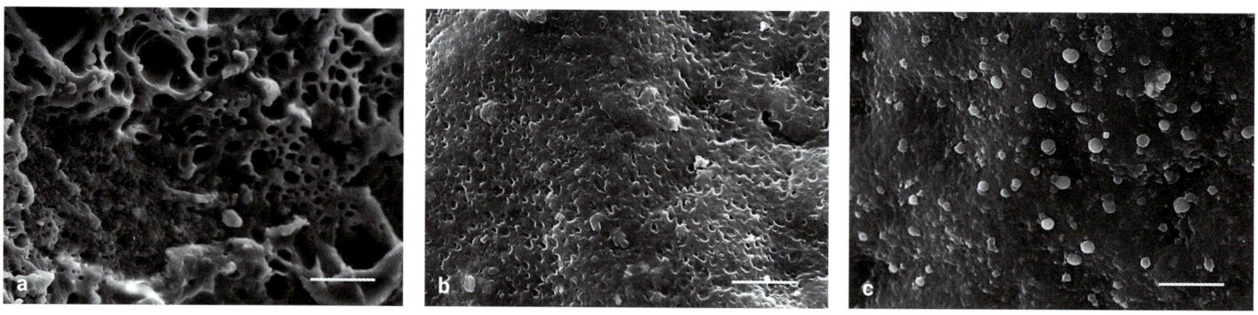

Fig 6-9 SEM images of PRF produced using tubes from the control group *(a)*, H_2SO_4 group *(b)*, and Piranha group *(c)*. Scale bar = 10 μm.

Study 5: Xu et al (2020)

Improving PRF membrane mechanical and bioactive properties by enhancing the activation of platelets through modification of tube hydrophilicity/chemical charge

Improving PRF tubes may be one area that requires the most critical and immediate attention for the optimization of PRF. It is obvious that a great deal of time, investigation, and funding has been performed to optimize PRF protocols and centrifugation devices (see chapters 2 and 3). That being said, very little research has paid attention to the actual tubes and their impact on the final size outcomes and quality of PRF.

In a study by Xu et al, the contact angles (hydrophilicity), contractile strength of PRF, and adhesion ability of platelets to glass was investigated following various glass-treatment modifications. First, to investigate the hydrophilic status of treated glass slides, the contact angle of five groups of boron silane glass slides were examined by contact angle measurements. Interestingly, it was found that the contact angles of H_2SO_4 (3.33 degrees) and Piranha 7:1 group (4 degrees) were significantly smaller than that of the control group (28.5 degrees), Piranha 7:3 group (15.3 degrees), and Piranha 4:1 group (5.83 degrees; Fig 6-8).

Thereafter, the fibrin network density of PRF was evaluated via SEM (Fig 6-9). It can be observed that the fibrin network produced through hydrophilic treatment tubes presented denser fibrin-like structure when compared to control groups without hydrophilic treatment (see Fig 6-9a). The contractile characteristics including elastic modulus, contractile stress, and force of the PRF membrane measured by Instron are presented in Fig 6-10. The surface

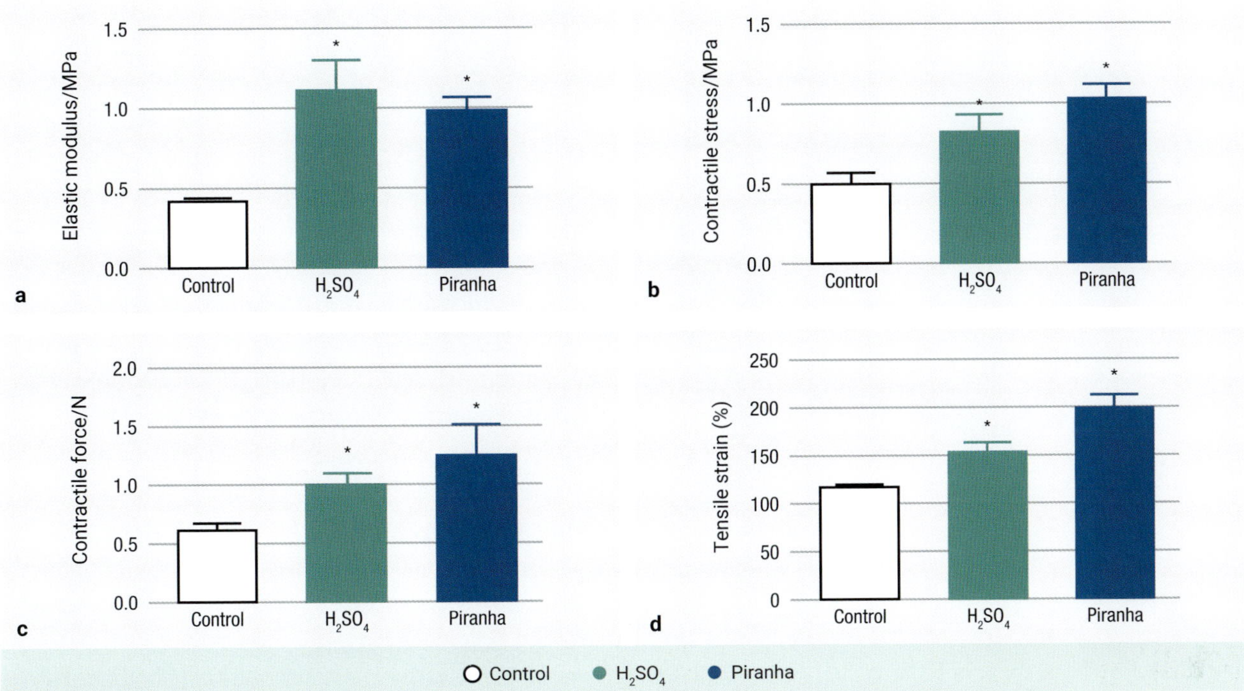

Fig 6-10 The contractile characteristics of PRF produced using tubes from the control group, H_2SO_4 group, and Piranha group. *(a)* Elastic modulus. *(b)* Contractile stress. *(c)* Contractile force. *(d)* Tensile strain. *$P < .05$.

treatments demonstrating the higher hydrophilicity including the H_2SO_4 group (1.115 MPa) and Piranha group (0.98 MPa) demonstrated over a twofold significantly higher elastic modulus when compared to the control group (0.415 MPa). A similar trend was also found for contractile stress, contractile force, and larger tensile strain, suggesting better use for clinical practice that favors stronger PRF membranes (see Fig 6-10).

Clinical significance

Chemical modification to tube surfaces resulted in faster fibrin clot formation and strong PRF clots for clinical use. Future research is underway to further optimize glass tube hydrophilic properties.

Study 6: Tunali et al (2014–2020)

Improving PRF membrane mechanical strength through change in material characterization and an introduction to titanium-PRF tubes

Titanium represents an extremely bioinert material with excellent hydrophilic properties. In addition, titanium is resistant to corrosion and has been utilized for dental implants because of its favorable biocompatibility. Little is known regarding compatibility of blood with metals; however, titanium is among the most thrombogenic materials, which may explain its outstanding ability to osseointegrate with bone.[27]

The coagulation activation caused by titanium is triggered by the intrinsic pathway from the generation of FXIIa-AT/C1

Fig 6-11 Clinical image of T-PRF tubes, which can also be utilized in clinical practice due to their excellent hydrophilic properties.

esterase inhibitor paralleled that of thrombin-antithrombin, and both thrombin-antithrombin complex and FXIIa-AT/C1 esterase inhibitor generation were abrogated by corn trypsin inhibitor, which is a specific inhibitor of FXIIa.[28] The binding of platelets is therefore increased on the titanium surface compared to other biomaterial surfaces (Fig 6-11). Furthermore, PDGF and other alpha-granule proteins (eg, TGF-β) are known to be potent promoters of osteogenesis, which suggests that the pronounced thrombogenic properties of titanium might contribute to the good osseointegration properties.[28]

Because of these properties, titanium-induced platelet aggregation is better than glass. Titanium-prepared PRF (T-PRF) therefore has stronger and thicker fibrin potential.[29] This stronger fibrin structure has also been demonstrated to extend the time for resorption of fibrin in vivo and increase the release time of growth factors.[30] Along with the experimental studies that reveal the properties of T-PRF, many successful clinical studies have also been conducted. T-PRF has been shown to be an important alternative to connective tissue grafts in the treatment of gingival recessions, for periodontal defects, sinus elevation applications, and crest preservation techniques.[31–33]

Clinical significance

Titanium represents an extremely bioinert material with excellent hydrophilic properties. T-PRF tubes have therefore been developed and utilized in clinical practice with excellent results. Future research is ongoing.

Discussion of Studies

Several studies have now been performed on this topic, highlighting the impact of silica/silicone coatings on either plastic or glass tubes and their potential toxic effect on human cells and/or inflammatory response when implanted in vivo. Generally, clinicians have for years been concerned with the choice of PRF centrifugation device, placing little importance on the impact of PRF tubes. With previously published studies over the past 1 to 2 years, it has convincingly been shown that PRF tubes are actually far more important than previously considered.

Unfortunately, much variability exists between centrifugation companies with respect to their tube quality, with virtually no research being performed on the topic to date. In 2018, a study by our group demonstrated convincingly that the PRF clots actually varied in size by as much as twofold between centrifugation tubes, even when all centrifugation protocols were carried out on the same machine from blood of the same patient at the same protocol.[24] This led several research groups from around the world to further investigate the impact of centrifugation tubes on the final outcomes of PRF clots.

A group of researchers led by Professor Tomoyuki Kawase in Japan obtained clear evidence that silica microparticles derived from commercially available blood collection tubes not only "leak" within PRF tubes[25] but also exert toxic effects on human periosteal cells by adsorbing on the plasma membrane and inducing apoptosis.[26] In addition, this cytotoxicity exceeded expectations, with silica microparticles contained in silica-coated tubes (eg, Neotube and Vacuette) completely disrupting cell growth and viability.

While in that study it was reported that amorphous silica is less toxic than crystalline silica, previous studies have demonstrated that amorphous silica is hazardous to health. In the group's study,[25] it was demonstrated that 5% to 30% silica microparticles, depending on tube brand, were included in the resulting PRF clots. The collective data support the prediction that PRF preparations using silica-coated tubes could be toxic to the surrounding cells at implantation sites. During and after preparation of a PRF matrix, silica microparticles may also overactivate or disrupt platelets[34] and other blood cells in the PRF matrix and reduce their therapeutic potency and efficacy. Future research on this topic is needed, as it was hypothesized that there may be ways to more safely produce PRF without chemical additive incorporation into PRF tubes.

In their study, Masuki et al[26] add the following to the discussion:

> To avoid misunderstanding, it must be noted that silica is different from silicone. In fact, a historical debate may have arisen because of this misunderstanding.[35,36] The website provided by Stream Peak International concisely summarizes the terminology regarding *silica*, *silicon*, and *silicone*. According to this website, silica, which is also known as *silicon dioxide*, is a compound that naturally forms in a reaction between oxygen and silicon. Silica is commonly used in the manufacturing of glass, ceramics, optical fiber, and cement. Silicon (Si) is the second most abundant element on Earth. However, it is rarely found in its original state as Si, as it readily reacts with oxygen to mainly form silicon dioxide. In contrast, silicone is a synthetic polymer created from the combination of silicon, oxygen, carbon, and/or hydrogen. Unlike natural materials that include silica and silicon, silicone is a manmade product that is manufactured in factories as a solid, a liquid, and a gel. Silicone is commonly used as sealant, electrical insulation, a component of cooking utensils, and a coating of test tubes. Therefore, even though silicone used for tube coatings may contain silica-like compounds, it cannot activate blood coagulation. Excess silicone coating actually delays coagulation. Furthermore, even if silicone has negative effects on the immune system and/or cells directly involved in tissue regeneration, these effects should be distinguished from those of silica. In any case, when platelet concentrates are prepared for use in regenerative therapy, we believe that real "plain" tubes that are approved by regulatory authorities of individual countries, regardless of their original materials, are better for clinical use.

Thus, it remains difficult to understand why companies would incorporate such chemical additives into PRF tubes.

Surprisingly, research conducted on the topic by our group has actually demonstrated that the addition of silicone even in plain glass tubes led to a drastic ~200% reduction in PRF clot sizes (see Fig 6-6). Furthermore, A-PRF tubes produced by certain manufacturers have undergone a surprising number of changes over the years in tube chemistry, glass coloration, and a recent addition of silicone, as depicted in Fig 6-6b. Strikingly, recommendations in extended protocol centrifugation times have even been proposed to compensate for the reduction in clotting efficacy due to the addition of silicone. Therefore, it is highly recommended that treating clinicians be aware of the great impact that centrifugation tubes play on the final outcomes of PRF-based matrices. When in doubt, the authors recommend filling tubes halfway with water and follow that with vigorous shaking to ensure no additives are present on the inner surface of PRF tubes (Videos 6-2 and 6-3).

Video 6-2 Video 6-3

Conclusion

This chapter highlighted the importance of PRF tubes with respect to the final outcomes of PRF clots. While many laboratory test tubes containing silica or silicone have been utilized, it is advised that clinicians wishing to utilize PRF for human clinical use select chemical-free tubes. Silica-coated plastic tubes, which are increasingly used as alternatives to glass tubes, produce a distinguishable type of fibrin matrix in terms of platelet distribution and contamination by silica particles that negatively impact cell survival and proliferation. Similarly, the addition of silicone to glass tubes has further been reported to cause increased inflammatory responses in humans with delays in clot formation following standard centrifugation protocols. Although there is vigorous debate among certain commercial entities on this topic, the authors caution that further investigation, particularly with chemical-free tubes, should be the focus of intensive research efforts over the coming years to substantially progress PRF therapy.

References

1. Alayan J, Vaquette C, Saifzadeh S, Hutmacher D, Ivanovski S. Comparison of early osseointegration of SLA and SLActive implants in maxillary sinus augmentation: A pilot study. Clin Oral Implants Res 2017;28:1325–1333.
2. Bosshardt DD, Salvi GE, Huynh-Ba G, Ivanovski S, Donos N, Lang NP. The role of bone debris in early healing adjacent to hydrophilic and hydrophobic implant surfaces in man. Clin Oral Implants Res 2011;22:357–364.
3. Donos N, Hamlet S, Lang NP, et al. Gene expression profile of osseointegration of a hydrophilic compared with a hydrophobic microrough implant surface. Clin Oral Implants Res 2011;22:365–372.
4. Hamlet SM, Lee RSB, Moon HJ, Alfarsi MA, Ivanovski S. Hydrophilic titanium surface-induced macrophage modulation promotes pro-osteogenic signalling. Clin Oral Implants Res 2019;30:1085–1096.
5. Lang NP, Salvi GE, Huynh-Ba G, Ivanovski S, Donos N, Bosshardt DD. Early osseointegration to hydrophilic and hydrophobic implant surfaces in humans. Clin Oral Implants Res 2011;22:349–356.

6. Nicolau P, Guerra F, Reis R, Krafft T, Benz K, Jackowski J. 10-year outcomes with immediate and early loaded implants with a chemically modified SLA surface. Quintessence Int 2019;50:114–124.
7. Wennerberg A, Galli S, Albrektsson T. Current knowledge about the hydrophilic and nanostructured SLActive surface. Clin Cosmet Investig Dent 2011;3:59–67.
8. Buser D, Broggini N, Wieland M, et al. Enhanced bone apposition to a chemically modified SLA titanium surface. J Dent Res 2004;83:529–533.
9. Miron RJ, Dham A, Dham U, Zhang Y, Pikos MA, Sculean A. The effect of age, gender, and time between blood draw and start of centrifugation on the size outcomes of platelet-rich fibrin (PRF) membranes. Clin Oral investig 2019;23:2179–2185.
10. Minors DS. Haemostasis, blood platelets and coagulation. Anaesth Intensive Care Med 2007;8:214–216.
11. Ruggeri ZM, Mendolicchio GL. Adhesion mechanisms in platelet function. Circulation Res 2007;100:1673–1685.
12. Bowen RA, Remaley AT. Interferences from blood collection tube components on clinical chemistry assays. Biochemia Medica 2014;24:31–44.
13. Cras J, Rowe-Taitt C, Nivens D, Ligler F. Comparison of chemical cleaning methods of glass in preparation for silanization. Biosensors Bioelectronics 1999;14:683–688.
14. Canning J, Petermann I, Cook K. Surface treatment of silicate based glass: Base piranha treatment versus 193nm laser processing. Presented at the Third Asia Pacific Optical Sensors Conference, International Society for Optics and Photonics, 2012.
15. Nakamura K, Morooka H, Tanabe Y, Yasuda E, Akatsu T, Shindo H. Surface oxidation and/or corrosion behavior of glass-like carbon in sulfuric and nitric acids, and in aqueous hydrogen peroxide. Corrosion Sci 2011;53:4010–4013.
16. Russell J. Spectrometric studies of γ-irradiated sulfuric acid glasses. Can J Chem 1967;45:839–845.
17. Vogler EA, Siedlecki CA. Contact activation of blood-plasma coagulation. Biomaterials 2009;30:1857–1869.
18. Gorbet MB, Sefton MV. Biomaterial-associated thrombosis: Roles of coagulation factors, complement, platelets and leukocytes. Biomaterials 2004;25:5681–5703.
19. Zhuo R, Siedlecki CA, Vogler EA. Autoactivation of blood factor XII at hydrophilic and hydrophobic surfaces. Biomaterials 2006;27: 4325–4332.
20. Zhuo R, Siedlecki CA, Vogler EA. Competitive-protein adsorption in contact activation of blood factor XII. Biomaterials 2007;28: 4355–4369.
21. Zarka R, Horev MB, Volberg T, et al. Differential modulation of platelet adhesion and spreading by adhesive ligand density. Nano Letters 2019;19:1418–1427.
22. Tutwiler V, Wang H, Litvinov RI, Weisel JW, Shenoy VB. Interplay of platelet contractility and elasticity of fibrin/erythrocytes in blood clot retraction. Biophys J 2017;112:714–723.
23. Thon JN, Italiano JE. Platelets: Production, morphology and ultrastructure. Antiplatelet Agents 2012:3–22.
24. Miron RJ, Xu H, Chai J, et al. Comparison of platelet-rich fibrin (PRF) produced using 3 commercially available centrifuges at both high (~ 700 g) and low (~ 200 g) relative centrifugation forces. Clin Oral Investig 2020;24:1171–1182.
25. Tsujino T, Takahashi A, Yamaguchi S, et al. Evidence for contamination of silica microparticles in advanced platelet-rich fibrin matrices prepared using silica-coated plastic tubes. Biomedicines 2019;7:45.
26. Masuki H, Isobe K, Kawabata H, et al. Acute cytotoxic effects of silica microparticles used for coating of plastic blood-collection tubes on human periosteal cells. Odontology 2020;108:545–552.
27. Hong J, Azens A, Ekdahl KN, Granqvist CG, Nilsson B. Material-specific thrombin generation following contact between metal surfaces and whole blood. Biomaterials 2005;26:1397–1403.
28. Hong J, Andersson J, Ekdahl KN, et al. Titanium is a highly thrombogenic biomaterial: Possible implications for osteogenesis. Thromb Haemost 1999;82:58–64.
29. Tunalı M, Özdemir H, Küçükodacı Z, et al. A novel platelet concentrate: Titanium-prepared platelet-rich fibrin. Biomed Res Int 2014;2014:209548.
30. Tunalı M, Özdemir H, Küçükodacı Z, Akman S, Fıratlı E. In vivo evaluation of titanium-prepared platelet-rich fibrin (T-PRF): A new platelet concentrate. Br J Oral Maxillofac Surg 2013;51:438–443.
31. Uzun BC, Ercan E, Tunalı M. Effectiveness and predictability of titanium-prepared platelet-rich fibrin for the management of multiple gingival recessions. Clin Oral Investig 2018;22:1345–1354.
32. Ustaoğlu G, Aydin ZU, Özelçi F. Comparison of GTR, T-PRF and open-flap debridement in the treatment of intrabony defects with endo-perio lesions: A randomized controlled trial. Med Oral Patol Oral Cir Bucal 2020;25:e117–e123.
33. Olgun E, Ozkan SY, Atmaca HT, Yalim M, Hendek MK. Comparison of the clinical, radiographic, and histological effects of titanium-prepared platelet rich fibrin to allograft materials in sinus-lifting procedures. J Investig Clin Dent 2018;9:e12347.
34. Komosa A, Rzymski P, Perek B, et al. Platelets redox balance assessment: Current evidence and methodological considerations. Vascul Pharmacol 2017;93:6–13.
35. O'Connell SM. Safety issues associated with platelet-rich fibrin method. Oral Surg Oral Med Oral Oral Pathol Oral Radiol Endod 2007;103:587.
36. Dohan DM, Del Corso M, Charrier J. Cytotoxicity analyses of Choukroun's platelet-rich fibrin (PRF) on a wide range of human cells: The answer to a commercial controversy. Oral Surg Oral Med Oral Pathol Oral Radiol Endod 2007;5:587–593.

7

Biologic Characterization of e-PRF Membranes

Contributors

Masako Fujioka-Kobayashi
Ezio Gheno
Tomoyuki Kawase
Anton Sculean
Yufeng Zhang
Richard J. Miron

Chapter Highlights

- *Extending the working properties of PRF from 2–3 weeks to 4–6 months (e-PRF)*
- *Protocol and technique to create e-PRF with the Bio-Heat device*
- *Clinical indications for use of e-PRF*

Video 7-1

One of the main reported drawbacks of PRF use is its rather fast resorption properties characterized within a 2- to 3-week time frame. However, recent studies have demonstrated that by heating a liquid platelet-poor plasma (PPP) layer, the resorption properties of heated albumin (albumin gel) can be significantly extended from 2 weeks to more than 4 months; this is called *extended-PRF*, or *e-PRF*. This chapter presents the biologic properties of this novel regenerative modality and includes animal data demonstrating its superior resorption properties. Basic research investigation from our laboratory has convincingly shown the ability for e-PRF to secrete several important growth factors found in standard PRF as well as the ability to improve gingival cell migration, proliferation, and collagen synthesis. Furthermore, while animal data found that while both L-PRF and H-PRF membranes demonstrated complete resorption by 21 days, the e-PRF membrane remained volume stable throughout the study duration with minimal resorption. This study demonstrates for the first time the marked improvement in membrane stability of e-PRF with future potential use as a biologic barrier membrane for GBR procedures with a longer-lasting half-life or as a biologic filler material in facial esthetic applications.

One of the main limitations of PRF has been its short in vivo turnover rate. In regenerative dentistry, it provides only a limited ability to act as a true "barrier membrane" because its use is associated with a typical 10- to 14-day resorption period in vivo.[1] An interesting attempt using the heat-compression technique with PRF membranes was introduced by Kawase et al, which aimed to use PRF membranes for guided tissue regeneration (GTR) treatment with extended working properties.[1] This technique extended the degradation properties of PRF to at least 3 weeks, whereas the standard PRF membranes were completely resorbed within 2 weeks.[1] This led to the hypothesis that heating PRF could favor a slower degradation rate of PRF when compared to standard PRF.

Kawase et al introduced the heat-compression technique of PRF, which extended its resorption properties.

In facial esthetics and plastic surgery, to overcome the rapid degradation of plasma and maintain volume stability, a novel technique was developed whereby platelet-poor plasma (PPP) containing roughly 60% albumin is heated to 75°C for 10 minutes to allow for the denaturation and breaking of many of the weak linkages or bonds (eg, hydrogen bonds) within its protein molecule. Following this process, the proteins are then restructured in a more densely organized protein structure with extended resorption properties of up to 4 to 6 months.[2–5] Because no cells or growth factors in the heat-treated PRF/PPP are capable of enduring the denaturation process, the platelet-rich layer from the buffy coat of C-PRF is mixed back into the heated PPP (albumin gel) after cooling. The resulting liquid biomaterial is termed *Alb-PRF* or *e-PRF*.[6]

In view of previous research to date on this topic, our research group has characterized the biologic properties of e-PRF on cell behavior, which may represent a potential improvement in PRF-based clinical applications.[6] This chapter highlights both the preclinical and clinical findings made thus far on this topic.

Heating of Plasma: Effect on Degradation Properties

Since the use of platelet concentrates first began three decades ago, the goal has always been to collect peripheral blood and utilize a centrifugation device to concentrate natural autologous growth factors. This easy-to-obtain technique allowed for a concentration of platelets, leukocytes, and growth factors that favored the future revascularization of tissues, cell recruitment, and cell proliferation—all favoring tissue regeneration.[7,8]

With the advent of new techniques and the search for improved materials, interest in this field quickly gravitated toward many avenues of esthetic medicine. For years, however, all improvements and techniques related to either PRP or PRF were focused on techniques that favored higher or better concentrations of collected growth factors and/or cells (see chapters 2 and 3). Among all of these blood-derived growth factors, the most researched in the esthetic area has been PRP. PRP provided documented evidence for the stimulation of cell recruitment and proliferation as well as increasing the production of type 1 and type 3 collagen. It has also been utilized in infiltrations, via microneedling, and as a facial topical overlaying the tissue to be regenerated.[9,10]

PRF, discovered almost 20 years ago, provided a new advantage in that a fibrin mesh could be obtained following centrifugation, thereby favoring a longer growth factor delivery vehicle, all via completely natural approaches (ie, no anticoagulants or other additives). Protocols and research in this field focused primarily on increasing the cellularity within the fibrin mesh or attempting to improve the distribution and production of growth factors within their formulation. However, while different protocols for autologous platelet concentrates are often suitable for specific clinical applications, limitations in the stability of this fibrin mesh may compromise their applicability in various clinical applications that demand improved stability.[11,12]

Recently, a novel technique was developed with these blood byproducts, which has now been utilized in various fields of medicine, including facial esthetics.[13] This product consists of the denaturized serum albumin with an extended working time (e-PRF). Simply stated, by heating and denaturing albumin, a modification in the secondary structure after the heating transforms the matter into a tridimensional structure. During this heating process,

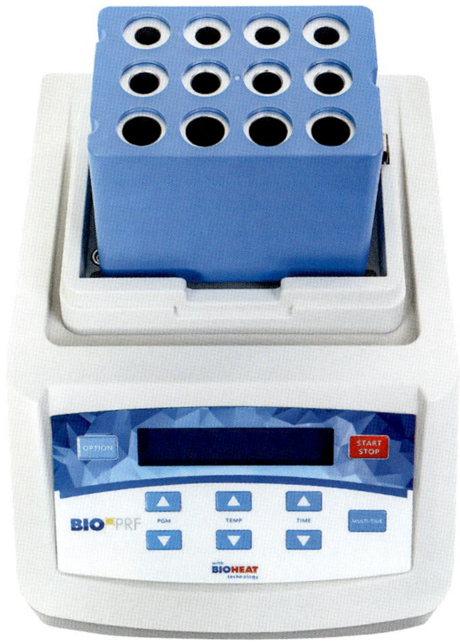

Fig 7-1 Image of the Bio-Heat medical apparatus for the production of e-PRF. Syringes of various sizes may be loaded into the upper compartment, and thereafter denaturing of albumin will occur at 75°C.

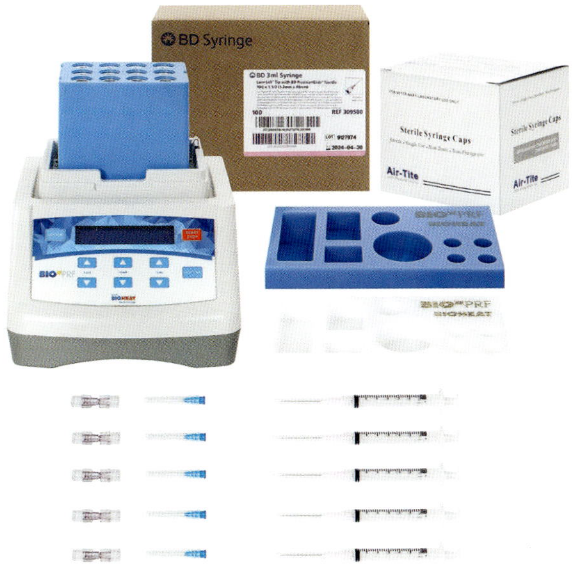

Fig 7-2 The Bio-Heat complete kit includes the blue syringe caps and female-female luer-lock connectors that are needed for the production of e-PRF.

new hydrogen and disulfide ligations in the enzymes are created that favor a larger tridimensional structure with drastic changes in its resorption properties and improve its stability over time. This in turn creates a "filler" derived from 100% whole blood, with extended resorption properties that last up to 6 months as opposed to 2 weeks.[13] While albumin is the most abundant human plasma protein, responsible for more than 50% of the total protein present in the bloodstream, it is important to note that during denaturation, collected growth factors and cells also lose their activity and undergo apoptosis at high temperatures. Therefore, a new protocol had to be developed following heating to reintroduce cells and growth factors back into the e-PRF as presented next.

During the heating process, new hydrogen and disulfide ligations in the enzymes are created that favor a larger tridimensional structure with drastic changes in its resorption properties and improve its stability over time.

Protocol to Produce e-PRF

In order to utilize this new technology, a specialized heating device (Bio-Heat, Bio-PRF) is required (Fig 7-1). A standard kit includes a variety of equipment, including 18G needles for blood collection, blue syringe caps for use in the Bio-Heat device, female-female luer-lock connectors, and Bio-Heat trays (Fig 7-2). The e-PRF is produced by collecting peripheral blood using 9- to 10-mL tubes, without adding any additives (Fig 7-3).

In a first step, peripheral blood is collected (see Fig 7-3a) and then placed in a horizontal centrifuge (Bio-PRF) utilizing a 700g or 2000g protocol for 8 minutes (see Fig 7-3b). After processing, it is possible to observe separation of blood layers into plasma and the remaining decanted red cells. PPP (2–4 mL) is then collected with a syringe (see Fig 7-3e) and placed into the Bio-Heat device (see Fig 7-3f), while the other blood portions (buffy coat, liquid-PRF, and red blood cells) are placed in the Bio-Cool device (see Fig 7-3g). The syringes containing PPP are then inserted into a heating device (Bio-Heat) for human serum albumin denaturation plasma to produce the albumin gel (see Fig 7-3f).

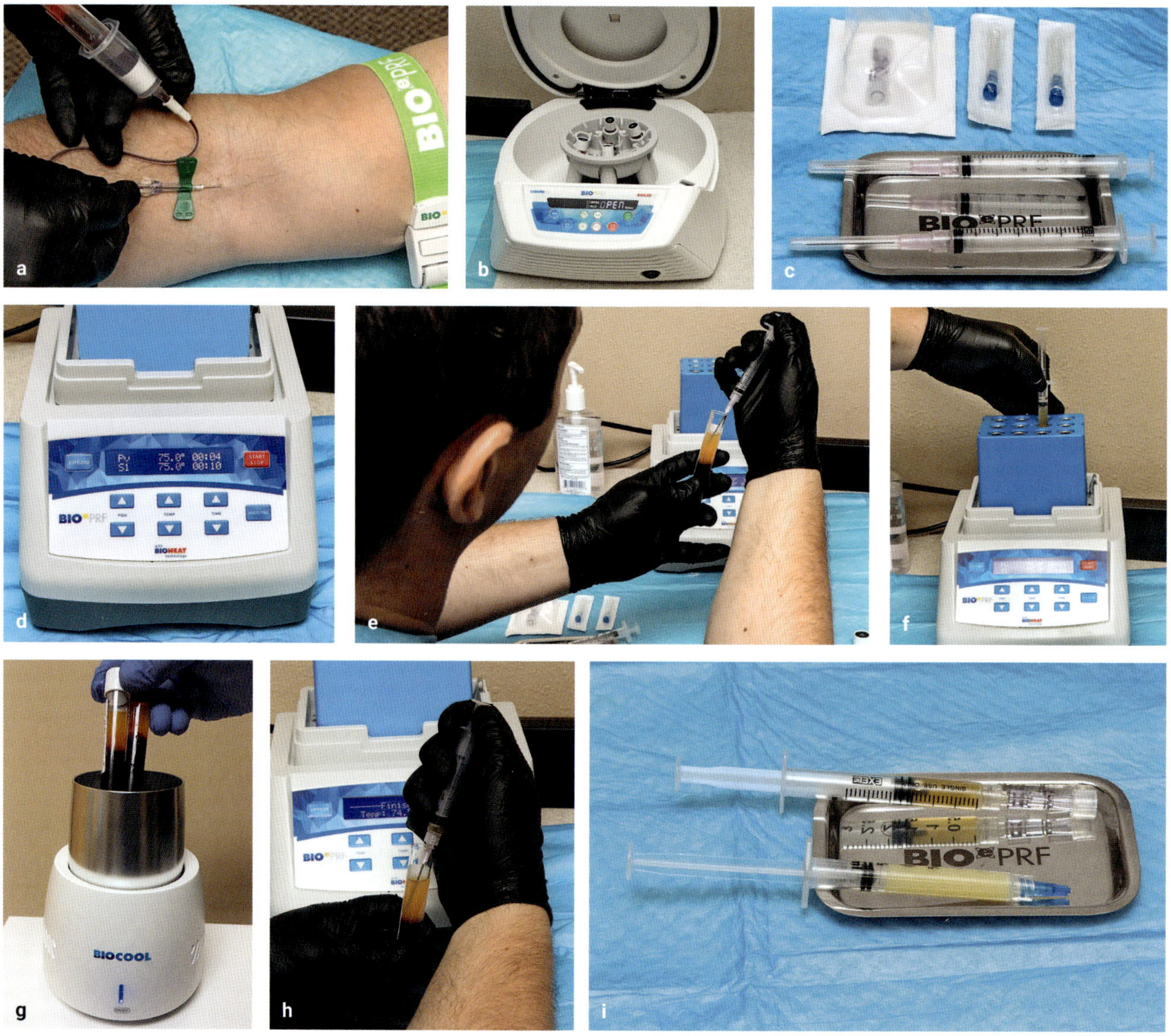

Fig 7-3 Step-by-step clinical demonstration of the production of e-PRF. *(a)* Venipuncture and blood collection. *(b)* Centrifugation. *(c)* Required materials for e-PRF production. *(d)* Bio-Heat medical device to heat the serum and PPP. Note that the machine must be preheated prior to its use. *(e)* Following centrifugation, collection of the serum plus PPP. *(f)* Introduction of the PPP in the Bio-Heat device at 75°C (10 minutes). *(g)* The remaining platelet-rich layer is kept in the Bio-Cool device to extend clotting time. *(h)* After 10 minutes, collection of the liquid-PRF + buffy coat zone. *(i)* Following 10 minutes, clinical differences in color are observed between the liquid-PRF *(top)* and the albumin gel.

After 10 minutes at an operating temperature of 75°C, the syringes are then removed and allowed to cool within the Bio-Cool device for 2 minutes (see Fig 7-3g). Liquid-PRF (preferably C-PRF) is then collected (see Fig 7-3h). Figure 7-3i demonstrates the noticeable color difference between the albumin gel and standard liquid-PRF. Thereafter, the albumin gel and the liquid-PRF are mixed together between syringes by passing back and forth using a female-female luer-lock connector (see Figs 7-3i to 7-3m; Video 7-2).

Video 7-2

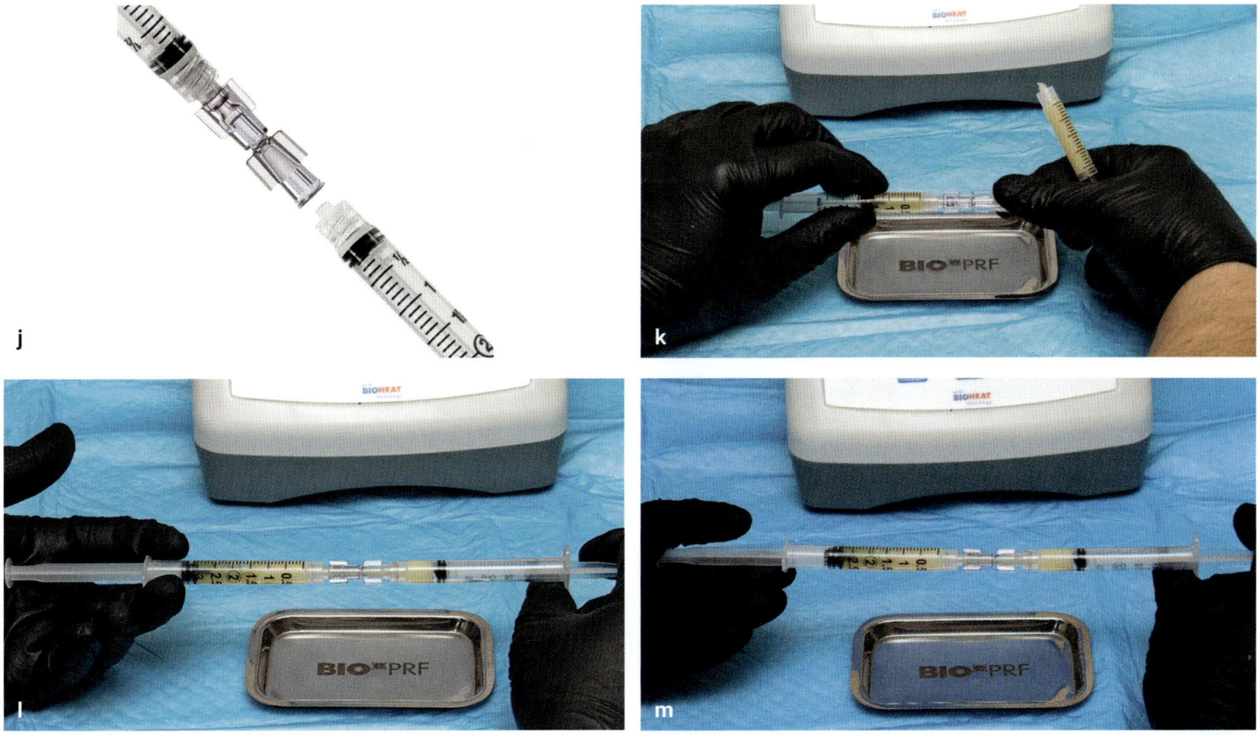

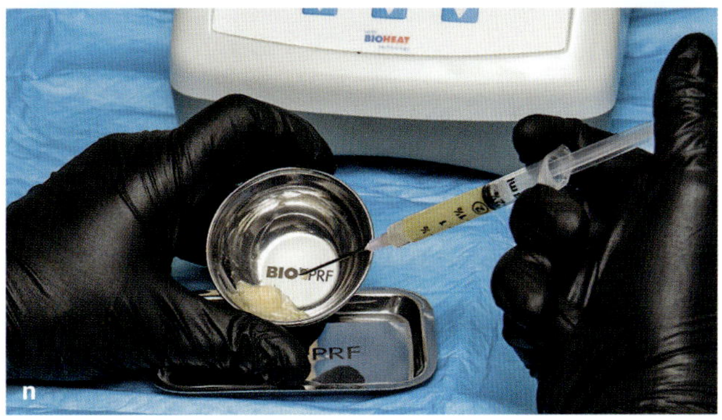

Fig 7-3 *(cont)* *(j and k)* A luer-lock mixer device is attached to both the liquid-PRF and albumin gel. *(l and m)* Mixing of the liquid-PRF and albumin gel back and forth to create e-PRF. *(n)* e-PRF ready for use. Note the ability to inject out of a syringe following adequate mixing. (Reprinted with permission from Davies and Miron.[13])

This should be passed back and forth between syringes a minimum of 10 times to allow for adequate mixing. Thereafter, the e-PRF can be utilized as an injecting autologous concentration of growth factors, cells, and heated albumin (see Fig 7-3n). As an injectable filler, a 25G needle is recommended. Smaller needles are difficult to allow for subcutaneous injections.

The preclinical research to date on this topic is presented below. A step-by-step flowchart of the technique is presented in Fig 7-4.

The Bio-Cool is a device set at a colder temperature that allows the liquid-PRF to remain liquid longer without converting to fibrin. This is essential during this protocol, because it may last upward of 20 minutes.

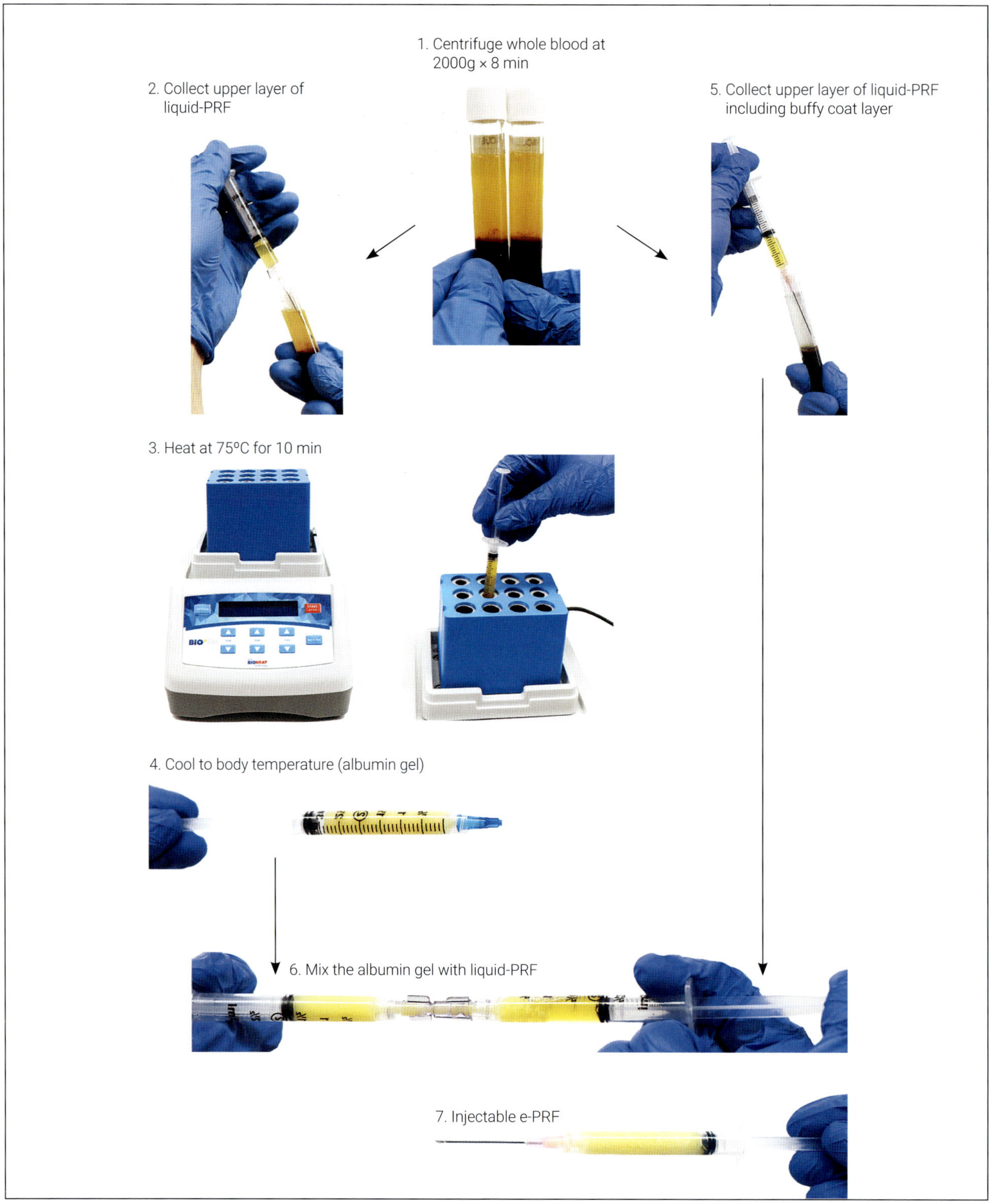

Fig 7-4 e-PRF preparation protocol. (1) Whole blood is centrifuged at 2000g for 8 minutes. The upper layer (yellow layer) shows the liquid plasma layer. (2) The uppermost layer of PPP is collected in a syringe. (3) The collected PPP is heated in a Bio-Heat device at 75°C for 10 minutes and thereafter (4) cooled to room temperature for approximately 10 minutes. An injectable albumin gel is then prepared. (5) The liquid platelet-rich layer (liquid-PRF), including the buffy coat layer with accumulated platelets and leukocytes, is collected in a separate syringe. (6) The albumin gel and native liquid-PRF are then thoroughly mixed by utilizing a female-female luer-lock connector. (7) Injectable e-PRF in final ready form. (Reprinted with permission from Fujioka-Kobayashi et al.[6])

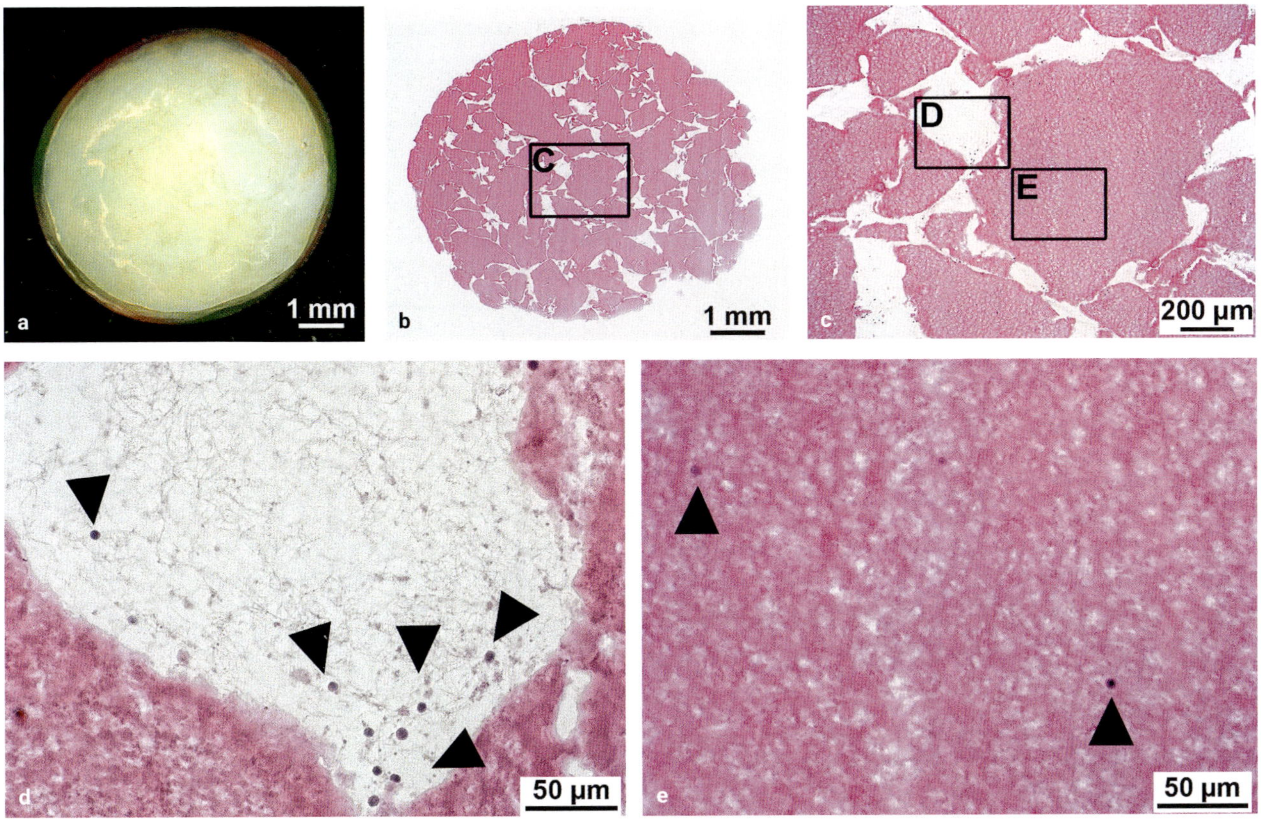

Fig 7-5 Microscopic observation of the e-PRF membrane. *(a)* A trimmed 8-mm e-PRF membrane sized with a biopsy punch. *(b)* H&E staining of the e-PRF section. *(c)* High-magnification view of the image shown in *b*. Two components, eosin-stained filler particle-like structure and matrix, are observed. *(d)* High-magnification view of the native liquid-PRF portion shown in *c*. Leukocytes are observed in the fibrin matrix. *(e)* High-magnification view of the denatured liquid PPP (albumin gel) shown in *c*. A dense fiber network is observed with few leukocytes. (Reprinted with permission from Fujioka-Kobayashi et al.[6])

Biologic Characterization of e-PRF

At the microscopic level, the frozen sections of e-PRF showed two composite structures with denatured PPP (albumin gel) particles and gelated liquid-PRF (Figs 7-5a to 7-5c). The loose matrix structure presented gelated liquid-PRF, including several leukocytes entrapped within the fibrin fibers (Fig 7-5d). The eosin-stained dense structures showed albumin gel, namely denatured plasma components, including few leukocytes that were most probably dead due to the heat treatment (Fig 7-5e).

The release kinetics of growth factors from e-PRF were investigated by ELISA, including PDGF-AA, PDGF-AB, PDGF-BB, TGF-β1, VEGF, EGF, and IGF-1 (Fig 7-6). e-PRF demonstrated a continuing release of growth factors up to 10 days (see Fig 7-6a). The total release of growth factor calculations revealed that TGF-β1 was released at the highest quantities, followed by PDGF-AA and PDGF-AB (see Fig 7-6b).

e-PRF was then investigated on gingival fibroblasts for the potential of cell viability, migration, proliferation, and expression of *TGF-β* and *Col1* (Fig 7-7). A proliferation assay demonstrated that e-PRF induced significantly higher cell numbers at 5 days when compared to those of the control (see Fig 7-7d). Investigation into mRNA levels revealed that e-PRF was able to significantly increase *TGF-β* expression at 3 and 7 days and *Col1a2* at 7 days postseeding when compared to expression in the control (see Fig 7-7e).

Thus, it was concluded that e-PRF is highly biocompatible, has extended ability to release growth factors over time, and also possesses the ability to improve gingival fibroblast cell behavior.

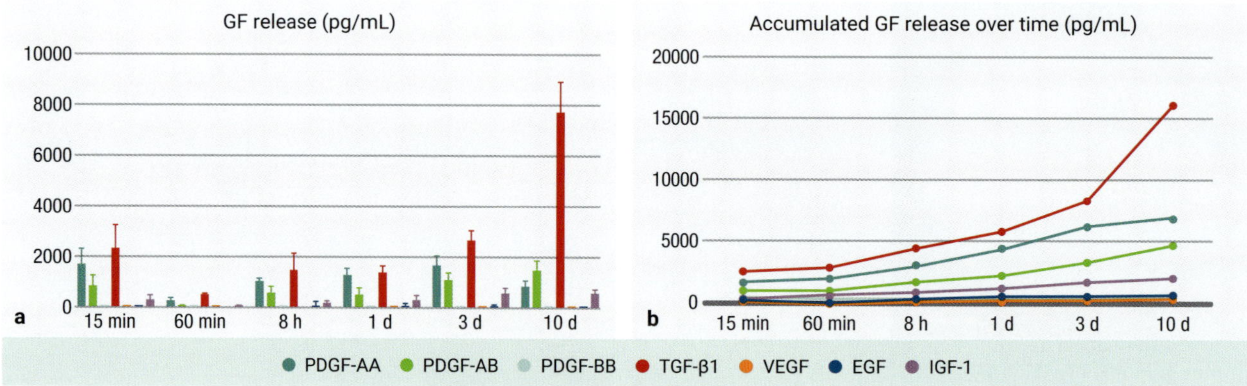

Fig 7-6 *(a)* ELISA protein quantification at each time point of PDGF-AA, PDGF-AB, PDGF-BB, TGF-β1, VEGF, EGF, and IGF-1 over a 10-day period. *(b)* Total accumulated growth factor released over a 10-day period for PDGF-AA, PDGF-AB, PDGF-BB, TGF-β1, VEGF, EGF, and IGF-1. (Reprinted with permission from Fujioka-Kobayashi et al.[6])

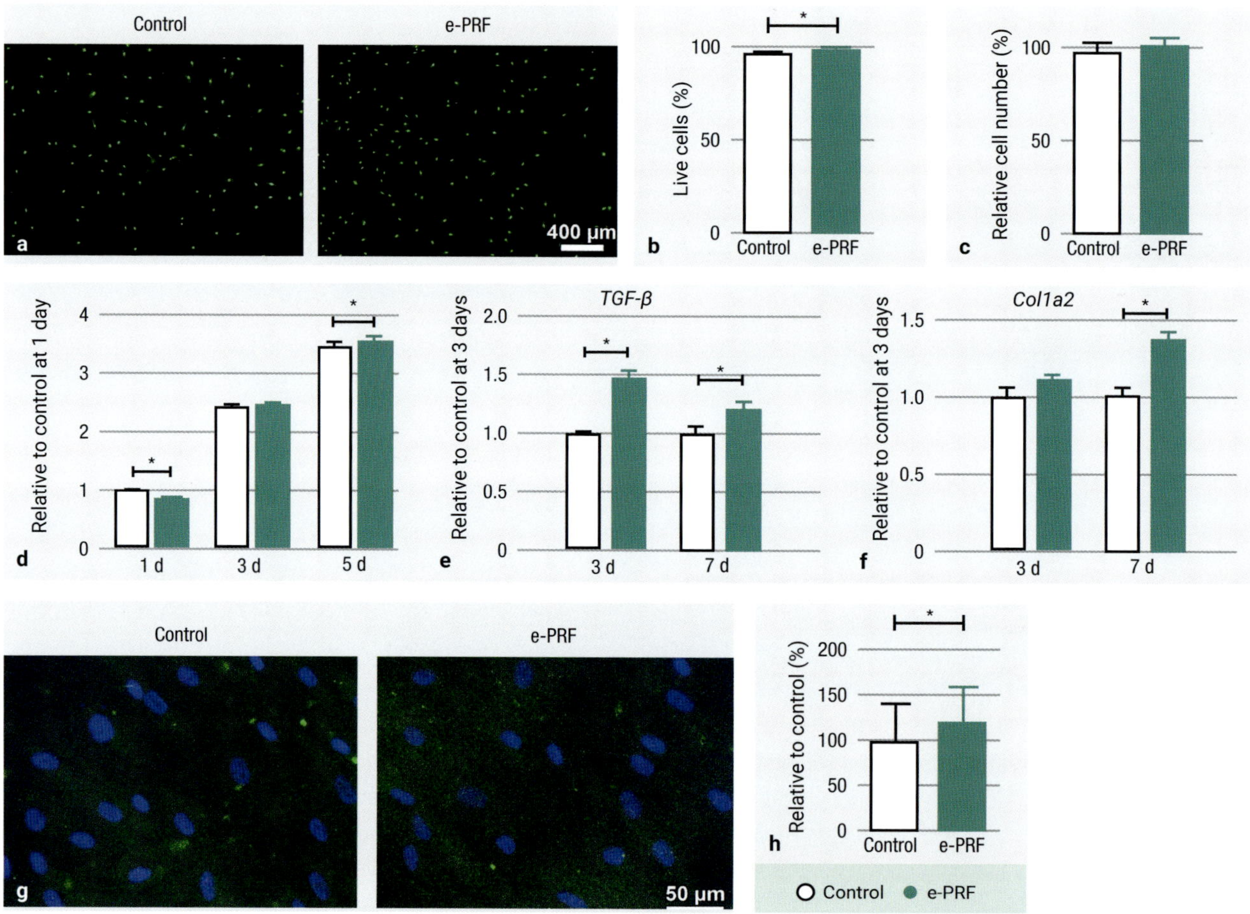

Fig 7-7 Cell behavior when stimulated with e-PRF. *(a and b)* Live/dead assay at 24 hours of human gingival fibroblasts treated with e-PRF. *(a)* The merged fluorescent images of live/dead staining with viable cells appearing in *green* and dead cells in *red*. *(b)* Cell viability was quantified as the percentage of living cells. *(c and d)* Effects of e-PRF on human gingival fibroblast cell migration *(c)* at 24 hours and cell proliferation *(d)* at 1, 3, and 5 days. *(e and f)* RT-PCR of human gingival fibroblasts cultured with e-PRF at 3 and 7 days for mRNA levels of *(e) TGF-β* and *(f) Col1a2*. *(g and h)* Immunofluorescent collagen 1 (COL1) staining of human gingival fibroblasts treated with e-PRF at 14 days. *(g)* The merged fluorescent images of COL1 staining *(green)* with DAPI staining *(blue)*. *(h)* Quantified values of COL1 staining in comparison with control samples. An *asterisk* (*) denotes a significant difference. (Reprinted with permission from Fujioka-Kobayashi et al.[6])

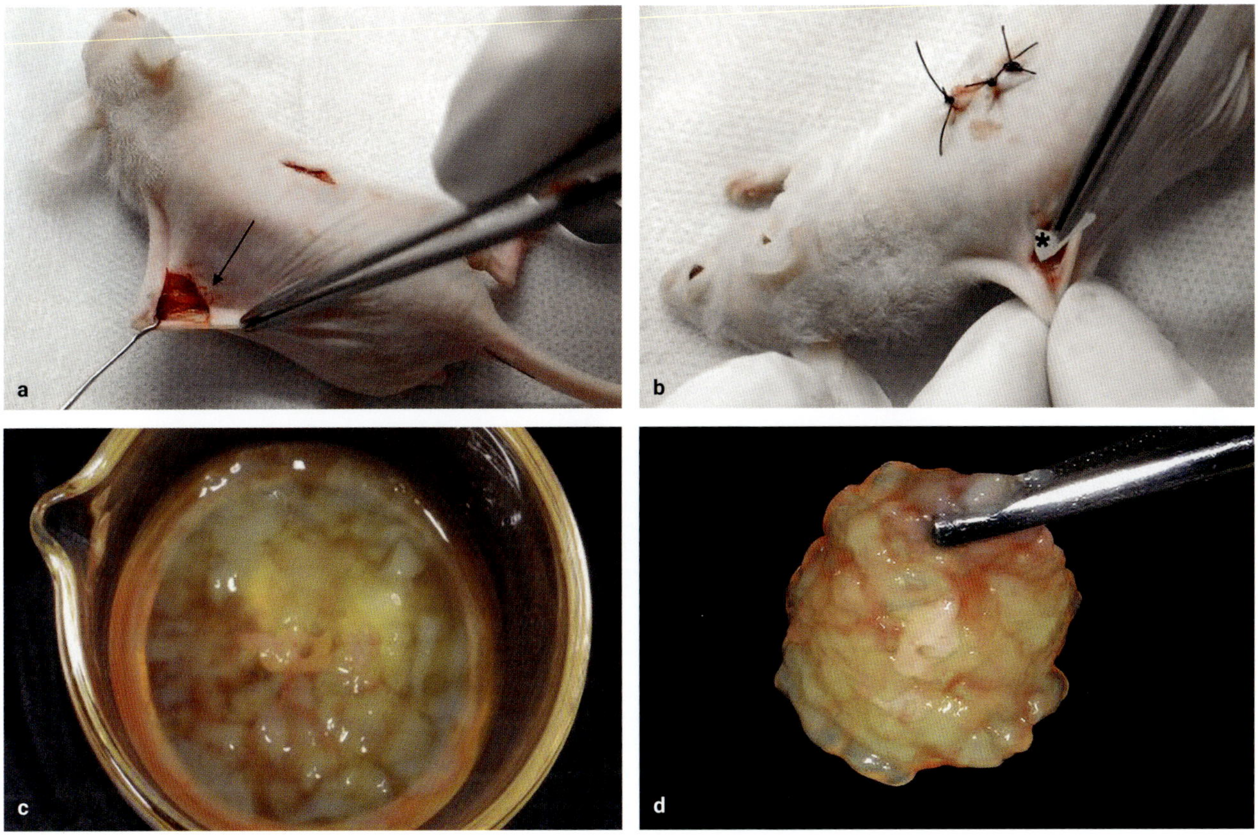

Fig 7-8 Surgical procedures for PRF implantation. *(a)* The *arrow* shows the skin incision and the subcutaneous pocket made by blunt dissection. *(b)* The *asterisk* (*) shows the fragment of the membrane being placed in the pocket. *(c)* Denatured albumin membrane combined with liquid-PRF in a borosilicate bowl. *(d)* Denatured albumin PRF combined with liquid-PRF following polymerization in solid e-PRF form. (Reprinted with permission from Gheno et al.[14])

In Vivo Evaluation of the Biocompatibility and Biodegradation of e-PRF

The inflammatory reaction, biocompatibility, and extended degradation properties of e-PRF were evaluated when compared to L-PRF and H-PRF at 7, 14, and 21 days in a nude mouse subcutaneous implantation model according to ISO 10993-6:2016.[14]

Figure 7-8 demonstrates an example surgical animal that was implanted with both L-PRF and e-PRF subcutaneously. Over time, the resorption properties of the implanted membranes were quantified and evaluated for histologic assessment. After 21 days, it was observed macroscopically that the majority of samples implanted with L-PRF or H-PRF demonstrated significant or complete resorption, whereas the e-PRF group demonstrated only a slight change in volume dimension (Fig 7-9). Of significance, Fig 7-9a demonstrates an animal that was subcutaneously implanted on either side of the animal midline with L-PRF and e-PRF. Note the complete resorption of the L-PRF side, whereas a noted bolus remained in the opposite e-PRF side. Figures 7-9b to 7-9d further demonstrated that the L-PRF and H-PRF groups demonstrated complete resorption at day 21. The e-PRF group remained with the presence of vascularization around the implanted biomaterial with noted volume stability over the entire study duration (see Fig 7-9d).

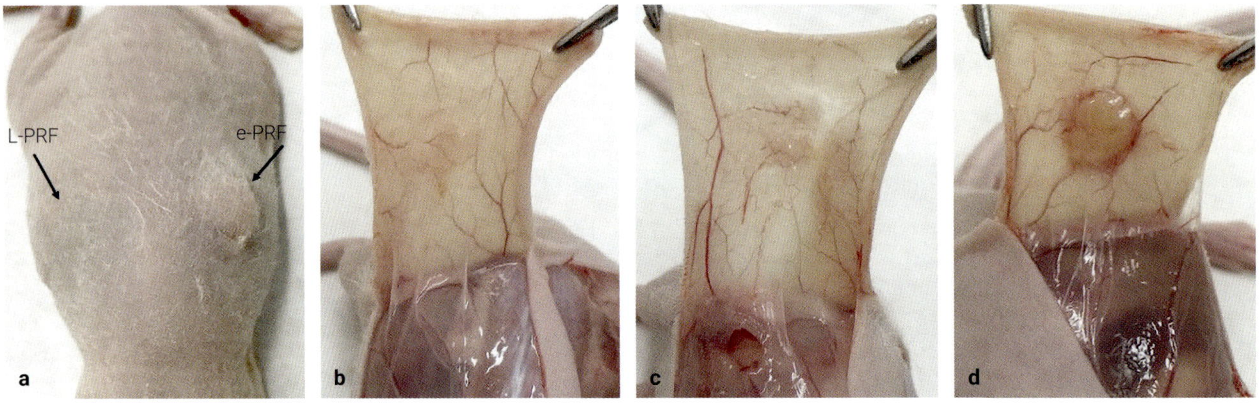

Fig 7-9 After 21 days of implantation, the e-PRF volume was observed macroscopically on the animal's back; it remained in place during all the experimental periods. *(a)* Animal's back with e-PRF volume *(arrow)*. *(b)* L-PRF group. *(c)* H-PRF. *(d)* e-PRF. (Reprinted with permission from Gheno et al.[14])

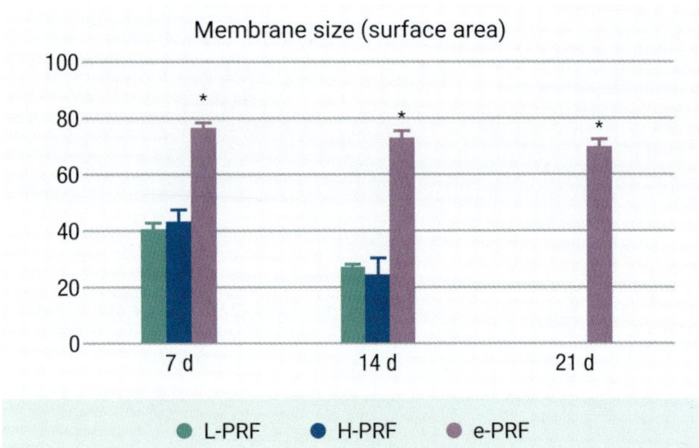

Fig 7-10 Quantification of the membrane surface area. Both the L-PRF and H-PRF groups noted complete resorption by day 21. e-PRF demonstrated superior volume stability over time. $P < .05$; an *asterisk* (*) denotes significantly greater surface area when compared to all other groups. (Reprinted with permission from Gheno et al.[14])

Quantification of the membrane size surface area demonstrated that both the L-PRF and H-PRF groups lost roughly 50% volume when compared to the e-PRF group by 7 days (Fig 7-10). Roughly 25% of the volume was lost by day 14, and complete resorption was noted by day 21. In contrast, from day 7 to day 21 the e-PRF group lost only 10% of its original volume, demonstrating superior volume stability over time (see Fig 7-10). Because of the autologous nature of both of these implanted biomaterials, all demonstrated excellent immune cell reactions at all time points (Figs 7-11 and 7-12).

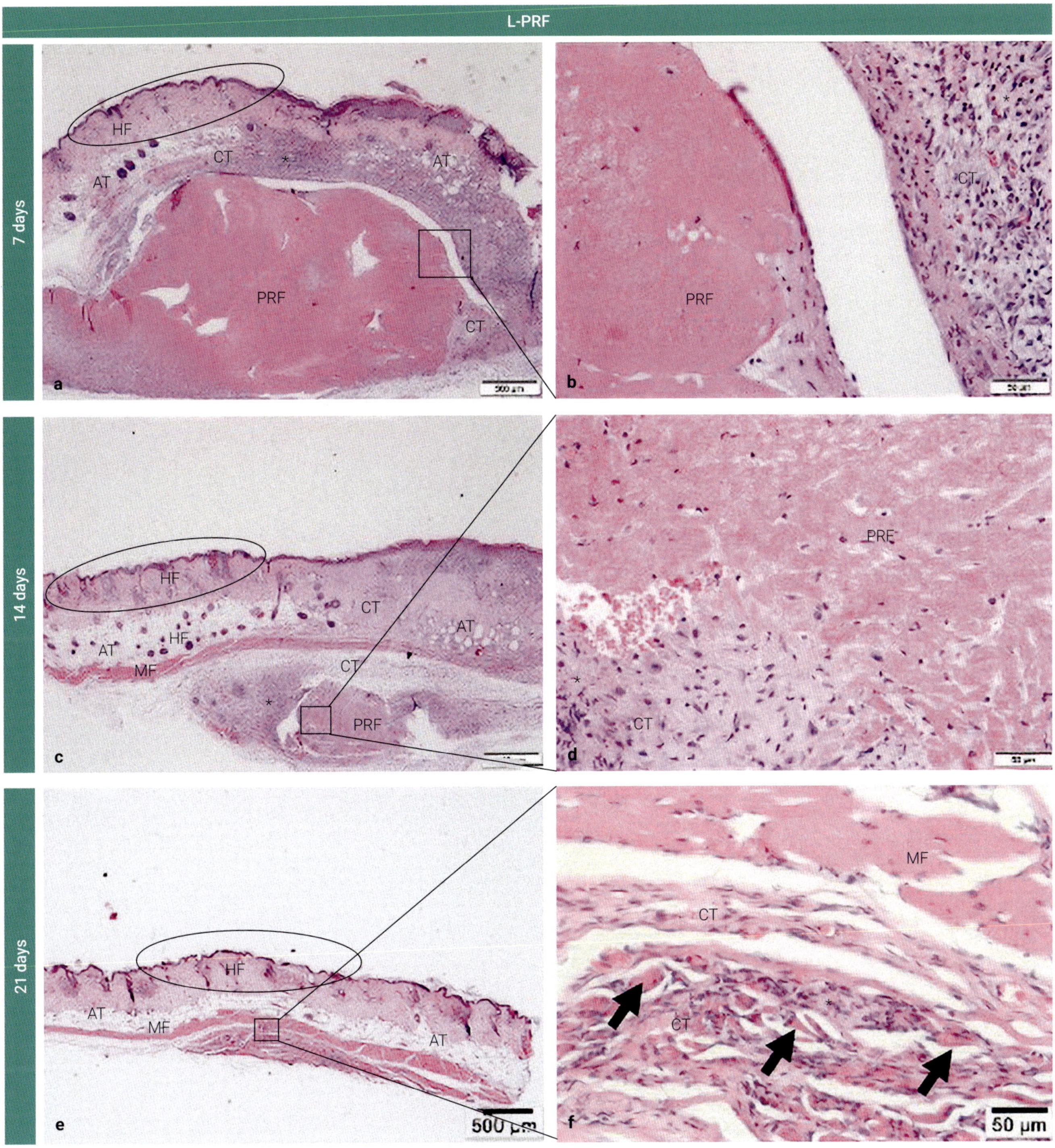

Fig 7-11 Photomicrographs of the L-PRF group at 7 days *(a and b)*, 14 days *(c and d)*, and 21 days *(f)*. *(a and b)* The epidermis and papillary dermis with hair follicle (HF), recovering connective tissue (CT) with moderate and diffuse inflammatory cells (*) surrounding the PRF, and adipocyte tissue (AT) are noted above the epithelium. *(c and d)* The epidermis and papillary dermis with HF and recovering CT with moderate and diffuse inflammatory cells (*) surrounding the PRF are noted. *(e and f)* The epidermis and papillary dermis with HF, recovering CT, rare inflammatory cells among muscle fibers (MF), and AT are noted. PRF was not observed in this period. (H&E stain; magnification ×4, scale bar = 500 µm for *a*, *c*, and *e*; ×40, scale bar = 50 µm for *b*, *d*, and *f*. (Reprinted with permission from Gheno et al.[14])

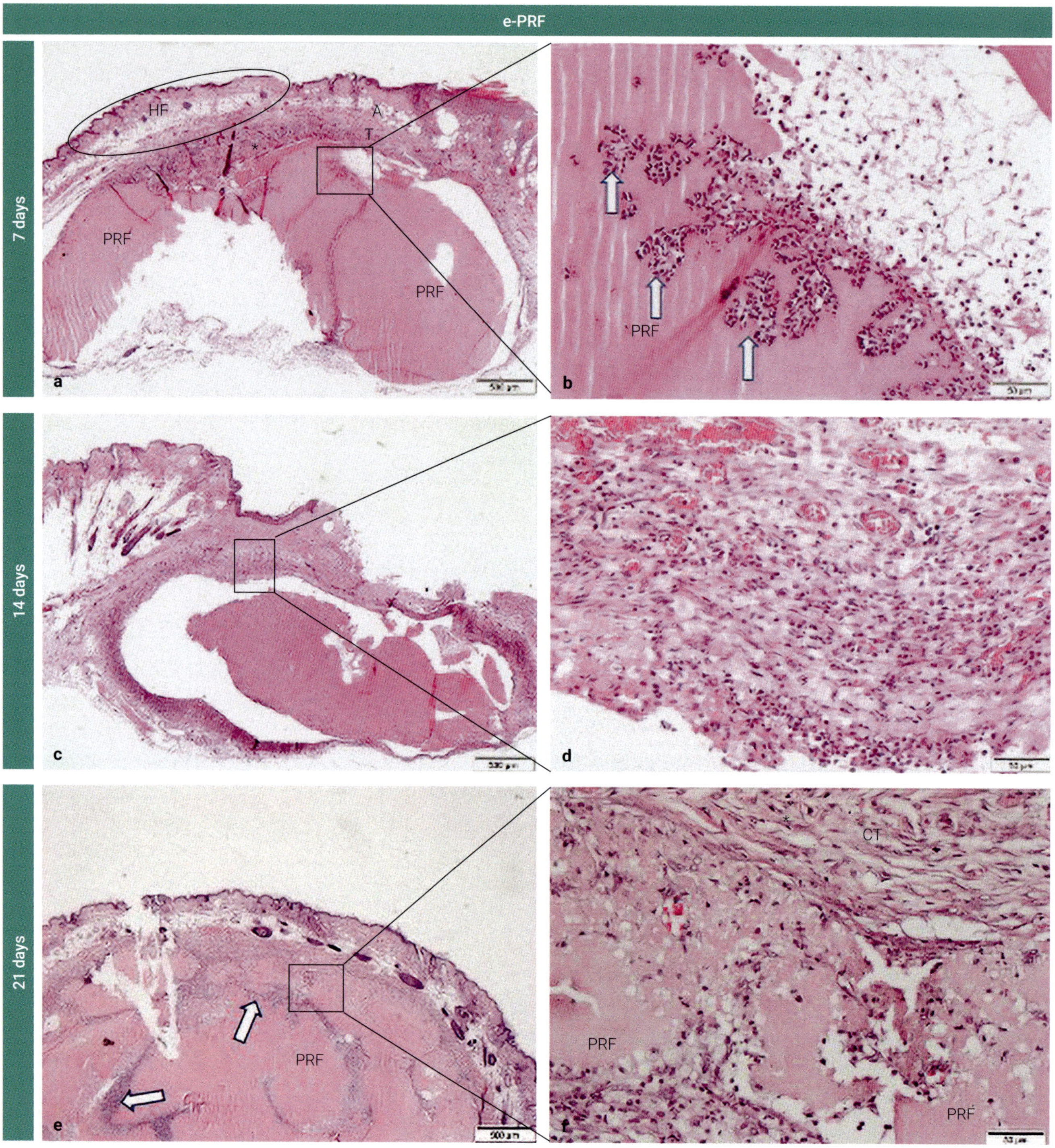

Fig 7-12 Photomicrographs of the e-PRF group at 7 days *(a and b)*, 14 days *(c and d)*, and 21 days *(f)*. *(a and b)* The epidermis and papillary dermis with hair follicle (HF) and recovering connective tissue (CT) with moderate and focal inflammatory cells (*) surrounding the PRF are noted, highlighting the presence of leukocyte groups inside the membrane *(white arrows)*. *(c and d)* The epidermis and papillary dermis with HF and recovering CT with moderate inflammatory cells (*) surrounding the PRF are noted. *(e and f)* The epidermis and papillary dermis with HF and recovering CT with disperse inflammatory cells (*) surrounding the PRF are noted, as well as the presence of leukocyte groups inside the membrane *(white arrows)*. (Reprinted with permission from Gheno et al.[14])

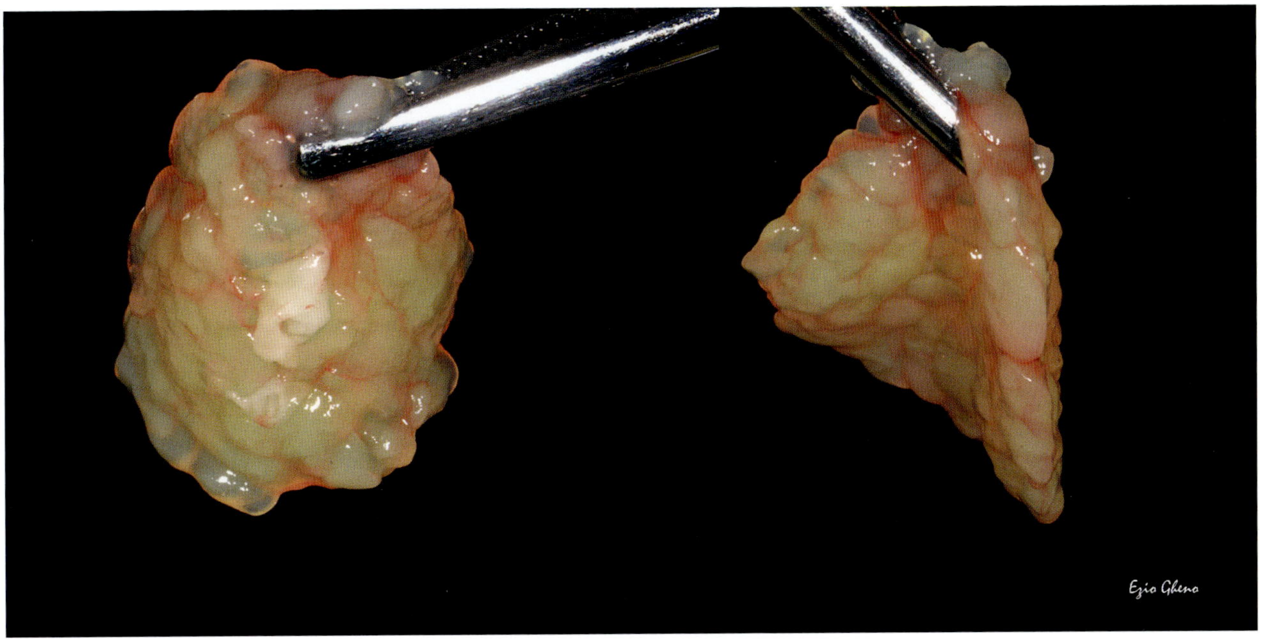

Fig 7-13 Clinical photograph of an e-PRF membrane utilized for recession coverage.

Discussion

More than 20 years have passed since it was shown that by heating and denaturing albumin, a modification in the secondary structure transforms the matter into a tridimensional structure whereby new hydrogen and disulfide ligations in the enzymes are created.[15] This favors a larger tridimensional structure with drastic changes in its resorption properties and improved stability over time.[15]

In 2015, Kawase et al investigated the degradation properties of PRF. By applying heat-treatment to standard PRF membranes through membrane compression, the resorption properties of PRF membranes were dramatically extended, opening several possibilities.[1] This chapter showed that e-PRF retains all the necessary regenerative cells and ability to release growth factors over time following reintroduction of C-PRF layer back into the albumin gel following heating/cooling and that e-PRF carries a dramatic improvement in membrane stability (less than 10% loss in the animal model). The data correlates well with human data because it is actually expected that more volume loss should occur in animals due to the faster metabolic rates of rodents.

While this change in protocol adds approximately 10 minutes to standard PRF membranes, it presents clear clinical benefits. It therefore becomes possible to create a true "barrier" or "filler" biomaterial derived from 100% autologous whole blood with drastically extended resorption properties. This has been utilized for various clinical indications, such as recession coverage cases as a membrane (Fig 7-13). It has also been injected for black triangles to improve papilla volume, with 5-year data now showing stable results (Figs 7-14 and 7-15 and Video 7-3). Furthermore, it has been utilized as an injectable biologic filler material in facial esthetics (Fig 7-16 and Videos 7-4 and 7-5). These clinical indications are highlighted in the later clinical chapters.

In 2019, a Bio-Heat tray was developed with the aim of specifically customizing the shapes of e-PRF to create specific geometries for extraction sockets (smaller, circular), membranes (square, rectangular), or diabetic foot ulcers (larger, circular). This allows for the easy fabrication of a customized kit that is autoclavable (Fig 7-17). Future research in this field is continuously ongoing with much future potential regarding its clinical benefits (see Videos 7-3 to 7-5).

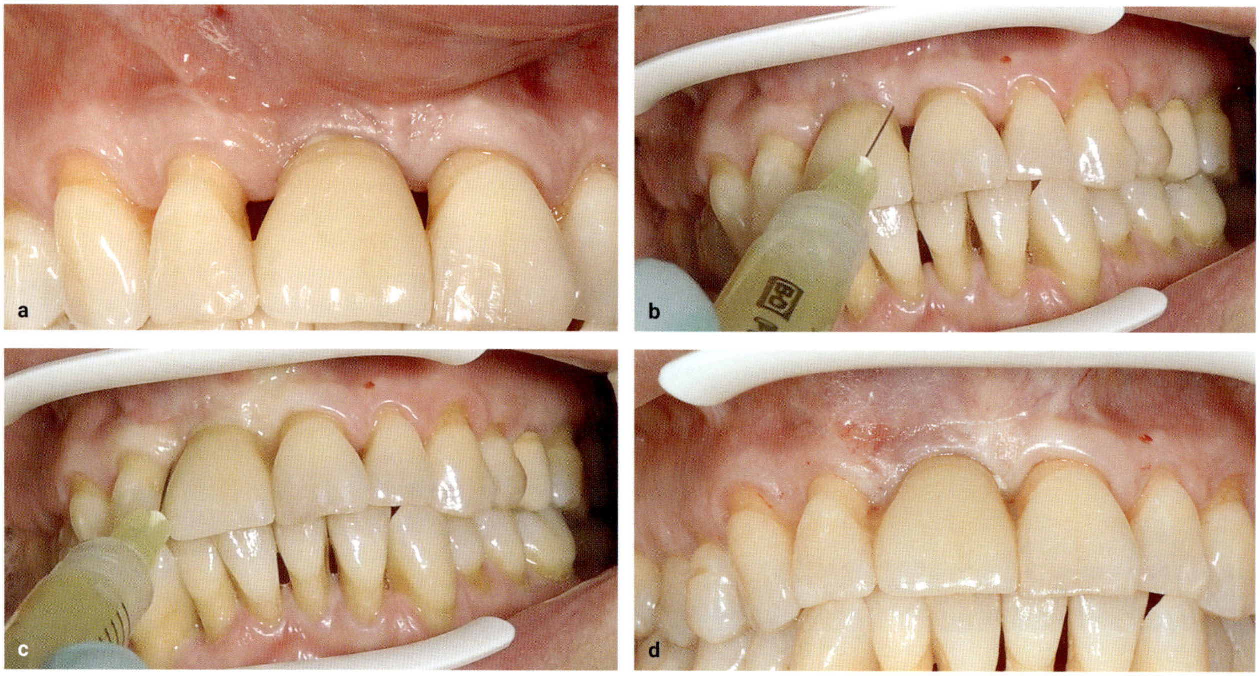

Fig 7-14 *(a)* Clinical photograph demonstrating maxillary anterior teeth with multiple black triangles. *(b and c)* Injections with e-PRF. *(d)* Final clinical photograph after injection of e-PRF (see Video 7-3). (Case performed by Dr Ezio Gheno.)

Video 7-3

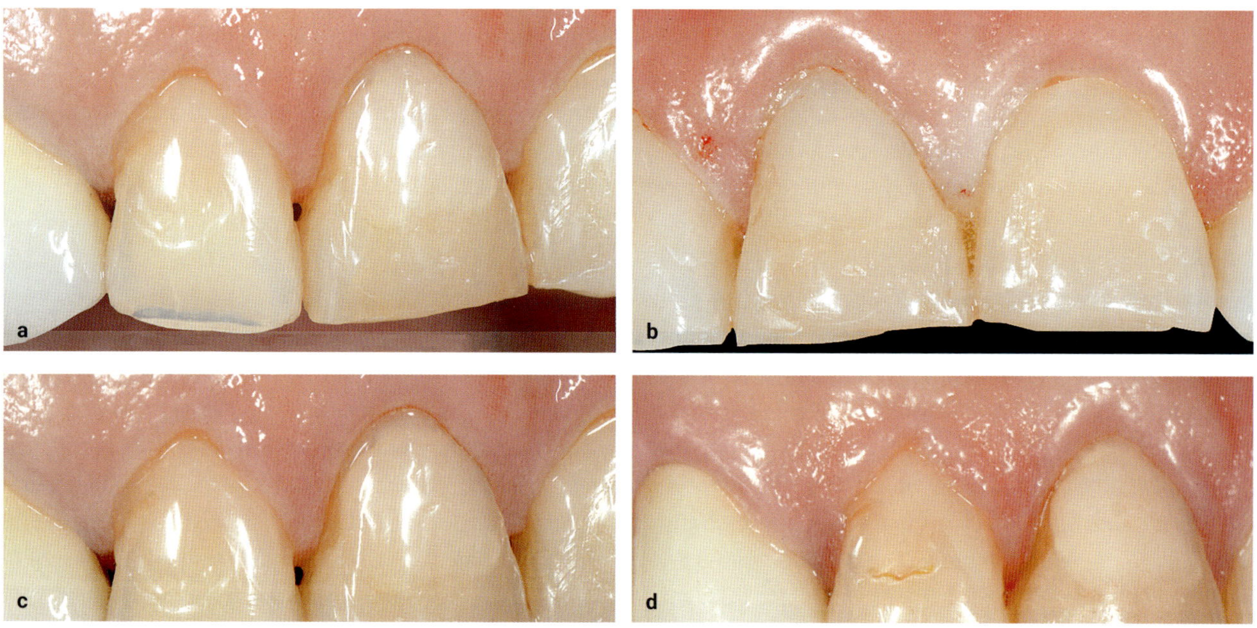

Fig 7-15 Cases demonstrating the use of e-PRF to augment lost papilla volume. *(a and b)* Papilla before and after injection with e-PRF, respectively. Note the excellent clinical outcome. *(c and d)* Another case with excellent volume stability 5 years postoperative. (Cases performed by Dr Ezio Gheno.)

Fig 7-16 *(a)* Clinical photograph of a woman in her mid 40s with pronounced marionette lines, deep nasolabial folds, and an overall aged facial appearance. *(b)* Final outcome following 3 treatments with 100% natural approaches including laser therapy (SmoothLase, Fotona), microneedling with PRF, and e-PRF injections. (Case performed by Dr Scott Delboccio; reprinted with permission from Davies and Miron.[13])

Video 7-4 Video 7-5

Fig 7-17 Custom Bio-Heat tray available to fabricate custom squares, rectangles, or circles of various sizes depending on the clinical application.

Conclusion

This chapter revealed the properties and protocol of e-PRF and introduced it as an autologous platelet scaffold with extended resorption properties. e-PRF may be used as an injectable biomaterial following mixing with a female-female luer-lock connector, or as a scaffold/membrane by preforming it in custom-sized trays to create specific and desirable shapes. Furthermore, e-PRF remains moldable upon either injection or during its shaping with a long-lasting growth factor release curve capable of stimulating tissue regeneration over extended periods of time. It possesses excellent biocompatibility and is sufficiently volume stable in preclinical animal models. It therefore offers much future potential use as a biologic barrier membrane for GBR procedures and root coverage procedures and as a biologic filler material in various esthetic medical applications. Future clinical research is ongoing and discussed throughout the remaining chapters of this textbook.

References

1. Kawase T, Kamiya M, Kobayashi M, et al. The heat-compression technique for the conversion of platelet-rich fibrin preparation to a barrier membrane with a reduced rate of biodegradation. J Biomed Mater Res B Appl Biomater 2015;103:825–831.
2. Jung SY, Kim HY, Oh HJ, Choi E, Cho MS, Kim HS. Feasibility of autologous plasma gel for tonsil-derived stem cell therapeutics in hypoparathyroidism. Sci Rep 2018;8:11896.
3. Doghaim NN, El-Tatawy RA, Neinaa YME. Assessment of the efficacy and safety of platelet poor plasma gel as autologous dermal filler for facial rejuvenation. J Cosmet Dermatol 2019;18:1271–1279.
4. Mourão CFAB, Gheno E, Lourenço ES, et al. Characterization of a new membrane from concentrated growth factors associated with denaturized Albumin (Alb-CGF) for clinical applications: A preliminary study. Int J Growth Factors Stem Cells Dent 2018;1:64.
5. Garg A. Autologous Blood Concentrates. Chicago: Quintessence, 2018:188.
6. Fujioka-Kobayashi M, Schaller B, Mourão CFAB, Zhang Y, Sculean A, Miron RJ. Biological characterization of an injectable platelet-rich fibrin mixture consisting of autologous albumin gel and liquid platelet-rich fibrin (E-PRF). Platelets 2020:1–8.
7. Miron RJ, Zucchelli G, Pikos MA, et al. Use of platelet-rich fibrin in regenerative dentistry: A systematic review. Clin Oral Investig 2017;21:1913–1927.
8. Miron RJ, Fujioka-Kobayashi M, Hernandez M, et al. Injectable platelet rich fibrin (i-PRF): Opportunities in regenerative dentistry? Clin Oral Investig 2017;21:2619–2627.
9. Wang X, Wang Y, Bosshardt DD, Miron RJ, Zhang Y. The role of macrophage polarization on fibroblast behavior—An in vitro investigation on titanium surfaces. Clin Oral Investig 2018;22:847–857.
10. Miron RJ, Dham A, Dham U, Zhang Y, Pikos MA, Sculean A. The effect of age, gender, and time between blood draw and start of centrifugation on the size outcomes of platelet-rich fibrin (PRF) membranes. Clin Oral Investig 2019;23:2179–2185.
11. Dohan DM, Choukroun J. PRP, cPRP, PRF, PRG, PRGF, FC... How to find your way in the jungle of platelet concentrates? Oral Surg Oral Med Oral Pathol Oral Radiol Endod 2007;103:305–306.
12. Foster TE, Puskas BL, Mandelbaum BR, Gerhardt MB, Rodeo SA. Platelet-rich plasma: From basic science to clinical applications. Am J Sports Med 2009;37:2259–2272.
13. Davies C, Miron RJ. PRF in Facial Esthetics. Chicago: Quintessence, 2020.
14. Gheno E, Mourão C, Mello-Machado RC, et al. In vivo evaluation of the biodegradation of a new denatured plasma membrane combined with liquid PRF (E-PRF). Platelets 2020:1–13.
15. Giancola C, De Sena C, Fessas D, Graziano G, Barone G. DSC studies on bovine serum albumin denaturation. Effects of ionic strength and SDS concentration. Int J Biol Macromol 1997;20:193–204.

8

Armamentarium in a PRF Kit

This chapter provides an overview of the armamentarium required to adequately introduce PRF into clinical practice. A variety of specialized instruments and equipment is needed in order to perform phlebotomy and be able to process PRF after centrifugation. Instruments required include a PRF centrifugation system, a PRF box, a variety of hand instruments, a PRF tube holder, a PRF bowl/tray, a tourniquet, butterfly needles, as well as both red solid-PRF and white/blue liquid-PRF tubes. This chapter also provides a video to learn the steps required into programming a PRF system.

Contributors

Richard J. Miron

Chapter Highlights

- *Armamentarium for PRF*
- *Tools required to successfully introduce PRF into your practice*
- *Programming a PRF centrifugation system*

Video 8-1

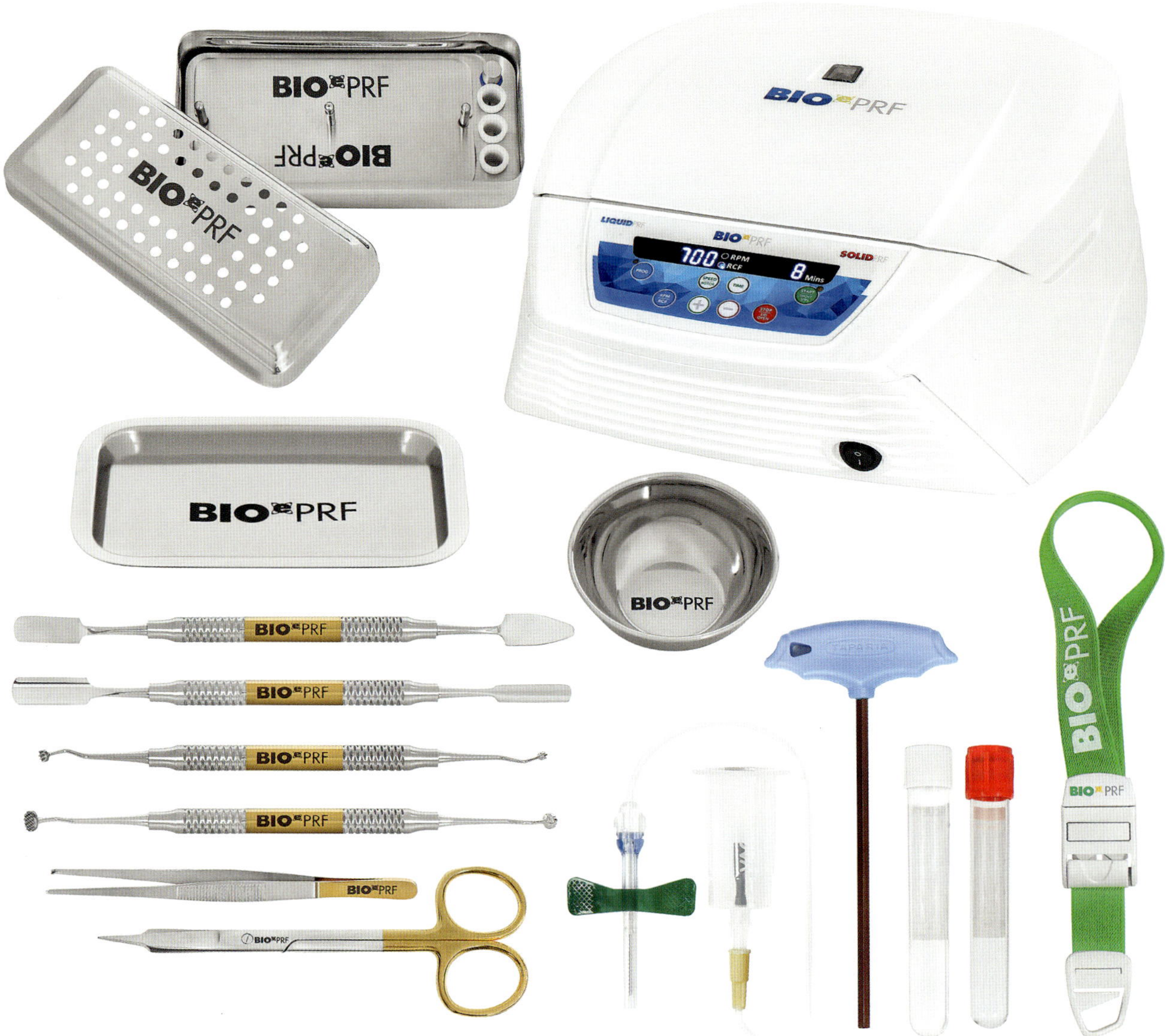

Fig 8-1 Example of a full PRF centrifugation kit.

This chapter provides the steps required to set up and program a new PRF system. A variety of centrifugation systems currently exist on the market, but they all contain very similar "starter packages" when it comes to their hardware (Fig 8-1). These typically include the following:

- **Bio-PRF centrifuge:** Horizontal centrifuge for the production of PRF up to 2220g (4000 rpm)
- **Bio-PRF box:** Utilized to compress membranes
- **Bio-PRF tube holder:** Utilized to hold PRF tubes following centrifugation
- **Bio-PRF tray:** Used to prepare membranes and separate the red blood cells from the PRF clot
- **Bio-PRF bowl:** Used to cut a Bio-PRF membrane to mix with particulate bone grafts
- **Bio-PRF scissors:** Used to cut PRF membranes and separate the red blood cell layer from the PRF clot
- **Bio-PRF forceps:** Used to remove PRF clots from the centrifugation tubes

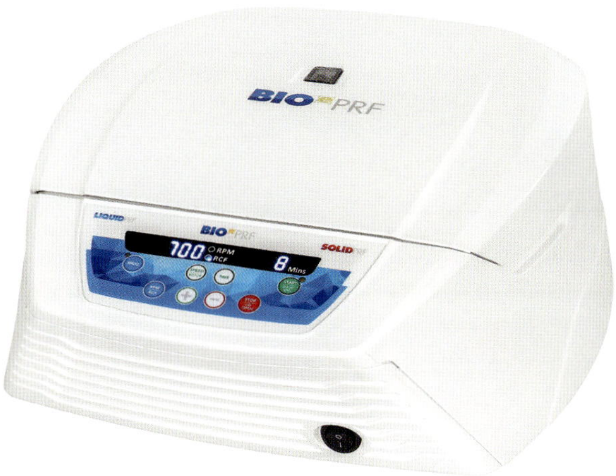

a

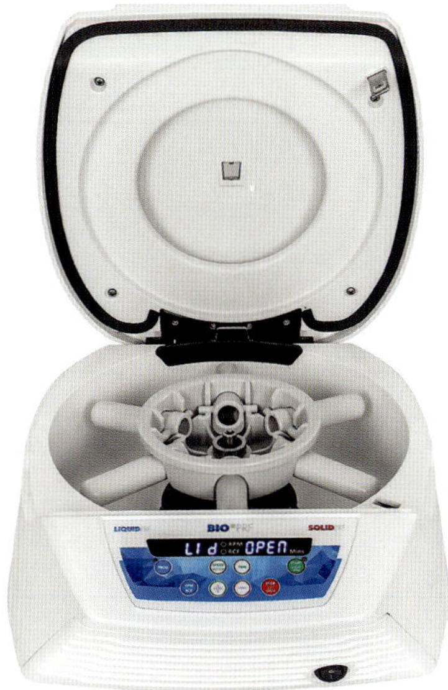

b

Fig 8-2 *(a)* Clinical illustration of a Bio-PRF centrifugation system. *(b)* Photograph demonstrating the ability for tubes to swing horizontally during the centrifugation process.

- **Bio-PRF pad:** Used to separate PRF membranes from the red clot and carry PRF membranes to the oral cavity
- **Bio-PRF small compactor:** Small instrument used to compact the PRF–bone graft complex into bone defects
- **Bio-PRF large compactor:** Large instrument used to compact the PRF–bone graft complex into bone defects
- **Bio-PRF double spoon:** Allows the carrying of bone graft material
- **Bio-PRF tourniquet:** Used to facilitate blood draws
- **Blood collection butterfly needles:** Used to collect blood via veins (box of 24)
- **Red cap tubes:** Used to make solid-PRF (box of 100)
- **White/blue cap tubes:** Used to make liquid-PRF (box of 100)

Bio-PRF Centrifuge

As reviewed in chapters 2 and 3, when selecting a new centrifugation system, preference is given to a horizontal centrifugation system when compared to fixed-angle due to its better ability to separate blood layers. The Bio-PRF system was the first commercially available horizontal centrifuge specifically designed for the production of PRF (Fig 8-2). All protocols by our laboratory have been optimized utilizing this system. This medical device is ideal and differs significantly from other branded centrifuges in that the device can be easily programmed.

When selecting a centrifugation system, it is important that an open platform be chosen and that the clinician is able to manually enter desired protocols. Otherwise, as research evolves, new centrifugation protocols necessitate the purchase of new centrifugation devices. It is much more cost-effective long-term to simply purchase a centrifugation system that may be easily programmable (Video 8-1).

When selecting a centrifugation system, it is best to select one with an open platform where protocols can be manually entered.

Video 8-1

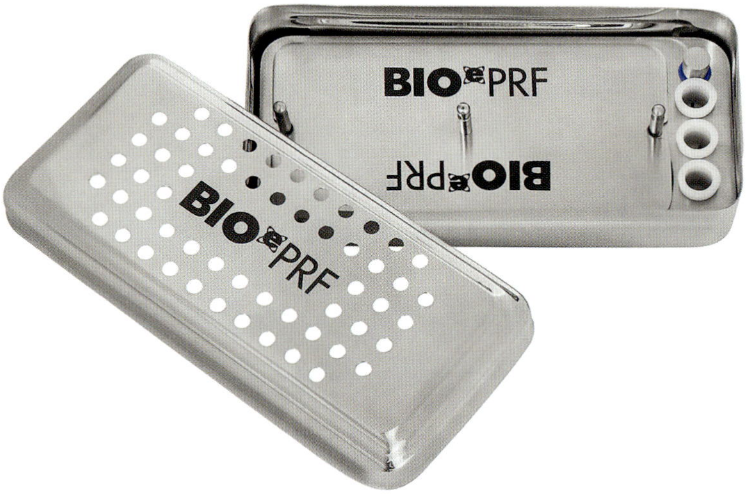

Fig 8-3 Bio-PRF box. The Bio-PRF box is utilized to fabricate PRF membranes and plugs. Following centrifugation, the PRF clots are removed and placed into this Bio-PRF box, where the clots are compressed into 1- to 2-mm-thick membranes.

Fig 8-4 *(a)* Bio-PRF tube holder. *(b)* Bio-PRF tray. *(c)* Bio-PRF bowl.

Bio-PRF Box

The Bio-PRF box allows for the compression of PRF clots into even-thickness membranes that can then be utilized as barrier membranes during GBR procedures (Fig 8-3). These PRF membranes contain a host a platelets and leukocytes that are capable of stimulating soft tissue wound healing and improving patient discomfort and pain. The box also contains slot cylinders to produce Bio-PRF plugs that can be utilized in extraction sockets. These protocols are reviewed in chapter 10.

Bio-PRF Tube Holder, Tray, and Bowl

Figure 8-4 illustrates the Bio-PRF tube holder, tray, and bowl. These are necessary equipment included in every system. The tube holders hold the tubes upright following centrifugation. This is particularly important following liquid-PRF protocols where the tubes remain liquid. Unless kept upright, the actual layers will mix and the liquid-PRF will be lost.

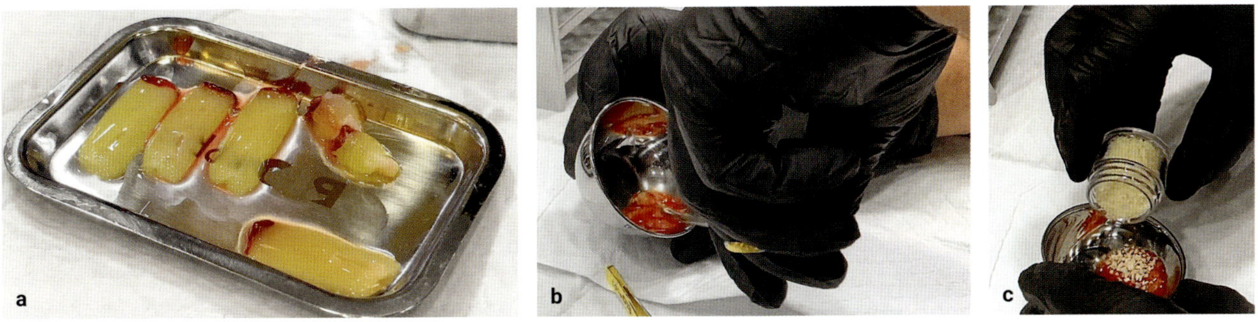

Fig 8-5 *(a)* Bio-PRF tray with several PRF membranes and PRF plugs demonstrated. *(b)* Bio-PRF bowl utilized for the cutting of PRF membranes into PRF fragments. *(c)* Introduction of allograft bone particles for the production of sticky bone.

Fig 8-6 Hand instruments included with standard PRF kits.

The Bio-PRF tray is utilized to gather PRF membranes and plugs on a plateau for clinical use (Fig 8-5a). The Bio-PRF bowl is utilized to cut PRF membranes into fragments for future mixing with bone graft particles to create sticky bone (Figs 8-5b and 8-5c), as later highlighted in chapter 10.

Bio-PRF Instrument Set

A variety of handling instruments are needed to remove the upper PRF clot from the tube following centrifugation as well as to handle PRF and utilize it for clinical applications. Typical sets include six instruments (Fig 8-6). The scissors are naturally utilized to cut PRF membranes and create PRF fragments, the forceps are utilized to remove the clots from tubes, and the Bio-PRF pad is used to separate the PRF membranes from the red clot layer and/or to carry the final PRF membranes to the oral cavity. Two compactors (small and large) are included in most kits and utilized in extraction sockets to pack the PRF/sticky bone into sockets. The double spoon is utilized for carrying bone graft particles and/or sticky bone toward the oral cavity.

Bio-PRF Tourniquet and Butterfly Needles

Naturally, all PRF kits should include a tourniquet (Fig 8-7) as well as butterfly needles (Fig 8-8). A tourniquet should be about 1 inch thick and is typically utilized with 21G needles. For children (eg, in endodontic vitality treatments) or for individuals with small veins, a smaller-gauge needle may be utilized (23G–25G). Note that as the needle gauge is reduced, blood flow is also reduced. As reviewed in chapter 2, PRF must be drawn relatively quickly (within 60–90 seconds), and therefore larger-gauge needles are generally preferred when possible. Most kits include 24 butterfly needles (21G).

Fig 8-7 Standard reusable tourniquet included in PRF kits.

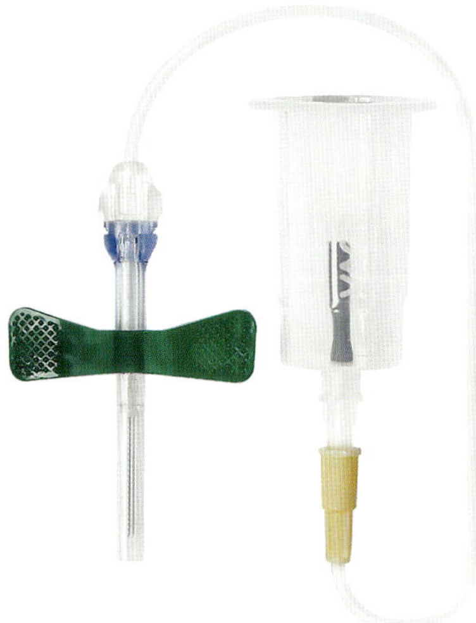

Fig 8-8 Standard butterfly needles utilized for the collection of peripheral blood. Most kits include 24 21G needles.

Solid-PRF and Liquid-PRF Tubes

Both hydrophilic solid-PRF tubes and hydrophobic liquid-PRF tubes are needed in standard PRF kits (Fig 8-9). Typically 100 tubes of each are provided in each kit. Because the red hydrophilic tubes are typically made of glass, they generally weigh more when compared to hydrophilic blue/white tubes made of PET. For this reason, red and white/blue tubes should never be balanced in a centrifuge. Instead, red tubes should balance with red tubes, and white tubes should balance with white tubes.

Video 8-2

Fig 8-9 *(a)* Box of 100 red hydrophilic solid-PRF tubes. *(b)* Box of 100 white/blue hydrophobic liquid-PRF tubes.

> ***Tip***
>
> It is wise to always keep one red tube filled with 10 mL of water as well as 1 white/blue tube filled with 10 mL of water within an office. If an odd number of tubes is drawn (for example five), a spare tube filled with water to the equivalent weight is rapidly available to balance the centrifugation system.

Conclusion

This chapter briefly highlighted the necessary armamentarium needed for the production of standard PRF. Most importantly, a horizontal centrifugation system should be selected with the ability to program centrifugation parameters within the system. Each kit will come with an array of hand instruments, a PRF box, tube holder, tray, bowl, and tourniquet. Furthermore, kits should include a minimum of 24 standard butterfly needles (21G), 100 solid-PRF tubes (red cap), and 100 liquid-PRF tubes (white/blue cap) to facilitate the incorporation of PRF into clinical practice.

9

Phlebotomy

Due to the need to rapidly begin centrifugation following blood draw, phlebotomy becomes an essential component for the production of PRF. Because no anticoagulants are utilized, centrifugation typically must be carried out within 60 to 90 seconds following blood draw, as reviewed in chapter 2. Therefore, office staff must be well equipped and prepared to perform the act of phlebotomy. This chapter highlights the essential components of phlebotomy and further provides standard tools and tips to collect blood samples. A 10-step protocol is provided with images and a video outlining the procedure.

Contributors

Richard J. Miron

Chapter Highlights

- *History of blood drawing*
- *Anatomy of the antecubital fossa area*
- *Equipment needed*
- *Step-by-step protocol*

Video 9-1

Since the Middle Ages, people have been fascinated by blood and its connection with well-being. In the Stone Age, it was actually common practice to take crude instruments, puncture blood vessels, and allow excess blood to drain out of the body to rid it of evil spirits and cleanse the body of impurities. Hippocrates, a Greek physician considered one of the "Fathers of Medicine," developed a theory that the body required a balance between the following four "humors": blood, phlegm, black bile, and yellow bile. If an excess of any of these was noted, it was believed that disease progression may result and that it was essential to remove that excess to restore health. Removing blood was called *bloodletting*. Techniques were fairly straightforward—just cut a vein with a sharp instrument and let the blood run.

It wasn't until the 17th and 18th centuries that phlebotomy was considered a major therapeutic option, where a lancet tool was used to withdraw approximately 10 mL of blood for diagnostic purposes. Naturally, phlebotomy has evolved considerably in modern medicine, where methods have dramatically improved and screening and diagnostic tests are utilized routinely for monitoring health.

Today, a variety of organizations and certification agencies exist to certify individuals in the act of phlebotomy. In America, these include the American Medical Technologists, the American Certification Agency, the American Society for Clinical Pathology, and the National Center for Competency Testing. In most dental offices, it is most common that an assistant become a certified phlebotomist and perform the act of phlebotomy.

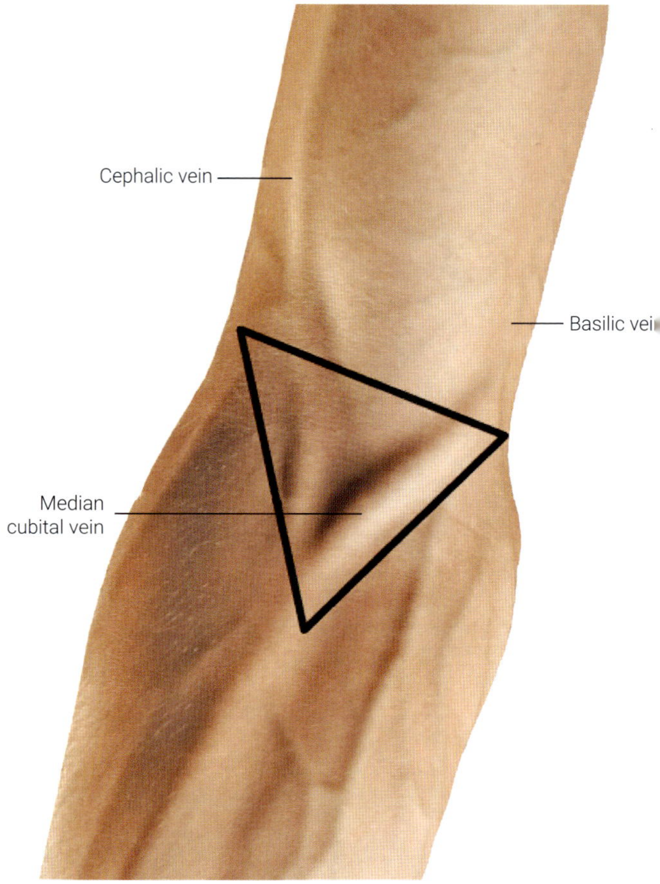

Fig 9-1 The antecubital fossa area of the arm. Note that the brachial artery and several major arm nerves pass through the antecubital fossa, so it is important to prioritize proper vein selection. (Reprinted with permission from McCall and Tankersley.[1])

Phlebotomy-Related Vascular Anatomy

Naturally, the phlebotomist must have adequate knowledge of the vascular anatomy of the arm, including its most appropriate veins for blood collection. As depicted in Fig 9-1, the antecubital fossa is the shallow triangular depression in the arm.[1] This is the first-choice location for venipuncture because of the number and size of veins located within this area superficially (ie, the veins are close to the surface). While it is important to note that anatomical differences exist from person to person, two very common distribution arrangements have been observed: the "H-shaped" (Fig 9-2) and "M-shaped" (Fig 9-3) pattern. These are typically easily identifiable, especially with a vein finder (vein light).

> ***CAUTION!***
>
> Take a close look at Figs 9-2 and 9-3 and notice the location of the brachial artery and several major nerves that pass through the antecubital fossa. While these nerves are generally found in deeper layers away from the superficial veins, ideally it is important to prioritize venipuncture away from the areas and at shallower depths to minimize the risk of accidental injury. The greatest chance of nerve injury is associated with venipuncture in the most medial and lateral portions of the antecubital fossa.

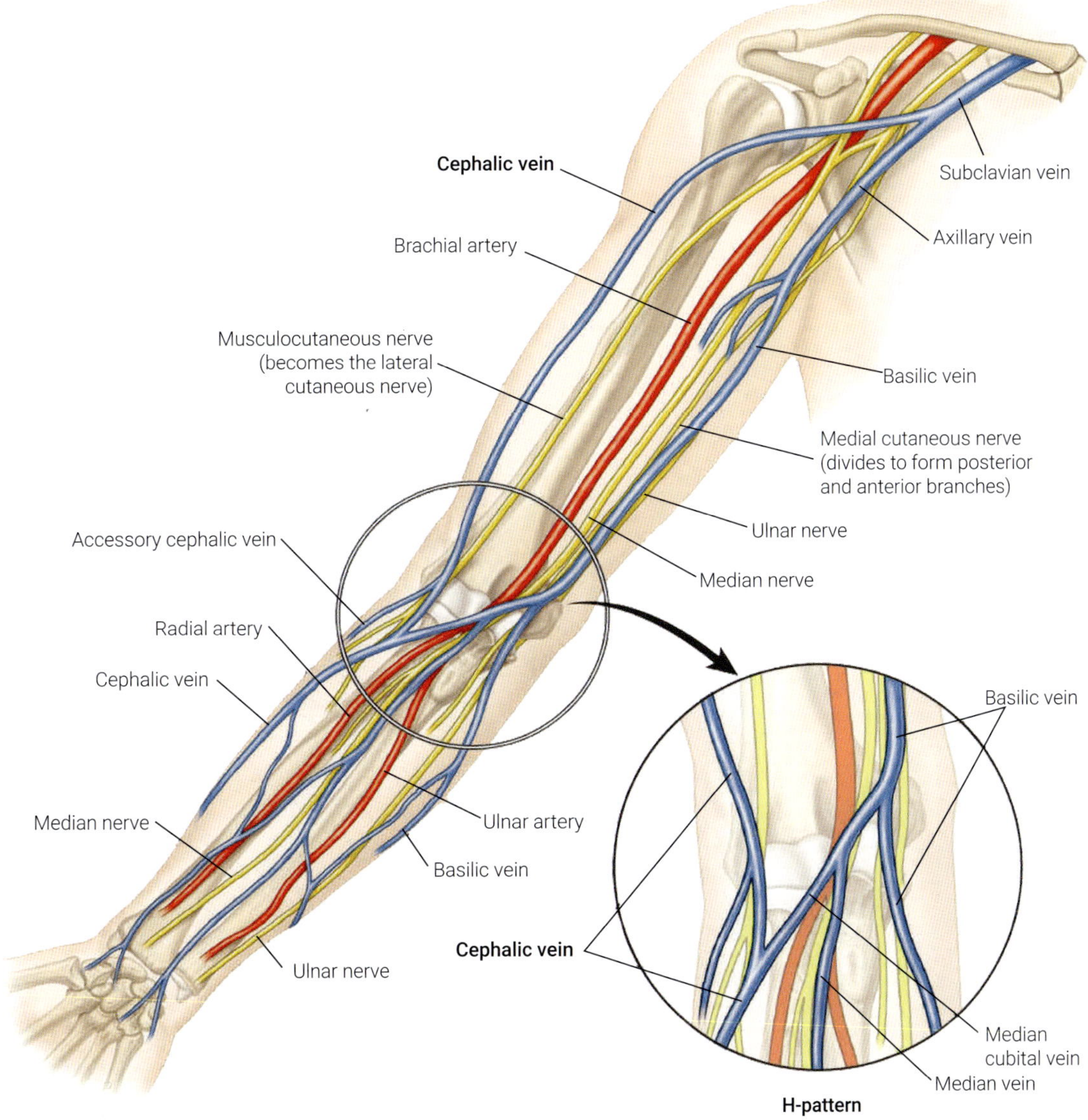

Fig 9-2 The principal veins of the right arm in anatomical position displaying the H-shaped pattern of antecubital veins. (Reprinted with permission from McCall and Tankersley.[1])

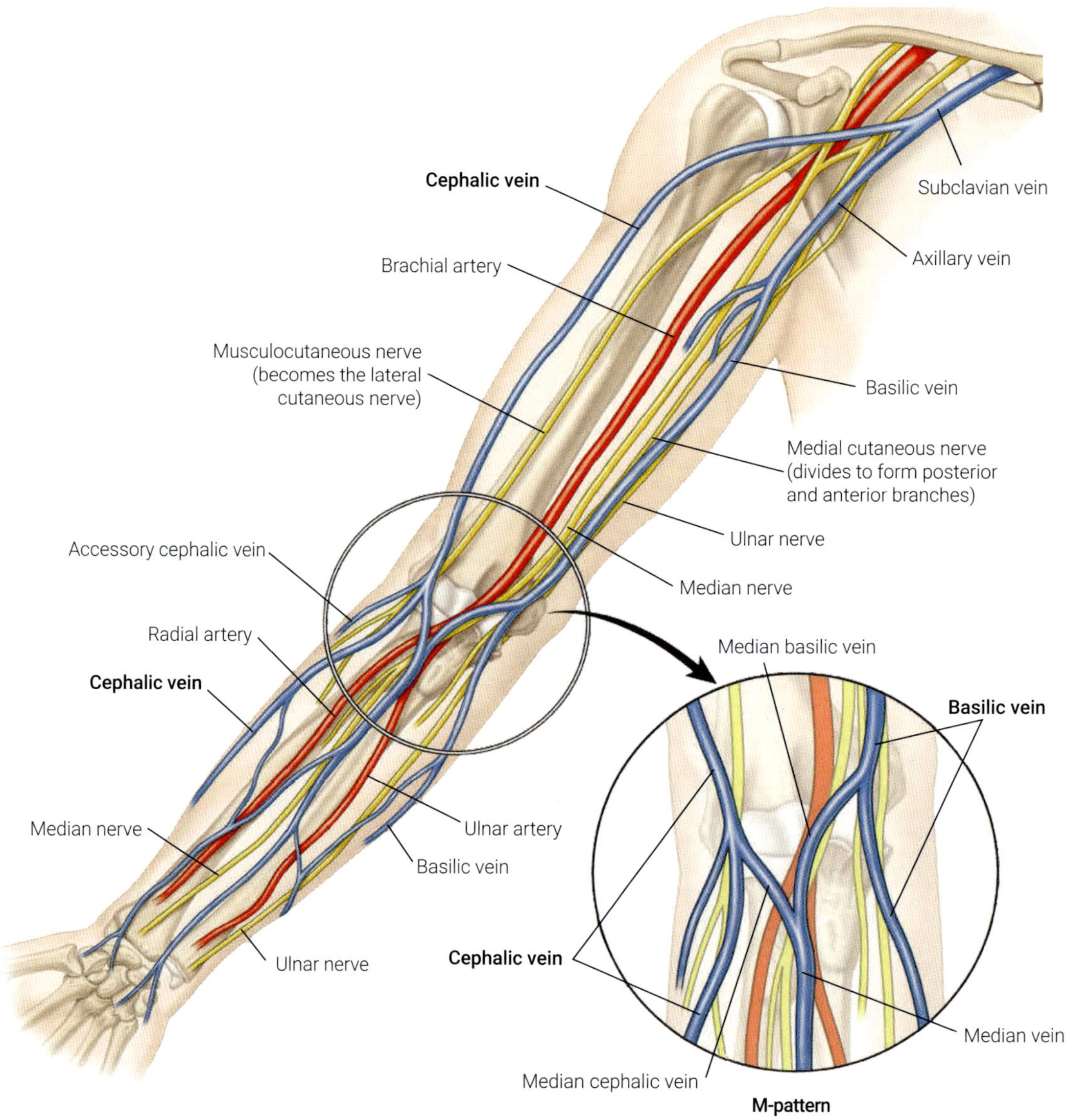

Fig 9-3 The principal veins of the right arm in anatomical position displaying the M-shaped pattern of antecubital veins. (Reprinted with permission from McCall and Tankersley.[1])

Understanding Vein Pattern Distributions

H-pattern antecubital veins

The H-pattern (see Fig 9-2) is found in the great majority of the population (approximately 70%) and includes arrangements of the median cubital vein, cephalic vein, and basilic vein as follows.

Median cubital vein

The median cubital vein is located near the center of the antecubital area and is the preferred choice for venipuncture in the H-shaped arrangement. It is typically larger, closer to the surface, more stationary, and away from other key anatomical structures (ie, arteries or nerves). It is also typically the easiest to locate, least painful to puncture, and least likely to bruise. Caution is advised for the most medial aspect (inner side) of this vein because it overlies the brachial artery and several nerves (should the phlebotomist accidentally penetrate too deeply).

Cephalic vein

The cephalic vein is located in the lateral aspect of the antecubital area and is the second-choice location for venipuncture in this pattern. Especially in obese patients, it is often easier to palpate when compared to the median cubital vein. It is preferable to avoid the most lateral portions of the cephalic vein and the accessory cephalic vein to prevent accidental injury to the lateral cutaneous nerve (carefully observe Fig 9-2).

Basilic vein

The basilic vein is the second-largest vein located on the medial aspect (inner side) of the antecubital area. It is, however, a last-option choice for venipuncture in both the H-pattern and M-pattern distributions. While it is generally easy to palpate, it is generally not as well anchored and as a result rolls more easily. A vein light is recommended to stabilize the vein. Furthermore, punctures to this vein tend to be more painful for patients and bruise more easily.

M-pattern antecubital veins

The distribution of veins is slightly different in the M-pattern (see Fig 9-3), where the cephalic vein, median vein, median cephalic vein, median basilic vein, and basilic vein are all venipuncture options. Preference is given as highlighted below.

Median vein

The median vein is located near the center of the antecubital area and is the first choice for venipuncture in the M-shaped pattern. It is furthest away from major nerves and arteries, well anchored, and tends to be less painful with reduced bruising.

Median cephalic vein

The median cephalic vein is located in the lateral aspect of the antecubital area and is the second choice for venipuncture in the M-pattern. It is also readily accessible and away from major nerves or arteries. As with the cephalic vein, avoiding the most lateral portions of the median cephalic vein is recommended to minimize accidental injury to the lateral cutaneous nerve (carefully observe Fig 9-3).

Median basilic vein

The median basilic vein is located in the medial aspect of the antecubital area and is a last option for venipuncture in the M-shaped pattern. It often appears larger and more accessible, but it is more painful to puncture and located closer to various nerves. It is utilized as a puncture site only when the median vein and median cephalic vein have been ruled out.

The difference between "medial" and "median" can be confusing, especially in the context of the location of antecubital veins. Medial means toward the midline of the body, while median means in the middle. It is important to understand these differences to better understand where to perform venipuncture.

Note

No perfect textbook drawing exists that can guarantee the exact location and pattern in every individual. Utilize your knowledge and experience as well as the available aids and tools (such as the vein light) to improve your success.

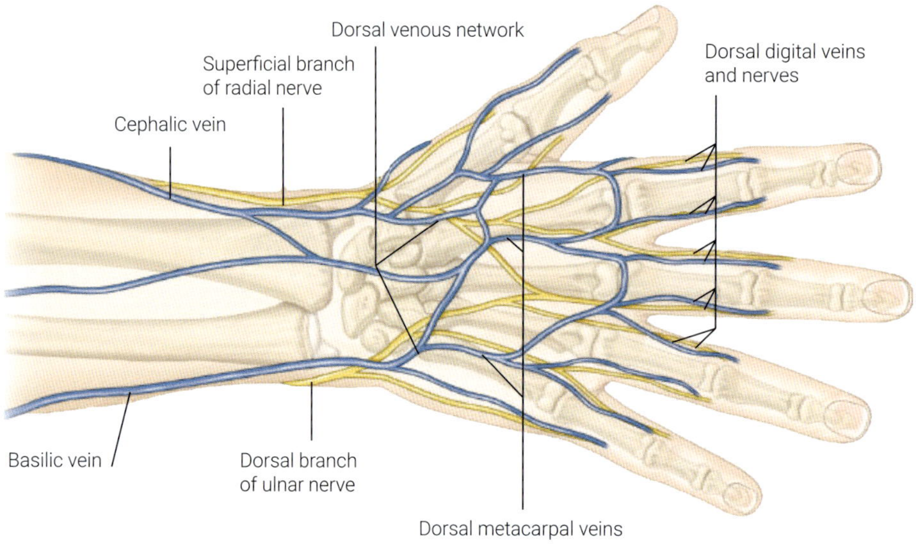

Fig 9-4 Right dorsal forearm, wrist, and hand veins. (Reprinted with permission from McCall and Tankersley.[1])

Table 9-1 Common venipuncture needle gauges with typical uses

Gauge	Needle type	Typical use
15–17	Special needle attached to collection bag	Collection of donor units, autologous blood donation, and therapeutic phlebotomy
20	Multisample hypodermic	Sometimes used when large-volume tubes are collected or large-volume syringes are used on patients with normal-size veins
21	Multisample hypodermic	Considered the standard venipuncture needle for routine venipuncture on patients with normal veins or for syringe blood culture collection
22	Multisample hypodermic	Used on older children and adult patients with small veins or for syringe draws on difficult veins
23	Blood collection	Used on the veins of infants and children and on difficult or hand veins of adults
25	Blood collection (thin-wall needle)	Used on the veins of infants and children and on difficult or hand veins of adults

Dorsal Forearm, Hand, and Wrist Veins

Larger veins are found in the antecubital fossa. However, if it is not possible to find a solid vein, or if one or two unsuccessful attempts have been made, the veins on the back of the hand and wrist are also acceptable for venipuncture (Fig 9-4). These veins are easily visualized but are typically more painful to draw from and often slower. In the case of hand venipuncture, a smaller 23G needle is recommended to accommodate the smaller veins. With these veins, it is not possible to draw six tubes within 90 seconds.

Equipment

A variety of equipment is needed to perform blood draws. While the tubes were discussed in great detail in chapter 6, this section focuses on the needles utilized for venipuncture for PRF.

Gauge needle type

The most common-sized needle utilized for collection of blood for production of PRF is a 21G butterfly needle. Table 9-1 summarizes different needle sizes and their typical uses. In children and in older females with small veins, it is recommended to use a smaller 23G needle to enhance success rates.

Needle length

Needles can come in various lengths but most commonly 1.0- and 1.5-inch needles are utilized for venipuncture. Many phlebotomists in regular laboratories prefer to use 1-inch needles because they are less intimidating to the patient. The advantage of a longer 1.5-inch needle is that it allows more accurate angulation visual.

Tube holders

Collection tubes are placed in the clear plastic, disposable cylinder. Its purpose is to facilitate tube placement and removal.

Note

Tube holders are available in several sizes, but this is not always obvious. Some manufacturers make smaller/larger-sized lids to specifically accommodate their tubes. As a result, some company needles cannot be utilized with other company tubes. Be aware of this!

Video 9-2

Vacuum

It is important to address tube vacuum because it is extremely relevant for blood draws, specifically for PRF because the draws must be completed within 90 seconds or less. Tubes fill with blood automatically because a negative pressure is added to the tube. This artificially creates a vacuum, allowing the pulling of air from the tube. We previously showed in a study that certain tubes (for example, BD) could take upward of 25 seconds to fill, whereas others would fill in 15 seconds. Because with PRF we are at a "race against time" during blood draws, naturally it is wiser to select a tube with a better vacuum.

Premature loss of vacuum can occur for any of the following reasons: improper storage, dropping the tube (small cracks can cause microleakage), advancing the tube too far onto the needle before venipuncture, needle bevel comes partially out of the skin during venipuncture.

Key point

The first tube will always underfill. This is as a result of the air in the tubing. Make sure to be aware of this as you are performing blood draws. Because your centrifugation device must be balanced, pay particular attention to this first tube and try to balance it equally with tube #2 (slightly less filled).

Venipuncture

By definition, *venipuncture* is the process of collecting or "drawing" blood from a vein. It is the most common way to collect blood specimens for laboratory testing and most frequently performed by a phlebotomist. A 10-step protocol is described next to ensure that timely and safe venipuncture is performed within a dental office (see Fig 9-7 and Video 9-1).

Step 1: Understand and be able to explain PRF

It is important that all staff be familiar with PRF and able to discuss this with patients. Most, if not all, patients have had a blood draw before for medical tests. Generally all that is required by the phlebotomist doing the procedure is a simple statement of intent to collect a few vials of blood for the production of PRF. Some patients, however, may want to know more and question you regarding its benefits. Therefore, knowledge of PRF by the phlebotomist is good practice to encourage a more professional working environment. This explanation can be in simple terms.

It is good practice to have all staff members familiar with PRF. Should a patient ask for additional information, general knowledge on PRF therapy will encourage a more professional working atmosphere.

During this step, it is also important to address needle phobia. It has been estimated that 10% of the population avoids medical tests for that reason. The patient will certainly know and may in fact inform you prior to your asking. This may cause more twitching by individuals during the blood draw, a heightened sensitivity to pain, pallor, profuse sweating, light-headedness, or even fainting. If such is the case where the phobia seems too great a risk, it may be recommended to avoid blood draws on such patients and utilize other recombinant growth factors when appropriate.

Step 2: Use appropriate PPE

When working with blood, proper hand sanitization as well as glove use is a requirement. It is important to understand the risks associated while working with blood and the absolute requirements to utilize gloves and other PPE appropriately.

Step 3: Apply a tourniquet and have the patient make a fist

It is important during venipuncture that the patient is positioned correctly to facilitate the procedure. A more comfortable phlebotomist will naturally have more successful venipuncture. For venipuncture in the antecubital area, always have the patient's arm extended downward in a straight line without bending at the elbow. Stability of this position minimizes the chances of rolling veins.

> ***Tip***
>
> Always have the arm fully extended to stabilize the veins and minimize vein rolling.

The tourniquet is applied 3 inches above the intended venipuncture site to allow more prominent visualization of veins by restricting venous blood flow (see Fig 9-7a). The aim of the tourniquet is to slow/stop blood flow of veins yet maintain blood from arteries. Therefore, the tourniquet should be applied tight enough to restrict the flow of blood back to the heart without preventing new blood from entering the arm. This allows more flow into the area and creates negative pressure, further facilitating the blood draw.

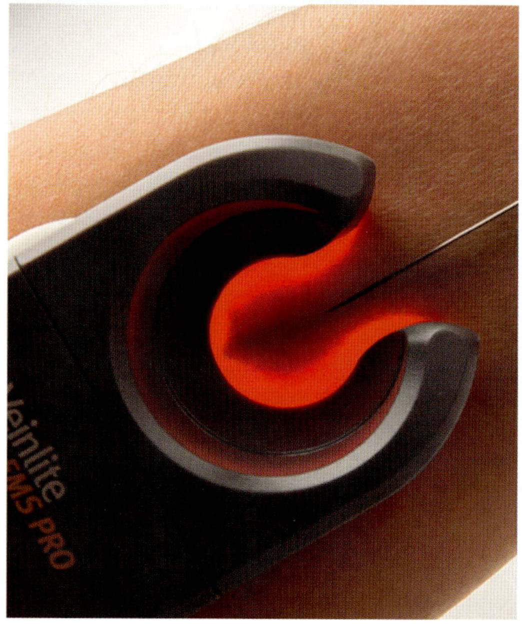

Fig 9-5 Clinical photograph of a transillumination Veinlite. (Provided by Veinlite.)

Once the tourniquet is placed 3 inches above the venipuncture site, ask the patient to clench and make a fist. This allows better visual/tactile sensation of the veins, facilitating venipuncture.

Step 4: Locate a vein

As previously discussed in the anatomy section of this chapter, the median cubital and median veins are normally best options because they are more superficial to the surface, more stationary, and less likely to result in nerve injury. They are the first choice for venipuncture, followed by the cephalic and median cephalic veins.

While an experienced phlebotomist can draw from either arm, patients generally have more prominent veins in their dominant arm. In previous times, palpation was the primary means to locate veins. Today, a transilluminator device (Veinlite) is definitely a tool useful for phlebotomists (Fig 9-5; see also Fig 9-7b). It allows for easy visualization of veins and is a highly recommended tool in dental offices. It is especially useful in obese patients, even for experienced phlebotomists.

Should difficulty be encountered when locating a vein, always remember that heating in general will result in the veins coming more toward the surface. Therefore, a gentle massage, tapping of the veins, or placing a warming pad to the area will all favor easier blood draws. Patient hydration is also key.

We recommend that beginners perform blood draws with the vein light in place. It is not wise to mark the area with a pen because this contaminates the site. It is also important to note that should too much time pass while searching for a vein (greater than 2 minutes), stasis (slowing or stopping of normal blood flow) will occur. Under these circumstances, remove the tourniquet and place it again after 2 minutes or so.

Step 5: Disinfect the area

It is also important to apply an antiseptic solution prior to venipuncture to minimize contamination (see Fig 9-7c). Routinely 70% isopropyl alcohol is utilized (either on a cotton roll, gauze pad, or prepackaged pad). Use circular motions, starting from the area and working your way outward. Allow the area to dry for a minimum of 30 seconds.

If the drying process is not complete prior to venipuncture, the patient will feel an additional burning/stinging sensation when the needle is inserted. Let the area dry naturally without fan waving or with another gauze pad.

Step 6: Prepare blood collection equipment

While the skin is drying after disinfection, the sterilized butterfly needle pack should be opened and the tubes prepared for blood draw. Remember that the entire blood draw should ideally be performed in less than 90 seconds (see chapter 2). If more than 2 minutes has passed from the time the tourniquet was first placed, it is generally advised to remove the tourniquet and replace it following an additional 120 seconds to minimize chance of stasis during the draw. The 15 to 20 seconds it takes for the skin to dry is also a good time to make sure that the centrifuge is set, open, and ready for use.

Step 7: Perform the venipuncture

First, observe that the needle bevel is facing upward and that no obvious damage to the needle is visible. Always hold the wing portion of the butterfly needle to facilitate venipuncture. At this point, ask the patient to remake a fist, anchor the vein (either with your hand or with the vein light), and insert the needle as highlighted below.

Needle insertion

With the bevel facing upward, line up the needle so that it is parallel with the vein. Your body should be positioned directly behind the needle so that you can clearly see a perfect path of insertion without "guesstimating" the correct path. At this point, notify the patient that they will feel a small poke.

It is very important that blood draws be performed typically with an insertion of roughly 30 degrees to minimize the chance of vein collapse (Fig 9-6a). Far too often, a beginner will insert the butterfly needle using a minimal angle (Fig 9-6b), which may actually increase the chance of vein collapse. Enter the needle at a steady pace (not too slowly) and you'll feel a slight "give" or decrease in resistance when you enter the vein. Phlebotomists refer to this feeling as a "pop," but it is a feeling and not a sound.

At this stage, when resistance has decreased, a small "flash" of blood will appear in the tubing (see Fig 9-7e). Stop advancing the needle, securely anchor the butterfly needle either with a Band-Aid or surgical tape (or by holding it in place during the procedure), secure the tube holder by holding it against the patient's arm, and begin collecting blood.

Step 8: Collect the vials of blood

At this stage, quick but smooth collection of blood is performed (see Figs 9-7g to 9-7i). The collection tube must always be advanced into the tube holder so that the tube is against the top portion of the tube holder. Often resistance will occur, so make sure staff is well trained to push the tube entirely into the top of the tube holder. Transition from tube to tube, making sure that your nondominant hand is always stabilized against the patient's arm. Blood should continue to fill until all tubes have been collected.

In the event that vein collapse does occur, attempts can be made to simply readjust the needle (there is no substitute here for experience and adequate training to fully master this technique, especially in the elderly population).

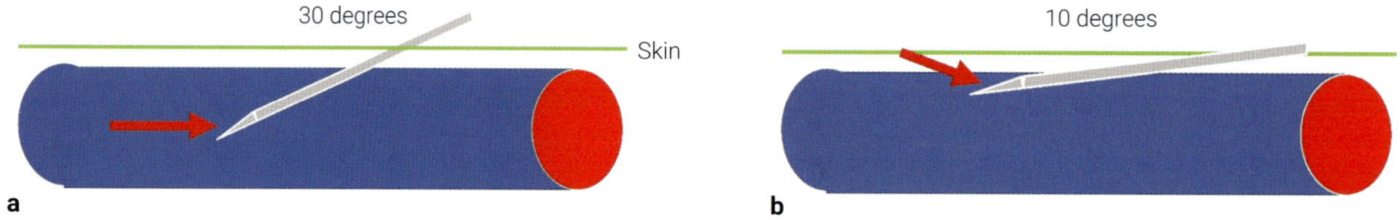

Fig 9-6 Angulation options for venipuncture. The bevel should always face upward. *(a)* Note that preference is given to a 15- to 30-degree angulation to allow for the bevel to face the opening. *(b)* Inexperienced phlebotomists will often enter the needle with too small of an angulation. This will in fact lead to a higher chance of vein collapse, whereby the vein wall will actually suction onto the opening of the needle.

In challenging patients, the tourniquet is typically left on until all tubes are filled. In a normal patient, the tourniquet may be removed while the last tube is filling.

You should never hear a hissing sound when performing a blood draw. If you do, it means the tube vacuum is escaping. This occurs either when the needle is not fully inserted into the vein or when the tube is not placed perfectly sealed within the tube holder.

Step 9: Withdraw the needle

When all tubes are collected, have another staff member bring the blood tubes to the centrifugation device and begin the spin cycle (see Fig 9-7l). Two important steps must take place prior to removing the needle. First, remove the tourniquet and wait at least 10 seconds to allow for pressure reduction. Second, if the patient remains with a clenched fist, make sure to advise the patient to relax their hand. Thereafter, the needle can be removed, and gauze is then applied to the area. Immediately after removal, activate the needle safety feature if present and discard it in a sharps container.

It is important not to place the gauze with pressure prior to needle removal. If pressure is applied prior to removing the needle, the needle removal is painful and may actually further slit the vein, causing more pain and bruising. Instead, apply the gauze pad over the puncture area without pressure, remove the needle fully, and then apply pressure with your nondominant hand thereafter.

It is acceptable to have the patient maintain pressure on the gauze while the phlebotomist discards the butterfly needle. Always remember to advise your patient to hold the arm straight and maintain pressure to the area to minimize the chance of bruising.

Remember

Do not bend the arm at the elbow. This additional pressure to the vein will increase the chance of bruising from platelets not being able to effectively form a plug.

Step 10: Check the patient's arm and apply a bandage

Once pressure has been held for a minute or so, examine the venipuncture site to determine if bleeding has stopped. Note that patients on anticoagulants will generally require more time to stop bleeding. Once the bleeding has stopped, apply a bandage to the puncture site (see Fig 9-7k).

Figure 9-7 demonstrates this step-by-step protocol for blood collection (see also Video 9-1).

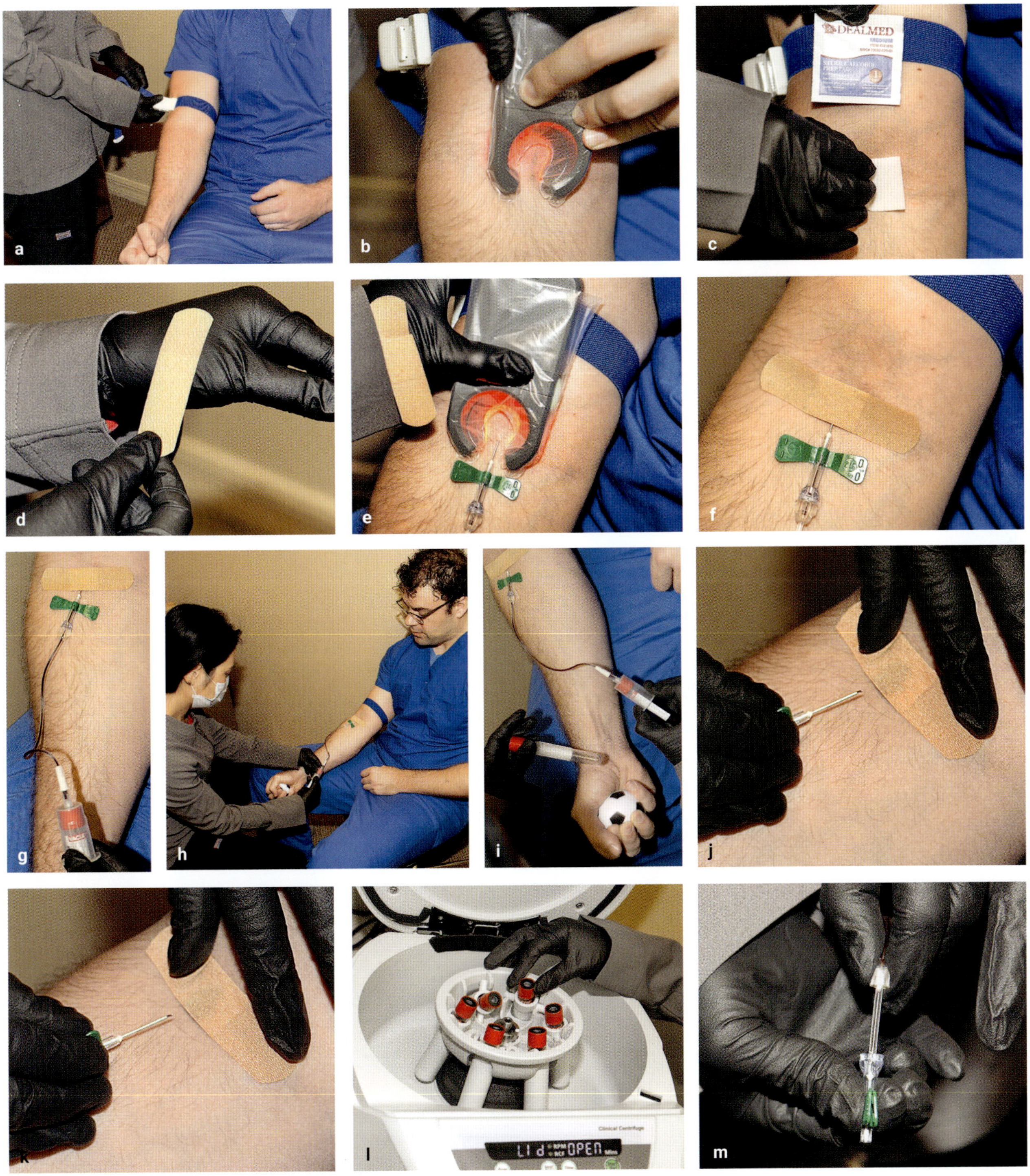

Fig 9-7 Illustration of blood collection for PRF. *(a)* First, a tourniquet is tied about 3 inches above the elbow. *(b)* A vein light is utilized to locate the vein. *(c)* Alcohol is used to disinfect the area. *(d)* A Band-Aid is typically attached to a nearby location (in this case the practitioner's glove) to speed use. *(e)* The butterfly needle is inserted into the vein at a 15- to 30-degree angle and parallel to the vein. *(f)* Backflow is observed within the butterfly needle. *(g to i)* The collection tubes are inserted, and vials of blood are collected. *(j)* Following blood draw, a Band-Aid is placed over the puncture site, and the butterfly needle is removed. *(k)* Compression is applied to the puncture site. *(l)* PRF tubes are placed in the centrifuge. *(m)* Many butterfly needles come with a safety feature locking the needle after use. (Reprinted with permission from Davies and Miron.[2])

Conclusion

This chapter reviewed the essentials of phlebotomy and the techniques to perform venipuncture in a routine dental practice. Proper equipment and a proper protocol are needed as well as a general overview of the anatomical structures in the antecubital fossa area. The step-by-step protocol outlined in this chapter will facilitate safe and effective venipuncture for the phlebotomist and staff.

References

1. McCall RE, Tankersley CM. Phlebotomy Essentials. Philadelphia: Lippincott Williams & Wilkins, 2008.
2. Davies C, Miron RJ. PRF in Facial Esthetics. Chicago: Quintessence, 2020.

10

Fabricating Various PRF Modalities

Contributors

Richard J. Miron

Chapter Highlights

- *Making PRF membranes and plugs*
- *Drawing liquid i-PRF and C-PRF*
- *Fabricating e-PRF membranes*
- *Fabricating "sticky bone," Bio-Graft, and Bio-Bone*

This chapter highlights the fabrication of various PRF modalities, including PRF membranes, PRF plugs, "sticky bone," e-PRF membranes, liquid i-PRF, and liquid C-PRF, as well as the production of newer protocols utilized to make Bio-Filler, Bio-Graft, and Bio-Bone in custom trays. A step-by-step protocol is provided for each harvesting technique, including the tube types utilized and fabrication time. Each is presented with images and videos detailing its production.

Video 10-0

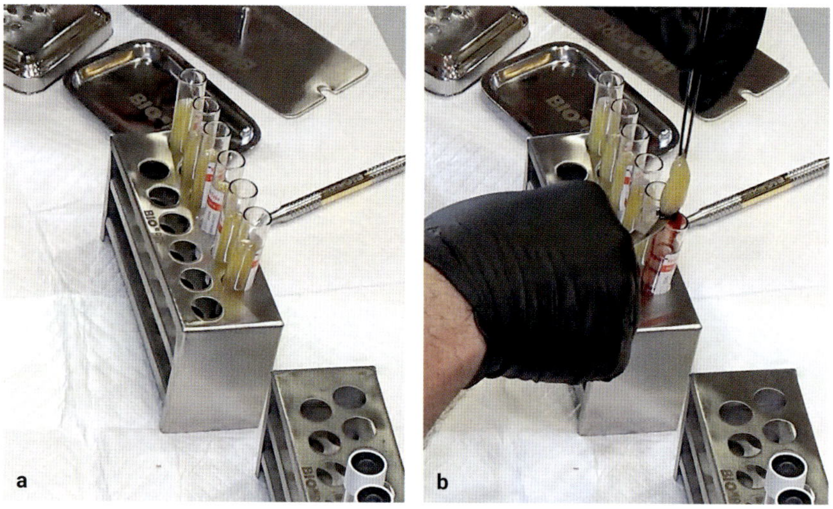

Fig 10-1 Method to remove the red corpuscle layer from the PRF layer. *(a)* After the centrifuge cycle, remove the red caps from the PRF tubes to provide oxygenation to the PRF clots (which favors a better fibrin clot). *(b)* Remove the excess red corpuscle layer. Do not cut the platelet-rich buffy coat zone!

This is a brief chapter that focuses on the end result of all the data and knowledge acquired from basic research studies presented to date in this book to fabricate an array of PRF modalities. The chapter begins by outlining the fabrication of PRF membranes and PRF plugs and thereafter focuses on differences between i-PRF and C-PRF liquid formulations. Lastly, sticky bone, Bio-Filler, Bio-Graft membranes, and Bio-Bone protocols are reviewed.

Solid-PRF Membranes and Plugs

The solid-PRF membrane or plug is created using the horizontal centrifuge and solid-PRF tubes (red cap). The spin time is 700g for 8 minutes, and a clot is then formed. Following the spin cycle, the lids may be removed and allowed to contact oxygen for 5 minutes to favor further clotting. Thereafter, the membranes are removed and the red layer is then gently removed either by peeling or with scissors. As reviewed in chapter 2, many platelets and leukocytes are found right at the buffy coat border, with many within the red cell layer. Do not cut the buffy coat zone rich in cells (Fig 10-1 and Video 10-1).

Video 10-1

The PRF clots are then transferred to the PRF box, and compression is applied (simple compression from the weight of the Bio-PRF lid). The PRF clots are dehydrated within 2 minutes and are then ready for use as 1- to 2-mm-thick PRF membranes (Fig 10-2 and Video 10-2).

Alternatively, if PRF plugs are desired, following removal of the PRF clots from the centrifugation tubes, the PRF clots may instead be placed inside the cylindric inserts and compressed until the piston is flush with the outer rim. Thereafter, the PRF plugs are removed and ready for use (Fig 10-3 and Video 10-3).

Solid-PRF protocol

- **Work schedule:** Phlebotomy 2 minutes | Centrifugation 8 minutes | Dehydration in PRF box 2 minutes | Ready for use
- **Total working time:** 12 minutes
- **Materials needed:** Bio-PRF centrifuge, solid PRF tubes, PRF box, hand instruments
- **Resorption profile:** 2–3 weeks

Video 10-2

Video 10-3

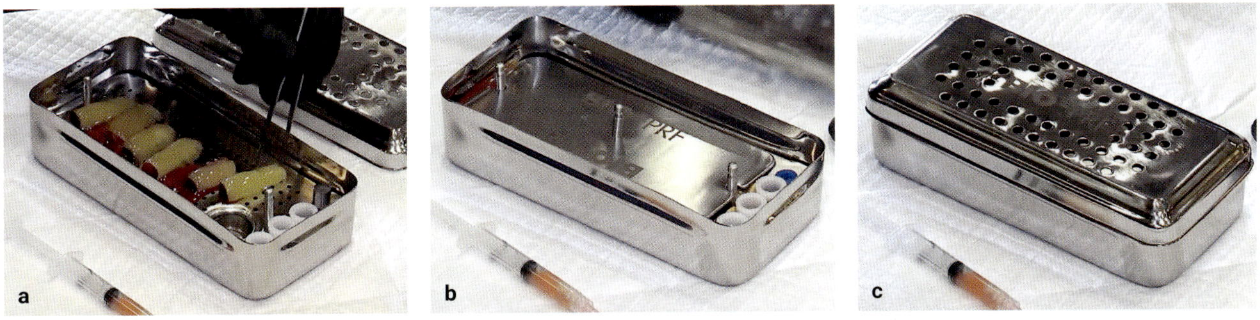

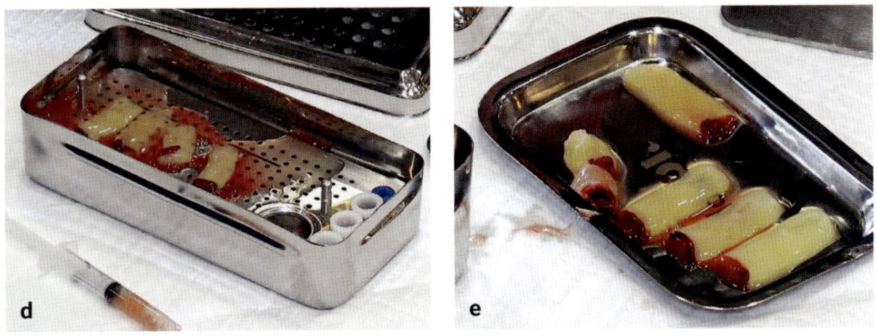

Fig 10-2 Use of the Bio-PRF box to create PRF membranes. *(a and b)* Once the PRF clots are placed in the Bio-PRF box, the Bio-PRF box lid is placed over the clots to compress the membranes. *(c)* The Bio-PRF box can be kept fully closed for up to 4 hours with the cells maintaining their viability. *(d)* When the PRF membranes are needed, the lid of the PRF box is removed, and the compressed 1- to 2-mm-thick PRF membranes are ready for use. *(e)* The PRF membranes may then be placed on the Bio-PRF tray and carried to the surgical site.

Fig 10-3 Use of the Bio-PRF cylinders to create PRF plugs. *(a)* The PRF clot is first placed within the cylinder. *(b and c)* Then the Bio-PRF piston is utilized to compress the PRF clot into a plug. *Note:* Do not overcompress the PRF plug because it may escape the bottom of the cylinder if excessive force is applied. *(d)* Final PRF plugs ready for use.

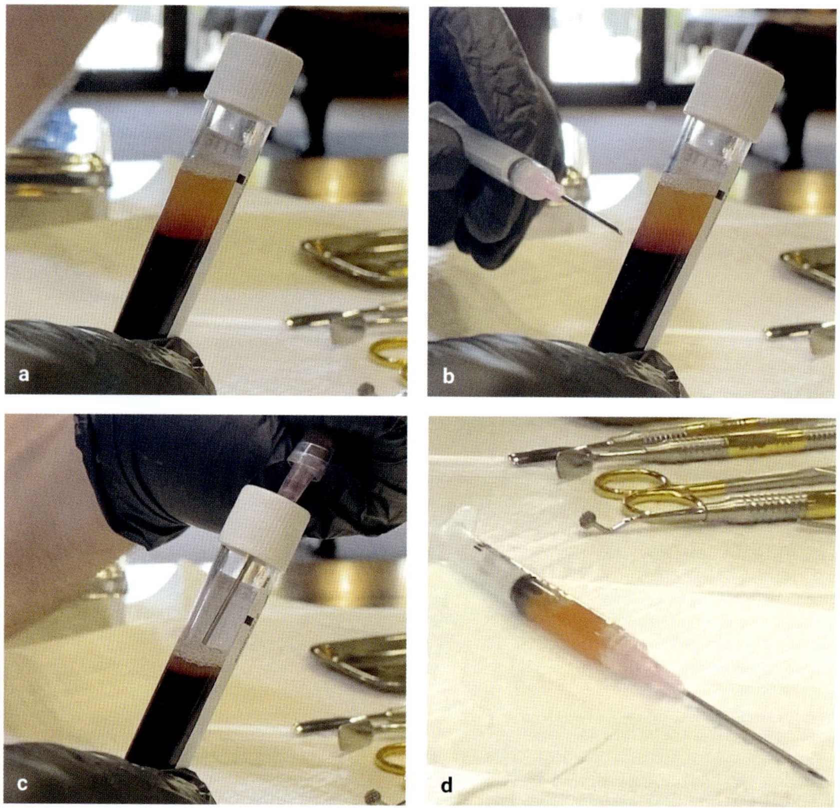

Fig 10-4 Protocol to draw liquid i-PRF. *(a)* Liquid-PRF is found in the upper 1- to 2-mL plasma zone following a 300g spin cycle for 5 minutes. Note that the lid is not removed to prevent oxygenation (clotting will occur more rapidly if removed). *(b)* Note that the cell-rich zone is found at the buffy coat. *(c)* The 18G needle is passed through the rubber portion of the lid, and liquid-PRF is drawn at the buffy coat zone. *(d)* Final liquid-PRF ready for use.

Liquid-PRF

Liquid-PRF is created using a horizontal centrifuge and liquid-PRF tubes (white/blue cap). The spin time is 300g for 5 minutes. Following the spin cycle, the lids should NOT be removed, because oxygen will speed clotting. Typically, an 18G, 1.5-inch syringe needle is utilized and penetrated through the lid to collect the liquid-PRF. As reviewed in chapter 2, many platelets and leukocytes are found right at the buffy coat zone, and therefore it is important to draw as near to the buffy coat zone as possible and even slightly within this layer (Fig 10-4 and Video 10-4). Liquid i-PRF is then ready for injection purposes.

Video 10-4

Liquid-PRF protocol

- **Work schedule:** Phlebotomy 2 minutes | Centrifugation 5 minutes | Blood collection 1 minute | Ready for use
- **Total working time:** 8 minutes
- **Materials needed:** Bio-PRF centrifuge, liquid-PRF tubes, syringe needle
- **Resorption profile:** 2–3 weeks

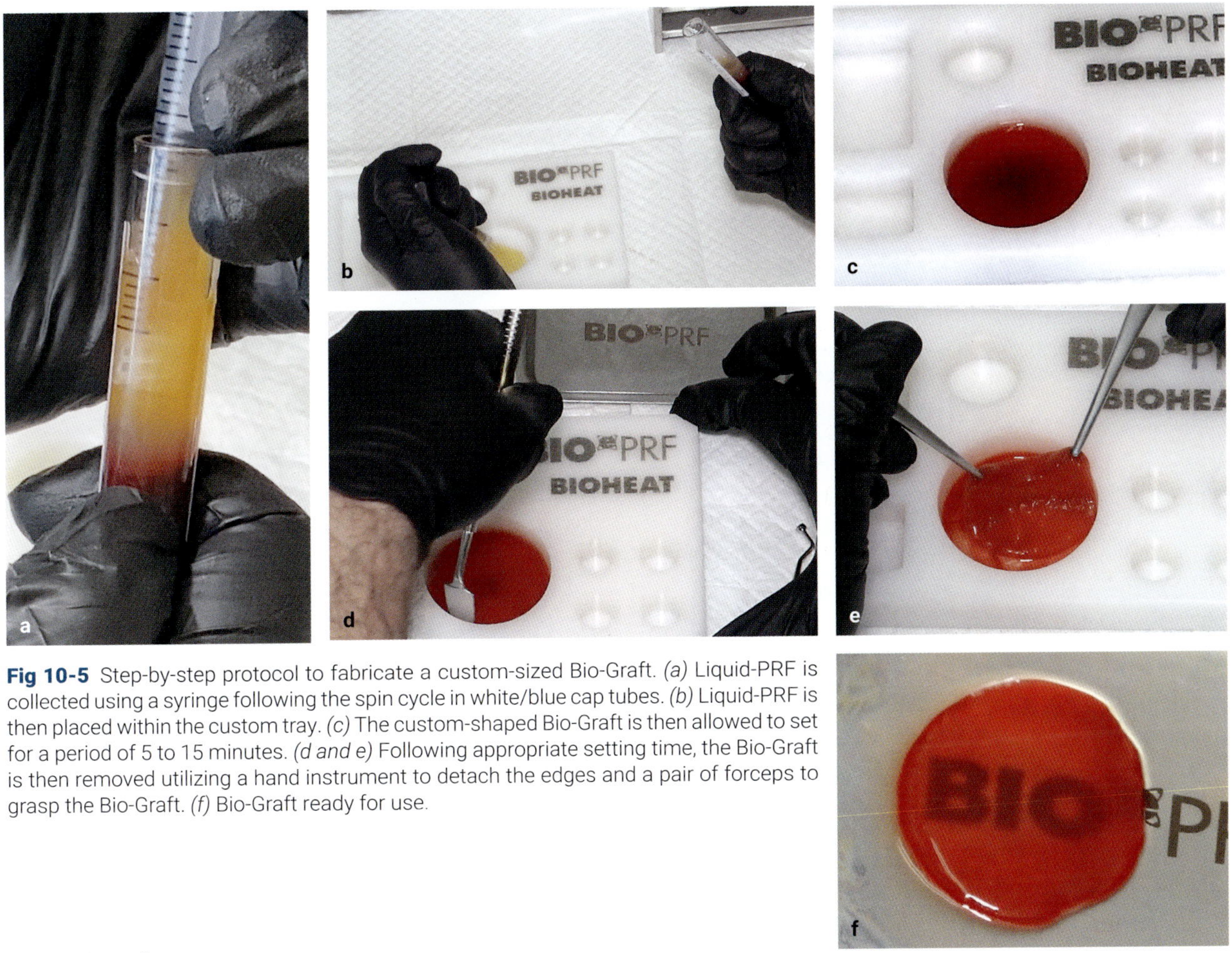

Fig 10-5 Step-by-step protocol to fabricate a custom-sized Bio-Graft. *(a)* Liquid-PRF is collected using a syringe following the spin cycle in white/blue cap tubes. *(b)* Liquid-PRF is then placed within the custom tray. *(c)* The custom-shaped Bio-Graft is then allowed to set for a period of 5 to 15 minutes. *(d and e)* Following appropriate setting time, the Bio-Graft is then removed utilizing a hand instrument to detach the edges and a pair of forceps to grasp the Bio-Graft. *(f)* Bio-Graft ready for use.

Bio-Graft

The Bio-Graft protocol is a biologic tissue graft containing a living custom shape designed for the treatment of various wounds such as diabetic ulcers. Briefly, two white/blue tubes are drawn and centrifuged in a horizontal centrifuge at 700g for 8 minutes. Following the spin cycle, the liquid-PRF is quickly removed from the tubes and placed into the Bio-Heat custom tray and allowed to set (Fig 10-5 and Video 10-5). During the solidification phase, additional growth factors, biomolecules, exosomes, stem cell extracts, or other materials may be added prior to fibrin clot formation, as reviewed in chapter 22. The Bio-Graft is then ready for use to cover wounds or as a larger membrane for GBR procedures.

Video 10-5

Bio-Graft protocol

- **Work schedule:** Phlebotomy 2 minutes | Centrifugation 8 minutes | Addition of liquid-PRF to the custom tray 1 minute | Setting 5 minutes | Ready for use
- **Total working time:** 16 minutes
- **Materials needed:** Bio-PRF centrifuge, solid-PRF tubes, syringe needle, custom tray, hand instruments
- **Resorption profile:** 2–3 weeks

Sticky Bone

Sticky bone is the term commonly given to the bone graft material with sticky/gummy consistency following the addition of PRF. Briefly, two white/blue tubes and two red tubes are drawn and centrifuged in a horizontal centrifuge

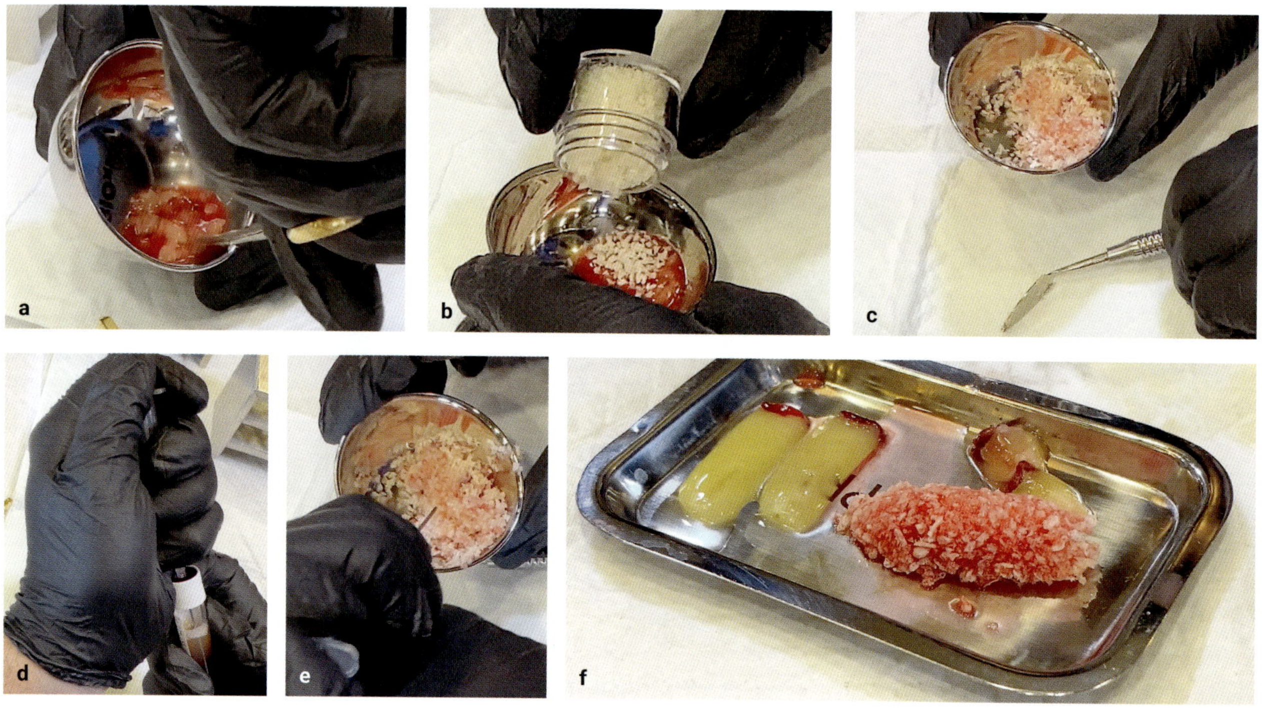

Fig 10-6 Step-by-step protocol to make sticky bone. *(a)* Two PRF membranes are cut up into small PRF fragments. *(b and c)* Bone allograft is then introduced and mixed accordingly. *(d and e)* Liquid-PRF is then collected and introduced within the bone graft complex, followed by appropriate shaping. *(f)* Within 2 minutes, sticky bone is ready for use.

at 700g for 8 minutes. Following the spin cycle, the red lids may be removed to allow oxygenation for 5 minutes if clotting has yet to occur.

The two PRF clots are removed and compressed using the Bio-PRF box to create PRF membranes under the standard protocol (see Fig 10-2). These are then transferred to the Bio-PRF bowl and cut into roughly 1-mm-sized PRF fragments. Thereafter, the bone graft is added, and gentle mixing occurs. Once the PRF fragments are evenly distributed within the bone graft, liquid-PRF is then introduced into the bone graft complex. Remember to take the liquid-PRF from the buffy coat layer because that area will generally be more concentrated with cells. Use as much liquid-PRF as needed to hydrate the entire graft (Fig 10-6 and Video 10-6). The sticky bone is then ready for the bone grafting procedure.

Video 10-6

A 1:1 ratio between bone graft and PRF fragments is generally recommended. Typically, allografts form a better sticky bone consistency because they incorporate collagen within their scaffolds, which induces better clot formation when compared to xenografts.

Sticky bone protocol

- **Work schedule:** Phlebotomy 2 minutes | Centrifugation 8 minutes | Dehydration in PRF box 2 minutes | Cutting PRF fragments 1 minute | Mixing with a bone graft 1 minute | Addition of liquid-PRF and bone graft for custom shaping 1 minute | Ready for use
- **Total working time:** 16 minutes
- **Materials needed:** Bio-PRF centrifuge, solid-PRF tubes, liquid-PRF tubes, allografts (preferred), PRF box, hand instruments
- **Resorption profile:** 2–3 weeks

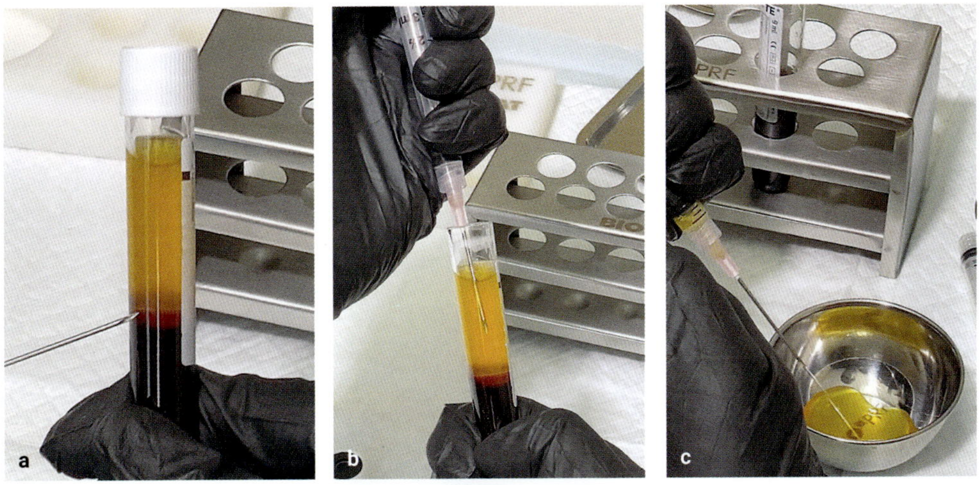

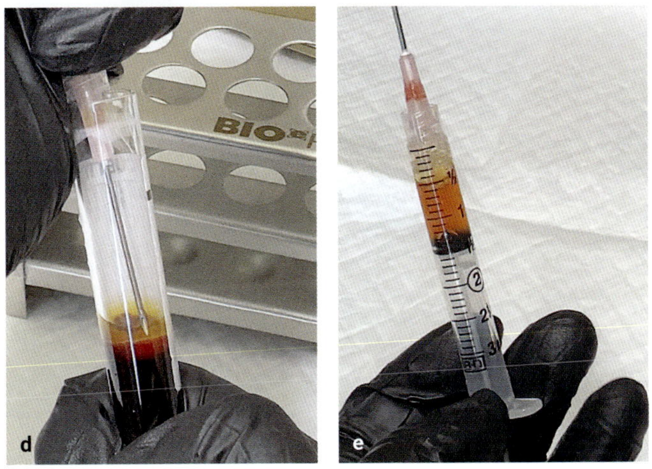

Fig 10-7 Protocol to draw liquid C-PRF. *(a)* Note that the cell-rich zone is found directly at the buffy coat zone, roughly 4 to 5 mL within the plasma zone. *(b and c)* It is important to discard the upper PPP layer first to favor more accurate collection of C-PRF. *(d)* Then the final 0.5- to 1-mL C-PRF layer can be harvested more easily. *(e)* C-PRF ready for use.

C-PRF

The C-PRF protocol is very similar to the liquid-PRF protocol, but it leads to a roughly 10-fold increase in cells and growth factors when compared to whole blood (see chapters 2 and 3). It is also created using the horizontal centrifuge and liquid-PRF tubes (white/blue cap). The spin time is much faster (2000g for 8 minutes). Following the spin cycle, the upper 3- to 4-mL layer is collected using an 18G, 1.5-inch needle and discarded. Thereafter, the remaining 0.5- to 1-mL layer just above the buffy coat is collected. As reviewed previously, many platelets and leukocytes are found directly within the buffy coat zone, so it is important to draw as near to the buffy coat zone as possible and even slightly within it (Fig 10-7 and Video 10-7). Using these higher spin protocols, many of the cells within the buffy coat will actually bunch together, and often a white/yellow layer of cells will form. Be sure to use the needle to break this layer up slightly during the collection of cells. The C-PRF is then ready for injection purposes.

C-PRF protocol

- **Work schedule:** Phlebotomy 2 minutes | Centrifugation 8 minutes | Blood collection 2 minutes | Ready for use
- **Total working time:** 12 minutes
- **Materials needed:** Bio-PRF centrifuge, liquid-PRF tubes, syringe needle
- **Resorption profile:** 2–3 weeks

Video 10-7

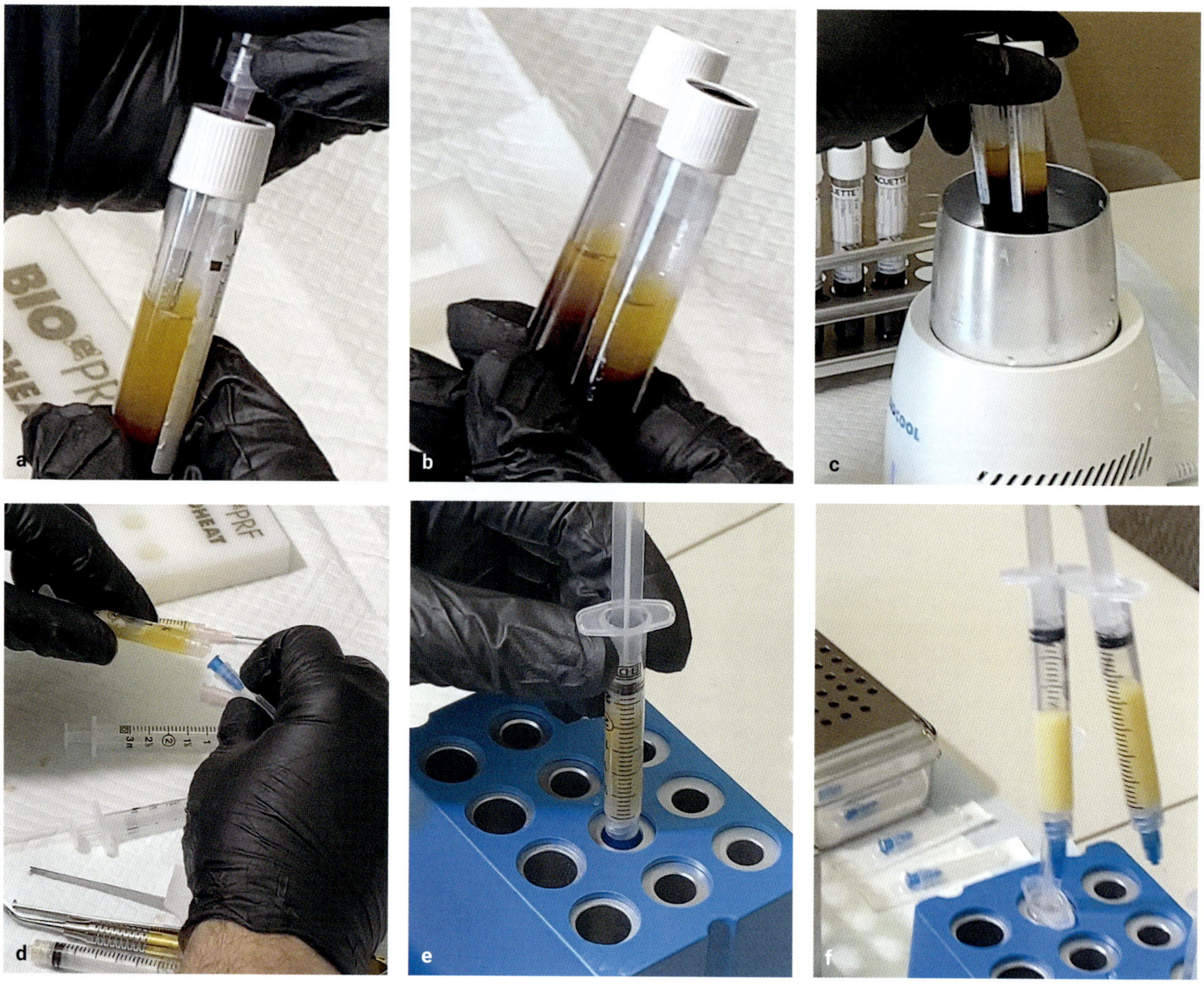

Fig 10-8 Step-by-step protocol to make the Bio-Filler ready for injectable purposes. *(a)* First, the upper 2-mL PPP layer is collected. *(b and c)* The remaining cell-rich layer near the buffy coat is then transferred to the Bio-Cool device to delay clotting. *(d and e)* The blue end caps are then attached to the syringes containing the PPP layer and immediately placed into the Bio-Heat device. *(f)* Following 10 minutes of heating at 75°C, note the difference in the consistency of the albumin gel.

Bio-Filler

Bio-Filler is fabricated using liquid-PRF tubes (white/blue cap) spun at 2000g for 8 minutes, followed by a Bio-Heat process.

From each liquid-PRF tube, the upper 2-mL layer of PPP is drawn into a 3-mL syringe. Each is placed in the Bio-Heat device for 10 minutes heated to 75°C. After 10 minutes, the syringes are placed in the Bio-Cool device for 1 to 2 minutes to cool. Meanwhile, the remaining C-PRF layer found at the buffy coat from each liquid-PRF tube is extracted into one syringe per tube. When the albumin gel has cooled down sufficiently in the Bio-Cool device, it is then mixed back together with the concentrated C-PRF layer using 3-mL syringes connected together via a specialized connector.

The result is a gel-like texture with extended resorption properties, or e-PRF (Fig 10-8 and Video 10-8). A 25G to 27G needle may then be utilized to collect the Bio-Filler for clinical use in facial esthetics.

Video 10-8

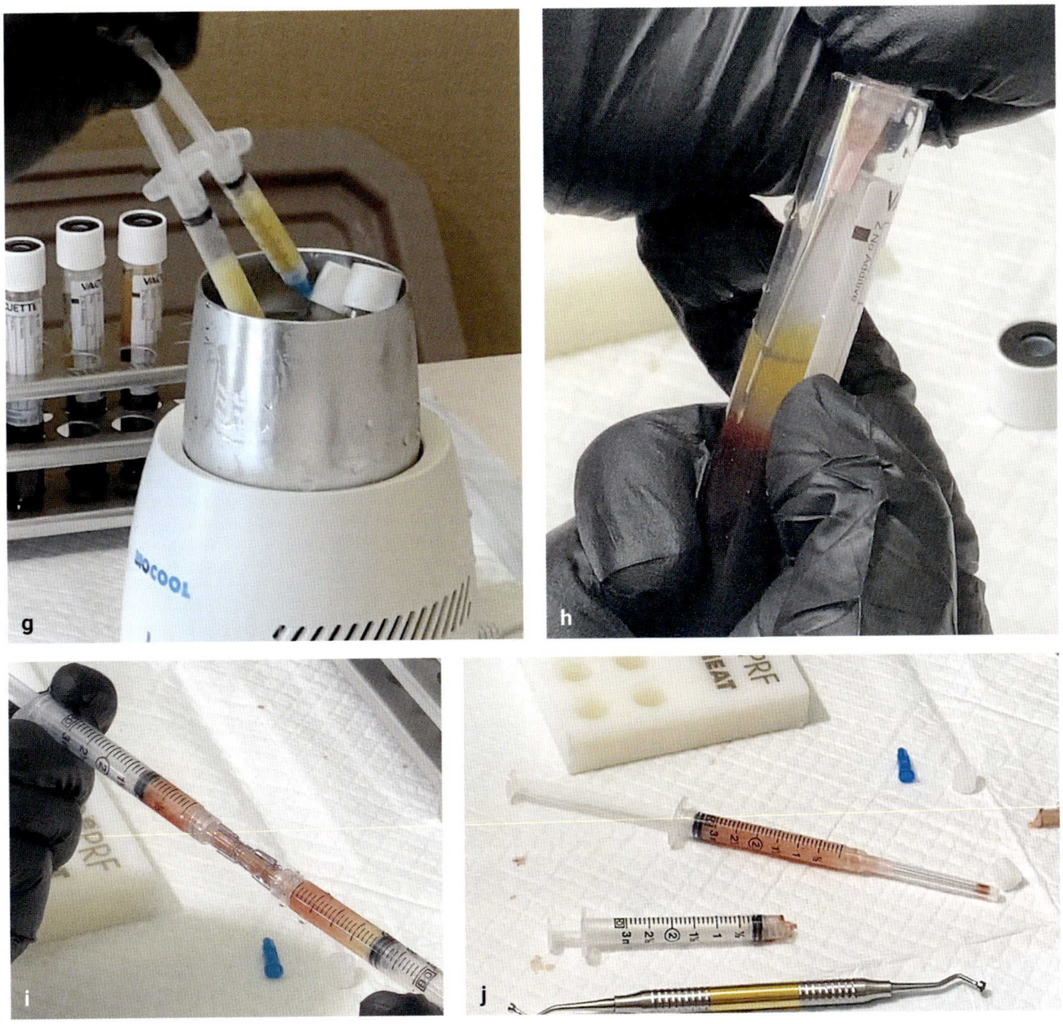

Fig 10-8 *(cont)* *(g)* The albumin gel is then transferred to the Bio-Cool device to allow for cooling. *(h)* The remaining C-PRF layer is collected from the buffy coat zone from the tubes. *(i)* The albumin gel is then mixed using a female-female luer-lock connector with the liquid C-PRF layer to form e-PRF (commercially termed *the Bio-Filler*). *(j)* Bio-Filler ready for use.

Bio-Filler protocol

- **Work schedule:** Phlebotomy 2 minutes | Centrifugation 8 minutes | Extraction of the PPP layer from liquid tubes 1 minute | Bio-Heat 10 minutes | Bio-Cool 1–2 minutes | Mixing 1 minute | Ready for use as an injectable in facial esthetics
- **Total working time:** 24 minutes
- **Materials needed:** Bio-PRF centrifuge, liquid-PRF tubes, Bio-Heat device, Bio-Cool device, syringe needle, special female-female connector
- **Resorption profile:** 4–6 months

e-PRF Membrane

The e-PRF membrane protocol is fabricated using liquid-PRF tubes (white/blue cap) spun using the Bio-PRF horizontal centrifuge at 700g for 8 minutes, followed by a Bio-Heat process. The slower protocol when compared to the Bio-Filler allows for more fibrinogen in the upper layer, favoring more stability of the membrane.

From each liquid-PRF tube, the upper 2-mL layer of PPP is drawn into a 3-mL syringe. Each is placed in the Bio-Heat device for 10 minutes heated to 75°C. After 10 minutes, the syringes are placed in the Bio-Cool device for 1 to 2 minutes to cool. Meanwhile, the remaining C-PRF layer found at the buffy coat from each liquid-PRF tube is extracted into one syringe per tube. When the albumin gel

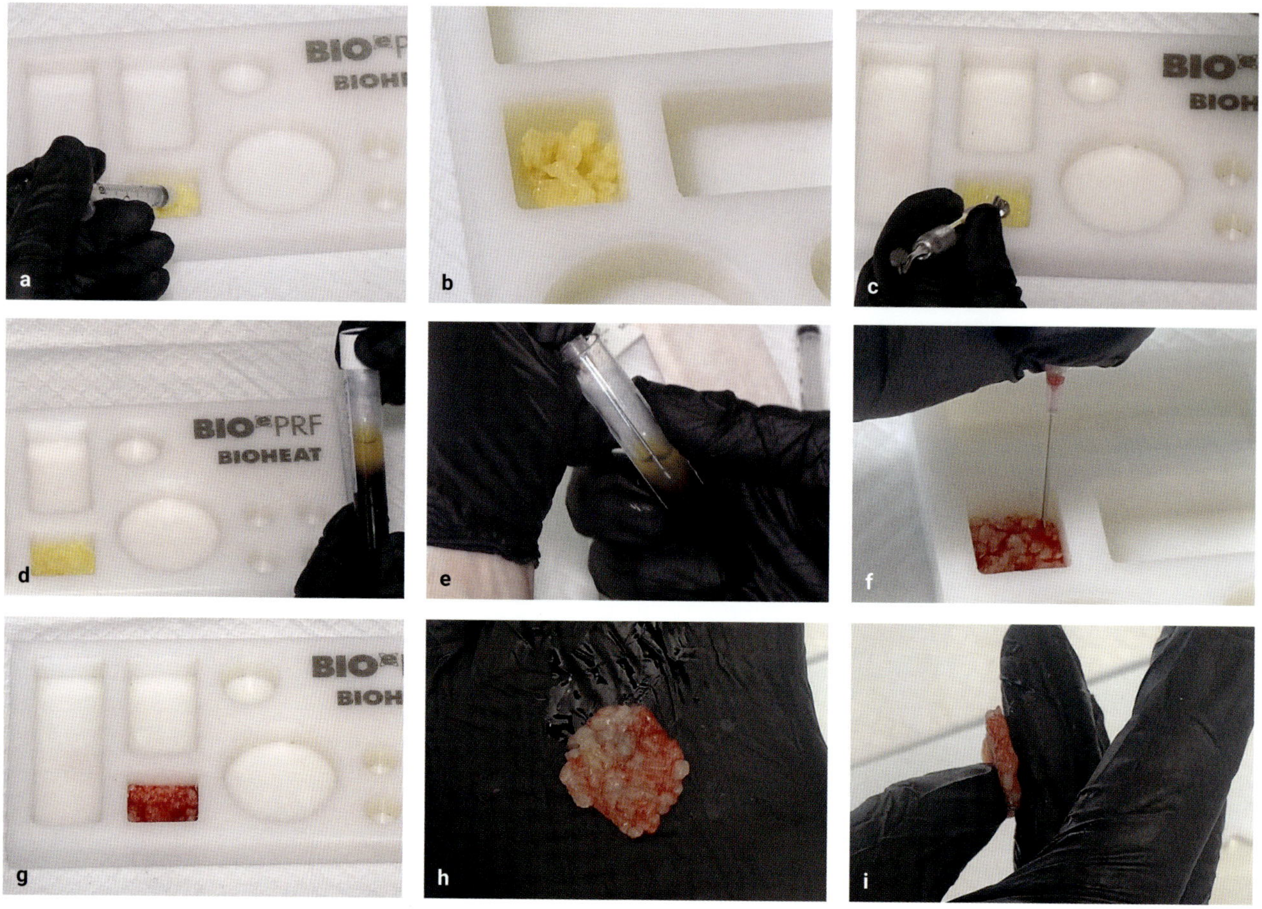

Fig 10-9 Step-by-step protocol to make e-PRF membranes. *(a)* The same steps from Figs 10-8a to 10-8h are applied. Instead of mixing the albumin gel and liquid C-PRF using the female-female luer-lock connector, the albumin gel is first placed directly into the Bio-Heat custom tray. *(b to d)* The albumin gel is then evenly laid flat within the custom tray. *(e and f)* The liquid C-PRF layer is then collected and mixed with the albumin gel in the custom tray. *(g)* The e-PRF is allowed to set for 15 minutes to allow adequate time for clotting. *(h and i)* The e-PRF membrane is then ready for use.

has cooled down, it is placed within the custom Bio-Heat tray and layered flat (Fig 10-9 and Video 10-9). Thereafter, the liquid C-PRF layer is applied to the albumin gel layer and allowed to set for 5 to -15 minutes. The e-PRF membrane is then ready for use with extended resorption properties.

Video 10-9

e-PRF membrane protocol

- **Work schedule:** Phlebotomy 2 minutes | Centrifugation 8 minutes | Extraction of the PPP layer from liquid tubes 1 minute | Bio-Heat 10 minutes | Bio-Cool 1–2 minutes | Addition of the albumin gel followed by liquid C-PRF into the custom tray 1 minute | Setting 5 minutes | Ready for use
- **Total working time:** 28 minutes
- **Materials needed:** Bio-PRF centrifuge, liquid-PRF tubes, Bio-Heat device, Bio-Cool device, custom tray, syringe needle, hand instruments
- **Resorption profile:** 4–6 months

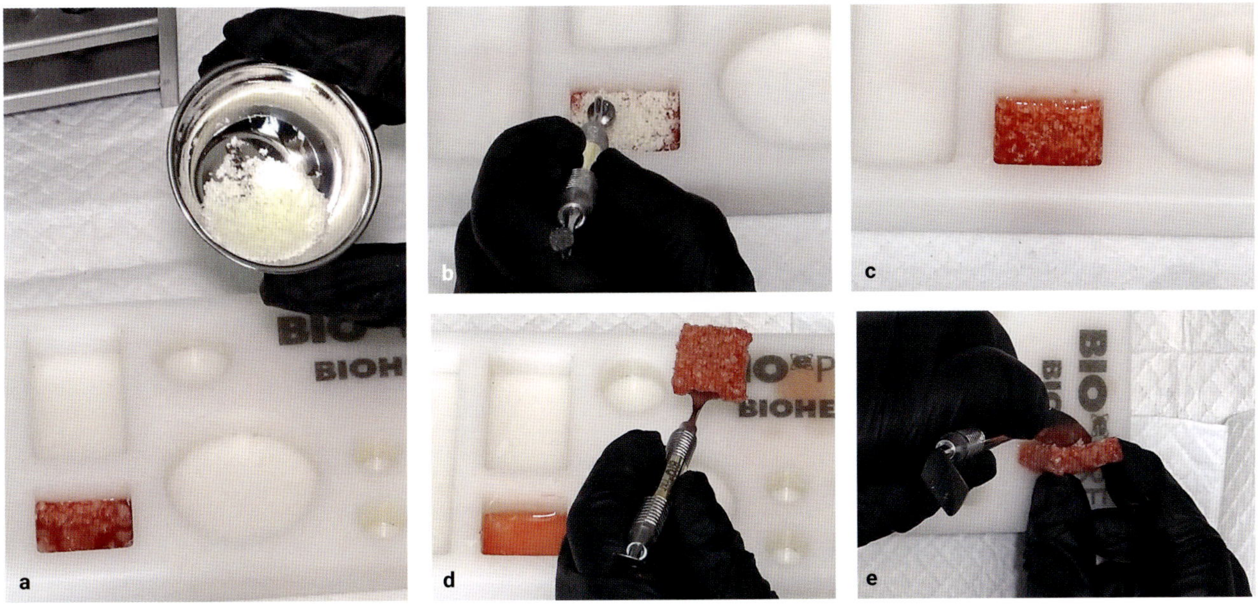

Fig 10-10 Step-by-step protocol to make Bio-Bone. *(a)* The Bio-Bone protocol follows the same protocol as the e-PRF protocol (see Fig 10-9), but prior to clotting, a bone allograft is introduced overtop of the e-PRF membrane. *(b)* The bone allograft is laid flat in a desired thickness depending on the required bone augmentation procedure. *(c)* Additional liquid-PRF is then introduced into the Bio-Bone complex to soak all bone allograft particles and is allowed to set for 15 minutes. *(d and e)* Bio-Bone ready for use. Note that the Bio-Bone is similar to traditional sticky bone but contains an outer e-PRF membrane with a 4- to 6-month resorption profile.

Bio-Bone

The Bio-Bone protocol is one that combines an e-PRF outer membrane with sticky bone. The aim is to create a ready-made bone graft with an outer e-PRF membrane that will harbor a 4- to 6-month resorption period. Briefly, four white/blue tubes are drawn and centrifuged in a horizontal centrifuge at 700g for 8 minutes.

From two liquid-PRF tubes, the upper 2-mL layer of PPP is drawn into a 3-mL syringe. Each is placed in the Bio-Heat device for 10 minutes heated to 75°C. After 10 minutes, the syringes are placed in the Bio-Cool device for 1 to 2 minutes to cool. Meanwhile, the remaining C-PRF layer found at the buffy coat from each liquid-PRF tube is extracted into a separate syringe. When the albumin gel has cooled down, it is placed within the custom Bio-Heat tray and layered flat. Thereafter, the liquid C-PRF layer is applied to the albumin gel layer. Prior to setting, an even layer of bone grafting material is added at the estimated thickness of the grafting procedure. Liquid-PRF from the remaining two tubes is then introduced into the bone graft complex (Fig 10-10 and Video 10-10). The graft complex is then allowed to set for 5 minutes, and the Bio-Bone is ready for use.

Bio-Bone protocol

- **Work schedule:** Phlebotomy 2 minutes | Centrifugation 8 minutes | Extraction of the PPP layer from liquid tubes 1 minute | Bio-Heat 10 minutes | Bio-Cool 1–2 minutes | Addition of the albumin gel followed by liquid C-PRF into the custom tray 1 minute | Addition of the bone graft to the custom tray 1 minute | Final layer of liquid PRF 1 minute | Ready for use
- **Total working time:** 25 minutes
- **Materials needed:** Bio-PRF centrifuge, liquid-PRF tubes, Bio-Heat device, Bio-Cool device, allografts (preferred), custom tray, syringe needle, hand instruments
- **Resorption profile:** 4–6 months

Video 10-10

Conclusion

This chapter highlighted the various protocols and options utilized to create a variety of grafting options with PRF, including PRF membranes, PRF plugs, sticky bone, liquid i-PRF, liquid C-PRF, and e-PRF membranes, as well as newer protocols utilized to make Bio-Filler, Bio-Graft, and Bio-Bone in custom trays. This field is rapidly evolving, and it is highly recommended to stay active with current research activities. For example, it is known that cooling PRF with the Bio-Cool will allow liquid-PRF to remain in liquid form longer, whereas heating PRF will allow it to clot faster and improve its stability (Video 10-11). In the final chapter of this textbook (see chapter 22), an incubator is utilized with a 3D printer to fabricate a stable-form custom 3D bone block that incorporates particulate allograft that are essentially "glued" together with heat-treated PRF. With the rapid advancements currently being made in digital technology, it is imperative that colleagues stay updated with new technologies to further improve patient treatment care and overall satisfaction at the dental office.

Video 10-11

11

Overview of Clinical Indications Using PRF

Contributors

Richard J. Miron

Chapter Highlights

- *Effects of PRF on hard tissues versus soft tissues*
- *Use of PRF for periodontal regeneration*
- *Use of PRF for bone regeneration*
- *Use of PRF in various other areas of dentistry and medicine, including facial esthetics*

Video 11-1

This chapter is dedicated to providing an overview of future clinical chapters on PRF with an accompanying video highlighting the widespread uses of PRF in regenerative dentistry. Over two decades have now passed since PRF was first utilized in clinical practice, and its use has exponentially increased over the past 5 years, with over 250 articles published each year on the topic. Many of these articles remain dedicated to its effect on the regeneration of tissues found in the oral cavity, and this chapter provides an overview of future chapters explaining just that.

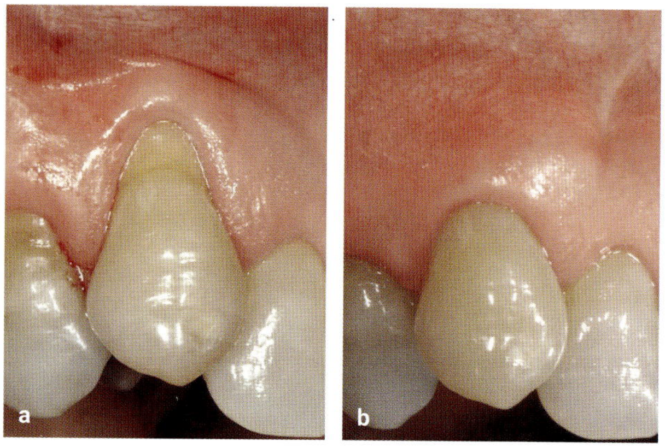

Fig 11-1 *(a and b)* Gingival recession treated with PRF alone. Notice the final esthetic outcome as well as the increased vascularization. Case reviewed in chapter 12. (Reprinted with permission from Miron and Choukroun.[1])

Regenerative medicine aims to regenerate lost or missing tissues as a result of disease or injury. While considered one of the fastest-growing fields in medicine, it remains difficult to fully regenerate tissues found in the oral cavity due to the fact that they are derived from various germ layers (mesoderm and ectoderm) with multiple cell types involved. Some tissues, such as the periodontal ligament, are further complicated because they contain Sharpey fibers spanning multiple tissue types including the cementum, periodontal ligament, and alveolar bone, which are overlaid by both connective tissue and epithelial cells. These various cell types (derived from different germ layers) consist of numerous cell populations in highly organized layers. Therefore, following their destruction, a complex regeneration process to correct their damage is necessary (but not yet predictably established).

Important for the regeneration of any tissue is angiogenesis, which permits the infusion of micronutrients and blood supply to the tissues. While various regenerative strategies have been proposed over the years for the regeneration of oral tissues, platelet concentrates have gained tremendous momentum as a regenerative agent capable of stimulating tissue healing due to their ability to rapidly increase neogenesis and cell recruitment to defective tissues. For these reasons, PRP/PRF has been utilized in virtually all fields of regenerative medicine.

Furthermore, unlike traditional PRP, which has been shown to facilitate the regeneration of numerous tissues, the use of PRF formulates a fibrin clot following its production (see chapter 2). This fibrin scaffold may then be utilized as a fibrin mesh with a better ability to release growth factors over time or as a scaffold for tissue regeneration. The purpose of this chapter is to overview the current literature to date on its use and outline future chapters investigating its use for either soft or hard tissue regeneration (see Video 11-1).

Recession Coverage and Periodontal Regeneration with PRF

PRF has most often been studied in randomized controlled trials (RCTs) in the field of periodontology. These studies include its use for root coverage procedures (see chapter 12) as well as for the regeneration of intrabony/furcation defects (see chapter 13). Overall, more than 40 RCTs are presented, with PRF clearly demonstrating superiority when used as an adjunct to coronally advanced flap (CAF) and also shown to significantly reduce pocket depths and improve clinical attachment levels of intrabony defects when compared to open flap debridement (OFD) alone.

It is now known that PRF can be utilized in an evidence-based manner instead of connective tissue grafts (CTGs) in Miller Class I and II defects with a thick tissue biotype; the use of PRF has been shown to improve vascularization, wound healing, and patient morbidity (Fig 11-1). Noteworthy is that limitations exist when keratinized tissue width is lacking (see chapter 12). In such cases, it remains essential to utilize a CTG (either alone or in combination with PRF) for patients with thin tissue biotypes or in Miller Class III recessions where the recession extends past the mucogingival junction.

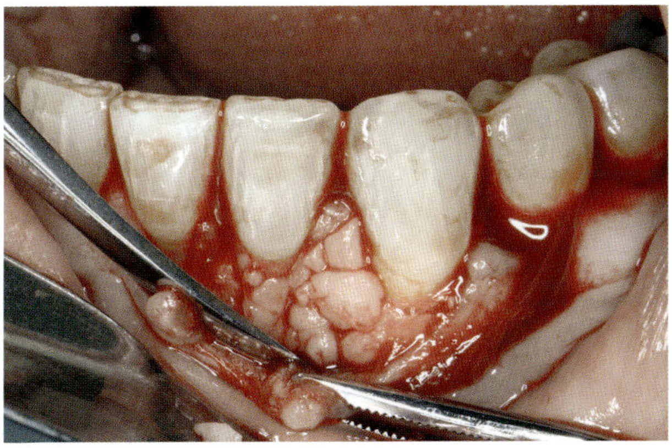

Fig 11-2 Use of PRF for intrabony defect regeneration. Case reviewed in chapter 13.

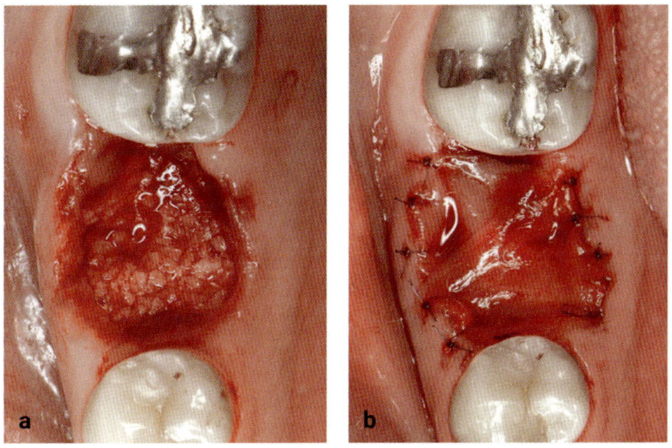

Fig 11-3 *(a)* Use of PRF with a bone grafting material during extraction site management. *(b)* The site was then covered with a PRF membrane. Case reviewed in chapter 14.

Thus, proper patient selection becomes crucial when PRF is utilized as a regenerative strategy for soft tissue grafting of gingival recessions, as highlighted in chapter 12. When utilized correctly, PRF has been shown to improve root coverage, with several RCTs demonstrating similar outcomes to CTGs, but proper case selection is essential.

Likewise, many RCTs have also investigated using PRF for the regeneration of periodontal intrabony and furcation defects. PRF has been shown to favor pocket depth reduction and clinical attachment level gains when compared to OFD alone (Fig 11-2). Over a dozen RCTs have also compared PRF to a bone grafting material, enamel matrix derivative, amniotic membranes, and in combination with other biomolecules and biomaterials. These are reviewed in chapter 13.

In summary, studies have now demonstrated the ability for PRF to stimulate soft tissue/periodontal repair with much greater evidence (in both number of RCTs as well as studies receiving positive outcomes) when compared to other fields of regenerative dentistry.

Use of PRF for Bone Regeneration

Prior to reviewing each individual chapter on PRF, it is important to note that PRF favors greater soft tissue regeneration when compared to bone formation. For this reason, the chapters dedicated to GBR and implant dentistry show how PRF may be utilized *(1)* to improve the handling of bone grafting materials and *(2)* as an outer barrier to facilitate soft tissue healing.

PRF favors greater soft tissue regeneration when compared to bone formation.

Chapter 14 is dedicated to extraction socket management; when utilized alone, PRF has demonstrated the ability to limit dimensional changes (though more variability in positive/negative outcomes exists in this field). More commonly, clinicians recommend its combination with bone grafting materials, where RCTs have demonstrated a better ability to maintain ridge volume when compared to utilizing PRF alone or bone grafting material alone (Fig 11-3; see chapter 14). Over 20 clinical trials have also demonstrated the positive outcomes utilizing PRF for the management of mandibular third molar extractions, where up to a 10-fold decrease in the rate of dry sockets has been observed (reviewed in detail in chapter 14).

Chapter 15 reviews the use of PRF in the field of implant dentistry. This includes using PRF to coat implant surfaces, the ability for PRF to influence osseointegration of implants, and its use for the management of peri-implantitis. Highlighted throughout this chapter is the use of PRF as an outer barrier membrane to minimize bacterial infiltration, especially during immediate implant placement (Fig 11-4).

With respect to using PRF as an adjunct to GBR procedures, it has been hypothesized that PRF could stimulate blood flow, lead to better grafting material handling/stability, and potentially improve new bone formation. Despite its use being widespread in the bone regeneration field and

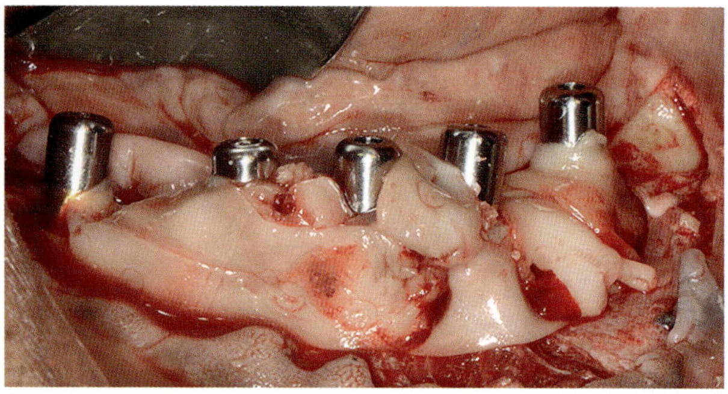

Fig 11-4 Use of PRF as an outer soft tissue barrier during implant placement. Case reviewed in chapter 15.

Fig 11-5 Use of PRF to improve the handling and stability of bone grafting particles. This concept is reviewed in chapter 16.

many colleagues exploring its use, very few studies have in fact investigated the use of PRF in GBR procedures, with limited data available. This topic is reviewed in chapter 16.

Similarly, PRF has also been utilized either alone or in combination with bone grafting materials for sinus elevation procedures (Fig 11-5). Chapter 17 reviews the literature and also sheds light on when to utilize PRF alone versus in combination with a bone grafting material for sinus grafting procedures. It will also review studies investigating the use of PRF as a barrier membrane for the closure of lateral windows as well as discuss its potential use to cover sinus membrane tears as opposed to utilizing a collagen membrane.

Lastly, PRF has also gained more widespread use in the field of oral and maxillofacial surgery. PRF has most commonly been studied in oral surgery for the management of osteonecrosis of the jaw, with over a dozen studies being reviewed in chapter 18.

Use of PRF in Various Other Fields of Dentistry and Medicine

The final chapters of this textbook focus on the use of PRF for new applications in regenerative dentistry and medicine. These cover the use of PRF in endodontics (chapter 19), the use of PRF in facial esthetics (chapter 20), as well as the use of PRF in medical applications (chapter 21). Specifically in medicine, liquid-PRF has been successfully utilized as an injectable and autologous biomaterial for osteoarthritic knees, temporomandibular joint disorders, as well as for diabetic ulcers and other large defect types by utilizing larger custom-shaped fibrin grafts (Bio-Graft).

Conclusion

PRF has now been widely utilized in regenerative dentistry for a variety of procedures for the repair and regeneration of either soft or hard tissues in the oral cavity. Initially, it was utilized in a similar fashion to PRP as a second-generation platelet concentrate with a better ability to more slowly and gradually deliver growth factors to defective tissues over an extended period of time. Over the years, the use of the actual fibrin scaffold has more commonly been utilized as a barrier to facilitate soft tissue healing as well as acting as a defense-fighting barrier to incoming pathogens. In the early years, several attempts were made utilizing PRF alone for various procedures. Today, it is much better understood when and where PRF is effective when utilized alone but also when a CTG, a bone grafting material, or a collagen barrier membrane may be favored. These topics are reviewed in the upcoming clinical chapters, which examine the evidence and offer clinical guidelines and recommendations for the use of PRF.

Reference

1. Miron RJ, Choukroun J (eds). Platelet Rich Fibrin in Regenerative Dentistry. Hoboken, NJ: Wiley, 2017.

12

Use of PRF for the Treatment of Gingival Recessions

Contributors

Richard J. Miron

Delia Tuttle

Vittorio Moraschini

Meizi Eliezer

Raluca Cosgarea

Søren Jepsen

Yoshinori Shirakata

Hom-Lay Wang

Anton Sculean

This first clinical chapter focuses on the use of PRF for root coverage procedures in comparison to other treatment modalities. In total, 17 RCTs are presented comparing the use of coronally advanced flap (CAF) with and without PRF versus connective tissue graft (CTG), enamel matrix derivative (EMD; Emdogain), and amniotic membrane. Reports on the ability of PRF to improve root coverage, clinical attachment level, keratinized mucosa width, and probing depth are evaluated. Thereafter, clinical case presentation as well as a step-by-step protocol utilizing PRF for root coverage via the vestibular incision subperiosteal tunnel access (VISTA) technique is presented. In summary, the present chapter demonstrates that PRF enhances root coverage and clinical attachment levels when compared to CAF, with little change in keratinized tissue width (KTW). Therefore, the use of PRF in conjunction with CAF represents a valid treatment modality for gingival recessions exhibiting adequate baseline KTW. In cases with limited baseline KTW, the use of CTGs alone or in combination with PRF may be preferred due to its greater ability to enhance KTW.

Chapter Highlights

- *Rationale for using PRF as a soft tissue graft substitute for periodontal root coverage*
- *Comparison of CAF with and without PRF versus CTG, EMD, and amniotic membrane for root coverage procedures*
- *Future research directions using PRF for the soft tissue coverage around peri-implant defects*

Video 12-1

Studies conducted in the United States have now demonstrated that approximately 90% of the population reports having at least one tooth with a 1-mm recession by age 60 years, with up to 40% of people displaying recessions greater than 3 mm.[1,2] The ultimate goal of root coverage procedures is total resolution of the defect by providing complete root coverage, with ideal keratinized and attached tissue and a seamless esthetic transition with neighboring tissues.[3] Multiple periodontal plastic surgical procedures with a variety of biomaterials have been proposed to correct these mucogingival deformities and thus rebuild the lost attachment apparatus. Traditionally, connective tissue grafts (CTGs) from the palate in combination with different flap designs have been utilized. Nevertheless, alternative treatment options, including various biomaterials and/or bioactive agents, have been proposed over the years with the aim of lowering patient morbidity.[3,4]

A variety of collagen-based membranes and dermal tissue derivatives from either allograft or xenograft origin have been brought to market for the management of gingival recessions.[5] While these substitute materials provide an excellent 3D matrix for the migration and proliferation of fibroblasts, reported disadvantages have included limited regenerative potential as well as a lack of long-term keratinization of tissues within the grafted regions.[6] In an attempt to increase the bioactivity of barrier membranes, a variety of commercial membranes derived from placental tissues (amnion) have also recently been brought to market, although long-term clinical data remain scarce.[7,8]

Similarly, another strategy has been the use of regenerative growth factors, either utilized alone or in combination with collagen membranes or CTGs, to stimulate the regenerative potential of fibroblasts within the defect area. A commonly utilized bioactive agent for the treatment of gingival recessions has been the use of enamel matrix derivative (EMD; Emdogain) derived from porcine origin. Results from animal and human studies have shown that EMD leads to positive clinical and histologic outcomes when combined with a CAF procedure.[9,10]

Over the past decade, PRF has been utilized because of its ability to particularly favor soft tissue healing over hard tissue healing.[11–13] This chapter provides information and clinical recommendations evaluating the use of PRF for the treatment of recession type RT1[14–16] (historically Miller class I or II) gingival recessions in comparison to other treatment options. Results from a recent systematic review and meta-analysis investigated the use of PRF in comparison to five categories, as follows: *(1)* CAF alone vs CAF/PRF, *(2)* CAF/CTG vs CAF/PRF, *(3)* CAF/EMD vs CAF/PRF, *(4)* CAF/AM (amniotic membrane) vs CAF/PRF, and *(5)* CAF/CTG vs CAF/CTG/PRF. The primary outcome evaluated was extent (in %) of relative root coverage (rRC). Secondary outcomes included clinical attachment level (CAL), keratinized tissue width (KTW) gain, and probing depths (PDs).

Evaluation of Clinical Studies Using PRF for Root Coverage

Table 12-1 demonstrates the results from a recent systematic review and meta-analysis of RCTs investigating the use of CAF/PRF for recession coverage compared to all other treatment modalities.[17–33] These findings were collected to more specifically address the clinical outcomes and recommendations for PRF with respect to its use in the treatment of RT1 (Miller class I and II)[15,16] gingival recessions compared to the use of other standard modalities currently utilized in the field. Overall, the majority of studies to date compared the use of CAF/PRF vs CAF alone or CAF/CTG (Figs 12-1 and 12-2). Furthermore, additional studies were gathered comparing CAF/EMD vs CAF/PRF, CAF/AM vs CAF/PRF, and CAF/CTG vs CAF/CTG/PRF. In the sections that follow, we highlight and discuss the summary of evidence from the current categories and further evaluate the strengths and limitations of each comparative analysis.

Study or Subgroup	PRF group Mean	PRF group SD	PRF group Total	Control group Mean	Control group SD	Control group Total	Weight	Mean Difference IV, Random, 95% CI	Year
1.1.1 CAF vs. CAF/PRF									
Aroca et al. 2009	80.7	14.7	67	91.5	11.4	67	7.2%	-10.80 [-15.25, -6.35]	2009
Padma et al. 2013	100	0.1	15	68.4	17.4	15	6.3%	31.60 [22.79, 40.41]	2013
Thamaraiselvan et al. 2015	74.1	29	10	65	44.4	10	1.9%	9.10 [-23.77, 41.97]	2015
Bozkurt Dogan et al. 2015	86.7	15.6	60	82.1	17.5	59	6.9%	4.60 [-1.36, 10.56]	2015
Gupta et al. 2015	91	19.9	15	86.6	23.8	15	4.5%	4.40 [-11.30, 20.10]	2015
Agarwal et al. 2016	56	36.7	15	33	40.4	15	2.4%	23.00 [-4.62, 50.62]	2016
Kumar et al. 2017	74.4	36.7	15	53.3	40.4	15	2.4%	21.10 [-6.52, 48.72]	2017
Dixit et al. 2018	82.8	5.8	12	79.5	5	12	7.3%	3.30 [-1.03, 7.63]	2018
Kuka et al. 2018	88.3	15.4	28	74.6	8.05	24	6.8%	13.70 [7.15, 20.25]	2018
Subtotal (95% CI)			**237**			**232**	**45.8%**	**9.60 [0.00, 19.19]**	
Heterogeneity: Tau² = 159.86; Chi² = 91.72, df = 8 (P < 0.00001); I² = 91%									
Test for overall effect: Z = 1.96 (P = 0.05)									
1.1.2 CAF/CTG vs. CAF/PRF									
Jankovic et al. 2012	88.7	10.7	15	92	15.5	15	6.1%	-3.30 [-12.83, 6.23]	2012
Eren and Atilla 2014	92.7	13.7	22	94.2	12.1	22	6.6%	-1.50 [-9.14, 6.14]	2014
Tunali et al. 2015	76.4	13.2	22	77.4	17.4	22	6.2%	-1.00 [-10.13, 8.13]	2015
Kumar et al. 2017	74.4	36.7	15	58.9	25.9	15	3.1%	15.50 [-7.23, 38.23]	2017
Öncü, 2017	77.1	14.2	30	84	16.3	30	6.5%	-6.90 [-14.64, 0.84]	2017
Mufti et al. 2017	51.1	36.9	16	64.7	37.8	16	2.6%	-13.60 [-39.48, 12.28]	2017
Culhaoglu et al. 2018	69.7	15.3	21	80.1	18.9	21	5.9%	-10.40 [-20.80, 0.00]	2018
Subtotal (95% CI)			**141**			**141**	**37.0%**	**-3.96 [-7.84, -0.08]**	
Heterogeneity: Tau² = 0.91; Chi² = 6.20, df = 6 (P = 0.40); I² = 3%									
Test for overall effect: Z = 2.00 (P = 0.05)									
1.1.3 CAF/EMD vs. CAF/PRF									
Jankovic et al. 2010	72.1	9.6	20	70.5	11.8	20	6.8%	1.60 [-5.07, 8.27]	2010
Subtotal (95% CI)			**20**			**20**	**6.8%**	**1.60 [-5.07, 8.27]**	
Heterogeneity: Not applicable									
Test for overall effect: Z = 0.47 (P = 0.64)									
1.1.4 CAF/AM vs. CAF/PRF									
Agarwal et al. 2016	56	36.7	15	36	25.9	15	3.1%	20.00 [-2.73, 42.73]	2016
Subtotal (95% CI)			**15**			**15**	**3.1%**	**20.00 [-2.73, 42.73]**	
Heterogeneity: Not applicable									
Test for overall effect: Z = 1.72 (P = 0.08)									
1.1.5 CAF/CTG vs. CAF/CTG/PRF									
Keceli et al. 2015	89.6	2.68	20	79.9	7.8	20	7.4%	9.70 [6.09, 13.31]	2015
Subtotal (95% CI)			**20**			**20**	**7.4%**	**9.70 [6.09, 13.31]**	
Heterogeneity: Not applicable									
Test for overall effect: Z = 5.26 (P < 0.00001)									
Total (95% CI)			**433**			**428**	**100.0%**	**4.28 [-0.93, 9.49]**	
Heterogeneity: Tau² = 92.47; Chi² = 126.38, df = 18 (P < 0.00001); I² = 86%									
Test for overall effect: Z = 1.61 (P = 0.11)									
Test for subgroup differences: Chi² = 29.37, df = 4 (P < 0.00001), I² = 86.4%									

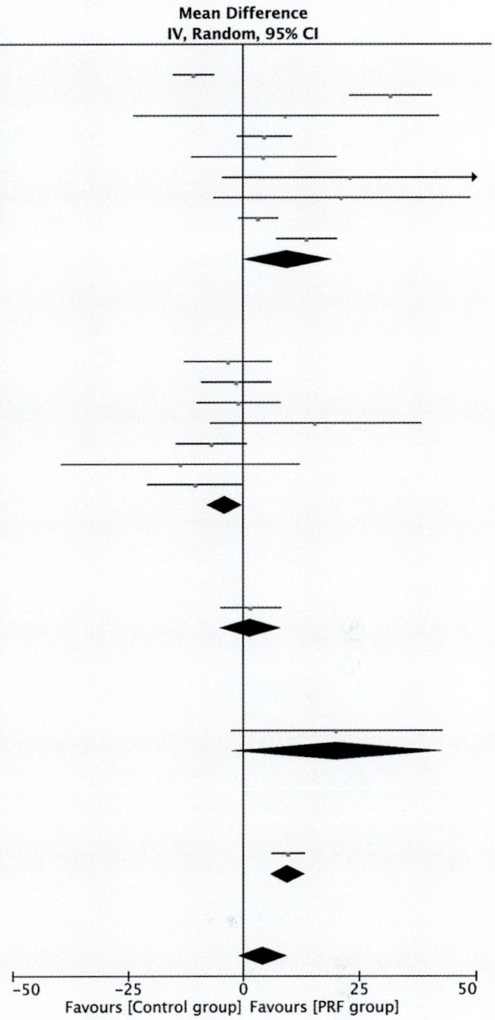

Fig 12-1 Forest plot for the event "relative root coverage" (reported as %rRC).

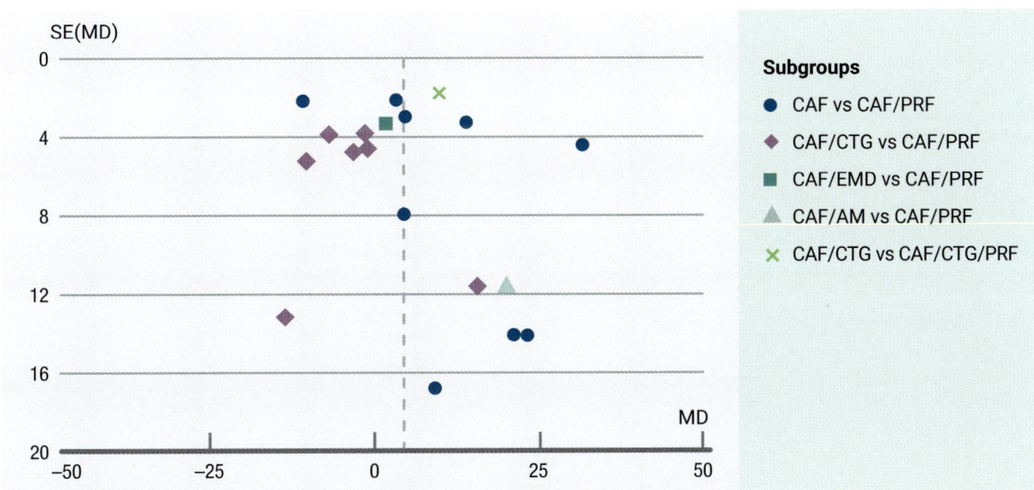

Fig 12-2 Funnel plot for the studies reporting the primary outcome "relative root coverage" (reported as %rRC).

Table 12-1 Main characteristics of the included studies

Authors (year)	Study design Follow-up	No. of treated recessions (per group) No. of subjects	Age range (mean) Sex	No. of smokers	Miller class	Site of recessions	Surgical technique
CAF vs CAF/PRF							
Aroca et al[17] (2009)	RCT (parallel) 6 months	67 (C) / 67 (T) 20	22–47 (31.7) M: 5 / F: 15	2	I and II	Maxillary and mandibular	CAF (C) CAF + PRF (T)
Padma et al[18] (2013)	RCT (split-mouth) 6 months	15 (C) / 15 (T) 15	18–35 (NR) NR	0	I and II	Maxillary and mandibular	CAF (C) CAF + PRF (T)
Bozkurt Doğan et al[19] (2015)	RCT (parallel) 6 months	59 (C) / 60 (T) 20	20–45 (37.1) M: 7 / F: 13	0	I and II	Maxillary	CAF (C) CAF + PRF (T)
Gupta et al[20] (2015)	RCT (split-mouth) 6 months	15 (C) / 15 (T) 26	20–50 (37.2) M: 16 / F: 10	0	I and II	Maxillary and mandibular	CAF (C) CAF + PRF (T)
Thamaraiselvan et al[21] (2015)	RCT (parallel) 6 months	10 (C) / 10 (T) 20	21–47 (NR) M: 18 / F: 2	0	I and II	Maxillary and mandibular	CAF (C) CAF + PRF (T)
Dixit et al[22] (2018)	RCT (parallel) 6 months	12 (C) / 12 (T) 12	18–50 (37.5) M: 7 / F: 5	0	I and II	Maxillary	CAF (C) CAF + PRF (T)
Kuka et al[23] (2018)	RCT (parallel) 12 months	24 (C) / 28 (T) 24	21–41 (32.3) M: 11 / F: 13	0	I	Maxillary and mandibular	CAF (C) CAF + PRF (T)
CAF/CTG vs CAF/PRF							
Jankovic et al[24] (2012)	RCT (split-mouth) 6 months	15 (C) / 15 (T) 15	19–47 (NR) M: 5 / F: 10	NR	I and II	Maxillary anterior	CAF + CTG (C) CAF + PRF (T)
Eren and Atilla[25] (2014)	RCT (split-mouth) 6 months	22 (C) / 22 (T) 22	18–52 (33.8) M: 9 / F: 13	0	I and II	Maxillary and mandibular	CAF + CTG (C) CAF + PRF (T)
Tunali et al[26] (2015)	RCT (parallel) 12 months	22 (C) / 22 (T) 10	25–52 (34.2) M: 4 / F: 6	0	I and II	Maxillary and mandibular	CAF + CTG (C) CAF + PRF (T)
Kumar et al[27] (2017)	RCT (parallel) 6 months	15 (C) / 15 (T1) / 15 (T2) 36	NR (33.2) M: 34 / F: 2	0	I and II	Maxillary	CAF (C) CAF + CTG (T1) CAF + PRF (T2)
Mufti et al[28] (2017)	RCT (parallel) 6 months	16 (C) / 16 (T) 32	NR (36.9) M: 16 / F: 16	0	I	Maxillary and mandibular	CAF + CTG (C) CAF + PRF (T)
Öncü[29] (2017)	RCT (split-mouth) 6 months	30 (C) / 30 (T) 20	20–60 (40) M: 9 / F: 11	0	I and II	Maxillary and mandibular	CAF + CTG (C) CAF + PRF (T)
Culhaoglu et al[30] (2018)	RCT (parallel) 6 months	21 (C) / 21 (T) 22	21–52 (37.7) M: 10 / F: 12	0	I	Maxillary and mandibular	CAF + CTG (C) CAF + PRF (T)
CAF/EMD vs CAF/PRF							
Jankovic et al[31] (2010)	RCT (split-mouth) 12 months	20 (C) / 20 (T) 20	21–48 (NR) M: 8 / F: 12	0	I and II	Maxillary anterior	CAF + EMD (C) CAF + PRF (T)
CAF/AM vs CAF/PRF							
Agarwal et al[32] (2016)	RCT (parallel) 6 months	15 (C) / 15 (T1) / 15 (T2) 30	> 18 (NR) M: 22 / F: 8	0	I and II	Maxillary	CAF (C) CAF + AM (T1) CAF + PRF (T2)
CAF/CTG vs CAF/CTG/PRF							
Keceli et al[33] (2015)	RCT (parallel) 6 months	20 (C) / 20 (T) 40	22–50 (40.7) M: 13 / F: 27	0	I and II	Maxillary and mandibular	CAF + CTG (C) CAF + CTG + PRF (T)

Table 12-1 ***(cont)***

Authors (year)	Mean difference in RC between baseline and final follow-up (%)	Mean difference in CAL between baseline and final follow-up (mm)	Mean difference in KMW between baseline and final follow-up (mm)	Mean difference in PD between baseline and final follow-up (mm)	Methods for PRF preparation		
					Centrifugation system	Volume of blood drawn	Centrifugation parameters speed (rpm) × time (min)
CAF vs CAF/PRF							
Aroca et al[17] (2009)	91.5 ± 11.4 (C) 80.7 ± 14.7 (T)	2.56 ± 1.56 (C) 2.47 ± 1.85 (T)	−0.48 ± 1.52 (C) −0.24 ± 0.77 (T)	0.30 ± 0.69 (C) 0.24 ± 0.77 (T)	EBA 20 (Hettich)	40 mL	3000 × 10
Padma et al[18] (2013)	68.4 ± 17.4 (C) 100 ± 0.0 (T)	2.69 ± 0.36 (C) 3.75 ± 1.9 (T)	2.19 ± 1.00 (C) 2.44 ± 0.90 (T)	0.87 ± 0.66 (C) 0.31 ± 0.45 (T)	NR	10 mL	3000 × 10
Bozkurt Doğan et al[19] (2015)	82.1 ± 17.5 (C) 86.7 ± 15.6 (T)	2.58 ± 0.62 (C) 2.83 ± 0.62 (T)	0.14 ± 0.63 (C) 0.58 ± 0.63 (T)	0.29 ± 0.46 (C) 0.37 ± 0.49 (T)	Medifuge	10 mL	2700 × 2 2400 × 4 2700 × 4 3000 × 3
Gupta et al[20] (2015)	86.6 ± 23.8 (C) 91.0 ± 19.9 (T)	2.46 ± 1.33 (C) 3.26 ± 0.87 (T)	1.40 ± 0.83 (C) 1.60 ± 0.67 (T)	0.40 ± 0.56 (C) 0.73 ± 0.46 (T)	REMI	10 mL	2700 × 12
Thamaraiselvan et al[21] (2015)	65.0 ± 44.4 (C) 74.1± 29.0 (T)	1.80 ± 0.91 (C) 2.50 ± 1.17 (T)	0.40 ± 0.69 (C) 0.40 ± 0.69 (T)	0.30 ± 0.48 (C) 0.40 ± 0.51 (T)	NR	10 mL	3000 × 10
Dixit et al[22] (2018)	79.5 ± 5 (C) 82.8 ± 5.8 (T)	2.50 ± 1.53 (C) 2.42 ± 1.20 (T)	NR	NR	NR	5 mL	2700 × 12
Kuka et al[23] (2018)	74.6 ± 8.05 (C) 88.3 ± 15.4 (T)	1.74 ± 0.24 (C) 2.10 ± 0.61 (T)	0.65 ± 0.47 (C) 0.70 ± 0.42 (T)	−0.78 ± 0.34 (C) −0.65 ± 0.24 (T)	EBA 20	10 mL	3000 × 10
CAF/CTG vs CAF/PRF							
Jankovic et al[24] (2012)	92.0 ± 15.5 (C) 88.7 ± 10.7 (T)	2.96 ± 0.42 (C) 2.87 ± 0.39 (T)	1.44 ± 0.63 (C) 0.88 ± 0.71 (T)	0.16 ± 0.09 (C) 0.21 ± 0.10 (T)	NR	10 mL	3000 × 10
Eren and Atilla[25] (2014)	94.2 ± 12.1 (C) 92.7 ± 13.7 (T)	2.09 ± 0.98 (C) 2.43 ± 0.89 (T)	1.22 ± 1.87 (C) 0.93 ± 1.87 (T)	−0.36 ± 0.64 (C) −0.02 ± 0.43 (T)	NF200 (Nüve Laboratory)	10 mL	400g × 12 (rpm NR)
Tunali et al[26] (2015)	77.4 ± 17.4 (C) 76.4 ± 13.2 (T)	3.04 ± 1.69 (C) 2.7 ± 2.17 (T)	0.6 ± 0.90 (C) 0.53 ± 0.89 (T)	0.31 ± 0.64 (C) 0.15 ± 0.68 (T)	Universal 320 (Hettich)	10 mL	2700 × 12
Kumar et al[27] (2017)	53.3 ± 40.4 (C) 58.9 ± 25.9 (T1) 74.4 ± 36.7 (T2)	1.00 ± 1.06 (C) 1.20 ± 0.94 (T1) 1.73 ± 0.88 (T2)	1.14 ± 1.31 (C) 1.20 ± 0.56 (T1) 1.14 ± 0.64 (T2)	−0.40 ± 0.29 (C) 0.00 ± 0.00 (T1) 0.34 ± 1.17 (T2)	PC-02 (Process)	10 mL	2700 × 12
Mufti et al[28] (2017)	64.7 ± 37.8 (C) 51.1 ± 36.9 (T)	0.32 ± 1.14 (C) 1.25 ± 1.01 (T)	0.32 ± 0.80 (C) 0.38 ± 1.93 (T)	NR	NR	10 mL	3000 × 10
Öncü[29] (2017)	84 ± 16.3 (C) 77.1 ± 14.2 (T)	3.76 ± 1.44 (C) 3.3 ± 1.59 (T)	1.73 ± 1.17 (C) 1.1 ± 1.16 (T)	0.16 ± 0.76 (C) 0.30 ± 0.63 (T)	PC-02	9 mL	2700 × 12
Culhaoglu et al[30] (2018)	80.1 ± 18.9 (C) 69.7 ± 15.3 (T)	2.31 ± 1.07 (C) 1.65 ± 0.97 (T)	2.24 ± 1.33 (C) 0.19 ± 1.61 (T)	0.14 ± 0.34 (C) 0.16 ± 0.38 (T)	PC-02	10 mL	2700 × 12
CAF/EMD vs CAF/PRF							
Jankovic et al[31] (2010)	70.5 ± 11.8 (C) 72.1 ± 9.6 (T)	2.65 ± 1.19 (C) 3.27 ± 1.19 (T)	0.60 ± 0.98 (C) 0.17 ± 0.90 (T)	0.1 ± 0.84 (C) 0.22 ± 0.83 (T)	NR	10 mL	3000 × 10
CAF/AM vs CAF/PRF							
Agarwal et al[32] (2016)	33 ± 40.4 (C) 36 ± 25.9 (T1) 56 ± 36.7 (T2)	0.86 ± 1.70 (C) 1.20 ± 1.75 (T1) 1.46 ± 2.49 (T2)	0.87 ± 1.32 (C) 0.93 ± 1.87 (T1) 1.20 ± 2.00 (T2)	0.14 ± 0.96 (C) 0.26 ± 0.43 (T1) 0.00 ± 0.53 (T2)	REMI	10 mL	2700 × 12
CAF/CTG vs CAF/CTG/PRF							
Keceli et al[33] (2015)	79.9 ± 7.8 (C) 89.6 ± 2.68 (T)	2.50 ± 1.00 (C) 3.1 ± 0.79 (T)	0.78 ± 1.71 (C) 1.18 ± 1.89 (T)	0.0 ± 0.31 (C) 0.15 ± 0.37 (T)	Mikro 22R (Hettich)	10 mL	NR

NR, not reported; C, control group; T, test group; M, male; F, female; RC, root coverage; KMW, keratinized mucosa width; PD, probing depth.

CAF alone vs CAF/PRF

In total, nine studies investigated the use of CAF/PRF vs CAF alone.[17–23] The meta-analysis demonstrated that the addition of PRF to CAF statistically significantly improved both rRC and CAL gains (Fig 12-3; see also Fig 12-1). No advantage was found between the groups for changes in either PD or KTW. Interestingly, one study demonstrated a statistically significant improvement in rRC in the control group, in which CAF alone was used, compared to the CAF/PRF group.[17] All other studies demonstrated better improvements in rRC in the CAF/PRF group (see Table 12-1). Overall, an approximate 10% to 25% increase in rRC was observed when PRF was added to CAF (see Table 12-1). Notably, however, the addition of PRF did not improve KTW (Fig 12-4).

Data from a recent systematic review found that PRF improved root coverage and CALs when compared to CAF but did not improve keratinized mucosa width.

Reported differences in final outcomes among studies may be due to the investigated parameters. Currently, there are no guidelines with respect to using PRF for treatment of RT1 (Miller class I and II) gingival recessions in terms of the ideal thickness of PRF, the number of PRF membranes needed per site, and/or the ideal surgical technique with respect to its use. Interestingly, it was noted that one study utilized only 5 mL of total blood volume collected yet maintained the use of the standard 2700-rpm protocol for 12 minutes.[22] Changes in total blood volume utilized during the centrifugation process influence not only the final cell concentration of PRF membranes but also the centrifugation g-forces produced at the actual PRF clot. The final platelet/leukocyte concentration that results following such modifications remains poorly investigated, which may explain the lack of improvement in rRC in certain studies. The proper use of centrifugation protocols and their respective centrifugation tubes is therefore of utmost importance to avoid potential unfavorable and/or inconsistent results.

CAF/CTG vs CAF/PRF

In total, seven studies investigated the use of CAF/CTG vs CAF/PRF.[24–30] Findings from the meta-analysis revealed that a statistically significant advantage was observed in the CAF/CTG group for rRC compared to the CAF/PRF group (see Fig 12-1). Furthermore, a number of studies concluded that the use of CTG led to statistically significant increases in keratinized tissue thickness compared to the use of PRF. Thus, it may be concluded that compared to the use of PRF, the use of CTG improves the keratinization of tissues, with an increase in both KTW and rRC. Jankovic et al,[24] Öncü,[29] and Culhaoglu et al[30] all demonstrated a statistically significant advantage in KTW in the CAF/CTG group when compared to the CAF/PRF group. In the later study, only a marginal increase in KTW was observed in the CAF/PRF group (0.19 mm), whereas the use of CAF/CTG led to a 2.24-mm increase in KTW compared to the findings in controls (over a 10-fold increase). Thus, for the treatment of gingival recessions associated with limited KTW or keratinized tissue thickness, the use of CTG led to better outcomes than the use of PRF.

For gingival recessions associated with limited keratinized tissues, CTG appears to lead to better outcomes than PRF.

CAF/CTG vs CAF/CTG/PRF

Despite the widespread use of PRF, only one study has investigated its use in combination with CTG.[33] In that study, a statistically significant increase in rRC and CAL was observed, although no change in PD or KTW was observed. Future clinical studies are therefore needed to further investigate whether the addition of PRF to CTG provides additional clinical benefit.

CAF/EMD vs CAF/PRF

Only one RCT has evaluated the use of CAF in combination with EMD when compared to PRF.[31] Following a 12-month healing period, no difference was observed in any of the investigated parameters.[31]

Fig 12-3 Gingival recession treated with PRF alone. *(a)* Preoperative view. *(b)* Situation after 3 months. Notice the final esthetic outcome as well as the increased vascularization when PRF is utilized. (Case performed by Dr Alexandre Amir Aalam; reprinted with permission from Miron and Choukroun.[34])

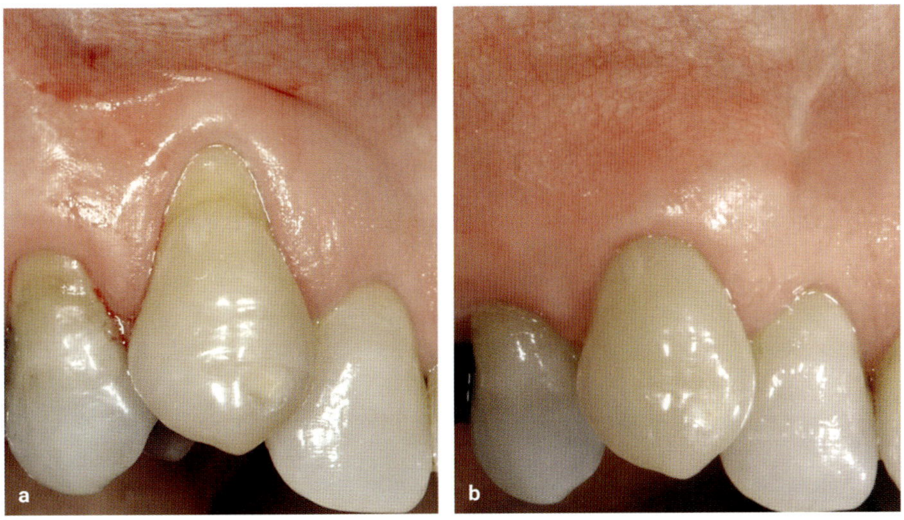

Study or Subgroup	PRF group Mean	PRF group SD	PRF group Total	Control group Mean	Control group SD	Control group Total	Weight	Mean Difference IV, Random, 95% CI	Year
1.3.1 CAF vs. CAF/PRF									
Aroca et al. 2009	0.24	0.77	67	0.48	1.52	67	7.8%	−0.24 [−0.65, 0.17]	2009
Padma et al. 2013	2.44	0.9	15	2.19	1	15	5.5%	0.25 [−0.43, 0.93]	2013
Gupta et al. 2015	1.6	0.67	15	1.4	0.83	15	6.6%	0.20 [−0.34, 0.74]	2015
Bozkurt Dogan et al. 2015	0.58	0.63	60	0.14	0.63	59	9.4%	0.44 [0.21, 0.67]	2015
Thamaraiselvan et al. 2015	0.4	0.69	10	0.4	0.69	10	6.1%	0.00 [−0.60, 0.60]	2015
Agarwal et al. 2016	1.2	2	15	0.87	1.32	15	2.7%	0.33 [−0.88, 1.54]	2016
Kumar et al. 2017	1.14	0.64	15	1.14	1.31	15	5.1%	0.00 [−0.74, 0.74]	2017
Kuka et al. 2018	0.7	0.42	28	0.65	0.47	24	9.2%	0.05 [−0.19, 0.29]	2018
Subtotal (95% CI)			**225**			**220**	**52.5%**	**0.14 [−0.06, 0.34]**	
Heterogeneity: Tau² = 0.03; Chi² = 10.90, df = 7 (P = 0.14); I² = 36%									
Test for overall effect: Z = 1.36 (P = 0.17)									
1.3.2 CAF/CTG vs. CAF/PRF									
Jankovic et al. 2012	0.88	0.71	15	1.44	0.63	15	7.2%	−0.56 [−1.04, −0.08]	2012
Eren and Atilla 2014	0.93	1.87	22	1.22	1.87	22	3.1%	−0.29 [−1.40, 0.82]	2014
Tunali et al. 2015	0.53	0.89	22	0.6	0.9	22	6.7%	−0.07 [−0.60, 0.46]	2015
Kumar et al. 2017	1.14	1.31	15	1.2	0.56	15	5.2%	−0.06 [−0.78, 0.66]	2017
Mufti et al. 2017	0.38	1.93	16	0.32	0.8	16	3.5%	0.06 [−0.96, 1.08]	2017
Öncü, 2017	1.1	1.16	30	1.73	1.17	30	6.2%	−0.63 [−1.22, −0.04]	2017
Culhaoglu et al. 2018	0.19	1.61	21	2.24	1.33	21	4.1%	−2.05 [−2.94, −1.16]	2018
Subtotal (95% CI)			**141**			**141**	**36.0%**	**−0.50 [−0.95, −0.05]**	
Heterogeneity: Tau² = 0.22; Chi² = 17.01, df = 6 (P = 0.009); I² = 65%									
Test for overall effect: Z = 2.19 (P = 0.03)									
1.3.3 CAF/EMD vs. CAF/PRF									
Jankovic et al. 2010	0.17	0.9	20	0.6	0.98	20	6.3%	−0.43 [−1.01, 0.15]	2010
Subtotal (95% CI)			**20**			**20**	**6.3%**	**−0.43 [−1.01, 0.15]**	
Heterogeneity: Not applicable									
Test for overall effect: Z = 1.45 (P = 0.15)									
1.3.4 CAF/AM vs. CAF/PRF									
Agarwal et al. 2016	1.2	2	15	0.93	1.87	15	2.2%	0.27 [−1.12, 1.66]	2016
Subtotal (95% CI)			**15**			**15**	**2.2%**	**0.27 [−1.12, 1.66]**	
Heterogeneity: Not applicable									
Test for overall effect: Z = 0.38 (P = 0.70)									
1.3.5 CAF/CTG vs. CAF/CTG/PRF									
Keceli et al. 2015	1.18	1.89	20	0.78	1.71	20	3.1%	0.40 [−0.72, 1.52]	2015
Subtotal (95% CI)			**20**			**20**	**3.1%**	**0.40 [−0.72, 1.52]**	
Heterogeneity: Not applicable									
Test for overall effect: Z = 0.70 (P = 0.48)									
Total (95% CI)			**421**			**416**	**100.0%**	**−0.12 [−0.36, 0.11]**	
Heterogeneity: Tau² = 0.14; Chi² = 50.19, df = 17 (P < 0.0001); I² = 66%									
Test for overall effect: Z = 1.04 (P = 0.30)									
Test for subgroup differences: Chi² = 9.36, df = 4 (P = 0.05), I² = 57.3%									

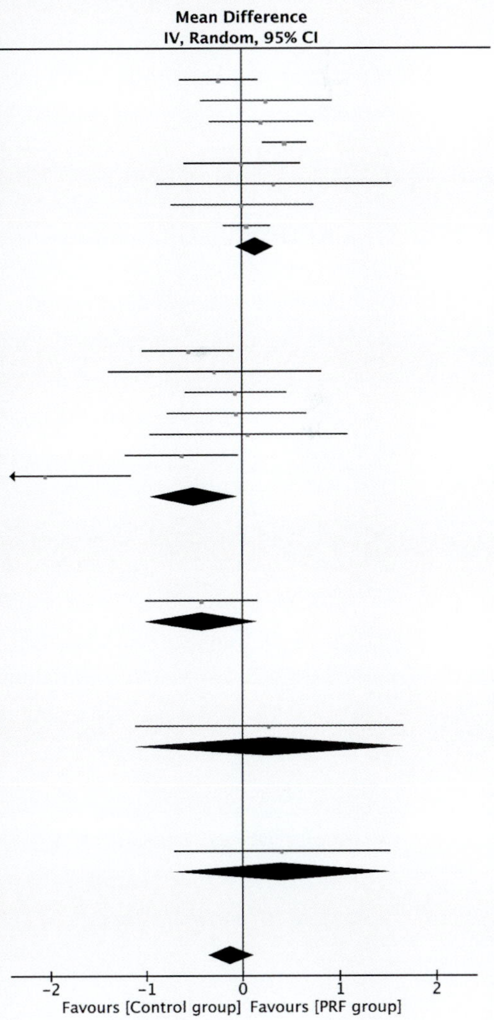

Fig 12-4 Forest plot for the event "keratinized tissue width" (reported in mm).

Table 12-2 Summary of the comparison of four outcomes between the PRF group and the other treatment modalities

The treatments that led to statistically better outcomes are shown in the boxes. All empty boxes signify no statistically significant difference between the groups.

	RC	CAL	KTW	PD
CAF vs PRF	PRF	PRF		
CTG vs PRF	CTG		CTG	
EMD vs PRF				
AM vs PRF				
CTG vs CTG/PRF	CTG/PRF	CTG/PRF		

CAF/AM vs CAF/PRF

One study compared the use of CAF/AM vs that of CAF/PRF.[32] Once again, although a 20% increase in rRC was observed in the CAF/PRF group, the large standard deviation combined with the limited number of clinical studies demonstrated no significant differences in any of the investigated parameters. Future research is needed.

Use of PRF in pain management

One interesting side finding reported in several studies was the highlighted decrease in patient morbidity/pain scores associated with the use of PRF. In total, five studies have reported advantages in the use of PRF for lowering pain scores compared to the use of CAF alone or CAF/CTG.[24,27,29,31,32] Furthermore, comparative RCTs utilizing visual analog scales have further demonstrated that the use of PRF placed within the donor site of the CTG leads to a statistically significant reduction in postoperative pain.[35–37] Therefore, the use of PRF provides an improvement in patient-reported morbidity at the donor site compared to the use of CAF alone, as well as during the harvesting of the CTG at the recipient site.

When CTGs are utilized, it has been shown that PRF placed within the donor site leads to a significant reduction in postoperative pain.

Clinical bottom line

1. The addition of CAF/PRF improved rRC and CAL when compared to CAF alone. Therefore, it should always be considered when adequate baseline KTW (2 mm) is present.
2. PRF did not improve KTW. Therefore, cases with inadequate baseline KTW should be treated with CAF/CTG or CAF/CTG/PRF.
3. There remains a lack of superiority between any of the following three groups for the treatment of recession coverage: PRF, EMD, and AM. Each of these has led to comparable results, with more studies needed to support the use of one versus the other.

See Table 12-2 for a summary of this data.

Clinical Procedure

Based on the findings presented here, the use of PRF is therefore supported when adequate baseline keratinized tissue is present. The most commonly utilized procedure with PRF is the vestibular incision subperiosteal tunnel access (VISTA) technique[38]; however, others have certainly demonstrated good results utilizing various other approaches (Fig 12-5).

Naturally, the use of a CTG harvested from the palate is utilized as a gold standard procedure due to the fact that

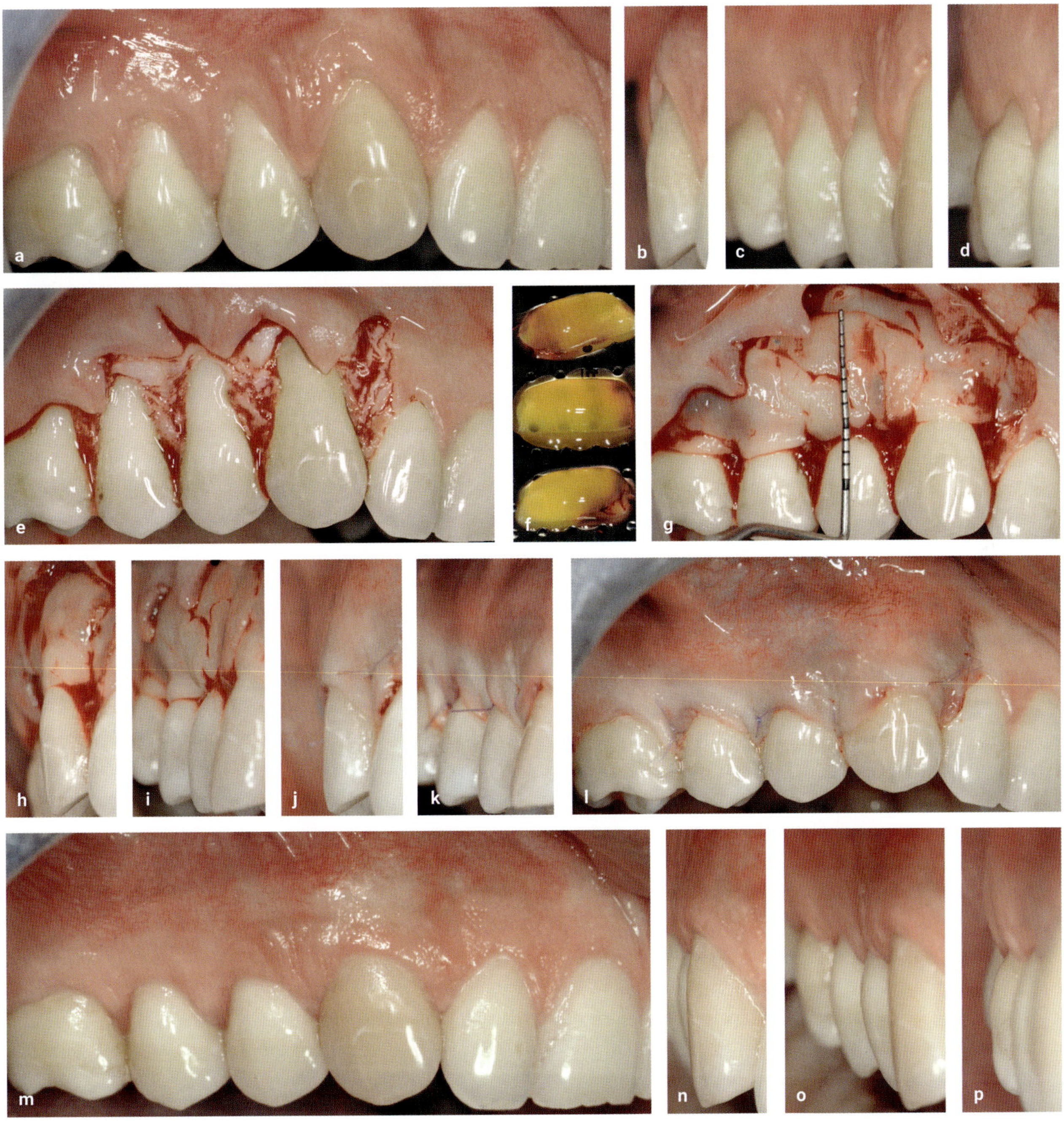

Fig 12-5 *(a to d)* Multiple gingival recessions from canine to first molar in the maxilla. *(e to l)* Surgical technique. *(e)* A flap for multiple gingival recessions has been elevated with a split-thickness approach. *(f)* A-PRF prepared. *(g)* The A-PRF has been applied to cover all teeth affected by gingival recessions. Multiple layers have been applied. *(h and i)* Lateral view showing the thickness of A-PRF material applied to the root exposures. *(j and k)* Lateral view showing the flap coronally advanced and completely covering the A-PRF material. *(l)* Frontal view showing the flap covering in excess all gingival recessions. *(m to p)* Six-month follow-up. *(m)* Complete root coverage with increase in keratinized tissue height has been achieved in all treated gingival recessions. *(n to p)* Lateral views showing the increase in gingival thickness at all teeth previously affected by gingival recessions. (Case performed by Dr Giovanni Zucchelli; reprinted with permission from Miron and Choukroun.[34])

the palate area is rich in keratinocytes/keratinized tissue. As demonstrated by Karring et al, when a CTG is transferred from the donor site to its recipient site, its genetic makeup will also be transplanted to the new area.[39] Following years of research with PRF, it is well understood today that PRF has no keratinocytes and no keratinized tissue within its fibrin matrix, and therefore the transfer of such tissues does not occur. Nevertheless, PRF does stimulate cell migration and cell proliferation. It is therefore critically important to understand that an adequate baseline KTW of 2 mm or greater is necessary when utilizing PRF alone to favor more predictable outcomes. If an existing band of KTW exists, PRF will act to further promote its angiogenesis and proliferation of cells, contributing to the potential further production of additional keratinized tissue. If, on the other hand, only a loose mucosal nonattached tissue is present, the formation of a similar poor-quality tissue should be expected. Far too often, clinicians anticipate PRF will help support root coverage procedures, but it is crucial that a basic understanding of its mechanism of action is present to lead to greater clinical success and predictability as well as long-term stability.

Step-by-step protocol

Figure 12-6 illustrates a typical step-by-step procedure utilized for full-arch cases. This protocol includes certain aspects of both the VISTA and other similar tunnel techniques.[38,40,41] A first step requires the placement of composite resin between the interproximal areas of treated teeth for sling suture attachment (Fig 12-6a). Thereafter, mechanical decontamination of the root surface is performed with both hand instruments and rotary instruments as needed (Fig 12-6b).

Root conditioning is then performed with standard materials such as EDTA or minocycline hydrochloride (Fig 12-6c). Following manufacturers' recommendations, the removal of the smear layer created by the root preparation will favor collagen fibers of the tome's tubule to be exposed and thus improve the quality of the type of attachment expected.

Vestibular access incisions are then performed 2 to 3 mm above the band of keratinized tissue, and a full-thickness flap is raised with specialized hand instruments (Fig 12-6d). The vertical mucosal incision allows for the future placement of PRF membranes and further aids in the relaxation and passive coronal displacement of tissue in future steps.

A full subperiosteal tunnel is then created to the cervical margins (Fig 12-6e). The use of either liquid-PRF or the plasma exudate from the PRF membranes is utilized to irrigate the tunnels and the roots (Fig 12-6f). Placement of PRF membranes is then introduced through the vestibular openings (Fig 12-6g). The PRF membranes are introduced and placed distally and then slowly filled mesially. At a minimum, two PRF membranes are required per pair of teeth, with three to four recommended. The addition of PRF membranes within the maxilla allows for the gradual physiologic displacement of the tissue more coronally, which will assist later with tension-free closure.

Apical mattress sutures are performed with tension-free closure, which is required[42] (Fig 12-6h). Tension-free closure provides better blood supply to the healing environment, and the apical mattress sutures avoid marginal suture tension on the PRF membranes.[43,44] Pini Prato et al demonstrated that a minimal flap tension closure ranging from 0.0 to 0.4 g favored less recession when compared to higher tension (4 to 7 g).[42] Interproximal vertical sling sutures are performed with the aim of coronally advancing the flap by roughly 3 mm above the CEJ (Fig 12-6i). Thereafter, any remaining liquid-PRF can be utilized and infiltrated within the areas to further stabilize the clot and minimize the motion/movement of PRF membranes (liquid-PRF will clot shortly upon injection; Fig 12-6j). It is recommended that a patient recall be performed 7 days postoperative (Fig 12-6k), and the sutures should be removed at day 15 (Fig 12-6l). A full case presentation is presented in Fig 12-7 and Video 12-2.

Video 12-2

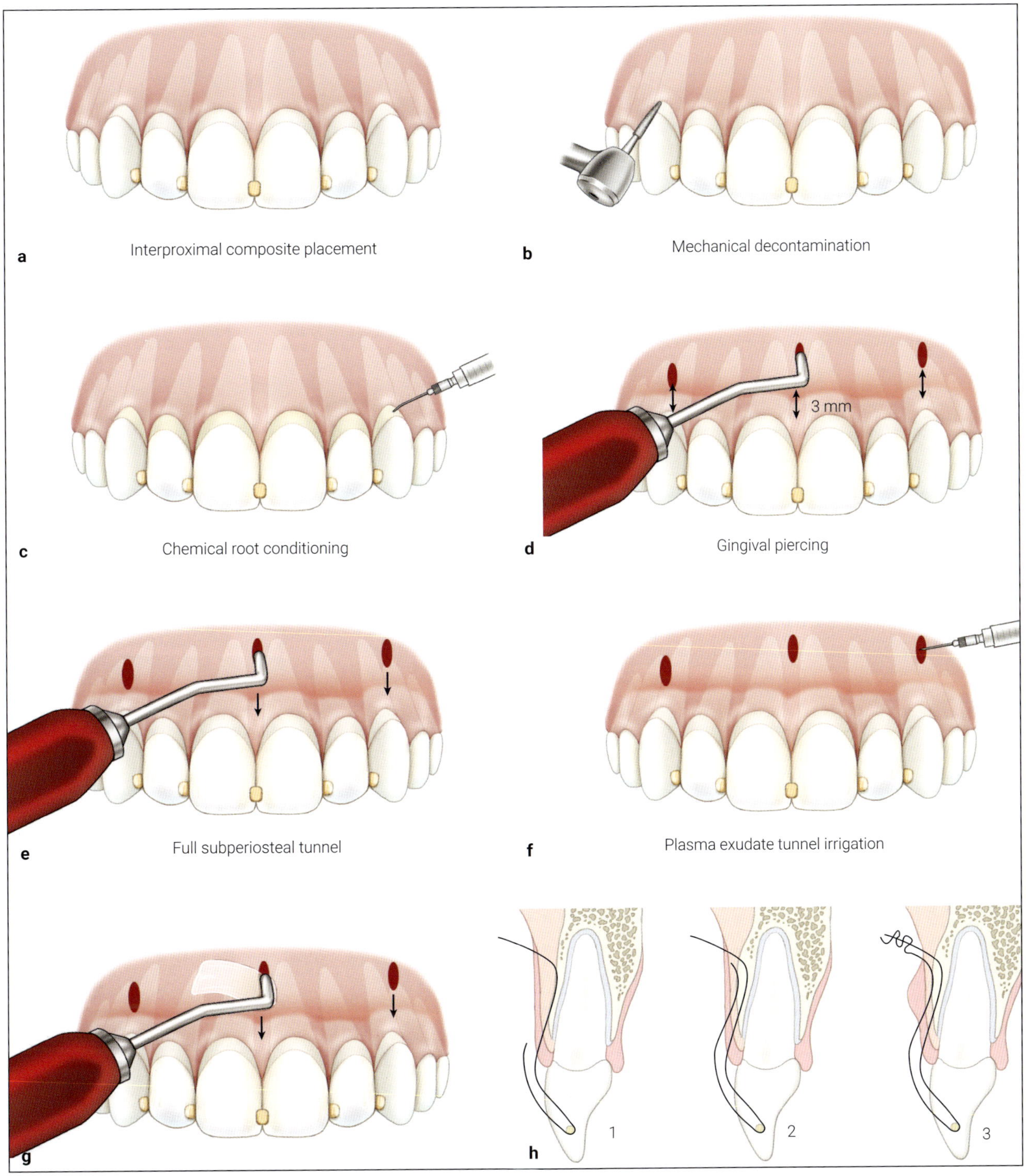

Fig 12-6 Step-by-step protocol describing the use of PRF in a CAF procedure. *(a)* Interproximal composite placement. *(b)* Mechanical decontamination of the root surface. *(c)* Chemical root conditioning. *(d)* Vestibular incision created 3 mm above the keratinized mucosa. *(e)* Creation of a full-thickness subperiosteal tunnel. *(f)* Liquid plasma irrigation of the tunnel. *(g)* Insertion of PRF membranes from a distal to mesial direction. *(h)* Apical mattress suturing technique.

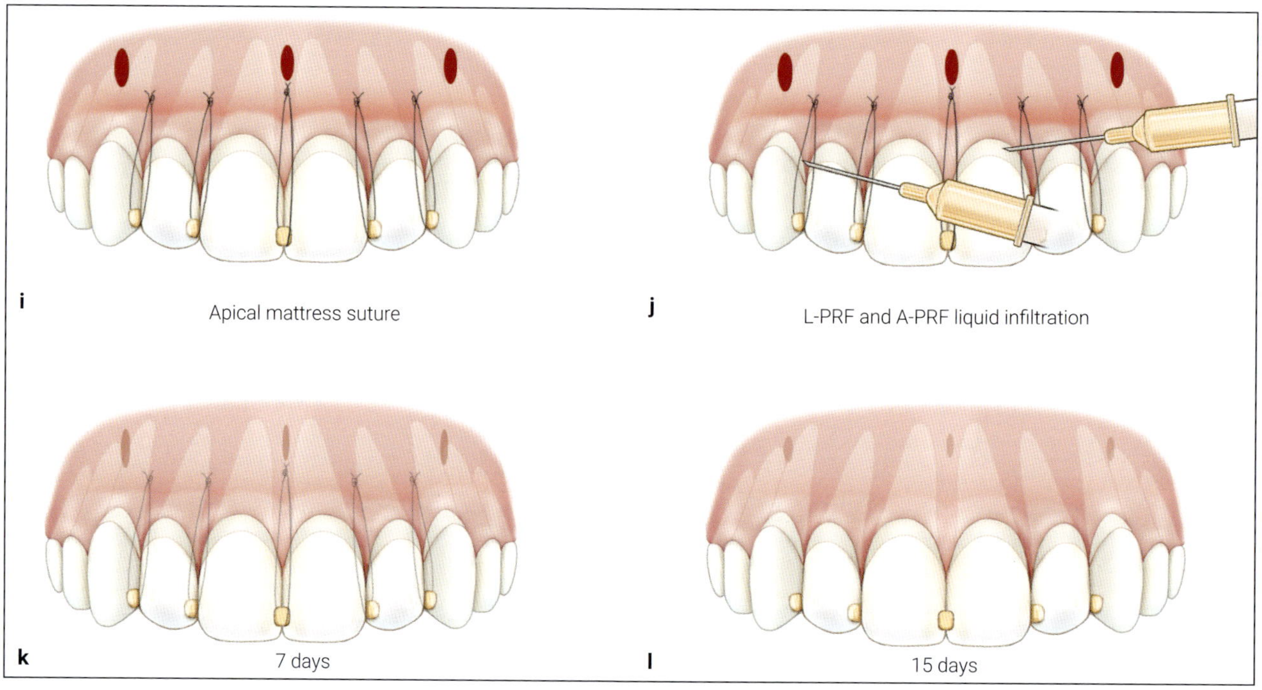

Fig 12-6 (cont) *(i)* Final apical mattress sutures with tension-free closure. *(j)* Liquid-PRF infiltration into the tunnel to stabilize the PRF membranes. *(k)* Postoperative recall at 7 days. *(l)* Suture removal at 15 days postoperative.

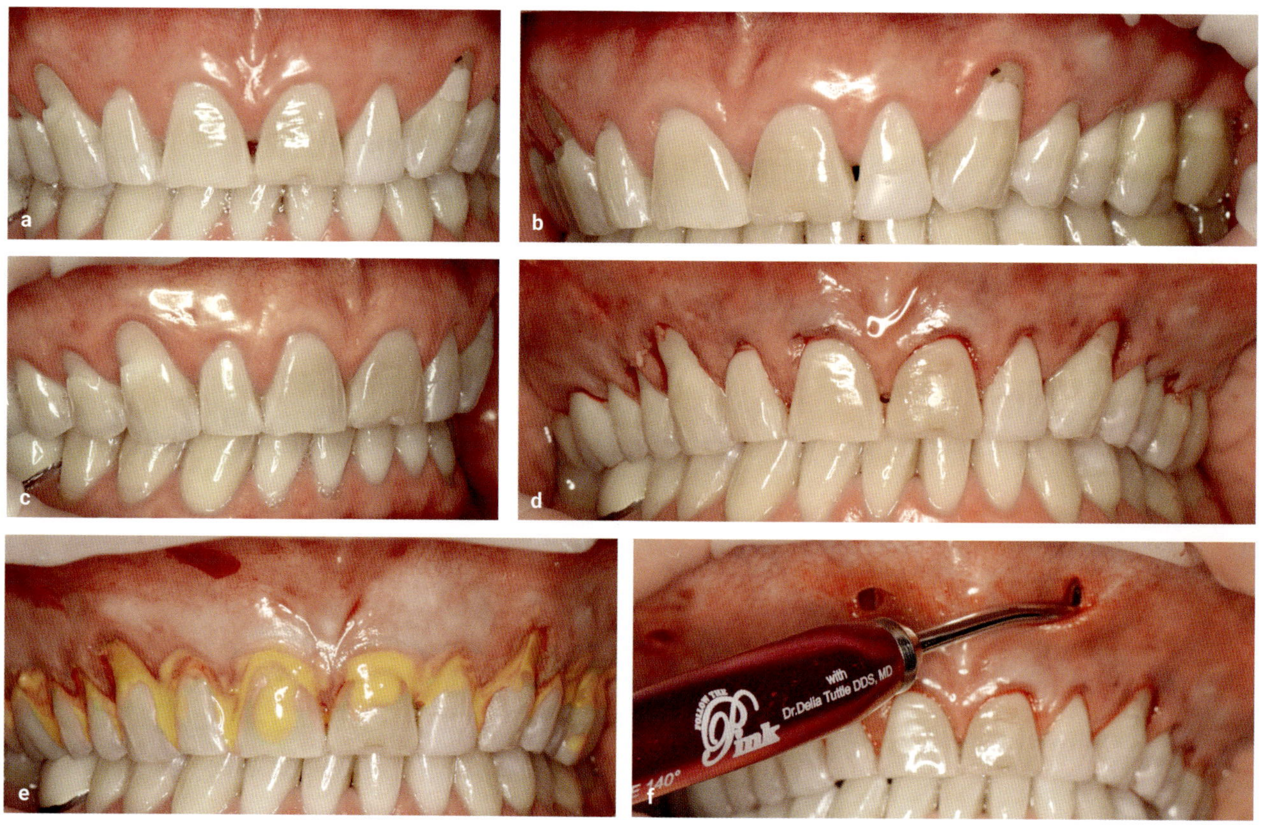

Fig 12-7 Step-by-step protocol demonstrating the clinical use of PRF in a CAF procedure. *(a to c)* Preoperative views. Notice the multiple gingival recessions in the maxillary arch. The patient elected not to have a CTG. *(d and e)* Root preparation following mechanical and chemical root conditioning, respectively. *(f)* Vestibular incision created 3 mm above the keratinized mucosa and creation of a full-thickness subperiosteal tunnel.

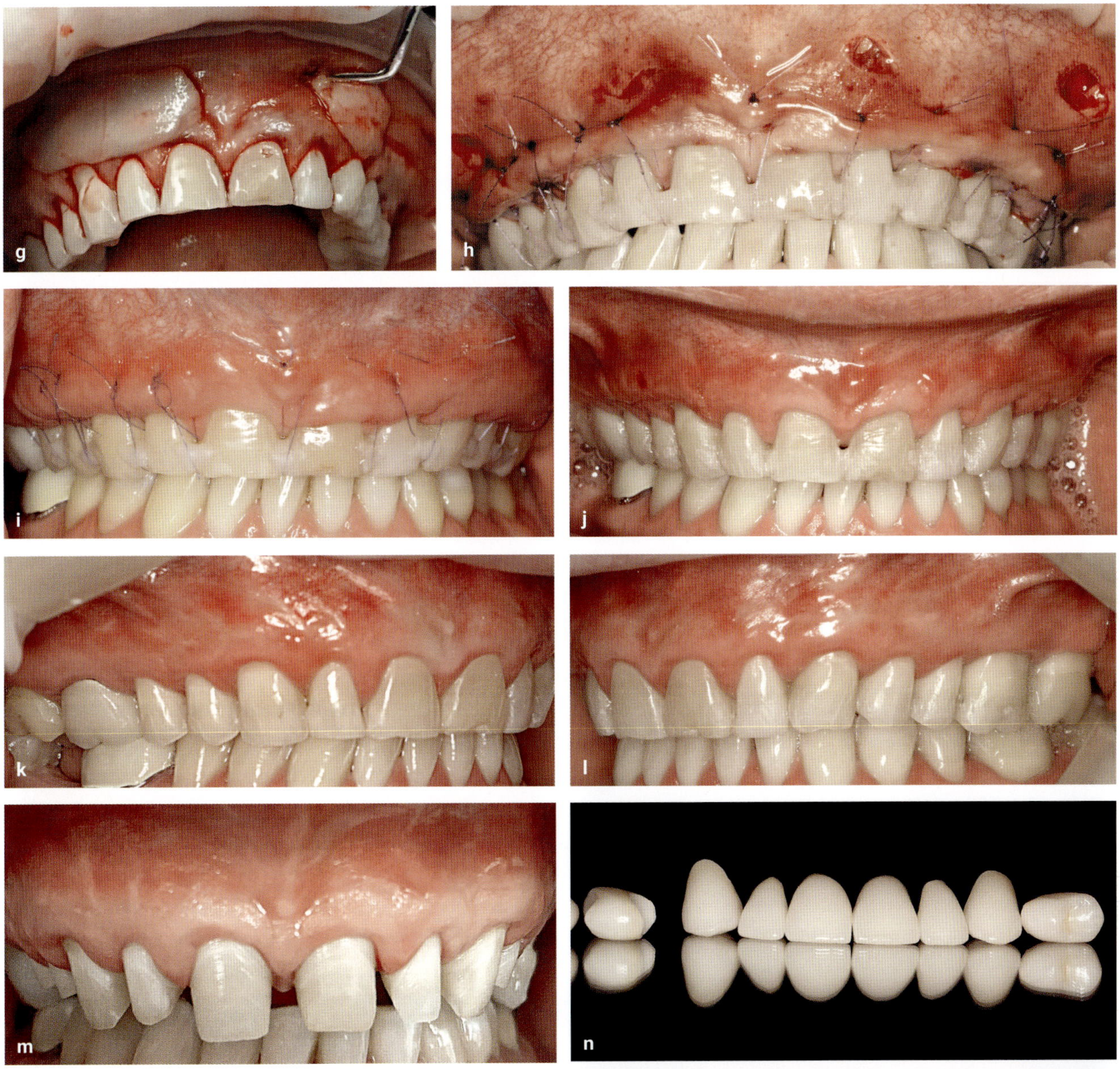

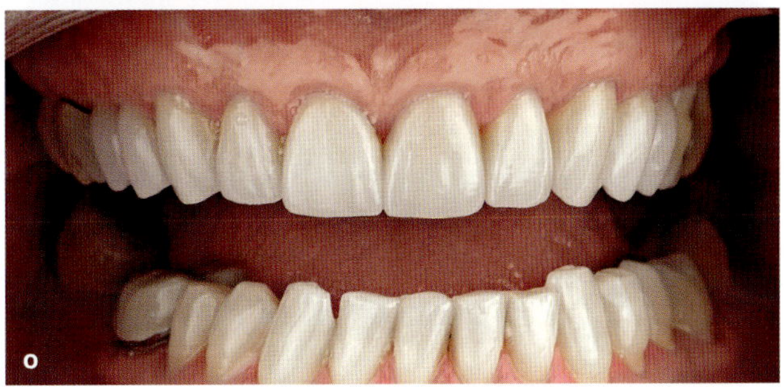

Fig 12-7 *(cont)* *(g)* Insertion of PRF membranes. *(h)* Final apical mattress sutures with tension-free closure. *(i)* Postoperative recall at 7 days. Note the excellent soft tissue wound healing. *(j)* Recall at 21 days postoperative. *(k and l)* Right and left maxillary quadrants at 9 months postoperative. *(m and n)* Crown restoration. *(o)* Final view. Note the excellent long-term maintenance at 3 years postoperative. (Case performed by Dr Delia Tuttle.)

Implications for Clinical Practice and Future Directions

In general, the use of PRF in combination with CAF procedures leads to statistically significant improvements in rRC and CAL gains compared to the use of CAF alone. Nevertheless, it is important to highlight the fact that relatively minimal change of KTW or tissue thickness was observed when PRF was utilized, which therefore highlights the fact that PRF alone is not sufficient to improve areas with deficiencies in keratinized tissues. All cases treated with PRF required adequate baseline KTW of at least 2 mm. Furthermore, it needs to be pointed out that at present there is no animal or human histologic evidence evaluating the healing type that may occur following the use of PRF in recession coverage procedures. Therefore, it is yet unknown to what extent the clinical improvements demonstrated after the use of PRF reflect a true periodontal regeneration. Additional studies are warranted to evaluate this important aspect.

All cases treated with PRF required adequate baseline KTW of at least 2 mm.

While the CTG has been the gold standard in the coverage of gingival recessions, alternative therapies are continuously desired. Although several RCTs highlighted within the present chapter demonstrated comparable results in rRC and CAL between PRF and CTG, it is important to note that the use of a CTG significantly improved KTW. Therefore, it is recommended to utilize CTGs for root coverage procedures in areas with minimal KTW.

Interestingly, there still exists no long-term data for sites grafted with PRF. Although RCTs on the topic have now been performed for over a decade, no single study has evaluated the long-term follow-up of patients beyond 18 months (see Table 12-1). This important missing data remains a priority for future research.

In addition, very rarely has surgical technique been discussed as a potential means of altering clinical outcomes. As such, variability in the results obtained has been observed, yet little discussion is generally provided by the authors regarding the surgical protocols/guidelines, which may further affect surgical outcomes when grafting with PRF. Questions as simple as "How many PRF membranes per tooth should be utilized for recession coverage?" remain vaguely answered, and substantial further research on the topic is needed. Similarly, surgical technique has been more recently discussed with respect to whether PRF should be utilized with CAF, tunneling, or vestibular incisions. While each clinical procedure offers advantages, studies performed over the upcoming decade should focus more specifically on technical guidelines for each technique while grafting with PRF.

Future studies comparing PRF with other commonly utilized biomaterials on the market are needed. For instance, it is surprising that although collagen membranes remain one of the most frequently utilized biomaterials in clinical practice to achieve root coverage, no single study has compared its clinical outcomes versus those associated with PRF. An array of studies on this topic would better address this missing information over the coming years. It has recently become common practice to harvest a liquid injectable-PRF that may be utilized to coat collagen membranes.[45] Future research aimed at investigating whether liquid-PRF improves the biocompatibility and/or regenerative potential of collagen membranes is therefore needed.

Future research comparing PRF with other regenerative agents such as EMD, AM, and rhPDGF would also be beneficial. Due to the shortage of RCTs on comparative growth factors, no differences have been reported to date among any of the above-mentioned groups. Nevertheless, being entirely autologous, PRF proves to be an alternative to commonly utilized synthetic or xenogeneic biomaterials and additionally favors greater rRC when compared to CAF alone.

Conclusion

Compared to the use of CAF alone, the use of CAF/PRF leads to statistically significantly higher rRC and CAL; however, it does not lead to substantial changes in KTW. When CAF/CTG was compared to CAF/PRF, no statistically significant difference in terms of PD and CAL were observed between the two approaches. It is important to note, however, that CAF/PRF led to a statistically significant improvement in the primary outcome of rRC as well as improvements in CAL. That being said, in cases with limited or lack of baseline KTW, the use of CTG should be preferred over PRF. No statistically significant differences in any of the investigated parameters were reported among the CAF/EMD, CAF/AM, and CAF/PRF groups. The use of PRF has been shown to improve patient-related outcomes such as postsurgical discomfort and pain. Future studies should aim to provide animal or human histologic evidence as well as the long-term effects of using PRF in recession coverage procedures.

References

1. Li R, Liu Y, Xu T, et al. The additional effect of autologous platelet concentrates to coronally advanced flap in the treatment of gingival recessions: A systematic review and meta-analysis. Biomed Res Int 2019;2019:2587245.
2. Meshram VS, Lambade PN, Meshram PV, Kadu A, Tiwari MS. The autologous platelet rich fibrin: A novel approach in osseous regeneration after cystic enucleation: A pilot study. Ind J Dent Res 2015;26:560.
3. Zucchelli G, De Sanctis M. Modified two-stage procedures for the treatment of gingival recession. Eur J Esthet Dent 2013;8:24–42.
4. Novaes AB Jr, Palioto DB. Experimental and clinical studies on plastic periodontal procedures. Periodontol 2000 2019;79:56–80.
5. Tavelli L, McGuire MK, Zucchelli G, et al. Biologics-based regenerative technologies for periodontal soft tissue engineering. J Periodontol 2020;91:147–154.
6. Jhaveri HM, Chavan MS, Tomar GB, Deshmukh VL, Wani MR, Miller PD Jr. Acellular dermal matrix seeded with autologous gingival fibroblasts for the treatment of gingival recession: A proof-of-concept study. J Periodontol 2010;81:616–625.
7. Shetty SS, Chatterjee A, Bose S. Bilateral multiple recession coverage with platelet-rich fibrin in comparison with amniotic membrane. J Indian Soc Periodontol 2014;18:102.
8. Sharma A, Yadav K. Amniotic membrane—A novel material for the root coverage: A case series. J Indian Soc Periodontol 2015;19:444.
9. Shirakata Y, Nakamura T, Shinohara Y, et al. Split-mouth evaluation of connective tissue graft with or without enamel matrix derivative for the treatment of isolated gingival recession defects in dogs. Clin Oral Investig 2019;23:3339–3349.
10. Cairo F, Nieri M, Pagliaro U. Efficacy of periodontal plastic surgery procedures in the treatment of localized facial gingival recessions. A systematic review. J Clin Periodontol 2014;41(suppl 15):S44–S62.
11. Miron RJ, Zucchelli G, Pikos MA, et al. Use of platelet-rich fibrin in regenerative dentistry: A systematic review. Clin Oral Investig 2017;21:1913–1927.
12. Najeeb S, Khurshid Z, Agwan MAS, Ansari SA, Zafar MS, Matinlinna JP. Regenerative potential of platelet rich fibrin (PRF) for curing intrabony periodontal defects: A systematic review of clinical studies. Tissue Eng Regen Med 2017;14:735–742.
13. Miron RJ, Fujioka-Kobayashi M, Bishara M, Zhang Y, Hernandez M, Choukroun J. Platelet-rich fibrin and soft tissue wound healing: A systematic review. Tissue Eng Part B Rev 2017;23:83–99.
14. Cairo F, Nieri M, Cincinelli S, Mervelt J, Pagliaro U. The interproximal clinical attachment level to classify gingival recessions and predict root coverage outcomes: An explorative and reliability study. J Clin Periodontol 2011;38:661–666.
15. Cortellini P, Bissada NF. Mucogingival conditions in the natural dentition: Narrative review, case definitions, and diagnostic considerations. J Clin Periodontol 2018;45(suppl 20):S190–S198.
16. Miller P Jr. A classification of marginal tissue recession. Int J Periodontics Restorative Dent 1985;5:8–13.
17. Aroca S, Keglevich T, Barbieri B, Gera I, Etienne D. Clinical evaluation of a modified coronally advanced flap alone or in combination with a platelet-rich fibrin membrane for the treatment of adjacent multiple gingival recessions: A 6-month study. J Periodontol 2009;80:244–252.
18. Padma R, Shilpa A, Kumar PA, Nagasri M, Kumar C, Sreedhar A. A split mouth randomized controlled study to evaluate the adjunctive effect of platelet-rich fibrin to coronally advanced flap in Miller's class I and II recession defects. J Indian Soc Periodontol 2013;17:631–636.
19. Bozkurt Doğan Ş, Ongoz Dede F, Balli U, Atalay EN, Durmuslar MC. Concentrated growth factor in the treatment of adjacent multiple gingival recessions: A split-mouth randomized clinical trial. J Clin Periodontol 2015;42:868–875.
20. Gupta S, Banthia R, Singh P, Banthia P, Raje S, Aggarwal N. Clinical evaluation and comparison of the efficacy of coronally advanced flap alone and in combination with platelet rich fibrin membrane in the treatment of Miller Class I and II gingival recessions. Contemp Clin Dent 2015;6:153–160.
21. Thamaraiselvan M, Elavarasu S, Thangakumaran S, Gadagi JS, Arthie T. Comparative clinical evaluation of coronally advanced flap with or without platelet rich fibrin membrane in the treatment of isolated gingival recession. J Indian Soc Periodontol 2015;19:66–71.
22. Dixit N, Lamba AK, Faraz F, Tandon S, Aggarwal K, Ahad A. Root coverage by modified coronally advanced flap with and without platelet-rich fibrin: A clinical study. Ind J Dent Res 2018;29:600–604.
23. Kuka S, Ipci SD, Cakar G, Yilmaz S. Clinical evaluation of coronally advanced flap with or without platelet-rich fibrin for the treatment of multiple gingival recessions. Clin Oral Investig 2018;22:1551–1558.
24. Jankovic S, Aleksic Z, Klokkevold P, et al. Use of platelet-rich fibrin membrane following treatment of gingival recession: A randomized clinical trial. Int J Periodontics Restorative Dent 2012;32:e41–e50.
25. Eren G, Atilla G. Platelet-rich fibrin in the treatment of localized gingival recessions: A split-mouth randomized clinical trial. Clin Oral Investig 2014;18:1941–1948.
26. Tunali M, Ozdemir H, Arabaciota T, Gurbuzer B, Pikdoken L, Firatli E. Clinical evaluation of autologous platelet-rich fibrin in the treatment of multiple adjacent gingival recession defects: A 12-month study. Int J Periodontics Restorative Dent 2015;35:105–114.
27. Kumar A, Bains VK, Jhingran R, Srivastava R, Madan R, Rizvi I. Patient-centered microsurgical management of gingival recession using coronally advanced flap with either platelet-rich fibrin or connective tissue graft: A comparative analysis. Contemp Clin Dent 2017;8:293–304.
28. Mufti S, Dadawala SM, Patel P, Shah M, Dave DH. Comparative evaluation of platelet-rich fibrin with connective tissue grafts in the treatment of Miller's class I gingival recessions. Contemp Clin Dent 2017;8:531–537.
29. Öncü E. The use of platelet-rich fibrin versus subepithelial connective tissue graft in treatment of multiple gingival recessions: A randomized clinical trial. Int J Periodontics Restorative Dent 2017;37:265–271.
30. Culhaoglu R, Taner L, Guler B. Evaluation of the effect of dose-dependent platelet-rich fibrin membrane on treatment of gingival recession: A randomized, controlled clinical trial. J Appl Oral Sci 2018;26:e20170278.
31. Jankovic S, Aleksic Z, Milinkovic I, Dimitrijevic B. The coronally advanced flap in combination with platelet-rich fibrin (PRF) and enamel matrix derivative in the treatment of gingival recession: A comparative study. Eur J Esthet Dent 2010;5:260–273.

32. Agarwal SK, Jhingran R, Bains VK, Srivastava R, Madan R, Rizvi I. Patient-centered evaluation of microsurgical management of gingival recession using coronally advanced flap with platelet-rich fibrin or amnion membrane: A comparative analysis. Eur J Dent 2016;10:121–133.
33. Keceli HG, Kamak G, Erdemir EO, Evginer MS, Dolgun A. The adjunctive effect of platelet-rich fibrin to connective tissue graft in the treatment of buccal recession defects: Results of a randomized, parallel-group controlled trial. J Periodontol 2015;86:1221–1230.
34. Miron RJ, Choukroun J (eds). Platelet Rich Fibrin in Regenerative Dentistry. Hoboken, NJ: Wiley, 2017.
35. Femminella B, Iaconi MC, Di Tullio M, et al. Clinical comparison of platelet-rich fibrin and a gelatin sponge in the management of palatal wounds after epithelialized free gingival graft harvest: A randomized clinical trial. J Periodontol 2016;87:103–113.
36. Lektemur Alpan A, Torumtay Cin G. PRF improves wound healing and postoperative discomfort after harvesting subepithelial connective tissue graft from palate: A randomized controlled trial. Clin Oral Investig 2020;24:425–436.
37. Ustaoglu G, Ercan E, Tunali M. The role of titanium-prepared platelet-rich fibrin in palatal mucosal wound healing and histoconduction. Acta Odontol Scand 2016;74:558–564.
38. Zadeh HH. Minimally invasive treatment of maxillary anterior gingival recession defects by vestibular incision subperiosteal tunnel access and platelet-derived growth factor BB. Int J Periodontics Restorative Dent 2011;31:653.
39. Karring T, Lang N, Löe H. The role of gingival connective tissue in determining epithelial differentiation. J Periodontal Res 1975;10:1–11.
40. Stimmelmayr M, Allen EP, Gernet W, et al. Treatment of gingival recession in the anterior mandible using the tunnel technique and a combination epithelialized-subepithelial connective tissue graft—A case series. Int J Periodontics Restorative Dent 2011;31:135–173.
41. Aalam AA, Aalam AK, Choukroun J. Novel technique in root coverage procedures: The fibrin assisted soft tissue promotion concept. Compendium 2017;38:220–225.
42. Pini Prato G, Pagliaro U, Baldi C, et al. Coronally advanced flap procedure for root coverage. Flap with tension versus flap without tension: A randomized controlled clinical study. J Periodontol 2000;71:188–201.
43. Mammoto T, Jiang A, Jiang E, Mammoto A. Platelet rich plasma extract promotes angiogenesis through the angiopoietin1-Tie2 pathway. Microvasc Res 2013;89:15–24.
44. Mammoto A, Connor KM, Mammoto T, et al. A mechanosensitive transcriptional mechanism that controls angiogenesis. Nature 2009;457:1103–1108.
45. Al-Maawi S, Herrera-Vizcaino C, Orlowska A, et al. Biologization of collagen-based biomaterials using liquid-platelet-rich fibrin: New insights into clinically applicable tissue engineering. Materials (Basel) 2019;12:3993.

13

Use of PRF for the Treatment of Intrabony and Furcation Defects

Contributors

Richard J. Miron
Vittorio Moraschini
Omid Moghaddas
Yoshinori Shirakata
Brian Mealey
Hom-Lay Wang
Anton Sculean

Chapter Highlights

- *Criteria and discussion of the therapeutic options for periodontal wound healing/regeneration*
- *Use of PRF alone for periodontal wound healing/regeneration*
- *Use of PRF in combination with bone grafts, barrier membranes, and other regenerative strategies*
- *Future research directions using PRF for the regeneration of intrabony and furcation defects*

Video 13-1

Over the past 20 years, an increasing trend has been observed in periodontology whereby the resolution(/regeneration) of intrabony and furcation defects has been accomplished by means of biologic agents and growth factors. Enamel matrix derivative (Emdogain) and rhPDGF have both been successfully utilized in regenerative periodontology due to their ability to regenerate the periodontal ligament, cementum, and alveolar bone with spanning Sharpey fibers. During periodontal regeneration, it has been well established that a stable blood clot remains a key necessity for its eventual replacement into periodontal tissues. With the advancements made in the field of PRF therapy, many years ago it was conceptualized that this "superclot" loaded with growth factors and supraphysiologic concentrations of platelets and leukocytes could be utilized as a stable fibrin clot serving as a space-making provisional matrix capable of further supporting angiogenesis. This chapter focuses on the nearly 40 RCTs investigating PRF for periodontal wound healing/regeneration. Discussion over the use of PRF alone versus in combination with bone grafts, growth factors, barrier membranes, as well as local therapeutic drugs such as metformin, statins, bisphosphonates, and antibiotics are explored as potential future use applications of PRF as a loading agent for biomolecule delivery.

The periodontium is a complex functional unit derived from several tissues responsible for the connection of teeth with their surrounding bone. This is accomplished by collagen Sharpey fibers that span from the root cementum through the periodontal ligament and attach to bundle alveolar bone[1,2] (Fig 13-1). The periodontium provides a flexible defense system with various host cells responsible for the maintenance of structural integrity of the tooth apparatus. It has been well described in the literature that failure to prevent infection by periodontal pathogens may cause gingivitis and, without treatment, may lead to the development of periodontitis and the eventual loss of teeth.[1]

Periodontal disease is one of the most prevalent diseases known to man. It begins as a superficial inflammation of the gingiva without attachment or bone loss (gingivitis) and later progresses to attachment loss with subsequent bone destruction (periodontitis). Results investigating the distribution of the disease from a national survey conducted in the United States found that over 47% of the adult population was affected.[3] Furthermore, it was found that 38.5% of the population had either a moderate or severe case (stage III or stage IV) of periodontitis. This finding is most alarming because the disease is characterized by an exponentially more difficult resolution and regeneration once advanced progression has taken place.

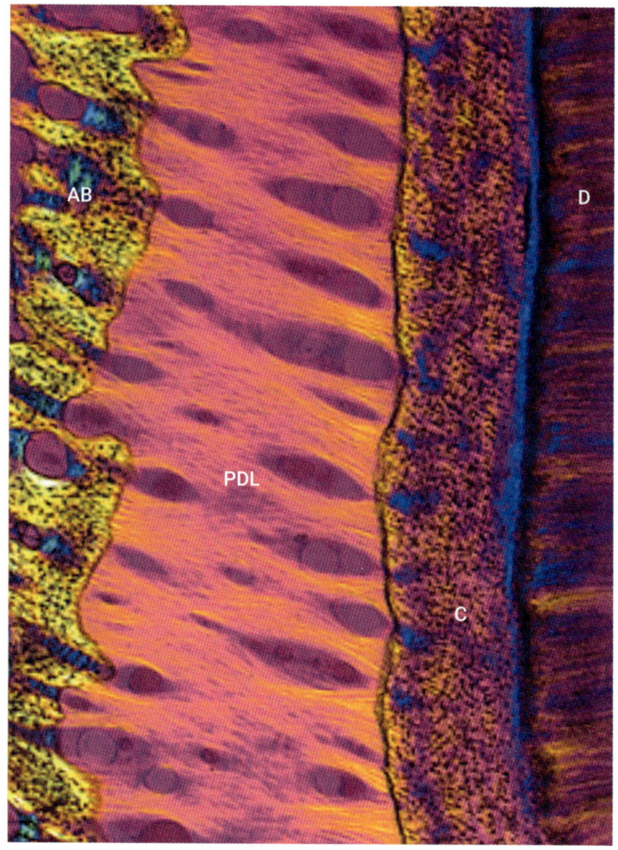

Fig 13-1 Micrograph illustrating the periodontal ligament (PDL) with its collagen fiber bundles spanning between the root covered with cementum (C) and the alveolar bone (AB). D, dentin. Undecalcified ground section, unstained and viewed under polarized light. (Reprinted with permission from Bosshardt et al.[2])

Surveys and epidemic studies have shown that periodontitis plays a pivotal role in a number of systemic diseases and conditions, including cardiovascular diseases (heart attack/stroke), Alzheimer disease, diabetes, obesity, and premature birth, among others.[4] It therefore becomes vital that dental practitioners and health care providers alike be aware of the characterization and disease progression, with more research placed on the regeneration of these tissues that have been lost due to periodontal disease.[5–7] Because true periodontal regeneration comprises regeneration of not only the periodontal ligament but also the surrounding alveolar bone, cementum, as well as the overlying soft tissues including new connective and epithelial tissues, complete periodontal regeneration remains complex, with much future research still needed to fulfill these criteria predictably. This chapter presents regenerative strategies that have been utilized over the years including the use of barrier membranes and bone grafting materials in guided tissue regeneration (GTR) procedures. We then highlight how growth factors and biologic agents have more recently become a preferred choice for intrabony and furcation defect resolution. Platelet concentrates are presented with a large focus on the growing number of RCTs that now support the use of PRF for the repair/regeneration of intrabony and furcation defects.

Treatment of periodontal disease is of utmost importance because it is linked to a number of systemic conditions, including heart attack, stroke, Alzheimer disease, diabetes, obesity, and premature birth, among others.

Role of Barrier Membranes in Periodontal Regeneration

The use of barrier membranes for GTR represents one of the first modalities adapted for periodontal regeneration. It was originally believed that in order to optimize periodontal regeneration, cells from periodontal ligament tissue and alveolar bone should be exclusively separated from the overlying soft tissues (epithelium or fibrous connective tissue) in order to allow repopulation of the periodontal defects. In 1982, Nyman et al first introduced this concept using a cellulose acetate laboratory filter by Millipore to act as a barrier separating periodontal structures from overlying soft tissues.[8] Today, a variety of barrier membranes derived from several sources have been developed that have biocompatibility with host tissues, space-making ability, mechanical strength, and ideal degradation properties, which are all requirements for GTR treatment of periodontal intrabony or furcation defects.

Although findings from initial studies utilizing barrier membranes confirmed greater clinical attachment levels (CALs),[9] a second surgery was required to remove the membrane and was shown to compromise some of the beneficial effects obtained during the regenerative procedure.[10] Because of these drawbacks, collagen resorbable membranes have since been more frequently utilized.[11] Today, a wide variety of resorbable barrier membranes are available, with various reported degradation rates and fabrication sources most suitable for various defects and cases. Still, complete periodontal regeneration is not possible utilizing barrier membranes alone.

Role of Bone Grafting Materials in Periodontal Regeneration

When intrabony/furcation defects reach greater sizes, often a bone grafting material is needed to prevent flap collapse and support the space maintenance requirements for regeneration of periodontal tissues. Originally, grafts were developed to serve as a passive, structural supporting network with the main criterion being biocompatibility, but advancements made in the field of tissue engineering have introduced a large array of materials, each possessing various advantages and disadvantages. Today, grafting procedures are extremely common and routine in implant dentistry, and as the population continues to age, conditions such as osteoporosis, arthritis, tumors, or trauma continue to necessitate the use of bone grafts.[12]

Bone grafts are typically characterized into four groups: autografts, allografts, xenografts, and alloplasts. While autogenous bone is considered the gold standard because of its osteoconductive, osteoinductive, and osteogenic potential,[13] its main drawback includes its limited supply, specifically for intrabony defect regeneration when small flaps are raised. Therefore, bone allograft harvested from another human cadaver has been the most common alternative. Fresh-frozen bone, freeze-dried bone allograft (FDBA), and demineralized freeze-dried bone allograft (DFDBA) have all been utilized successfully for the regeneration of intrabony/furcation defects.[14–17] Because certain countries do not permit the use of allografts, xenografts derived from various animal sources as well as laboratory-made synthetic alloplasts have also been widely used.[18]

Biologic Agents/Growth Factors for Periodontal Regeneration

The use of biologic agents such as growth factors to promote periodontal regeneration has increased tremendously over the past two decades. A variety of novel research in the mid 1990s focused on developing delivery systems for growth factors accordingly to support periodontal regeneration. Because the regeneration of periodontal tissues is much more complex than most tissues due mainly to the fact that it comprises many tissues/cell types from different embryonic origins, a variety of biologic agents including enamel matrix derivative (EMD), rhPDGF-BB, recombinant human fibroblast growth factor-2 (rhFGF-2), BMPs (BMP-2 and BMP-7), teriparatide parathyroid hormone (PTH), and growth differential factor-5 (GDF-5), PRP, and PRF have all been investigated for periodontal regeneration.[18]

The understanding of root cementum development and formation led to the development of EMD, where the proteins comprised within EMD, mainly amelogenins, were aimed at mimicking normal tissue development.[19–23] Similarly, rhPDGF was granted FDA approval for both medical and dental purposes and has been the first such growth factor of its kind.[24,25] Its main action was derived from the main regenerative properties of platelets (hence its name *platelet-derived growth factor*). As reviewed in chapter 2,

PDGF aims to promote the rapid cell migration to wound sites, and research has further shown its ability to promote intrabony defect resolution (or regeneration) in human RCTs.[26–37] Teriparatide PTH, rhFGF-2, and GDF-5 have also all been attempted for periodontal regeneration and shown some promising results, but future studies in this area are needed before these materials can be widely used.[38–40] Studies also examined the effect of BMP-2 and BMP-7 for periodontal regeneration, but due to their osteoinductive potential, they have not yet been utilized in humans.[41,42]

Evaluation of Clinical Studies Using PRF for Intrabony Defect Regeneration

Early experiments revealed that PRP could be utilized to facilitate periodontal regeneration of either intrabony or furcation defects. Nevertheless, PRF has two distinct advantages when compared to PRP.[43,44] First, because anticoagulants are not utilized, a fibrin scaffold is found with entrapment of cells and growth factors. This favors the slower and more gradual release of growth factors over time (see chapter 2). More importantly, because a stable clot is required for intrabony defect regeneration, PRF provides a ready-made "superclot" that can be implanted into periodontal defects. Second, PRF contains leukocytes and macrophages, known cell types implicated in immunity and host defense.[45,46] Because periodontal defects are the result of invading bacterial pathogens, the inclusion of white blood cells contained within PRF is hypothesized to further act as a bacteria-resistant matrix capable of fighting bacterial pathogens.

Because periodontal defects are the result of invading bacterial pathogens, the inclusion of white blood cells contained within PRF is hypothesized to further act as a bacteria-resistant matrix capable of fighting bacterial pathogens.

In total, 38 RCTs have investigated the use of PRF for either intrabony or furcation defect resolution (Tables 13-1 and 13-2).[47–85] These are characterized as either comparing open flap debridement (OFD) vs PRF, bone graft (BG) vs PRF, BG vs BG/PRF, collagen membrane vs PRF, PRP vs PRF, EMD vs PRF, EMD vs EMD/PRF, PRF vs PRF+metformin, PRF vs PRF+bisphosphonates, and PRF vs PRF+statins. Next, we highlight and discuss results from a recent systematic review with meta-analysis characterizing the strengths and limitations of utilizing PRF for periodontal wound healing (regeneration).

OFD alone vs with PRF

In total, 14 studies have evaluated the use of PRF as an adjunct to standard OFD when compared to OFD alone[47–54] (Table 13-1 and Figs 13-2 and 13-3). In summary, 13 of the 14 studies demonstrated significant improvements in mean probing depth (PD) reduction, with a mean difference of 1.26 mm (see Fig 13-2); 11 of the 14 studies demonstrated significant improvements in mean CAL gain, with a mean difference of −1.39 mm (see Fig 13-3); and all studies showed significant improvement in bone fill (data not shown). In summary, it was observed that on average the results from the 14 RCTs demonstrated a significant relative PD reduction of ~1.3 mm and CAL gain of ~1.5 mm when PRF was additionally filled into intrabony defects following OFD (see Figs 13-2 and 13-3).

Bone graft vs PRF

In a second series of studies, five studies evaluated the use of a BG versus PRF (see Figs 13-2 and 13-3).[55–59] In general, no significant difference was found between the two groups. One study demonstrated significantly better results for PRF,[59] and one study demonstrated significantly more favorable outcomes for BG.[58] The three remaining studies demonstrated no differences between the groups. The meta-analysis also demonstrated no differences in PD reduction, CAL gain, or bone fill (BF) between the two groups. Therefore, it may be concluded that PRF leads to similar outcomes when compared to BG for intrabony defect repair/regeneration.

Table 13-1 Characteristics of the RCTs evaluating intrabony defects

Authors (year)	Study design Follow-up	No. of participants Sex Mean age	Groups	Bone defect type	Smokers (Y, N)	Conclusions
OFD vs PRF						
Sharma and Pradeep[47] (2011)	RCT (parallel) 9 months	35 M: 24 / F: 18 35	C: 28, OFD T: 28, OFD+PRF	3 walls	N	There was greater PD reduction, CAL gain, and BF at sites treated with PRF+OFD compared to OFD alone.
Thorat et al[48] (2011)	RCT (parallel) 9 months	32 M: 20 / F: 12 31	C: 16, OFD T: 16, OFD+PRF	2 and 3 walls	N	There was greater reduction in PD, more CAL gain, and greater intrabony defect (IBD) fill at sites treated with PRF than the OFD alone.
Rosamma et al[49] (2014)	RCT (split-mouth) 12 months	15 M: 6 / F: 9 29	C: 15, OFD T: 15, OFD+PRF	3 walls	N	The use of PRF is more effective than OFD alone in the management of IBDs.
Ajwani et al[50] (2015)	RCT (split-mouth) 9 months	20 M: 10 / F: 10 31	C: 20, OFD T: 20, OFD+PRF	2 and 3 walls	N	Adjunctive use of PRF with OFD significantly improves defect fill when compared to OFD alone.
Bajaj et al[51] (2017)	RCT (parallel) 9 months	17 M: 9 / F: 8 29.7	C: 27, OFD T: 27, OFD+PRF	2 and 3 walls	N	There was greater BF at sites treated with PRF with conventional OFD than conventional OFD alone.
Patel et al[52] (2017)	RCT (split-mouth) 12 months	13 M: 4 / F: 9 44	C: 13, OFD T: 13, OFD+PRF	2 and 3 walls	N	The adjunctive use of PRF with conventional OFD may be potentially used in the treatment of IBDs.
Pradeep et al[53] (2017)	RCT (parallel) 9 months	62 M: 34 / F: 28 39.7	C: 18, OFD T1: 19, OFD+PRF T2: 20, OFD+PRF+BG	3 walls	N	Treatment of IBDs with PRF results in significant improvements of clinical parameters compared to baseline.
Thorat et al[54] (2017)	RCT (split-mouth) 12 months	15 M: 7 / F: 8 25	C: 15, OFD T: 15, OFD+PRF	3 walls	NR	Use of PRF significantly enhances the clinical and radiographic outcomes of OFD in the treatment of IBDs.
BG vs PRF						
Shah et al[55] (2014)	RCT (split-mouth) 6 months	20 NR NR	C: 20, OFD+DFDBA T: 20, OFD+PRF	2 and 3 walls	N	PRF has shown significant results after 6 months, which is comparable to DFDBA for periodontal regeneration.
Mathur et al[56] (2015)	RCT (parallel) 6 months	25 M: 14 / F: 11 NR	C: 19, OFD+ABG T: 19, OFD+PRF	2 and 3 walls	N	The use of either PRF or autogenous bone graft (ABG) were effective in the treatment of IBDs.
Chadwick et al[57] (2016)	RCT (parallel) 6 months	36 M: 20 / F: 16 55	C: 19, OFD+DFDBA T: 17, OFD+PRF	2 and 3 walls	Y	Treatment of IBDs with either DFDBA or PRF resulted in a significant gain in CAL as well as BF after 6 months of healing, with no significant difference.
Galav et al[58] (2016)	RCT (split-mouth) 9 months	20 NR 45	C: 20, OFD+ABG T: 20, OFD+PRF	2 and 3 walls	N	Both ABG and PRF can be used predictably to reconstruct lost periodontal structures.
Yajamanya et al[59] (2017)	RCT (parallel) 9 months	90 NR NR	C: 28, OFD T1: 28, OFD+BG T2: 28, OFD+PRF	2 and 3 walls	N	This study shows marked improvements in the clinical parameters and radiographic outcomes with both BG and PRF to treat periodontal IBDs as compared to OFD alone.
BG vs BG+PRF						
Bansal and Bharti[60] (2013)	RCT (split-mouth) 6 months	10 NR NR	C: 10, OFD+DFDBA T: 10, OFD+DFDBA+PRF	NR	NR	There was a significantly greater PD reduction and CAL when PRF was added to DFDBA.
Elgendy and Abo Shady[61] (2015)	RCT (split-mouth) 6 months	20 NR 44	C: 20, OFD+BG T: 20, OFD+BG+PRF	NR	Y	Both treatment groups showed a significant PD reduction and CAL gain 6 months after surgery. However, there was a significantly greater PD reduction and CAL gain when PRF was added to BG.

OFD, open flap debridement; BG, bone graft; NR, not reported; C, control group; T, test group, M, male; F, female; CAL, clinical attachment level; PD, probing depth; BF, bone fill.

Table 13-1 *(cont)* Characteristics of the RCTs evaluating intrabony defects

Authors (year)	Study design Follow-up	No. of participants Sex Mean age	Groups	Bone defect type	Smokers (Y, N)	Conclusions
Agarwal et al[62] (2016)	RCT (split-mouth) 12 months	30 M: 14 / F: 16 52	C: 30, OFD + DFDBA T: 30, OFD+DFDBA/PRF	2 and 3 walls	N	The combination of PRF and DFDBA is more effective than DFDBA alone.
Naqvi et al[63] (2017)	RCT (split-mouth) 9 months	10 M: 14 / F: 11 NR	C: 10, OFD+BG T: 10, OFD+BG+PRF	2 and 3 walls	N	The results of this study showed both BG putty alone and the combination of PRF and BG putty are effective in the treatment of IBDs.
Sezgin et al[64] (2017)	RCT (split-mouth) 6 months	15 M: 8 / F: 7 46	C: 15, OFD+ABBM T: 15, OFD+ABBM+PRF	2 and 3 walls	N	The results of this study indicate that both therapies are effective in the treatment of IBDs.
Bodhare et al[65] (2019)	RCT (split-mouth) 6 months	20 M: 11 / F: 9 36	C: 20, OFD+BG T: 20, OFD+BG+PRF	2 and 3 walls	N	BG when used in combination with PRF is found to be more effective in gain in CAL, reduction in PD, and achieving greater BF as compared to treatment with BG alone.
Barrier membrane vs PRF						
Panda et al[66] (2016)	RCT (split-mouth) 9 months	18 M: 10 / F: 8 38	C: 18, OFD+BM T: 18, OFD+PRF+BM	3 walls	N	The adjunctive use of PRF in combination with a barrier membrane is more effective in the treatment of IBDs in chronic periodontitis as compared with use of a barrier membrane alone.
PRP vs PRF						
Pradeep et al[67] (2012)	RCT (parallel) 9 months	54 M: 27 / F: 27 36.8	C: 17, OFD T1: 16, OFD+PRP T2: 17, OFD+PRF	3 walls	N	There was similar PD reduction, CAL gain, and BF at sites treated with PRF or PRP with conventional OFD.
EMD vs PRF						
Gupta et al[68] (2014)	RCT (parallel) 6 months	30 M: 15 / F: 15 NR	C: 22, OFD+EMD T: 19, OFD+PRF	3 walls	N	Both EMD and PRF were effective in the regeneration of IBDs.
EMD vs EMD+PRF						
Aydemir Turkal et al[69] (2016)	RCT (split-mouth) 6 months	28 M: 14 / F: 14 38.5	C: 24, OFD+EMD T: 25, OFD+EMD+PRF	2 and 3 walls	N	The addition of PRF did not improve the clinical and radiographic outcomes.
PRF vs PRF+MF (metformin)						
Pradeep et al[70] (2015)	RCT (parallel) 9 months	120 M: 60 / F: 60 41	C: 30, OFD T1: 30, OFD+1% MF T2: 30, OFD+PRF T3: 30, OFD+1% MF+PRF	3 walls	N	The study showed that the PRF group was more effective than OFD alone in the management of IBDs.
PRF vs PRF+ALN (alendronate)						
Kanoriya et al[71] (2016)	RCT (parallel) 9 months	90 M: 43 / F: 47 40	C: 30, OFD T1: 30, OFD+PRF T2: 30, OFD+PRF+ 1% ALN	2 and 3 walls	N	Combined approach therapy of PRF+1% ALN for IBDs treatment showed better clinical parameter outcomes compared with PRF and OFD alone.
PRF vs PRF+statins						
Martande et al[72] (2016)	RCT (parallel) 9 months	90 M: 48 / F: 48 37.6	C: 30, OFD T1: 30, OFD+PRF T2: 30, OFD+PRF+1.2% ATV	3 walls	N	PRF+1.2% ATV (atorvastatin) showed similar improvements in clinical parameters with a greater percentage radiographic defect depth reduction compared with PRF alone in treatment of IBDs.
Pradeep et al[73] (2016)	RCT (parallel) 9 months	90 M: 45 / F: 45 35	C: 30, OFD T1: 30, OFD+PRF T2: 30, OFD+PRF+1.2% RSV	2 and 3 walls	N	OFD with 1.2% RSV (rosuvastatin) and PRF results in significantly greater periodontal benefits compared with OFD alone or with PRF.

Table 13-1 *(cont)*

				Methods for PRF preparation		
Authors (year)	Mean difference in PD between baseline and final follow-up (mm)	Mean difference in CAL between baseline and final follow-up (mm)	Mean difference in BF between baseline and final follow-up (mm)	Centrifugation system	Volume of blood drawn	Centrifugation parameters speed (rpm) × time (min)
OFD vs PRF						
Sharma and Pradeep[47] (2011)	3.21 ± 1.64 (C) 4.55 ± 1.87 (T)	2.77 ± 1.44 (C) 3.31 ± 1.76 (T)	0.09 ± 0.11 (C) 2.5 ± 0.78 (T)	R-4C	10 mL	3000 × 10
Thorat et al[48] (2011)	3.56 ± 1.09 (C) 4.69 ± 1.45 (T)	2.13 ± 1.71 (C) 4.13 ± 1.63 (T)	1.24 ± 0.69 (C) 2.12 ± 0.69 (T)	NR	10 mL	2700 × 12
Rosamma et al[49] (2014)	2.40 ± 0.63 (C) 4.67 ± 0.90 (T)	1.40 ± 1.06 (C) 4.73 ± 0.88 (T)	0.64 ± 0.50 (C) 1.93 ± 1.07 (T)	KW-70 (Almicro Instruments)	10 mL	3000 × 10
Ajwani et al[50] (2015)	1.60 ± 0.84 (C) 1.90 ± 0.74 (T)	1.30 ± 0.68 (C) 1.80 ± 0.63 (T)	0.80 ± 0.35 (C) 1.45 ± 0.50 (T)	R-4C	10 mL	3000 × 10
Bajaj et al[51] (2017)	2.14 ± 1.26 (C) 3.14 ± 1.26 (T)	1.59 ± 1.01 (C) 2.66 ± 1.07 (T)	0.84 ± 0.99 (C) 2.24 ± 0.66 (T)	R-4C	10 mL	3000 × 10
Patel et al[52] (2017)	2.40 ± 0.84 (C) 4.20 ± 1.69 (T)	2.10 ± 0.74 (C) 3.70 ± 0.67 (T)	NR	R-8C (REMI)	15 mL	3000 × 10
Pradeep et al[53] (2017)	2.97 ± 0.93 (C) 3.90 ± 1.09 (T1) 4.27 ± 0.98 (T2)	2.67 ± 1.09 (C) 3.03 ± 1.16 (T1) 3.67 ± 1.03 (T2)	0.93 ± 0.83 (C) 3.20 ± 0.89 (T1) 3.87 ± 1.33 (T2)	R-4C	10 mL	3000 × 10
Thorat et al[54] (2017)	1.50 ± 0.34 (C) 4.00 ± 0.63 (T)	0.33 ± 1.21 (C) 4.00 ± 0.63 (T)	1.67 ± 0.06 (C) 3.09 ± 0.50 (T)	R-4C	10 mL	3000 × 12
BG vs PRF						
Shah et al[55] (2014)	3.70 ± 0.68 (C) 3.67 ± 0.69 (T)	2.97 ± 1.68 (C) 2.97 ± 1.56 (T)	0.32 ± 1.59 (C) 0.42 ± 1.38 (T)	NR	10 mL	3000 × 10
Mathur et al[56] (2015)	2.40 ± 1.06 (C) 2.67 ± 1.29 (T)	2.67 ± 1.63 (C) 2.53 ± 1.06 (T)	2.66 ± 1.84 (C) 2.93 ± 1.79 (T)	R-4C	NR	3000 × 10
Chadwick et al[57] (2016)	2.00 ± 1.37 (C) 2.12 ± 1.41 (T)	1.16 ± 1.33 (C) 1.03 ± 0.86 (T)	1.53 ± 1.64 (C) 1.35 ± 1.69 (T)	Centrifuge 1310 (Salvin)	10 mL	3000 × 10
Galav et al[58] (2016)	4.80 ± 0.57 (C) 4.10 ± 0.63 (T)	4.50 ± 0.52 (C) 3.90 ± 0.37 (T)	4.10 ± 0.47 (C) 4.59 ± 0.70 (T)	NR	10 mL	3000 × 10
Yajamanya et al[59] (2017)	3.68 ± 0.72 (C) 5.57 ± 1.10 (T1) 6.11 ± 0.92 (T2)	4.14 ± 0.76 (C) 6.57 ± 1.45 (T1) 6.74 ± 1.55 (T2)	NR	NR	10 mL	3000 × 10
BG vs BG+PRF						
Bansal and Bharti[60] (2013)	3.10 ± 0.74 (C) 4.00 ± 0.82 (T)	2.30 ± 0.70 (C) 3.40 ± 0.60 (T)	1.93 ± 1.21 (C) 2.13 ± 1.28 (T)	NR	10 mL	3000 × 10
Elgendy and Abo Shady[61] (2015)	3.33 ± 0.36 (C) 3.30 ± 0.18 (T)	3.55 ± 0.13 (C) 3.50 ± 0.06 (T)	NR	NR	10 mL	3000 × 10
Agarwal et al[62] (2016)	3.60 ± 0.51 (C) 4.15 ± 0.84 (T)	2.61 ± 0.68 (C) 3.73 ± 0.74 (T)	2.49 ± 0.64 (C) 3.50 ± 0.67 (T)	NR	10 mL	400g × 12

OFD, open flap debridement; BG, bone graft; NR, not reported; C, control group; T, test group, M, male; F, female; BM, barrier membrane; ABBM, anorganic bovine bone mineral; CAL, clinical attachment level; PD, probing depth; BF, bone fill.

Table 13-1 *(cont)* Characteristics of the RCTs evaluating intrabony defects

				Methods for PRF preparation		
Authors (year)	**Mean difference in PD between baseline and final follow-up (mm)**	**Mean difference in CAL between baseline and final follow-up (mm)**	**Mean difference in BF between baseline and final follow-up (mm)**	**Centrifugation system**	**Volume of blood drawn**	**Centrifugation parameters speed (rpm) × time (min)**
Naqvi et al[63] (2017)	3.15 ± 1.06 (C) 3.20 ± 2.30 (T)	3.15 ± 1.06 (C) 4.10 ± 1.73 (T)	5.70 ± 1.37 (C) 7.10 ± 1.37 (T)	NR	10 mL	3000 × 10
Sezgin et al[64] (2017)	4.21 ± 1.21 (C) 4.93 ± 1.22 (T)	3.27 ± 1.34 (C) 4.47 ± 1.60 (T)	1.98 ± 0.80 (C) 2.55 ± 1.15 (T)	PC-02 (Process)	10 mL	2700 × 12
Bodhare et al[65] (2019)	5.65 ± 1.66 (C) 5.75 ± 1.16 (T)	4.20 ± 1.70 (C) 5.05 ± 1.09 (T)	2.56 ± 0.95 (C) 3.51 ± 1.17 (T)	R-4C	10 mL	3000 × 10
Barrier membrane vs PRF						
Panda et al[66] (2016)	3.19 ± 1.33 (C) 3.88 ± 1.15 (T)	3.38 ± 1.45 (C) 4.44 ± 1.50 (T)	0.80 ± 0.28 (C) 2.10 ± 0.64 (T)	R-4C (REMI)	5 mL	3000 × 10
PRP vs PRF						
Pradeep et al[67] (2012)	2.97 ± 0.93 (C) 3.77 ± 1.07 (T1) 3.77 ± 1.19 (T2)	2.83 ± 0.91 (C) 2.93 ± 1.08 (T1) 3.17 ± 1.29 (T2)	0.13 ± 1.46 (C) 2.70 ± 0.79 (T1) 2.80 ± 0.89 (T2)	R-4C	3 × 10 mL	3000 × 10
EMD vs PRF						
Gupta et al[68] (2014)	1.80 ± 0.56 (C) 1.80 ± 0.77 (T)	2.00 ± 0.54 (C) 1.87 ± 0.91 (T)	2.08 ± 0.77 (C) 1.67 ± 1.17 (T)	R-4C	10 mL	3000 × 12
EMD vs EMD+PRF						
Aydemir Turkal et al[69] (2016)	3.88 ± 1.26 (C) 4.00 ± 1.38 (T)	3.29 ± 1.30 (C) 3.42 ± 1.28 (T)	1.13 ± 0.50 (C) 1.21 ± 0.68 (T)	Mikro 22 R (Hettich)	10 mL	400 × 10
PRF vs PRF+MF						
Pradeep et al[70] (2015)	3.00 ± 0.18 (C) 3.93 ± 0.25 (T1) 4.00 ± 0.18 (T2) 4.90 ± 0.30 (T3)	2.96 ± 0.18 (C) 3.93 ± 0.25 (T1) 4.03 ± 0.18 (T2) 4.90 ± 0.30 (T3)	0.49 ± 0.27 (C) 2.56 ± 0.28 (T1) 2.53 ± 0.30 (T2) 2.77 ± 0.30 (T3)	R-4C	10 mL	3000 × 10
PRF vs PRF+ALN						
Kanoriya et al[71] (2016)	2.86 ± 0.68 (C) 3.70 ± 0.91 (T1) 4.53 ± 0.81 (T2)	3.03 ± 0.18 (C) 4.20 ± 0.66 (T1) 5.16 ± 0.46 (T2)	0.38 ± 0.26 (C) 2.42 ± 0.21 (T1) 2.84 ± 0.26 (T2)	R-4C	10 mL	3000 × 10
PRF vs PRF+statins						
Martande et al[72] (2016)	2.76 ± 1.43 (C) 3.76 ± 1.12 (T1) 4.06 ± 1.22 (T2)	2.50 ± 1.33 (C) 3.40 ± 1.13 (T1) 3.66 ± 1.42 (T2)	0.27 ± 0.19 (C) 2.46 ± 0.33 (T1) 2.58 ± 0.36 (T2)	R-4C	10–15 mL	3000 × 12–14
Pradeep et al[73] (2016)	3.10 ± 0.30 (C) 4.03 ± 0.18 (T1) 4.90 ± 0.31 (T2)	2.47 ± 0.77 (C) 3.30 ± 0.65 (T1) 3.93 ± 0.78 (T2)	1.43 ± 0.50 (C) 3.17 ± 0.65 (T1) 3.63 ± 0.67 (T2)	R-4C	10 mL	3000 × 10

Study or Subgroup	PRF Mean	PRF SD	PRF Total	Control Mean	Control SD	Control Total	Weight	Mean Difference IV, Random, 95% CI
1.1.1 OFD vs. PRF								
Sharma and Pradeep (2011)	4,55	1,87	28	3,21	1,64	28	2,2%	1,34 [0,42, 2,26]
Thorat et al. (2011)	4,69	1,45	16	3,56	1,09	16	2,2%	1,13 [0,24, 2,02]
Pradeep et al. (2012)	3,77	1,19	17	2,97	0,93	17	2,7%	0,80 [0,08, 1,52]
Rosamma et al. (2014)	4,67	0,9	15	2,4	0,63	15	3,1%	2,27 [1,71, 2,83]
Ajwani et al. (2015)	1,9	0,74	20	1,6	0,84	20	3,3%	0,30 [−0,19, 0,79]
Pradeep et al. (2015)	4	0,18	30	3	0,18	30	4,2%	1,00 [0,91, 1,09]
Kanoriya et al. (2016)	3,7	0,91	30	2,86	0,68	30	3,6%	0,84 [0,43, 1,25]
Martande et al. (2016)	3,76	1,12	30	2,76	1,43	30	2,9%	1,00 [0,35, 1,65]
Pradeep et al. (2016)	4,03	0,18	30	3,1	0,3	30	4,1%	0,93 [0,80, 1,06]
Bajaj et al. (2017)	3,14	1,26	27	2,14	1,26	27	2,8%	1,00 [0,33, 1,67]
Patel et al. (2017)	4,2	1,69	13	2,4	0,84	13	1,9%	1,80 [0,77, 2,83]
Pradeep et al. (2017)	3,9	1,09	19	2,97	0,93	18	2,9%	0,93 [0,28, 1,58]
Thorat et al. (2017)	4	0,63	15	1,5	0,34	15	3,7%	2,50 [2,14, 2,86]
Yajamanya et al. (2017)	5,57	1,1	28	3,68	0,72	28	3,3%	1,89 [1,40, 2,38]
Subtotal (95% CI)			**318**			**317**	**42,8%**	**1,26 [0,99, 1,53]**
Heterogeneity: Tau² = 0,18; Chi² = 109,25, df = 13 (P < 0.00001); I² = 88%								
Test for overall effect: Z = 9,10 (P < 0.00001)								
1.1.2 BG vs. PRF								
Mathur et al. (2015)	2,67	1,29	19	2,4	1,06	19	2,6%	0,27 [−0,48, 1,02]
Shah et al. (2015)	3,67	0,69	20	3,7	0,68	20	3,5%	−0,03 [−0,45, 0,39]
Chadwick et al. (2016)	2,12	1,41	17	2	1,37	19	2,2%	0,12 [−0,79, 1,03]
Galav et al. (2016)	4,1	0,63	20	4,8	0,57	20	3,6%	−0,70 [−1,07, −0,33]
Yajamanya et al. (2017)	6,11	0,92	28	5,57	1,1	28	3,2%	0,54 [0,01, 1,07]
Subtotal (95% CI)			**104**			**106**	**15,1%**	**0,00 [−0,50, 0,50]**
Heterogeneity: Tau² = 0,23; Chi² = 16,70, df = 4 (P = 0,002); I² = 76%								
Test for overall effect: Z = 0,01 (P = 0,99)								
1.1.3 BG vs. BG/PRF								
Bansal and Bharti (2013)	4	0,82	10	3,1	0,74	10	2,8%	0,90 [0,22, 1,58]
Elgendy and Abo Shady (2015)	3,3	0,18	20	3,33	0,36	20	4,1%	−0,03 [−0,21, 0,15]
Agarwal et al. (2016)	4,15	0,84	30	3,6	0,51	30	3,7%	0,55 [0,20, 0,90]
Naqvi et al. (2017)	3,2	2,3	10	3,15	1,06	10	1,1%	0,05 [−1,52, 1,62]
Sezgin et al. (2017)	4,93	1,22	15	4,21	1,21	15	2,3%	0,72 [−0,15, 1,59]
Bodhare et al. (2019)	5,75	1,16	20	5,65	1,66	20	2,2%	0,10 [−0,79, 0,99]
Subtotal (95% CI)			**105**			**105**	**16,2%**	**0,38 [−0,02, 0,77]**
Heterogeneity: Tau² = 0,13; Chi² = 15,25, df = 5 (P = 0,009); I² = 67%								
Test for overall effect: Z = 1,88 (P = 0,06)								
1.1.4 Barrier membrane vs. PRF								
Panda et al. (2016)	3,88	1,15	18	3,19	1,33	18	2,4%	0,69 [−0,12, 1,50]
Subtotal (95% CI)			**18**			**18**	**2,4%**	**0,69 [−0,12, 1,50]**
Heterogeneity: Not applicable								
Test for overall effect: Z = 1,66 (P = 0,10)								
1.1.5 PRP vs. PRF								
Pradeep et al. (2012)	3,77	1,19	17	3,77	1,07	16	2,5%	0,00 [−0,77, 0,77]
Subtotal (95% CI)			**17**			**16**	**2,5%**	**0,00 [−0,77, 0,77]**
Heterogeneity: Not applicable								
Test for overall effect: Z = 0,00 (P = 1,00)								
1.1.6 EMD vs. PRF								
Gupta et al. (2014)	1,8	0,77	19	1,8	0,56	22	3,5%	0,00 [−0,42, 0,42]
Subtotal (95% CI)			**19**			**22**	**3,5%**	**0,00 [−0,42, 0,42]**
Heterogeneity: Not applicable								
Test for overall effect: Z = 0,00 (P = 1,00)								
1.1.7 EMD vs. EMD + PRF								
Aydemir Turkal et al. (2016)	4	1,38	25	3,88	1,26	24	2,6%	0,12 [−0,62, 0,86]
Subtotal (95% CI)			**25**			**24**	**2,6%**	**0,12 [−0,62, 0,86]**
Heterogeneity: Not applicable								
Test for overall effect: Z = 0,32 (P = 0,75)								
1.1.8 PRF vs. PRF + Metformin								
Pradeep et al. (2015)	4,9	0,3	30	4	0,18	30	4,1%	0,90 [0,77, 1,03]
Subtotal (95% CI)			**30**			**30**	**4,1%**	**0,90 [0,77, 1,03]**
Heterogeneity: Not applicable								
Test for overall effect: Z = 14,09 (P < 0.00001)								
1.1.9 PRF vs. PRF + Bisphosphonates								
Kanoriya et al. (2016)	4,53	0,81	30	3,7	0,91	30	3,5%	0,83 [0,39, 1,27]
Subtotal (95% CI)			**30**			**30**	**3,5%**	**0,83 [0,39, 1,27]**
Heterogeneity: Not applicable								
Test for overall effect: Z = 3,73 (P = 0,0002)								
1.1.10 PRF vs. PRF + Statins								
Martande et al. (2016)	4,06	1,22	30	3,76	1,12	30	3,0%	0,30 [−0,29, 0,89]
Pradeep et al. (2016)	4,9	0,31	30	4,03	0,18	30	4,1%	0,87 [0,74, 1,00]
Subtotal (95% CI)			**60**			**60**	**7,1%**	**0,66 [0,12, 1,20]**
Heterogeneity: Tau² = 0,11; Chi² = 3,39, df = 1 (P = 0,07); I² = 71%								
Test for overall effect: Z = 2,41 (P = 0,02)								
Total (95% CI)			**726**			**728**	**100,0%**	**0,73 [0,54, 0,93]**
Heterogeneity: Tau² = 0,24; Chi² = 363,05, df = 32 (P < 0.00001); I² = 91%								
Test for overall effect: Z = 7,37 (P < 0.00001)								
Test for subgroup differences: Chi² = 49,48, df = 9 (P < 0.00001), I² = 81,8%								

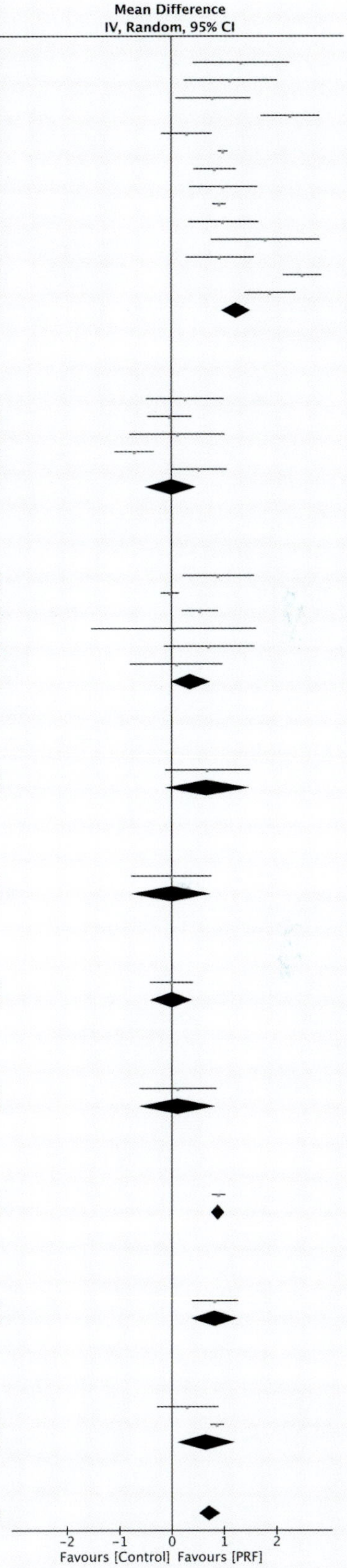

Fig 13-2 Forest plot for the event periodontal "pocket depth" reduction of intrabony defects (reported as PD in mm).

	PRF			Control				Mean Difference
Study or Subgroup	**Mean**	**SD**	**Total**	**Mean**	**SD**	**Total**	**Weight**	**IV, Random, 95% CI**
2.1.1 OFD vs. PRF								
Sharma and Pradeep (2011)	3,31	1,76	28	2,77	1,44	28	2,7%	0,54 [-0,30, 1,38]
Thorat et al. (2011)	4,13	1,63	16	2,13	1,71	16	2,2%	2,00 [0,84, 3,16]
Pradeep et al. (2012)	3,17	1,29	17	2,83	0,91	17	2,9%	0,34 [-0,41, 1,09]
Rosamma et al. (2014)	4,73	0,88	15	1,4	1,06	15	3,0%	3,33 [2,63, 4,03]
Ajwani et al. (2015)	1,8	0,63	20	1,3	0,68	20	3,4%	0,50 [0,09, 0,91]
Pradeep et al. (2015)	4,03	0,18	30	2,96	0,18	30	3,7%	1,07 [0,98, 1,16]
Kanoriya et al. (2016)	4,2	0,66	30	3,03	0,18	30	3,6%	1,17 [0,93, 1,41]
Martande et al. (2016)	3,4	1,13	30	2,5	1,33	30	3,1%	0,90 [0,28, 1,52]
Pradeep et al. (2016)	3,3	0,65	30	2,47	0,77	30	3,5%	0,83 [0,47, 1,19]
Bajaj et al. (2017)	2,66	1,07	27	1,59	1,01	27	3,2%	1,07 [0,51, 1,63]
Patel et al. (2017)	3,7	0,67	13	2,1	0,74	13	3,2%	1,60 [1,06, 2,14]
Pradeep et al. (2017)	3,03	1,16	19	2,67	1,09	18	2,9%	0,36 [-0,36, 1,08]
Thorat et al. (2017)	4	0,63	15	0,33	1,21	15	3,0%	3,67 [2,98, 4,36]
Yajamanya et al. (2017)	6,74	1,55	28	4,14	0,76	28	3,0%	2,60 [1,96, 3,24]
Subtotal (95% CI)			**318**			**317**	**43,2%**	**1,39 [1,03, 1,76]**
Heterogeneity: Tau² = 0,39; Chi² = 140,27, df = 13 (P < 0.00001); I² = 91%								
Test for overall effect: Z = 7,51 (P < 0.00001)								
2.1.2 BG vs. PRF								
Mathur et al. (2015)	2,53	1,06	19	2,67	1,63	19	2,7%	-0,14 [-1,01, 0,73]
Shah et al. (2015)	2,97	1,56	20	2,97	1,68	20	2,4%	0,00 [-1,00, 1,00]
Chadwick et al. (2016)	1,03	0,86	17	1,16	1,33	19	2,9%	-0,13 [-0,85, 0,59]
Galav et al. (2016)	3,9	0,37	20	4,5	0,52	20	3,5%	-0,60 [-0,88, -0,32]
Yajamanya et al. (2017)	6,57	1,45	28	6,74	1,55	28	2,8%	-0,17 [-0,96, 0,62]
Subtotal (95% CI)			**104**			**106**	**14,3%**	**-0,45 [-0,68, -0,22]**
Heterogeneity: Tau² = 0,00; Chi² = 3,60, df = 4 (P = 0,46); I² = 0%								
Test for overall effect: Z = 3,81 (P = 0,0001)								
2.1.3 BG vs. BG/PRF								
Bansal and Bharti (2013)	3,4	0,6	10	2,3	0,7	10	3,2%	1,10 [0,53, 1,67]
Elgendy and Abo Shady (2015)	3,5	0,06	20	3,55	0,13	20	3,7%	-0,05 [-0,11, 0,01]
Agarwal et al. (2016)	3,73	0,74	30	2,61	0,68	30	3,5%	1,12 [0,76, 1,48]
Naqvi et al. (2017)	4,1	1,73	10	3,15	1,06	10	2,0%	0,95 [-0,31, 2,21]
Sezgin et al. (2017)	4,47	1,6	15	3,27	1,34	15	2,4%	1,20 [0,14, 2,26]
Bodhare et al. (2019)	5,05	1,09	20	4,2	1,7	20	2,6%	0,85 [-0,04, 1,74]
Subtotal (95% CI)			**105**			**105**	**17,3%**	**0,82 [0,14, 1,51]**
Heterogeneity: Tau² = 0,59; Chi² = 64,57, df = 5 (P < 0.00001); I² = 92%								
Test for overall effect: Z = 2,35 (P = 0,02)								
2.1.4 Barrier membrane vs. PRF								
Panda et al. (2016)	4,44	1,5	18	3,38	1,45	18	2,5%	1,06 [0,10, 2,02]
Subtotal (95% CI)			**18**			**18**	**2,5%**	**1,06 [0,10, 2,02]**
Heterogeneity: Not applicable								
Test for overall effect: Z = 2,16 (P = 0,03)								
2.1.5 PRP vs. PRF								
Pradeep et al. (2012)	3,17	1,29	17	2,93	1,08	16	2,8%	0,24 [-0,57, 1,05]
Subtotal (95% CI)			**17**			**16**	**2,8%**	**0,24 [-0,57, 1,05]**
Heterogeneity: Not applicable								
Test for overall effect: Z = 0,58 (P = 0,56)								
2.1.6 EMD vs. PRF								
Gupta et al. (2014)	1,87	0,91	19	2	0,54	22	3,3%	-0,13 [-0,60, 0,34]
Subtotal (95% CI)			**19**			**22**	**3,3%**	**-0,13 [-0,60, 0,34]**
Heterogeneity: Not applicable								
Test for overall effect: Z = 0,55 (P = 0,59)								
2.1.7 EMD vs. EMD + PRF								
Aydemir Turkal et al. (2016)	3,42	1,28	25	3,29	1,3	24	2,9%	0,13 [-0,59, 0,85]
Subtotal (95% CI)			**25**			**24**	**2,9%**	**0,13 [-0,59, 0,85]**
Heterogeneity: Not applicable								
Test for overall effect: Z = 0,35 (P = 0,72)								
2.1.8 PRF vs. PRF + Metformin								
Pradeep et al. (2015)	4,9	0,3	30	4,03	0,18	30	3,6%	0,87 [0,74, 1,00]
Subtotal (95% CI)			**30**			**30**	**3,6%**	**0,87 [0,74, 1,00]**
Heterogeneity: Not applicable								
Test for overall effect: Z = 13,62 (P < 0.00001)								
2.1.9 PRF vs. PRF + Bisphosphonates								
Kanoriya et al. (2016)	5,16	0,46	30	4,2	0,66	30	3,5%	0,96 [0,67, 1,25]
Subtotal (95% CI)			**30**			**30**	**3,5%**	**0,96 [0,67, 1,25]**
Heterogeneity: Not applicable								
Test for overall effect: Z = 6,54 (P < 0.00001)								
2.1.10 PRF vs. PRF + Statins								
Martande et al. (2016)	3,66	1,42	30	3,4	1,13	30	3,0%	0,26 [-0,39, 0,91]
Pradeep et al. (2016)	3,93	0,79	30	3,3	0,65	30	3,4%	0,63 [0,26, 1,00]
Subtotal (95% CI)			**60**			**60**	**6,5%**	**0,54 [0,22, 0,86]**
Heterogeneity: Tau² = 0,00; Chi² = 0,95, df = 1 (P = 0,33); I² = 0%								
Test for overall effect: Z = 3,32 (P = 0,0009)								
Total (95% CI)			**726**			**728**	**100,0%**	**0,84 [0,57, 1,11]**
Heterogeneity: Tau² = 0,52; Chi² = 830,29, df = 32 (P < 0.00001); I² = 96%								
Test for overall effect: Z = 6,10 (P < 0.00001)								
Test for subgroup differences: Chi² = 133,72, df = 9 (P < 0.00001), I² = 93,3%								

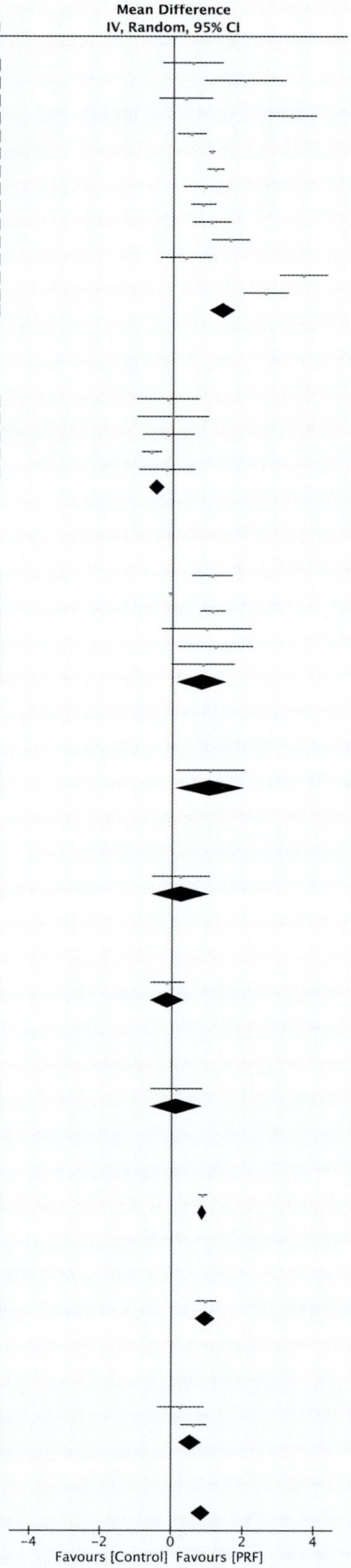

Fig 13-3 Forest plot for the event periodontal "clinical attachment level" gains of intrabony defects (reported as CAL in mm).

Bone graft vs bone graft with PRF

In a third series of investigated studies, six studies evaluated the additional use of PRF to BG when compared to BG alone (see Figs 13-2 and 13-3).[60–65] Of the six studies, two demonstrated a significant improvement in PD and CAL gain when compared to BG alone,[60,62] while the other four studies demonstrated no significant difference.[61,63–65] A final reported nonsignificant reduction in PD (~0.5 mm) was observed. When investigating CAL gain, two studies demonstrated a significant advantage, whereas three others demonstrated nonsignificant improvements in CAL. Following meta-analysis, it was found that the additional use of PRF to BG led to a statistically significant 1-mm gain in CAL when compared to BG alone and also significant improvements in BF. Therefore, some clinical benefit was observed when PRF was combined with a bone grafting material.

Additional RCTs evaluating PRF

An additional eight studies evaluated the use of PRF in various RCTs as highlighted herein.[66–73] No meta-analysis could be performed, but general trends were reported (see Figs 13-2 and 13-3). The comparison investigating PRF vs a collagen barrier membrane led to no significant difference in PD; however, significant improvements were observed for CAL and BF favoring the PRF group when compared to collagen membranes.[66] Interestingly, no differences in any of the investigated parameters were observed for single RCTs investigating *(1)* PRP vs PRF,[67] *(2)* EMD vs PRF,[68] or *(3)* EMD vs EMD+PRF,[69] so no advantages were noted. Future studies are needed in this area.

Lastly, four studies have investigated PRF in combination with either metformin,[70] bisphosphonates,[71] or statins.[72,73] There was a significant advantage in PD reduction, CAL gain, and BF for the combined use of PRF with each of the above modalities when compared to PRF alone. Therefore, it appears that future research investigating PRF as a drug-delivery system for various local therapeutic agents/biomolecules may provide further clinical benefit. At present, however, the above trends are simply reported in single RCTs with no meta-analysis being performed due to the low number of studies.

PRF may provide benefit as a drug-delivery system for various local therapeutic agents and biomolecules.

Clinical bottom line

1. PRF significantly improved PD, CAL gain, and BF when compared to OFD alone.
2. No difference was reported when comparing PRF to a BG for intrabony defect resolution.
3. Mild improvements were observed when PRF was added to a BG, including significant improvements in CAL gain and BF.
4. No difference was observed when comparing PRF to EMD either utilized as individual strategies or in combination.
5. The addition of various drugs such as metformin, bisphosphonates, and statins to PRF all demonstrated significant advantages when compared to PRF alone.

Clinical Procedure

The general grafting procedure utilizing PRF for periodontal wound healing/regeneration has been favored primarily for narrow vertical intrabony defects when space maintenance is not a key requirement (Fig 13-4). When defects are significantly larger, typically a bone grafting material may be utilized in combination with PRF, favoring better CAL gain and BF to prevent flap collapse.

When PRF is to be utilized alone, typically a flap is reflected followed by defect debridement (see Fig 13-4c). A PRF membrane is then cut into small portions and placed within the defect with slight overfill (see Fig 13-4d). Note that the bottom portion of the PRF membrane should be utilized because it incorporates more cells from within the buffy coat region (see chapter 2). Once applied, PRF is then typically compressed with a moist gauze. Thereafter, flaps are sutured closed (see Fig 13-4e). Following 6 months of healing, the defect healing will generally resemble that of a demineralized bone allograft as observed from multiple RCTs with meta-analysis, as presented earlier in this chapter (see Figs 13-4f and 13-4g).

The bottom portion of a PRF membrane should be utilized because it incorporates more cells from within the buffy coat region.

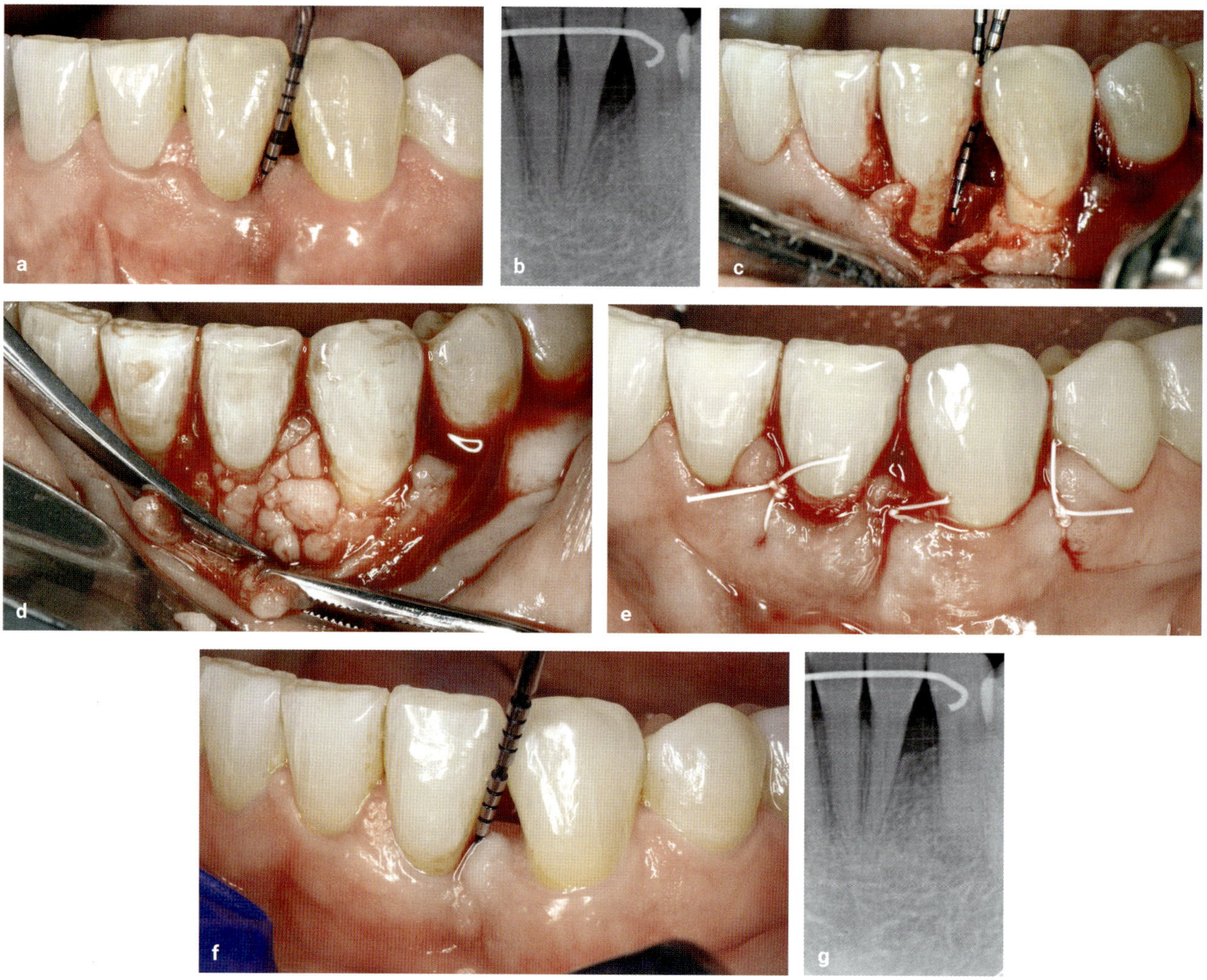

Fig 13-4 Narrow two- to three-wall intrabony defect treated with PRF alone. *(a)* Preoperative PD of a mandibular left lateral incisor. *(b)* Preoperative radiograph with Fixott-Everett grid. *(c)* Periodontal probing in the defect after flap reflection and debridement. *(d)* PRF placed into the defect and overfilled. (PRF was then compressed with a moist gauze.) *(e)* Flaps sutured. *(f)* Six-month postoperative probing depth and bone sounding. *(g)* Six-month postoperative radiograph with Fixott-Everett grid. Notice the defect fill when compared to *b*. (Case performed by resident Dr Jane Chadwick under supervision from Dr Brian Mealey; reprinted with permission from Miron and Choukroun.[74])

When the treating clinician fears that space maintenance may be an issue, it is generally recommended to combine PRF with a bone grafting material; results from a recent systematic review and meta-analysis demonstrated greater CAL gains and BF when utilizing this combination approach. Following flap elevation, PRF is generally premixed with a bone grafting material and placed within the bone defect, once again with slight overfill (Fig 13-5). The defect may

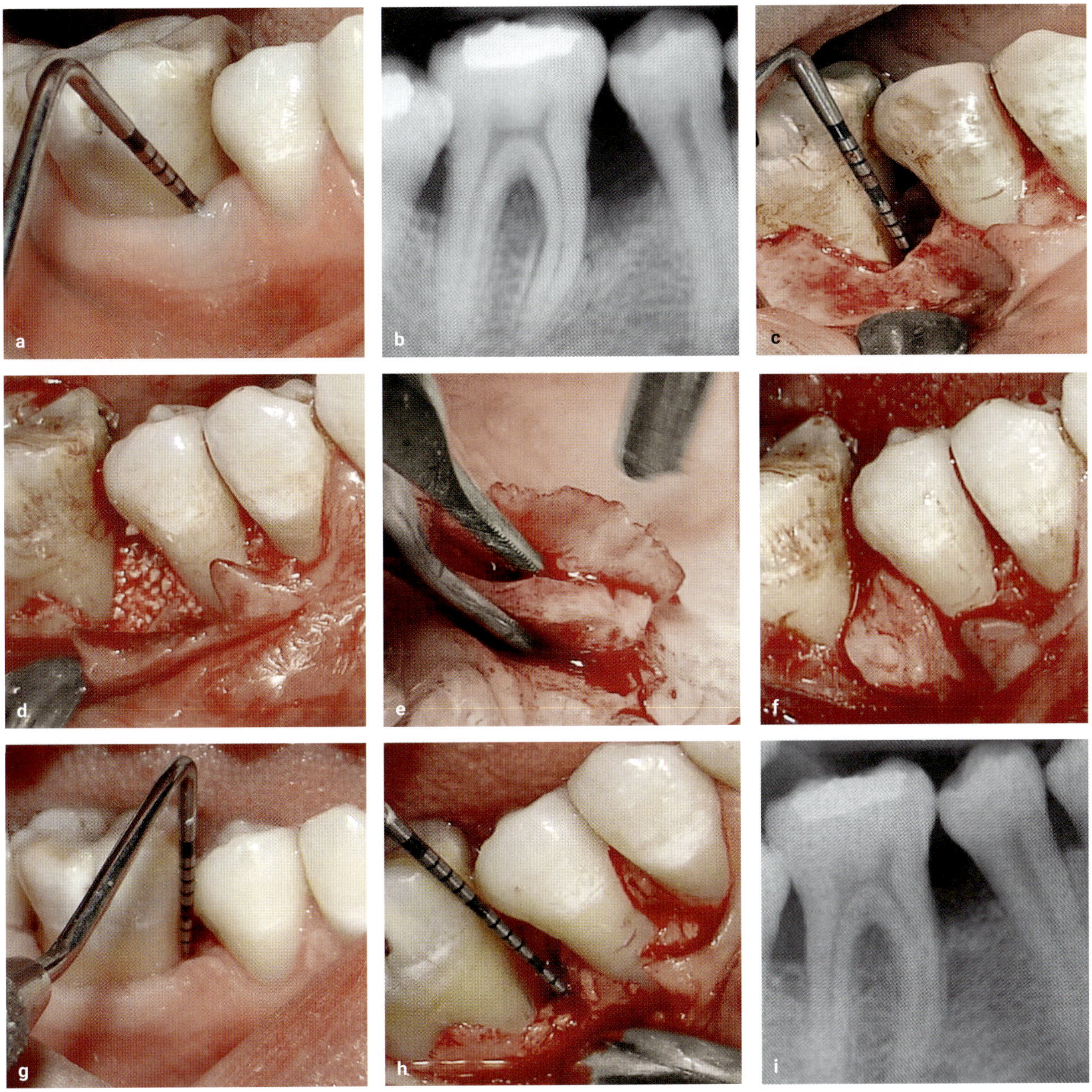

Fig 13-5 Wide three-wall intrabony defect treated with DBBM+PRF with a connective tissue graft (CTG). *(a)* PD of 9 mm. *(b)* Periapical radiograph demonstrating large intrabony defect. *(c)* Visual representation of a 6-mm intrabony defect. *(d)* The intrabony defect was filled with DBBM+PRF. *(e)* CTG harvested from the palate. The CTG was trimmed to 1 mm following harvest. *(f)* Placement and adaptation of the CTG on top of the grafted site. *(g)* Visual representation of the PD reduced to 6 mm. *(h)* Flap elevation demonstrating bone formation within the defect. *(i)* Six-month postoperative radiograph demonstrating defect fill. (Case performed by Dr Omid Moghaddas.)

also be covered with a thin 1-mm connective tissue graft to prevent soft tissue infiltration and improve keratinized tissue[87] (see Fig 13-5f). Note that in Figs 13-5 and 13-6, substantial bone gain was observed. Following flap elevation (study protocol), it was also noted that the intrabony defect had adequately regenerated with real bone following 6 months of healing utilizing this approach.

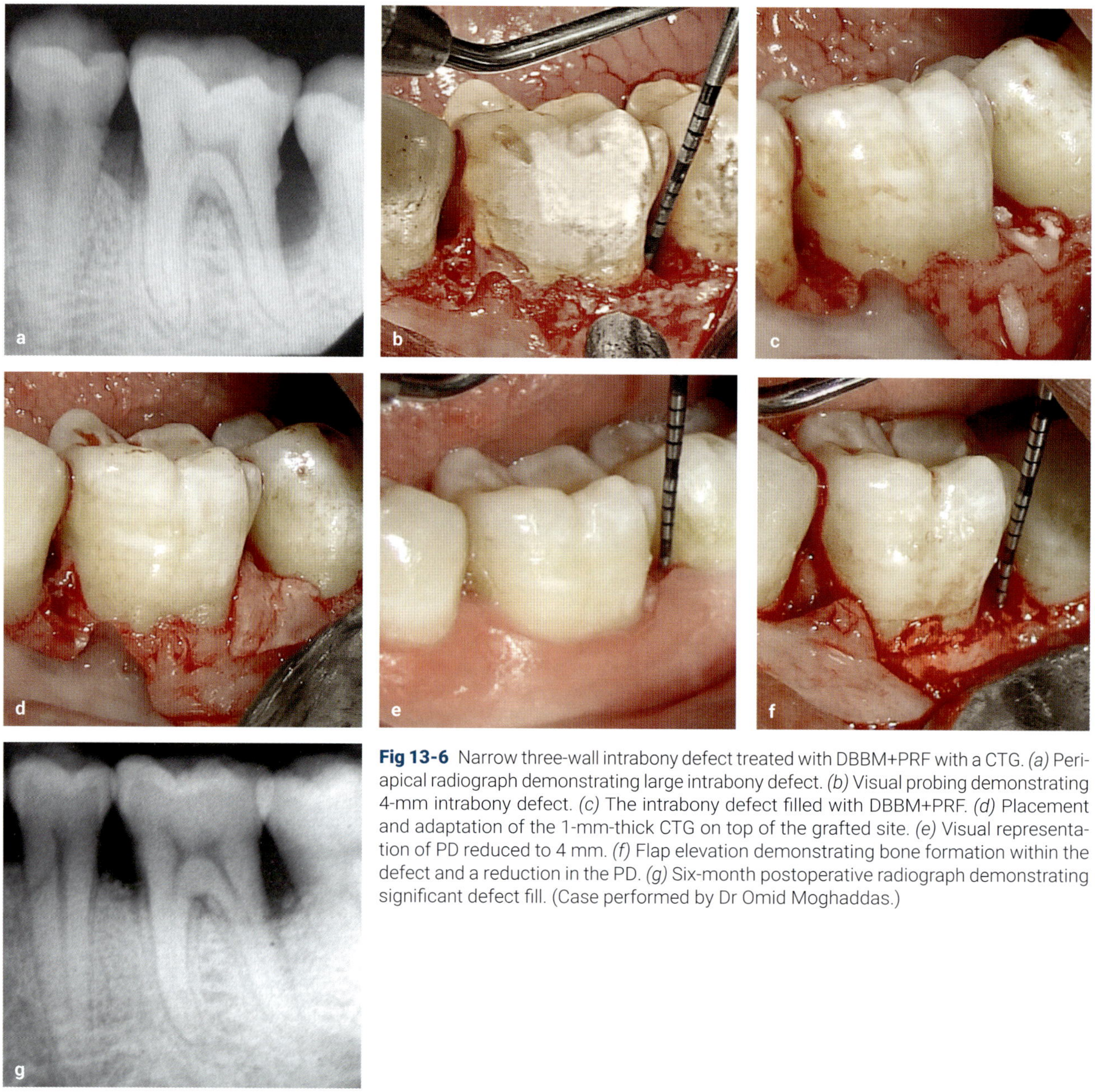

Fig 13-6 Narrow three-wall intrabony defect treated with DBBM+PRF with a CTG. *(a)* Periapical radiograph demonstrating large intrabony defect. *(b)* Visual probing demonstrating 4-mm intrabony defect. *(c)* The intrabony defect filled with DBBM+PRF. *(d)* Placement and adaptation of the 1-mm-thick CTG on top of the grafted site. *(e)* Visual representation of PD reduced to 4 mm. *(f)* Flap elevation demonstrating bone formation within the defect and a reduction in the PD. *(g)* Six-month postoperative radiograph demonstrating significant defect fill. (Case performed by Dr Omid Moghaddas.)

Evaluation of Clinical Studies Using PRF for Furcation Defect Regeneration

In total, 12 RCTs have investigated PRF for the regeneration of furcation defects[75–86] (Fig 13-7 and Table 13-2). In general, five studies evaluated the use of PRF vs OFD and also found additional PD reduction (~1.5 mm) and CAL gain when PRF was used as an adjunctive therapy to OFD.[75–86] Once again, the comparison between PRF alone vs BG alone led to similar outcomes for all investigated parameters, including PD reduction, CAL gain, and BF (three studies).[76–78] Two studies compared the use of BG vs BG plus PRF.[79,80] Here once again significant advantages in PD reduction and BF were observed. No differences were reported comparing PRP to PRF,[82] whereas the additional incorporation of various drugs including either metformin, bisphosphonates, or statins all led to significant improvements in PD reduction, CAL gain, and BF.[83–86]

	PRF			Control				Mean Difference
Study or Subgroup	**Mean**	**SD**	**Total**	**Mean**	**SD**	**Total**	**Weight**	**IV, Random, 95% CI**
1.1.1 OFD vs. PRF								
Sharma et al. (2011)	4,06	0,42	18	2,89	0,68	18	6,6%	1,17 [0,80, 1,54]
Bajaj et al. (2013)	4,29	1,04	24	1,58	1,02	23	6,1%	2,71 [2,12, 3,30]
Pradeep et al. (2016)	4,03	0,18	30	3,1	0,3	30	7,0%	0,93 [0,80, 1,06]
Siddiqui et al. (2016)	2,27	1,1	15	1,03	0,67	15	6,0%	1,24 [0,59, 1,89]
Kanoriya et al. (2017)	3,69	0,76	24	2,41	0,77	23	6,5%	1,28 [0,84, 1,72]
Subtotal (95% CI)			**111**			**109**	**32,2%**	**1,42 [0,92, 1,92]**
Heterogeneity: Tau² = 0,27; Chi² = 35,83, df = 4 (P < 0.00001); I² = 89%								
Test for overall effect: Z = 5,59 (P < 0.00001)								
1.1.2 BG vs. PRF								
Biswas et al. (2016)	3,1	0,22	10	3,7	0,11	10	6,9%	−0,60 [−0,75, −0,45]
Siddiqui et al. (2016)	2,27	1,1	15	2,47	1,15	15	5,5%	−0,20 [−1,01, 0,61]
Basireddy et al. (2019)	2,5	0,52	14	2,36	0,5	14	6,6%	0,14 [−0,24, 0,52]
Subtotal (95% CI)			**39**			**39**	**19,1%**	**−0,24 [−0,81, 0,33]**
Heterogeneity: Tau² = 0,20; Chi² = 13,17, df = 2 (P = 0,001); I² = 85%								
Test for overall effect: Z = 0,84 (P = 0,40)								
1.1.3 BG vs. BG + PRF								
Lohi et al. (2017)	3,37	1,06	10	2,4	0,52	8	5,7%	0,97 [0,22, 1,72]
Rani et al. (2018)	3,5	2,27	10	2,8	1,93	10	2,9%	0,70 [−1,15, 2,55]
Subtotal (95% CI)			**20**			**18**	**8,6%**	**0,93 [0,24, 1,63]**
Heterogeneity: Tau² = 0,00; Chi² = 0,07, df = 1 (P = 0,79); I² = 0%								
Test for overall effect: Z = 2,63 (P = 0,009)								
1.1.4 Collagen membrane vs. PRF								
Mehta et al. (2018)	2,45	0,4	18	2,84	0,3	18	6,9%	−0,39 [−0,62, −0,16]
Subtotal (95% CI)			**18**			**18**	**6,9%**	**−0,39 [−0,62, −0,16]**
Heterogeneity: Not applicable								
Test for overall effect: Z = 3,31 (P = 0,0009)								
1.1.5 PRP vs. PRF								
Bajaj et al. (2013)	4,29	1,04	24	3,92	0,93	25	6,2%	0,37 [−0,18, 0,92]
Subtotal (95% CI)			**24**			**25**	**6,2%**	**0,37 [−0,18, 0,92]**
Heterogeneity: Not applicable								
Test for overall effect: Z = 1,31 (P = 0,19)								
1.1.6 PRF vs. PRF + Metformin								
Sharma et al. (2017)	3,2	0,24	15	1,64	0,24	15	6,9%	1,56 [1,39, 1,73]
Subtotal (95% CI)			**15**			**15**	**6,9%**	**1,56 [1,39, 1,73]**
Heterogeneity: Not applicable								
Test for overall effect: Z = 17,80 (P < 0.00001)								
1.1.7 PRF vs. PRF + Bisphosphonates								
Kanoriya et al. (2017)	4,4	0,57	25	3,69	0,76	24	6,6%	0,71 [0,33, 1,09]
Wanikar et al. (2019)	2,85	0,88	20	1,85	0,59	20	6,4%	1,00 [0,54, 1,46]
Subtotal (95% CI)			**45**			**44**	**13,1%**	**0,83 [0,53, 1,12]**
Heterogeneity: Tau² = 0,00; Chi² = 0,90, df = 1 (P = 0,34); I² = 0%								
Test for overall effect: Z = 5,52 (P < 0.00001)								
1.1.8 PRF vs. PRF + Statins								
Pradeep et al. (2016)	4,9	0,31	30	4,03	0,18	30	7,0%	0,87 [0,74, 1,00]
Subtotal (95% CI)			**30**			**30**	**7,0%**	**0,87 [0,74, 1,00]**
Heterogeneity: Not applicable								
Test for overall effect: Z = 13,29 (P < 0.00001)								
Total (95% CI)			**302**			**298**	**100,0%**	**0,77 [0,36, 1,19]**
Heterogeneity: Tau² = 0,64; Chi² = 555,45, df = 15 (P < 0.00001); I² = 97%								
Test for overall effect: Z = 3,66 (P = 0,0002)								
Test for subgroup differences: Chi² = 198,66, df = 7 (P < 0.00001), I² = 96,5%								

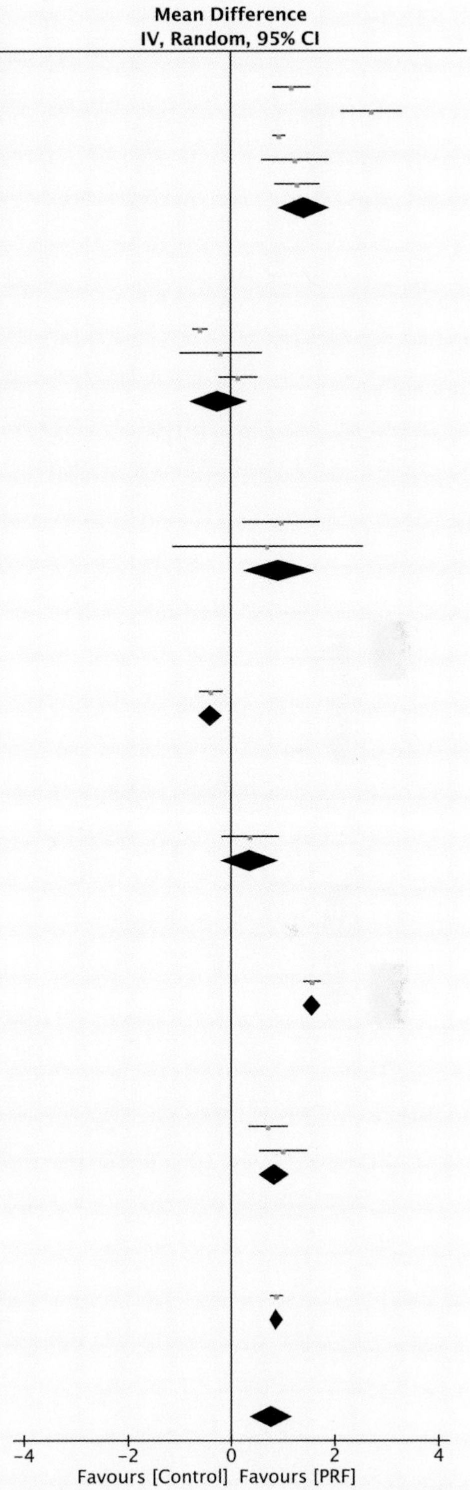

Fig 13-7 Forest plot for the event periodontal "probing depth" reduction of furcation defects (reported as PD in mm).

Table 13-2 Characteristics of the RCTs evaluating furcation defects

Authors (year)	Study design Follow-up	No. of participants Sex Mean age	Groups	Furcation degree	Smokers (Y, N)	Conclusions
OFD vs PRF						
Sharma and Pradeep[75] (2011)	RCT (split-mouth) 9 months	18 M: 10 / F: 8 34	C: 18, OFD T: 18, OFD+PRF	II	N	Significant improvement with autologous PRF implies its role as a regenerative material in the treatment of FD.
BG vs PRF						
Biswas et al[76] (2016)	RCT (parallel) 6 months	15 M: 10 / F: 5 38	C: 10, OFD+DFDBA T: 10, OFD+DFDBA+PRF	II	N	PRF seems to favor soft tissue healing but has no additional benefit in bone regeneration when used in combination with DFDBA.
Siddiqui et al[77] (2016)	RCT (parallel) 6 months	31 M: 24 / F: 7 40.9	C: 15, OFD T1: 15, OFD+BG T2:15, OFD+PRF	II	N	For both test and control groups, there was statistically significant improvement at 6 months follow-up from baseline values.
Basireddy et al[78] (2019)	RCT (split-mouth) 6 months	14 NR NR	C: 14, OFD+BG T: 14, OFD+PRF	II	N	Use of bioactive glass yields superior clinical results, including increased pocket depth reduction, as compared to an autologous platelet concentrate.
BG vs BG+PRF						
Lohi et al[79] (2017)	RCT (parallel) 6 months	14 M: 10 / F: 4 42.3	C: 8, OFD+BG T: 10, OFD+BG+PRF	II	N	Adjunctive use of PRF with bone graft may be a more effective treatment.
Rani et al[80] (2018)	RCT (parallel) 6 months	20 NR NR	C: 10, OFD+BTCP T: 10, OFD+BTCP+PRF	II	N	Adjunctive use of PRF with bone graft may be a more effective treatment.
Collagen membrane vs PRF						
Mehta et al[81] (2018)	RCT (split-mouth) 6 months	18 NR NR	C: 18, OFD+DFDBA+CM T: 18, OFD+DFDBA+PRF	II	N	Both groups showed statistically significant outcomes in intragroup comparison from baseline to 3 and 6 months. However, no statistical difference between PRF and collagen membrane groups were reported.
PRP vs PRF						
Bajaj et al[82] (2013)	RCT (parallel) 9 months	42 M: 22 / F: 20 39	C: 23, OFD T1: 25, OFD+PRP T2: 24, OFD+PRF	II	NR	The use of autologous PRF and PRP were both effective in the treatment of FD with uneventful healing of sites.
PRF vs PRF+MF (metformin)						
Sharma et al[83] (2017)	RCT (parallel) 6 months	22 M: 12 / F: 10 41	C: 15, OFD+PRF T: 15, OFD+PRF+MF	II	NR	PRF when combined with a potential osteogenic agent like MF can provide a better therapeutic benefit to a furcation-involved tooth.
PRF vs PRF+ALN (alendronate)						
Kanoriya et al[84] (2017)	RCT (parallel) 9 months	72 M: 36 / F: 36 38	C: 23, OFD T1:24, OFD+PRF T2: 25, OFD+PRF+1%ALN	II	N	FD treatment with PRF combined with 1% ALN gel results in significant therapeutic outcomes when compared with PRF and OFD alone.
Wanikar et al[85] (2019)	RCT (split-mouth) 6 months	20 M: 6 / F: 14 48	C: 20, OFD+PRF T: 20, OFD+PRF+1%ALN	II	N	PRF+ALN–treated defects exhibited better clinical and radiographic outcomes suggestive of enhanced periodontal regeneration when compared to sites treated with PRF alone.
PRF vs PRF+RSV (rosuvastatin)						
Pradeep et al[86] (2016)	RCT (parallel) 9 months	90 M: 45 / F: 45 35	C: 30, OFD T1:30 OFD+PRF T2: 30, OFD+PRF+1.2% RSV	II	N	OFD with RSV (1.2%) and PRF results in significantly greater periodontal benefits compared with OFD alone or with PRF.

Table 13-2 *(cont)*

				Methods for PRF preparation		
Authors (year)	**Mean difference in PD between baseline and final follow-up (mm)**	**Mean difference in CAL between baseline and final follow-up (mm)**	**Mean difference in BF between baseline and final follow-up (mm)**	**Centrifugation system**	**Volume of blood drawn**	**Centrifugation parameters speed (rpm) × time (min)**
OFD vs PRF						
Sharma and Pradeep[75] (2011)	2.89 ± 0.68 (C) 4.06 ± 0.42 (T)	1.28 ± 0.46 (C) 2.33 ± 0.49 (T)	0.62 ± 0.22 (C) 2.01 ± 0.16 (T)	R-4C	10 mL	3000 × 10
BG vs PRF						
Biswas et al[76] (2016)	3.70 ± 0.11 (C) 3.10 ± 0.22 (T)	3.70 ± 0.10 (C) 2.90 ± 0.12 (T)	NR	NR	NR	NR
Siddiqui et al[77] (2016)	1.03 ± 0.67 (C) 2.47 ± 1.51 (T1) 2.27 ± 1.10 (T2)	0.93 ± 0.46 (C) 2.53 ± 0.83 (T1) 2.40 ± 0.91 (T2)	0.73 ± 0.46 (C) 2.20 ± 0.77 (T1) 1.93 ± 0.59 (T2)	R-4C	NR	3000 × 10
Basireddy et al[78] (2019)	2.36 ± 0.50 (C) 2.50 ± 0.52 (T)	1.79 ± 0.80 (C) 2.36 ± 0.50 (T)	1.01 ± 0.16 (C) 1.19 ± 0.08 (T)	R-8C	10 mL	3000 × 10
BG vs BG+PRF						
Lohi et al[79] (2017)	2.40 ± 0.52 (C) 3.37 ± 1.06 (T)	1.90 ± 0.57 (C) 3.00 ± 0.93 (T)	0.60 ± 0.70 (C) 1.38 ± 0.52 (T)	NR	10 mL	3000 × 10
Rani et al[80] (2018)	2.80 ± 1.93 (C) 3.50 ± 2.27 (T)	3.00 ± 1.49 (C) 2.80 ± 1.40 (T)	NR	Systonic Lab and Scientific Instruments	15 mL	3000 × 12
Collagen membrane vs PRF						
Mehta et al[81] (2018)	2.84 ± 0.30 (C) 2.45 ± 0.40 (T)	3.00 ± 0.40 (C) 2.80 ± 0.20 (T)	NR	NR	10 mL	3000 × 10
PRP vs PRF						
Bajaj et al[82] (2013)	1.58 ± 1.02 (C) 3.92 ± 0.93 (T1) 4.29 ± 1.04 (T2)	1.37 ± 0.58 (C) 2.71 ± 1.04 (T1) 2.87 ± 0.85 (T2)	0.11 ± 0.03 (C) 1.77 ± 0.52 (T1) 1.85 ± 0.49 (T2)	R-4C	10 mL	400g × 10
PRF vs PRF+MF						
Sharma et al[83] (2017)	1.64 ± 0.24 (C) 3.20 ± 0.24 (T)	2.07 ± 0.21 (C) 2.94 ± 0.22 (T)	0.21 ± 0.15 (C) 0.66 ± 0.40 (T)	R-4C	NR	3000 × 10
PRF vs PRF+ALN						
Kanoriya et al[84] (2017)	2.41 ± 0.77 (C) 3.69 ± 0.76 (T1) 4.40 ± 0.57 (T2)	2.33 ± 0.48 (C) 3.39 ± 0.49 (T1) 4.12 ± 0.60 (T2)	0.52 ± 0.19 (C) 2.59 ± 0.32 (T1) 2.92 ± 0.25 (T2)	R-4C	10 mL	3000 × 10
Wanikar et al[85] (2019)	1.85 ± 0.59 (C) 2.85 ± 0.88 (T)	1.90 ± 0.64 (C) 3.05 ± 0.98 (T)	1.70 ± 0.73 (C) 2.30 ± 0.73 (T)	R-4C	NR	3000 × 10
PRF vs PRF+RSV						
Pradeep et al[86] (2016)	3.10 ± 0.30 (C) 4.03 ± 0.18 (T1) 4.90 ± 0.31 (T2)	2.47 ± 0.77 (C) 3.30 ± 0.65 (T1) 3.93 ± 0.78 (T2)	1.43 ± 0.50 (C) 3.17 ± 0.65 (T1) 3.63 ± 0.67 (T2)	R-4C	10 mL	3000 × 10

NR, not reported; C, control group; T, test group, M, male; F, female; CAL, clinical attachment level; PD, probing depth; BF, bone fill.

Discussion and Future Research

Despite the fact that PRF is only beginning to be utilized more commonly in routine clinical practice for the treatment of intrabony/furcation defects, it remains interesting to note that nearly 40 RCTs have thus far evaluated its potential for periodontal wound regeneration. The formation of a blood clot alone has been shown to be one of the key necessary features in order for periodontal regeneration to take place, as long as bacterial pathogens have been completely eliminated. Evidence from the literature suggests that blood clot formation alone is enough to treat a number of intrabony defects where space maintenance is not an issue.[88] PRF therefore acts in a similar fashion, whereby the fibrin scaffold can be inserted into the periodontal pocket acting as a stable clot, with significant increases in platelets, leukocytes, and growth factors. Because periodontal regeneration remains complex due to the number of tissues needed to be regenerated (new cementum, periodontal ligament, and alveolar bone), as well as the fact that Sharpey fibers need to be oriented functionally to support the tooth apparatus, it remains difficult to assess whether PRF actually leads to true periodontal regeneration; no human histologic evidence exists to date on the topic (despite the nearly 50 RCTs having been performed).

Nevertheless, it is known that periodontal disease is caused by bacterial pathogens, and an increase in regenerative growth factors and cells, as well as its incorporation of defense-fighting leukocytes, is certainly hypothesized to favor defect resolution and potentially mitigate tissue inflammation. As highlighted in chapter 2 from basic research studies, PRF has the ability to favor M2 macrophage polarization and also decrease tissue inflammation.[89,90] Furthermore, angiogenesis is an important factor for tissue regeneration, and PRF releases a number of proangiogenic and profibrotic agents capable of further speeding periodontal tissue repopulation.[7,88]

The biologic advantages of PRF have been shown to act locally by quickly stimulating a large number of cell types—including endothelial cells, gingival fibroblasts, chondrocytes, and osteoblasts—by influencing their recruitment, proliferation, and/or differentiation.[91,92] Thus, PRF may prove beneficial for the regeneration of specific tissues such as the periodontium because several cell types and tissue types require regeneration. While it is known that the beneficial effects of PRF may partially be due to the large number of secreted autologous blood-derived growth factors, rhPDGF has been one of the main recombinant growth factors sold to date in North America for the regeneration of periodontal tissues.[24,25,93,94] Although recombinant proteins have a regenerative potential well documented in the literature,[95–97] their associated costs and other secondary adverse effects including biocompatibility, lower stability, and potential swelling may favor the use of autologous PRF.[98,99] Future comparative studies including a cost-benefit analysis between both modalities remain necessary.

Several research topics also remain at the forefront of needed research in this space. As previously mentioned, no single study has characterized PRF at the histologic level in a well-characterized human study. It has already been well established in the literature that PRF favors soft tissue wound healing when compared to hard tissues.[100] Because periodontitis is characterized not only by periodontal ligament breakdown but also by breakdown of cementum and alveolar bone, the regenerative potential of each of these tissues needs to be further characterized via histologic evaluation, ideally in human studies. Chapter 22 presents an animal (dog) model investigating various forms of PRF (L-PRF vs H-PRF vs e-PRF) in intrabony defects. Human histologic studies are planned and needed to further evaluate the regenerative potential of PRF.

It is also noteworthy that the use of PRF has been shown to decrease the rate of localized alveolar osteitis 9.5-fold following third molar extractions, suggesting a beneficial wound healing effect with potential defense fighting of PRF in extraction sites.[101] This may be due to the presence of leukocytes contained within PRF. It would be of interest to determine the change in secreted inflammatory markers and population of periodontal pathogens following therapy with PRF. Limited data is available to date on this topic.

Furthermore, more recent research has investigated photodynamic therapy or Nd:YAG/Er:YAG lasers for periodontal decontamination during microsurgical debridement of pockets[102,103] (Fig 13-8). The goal of laser therapy is to eliminate periodontal pathogens and stimulate periodontal tissues, favoring new attachment gain specifically by nonsurgical periodontal modalities. As reviewed in Fig 13-8, PRF may be utilized following scaling and root planing plus laser therapy and introduced/tucked within pockets and intrabony defects without flap elevation. While this area of

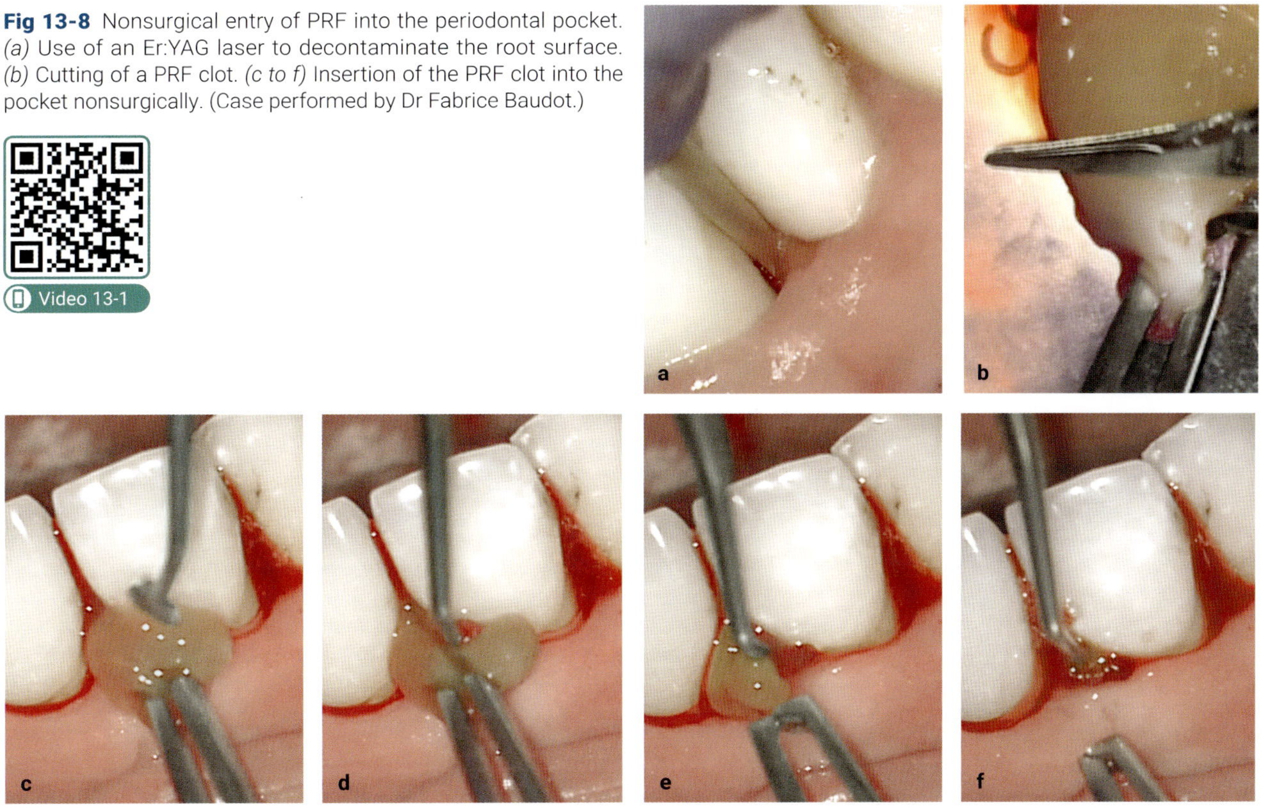

Fig 13-8 Nonsurgical entry of PRF into the periodontal pocket. *(a)* Use of an Er:YAG laser to decontaminate the root surface. *(b)* Cutting of a PRF clot. *(c to f)* Insertion of the PRF clot into the pocket nonsurgically. (Case performed by Dr Fabrice Baudot.)

Video 13-1

research remains preliminary, future investigation aims to reveal the potential of this combination approach for regeneration of periodontal and peri-implant defects. Furthermore, advancements made utilizing concentrated forms of PRF as well as e-PRF protocols extending resorption profiles from 2–3 weeks to 4–6 months show the potential for various forms of PRF to favor periodontal regeneration.

Conclusion

The data to date points to the fact that PRF leads to significant improvements in PD reduction, CAL gain, and BF when compared to OFD alone. Furthermore, data has now demonstrated comparable results when intrabony defects are filled with either PRF vs a BG material, and significant improvements in CAL and BF were observed when PRF was combined with BGs. Future research evaluating the use of PRF in combination with various additional drugs such as metformin, bisphosphonates, statins, and antibiotics may additionally improve clinical outcomes, but future RCTs are needed on each of these specific topics. In addition, future human histologic evidence is needed to verify if PRF actually leads to true periodontal regeneration in humans.

References

1. Bosshardt DD, Sculean A. Does periodontal tissue regeneration really work? Periodontol 2000 2009;51:208–219.
2. Bosshardt DD, Stadlinger B, Terheyden H. Cell-to-cell communication—Periodontal regeneration. Clin Oral Implants Res 2015;26:229–239.
3. Eke PI, Dye BA, Wei L, Thornton-Evans GO, Genco RJ; CDC Periodontal Disease Surveillance workgroup. Prevalence of periodontitis in adults in the United States: 2009 and 2010. J Dent Res 2012;91:914–920.
4. Cullinan MP, Seymour GJ. Periodontal disease and systemic illness: Will the evidence ever be enough? Periodontol 2000 2013;62:271–286.
5. Grzesik WJ, Narayanan AS. Cementum and periodontal wound healing and regeneration. Crit Rev Oral Biol Med 2002;13:474–484.
6. Wikesjö UM, Selvig KA. Periodontal wound healing and regeneration. Periodontol 2000 1999;19:21–39.
7. Wang HL, Greenwell H, Fiorellini J, et al. Periodontal regeneration. J Periodontol 2005;76:1601–1622.

8. Nyman S, Lindhe J, Karring T, Rylander H. New attachment following surgical treatment of human periodontal disease. J Clin Periodontol 1982;9:290–296.
9. Pihlstrom BL, McHugh RB, Oliphant TH, Ortiz-Campos C. Comparison of surgical and nonsurgical treatment of periodontal disease. A review of current studies and additional results after 6½ years. J Clin Periodontol 1983;10:524–541.
10. AlGhamdi AS, Ciancio SG. Guided tissue regeneration membranes for periodontal regeneration—A literature review. J Int Acad Periodontol 2009;11:226–231.
11. Zhang Y, Zhang X, Shi B, Miron. R. Membranes for guided tissue and bone regeneration. Ann Oral Maxillofac Surg 2013;1:10.
12. Place ES, Evans ND, Stevens MM. Complexity in biomaterials for tissue engineering. Nat Mater 2009;8:457–470.
13. Miron RJ, Sculean A, Shuang Y, et al. Osteoinductive potential of a novel biphasic calcium phosphate bone graft in comparison with autographs, xenografts, and DFDBA. Clin Oral Implants Res 2016;27:668–675.
14. Fucini SE, Quintero G, Gher ME, Black BS, Richardson AC. Small versus large particles of demineralized freeze-dried bone allografts in human intrabony periodontal defects. J Periodontol 1993;64:844–847.
15. Harasty LA, Brownstein CN, Deasy MJ. Regeneration of intrabony defects: Comparing e-PTFE membrane vs. decalcified freeze dried bone allograft—A pilot study. Periodontal Clin Investig 1999;21:10–17.
16. Parashis A, Andronikaki-Faldami A, Tsiklakis K. Comparison of 2 regenerative procedures—Guided tissue regeneration and demineralized freeze-dried bone allograft in the treatment of intrabony defects: A clinical and radiographic study. J Periodontol 1998;69:751–758.
17. Reynolds MA, Bowers GM. Fate of demineralized freeze-dried bone allografts in human intrabony defects. J Periodontol 1996;67:150–157.
18. Miron RJ, Zhang Y (eds). Next-Generation Biomaterials for Bone & Periodontal Regeneration. Chicago: Quintessence, 2019.
19. Gestrelius S, Andersson C, Johansson AC, et al. Formulation of enamel matrix derivative for surface coating. Kinetics and cell colonization. J Clin Periodontol 1997;24:678–684.
20. Gestrelius S, Andersson C, Lidstrom D, Hammarstrom L, Somerman M. In vitro studies on periodontal ligament cells and enamel matrix derivative. J Clin Periodontol 1997;24:685–692.
21. Heijl L. Periodontal regeneration with enamel matrix derivative in one human experimental defect. A case report. J Clin Periodontol 1997;24:693–696.
22. Heijl L, Heden G, Svardstrom G, Ostgren A. Enamel matrix derivative (Emdogain) in the treatment of intrabony periodontal defects. J Clin Periodontol 1997;24:705–714.
23. Zetterstrom O, Andersson C, Eriksson L, et al. Clinical safety of enamel matrix derivative (Emdogain) in the treatment of periodontal defects. J Clin Periodontol 1997;24:697–704.
24. Steed DL, Donohoe D, Webster MW, Lindsley L. Effect of extensive debridement and treatment on the healing of diabetic foot ulcers. Diabetic Ulcer Study Group. J Am Coll Surg 1996;183:61–64.
25. Wieman TJ, Smiell JM, Su Y. Efficacy and safety of a topical gel formulation of recombinant human platelet-derived growth factor-BB (becaplermin) in patients with chronic neuropathic diabetic ulcers. A phase III randomized placebo-controlled double-blind study. Diabetes Care 1998;21:822–827.
26. Ronnstrand L, Heldin CH. Mechanisms of platelet-derived growth factor-induced chemotaxis. Int J Cancer 2001;91:757–762.
27. Nevins M, Camelo M, Nevins ML, Schenk RK, Lynch SE. Periodontal regeneration in humans using recombinant human platelet-derived growth factor-BB (rhPDGF-BB) and allogenic bone. J Periodontol 2003;74:1282–1292.
28. Camelo M, Nevins ML, Schenk RK, Lynch SE, Nevins M. Periodontal regeneration in human Class II furcations using purified recombinant human platelet-derived growth factor-BB (rhPDGF-BB) with bone allograft. Int J Periodontics Restorative Dent 2003; 23:213–225.
29. Rosen PS, Toscano N, Holzclaw D, Reynolds MA. A retrospective consecutive case series using mineralized allograft combined with recombinant human platelet-derived growth factor BB to treat moderate to severe osseous lesions. Int J Periodontics Restorative Dent 2011;31:335–342.
30. Khoshkam V, Chan HL, Lin GH, et al. Outcomes of regenerative treatment with rhPDGF-BB and rhFGF-2 for periodontal intrabony defects: A systematic review and meta-analysis. J Clin Periodontol 2015;42:272–280.
31. Nevins M, Hanratty J, Lynch SE. Clinical results using recombinant human platelet-derived growth factor and mineralized freeze-dried bone allograft in periodontal defects. Int J Periodontics Restorative Dent 2007;27:421–427.
32. Nevins M, Nevins ML, Karimbux N, Kim SW, Schupbach P, Kim DM. The combination of purified recombinant human platelet-derived growth factor-BB and equine particulate bone graft for periodontal regeneration. J Periodontol 2012;83:565–573.
33. McGuire MK, Kao RT, Nevins M, Lynch SE. rhPDGF-BB promotes healing of periodontal defects: 24-month clinical and radiographic observations. Int J Periodontics Restorative Dent 2006;26: 223–231.
34. McGuire MK, Scheyer ET, Schupbach P. Growth factor-mediated treatment of recession defects: A randomized controlled trial and histologic and microcomputed tomography examination. J Periodontol 2009;80:550–564.
35. McGuire MK, Scheyer T, Nevins M, Schupbach P. Evaluation of human recession defects treated with coronally advanced flaps and either purified recombinant human platelet-derived growth factor-BB with beta tricalcium phosphate or connective tissue: A histologic and microcomputed tomographic examination. Int J Periodontics Restorative Dent 2009;29:7–21.
36. Nevins M, Kao RT, McGuire MK, et al. Platelet-derived growth factor promotes periodontal regeneration in localized osseous defects: 36-month extension results from a randomized, controlled, double-masked clinical trial. J Periodontol 2013;84:456–464.
37. Rosen PS. Using recombinant platelet-derived growth factor to facilitate wound healing. Compend Contin Educ Dent 2006;27: 520–525.
38. Kitamura M, Akamatsu M, Kawanami M, et al. Randomized placebo-controlled and controlled non-inferiority phase iii trials comparing trafermin, a recombinant human fibroblast growth factor 2, and enamel matrix derivative in periodontal regeneration in intrabony defects. J Bone Miner Res 2016;31:806–814.
39. Bashutski JD, Eber RM, Kinney JS, et al. Teriparatide and osseous regeneration in the oral cavity. N Engl J Med 2010;363:2396–2405.
40. Stavropoulos A, Windisch P, Gera I, Capsius B, Sculean A, Wikesjö UM. A phase II a randomized controlled clinical and histological pilot study evaluating rhGDF-5/β-TCP for periodontal regeneration. J Clin Periodontol 2011;38:1044–1054.
41. Reddi AH, Cunningham NS. Initiation and promotion of bone differentiation by bone morphogenetic proteins. J Bone Miner Res 1993;8(suppl 2):S499–S502.
42. Rao SM, Ugale GM, Warad SB. Bone morphogenetic proteins: Periodontal regeneration. N Am J Med Sci 2013;5:161–168.
43. Saleem M, Pisani F, Zahid FM, et al. Adjunctive platelet-rich plasma (PRP) in infrabony regenerative treatment: A systematic review and RCT's meta-analysis. Stem Cells Int 2018;2018.

44. Lekovic V, Camargo PM, Weinlaender M, Vasilic N, Kenney EB. Comparison of platelet-rich plasma, bovine porous bone mineral, and guided tissue regeneration versus platelet-rich plasma and bovine porous bone mineral in the treatment of intrabony defects: A reentry study. J Periodontol 2002;73:198–205.
45. Clark RA. Fibrin and wound healing. Ann N Y Acad Sci 2001;936: 355–367.
46. Choukroun J, Diss A, Simonpieri A, et al. Platelet-rich fibrin (PRF): A second-generation platelet concentrate. Part IV: Clinical effects on tissue healing. Oral Surg Oral Med Oral Pathol Oral Radiol Endod 2006;101:e56–e60.
47. Sharma A, Pradeep A. Treatment of 3-wall intrabony defects in patients with chronic periodontitis with autologous platelet-rich fibrin: A randomized controlled clinical trial. J Periodontol 2011;82:1705–1712.
48. Thorat M, Pradeep A, Pallavi B. Clinical effect of autologous platelet-rich fibrin in the treatment of intrabony defects: A controlled clinical trial. J Clin Periodontol 2011;38:925–932.
49. Rosamma Joseph V, Sam G, Amol NV. Clinical evaluation of autologous platelet rich fibrin in horizontal alveolar bony defects. J Clin Diagn Res 2014;8:ZC43.
50. Ajwani H, Shetty S, Gopalakrishnan D, et al. Comparative evaluation of platelet-rich fibrin biomaterial and open flap debridement in the treatment of two and three wall intrabony defects. J Int Oral Health 2015;7:32–37.
51. Bajaj P, Agarwal E, Rao NS, et al. Autologous platelet-rich fibrin in the treatment of 3-wall intrabony defects in aggressive periodontitis: A randomized controlled clinical trial. J Periodontol 2017;88:1186–1191.
52. Patel GK, Gaekwad SS, Gujjari SK, S CV. Platelet-rich fibrin in regeneration of intrabony defects: A randomized controlled trial. J Periodontol 2017;88:1192–1199.
53. Pradeep AR, Bajaj P, Rao NS, Agarwal E, Naik SB. Platelet-rich fibrin combined with a porous hydroxyapatite graft for the treatment of 3-wall intrabony defects in chronic periodontitis: A randomized controlled clinical trial. J Periodontol 2017;88: 1288–1296.
54. Thorat M, Baghele ON, S RP. Adjunctive effect of autologous platelet-rich fibrin in the treatment of intrabony defects in localized aggressive periodontitis patients: A randomized controlled split-mouth clinical trial. Int J Periodontics Restorative Dent 2017;37:e302–e309.
55. Shah M, Deshpande N, Bharwani A, Nadig P, Doshi V, Dave D. Effectiveness of autologous platelet-rich fibrin in the treatment of intra-bony defects: A systematic review and meta-analysis. J Indian Soc Periodontol 2014;18:698–704.
56. Mathur A, Bains VK, Gupta V, Jhingran R, Singh GP. Evaluation of intrabony defects treated with platelet-rich fibrin or autogenous bone graft: A comparative analysis. Eur J Dent 2015;9:100–108.
57. Chadwick JK, Mills MP, Mealey BL. Clinical and radiographic evaluation of demineralized freeze-dried bone allograft versus platelet-rich fibrin for the treatment of periodontal intrabony defects in humans. J Periodontol 2016;87:1253–1260.
58. Galav S, Chandrashekar KT, Mishra R, Tripathi V, Agarwal R, Galav A. Comparative evaluation of platelet-rich fibrin and autogenous bone graft for the treatment of infrabony defects in chronic periodontitis: Clinical, radiological, and surgical reentry. Indian J Dent Res 2016;27:502–507.
59. Yajamanya SR, Chatterjee A, Hussain A, Coutinho A, Das S, Subbaiah S. Bioactive glass versus autologous platelet-rich fibrin for treating periodontal intrabony defects: A comparative clinical study. J Indian Soc Periodontol 2017;21:32–36.
60. Bansal C, Bharti V. Evaluation of efficacy of autologous platelet-rich fibrin with demineralized-freeze dried bone allograft in the treatment of periodontal intrabony defects. J Indian Soc Periodontol 2013;17:361–366.
61. Elgendy EA, Abo Shady TE. Clinical and radiographic evaluation of nanocrystalline hydroxyapatite with or without platelet-rich fibrin membrane in the treatment of periodontal intrabony defects. J Indian Soc Periodontol 2015;19:61–65.
62. Agarwal A, Gupta ND, Jain A. Platelet rich fibrin combined with decalcified freeze-dried bone allograft for the treatment of human intrabony periodontal defects: A randomized split mouth clinical trail. Acta Odontol Scand 2016;74:36–43.
63. Naqvi A, Gopalakrishnan D, Bhasin MT, Sharma N, Haider K, Martande S. Comparative evaluation of bioactive glass putty and platelet rich fibrin in the treatment of human periodontal intrabony defects: A randomized control trial. J Clin Diagn Res 2017;11:Zc09–Zc13.
64. Sezgin Y, Uraz A, Taner IL, Çulhaoğlu R. Effects of platelet-rich fibrin on healing of intra-bony defects treated with anorganic bovine bone mineral. Braz Oral Res 2017;31:e15.
65. Bodhare GH, Kolte AP, Kolte RA, Shirke PY. Clinical and radiographic evaluation and comparison of bioactive bone alloplast morsels when used alone and in combination with platelet-rich fibrin in the treatment of periodontal intrabony defects—A randomized controlled trial. J Periodontol 2019;90:584–594.
66. Panda S, Sankari M, Satpathy A, et al. Adjunctive effect of autologous platelet-rich fibrin to barrier membrane in the treatment of periodontal intrabony defects. J Craniofac Surg 2016;27:691–696.
67. Pradeep AR, Rao NS, Agarwal E, Bajaj P, Kumari M, Naik SB. Comparative evaluation of autologous platelet-rich fibrin and platelet-rich plasma in the treatment of 3-wall intrabony defects in chronic periodontitis: A randomized controlled clinical trial. J Periodontol 2012;83:1499–1507.
68. Gupta SJ, Jhingran R, Gupta V, Bains VK, Madan R, Rizvi I. Efficacy of platelet-rich fibrin vs. enamel matrix derivative in the treatment of periodontal intrabony defects: A clinical and cone beam computed tomography study. J Int Acad Periodontol 2014;16:86–96.
69. Aydemir Turkal H, Demirer S, Dolgun A, Keceli HG. Evaluation of the adjunctive effect of platelet-rich fibrin to enamel matrix derivative in the treatment of intrabony defects: Six-month results of a randomized, split-mouth, controlled clinical study. J Clin Periodontol 2016;43:955–964.
70. Pradeep AR, Nagpal K, Karvekar S, Patnaik K, Naik SB, Guruprasad CN. Platelet-rich fibrin with 1% metformin for the treatment of intrabony defects in chronic periodontitis: A randomized controlled clinical trial. J Periodontol 2015;86:729–737.
71. Kanoriya D, Pradeep AR, Singhal S, Garg V, Guruprasad CN. Synergistic approach using platelet-rich fibrin and 1% alendronate for intrabony defect treatment in chronic periodontitis: A randomized clinical trial. J Periodontol 2016;87:1427–1435.
72. Martande SS, Kumari M, Pradeep AR, Singh SP, Suke DK, Guruprasad CN. Platelet-rich fibrin combined with 1.2% atorvastatin for treatment of intrabony defects in chronic periodontitis: A randomized controlled clinical trial. J Periodontol 2016;87:1039–1046.
73. Pradeep AR, Garg V, Kanoriya D, Singhal S. Platelet-rich fibrin with 1.2% rosuvastatin for treatment of intrabony defects in chronic periodontitis: A randomized controlled clinical trial. J Periodontol 2016;87:1468–1473.
74. Miron RJ, Choukroun J (eds). Platelet Rich Fibrin in Regenerative Dentistry. Hoboken, NJ: Wiley, 2017.
75. Sharma A, Pradeep AR. Autologous platelet-rich fibrin in the treatment of mandibular degree II furcation defects: A randomized clinical trial. J Periodontol 2011;82:1396–1403.

76. Biswas S, Sambashivaiah S, Kulal R, Bilichodmath S, Kurtzman GM. Comparative evaluation of bioactive glass (putty) and platelet rich fibrin in treating furcation defects. J Oral Implantol 2016;42: 411–415.
77. Siddiqui ZR, Jhingran R, Bains VK, Srivastava R, Madan R, Rizvi I. Comparative evaluation of platelet-rich fibrin versus beta-tri-calcium phosphate in the treatment of grade II mandibular furcation defects using cone-beam computed tomography. Eur J Dent 2016;10:496–506.
78. Basireddy A, Prathypaty SK, Yendluri DB, Potharaju SP. Demineralized freeze-dried bone allograft with or without platelet-rich fibrin in the treatment of mandibular degree II furcation defects: A clinical and cone beam computed tomography study. J Indian Soc Periodontol 2019;23:242–248.
79. Lohi HS, Nayak DG, Uppoor AS. Comparative evaluation of the efficacy of bioactive ceramic composite granules alone and in combination with platelet rich fibrin in the treatment of mandibular Class II furcation defects: A clinical and radiographic study. J Clin Diagn Res 2017;11:Zc76–Zc80.
80. Rani N, Kaushal S, Singh S, Nandlal, Khan MA, Pathak AK. Evaluation of the relative efficacy of autologous platelet-rich fibrin membrane in combination with β-tricalcium phosphate (Septodont-resorbable tissue replacement) alloplast versus β-TCP alloplast alone in the treatment of grade II furcation defects. Natl J Maxillofac Surg 2018;9:196–204.
81. Mehta DB, Deshpande NC, Dandekar SA. Comparative evaluation of platelet-rich fibrin membrane and collagen membrane along with demineralized freeze-dried bone allograft in grade II furcation defects: A randomized controlled study. J Indian Soc Periodontol 2018;22:322–327.
82. Bajaj P, Pradeep AR, Agarwal E, et al. Comparative evaluation of autologous platelet-rich fibrin and platelet-rich plasma in the treatment of mandibular degree II furcation defects: A randomized controlled clinical trial. J Periodontal Res 2013;48:573–581.
83. Sharma P, Grover HS, Masamatti SS, Saksena N. A clinicoradiographic assessment of 1% metformin gel with platelet-rich fibrin in the treatment of mandibular grade II furcation defects. J Indian Soc Periodontol 2017;21:303–308.
84. Kanoriya D, Pradeep AR, Garg V, Singhal S. Mandibular degree II furcation defects treatment with platelet-rich fibrin and 1% alendronate gel combination: A randomized controlled clinical trial. J Periodontol 2017;88:250–258.
85. Wanikar I, Rathod S, Kolte AP. Clinico-radiographic evaluation of 1% alendronate gel as an adjunct and smart blood derivative platelet rich fibrin in grade II furcation defects. J Periodontol 2019;90:52–60.
86. Pradeep AR, Karvekar S, Nagpal K, Patnaik K, Raju A, Singh P. Rosuvastatin 1.2 mg in situ gel combined with 1:1 mixture of autologous platelet-rich fibrin and porous hydroxyapatite bone graft in surgical treatment of mandibular Class II furcation defects: A randomized clinical control trial. J Periodontol 2016; 87:5–13.
87. Moghaddas H, Soltani L, Moghaddas O. Efficacy of palatal connective tissue graft as a membrane in the treatment of intrabony defects. J Periodontol Implant Dent 2011;2:70–76.
88. Wang HL, Boyapati L. "PASS" principles for predictable bone regeneration. Implant Dent 2006;15:8–17.
89. Nasirzade J, Kargarpour Z, Hasannia S, Strauss FJ, Gruber R. Platelet-rich fibrin elicits an anti-inflammatory response in macrophages in vitro. J Periodontol 2020;91:244–252.
90. Zhang J, Yin C, Zhao Q, et al. Anti-inflammation effects of injectable platelet-rich fibrin via macrophages and dendritic cells. J Biomed Mater Res A 2020;108:61–68.
91. Roy S, Driggs J, Elgharably H, et al. Platelet-rich fibrin matrix improves wound angiogenesis via inducing endothelial cell proliferation. Wound Repair Regen 2011;19:753–766.
92. Chen FM, Wu LA, Zhang M, Zhang R, Sun HH. Homing of endogenous stem/progenitor cells for in situ tissue regeneration: Promises, strategies, and translational perspectives. Biomaterials 2011;32:3189–3209.
93. White AP, Vaccaro AR, Hall JA, Whang PG, Friel BC, McKee MD. Clinical applications of BMP-7/OP-1 in fractures, nonunions and spinal fusion. Int Orthop 2007;31:735–741.
94. Miron RJ, Zhang YF. Osteoinduction: A review of old concepts with new standards. J Dent Res 2012;91:736–744.
95. Young CS, Ladd PA, Browning CF, et al. Release, biological potency, and biochemical integrity of recombinant human platelet-derived growth factor-BB (rhPDGF-BB) combined with Augment Bone Graft or *GEM 21S* beta-tricalcium phosphate (β-TCP). J Control Release 2009;140:250–255.
96. Park YJ, Lee YM, Lee JY, Seol YJ, Chung CP, Lee SJ. Controlled release of platelet-derived growth factor-BB from chondroitin sulfate-chitosan sponge for guided bone regeneration. J Control Release 2000;67:385–394.
97. Wissink MJ, Beernink R, Poot AA, et al. Improved endothelialization of vascular grafts by local release of growth factor from heparinized collagen matrices. J Control Release 2000;64:103–114.
98. Delgado JJ, Evora C, Sanchez E, Baro M, Delgado A. Validation of a method for non-invasive in vivo measurement of growth factor release from a local delivery system in bone. J Control Release 2006;114:223–229.
99. Oe S, Fukunaka Y, Hirose T, Yamaoka Y, Tabata Y. A trial on regeneration therapy of rat liver cirrhosis by controlled release of hepatocyte growth factor. J Control Release 2003;88:193–200.
100. Miron RJ, Fujioka-Kobayashi M, Bishara M, Zhang Y, Hernandez M, Choukroun J. Platelet-rich fibrin and soft tissue wound healing: A systematic review. Tissue Eng Part B Rev 2017;23:83–99.
101. Hoaglin DR, Lines GK. Prevention of localized osteitis in mandibular third-molar sites using platelet-rich fibrin. Int J Dent 2013; 2013:875380.
102. Sculean A, Aoki A, Romanos G, Schwarz F, Miron RJ, Cosgarea R. Is photodynamic therapy an effective treatment for periodontal and peri-implant infections? Dent Clin North Am 2015;59:831–858.
103. Grzech-Leśniak K, Sculean A, Gašpirc B. Laser reduction of specific microorganisms in the periodontal pocket using Er:YAG and Nd:YAG lasers: A randomized controlled clinical study. Lasers Med Sci 2018;33:1461–1470.

14

Use of PRF for Extraction Site Management

Contributors

Richard J. Miron

Michael A. Pikos

Jonathan Du Toit

Mark Bishara

Senthil Selvan

Alfonso Gil

Massimo Del Fabbro

PRF has been successfully utilized for the management of soft/periodontal tissues, as reviewed in previous chapters. This chapter highlights its use for bone regeneration in extraction site management. Specific to socket grafting, PRF has been investigated in over 20 clinical trials when utilized either alone or in combination with a bone grafting material. Research over the past 15 years has clearly demonstrated that major bone loss occurs postextraction due to loss of primary blood supply (ie, the periodontal ligament). Strategies have therefore been employed whereby PRF is utilized either alone or in combination with a bone graft to limit dimensional changes postextraction by providing the necessary growth factors and regenerative cells favoring angiogenesis. It has also been utilized in over 20 clinical trials for the management of third molar extraction, where it has been shown to decrease the rate of dry sockets by as much as 10-fold. This chapter provides evidence from RCTs on this topic and also overviews human histologic samples grafted with and without PRF for the management of extraction sites.

Chapter Highlights

- *Dimensional changes occurring postextraction*
- *Clinical uses of PRF for the management of extraction sockets either alone or with a bone grafting material*
- *Use of PRF to facilitate soft tissue healing over extraction sites*
- *Use of PRF for third molar extractions*

Tooth loss is caused by a variety of conditions affecting the periodontium, including periodontal disease, trauma, tooth fracture, and tooth decay. In the United States alone, over 36 million people are completely edentulous, and 120 million people are missing at least one tooth.[1] It is now well understood that following tooth loss, marked alterations in the bone alveolar structure also occur.[2,3] In fact, over 50 years ago one of the first evidenced articles on this topic was published in the *Australian Dental Journal*.[4] While numerous advancements have been made in biomaterials over the past several decades, it is essential that we understand and are able to characterize these changes with the aim of minimizing bone loss over time. As such, a great deal of research utilizing bone grafts, different types of barrier membranes, as well as biologic agents and growth factors including PRF have been investigated to prevent dimensional changes in the bone morphology and architecture resulting from tooth loss.[5]

The most commonly utilized biomaterial to limit dimensional changes postextraction is a bone grafting material. This chapter aims to summarize the current literature on this topic and presents evidence from over 40 RCTs investigating the use of PRF either alone or with a bone graft for extraction site management. First, however, a basic understanding regarding the biologic events taking place following tooth extraction is presented from both animal and human studies.

Dimensional Changes Occurring Postextraction

Ridge preservation is typically described as a process by which a socket is grafted with nonviable grafts (ie, bone grafts, collagen sponges, etc).[6] These materials represent the matrix on which cell adhesion molecules from plasma (fibrin and fibronectin) and from platelets (vitronectin) will adhere, connecting the graft particles together within the socket. Signaling molecules are derived from growth factors when platelets degranulate in the blood clot, allowing progenitor cells to migrate from open marrow spaces of the socket itself during ridge preservation or from the bony walls of the sinus cavity in sinus augmentations. Similar to the mechanism of osseointegration, osteoprogenitor cells divide, and daughter cells differentiate toward osteoblasts via adhesion molecules and signaling factors.

Healing process

The following six events take place during the healing process (Fig 14-1)[6,7]:

1. The socket fills with blood from the severed vessels, which contain proteins and growth factors. Cells initiate a series of events that lead to the formation of a fibrin network along with platelets, forming a blood clot or coagulum within the first 24 hours.[8]
2. The coagulum, acting as a physical matrix, directs the migration of cells. Neutrophils and later macrophages enter the wound site and digest bacteria and tissue debris within the wound. They release growth factors and cytokines that induce and amplify the migration of mesenchymal cells and their activity within the coagulum.[9]
3. Within days, the blood clot begins to break down (fibrinolysis). The proliferation of mesenchymal cells leads to gradual replacement of the coagulum by granulation tissue (2–4 days).[10]
4. By end of week 1, a vascular network is formed, and by week 2, the marginal portion of the extraction socket is covered with young connective tissue rich in vessels and inflammatory cells.[11]
5. At 4 to 6 weeks, most of the alveolus is filled with woven bone, while the soft tissue becomes keratinized.[9,10]
6. At 4 to 6 months, the mineral tissue within the original socket is reinforced with layers of lamellar bone that is deposited onto the previously formed woven bone.[9,10]

Natural dimensional changes occurring postextraction

The dimensional changes that occur following tooth extraction remain inevitable, even if biomaterials are utilized. As already mentioned, more than 50 years have passed since initial clinical investigations began to notice significant bony changes.[4] While these preliminary findings were scarce, it fueled a great deal of research into this topic over the following decades to investigate more precisely the characterization of bone loss following tooth loss. In one of the most-often cited preclinical studies on the topic, Araújo and Lindhe studied dimensional changes in a canine model following tooth loss at 1, 2, 4, and 8 weeks and investigated histologically the dimensional changes

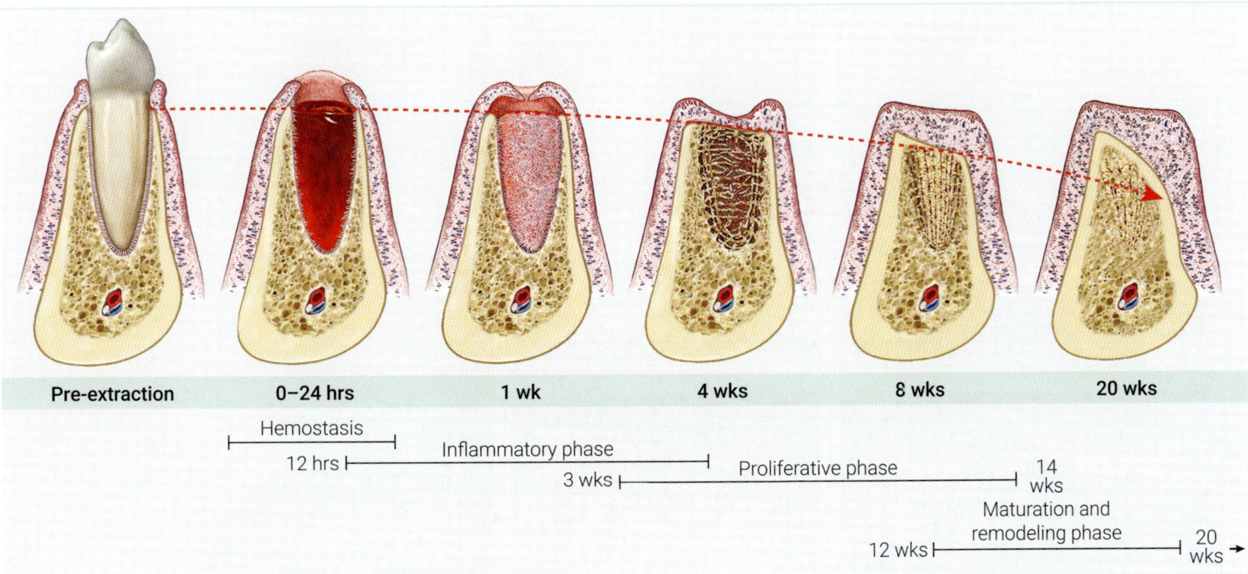

Fig 14-1 Healing process following tooth extraction. (Reprinted with permission from Avila-Ortiz and Zadeh.[6])

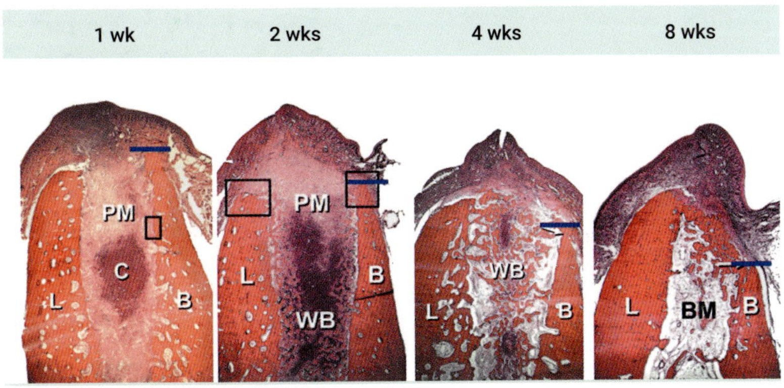

Fig 14-2 Overview of the extraction site after 1, 2, 4, and 8 weeks of healing. Note the large amounts of provisional matrix and, in the center of the socket, remaining blood clot. The buccal bone crest is made exclusively of bundle bone while the lingual crest is comprised of a mixture of cortical bone and bundle bone. Note the amount of bone loss occurring within an 8-week period, especially on the buccal surface. C, blood clot; B, buccal; L, lingual; PM, provisional matrix; WB, woven bone; BM, bone marrow. H&E staining; original magnification ×16. (Modified with permission from Araújo and Lindhe.[10])

over time.[10] They showed convincingly that the dimensional changes and bone loss following tooth extraction occurred rapidly and within as little as an 8-week healing period (Fig 14-2). It was observed histologically that within 2 weeks, a large number of bone-resorbing multinucleated osteoclasts were found on the buccal bone surface, responsible for the future resorption that occurs shortly thereafter.[10] By 8 weeks, a pronounced loss of bone and ridge dimension was observed, most notably on the thin buccal surface (see Fig 14-2).[10] The reason for the faster bone resorption on the buccal bone wall was hypothesized to be due to it being comprised primarily of bundle bone, and therefore modeling resulted in substantial vertical reduction of the buccal crest.[10] In 2014, Scala et al carried out a similar study in monkeys.[12] The primate model, which is deemed more relevant because it bears more similarities to humans, found that half the socket's bundle bone was lost by day 30 and 90% by 3 months.[12]

Human studies have also been performed using a variety of techniques and analysis methods. In general, an average of 40% to 60% of original height and width can be expected to be lost postextraction (especially in the esthetic zone), with most of that loss occurring within the first 2 years.[13–15] Schropp et al assessed bone formation in

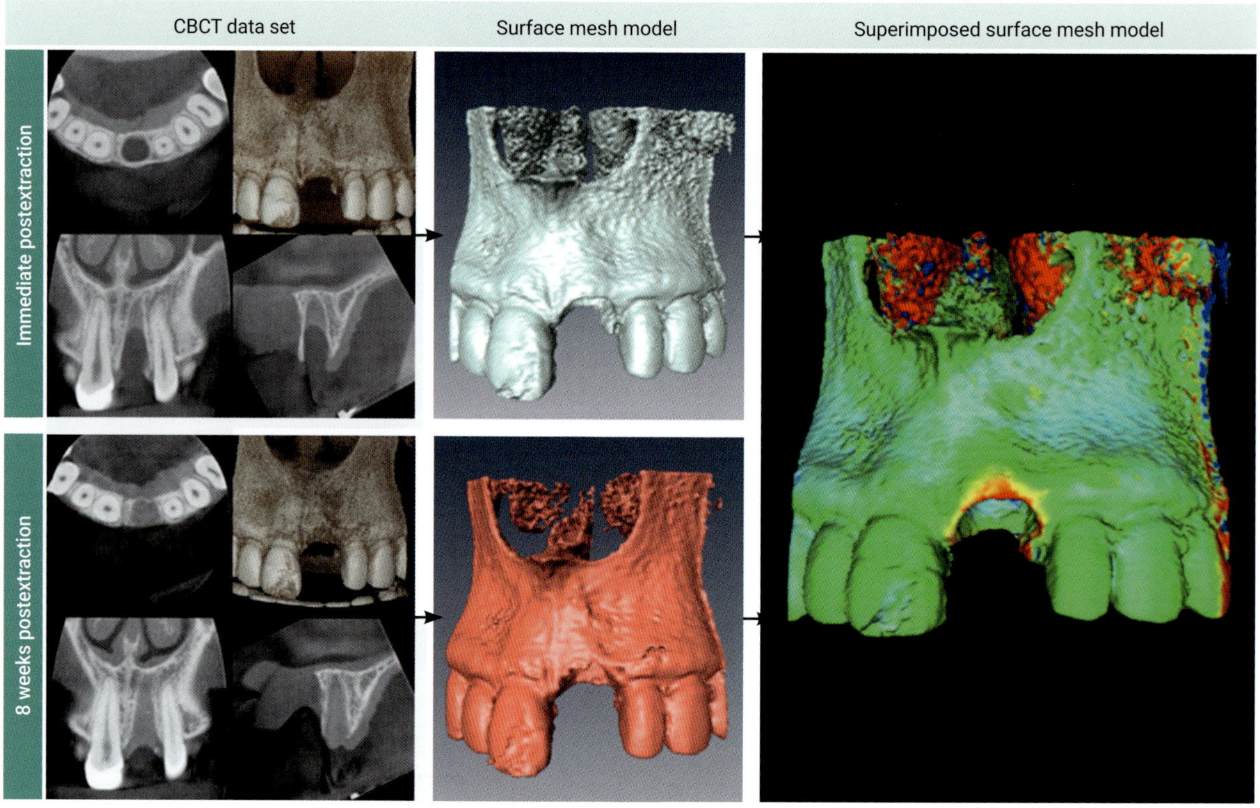

Fig 14-3 Radiographic examination following tooth extraction. The DICOM files of the obtained CBCT data sets, immediately postextraction and following 8 weeks of healing, were converted into a surface mesh model with digital imaging software. The two surface mesh models were superimposed and rigidly aligned with anatomical landmarks. The distance between the two surface meshes was presented as color-coded figures to identify zones of facial bone resorption. (Reprinted with permission from Chappuis et al.[19])

the alveolus and contour changes of the alveolar process after tooth extraction. It was found that the maximum loss of tissue contour occurs during the first few months after tooth extraction.[16] The width of the alveolar ridge was reduced by 50% during the observation period of 1 year, which is in agreement with earlier studies.[4,17,18] Approximately two-thirds of the alveolar bone loss occurred within the first 3 months. Additionally, the loss of crestal bone height mainly occurred within the 3-month period after tooth extraction, with reorganization of the lamina dura occurring during the remaining healing period.[16]

Interestingly, CBCT has made possible the visualization of dimensional changes in humans by superimposing CBCT images over various healing periods. Chappuis et al clearly demonstrated the marked postextraction dimensional changes that occur following an 8-week healing period (Fig 14-3).[19] The aims of this study were *(1)* to characterize the thickness of the buccal bone wall specifically in the esthetic zone and *(2)* to evaluate both horizontal and vertical dimensional changes that occur following an 8-week healing period. It was found that 69% of human cases presented with a buccal wall surface thinner than 1 mm. Furthermore, it was observed that after an 8-week healing period, an average vertical bone loss of 5.2 mm was reported (Fig 14-4).[19] These changes were deemed to be 2.5 to 3 times more severe than those previously reported by Araújo and Lindhe, and this was hypothesized to be caused by the animal's ability to regenerate at higher rates as well as being utilized at surgery at a young age. Nevertheless, Fig 14-5 demonstrates the marked impact of tooth extraction on wound healing after only an 8-week period, thereby requiring substantial new bone regeneration prior to implant placement in such cases.

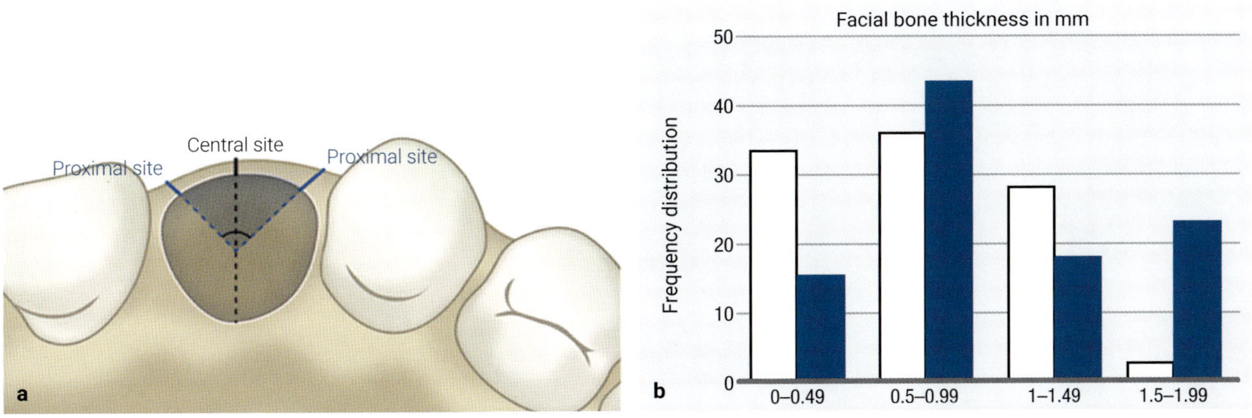

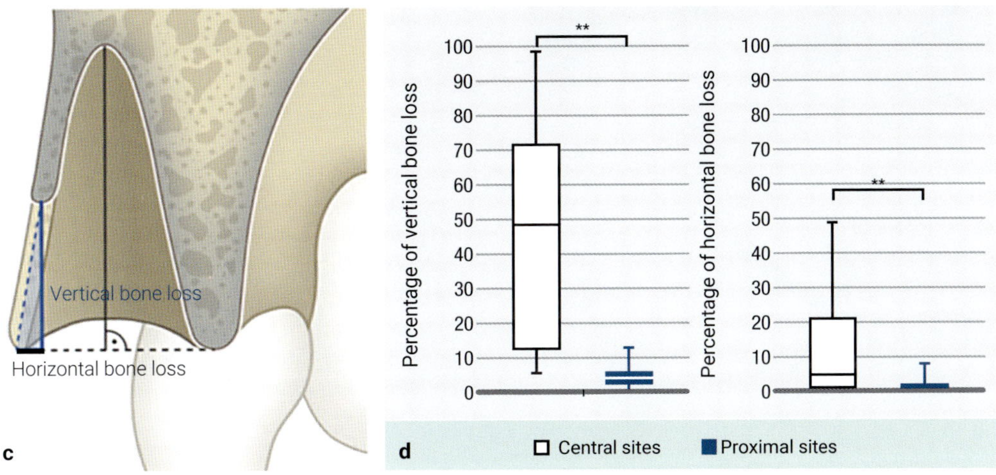

Fig 14-4 Baseline measurements and dimensional and vertical bone loss after 8 weeks of healing. *(a)* The analysis was performed in central and proximal sites oriented at a 45-degree angle with the tooth axis as a reference. *(b)* Frequency distribution of facial bone wall thickness in central and proximal sites. *(c)* A horizontal reference line was traced connecting the facial and palatal bone wall for standardized measurements. The point-to-point distance between the two surface meshes with the respective angle to the reference line was obtained for each sample, and the vertical and horizontal bone losses were calculated accordingly. *(d)* Percentage of vertical bone loss in central and proximal sites. *(e)* Percentage of horizontal bone loss in central and proximal sites. $^{**}P < .0001$. (Reprinted with permission from Chappuis et al.[19])

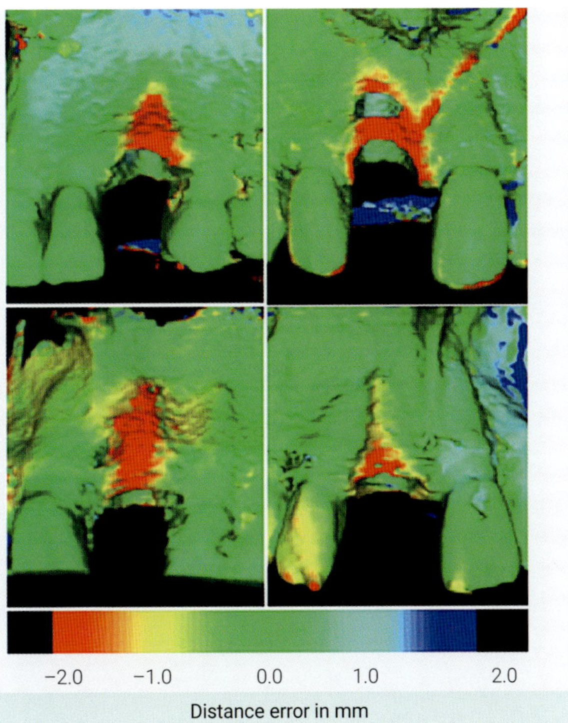

Fig 14-5 Characteristic bone resorption patterns. A thin-wall phenotype showed a facial bone wall thickness of ≤ 1 mm and revealed a progressive bone resorption pattern after 8 weeks of healing. (Reprinted with permission from Chappuis et al.[19])

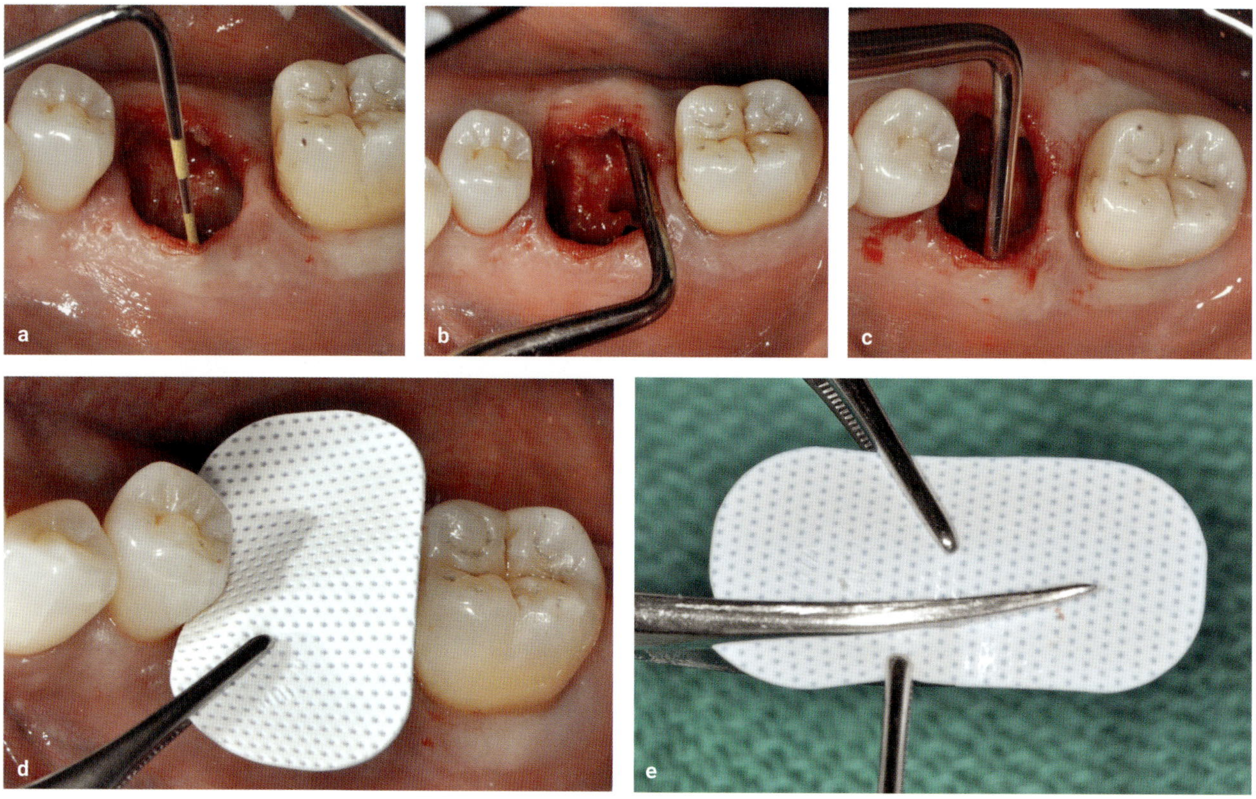

Fig 14-6 Use of a dPTFE membrane (Cytoplast) as a barrier to facilitate GBR. *(a)* Defect exploration. *(b and c)* Buccal and lingual subperiosteal pouches are created to facilitate PTFE membrane placement. *(d)* Correct membrane sizing for extraction socket. *(e)* Trimming of the membrane to accommodate optimal placement.

Conventional socket grafting and ridge preservation techniques

It has been well demonstrated in the literature in numerous RCTs and systematic reviews that socket grafting can limit dimensional changes; however, to date no single therapy can predictably maintain 100% of the ridge dimension.[5,20–24] Furthermore, without a consistent and constant blood supply, bone cannot exist.[25] While bone grafting materials are most commonly utilized and certainly limit dimensional changes postextraction, a resulting 0.5 to 1 mm of horizontal and vertical loss occurs mainly on the buccal surface.[5] Figure 14-6 demonstrates a step-by-step routine procedure highlighting standard grafting with a bone allograft covered by a PTFE membrane.[7]

Without a consistent and constant blood supply, bone cannot exist.

Over the past 10 years, research has primarily focused on minimizing the dimensional changes occurring postextraction by favoring the use of bone biomaterials. These include barrier membranes,[26–29] bone grafting materials,[27,29–31] and growth factor therapies.[32–35] The most widely utilized bone grafting material for extraction site management is a mineralized allograft with a 50:50 corticocancellous mixture of freeze-dried bone allograft (FDBA). Barrier membranes have frequently been utilized to prevent soft tissue infiltration. Despite the use of biomaterials, no single

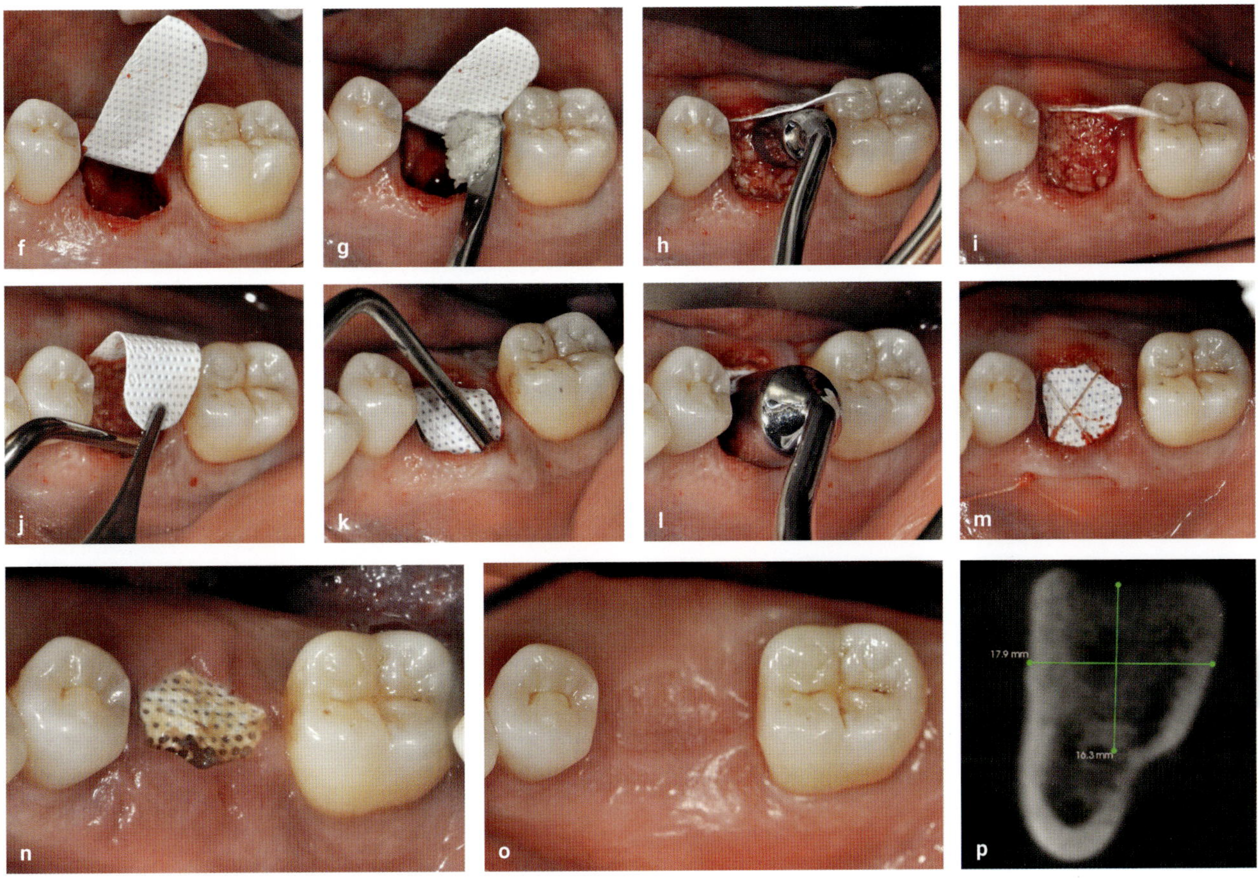

Fig 14-6 *(cont)* *(f to i)* The membrane is first tucked into the lingual pouch, and FDBA (MinerOss, 0.6–1.2 mm) is then placed and condensed into the socket. The graft must be moderately condensed. *(j to m)* The membrane is tucked into the buccal pouch, compressed, and sutured with a horizontal crisscross suture technique. *(n)* Four weeks postoperative, the sutures and membrane are removed. *(o)* The soft tissue has healed uneventfully at 4 months. *(p)* Notice the excellent dimensional stability demonstrated radiographically in both the horizontal and vertical dimensions at 4 months. (Reprinted with permission from Pikos and Miron.[7])

therapy (either utilized alone or in combination) can predictably prevent some dimensional change. It has also been demonstrated that the principle of GBR for extraction site management (using a barrier to exclude the soft tissue) favors new bone growth within an extraction socket. In a 10-case study by Lekovic et al comparing the outcome of alveolar ridge preservation postextraction using absorbable barrier membranes alone, it was found that after 6 months, significantly less crestal bone loss (0.38 vs 1.50 mm), more internal socket fill (5.81 vs 3.94 mm), and less horizontal ridge resorption (1.31 vs 4.56 mm) was found in the membrane group when compared to the empty control group.[36] Therefore, the concept of providing a barrier membrane on the outer layer of the socket was enough to prevent collapse and facilitate bone regrowth.[36]

In summary, two key parameters are noted[7]:

- To date, there are currently no available options to completely prevent dimensional changes following tooth extraction. Resorption of bundle bone will occur regardless of the alveolar ridge preservation technique utilized.
- There exists no ideal or favored method to preserve dimensional changes of the alveolar ridge including using GBR techniques, socket fillers, socket seals, or combinations of the above-mentioned techniques.

Table 14-1 Use of PRF versus natural healing for extraction site management

Authors (year)	Study design Follow-up	No. of participants Sex Mean age	Groups	Smokers (Y, N)	Conclusions
Use of PRF for extraction socket healing (bone)					
Hauser et al[37] (2013)	RCT evaluating microarchitecture and intrinsic bone tissue quality of the alveolar bone after premolar extraction therapy. 8 weeks	23 (23 sites) M: 9 / F: 14 47.0	C: 8, clot T1: 9, PRF T2: 6, mucosal flap+PRF	NR	Analysis by microcomputed tomography showed better bone healing with improvement of the microarchitecture (*P* < .05) in the PRF group without flap. An invasive surgical procedure with a mucosal flap appeared to completely neutralize the advantages of the PRF.
Suttapreyasri and Leepong[38] (2013)	RCT evaluating alveolar ridge contour changes and crestal bone resorption in dental casts and periapical radiographs. Up to 8 weeks	8 (20 sites) M: 3 / F: 5 22.6	C: 10, clot T: 10, PRF	NR	PRF clinically showed early healing of soft tissue covering socket orifices in the first 4 weeks. Neither better alveolar ridge preservation nor enhanced bone formation of PRF was observed in the extraction socket.
Temmerman et al[39] (2016)	RCT (split-mouth single bilateral and closely symmetric tooth extractions) Mean ridge width differences between time points were measured at three levels below the crest on both the buccal and lingual sides.	22 (44 sites) M: 15 / F: 7 54	C: 22, clot T: 22, PRF	N	The use of PRF as a socket filling material to achieve preservation of horizontal and vertical ridge dimension at 3 months after tooth extraction is beneficial.
Du Toit et al[40] (2016)	RCT (split-mouth) Human bone biopsy specimens from extraction sockets were treated with PRF versus natural healing at 3 months.	4 (8 sites) M: 3 / F: 1 39.5	C: 4, clot T: 4, PRF	N	Bone derived from PRF histologically did not differ from bone that healed without intervention.

Evaluation of Clinical Studies Using PRF for Extraction Sockets

PRF versus natural wound healing

As mentioned, alveolar bone is largely dependent on the supporting tooth structure and will rapidly undergo remodeling following extraction as a result of loss of the periodontal ligament's blood supply. Interestingly, PRF as a biomaterial for extraction socket management and healing has become a frequent topic of research in recent years (Table 14-1).[37–47] Many years ago, it was hypothesized that ridge preservation might be achieved by using PRF with the rationale of applying proangiogenic and regenerative growth factors to stimulate healing in extraction sockets. Table 14-1 presents the current studies evaluating dimensional changes postextraction with PRF in comparison to a control or to a bone grafting material. In total, 11 studies have now reported the effects of PRF versus natural wound healing.[37–47]

Hauser et al were one of the first to show that PRF was capable of inducing new bone formation in extraction sockets when compared to controls.[37] Analysis by microcomputed tomography showed better bone healing with improvements in the microarchitecture in the group treated with PRF. It was also shown that PRF had a significant effect on intrinsic bone tissue quality and preservation of the alveolar width. Interestingly, an invasive surgical procedure with a mucosal flap appeared to completely neutralize the advantages of PRF.[37] For these reasons, it is strongly recommended not to raise a flap during routine tooth extraction, as others have also shown in similar studies.

Suttapreyasri et al then demonstrated that PRF showed better early healing of soft tissue covering socket orifices in the first 4 weeks.[38] However, neither better alveolar

Table 14-1 *(cont)*

Authors (year)	Study design Follow-up	No. of participants Sex Mean age	Groups	Smokers (Y, N)	Conclusions
Alzahrani et al[41] (2017)	RCT Ridge width was assessed using cast analysis with the help of an acrylic stent and a pair of calipers. Radiographic analysis of the socket surface area was performed using computer graphic software program at 1, 4, and 8 weeks.	24 (24 sites) M: 39 / F: 5 37.8	C: 12, clot T: 12, PRF	N	Ridge width proportions were significantly higher in the test group as compared to the control group between baseline and 4 and 8 weeks, respectively. The mean radiographic bone fill was significantly higher in the test group than the control group at all time intervals.
Zhang et al[42] (2018)	RCT CBCT was compared at 3 months to assess alveolar ridge height, width, bone mineral density, and histomorphometry.	28 (28 sites) M: 14 / F: 14 33.9	C: 14, clot T: 14, PRF	N	Histologic analysis of novel bone formation confirmed that PRF increased the quality of the novel bone and enhanced the rate of bone formation; however, the effect of PRF was not significant to reduce alveolar bone resorption in the extraction socket alone.
Srinivas et al[43] (2018)	RCT (split-mouth) evaluating bone height and density.	30 (60 sites) NR 20–50	C: 30, clot T: 30, PRF	NR	Patients in the PRF group had better Healing Index compared to the control group. Use of PRF showed a comparable increase in bone density too.
Areewong et al[44] (2019)	RCT comparing new bone formation with PRF versus with normal healing via histomorphometric analysis.	33 (36 sites) M: 15 / F: 21 50.7	C: 18, clot T: 18, PRF	Y	It may be concluded that the use of PRF in ARP does not statistically significantly enhance new bone formation after tooth extraction compared to normal wound healing.
Canellas et al[45] (2020)	RCT CBCT imaging was performed immediately after tooth extraction and after 3 months, prior to implant surgery.	48 (48 sites) M: 21 / F: 27 44.8	C: 24, clot T: 24, PRF	N	Histomorphometric analysis showed a higher percentage of new bone formation in the PRF group compared with the control group.
Use of PRF for extraction socket healing (soft tissue and pain scores)					
Marenzi et al[46] (2015)	RCT Pain was assessed using VAS (1 to 10) at intervals of 1, 2, 3, and 4 days. The soft tissue healing was evaluated at 3, 7, 14, and 21 days using the modified Healing Index (4 to 12).	26 (108 sites) M: 9 / F: 17 53	C: 54, clot T: 54, PRF	N	PRF improved the postoperative pain, promoted the soft tissue healing process, and reduced the early adverse effects of the inflammation.
Temmerman et al[39] (2016)	RCT (split-mouth) VAS score	22 (44 sites) M: 15 / F: 7 54	C: 22, clot T: 22, PRF	N	In this study, a significantly reduced level of postoperative pain could be seen at sites treated with PRF during the early healing phases.
de Almeida Barros Mourão et al[47] (2020)	One week after extraction, an assessment of soft tissue healing around the sockets was performed using the Healing Index.	32 (32 sites) M: 13 / F: 19 37	C: 16, clot T: 16, PRF	N	The results of the present study demonstrate that whenever improved healing of the extraction socket is needed, the use of PRF should be considered. In addition, the use of PRF decreases postoperative pain and discomfort.

NR, not reported; C, control group; T, test group; M, male; F, female; VAS, visual analog scale; ARP, alveolar ridge preservation.

Table 14-1 ***(cont)*** Use of PRF versus natural healing for extraction site management

				Methods for PRF preparation		
Authors (year)	**Horizontal bone loss (mm)**	**Vertical bone loss (mm)**	**Bone density**	**Centrifugation system**	**Volume of tubes for blood drawn**	**Centrifugation parameters speed (rpm) × time (min)**
Use of PRF in sinus augmentation procedure						
Hauser et al[37] (2013)	0.43 ± 0.18 (C) 0.06 ± 0.05 (T1) 0.42 ± 0.2 (T2)	M: 0.77 ± 0.17 (C) D: 2.07 ± 0.81 (C) M: 1.21 ± 0.40 (T1) D: 0.76 ± 0.25 (T1) M: 0.86 ± 0.34 (T2) D: 2.15 ± 1.05 (T2)	*CBCT values:* 780 ± 10 (C) 820 ± 23 (T1) 832 ± 18 (T2)	NR	8-mL tubes (32 mL total)	2700 × 12
Sutta-preyasri and Leepong[38] (2013)	B: 1.61 ± 0.43 (C) L: 1.78 ± 0.47 (C) B: 1.79 ± 0.9 (T) L: 0.42 ± 0.39 (T)	M: 1.33 (C) D: 1.14 (C) M: 0.7 (T) D: 1.23 (T)	NR	EBA 20 (Hettich)	10-mL glass tubes	3000 × 10
Temmer-man et al[39] (2016)	51.92% ± 40.31 (C) 22.84% ± 24.28 (T)	1.5 ± 1.3 (C) 0.5 ± 2.3 (T)	63.3% ± 31.9% (C) 94.7% ± 26.9% (T)	IntraSpin	10-mL tubes (20–50 mL total)	2700 × 12
Du Toit et al[40] (2016)	NR	NR	47.9% ± 18.1% (C) 50.7% ± 13.3% (T)	EBA 20	10-mL tubes (20 mL total)	3000 × 10
Alzahrani et al[41] (2017)	13.54% (C) 8.58% (T)	NR	80.3% ± 2.6% (C) 88.8% ± 1.5% (T)	IntraSpin	10-mL tubes (20–50 mL total)	2700 × 12
Zhang et al[42] (2018)	2.08 ± 1.67 (C) 1.05 ± 1.78 (T)	B: 2.80 ± 1.81 (C) 1.60 ± 1.46 (T) L: 2.05 ± 1.29 (C) 1.00 ± 0.71 (T)	*Osteoid area (%):* 2.81 ± 1.21 (C) 9.76 + 4.01 (T)	Universal 320 (Hettich)	9-mL tubes (18 mL total)	400g × 10 (rpm NR)
Srinivas et al[43] (2018)	NR	1.90 ± 0.50 1.65 ± 0.28	*CBCT values:* 295.87 ± 87.21 (C) 564.76 ± 94.86 (T)	NR	10-mL tubes	3000 × 10
Areewong et al[44] (2019)	NR	NR	26.33% ± 19.63% (C) 31.33% ± 18.00% (T)	IntraSpin	NR	2700 × 12
Canellas et al[45] (2020)	22.27 ± 1.2 (C) 0.93 ± 0.9 (T)	1.39 ± 1.2 (C) 0.70 ± 0.7 (T)	39.69% ± 11.13% (C) 55.96% ± 11.97% (T)	IntraSpin	9-mL tubes (up to 54 mL total)	2700 × 12

Use of PRF for extraction socket healing (soft tissue and pain scores)

Authors (year)	**VAS score (0–1)**	**Healing Index (4–12)**	**Analgesics taken**	**Centrifugation system**	**Volume of tubes for blood drawn**	**Centrifugation parameters speed (rpm) × time (min)**
Marenzi et al[46] (2015)	*At 3 days:* 4.5 ± 0.7 (C) 3.2 ± 0.3 (T) (sig)	*At 7 days:* 4.9 ± 0.3 (C) 4.5 ± 0.5 (T)	NR	IntraSpin	9-mL tubes (18–54 mL total)	2700 × 12
Temmer-man et al[39] (2016)	*At 3 days:* 2.45 (0.09–4.64) (C) 1.65 (0.09–3.12) (T)	NR	NR	IntraSpin	10-mL tubes (20–50 mL total)	2700 × 12
de Almeida Barros Mourão et al[47] (2020)	*At 7 days:* 4.00 ± 1.15 5.12 ± 1.08	*At 7 days:* 3.18 ± 0.65 (C) 3.81 ± 0.54 (T)	*At 7 days:* 1.75 ± 0.85 (C) 1 ± 1.15 (T)	IntraSpin	10-mL tubes	2700 × 12

NR, not reported; C, control group; T, test group; M, male; F, female; VAS, visual analog scale.

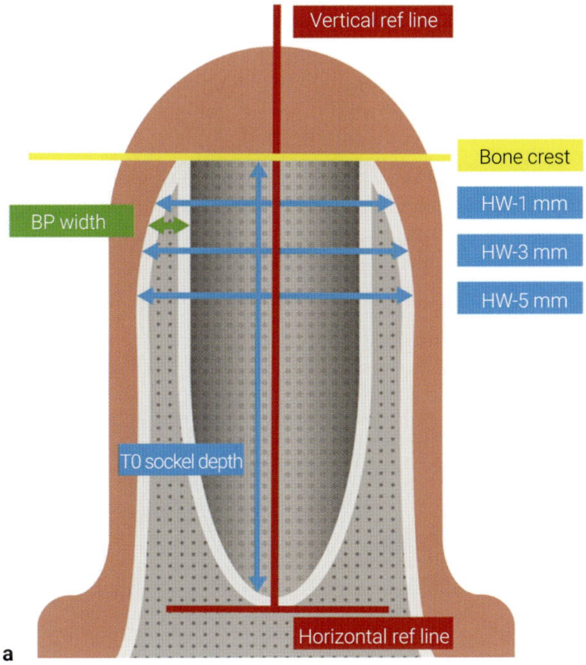

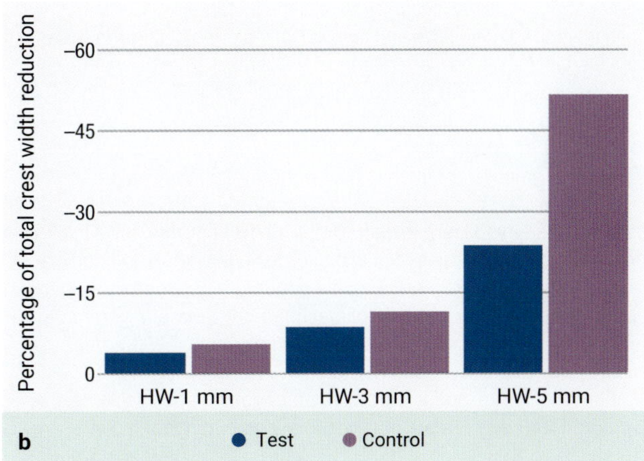

Fig 14-7 *(a)* Cross-sectional slice of a test/control site at baseline (immediately after extraction). HW-1 mm, HW-3 mm, HW-5 mm represent the measurements performed at three levels below the bone crest. The width of the buccal plate (BP width) was measured 1 mm below the crest. The depth of the socket was measured as the deepest point of the socket to the bone crest. *(b)* Changes (expressed in %) in ridge height and width between control (unfilled sockets) and test (sockets filled with PRF) at 3 months based on CBCT measurements. (Reprinted with permission from Temmerman et al.[39])

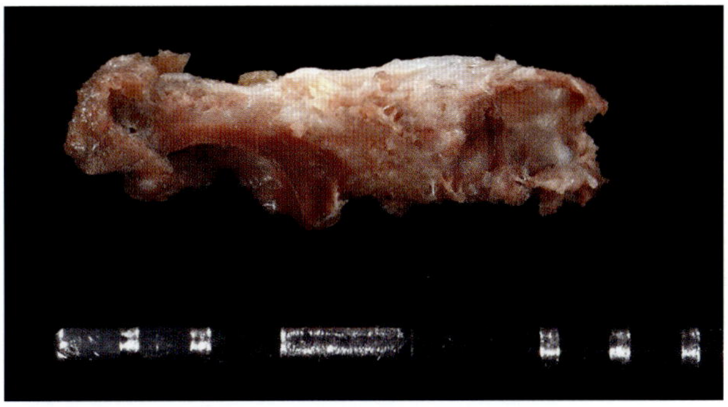

Fig 14-8 A 2 × 7–mm trephined bone biopsy from a human extraction socket filled with PRF plugs, retrieved after 90 days of healing. (Reprinted with permission from Du Toit et al.[40])

ridge preservation nor enhanced bone formation in the PRF group was observed.[38]

Temmerman et al demonstrated that the use of PRF as a socket filling material to achieve preservation of horizontal and vertical ridge dimension at 3 months after tooth extraction is beneficial (Fig 14-7a).[39] In that study, as much as a twofold increase (from 53% to 22%) of the ridge width was better maintained when PRF was utilized postextraction (Fig 14-7b).[39]

Du Toit et al investigated histologically in a split-mouth study of human bone biopsy specimens from extraction sockets treated with PRF versus natural healing at 3 months (Fig 14-8).[40] At the time of implant placement, a trephined bone core was retrieved and processed for histologic evaluation (Fig 14-9). The findings resulted in a 9.9% ± 5.9% gain in newly formed osteoid in the PRF group versus 4% ± 2.1% for specimens derived from the control sites. Due to the low sample size, this was not deemed significant (P = .089; Fig 14-10).[40]

Alzahrani et al investigated the ridge width assessed using cast analysis with the help of an acrylic stent and a pair of calipers.[41] Radiographic analysis of socket surface

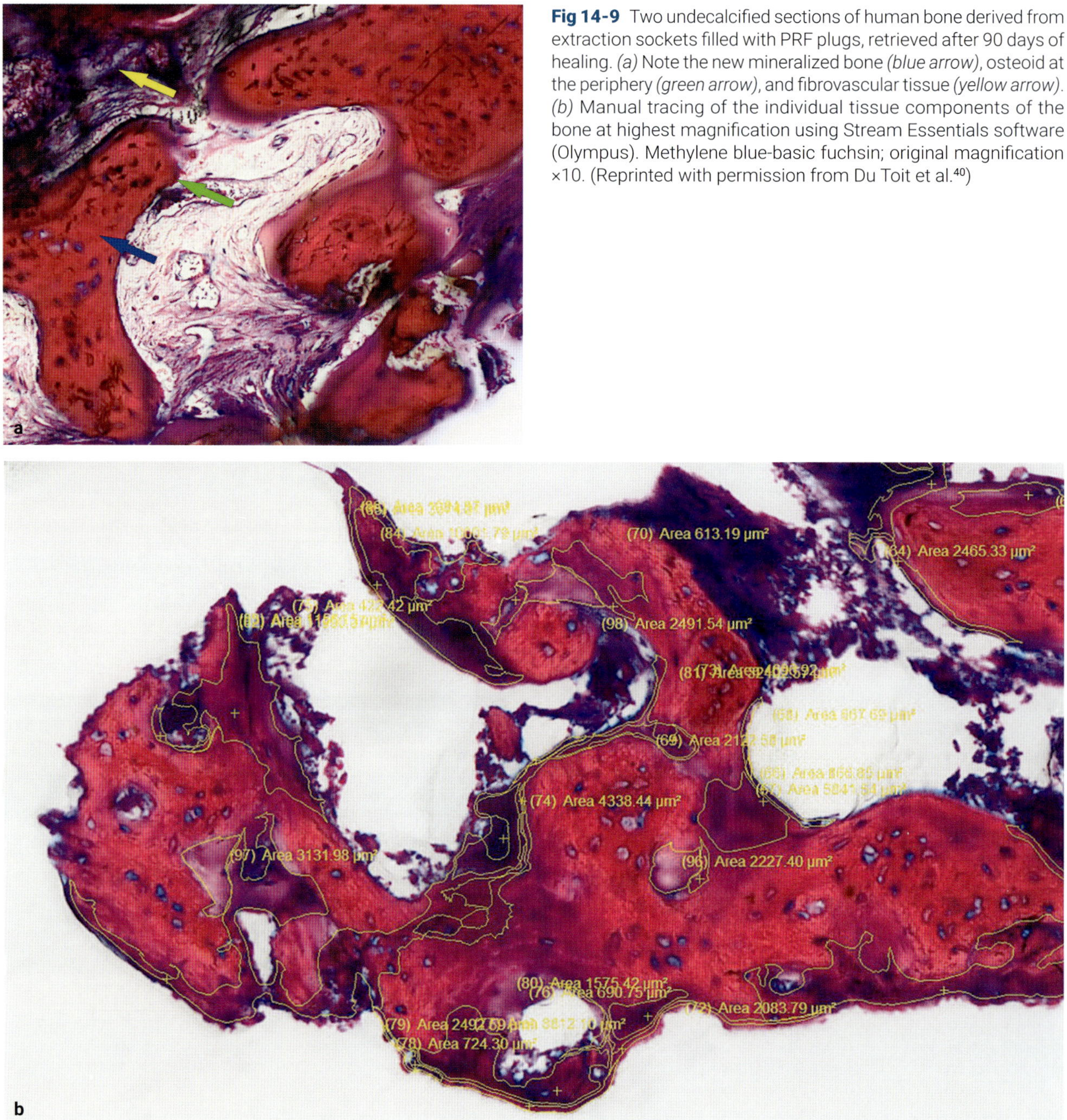

Fig 14-9 Two undecalcified sections of human bone derived from extraction sockets filled with PRF plugs, retrieved after 90 days of healing. *(a)* Note the new mineralized bone *(blue arrow)*, osteoid at the periphery *(green arrow)*, and fibrovascular tissue *(yellow arrow)*. *(b)* Manual tracing of the individual tissue components of the bone at highest magnification using Stream Essentials software (Olympus). Methylene blue-basic fuchsin; original magnification ×10. (Reprinted with permission from Du Toit et al.[40])

area was performed using computer graphic software at 1, 4, and 8 weeks. It was concluded that the mean radiographic bone fill was significantly higher in the PRF group when compared to the control group at all time intervals.[41]

Zhang et al performed CBCT and investigated alveolar ridge height, alveolar ridge width, bone mineral density, and histomorphometry at 3 months.[42] Histologic analysis of new bone formation confirmed that PRF increased the quality and enhanced the rate of bone formation; however, the effect of PRF was not significant enough to reduce alveolar bone resorption in the extraction socket alone.[42]

In 2018, Srinivas et al investigated bone height and density in a split-mouth study using PRF versus natural wound healing.[43] Patients in the PRF group had better Healing Index scores when compared to the control group without PRF and also showed comparable improvements in bone density.[43]

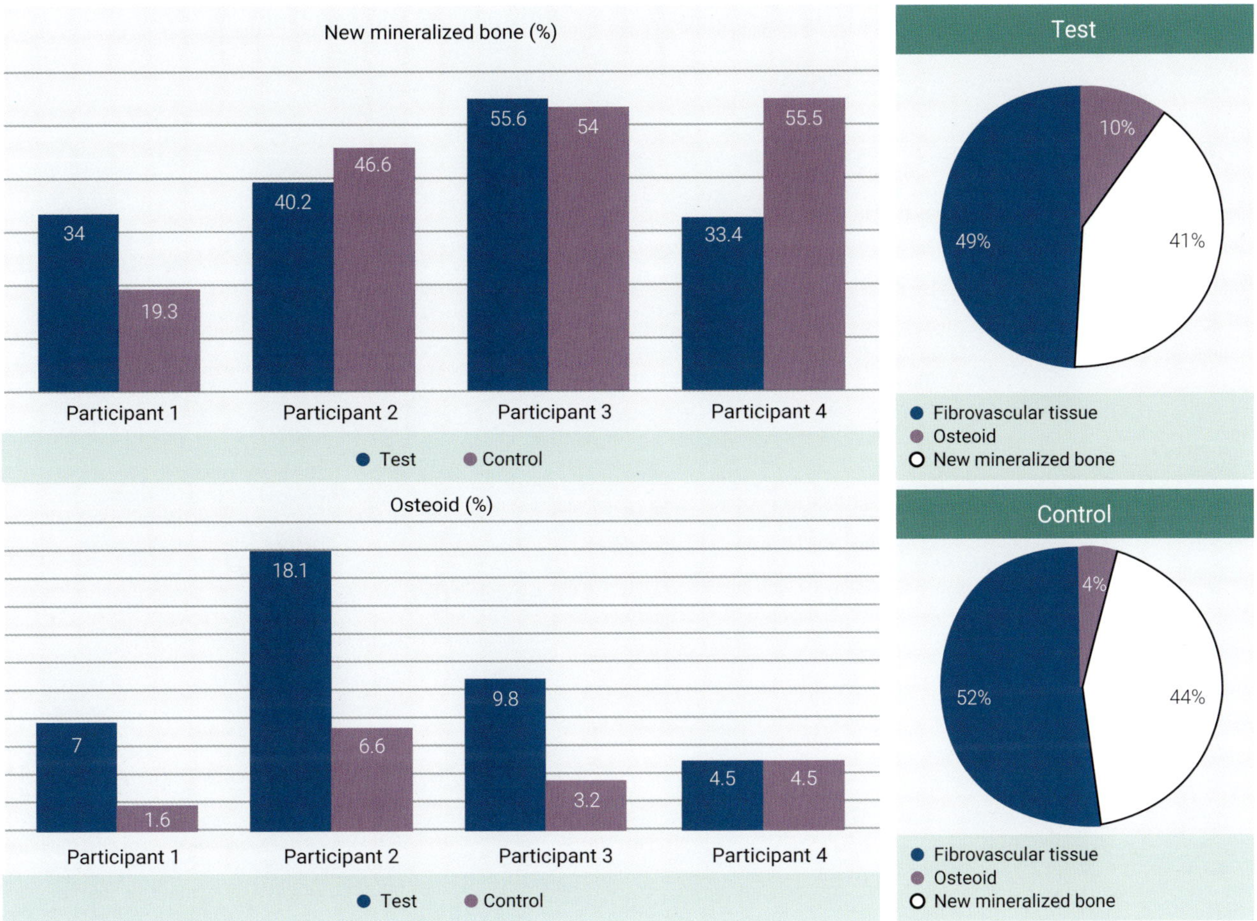

Fig 14-10 Comparative results for PRF and control sites for mineralized bone and osteoid measured (bar graphs) and newly formed bone to fibrovascular tissue ratios for PRF and control sites (pie charts). (Reprinted with permission from Du Toit et al.[40])

In 2019, Areewong et al used histomorphometric analysis to compare new bone formation using PRF versus normal healing.[44] It was concluded that the use of PRF in alveolar ridge preservation does not statistically significantly enhance new bone formation after tooth extraction compared to normal wound healing.[44]

In 2020, Canellas et al performed CBCT immediately after and 3 months after tooth extraction (prior to implant surgery) and compared bone density.[45] A higher percentage of new bone formation in the PRF group was observed compared with the control group.[45]

Furthermore, three studies investigated the effects of PRF on soft tissue healing when compared to controls.[39,46,47] In each of the studies, PRF improved the postoperative pain, promoted the soft tissue healing process, and reduced the early adverse effects of the inflammation.[39,46,47]

In summary, the majority of studies have demonstrated the ability for PRF to limit dimensional changes postextraction. Furthermore, 100% of the studies have shown some benefit for soft tissue wound healing. The results from a recent systematic review (2019) found that among the 588 eligible articles found in the initial search (7 published studies from 2012 through 2019), the authors' qualitative analysis showed that PRF played a positive role in reducing postoperative pain and ridge dimension changes following tooth extraction. Results from their meta-analysis showed that smaller mesial bone height changes and a greater percentage of bone fill were observed in the PRF group.[48] Parallel to those findings, a recent systematic review found that plasma rich in growth factors may bring advantages in some relevant clinical and radiographic outcomes, such as bone density and soft tissue healing, after tooth extraction and may also reduce postoperative adverse events, complications, and patient discomfort.[49]

Table 14-2 Use of PRF versus with a bone grafting material for extraction site management

Authors (year)	Study design Follow-up	No. of participants Sex Mean age	Groups	Smokers (Y, N)	Conclusions
Das et al[50] (2016)	RCT evaluating clinicoradiographic efficacy of PRF and β-TCP-Cl in preserving extraction sockets at 6 months.	26 (30 sites) M: 13 / F: 13 31.2	C: 15, β-TCP-Cl T: 15, PRF	N	The use of either autologous PRF or β-TCP-Cl was effective in socket preservation. Results obtained from PRF were similar to β-TCP-Cl.
Thakkar et al[51] (2016)	RCT evaluating clinical and radiographic bone fill in extraction sockets at 3 and 6 months.	36 (36 sites) M: 22 / F: 14 20–55	C: 18, DFDBA T: 18, DFDBA+PRF	N	The addition of PRF to DFDBA favored the prevention of ridge width reduction, while there was no statistical difference in reduction of ridge height among the groups at different intervals.
Girish Kumar et al[52] (2018)	RCT: PRF was used as a socket plug with or without use of plaster of Paris (POP) as a bone substitute to preserve the alveolar ridge postextraction at 6 months.	48 (90 sites) M: 15 / F: 33 44.4	C1: 30, clot C2: 30, PRF T2: 30, PRF+POP	NR	Sockets that were grafted with POP showed better ridge preservation and postoperative comfort even though the difference in ridge resorption between the three groups was not statistically significant.
Clark et al[53] (2018)	RCT evaluating the efficacy of PRF alone or with FDBA in improving vital bone formation and alveolar dimensional stability during ridge preservation at 15 weeks.	40 (40 sites) NR NR	C: 10, clot C1: 10, PRF T1: 10, FDBA T2: 10, FDBA+PRF	N	Significantly greater loss of ridge height was noted in the blood clot group (3.8 ± 2.0 mm) compared to the PRF (1.8 ± 2.1 mm) and PRF+FDBA (1.0 ± 2.3 mm) groups ($P < .05$). Significantly more vital bone was present in the PRF group (46% ± 18%) compared to the FDBA group (29% ± 14%; $P < .05$).
Mendoza-Azpur et al[54] (2019)	RCT evaluating clinical and histologic differences of using a combination of alloplastic β-TCP and a cross-linked collagen membrane versus autologous L-PRF in ridge preservation after dental extraction and healing at 4 months.	51 (51 sites) M: 21 / F: 30 47.7	C: 25, β-TCP T: 26, PRF	N	PRF concentrate accelerates wound healing in postextraction sockets in terms of new mineralized tissue component. However, the use of β-TCP appears to be superior to maintain buccolingual volume and the final position of the mucogingival junction.
de Angelis et al[55] (2019)	RCT comparing the clinical and radiographic outcomes of different ridge preservation procedures based on the use of PRF versus xenograft at 6 months.	45 (45 sites) M: 19 / F: 26 51.2	C: 15, PRF T1: 15, xenograft T2: 15, xenograft+PRF	N	The PRF group experienced significantly greater horizontal and vertical bone resorption. The L-PRF+xenograft group had less vertical and horizontal bone resorption than the xenograft alone group. Statistically significant differences in postoperative pain and wound healing were observed, with the xenograft group in particular having higher values for pain and experiencing delayed wound healing.
Ouyyamwongs et al[56] (2019)	RCT to assess the potential of using autologous demineralized tooth matrix (aDTM) in combination with PRF membrane (aDTM/PRF) or PRF membrane alone (control) at 2, 4, and 8 weeks.	12 (40 sites) M: 2 / F: 10 20.5	C: 20, PRF T: 20, PRF+dentin	NR	Application of aDTM with PRF membrane is useful for ridge preservation by reducing the horizontal ridge collapse and promoting bone healing as shown clinically and radiographically.

PRF versus a bone grafting material

Table 14-2 demonstrates a total of seven RCTs that have investigated PRF versus or in combination with a bone grafting material.[50–56]

Das et al were one of the first to investigate the use of PRF in comparison to β-tricalcium phosphate with collagen (β-TCP-Cl) in preserving extraction sockets at 6 months.[50] It was found that the use of either autologous PRF or β-TCP-Cl was effective in socket preservation. Results obtained with

Table 14-2 *(cont)*

				Methods for PRF preparation		
Authors (year)	**Horizontal bone loss (mm)**	**Vertical bone loss (mm)**	**Bone density**	**Centrifugation system**	**Volume of tubes for blood drawn**	**Centrifugation parameters speed (rpm) × time (min)**
Das et al[50] (2016)	−1.52 ± 1.22 −0.86 ± 1.76	−1.17 ± 5.96 −0.35 ± 6.52	703.09 ± 76.39 −54.83 ± 104.98	REMI tabletop	10-mL tubes	NR
Thakkar et al[51] (2016)	−1.36 ± 0.70 −1.06 ± 0.38	−1.39 ± 0.5 −1.08 ± 0.43	NR	NR	10-mL glass tubes	3000 × 10
Girish Kumar et al[52] (2018)	3 ± 0.83 3 ± 0.64 2.9 ± 0.8	3.3 ± 0.61 3 ± 0.8 2.8 ± 0.46	63.3% ± 31.9% (C) 94.7% ± 26.9% (T)	NR	10-mL tubes	3000 × 10
Clark et al[53] (2018)	2.9 ± 1.7 2.8 ± 1.2 2.5 ± 1.1 1.9 ± 1.1	3.8 ± 2.0 1.8 ± 2.1 2.2 ± 1.8 1.0 ± 2.3	487 ± 64 493 ± 70 551 ± 58 521 ± 58	Process for PRF	10-mL glass tubes	1300 × 8 (200g)
Mendoza-Azpur et al[54] (2019)	1.16 ± 0.55 2.19 ± 0.80	NR	26.14% ± 7.49% 77.33% ± 9.80%	NR	10-mL tubes	3000 × 10
de Angelis et al[55] (2019)	2.8 ± 0.31 1.12 ± 0.28 1.05 ± 0.23	2.24 ± 0.66 0.58 ± 0.25 0.42 ± 0.48	*Osteoid area (%):* 2.81 ± 1.21 (C) 9.76 + 4.01 (T)	IntraSpin	9-mL tubes	2700 × 12
Ouyyamwongs et al[56] (2019)	1.34 ± 0.37 1.04 ± 0.40	0.70 ± 0.28 0.79 ± 0.47	35.85% ± 15.15% 44.84% ± 9.12%	EBA 20	10-mL tubes	3000 × 10

NR, not reported; C, control group; T, test group; M, male; F, female; DFDBA, demineralized freeze-dried bone allograft; FDBA, freeze-dried bone allograft.

PRF were similar to those obtained with β-TCP-Cl. Thus, it was concluded that the use of this synthetic bone graft was slightly (yet nonsignificantly) superior to PRF.[50]

In a 2016 study by Thakkar et al, DFDBA was utilized for extraction site management with and without PRF and evaluated at 3 and 6 months.[51] It was concluded that the addition of PRF to DFDBA favored the prevention of ridge width reduction, while there was no statistically significant difference in reduction of ridge height among groups at different intervals.[51]

Girish Kumar et al investigated socket fill with PRF alone versus with a plaster of Paris (POP) bone substitute versus that of natural healing (control) at 6 months.[52] It was concluded that sockets that were grafted with POP showed better ridge preservation and postoperative comfort, even though the difference in ridge resorption between the three groups was not statistically significant.[52]

In 2018, Clark et al performed one of the largest studies on the topic in which four groups were evaluated for vital bone formation and alveolar dimensional stability during ridge preservation at 15 weeks: *(1)* natural wound healing, *(2)* PRF, *(3)* FDBA, and *(4)* FDBA+PRF.[53] It was concluded that significantly greater loss of ridge height was noted in the natural healing group (3.8 ± 2.0 mm) compared to the PRF (1.8 ± 2.1 mm) and PRF+FDBA (1.0 ± 2.3 mm) groups ($P < .05$).[53] Significantly more vital bone was present in the PRF group (46% ± 18%) compared to the FDBA group (29% ± 14%; $P < .05$). Overall, a consistent trend was observed that PRF was able to increase vital bone formation, whereas the bone allograft was better able to maintain the ridge width. The combination approach led to the greatest ability to maintain the ridge.[53]

In 2019, Mendoza-Azpur et al investigated β-TCP versus PRF following 4 months of healing.[54] It was concluded once again that PRF accelerated wound healing in postextraction sockets in terms of new mineralized tissue component, but the use of β-TCP biomaterial was superior at maintaining the buccolingual volume and the final position of the mucogingival junction.[54]

In a study by de Angelis in 2019, three groups were compared at 6 months: *(1)* PRF, *(2)* xenograft, and *(3)* xenograft+PRF.[55] It was concluded that the PRF group experienced significantly greater horizontal and vertical bone resorption. The L-PRF+xenograft group had less vertical

Table 14-3 Use of PRF for third molar extractions

Authors (year)	Study design Follow-up	No. of participants Sex Mean age	Groups	Smokers (Y, N)	Conclusions
Gürbüzer et al[57] (2010)	Four weeks after surgery, scintigrams were obtained to evaluate scintigraphic differences between PRF-treated and non–PRF-treated sockets.	14 (14 sites) M: 7 / F: 7 24.9	C: 14, clot T: 14, PRF	N	PRF might not lead to enhanced bone healing in soft tissue impacted mandibular third molar extraction sockets 4 weeks after surgery.
Ruga et al[58] (2011)	Prospective split-mouth study to assess the outcomes of a surgical protocol to remove third molars including the use of ultrasound bone surgery devices and PRF as a grafting material.	14 (28 sites) M: 6 / F: 8 29.5	C: 14, clot T: 14, PRF	Y	Combined action of PRF and piezoelectric surgery can be considered a safe and fine technique for third molar surgery and alveolar socket healing.
Hoaglin and Lines[59] (2013)	Split-mouth study evaluating localized osteitis within 7–10 days of surgery.	100 (200 sites) M: 46 / F: 54 NR	C: 100, clot T: 100, PRF	Y	PRF placement within the extraction sites resulted in 1% occurrence of dry socket versus 9.5% in the control sites.
Eshghpour et al[60] (2014)	Double-blinded split-mouth study wherein bilateral impacted mandibular third molars were extracted; one socket received PRF and the other served as a control.	78 (156 sites) M: 33 / F: 45 25	C: 78, clot T: 78, PRF	N	The overall frequency of alveolar osteitis (AO) was 14.74% for all surgeries. The frequency of AO in the PRF group was significantly lower than in the non-PRF group.
Kumar et al[61] (2015)	RCT The outcome variables were pain, swelling, maximum mouth opening, periodontal pocket depth, and bone formation, with a follow-up period of 3 months.	31 (31 sites) NR 26.1	C: 15, clot T: 16, PRF	NR	The application of PRF lessens the severity of immediate postoperative sequelae, decreases preoperative pocket depth, and hastens bone formation.
Yelamali and Saikrishna[62] (2015)	Split-mouth study of 20 patients who underwent bilateral extraction of impacted third molars with PRP vs PRF. Both soft tissue at 1 week and bone at 4 months were compared.	20 (40 sites) M: 12 / F: 8 25.5	C: 20, PRP T: 20, PRF	NR	PRF is significantly better in promoting soft tissue healing and also faster regeneration of bone after third molar extraction in comparison with PRP.

and horizontal bone resorption than the xenograft alone group. Statistically significant differences in postoperative pain and wound healing were observed in the xenograft alone group, which had higher values for pain and delayed wound healing.[55]

In a final study by Ouyyamwongs et al in 2019, the use of PRF was compared to PRF with a dentin graft. It was concluded that application of the autologous dentin with PRF was useful for ridge preservation by reducing the horizontal ridge collapse and promoting bone healing as shown clinically and radiographically.[56]

In conclusion, it has been well established that the use of a bone grafting material is better able to maintain ridge dimension (especially ridge width), whereas the additional use of PRF in combination with the bone grafting material may favorably improve new vital bone formation in the area.

PRF for mandibular third molar extractions

Table 14-3 demonstrates the data from 23 RCTs evaluating the use of PRF after extraction of mandibular third molars when compared to natural wound healing.[57–79] In nearly all the studies, PRF proved to add value in either lowering the rates of dry sockets, improving soft tissue healing, and/or decreasing postoperative pain and use of analgesics (Table 14-3).

In one of the first comparative studies investigating PRF for mandibular third molars, Hoaglin and Lines investigated in a split-mouth study the number of alveolar infections (dry

Table 14-3 *(cont)*

Authors (year)	Study design Follow-up	No. of participants Sex Mean age	Groups	Smokers (Y, N)	Conclusions
Uyanik et al[63] (2015)	Bilateral third molar extraction. Parameters assessed at baseline for each patient included pain, the number of analgesics taken, trismus, and cheek swelling. These variables were also assessed on postoperative days 1, 2, 3, and 7.	20 (40 sites) M: 10 / F: 10 22.5	C1: 10 T1: 10, PRF T2: 10, Piezosurgery alone T3: 10, Piezosurgery+PRF	N	Patients in the PRF group had better Healing Index scores compared to groups without PRF. Use of PRF showed a comparable increase in bone density too.
Ozgul et al[64] (2015)	RCT The pain was evaluated in the postoperative period using a VAS of 100.	56 (112 sites) M: 23 / F: 33 50.7	C: 56, clot T: 56, PRF	NR	Horizontal and vertical measurements showed more swelling at the control side (without PRF) on the 3rd day postoperatively ($P < .05$). There were no statistically significant differences regarding pain among the groups.
Baslarli et al[65] (2015)	At days 30 and 90 postoperative, panoramic images and bone scintigrams were taken to evaluate the bone healing between PRF-treated and non–PRF-treated sockets.	20 (40 sites) M: 7 / F: 13 23.9	C: 20, clot T: 20, PRF	N	PRF might not lead to enhanced bone healing in impacted mandibular third molar extraction sockets 30 and 90 days postsurgery.
Dutta et al[66] (2016)	RCT Pain was assessed using VAS (1 to 10) at intervals of 1, 2, 3, and 4 days. The soft tissue healing was evaluated at 3, 7, 14, and 21 days using the modified Healing Index (4 to 12).	26 (108 sites) M: 9 / F: 17 53	C: 54, clot T: 54, PRF	N	PRF improved the postoperative pain, promoted the soft tissue healing process, and reduced the early adverse effects of the inflammation.
Bilginaylar et al[67] (2016)	RCT (split-mouth) VAS score	22 (44 sites) M: 5 / F: 7 54	C: 22, clot T: 22, PRF	N	In this study, a significantly reduced level of postoperative pain could be seen at sites treated with PRF during the early healing phases.
Varghese et al[68] (2017)	One week after extraction, an assessment of soft tissue healing around the sockets was performed using the Healing Index. Also, postoperative pain by VAS and the number of consumed analgesic tablets were recorded.	32 (32 sites) M: 13 / F: 19 37	C: 16, clot T: 16, PRF	N	The results of the present study demonstrate that whenever improved healing of the extraction socket is needed, the use of PRF should be considered. In addition, the use of PRF decreases postoperative pain and discomfort.
Sharma et al[69] (2017)	Single-arm clinical trial assessing the efficacy of PRF on the pain and healing of the extraction socket related with established AO.	100 (100 sites) M: 79 / F: 21 18–40	C: 50, clot T: 50, PRF	NR	There was significant reduction in pain associated with AO 3 and 7 days after PRF placement day in the extraction socket, along with marked decrease in the degree of inflammation on day 3. There was better wound healing by the end of the second week.
Gülşen et al[70] (2017)	Prospective split-mouth study Postoperative edema was measured with a flexible tape measure by calculating the distance between several facial landmarks on postoperative days 2 and 7. Postoperative pain was evaluated with a line-type VAS and a verbal scale.	30 (60 sites) M: 21 / F: 9 20.0	C: 30, clot T: 30, PRF	NR	Both groups recorded significant improvement compared to the baseline levels in almost all of the outcome variables. There was no statistically significant difference between the study and control groups.

NR, not reported; C, control group; T, test group; M, male; F, female.

Table 14-3 *(cont)* Use of PRF for third molar extractions

Authors (year)	Study design Follow-up	No. of participants Sex Mean age	Groups	Smokers (Y, N)	Conclusions
Asutay et al[71] (2017)	Double-blinded, split-mouth randomized study evaluating pain, swelling, and trismus levels following third molar surgery.	30 (60 sites) M: 24 / F: 6 20.32	C: 30, clot T: 30, PRF	N	No significant differences were reported between the control and study groups regarding postoperative pain, swelling, and trismus. In the control group, 10% reported dry sockets versus 3.3% in the PRF group.
Unsal et al[72] (2018)	Split-mouth randomized study evaluating postoperative pain levels and periodontal probing depth following extraction of bilaterally symmetric partially erupted mandibular third molars.	50 (100 sites) M: 33 / F: 17 23.96	C: 50, clot T: 50, PRF	Y	None of the smokers in the PRF group and 37.5% of smokers in the control group were diagnosed with AO.
Jeyaraj and Chakra-narayan[73] (2018)	Parallel, prospective study evaluating differences in soft tissue healing and bony regeneration of impacted mandibular third molar extraction sites with and without PRF.	60 (60 sites) NR NR	C: 30, clot T: 30, PRF	NR	It was found that the study group in which autologous PRF had been incorporated into the operative site exhibited quick and complication-free soft tissue healing as well as a much quicker reossification and bone fill of the extraction socket.
Dar et al[74] (2018)	Prospective clinical split-mouth study evaluating the effectiveness of PRF on soft tissue healing and bone tissue healing in terms of postoperative pain, postoperative swelling, soft tissue healing, and the quality of bone healing at the mandibular third molar socket.	30 (60 sites) M: 33 / F: 17 23.6	C: 30, clot T: 30, PRF	N	The mean postoperative pain score (VAS) and swelling were lower for the PRF group at all time points.
Daugela et al[75] (2018)	Split-mouth RCT evaluating wound healing, patient postoperative discomfort, and incidence of AO following third molar extraction.	34 (68 sites) M: 20 / F: 14 22.8	C: 30, clot T: 30, PRF	N	Sites treated with PRF resulted in improved pain VAS scores in the first postoperative week and significant reduction in facial swelling. In the control group, 11.8% reported dry sockets versus 0% in the PRF group.
Zahid and Nadershah[76] (2019)	Split-mouth, randomized, double-blind clinical trial evaluating pain, swelling, and healing at day 7 following third molar extraction.	10 (20 sites) M: 10 / F: 10 24	C: 10, clot T: 10, PRF	N	The findings of this study demonstrate PRF as a potential biomaterial for lessening the severity of pain and swelling after third molar surgery.
Ritto et al[77] (2019)	Prospective, double-blind, split-mouth study assessing postoperative bone regeneration, pain, and soft tissue healing.	17 (34 sites) M: 7 / F:10 21.8	C: 17, clot T: 17, PRF	N	The application of PRF improved bone density, which was higher in the test group ($P = .007$). There was no statistical difference related to pain or soft tissue healing between the groups ($P > .05$).
Yüce and Kömerik[78] (2019)	Parallel study evaluating pain, soft tissue healing, and bone density.	40 (40 sites) M: 22 / F: 18 31.2	C: 20, saline rinse T: 20, PRF	N	PRF application demonstrated rapidly and continually reduced pain intensity at each respective time point in comparison to the control and faster healing rates of epithelium and hard tissue.
Afat et al[79] (2019)	Parallel controlled study The primary outcome variable was healing score for the mucosa over the extraction socket on days 7, 14, and 21. Secondary outcome variables were frequency of postoperative complications, hemorrhagic complications, AO, and postoperative wound infection.	60 (60 sites) M: 38 / F: 22 22.4	C: 20, clot T1: 20, PRF T2: 20, PRF+HA (hyaluronic acid)	N	Mean healing scores for the mucosa on days 7, 14, and 21 were significantly better in the PRF+HA group than the control group. No dry sockets were reported in either of the test groups with PRF.

Table 14-3 *(cont)*

				Methods for PRF preparation		
Authors (year)	**Dry sockets (%)**	**VAS scores/pain**	**Soft and hard tissue healing**	**Centrifugation system**	**Volume of tubes for blood drawn**	**Centrifugation parameters speed (rpm) × time (min)**
Gürbüzer et al[57] (2010)	NR	NR	*Scintigraphic analysis:* 4.61 ± 1.02 4.54 ± 1.03	Universal 320 (Hettich)	10-mL tubes	400g (2030 rpm) × 10
Ruga et al[58] (2011)	NR	5.42 (C) 4.92 (T)	*Insufficient cases (healing):* 3 of 14 cases (C) 0 of 14 cases (T)	Process for PRF	9-mL tubes (18–54 mL total)	2700 × 12
Hoaglin and Lines[59] (2013)	9.5% (C) 1% (T)	NR	NR	NR	8.5–10 mL	2700 × 10–12
Eshghpour et al[60] (2014)	20.5% (C) 9.0% (T)	NR	NR	Labofuge 400R (Heraeus)	10-mL tubes	3000 × 10
Kumar et al[61] (2015)	NR	*Slight/severe pain:* 60% (C) 12.5% (T) *Slight/severe swelling:* 53.3% (C) 18.8% (T)	NR	NR	5 mL	3000 × 10
Yelamali and Saikrishna[62] (2015)	NR	NR	*Soft Tissue Index:* 4.10 ± 0.97 (C) 4.60 ± 0.60 (T) *Bone Index:* 145.21 ± 11.38 (C) 156.11 ± 9.75 (T)	NR	6 mL	3000 × 10
Uyanik et al[63] (2015)	NR	*Pain scores:* 74.60 ± 35.21 (C) 25.00 ± 18.99 (T1) 48.51 ± 44.15 (T2) 24.45 ± 14.95 (T3) *No. of analgesics taken:* 9.4 ± 4.81 (C) 5.6 ± 3.02 (T1) 9.5 ± 6.11 (T2) 4.3 ± 2.94 (T3)	NR	Elektro-mag M 415 P	10-mL tubes	3000 × 10
Ozgul et al[64] (2015)	NR	*Swelling:* 3.62 ± 3.51 (C) 1.83 ± 2.52 (T) *Pain:* 26.48 ± 30.36 (C) 25.50 ± 29.95 (T)	NR	NR	10-mL tubes	3000 × 10
Baslarli et al[65] (2015)	NR	NR	*Bone healing (3 months):* 3.96 ± 1.00 (C) 4.10 ± 1.10 (T)	NR	9 mL	3000 × 10
Dutta et al[66] (2016)	NR	NR	*Bone healing (3 months):* 3.96 ± 1.00 (C) 4.10 ± 1.10 (T)	NR	9 mL	3000 × 10

NR, not reported; C, control group; T, test group; M, male; F, female.

Table 14-3 *(cont)* Use of PRF for third molar extractions

				Methods for PRF preparation		
Authors (year)	**Dry sockets (%)**	**VAS scores/pain**	**Soft and hard tissue healing**	**Centrifugation system**	**Volume of tubes for blood drawn**	**Centrifugation parameters speed (rpm) × time (min)**
Bilginaylar et al[67] (2016)	*At 3 days:* 4.5 ± 0.7 (C) 3.2 ± 0.3 (T) (sig)	*At 7 days:* 4.9 ± 0.3 (C) 4.5 ± 0.5 (T)	NR	IntraSpin	9-mL tubes (18–54 mL total)	2700 × 12
Varghese et al[68] (2017)	*At 3 days:* 2.45 (0.09–4.64) (C) 1.65 (0.09–3.12) (T)	NR	NR	IntraSpin	10-mL tubes (20–50 mL total)	2700 × 12
Sharma et al[69] (2017)	NR	*At 3 days:* 2.38 ± 0.48 (C) 1.03 ± 0.17 (T)	NR	R-8C BL (REMI)	10-mL tubes	3000 × 12
Gülşen et al[70] (2017)	NR	*At 3 days:* 8.0 ± 12.3 (C) 7.9 ± 12.1 (T)	NR	NF 200 (Nüve)	10-mL tubes (30 mL total)	3000 × 10
Asutay et al[71] (2017)	3/30 = 10% (C) 1/30 = 3.33% (T)	*At 3 days:* 17.73 ± 24.90 (C) 22.00 ± 23.77 (T)	NR	NR	10-mL tubes	2700 × 12
Unsal et al[72] (2018)	9/50 = 18% (C) 4/50 = 8% (T) *In smokers:* 6/16 = 37.5% (C) 0/16 = 0% (T)	*At 3 days:* 1.40 (C) 0.88 (T)	NR	NF 200	10-mL tubes	3000 × 10
Jeyaraj and Chakra-narayan[73] (2018)	NR	*At 3 days:* 3.30 ± 1.18 (C) 1.50 ± 1.01 (T)	*Bone density:* 0.87 ± 0.73 (C) 2.53 ± 0.57 (T)	NR	10-mL tubes (20 mL total)	2700 × 12
Dar et al[74] (2018)	NR	*At 3 days:* 2.20 ± 1.86 (C) 0.50 ± 0.94 (T)	*Soft tissue healing:* 2.50 ± 0.57 (C) 2.83 ± 0.65 (T)	NR	10 mL tubes	3000 × 10
Daugela et al[75] (2018)	4/34 = 11.8% (C) 0/34 = 0 % (T)	*At 3 days:* 3.13 ± 1.28 (C) 1.67 ± 0.88 (T)	NR	EBA 20	9-mL tubes (20 mL total)	2800 × 12
Zahid and Nader-shah[76] (2019)	NR	5 (4.25–6.75) (C) 2.5 (2–4.75) (T)	NR	Duo Quattro	10-mL tubes (20 mL total)	1300 × 13
Ritto et al[77] (2019)	NR	*Pain at 3 days:* 3.11 ± 2.61 (C) 2.85 ± 2.17 (T)	*Bone density:* 522.5 ± 352.3 (C) 954.1 ± 500.8 (T)	NR	10-mL tubes	2700 × 12
Yüce and Kömerik[78] (2019)	NR	3.6 ± 1.19 (C) 0.45 ± 0.51 (T) *Use of analgesics:* 13.05 ± 1.32 (C) 3.6 ± 1.19 (T)	*Epithelial healing:* 3.55 ± 0.6 (C) 4.8 ± 0.41 (T) *Bone density:* 95.8 ± 6.39 (C) 121.2 ± 6.35 (T)	Duo Quattro	10-mL tubes	1300 × 8
Afat et al[79] (2019)	1/20 = 5% (C) 0/20 = 0% (T1) 0/20 = 0% (T2)	*At 7 days:* 2.75 ± 0.55 (C) 1.6 ± 0.5 (T1) 1.5 ± 0.51 (T2)	NR	Elektro-mag M 615 P	10-mL tubes (20 mL total)	3000 × 10

NR, not reported; C, control group; T, test group; M, male; F, female.

sockets) resulting following natural healing versus healing with PRF.[59] It was found that placement of PRF within the extraction sites was associated with a 1% rate of dry sockets versus 9.5% in the control sites (nearly a 10-fold reduction).[59] Eshghpour et al also found that the frequency of alveolar osteitis in the PRF group was significantly lower than in the non-PRF group.[60] In a final study carried out by a split-mouth randomized design, 78 mandibular third molar sockets treated with PRF were compared to 78 controls.[60] Overall, alveolar osteitis occurred more in their study, but once again it was concluded that treating sockets with PRF reduced the incidence of this complication by about half (PRF 9% vs control 20.5%).[60]

PRF has more frequently been investigated for its ability to decrease postoperative pain and reduce the number of analgesics taken by the patient. Pain following third molar extractions typically arises largely from soft tissue incision, trauma, and stretching of tissues. In a study by Bilginaylar and Uyanik, patients were assessed for pain, the number of analgesics taken, trismus, and swelling on days 1, 2, 3, and 7 following removal of impacted third molars.[67] A significant reduction in pain on days 1, 2, and 3, and in the number of analgesics taken on days 2 and 3, was observed in the PRF group. The findings from this study indicated that *(1)* PRF was able to reduce postoperative pain largely due to the improvements in soft tissue healing, while reducing infections due to the presence of microbial-fighting immune cells (leukocytes), and *(2)* PRF led to a reduction in the number of analgesics required by these patients to manage their pain level.[67] The ability for PRF to decrease postoperative pain has been reported in over a dozen studies (see Table 14-3).

Interestingly, another study proposed using PRF to prevent postoperative infection in extraction sockets of mandibular third molars.[59] The rationale is based on the local delivery of concentrated leukocytes, which are able to participate in an inflammatory response well understood to ward off infective pathogens. This retrospective report compared 200 mandibular third molar sites treated with versus without PRF. The results showed only two cases of localized osteitis with PRF treatment versus 19 cases in the control group (1% vs 9.5% respectively).[59] While the etiologies of localized osteitis are multiple and not well understood, the authors also point to the fact that the non-PRF treatment group (controls) required an additional 6.5 hours of clinical time to manage local infections, necessitating extra surgical time and costs to resolve these problems. These authors demonstrate that preventive treatment of localized osteitis can be accomplished using a low-cost, autogenous, soluble, biologic material, and that PRF enhanced third molar socket healing/clot retention and greatly decreased the clinical time required for postoperative management of infection.[59]

Clinical bottom line

1. The use of PRF alone has been shown to limit dimensional changes when compared to natural healing.
2. The use of a bone grafting material has more favorably been shown to limit dimensional changes when compared to PRF. When considering bone graft versus PRF, it is important to note that more horizontal bone loss will occur in the PRF alone group.
3. The use of PRF in combination with a bone grafting material has been shown to favor significantly more vital bone formation when compared to bone graft alone. Thus, there may potentially be advantages of combining bone grafts with PRF, though data remains inconclusive.
4. The use of PRF has been shown to favor better soft tissue healing when utilized as an outer barrier membrane, with improvements in VAS scores and reported patient pain scores as well as reduced number of analgesics taken.
5. Several studies have now shown that PRF significantly lowers the rate of dry sockets and postoperative discomfort following third molar tooth extraction.

Clinical Procedure

Based on the evidence from the literature, it is known that the use of PRF alone for extraction site management will limit the dimensional changes. Furthermore, its use has been shown to better maintain ridge dimensions, as presented in Fig 14-11, when compared to natural healing. Typically for such procedures, compressed PRF plugs are formed (reviewed in chapter 10) and placed within the extraction site. Ideally, it is advised to use as many as possible. Following their incorporation into the extraction

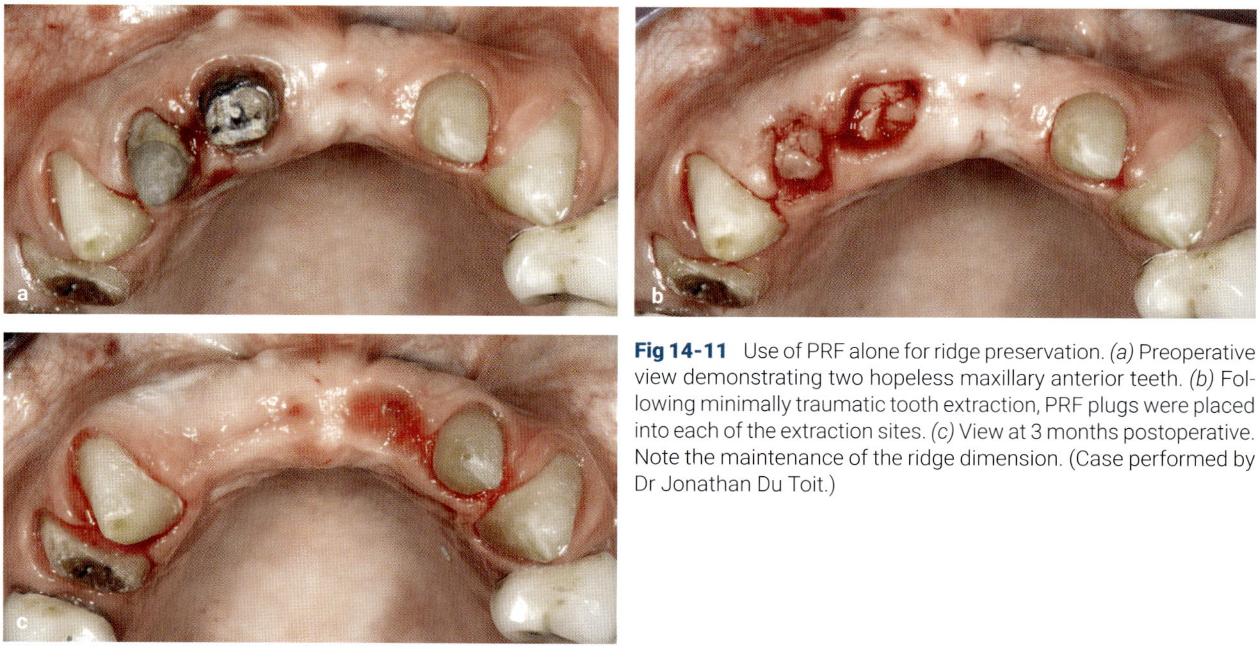

Fig 14-11 Use of PRF alone for ridge preservation. *(a)* Preoperative view demonstrating two hopeless maxillary anterior teeth. *(b)* Following minimally traumatic tooth extraction, PRF plugs were placed into each of the extraction sites. *(c)* View at 3 months postoperative. Note the maintenance of the ridge dimension. (Case performed by Dr Jonathan Du Toit.)

site, a typical cross-suture is utilized. PRF may be left exposed to the oral cavity without difficulty. Within a 3- to 4-month period, implant placement is then advised.

> ***Note***
>
> If PRF alone is utilized, it is important NOT to wait much longer than 4 months after grafting for implant placement because the ridge will continue to collapse should implant placement not be performed. Remember that the PRF scaffold lasts only a 2-week period, and therefore the turnover rate is extremely fast compared to a bone grafting material.

A similar approach is taken when third molars (especially mandibular third molars) are extracted. Typically, compression of PRF clots into PRF plugs is not needed because the goal is simply to fill the extraction socket with a fibrin network to avoid the rate of dry sockets and/or complications as well as to improve patient-perceived pain and reduce the number of analgesics taken. In such cases, a full-size PRF clot is placed into the extraction, as demonstrated in Fig 14-12.

The data from a number of RCTs now point to the fact that the use of a bone grafting material better maintains dimensional stability of the ridge postextraction (see Table 14-2). Therefore, it is highly advised to use a bone grafting material for such cases, preferably a bone allograft.[7] In such cases, the use of PRF can be mixed with a bone allograft (Fig 14-13) in a sticky bone type consistency (see chapter 10). Thereafter, a PRF membrane that has been flattened can be utilized overtop the extraction site to improve soft tissue healing.

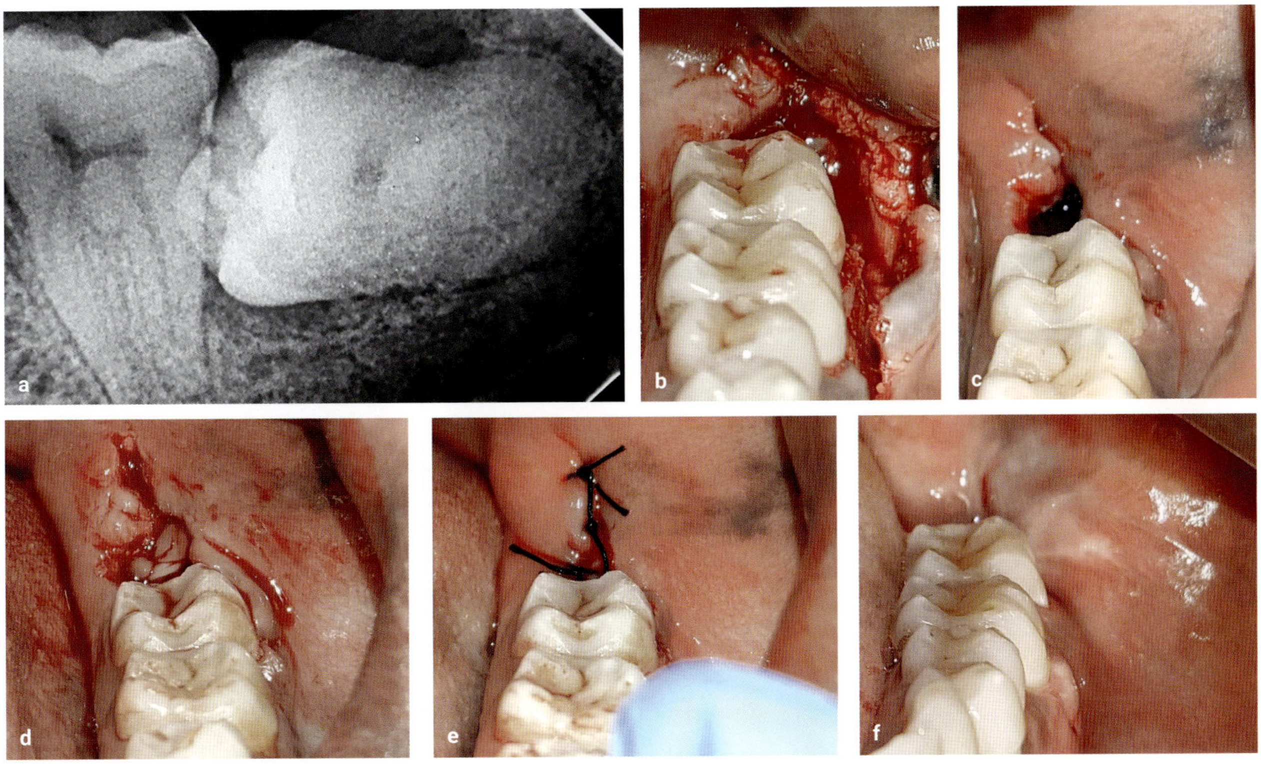

Fig 14-12 Use of PRF for third molar extraction. *(a)* Teenage patient presents with horizontally impacted third molar requiring surgical extraction. *(b)* Following flap elevation, tooth extraction was performed. *(c)* Following tooth removal, note the large soft tissue defect. *(d)* PRF placement within the extraction site to favor wound healing. *(e)* Soft tissue closure. *(f)* At 14 days postoperative, note the soft tissue healing. This case demonstrates the successful use of PRF to minimize the risk of dry sockets following third molar removal. (Case performed by Dr Senthil Selvan.)

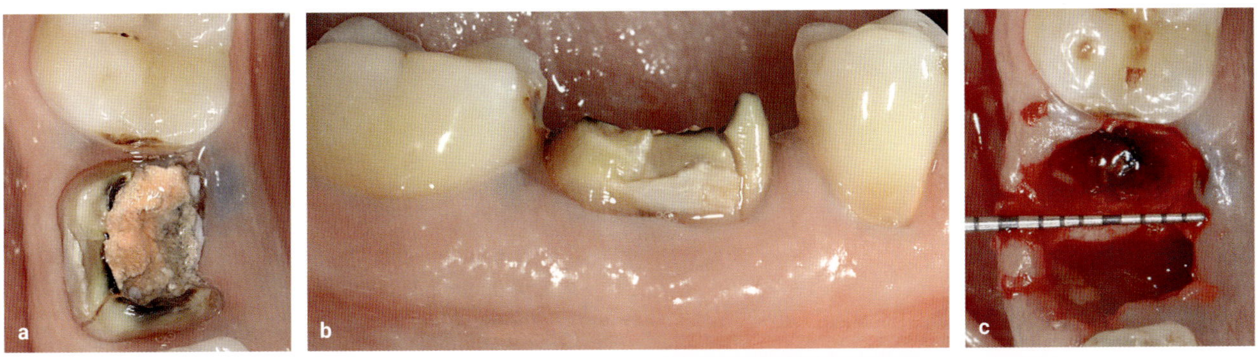

Fig 14-13 Use of PRF in combination with a bone allograft for ridge preservation. *(a and b)* Occlusal and buccal views of a hopeless mandibular first molar. *(c)* Occlusal view following minimally traumatic tooth extraction. *(d)* Mixing of liquid-PRF with a bone allograft. *(e)* Placement of the sticky bone complex into the extraction site.

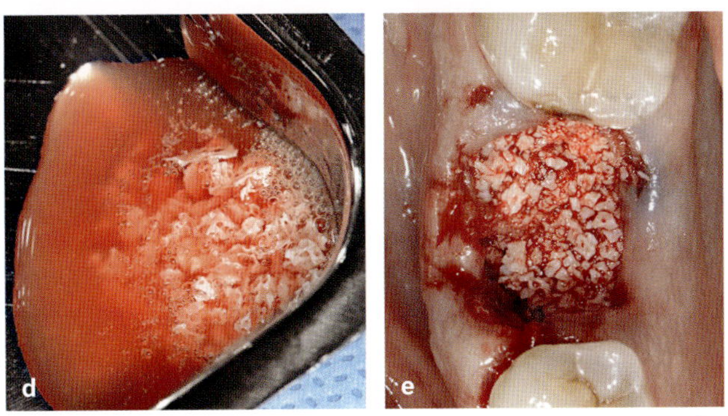

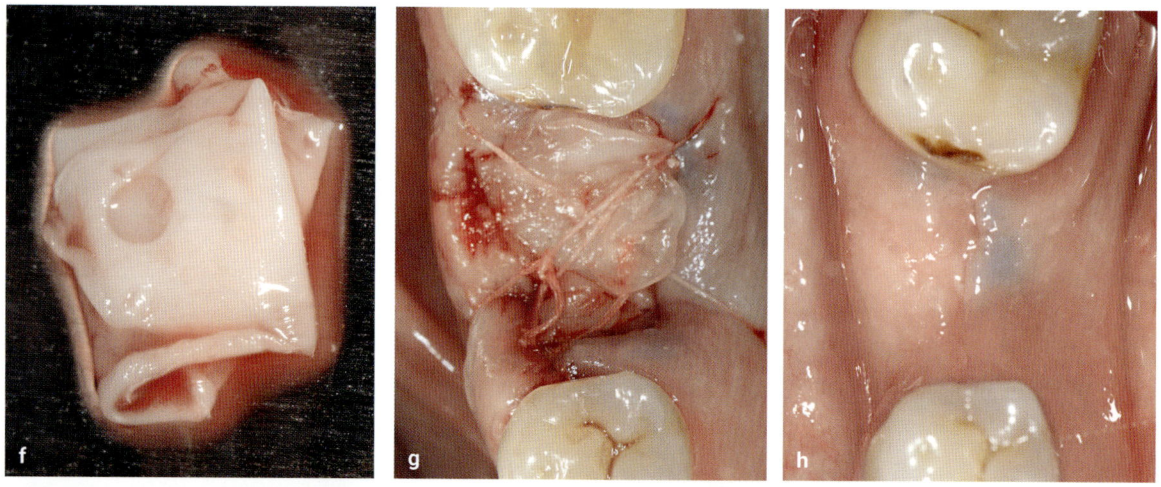

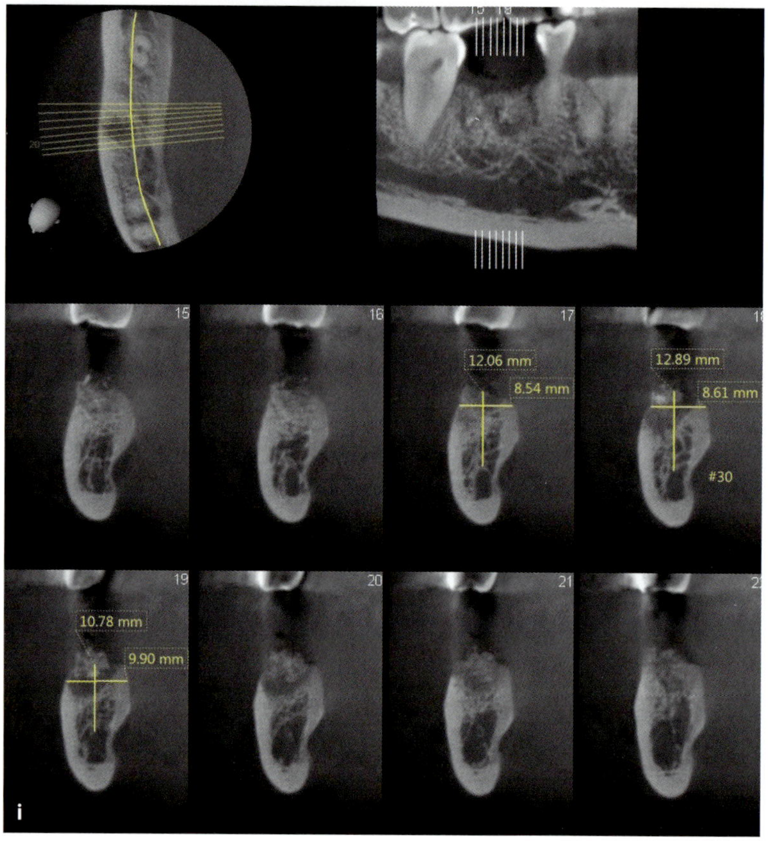

Fig 14-13 *(cont)* *(f)* A PRF membrane was created and folded to be placed over the extraction socket. *(g)* Final occlusal view after tooth extraction and socket grafting with the PRF membrane placed overtop the bone graft complex followed by a final cross-suture. *(h)* At 4 months postoperative, note the excellent maintenance of the ridge. *(i)* CBCT visualization of the maintenance of the ridge in both width and height. (Case performed by Dr Alfonso Gil.)

Figure 14-14 shows a similar case whereby PRF was mixed with a synthetic bone grafting material in a country where bone allografts are not permitted. A similar approach can be utilized favoring a better ability to maintain ridge width and height by incorporating a bone graft within the grafted area.

Implications for Clinical Practice and Future Directions

Over the past decade, much research has focused on minimizing the dimensional changes occurring postextraction by utilizing a variety of bone biomaterials. Most frequently,

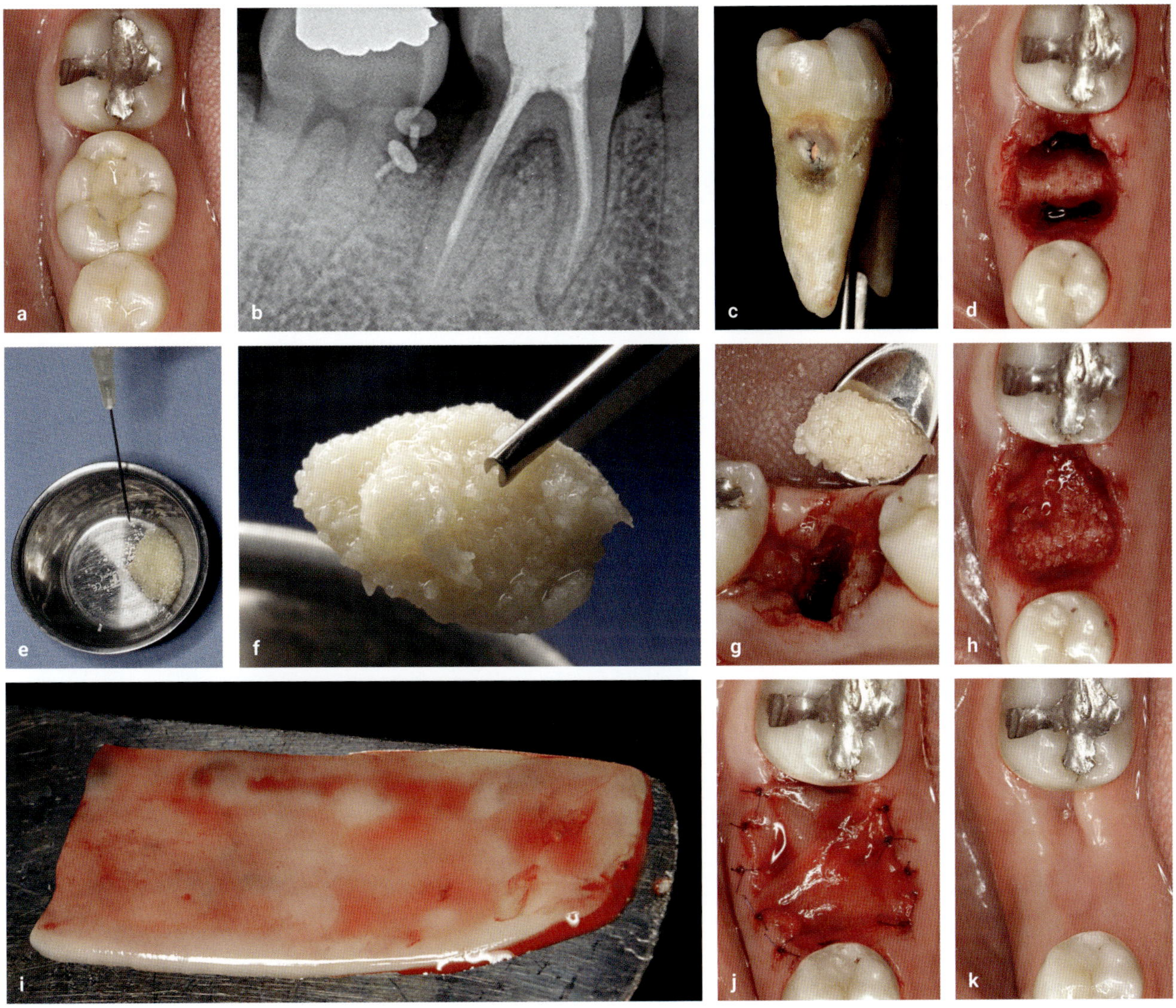

Fig 14-14 Use of PRF in combination with a synthetic bone grafting material for ridge preservation. *(a and b)* Clinical and radiographic views of a hopeless mandibular first molar. *(c)* Following tooth extraction, note the large lesion. *(d)* Occlusal view following tooth removal. *(e)* Addition of liquid-PRF to a synthetic bone grafting material. *(f)* Final sticky bone complex. *(g and h)* Filling of the extraction socket with this synthetic sticky bone. *(i and j)* Use of a PRF membrane for soft tissue coverage. *(k)* Healing at 8 weeks postoperative. (Case performed by Dr Jonathan Du Toit.)

bone grafting materials have been utilized because of their better ability to maintain ridge dimension. Furthermore, various barrier membranes, including collagen, PTFE, and more recently PRF, have frequently been utilized to either prevent soft tissue infiltration or promote faster soft tissue healing, both favoring new bone formation. One of the limitations to PRF is the fact that it resorbs within a 2-week healing period and therefore is not necessarily the "ideal" biomaterial when it comes to excluding soft tissue infiltration. Nevertheless, numerous studies have demonstrated that when utilized above a bone grafting material, the grafted site is able to maintain both ridge height and width similar to other barriers (see Table 14-2). Interestingly, the advancements made with respect to the Bio-Heat technology of PRF brings new avenues of possibilities with the ability to extend the working properties of PRF from 2–3 weeks to 4–6 months.

It is also well known today from various systematic reviews that, on average, dimensional changes postextraction are in the 0.5- to 1-mm range when utilizing a bone grafting material. Most RCTs investigating the use of PRF fall more frequently in the 1.5- to 2.5-mm range, with some studies demonstrating no difference when compared to simple natural healing (see Table 14-1). Thus, even in well-conducted studies such as that presented in Fig 14-7, which demonstrated a 23% reduction in ridge volume when compared to 52% during natural wound healing, the 22% still represents roughly a 1.5-mm ridge width loss, which remains more than the data found in various systematic reviews on the topic when utilizing bone grafting materials alone, as well as when compared to combining PRF with a bone grafting material, as highlighted in Table 14-3. This important finding highlights the need for considering a bone grafting material, especially when the buccal and/or lingual plate is compromised.

Furthermore, over the past 15 years, research has also convincingly demonstrated the negative impact of raising flaps during tooth extraction. Because the periodontal ligament (the main source of blood supply) is removed following tooth loss, it has been the focus of much further research to evaluate the effect of flap elevation during routine extractions (thereby also removing the blood supply from the periosteum).[80] Today, it is highly recommended that tooth extraction be performed as atraumatically as possible without the use of flap elevation when possible (the exception, of course, being impacted third molars).

The effects of tooth loss on dimensional changes have now been extensively reported in the literature.[5,10,19–24] Due to considerable variations in humans, clinicians need to be aware that not all sockets resorb equally, not all patients are genetically predisposed to resorb equally, and PRF may contribute to ridge preservation differently between sites and patients. For instance, Chappuis et al showed convincingly that facial bone thickness in the esthetic zone is a critical factor affecting potential facial bone resorption.[19]

One benefit of socket grafting with PRF is the improvements in soft tissue wound healing.[81] Together with the findings reported in chapters 12 and 13 on gingival recession coverage and periodontal regeneration, it becomes increasingly clear that PRF acts to primarily promote soft tissue wound healing. There are additional studies now found in the literature that clearly point to the fact that *(1)* soft tissue heals faster when PRF is utilized as an outer barrier layer and *(2)* specifically in third molars, PRF dramatically reduces the rate of dry sockets nearly 10-fold. These improvements lead to fewer postoperative complications, reduced postoperative patient pain, as well as reduced number of analgesics taken by patients.

One patient population that particularly may benefit from PRF use is patients that are currently taking various bisphosphonates or antiresorptive medications. These patients are naturally at a higher risk for dry sockets, reduced blood flow, and developing osteonecrosis of the jaw (ONJ). It is advised that by simply utilizing PRF in extraction sites in these patients, the rate and chance of developing ONJ (even in low-dose bisphosphonate users for osteoporosis) will be dramatically lowered worldwide, and it is highly recommended (see chapter 18).

Conclusion

This chapter presented the literature on the topic of extraction site management and drew some conclusions based on best practices from evidence-based findings. It is well known in the literature that PRF will limit dimensional changes postextraction when utilized alone, but better results are obtained with a bone grafting material. Thus, the combination approach should be considered as a gold standard approach, with clinicians favoring *(1)* bone grafting material+PRF, *(2)* bone grafting material alone, and *(3)* PRF alone (in this order). PRF has been shown to favor better soft tissue wound healing, particularly in third molar sites where the risk of dry socket exists in far greater percentages. PRF should therefore be considered in all cases. PRF contains leukocytes that fight against incoming pathogens and therefore has been shown to reduce postoperative pain, infection, and the total number of analgesics taken postoperatively. Future research aimed at investigating newer technologies such as e-PRF plugs lasting 4 to 6 months as well as production of PRF utilizing horizontal centrifugation are needed to evaluate if an increase in leukocyte number and growth factor release from PRF scaffolds may translate into further clinical benefit.

References

1. Dye BA, Thornton-Evans G, Li X, Iafolla T. Dental caries and tooth loss in adults in the United States, 2011–2012. Washington, DC: US Department of Health and Human Services, Centers for Disease Control and Prevention, National Center for Health Statistics, 2015.
2. Schropp L, Kostopoulos L, Wenzel A. Bone healing following immediate versus delayed placement of titanium implants into extraction sockets: A prospective clinical study. Int J Oral Maxillofac Implants 2003;18:189–199.
3. Botticelli D, Persson LG, Lindhe J, Berglundh T. Bone tissue formation adjacent to implants placed in fresh extraction sockets: An experimental study in dogs. Clin Oral Implants Res 2006;17:351–358.
4. Johnson K. A study of the dimensional changes occurring in the maxilla after tooth extraction.—Part I. Normal healing. Aust Dent J 1963;8:428–433.
5. Morjaria KR, Wilson R, Palmer RM. Bone healing after tooth extraction with or without an intervention: A systematic review of randomized controlled trials. Clin Implant Dent Relat Res 2014;16:1–20.
6. Avila-Ortiz G, Zadeh HH. Management of the extraction site: Socket grafting. In: Nevins M, Wang HL (eds). Implant Therapy: Clinical Approaches and Evidence of Success, ed 2. Chicago: Quintessence, 2019:127–147.
7. Pikos MA, Miron RJ. Bone Augmentation in Implant Dentistry: A Step-by-Step Guide to Predictable Alveolar Ridge and Sinus Grafting. Chicago: Quintessence, 2019.
8. Amler MH. The time sequence of tissue regeneration in human extraction wounds. Oral Surg Oral Med Oral Pathol 1969;27: 309–318.
9. Lin WL, McCulloch CA, Cho MI. Differentiation of periodontal ligament fibroblasts into osteoblasts during socket healing after tooth extraction in the rat. Anat Rec 1994;240:492–506.
10. Araújo MG, Lindhe J. Dimensional ridge alterations following tooth extraction. An experimental study in the dog. J Clin Periodontol 2005;32:212–218.
11. Cardaropoli G, Araújo M, Lindhe J. Dynamics of bone tissue formation in tooth extraction sites: An experimental study in dogs. J Clin Periodontol 2003;30:809–818.
12. Scala A, Lang NP, Schweikert MT, Oliveira JA, Rangel-Garcia I, Botticelli D. Sequential healing of open extraction sockets. An experimental study in monkeys. Clin Oral Implants Res 2014;25:288–295.
13. Werbitt MJ, Goldberg PV. Immediate implantation. Preservation of bone volume and osseous regeneration [in French]. J Parodontol 1991;10:157–166.
14. Werbitt MJ, Goldberg PV. The immediate implant: Bone preservation and bone regeneration. Int J Periodontics Restorative Dent 1992;12:206–217.
15. Polizzi G, Grunder U, Goene R, et al. Immediate and delayed implant placement into extraction sockets: A 5-year report. Clin Implant Dent Relat Res 2000;2:93–99.
16. Schropp L, Wenzel A, Kostopoulos L, Karring T. Bone healing and soft tissue contour changes following single-tooth extraction: A clinical and radiographic 12-month prospective study. Int J Periodontics Restorative Dent 2003;23:313–323.
17. Pietrokovski J, Massler M. Alveolar ridge resorption following tooth extraction. J Prosthet Dent 1967;17:21–27.
18. Johnson K. A study of the dimensional changes occurring in the maxilla following tooth extraction. Aust Dent J 1969;14:241–244.
19. Chappuis V, Engel O, Reyes M, Shahim K, Nolte LP, Buser D. Ridge alterations post-extraction in the esthetic zone: A 3D analysis with CBCT. J Dent Res 2013;92(12 suppl):195S–201S.
20. De Risi V, Clementini M, Vittorini G, Mannocci A, De Sanctis M. Alveolar ridge preservation techniques: A systematic review and meta-analysis of histological and histomorphometrical data. Clin Oral Implants Res 2015;26:50–68.
21. Jambhekar S, Kernen F, Bidra AS. Clinical and histologic outcomes of socket grafting after flapless tooth extraction: A systematic review of randomized controlled clinical trials. J Prosthet Dent 2015;113:371–382.
22. Moraschini V, Barboza ED. Quality assessment of systematic reviews on alveolar socket preservation. Int J Oral Maxillofac Surg 2016;45:1126–1134.
23. Spagnoli D, Choi C. Extraction socket grafting and buccal wall regeneration with recombinant human bone morphogenetic protein-2 and acellular collagen sponge. Atlas Oral Maxillofac Surg Clin North Am 2013;21:175–183.
24. Tan WL, Wong TL, Wong MC, Lang NP. A systematic review of post-extractional alveolar hard and soft tissue dimensional changes in humans. Clin Oral Implants Res 2012;23(suppl 5):1–21.
25. Mammoto A, Connor KM, Mammoto T, et al. A mechanosensitive transcriptional mechanism that controls angiogenesis. Nature 2009;457:1103–1108.
26. Brkovic BM, Prasad HS, Konandreas G, et al. Simple preservation of a maxillary extraction socket using beta-tricalcium phosphate with type I collagen: Preliminary clinical and histomorphometric observations. J Can Dent Assoc 2008;74:523–528.
27. Brkovic BM, Prasad HS, Rohrer MD, et al. Beta-tricalcium phosphate/type I collagen cones with or without a barrier membrane in human extraction socket healing: Clinical, histologic, histomorphometric, and immunohistochemical evaluation. Clin Oral Investig 2012;16:581–590.
28. Mardas N, Chadha V, Donos N. Alveolar ridge preservation with guided bone regeneration and a synthetic bone substitute or a bovine-derived xenograft: A randomized, controlled clinical trial. Clin Oral Implants Res 2010;21:688–698.
29. Wallace S. Histomorphometric and 3D cone-beam computerized tomographic evaluation of socket preservation in molar extraction sites using human particulate mineralized cancellous allograft bone with a porcine collagen xenograft barrier: A case series. J Oral Implantol 2015;41:291–297.
30. Bayat M, Momen Heravi F, Mahmoudi M, Bahrami N. Bone reconstruction following application of bone matrix gelatin to alveolar defects: A randomized clinical trial. Int J Organ Transplant Med 2015;6:176–181.
31. Mardas N, D'Aiuto F, Mezzomo L, Arzoumanidi M, Donos N. Radiographic alveolar bone changes following ridge preservation with two different biomaterials. Clin Oral Implants Res 2011;22: 416–423.
32. Coomes AM, Mealey BL, Huynh-Ba G, Barboza-Arguello C, Moore WS, Cochran DL. Buccal bone formation after flapless extraction: A randomized, controlled clinical trial comparing recombinant human bone morphogenetic protein 2/absorbable collagen carrier and collagen sponge alone. J Periodontol 2014;85:525–535.
33. Fiorellini JP, Howell TH, Cochran D, et al. Randomized study evaluating recombinant human bone morphogenetic protein-2 for extraction socket augmentation. J Periodontol 2005;76:605–613.
34. Misch CM. The use of recombinant human bone morphogenetic protein-2 for the repair of extraction socket defects: A technical modification and case series report. Int J Oral Maxillofac Implants 2010;25:1246–1252.
35. Wallace SC, Pikos MA, Prasad H. De novo bone regeneration in human extraction sites using recombinant human bone morphogenetic protein-2/ACS: A clinical, histomorphometric, densitometric, and 3-dimensional cone-beam computerized tomographic scan evaluation. Implant Dent 2014;23:132–137.
36. Lekovic V, Kenney E, Weinlaender M, et al. A bone regenerative approach to alveolar ridge maintenance following tooth extraction. Report of 10 cases. J Periodontol 1997;68:563–570.
37. Hauser F, Gaydarov N, Badoud I, Vazquez L, Bernard JP, Ammann P. Clinical and histological evaluation of postextraction platelet-rich fibrin socket filling: A prospective randomized controlled study. Implant Dent 2013;22:295–303.

38. Suttapreyasri S, Leepong N. Influence of platelet-rich fibrin on alveolar ridge preservation. J Craniofac Surg 2013;24:1088–1094.
39. Temmerman A, Vandessel J, Castro A, et al. The use of leucocyte and platelet-rich fibrin in socket management and ridge preservation: A split-mouth, randomized, controlled clinical trial. J Clin Periodontol 2016;43:990–999.
40. Du Toit J, Siebold A, Dreyer A, Gluckman H. Choukroun platelet-rich fibrin as an autogenous graft biomaterial in preimplant surgery: Results of a preliminary randomized, human histomorphometric, split-mouth study. Int J Periodontics Restorative Dent 2016;36(suppl):S75–S86.
41. Alzahrani AA, Murriky A, Shafik S. Influence of platelet rich fibrin on post-extraction socket healing: A clinical and radiographic study. Saudi Dent J 2017;29:149–155.
42. Zhang Y, Ruan Z, Shen M, et al. Clinical effect of platelet-rich fibrin on the preservation of the alveolar ridge following tooth extraction. Exp Ther Med 2018;15:2277–2286.
43. Srinivas B, Das P, Rana MM, Qureshi AQ, Vaidya KC, Ahmed Raziuddin SJ. Wound healing and bone regeneration in postextraction sockets with and without platelet-rich fibrin. Ann Maxillofac Surg 2018;8:28–34.
44. Areewong K, Chantaramungkorn M, Khongkhunthian P. Platelet-rich fibrin to preserve alveolar bone sockets following tooth extraction: A randomized controlled trial. Clin Implant Dent Relat Res 2019;21:1156–1163.
45. Canellas J, da Costa RC, Breves RC, et al. Tomographic and histomorphometric evaluation of socket healing after tooth extraction using leukocyte- and platelet-rich fibrin: A randomized, single-blind, controlled clinical trial. J Craniomaxillofac Surg 2020;48:24–32.
46. Marenzi G, Riccitiello F, Tia M, di Lauro A, Sammartino G. Influence of leukocyte- and platelet-rich fibrin (L-PRF) in the healing of simple postextraction sockets: A split-mouth study. Biomed Res Int 2015;2015:369273.
47. de Almeida Barros Mourão CF, de Mello-Machado RC, Javid K, Moraschini V. The use of leukocyte- and platelet-rich fibrin in the management of soft tissue healing and pain in post-extraction sockets: A randomized clinical trial. J Craniomaxillofac Surg 2020;48:452–457.
48. Pan J, Xu Q, Hou J, et al. Effect of platelet-rich fibrin on alveolar ridge preservation: A systematic review. J Am Dent Assoc 2019;150:766–778.
49. Del Fabbro M, Panda S, Taschieri S. Adjunctive use of plasma rich in growth factors for improving alveolar socket healing: A systematic review. J Evid Based Dent Pract 2019;19:166–176.
50. Das S, Jhingran R, Bains VK, Madan R, Srivastava R, Rizvi I. Socket preservation by beta-tri-calcium phosphate with collagen compared to platelet-rich fibrin: A clinico-radiographic study. Eur J Dent 2016;10:264–276.
51. Thakkar DJ, Deshpande NC, Dave DH, Narayankar SD. A comparative evaluation of extraction socket preservation with demineralized freeze-dried bone allograft alone and along with platelet-rich fibrin: A clinical and radiographic study. Contemp Clin Dent 2016;7:371–376.
52. Girish Kumar N, Chaudhary R, Kumar I, Arora SS, Kumar N, Singh H. To assess the efficacy of socket plug technique using platelet rich fibrin with or without the use of bone substitute in alveolar ridge preservation: A prospective randomised controlled study. Oral Maxillofac Surg 2018;22:135–142.
53. Clark D, Rajendran Y, Paydar S, et al. Advanced platelet-rich fibrin and freeze-dried bone allograft for ridge preservation: A randomized controlled clinical trial. J Periodontol 2018;89:379–387.
54. Mendoza-Azpur G, Olaechea A, Padial-Molina M, et al. Composite alloplastic biomaterial vs. autologous platelet-rich fibrin in ridge preservation. J Clin Med 2019;8:223.
55. De Angelis P, De Angelis S, Passarelli PC, Liguori MG, Manicone PF, D'Addona A. Hard and soft tissue evaluation of different socket preservation procedures using leukocyte and platelet-rich fibrin: A retrospective clinical and volumetric analysis. J Oral Maxillofac Surg 2019;77:1807–1815.
56. Ouyyamwongs W, Leepong N, Suttapreyasri S. Alveolar ridge preservation using autologous demineralized tooth matrix and platelet-rich fibrin versus platelet-rich fibrin alone: A split-mouth randomized controlled clinical trial. Implant Dent 2019;28:455–462.
57. Gürbüzer B, Pikdöken L, Tunali M, Urhan M, Küçükodaci Z, Ercan F. Scintigraphic evaluation of osteoblastic activity in extraction sockets treated with platelet-rich fibrin. J Oral Maxillofac Surg 2010;68:980–989.
58. Ruga E, Gallesio C, Boffano P. Platelet-rich fibrin and piezoelectric surgery: A safe technique for the prevention of periodontal complications in third molar surgery. J Craniofac Surg 2011;22:1951–1955.
59. Hoaglin DR, Lines GK. Prevention of localized osteitis in mandibular third-molar sites using platelet-rich fibrin. Int J Dent 2013;2013:875380.
60. Eshghpour M, Dastmalchi P, Nekooei AH, Nejat A. Effect of platelet-rich fibrin on frequency of alveolar osteitis following mandibular third molar surgery: A double-blinded randomized clinical trial. J Oral Maxillofac Surg 2014;72:1463–1467.
61. Kumar N, Prasad K, Ramanujam L, K R, Dexith J, Chauhan A. Evaluation of treatment outcome after impacted mandibular third molar surgery with the use of autologous platelet-rich fibrin: A randomized controlled clinical study. J Oral Maxillofac Surg 2015;73:1042–1049.
62. Yelamali T, Saikrishna D. Role of platelet rich fibrin and platelet rich plasma in wound healing of extracted third molar sockets: A comparative study. J Maxillofac Oral Surg 2015;14:410–416.
63. Uyanık LO, Bilginaylar K, Etikan İ. Effects of platelet-rich fibrin and piezosurgery on impacted mandibular third molar surgery outcomes. Head Face Med 2015;11:25.
64. Ozgul O, Senses F, Er N, et al. Efficacy of platelet rich fibrin in the reduction of the pain and swelling after impacted third molar surgery: Randomized multicenter split-mouth clinical trial. Head Face Med 2015;11:37.
65. Baslarli O, Tumer C, Ugur O, Vatankulu B. Evaluation of osteoblastic activity in extraction sockets treated with platelet-rich fibrin. Med Oral Patol Oral Cir Bucal 2015;20:e111–e116.
66. Dutta SR, Passi D, Singh P, Sharma S, Singh M, Srivastava D. A randomized comparative prospective study of platelet-rich plasma, platelet-rich fibrin, and hydroxyapatite as a graft material for mandibular third molar extraction socket healing. Natl J Maxillofac Surg 2016;7:45–51.
67. Bilginaylar K, Uyanik LO. Evaluation of the effects of platelet-rich fibrin and piezosurgery on outcomes after removal of impacted mandibular third molars. Br J Oral Maxillofac Surg 2016;54:629–633.
68. Varghese MP, Manuel S, Kumar LKS. Potential for osseous regeneration of platelet-rich fibrin—A comparative study in mandibular third molar impaction sockets. J Oral Maxillofac Surg 2017;75:1322–1329.
69. Sharma A, Aggarwal N, Rastogi S, Choudhury R, Tripathi S. Effectiveness of platelet-rich fibrin in the management of pain and delayed wound healing associated with established alveolar osteitis (dry socket). Eur J Dent 2017;11:508–513.
70. Gülşen U, Şentürk MF. Effect of platelet rich fibrin on edema and pain following third molar surgery: A split mouth control study. BMC Oral Health 2017;17:79.
71. Asutay F, Yolcu Ü, Geçör O, Acar AH, Öztürk SA, Malkoç S. An evaluation of effects of platelet-rich-fibrin on postoperative morbidities after lower third molar surgery. Niger J Clin Pract 2017;20:1531–1536.

72. Unsal H, Erbasar GNHE. Evaluation of the effect of platelet-rich fibrin on the alveolar osteitis incidence and periodontal probing depth after extracting partially erupted mandibular third molars extraction. Niger J Clin Pract 2018;21:201–205.
73. Jeyaraj PE, Chakranarayan A. Soft tissue healing and bony regeneration of impacted mandibular third molar extraction sockets, following postoperative incorporation of platelet-rich fibrin. Ann Maxillofac Surg 2018;8:10–18.
74. Dar MM, Shah AA, Najar AL, Younis M, Kapoor M, Dar JI. Healing potential of platelet rich fibrin in impacted mandibular third molar extraction sockets. Ann Maxillofac Surg 2018;8:206–213.
75. Daugela P, Grimuta V, Sakavicius D, Jonaitis J, Juodzbalys G. Influence of leukocyte- and platelet-rich fibrin (L-PRF) on the outcomes of impacted mandibular third molar removal surgery: A split-mouth randomized clinical trial. Quintessence Int 2018;49:377–388.
76. Zahid TM, Nadershah M. Effect of advanced platelet-rich fibrin on wound healing after third molar extraction: A split-mouth randomized double-blind study. J Contemp Dent Pract 2019;20:1164–1170.
77. Ritto FG, Pimentel T, Canellas JVS, Junger B, Cruz M, Medeiros PJ. Randomized double-blind clinical trial evaluation of bone healing after third molar surgery with the use of leukocyte- and platelet-rich fibrin. Int J Oral Maxillofac Surg 2019;48:1088–1093.
78. Yüce E, Kömerik N. Potential effects of advanced platelet rich fibrin as a wound-healing accelerator in the management of alveolar osteitis: A randomized clinical trial. Niger J Clin Pract 2019;22:1189–1195.
79. Afat IM, Akdoğan ET, Gönül O. Effects of leukocyte- and platelet-rich fibrin alone and combined with hyaluronic acid on early soft tissue healing after surgical extraction of impacted mandibular third molars: A prospective clinical study. J Craniomaxillofac Surg 2019;47:280–286.
80. Kotsakis G, Chrepa V, Marcou N, Prasad H, Hinrichs J. Flapless alveolar ridge preservation utilizing the "socket-plug" technique: Clinical technique and review of the literature. J Oral Implantol 2014;40:690–698.
81. Miron RJ, Fujioka-Kobayashi M, Bishara M, Zhang Y, Hernandez M, Choukroun J. Platelet-rich fibrin and soft tissue wound healing: A systematic review. Tissue Eng Part B Rev 2017;23:83–89.

15

Use of PRF as an Adjunct Therapy to Implant Dentistry

Contributors

Howard Gluckman
Michael A. Pikos
Luis Bessa
Richard J. Martin
Luigi Canullo
Richard J. Miron

Chapter Highlights

- *Overview of the background RCTs evaluating PRF in implant dentistry*
- *Reported clinical scenarios that apply PRF during implant placement*
- *The growing use of PRF for soft tissue healing following immediate implant placement*
- *The future of PRF at implant sites*

Dental implants have gradually become the standard of care for the treatment of missing teeth. While they are generally considered a highly predictable modality, proper knowledge of their placement is critical for predictable results. The symbiotic relationship between bone and soft tissue is increasingly understood to be integral for long-term implant success. Therefore, it is essential to ensure the quality and quantity of bone and soft tissues surrounding the implants for their long-term success and maintenance.

This chapter reviews the use PRF as an adjunct therapy to implant dentistry. It reports on the ability of PRF to impact implant osseointegration and its effect on soft tissue integration. A review of the literature investigating bone-to-implant contact (BIC) and implant stability quotient (ISQ) values with and without PRF is presented. We also highlight recommended clinical protocols and discuss the growing avenue of treatment modalities for the management of peri-implantitis. The chapter sheds light on the advantages of PRF for soft tissue healing around implants, especially for immediate implant placement, and provides a better understanding of the symbiotic relationship for long-term health and esthetics of implants. Lastly, it deals with the issues of tissue healing around implants using PRF, with a focus on bone healing during osseointegration as well as soft tissue enhancement around implants.

The scientific data is clear: Tooth loss leads to a significant reduction in ridge dimensions (see chapter 14). The most apparent benefit of PRF as a biomaterial is that it may accelerate growth and healing during the augmentation and reconstruction of those sites.[1] One of the first instances where the clinician may use PRF is within extraction sockets to limit dimensional changes postextraction, presumably to place implants at a later stage. The previous chapter addressed the topic of socket management with PRF in detail. This chapter builds from there and addresses of the use of PRF at implant sites. Numerous questions arise when discussing PRF in the context of implants: Can I or should I use PRF when I place implants? What are the benefits? When or how should I use PRF?

Unfortunately, limited data and literature currently exist on this topic, as addressed later in this chapter. What is certain is that the loss of ridge architecture following tooth extraction often demands augmentation to provide adequate, healthy hard and soft tissue to accommodate the future placement of dental implants.[2,3] The literature is abundant with sound data that prescribes the quantity of healthy bone and soft tissue required.[4,5]

A second distinction can also be made between PRF's contribution to hard tissue healing and soft tissue healing. Soft tissue volume has become the focus of research recently.[4,6,7] Sufficient, stable bone is key to the long-term success of implants, and the difficulty in predictably regenerating this bone when lost remains a significant challenge.[2] It has been shown that a minimum of 2 to 4 mm of bone around the implant is essential to maintain the soft tissue's stability. Vice versa, it is vital that a minimum of 2 to 3 mm of attached soft tissue thickness in both the vertical and horizontal dimensions be present to protect bone from resorption.[4,5] Far too often, one or the other of these components is neglected, which leads to potential long-term breakdown.[4,8,9] As such, the clinician may seek to better understand what effects PRF may provide, not only during osseointegration but also for the healing of soft and hard tissues during implant placement.

Hereafter, the topic of PRF's use with dental implants is explored, describing the contributions of this biomaterial's growth-promoting factors that may contribute to the success and performance of these biomedical devices. Its use during implant placement may include the treatment of peri-implant defects, osseointegration, and soft tissue healing at implant sites.

Evaluation of Clinical Studies Using PRF for Implant Dentistry

The use of PRF has been investigated in six RCTs in implant dentistry, as reviewed in a recent systematic review[10–15] (Table 15-1). The majority assessed either implant stability quotient (ISQ) values or bone level changes after implant placement when PRF was utilized.[10–13] One study assessed soft tissue healing and mucosal thickness changes when PRF was being used during implant therapy,[14] and another investigated PRF for the management of peri-implantitis.[15]

PRF and implant stability

A prospective study by Boora et al (2015) investigated the effect of PRF on peri-implant tissue response following one-stage implant placement with nonfunctional immediate provisionalization in the maxillary anterior region with 20 patients.[10] The mean marginal bone changes were observed from baseline to 3 months, with slightly lesser changes observed in the PRF group (0.5–0.6 mm of bone change in the control group vs 0.2 mm in the PRF group). No significant differences in probing depth (PD) and bleeding on probing (BoP) were noted during follow-up.[10]

A second study by Öncü and Alaaddinoğlu (2015) investigated mean ISQ values at 1 week and 1 month postoperatively.[11] It was found that the PRF groups demonstrated ISQ values of 69.3 ± 10.5, whereas the control group demonstrated values of 64.5 ± 12.2 at the end of the first week. The mean ISQs at 1 month postoperatively were 77.1 ± 7.1 for the PRF group and 70.5 ± 7.7 for the control group.[11]

A third study by Tabrizi et al (2018) evaluated the stability of implants placed in the posterior maxilla, with or without the use of PRF, in a split-mouth RCT.[12] Twenty patients with missing teeth in the molar regions of the maxilla requiring bilateral implants were included (control group implant placement alone vs test group with PRF). Implant stability was assessed by resonance frequency analysis (RFA) at 2, 4, and 6 weeks after placement. At 2 weeks, the mean ISQ was 60.60 ± 3.42 in the test group and 58.25 ± 3.64 in the control group; at 4 weeks, it was 70.30 ± 3.36 and 67.15 ± 4.33, respectively; and at 6 weeks, it was 78.45 ± 3.36 and 76.15 ± 2.94, respectively. Significant differences in RFA were found between the groups at 2 weeks ($P = .04$), 4 weeks ($P = .014$), and 6 weeks ($P = .027$) after placement. The study suggests that PRF's use may enhance

Table 15-1 Characteristics of RCTs evaluating the use of PRF for implant dentistry

Authors (year)	Study design Follow-up	No. of participants Sex Mean age	Groups	Smokers (Y, N)	Conclusions
Use of PRF for implant placement (osseointegration)					
Boora et al[10] (2015)	RCT (one-stage, nonfunctional immediate implant with or without PRF) 3 months	20 (20 implants) M: 15 / F: 5 24.6	C: 10, implants alone T: 10, implants+PRF	NR	At 3 months, all implants remained osseointegrated. The mean marginal bone changes were observed from baseline to 3 months in both groups with lesser changes observed in the PRF group. No significant differences in PD and BoP were noted during follow-up.
Öncü and Alaaddinoğlu[11] (2015)	RCT (implants coated with or without PRF and investigated for ISQ) 1 month	20 (64 implants) M: 14 / F: 6 44.2	C: 33, implants alone T: 31, implants+PRF	Y	Mean ISQs of the PRF+implants group were 69.3 ± 10.5, and mean ISQs for the implants alone group were 64.5 ± 12.2 at the end of the first week. The mean ISQs at 4 weeks postoperatively were 77.1 ± 7.1 for the PRF+implants group and 70.5 ± 7.7 for the implants alone group.
Tabrizi et al[12] (2018)	RCT (split-mouth bilateral implant placement with or without PRF) 6 weeks	20 (40 implants) M: 14 / F: 6 39.6	C: 20, implants alone T: 20, implants+PRF	NR	At 6 weeks, implant ISQ values were 76.15 ± 2.94 and 78.45 ± 3.36 in the control and test groups, respectively.
Diana et al[13] (2018)	RCT (implants coated with or without PRF and investigated for ISQ) 3 months	31 (41 implants) M: 18 / F: 13 28.5	C: 20, implants alone T: 21, implants+PRF	Y	No significant difference was observed between the groups in terms of implant stability over the 3-month period, with ISQ values of 56.58 ± 18.81 to 71.32 ± 7.82 in the PRF group and 60.61 ± 11.49 to 70.06 ± 8.96 in the control group.
Use of PRF for implant placement (soft tissue management)					
Hehn et al[14] (2016)	RCT (implant placement with or without soft tissue augmentation with PRF) 6 months	31 (31 implants) M: 16 / F: 15 53.8	C: 21, implants alone T: 10, implants+PRF	N	Within the limits of this study and the early determination of the test group, this study concludes that soft tissue augmentation with PRF performed with a split-flap technique cannot be recommended for thickening thin mucosa.
Use of PRF for the management of peri-implantitis					
Hamzacebi et al[15] (2015)	RCT (conventional flap surgery for the treatment of peri-implant bone loss with or without PRF) 6 months	19 (38 implants) M: 11 / F: 8 60.9	C: 19, OFD T: 19, OFD+PRF	NR	At 3 and 6 months postsurgery, the PRF group demonstrated higher mean PD reductions and CAL gains compared with the control group. In addition, the increase in the amount of keratinized mucosa from baseline to 6 months postoperatively was statistically significant for the PRF group.

NR, not reported; C, control group; T, test group; M, male; F, female; PD, probing depth; BoP, bleeding on probing; ISQ, implant stability quotient; CAL, clinical attachment level.

the postinsertion stability of dental implants placed in the posterior maxilla during the healing period.[12]

Lastly, Diana et al (2018) performed an RCT to investigate immediate implants grafted with PRF.[13] PRF was placed in the peri-implant region of the study group (n = 21). In contrast, no augmentation was done in the control group (n = 20). A significant increase in implant stability was noted in both groups over time (ISQ: study group, 56.58 ± 18.81 to 71.32 ± 7.82; control group, 60.61 ± 11.49 to 70.06 ± 8.96; *P* = .01), but no significant difference was observed between the groups. It was therefore concluded that PRF did not influence the primary stability of immediate implants.[13]

Table 15-1 *(cont)* Characteristics of RCTs evaluating the use of PRF for implant dentistry

				Methods for PRF preparation		
Authors (year)	**ISQ values**	**Bone level changes (mm)**	**Other changes**	**Centrifugation system**	**Volume of tubes for blood drawn**	**Centrifugation parameters speed (rpm) × time (min)**
Use of PRF for implant placement (osseointegration)						
Boora et al[10] (2015)	NR	Mesial: 0.50 ± 0.2 (C) 0.2 ± 0.0 (T) Distal: 0.60 ± (C) 0.2 ± 0.0 (T)	*PD at 3 months:* 3.45 ± 1.17 (C) 3.3 ± 1.39 (T) *BoP at 3 months:* 30% (C) vs 20% (T)	NR	10-mL test tubes	3000 × 10–12
Öncü and Alaaddinoğlu[11] (2015)	70.4 ± 7.7 (C) 77.1 ± 6.0 (T)	NR	NR	IntraSpin	9-mL glass-coated plastic tubes	2700 × 12
Tabrizi et al[12] (2018)	76.15 ± 2.94 (C) 78.45 ± 3.36 (T)	NR	NR	IntraSpin	10-mL tubes (NR)	28000 × 12*
Diana et al[13] (2018)	70.06 ± 8.96 (C) 71.32 ± 7.82 (T)	Mesial: 0.85 ± 0.76 (C) 1.17 ± 1.14 (T) Distal: 0.92 ± 0.34 (C) 1.15 ± 0.96 (T)	*PD at 1 year:* 2.60 ± 0.68 (C) 2.01 ± 0.62 (T)	NR	NR	NR
Use of PRF for implant placement (soft tissue management)						
Hehn et al[14] (2016)	*Mucosal thickness changes (mm):* At crest: 2.6 ± 0.4 to 2.6 ± 0.6 (C) 2.2 ± 0.4 to 0.9 ± 1 (T) Buccal and lingual: No significant changes	*Bone loss (defect depth/defect width in mm):* Mesial: 0.7 ± 0.6 / 0.5 ± 0.4 (C) 0.7 ± 0.4 / 0.5 ± 0.4 (T) Distal: 0.8 ± 0.7 / 0.5 ± 0.5 (C) 0.8 ± 0.4 / 0.6 ± 0.3 (T)	NR	IntraSpin	NR	NR
Use of PRF for the management of peri-implantitis						
Hamzacebi et al[15] (2015)	*PD reduction (mm):* 2.05 ± 0.77 (C) 2.82 ± 1.03 (T)	*CAL gains (mm):* 1.84 ± 0.81 (C) 3.31 ± 1.08 (T)	*Change in keratinized mucosa (mm):* 0.62 ± 0.48 (C) −0.05 ± 0.15 (T)	NR	NR	NR

NR, not reported; C, control group; T, test group; M, male; F, female; PD, probing depth; BoP, bleeding on probing; ISQ, implant stability quotient; CAL, clinical attachment level. *Reported like this.

Therefore, in summary, of the four controlled studies investigating the impact of PRF on implants' primary stability, three demonstrated some benefit for using PRF, though only minimal improvements were reported.

PRF and soft tissue

An RCT investigating the effects of PRF on soft tissue healing around implants performed by Hehn et al evaluated soft tissue thickness and initial marginal bone loss around implants using a split-flap technique.[14] Tissue thickness was measured at the point of implant insertion (baseline) and at the time of reentry after 3 months. Standardized digital radiographs were obtained for evaluation at the time of implant placement, reentry after 3 months, and at a 6-month follow-up. It was observed that soft tissue augmentation with PRF led to an actual decrease in tissue volume and a significant tissue loss. This study concluded that soft tissue augmentation with PRF performed using a split-flap technique cannot be recommended for thickening thin mucosa.[14]

PRF and the treatment of peri-implantitis

In a study investigating the effect of PRF on peri-implantitis, conventional flap surgery was utilized to manage peri-implantitis cases with or without PRF.[15] At the surgical phase, full-thickness mucoperiosteal flaps were raised to gain access to the implant surface, and the inflammatory tissue was removed using hand instruments.[15] The implant surface was decontaminated using a PeriBrush and 4% pH 1 citric acid for 3 minutes or tetracycline hydrochloride (HCl) solution (1 g) in 20 mL of sterile saline. In the control group, the defect was rinsed and closed. In the test group, the defect was filled with PRF membranes, and PRF plugs were placed over the suprabony component of the defect.[15]

At 3 and 6 months after surgery, respectively, the PRF group demonstrated better mean PD reductions (2.41 ± 1.06 and 2.82 ± 1.03 mm vs 1.65 ± 1.02 and 2.05 ± 0.77 mm) and more clinical attachment level (CAL) gains (2.89 ± 1.01 and 3.31 ± 1.08 mm vs 1.43 ± 1.08 and 1.84 ± 0.81 mm) compared to the control group. The increase in the amount of keratinized mucosa from baseline to 6 months postoperatively was statistically significant for the PRF group ($P < .001$). Hence, the current study's data led to the conclusion that PRF application in peri-implant bone loss provided better clinical results than conventional flap surgery alone.[15]

Clinical bottom line

1. PRF may enhance primary implant stability, though inconclusive/minimal data was found following assessment using ISQ and RFA parameters. To date, no histologic data evaluating bone-to-implant contact (BIC) exists on the topic.
2. The data also recommends avoiding soft tissue augmentation with PRF using a split-flap technique. Full-thickness flaps are therefore recommended.
3. While only one study investigated the use of PRF for the management of peri-implantitis, it was generally found that its application led to significant improvements in PD reduction, CAL gain, and the amount of keratinized mucosa from baseline to 6 months.

Clinical Procedures

PRF treatment of peri-implant defects

The two peri-implant defect types receiving the most attention are coronal bone loss seen in peri-implantitis and the buccal gap seen with immediate implant placement. It is widely known that immediate implant placement dictates a more palatal or lingual approach and a deeper placement to establish primary stability. The buccal gap is observed as the implant is placed away from the buccal plate[5] (Fig 15-1a). The literature reports on a variety of approaches to manage this gap, although no intervention is recommended other than allowing a whole blood clot to organize into bone or grafting this space with an assortment of materials.[5,16,17] It is apparent that PRF may contribute its leukocyte cytokines and growth factors to this process, which may have a positive influence on the healing of this bony defect (Figs 15-1b to 15-1d).

Lee et al created buccal gap defects during implant placement to experimentally simulate this in an animal model.[18] Positive results were demonstrated with an increase in bone volume in the defect area and in the interthread spaces when augmented with PRF. Additional animal studies have shown that PRF alone or with particulate bone material in noninfective peri-implant defects showed high BIC of 61% and 73%, respectively.[19,20] As such, the literature seems to support the use of PRF as beneficial during the filling within this buccal gap during immediate placement, or in combination with a bone biomaterial. Noteworthy and as demonstrated through various case presentations, PRF is most commonly utilized at the coronal interface to promote soft tissue healing and prevent bacterial invasion. Figure 15-2 represents a grafted site with PRF utilized in the coronal portion to facilitate soft tissue wound healing around an immediate implant case with a slight gingival recession. Note that while PRF is utilized in such immediate implant cases to favor soft tissue healing, it also minimizes bacterial infiltration toward the roughened portion of the implant surface. Figure 15-3 presents a second case following a similar approach, however starting with a baseline buccal gingival recession on a hopeless tooth. During implant placement, the coronal portion of the grafted site was treated with PRF alone. In this case, PRF helped promote soft tissue healing, and the gingival recession was no longer present at 1-year follow-up (see Fig 15-3l).

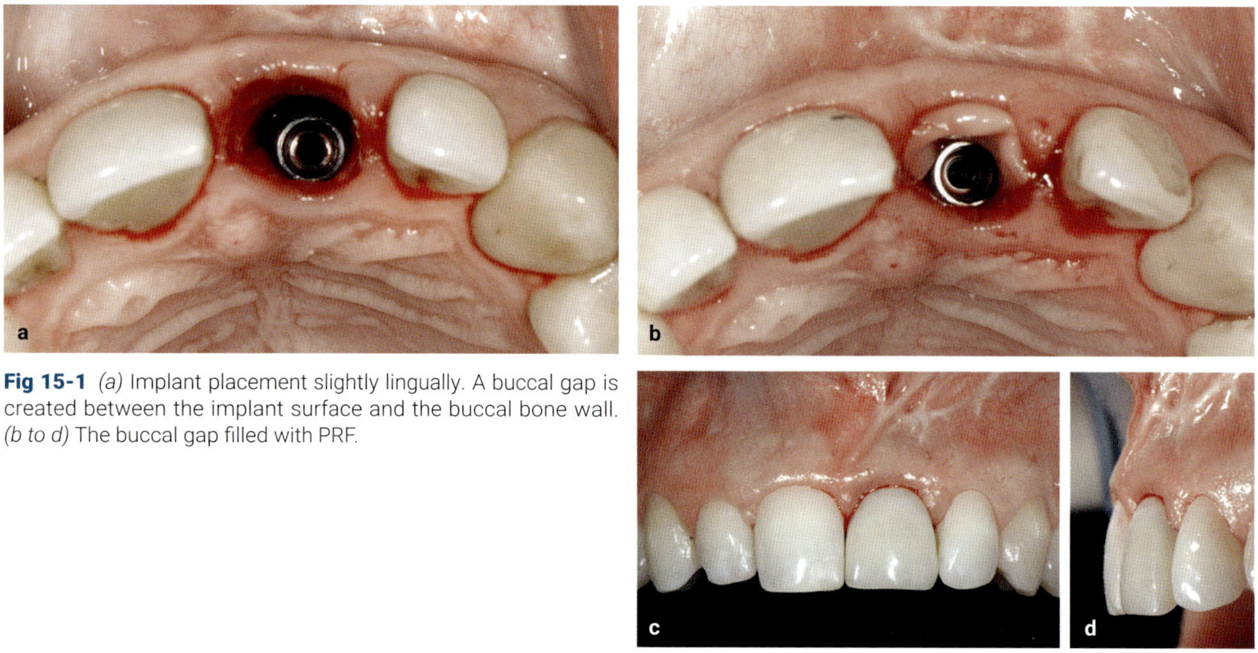

Fig 15-1 *(a)* Implant placement slightly lingually. A buccal gap is created between the implant surface and the buccal bone wall. *(b to d)* The buccal gap filled with PRF.

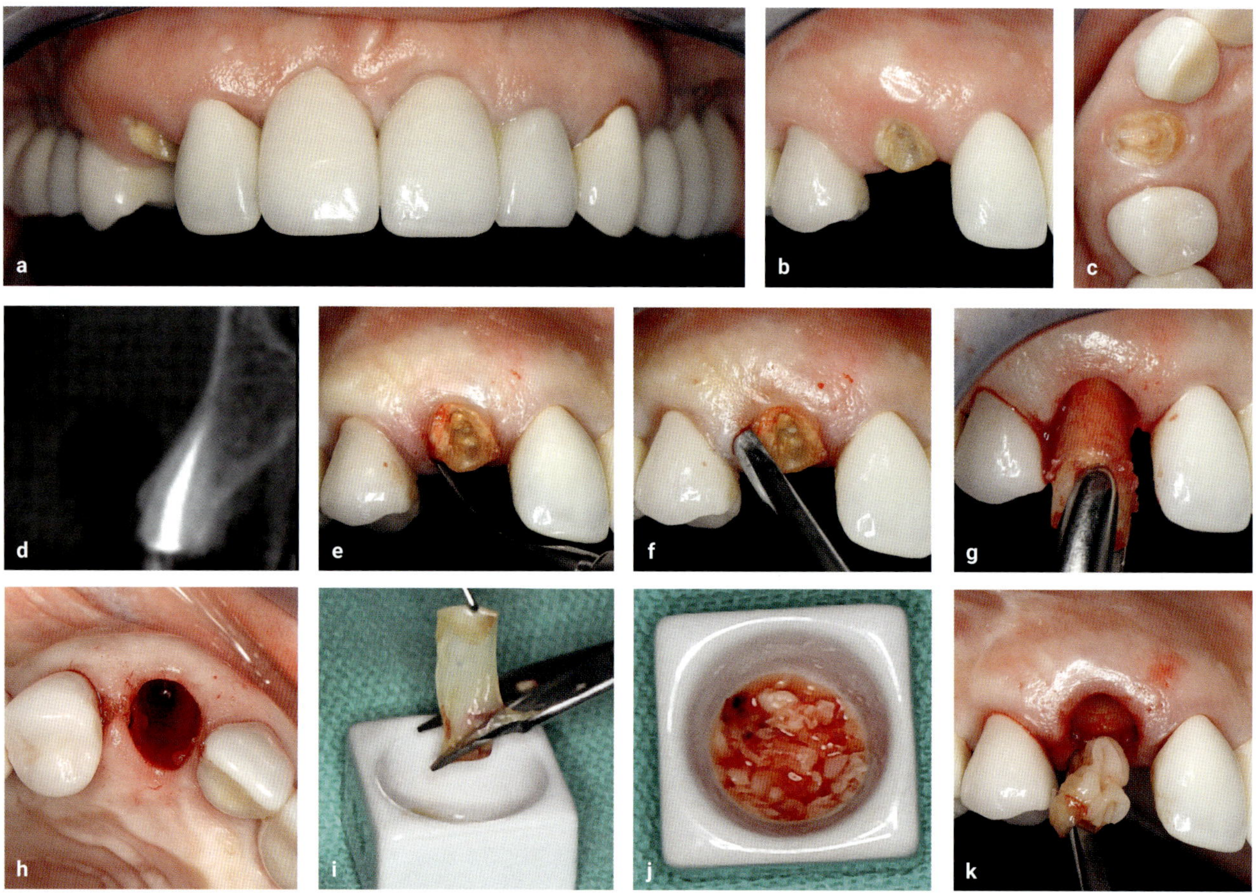

Fig 15-2 Flapless extraction and immediate implant placement of a maxillary right canine. *(a to c)* Labial and occlusal views of a fractured maxillary right canine. *(d)* CBCT image demonstrating thin buccal bone. *(e to g)* Flapless tooth extraction. *(h)* Occlusal view following tooth extraction. *(i to k)* PRF membranes were cut into fragments and inserted into the extraction site along the buccal surface.

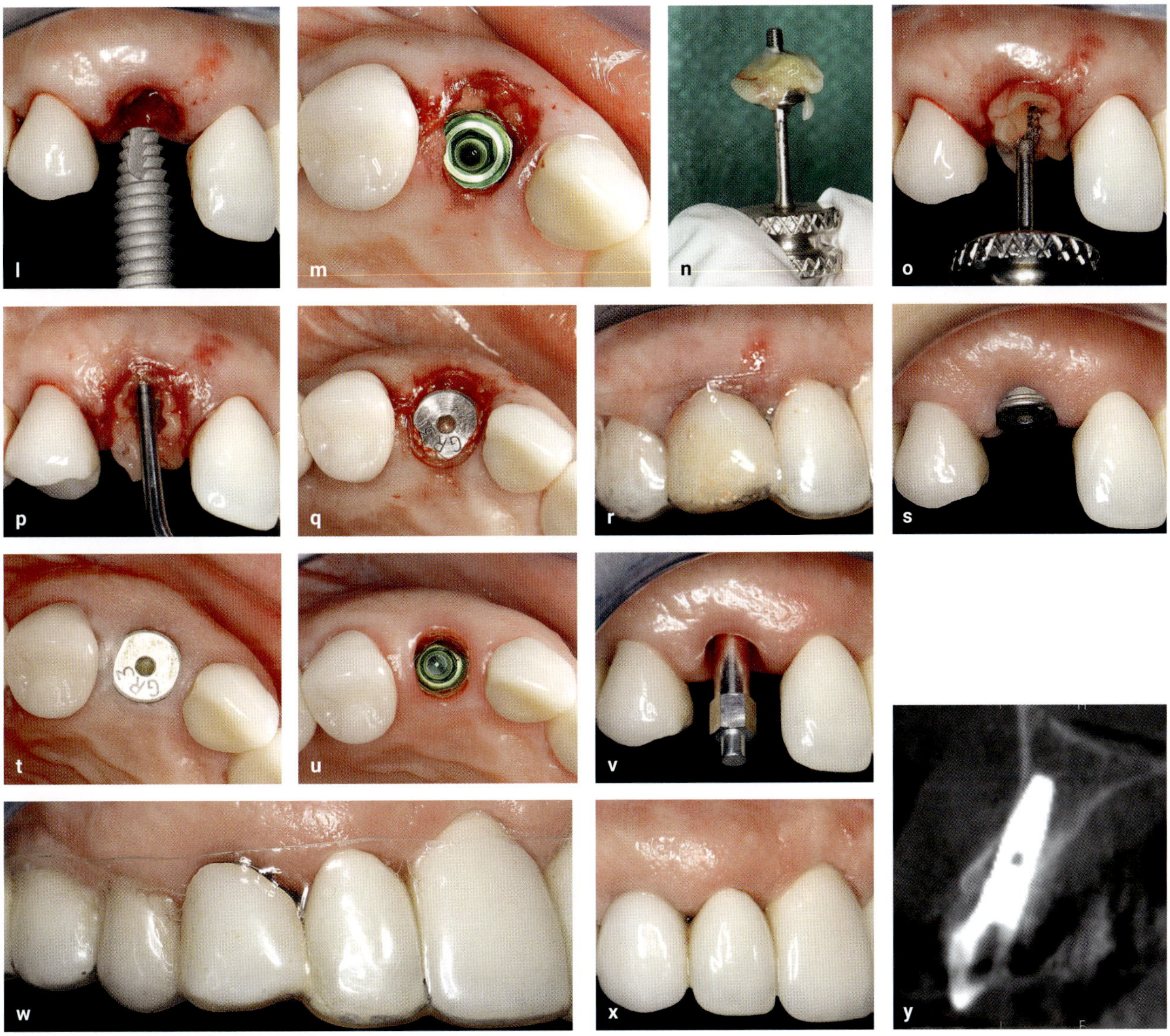

Fig 15-2 *(cont)* *(l)* Implant placement. *(m)* PRF fragments further inserted into the buccal gap. *(n and o)* A PRF membrane was wrapped around the healing abutment. *(p)* A PRF membrane was further packed into the gaps around the coronal portion of the healing abutment. *(q)* Occlusal view of immediate implant. *(r)* Provisional crown. *(s and t)* Labial and occlusal views at 3 weeks. *(u)* Tissue at 3 months. Notice the excellent soft tissue healing. *(v)* ISQ of 83, demonstrating excellent stability of the implant. *(w)* Final restoration. *(x)* Final restoration at 3 years. *(y)* CBCT image at 3 years demonstrating excellent buccal bone stability. (Case performed by Dr Michael A. Pikos; reprinted with permission from Pikos and Miron.[21])

Contrary to the augmentation of the buccal gap, treating peri-implant defects as a result of peri-implantitis is far more complex.[22] This topic remains mostly unresolved, and its exact etiology with reliable treatment options is poorly understood.[23] That said, clinical studies have now begun to investigate the use of PRF in human studies.[24] In the study by Guler et al, the implants had full-thickness flaps raised and decontaminated. The experimental group additionally utilized PRF placed within the bone defect before the flap was closed. When healing occurred with PRF, a difference in PD reduction was observed.[24] Furthermore, CALs seemed to benefit, and an increase in keratinized mucosa was reported. These results suggest that PRF may be beneficial for the treatment of peri-implantitis defects; however, much further research remains necessary to validate these preliminary findings. The clinician should be aware that the treatment of peri-implantitis is currently unpredictable, with significant variations in bone defects and diverse responses to treatment.[25] The evidence remains extremely limited and scarce, supporting the use of PRF for the treatment of peri-implantitis with future studies needed.

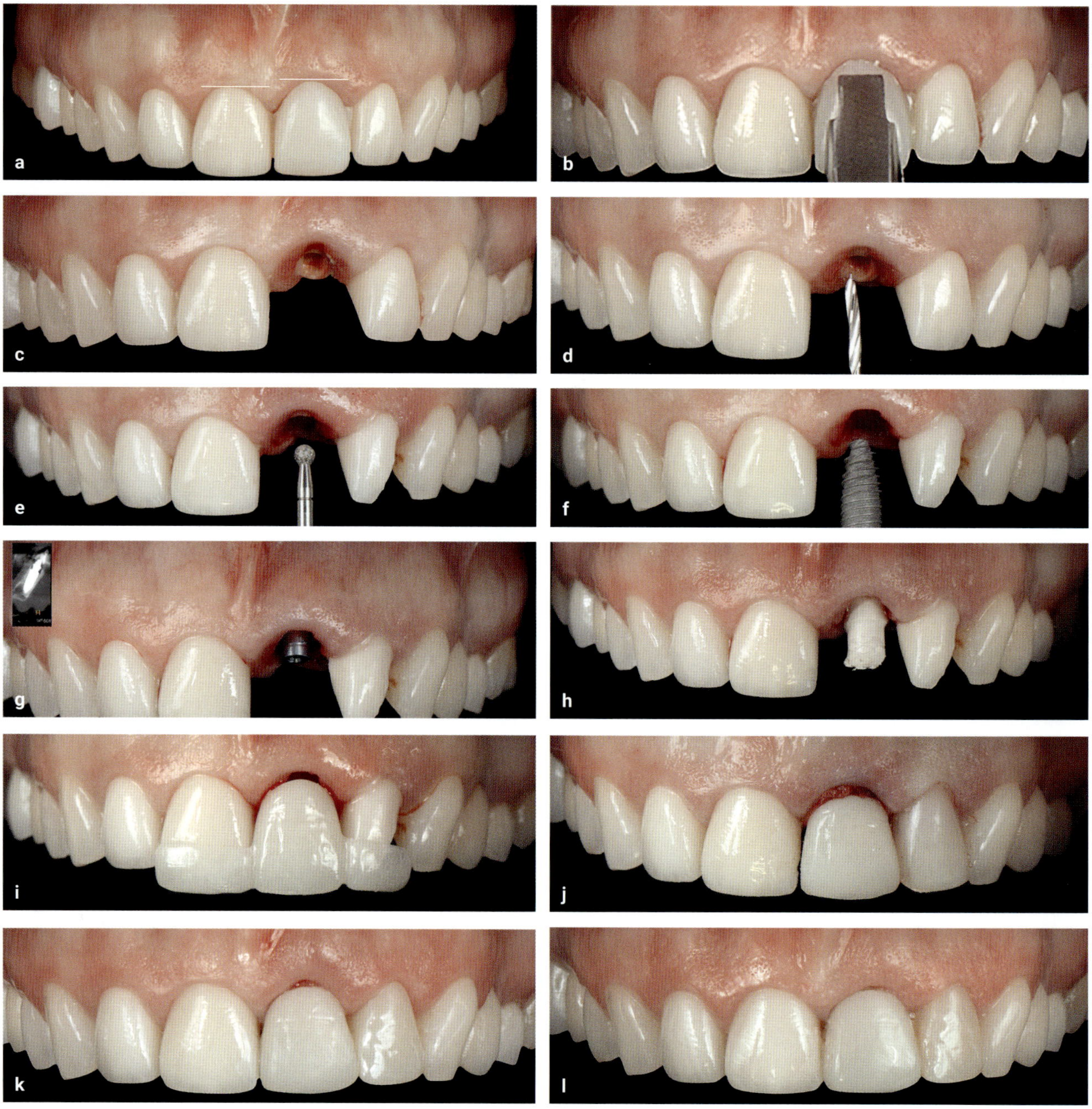

Fig 15-3 Implant placement utilizing a socket shield approach with PRF to facilitate soft tissue healing and improvements in gingival contouring. *(a)* Initial baseline clinical image demonstrating a hopeless tooth with an obvious associated gingival recession. This case was planned for immediate implant placement utilizing a partial extraction therapy. *(b and c)* Following crown removal. *(d and e)* The palatal surface of the root was removed. *(f and g)* Implant placement. *(h and i)* Provisional crown fabrication. *(j)* PRF was then placed within the soft tissue gap to facilitate soft tissue healing followed by final seating of the provisional crown. *(k)* Note the healing 2 weeks postoperative. *(l)* Note the correction of the gingival recession at 1 year utilizing this approach. (Case performed by Dr Luis Bessa.)

PRF and soft tissue healing at implants

It has been shown that thick soft tissue favors coronal peri-implant bone stability.[4–6] As mentioned previously, a symbiotic relationship between bone and soft tissue is necessary to maintain the implant's stability and integrity. Bone and its augmentation in numerous reconstruction techniques traditionally has been a keen topic in implant dentistry.[2] Current understanding now stresses the importance of bone to maintain the soft tissue and the soft tissue to reinforce bone stability. The lack of adequate soft and hard tissue development could be one of the reasons why high levels of peri-implantitis are observed in the current literature.[26] Because PRF has been shown to act on soft

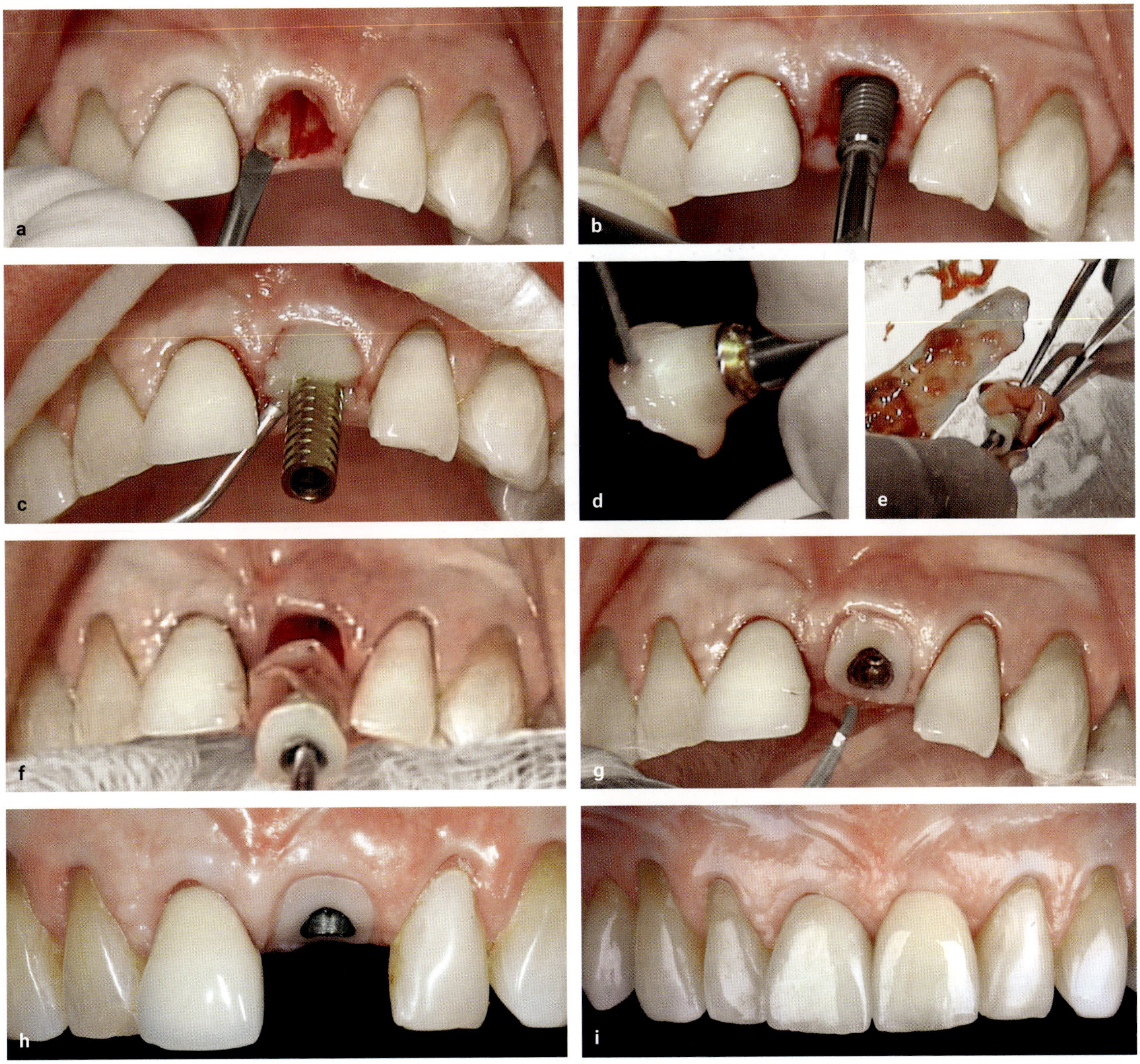

Fig 15-4 Use of PRF for soft tissue healing around a custom-fabricated abutment during immediate implant placement. *(a)* Partial extraction of the lingual portion of the root. *(b)* Implant placement. *(c)* Fabrication of a custom healing abutment. *(d)* Finishing fabrication of the custom healing abutment extraorally. *(e)* Placement of a PRF membrane folded in half through the custom healing abutment (poncho technique). *(f and g)* Placement and fixation of the custom healing abutment with a PRF membrane to facilitate soft tissue healing. *(h)* Healing of soft tissues around custom abutment. Note the excellent maintenance of soft tissues. *(i)* Final crown placement. See Video 15-1 for complete case presentation. (Case performed by Dr Howard Gluckman.)

tissue wound healing primarily, its use in implant dentistry, when placed immediately or delayed, has been heavily utilized adjacent to soft tissues, as presented in the following cases. The concept takes advantage of the local application of an autogenous biomaterial rich in growth factors that stimulate neoangiogenesis and collagen formation within the soft tissue flap above an implant. Moreover, when compressed into membranes, PRF can maintain the integrity of an augmentation procedure, enhancing site protection when used in conjunction with other barrier membranes and contributing to the healing of the overlying flap.[27,28] Figure 15-4 depicts this concept, whereby PRF is utilized with a custom provisional abutment to facilitate soft tissue regeneration and attachment around the abutment and minimize potential bacterial invasion toward the implant surface.

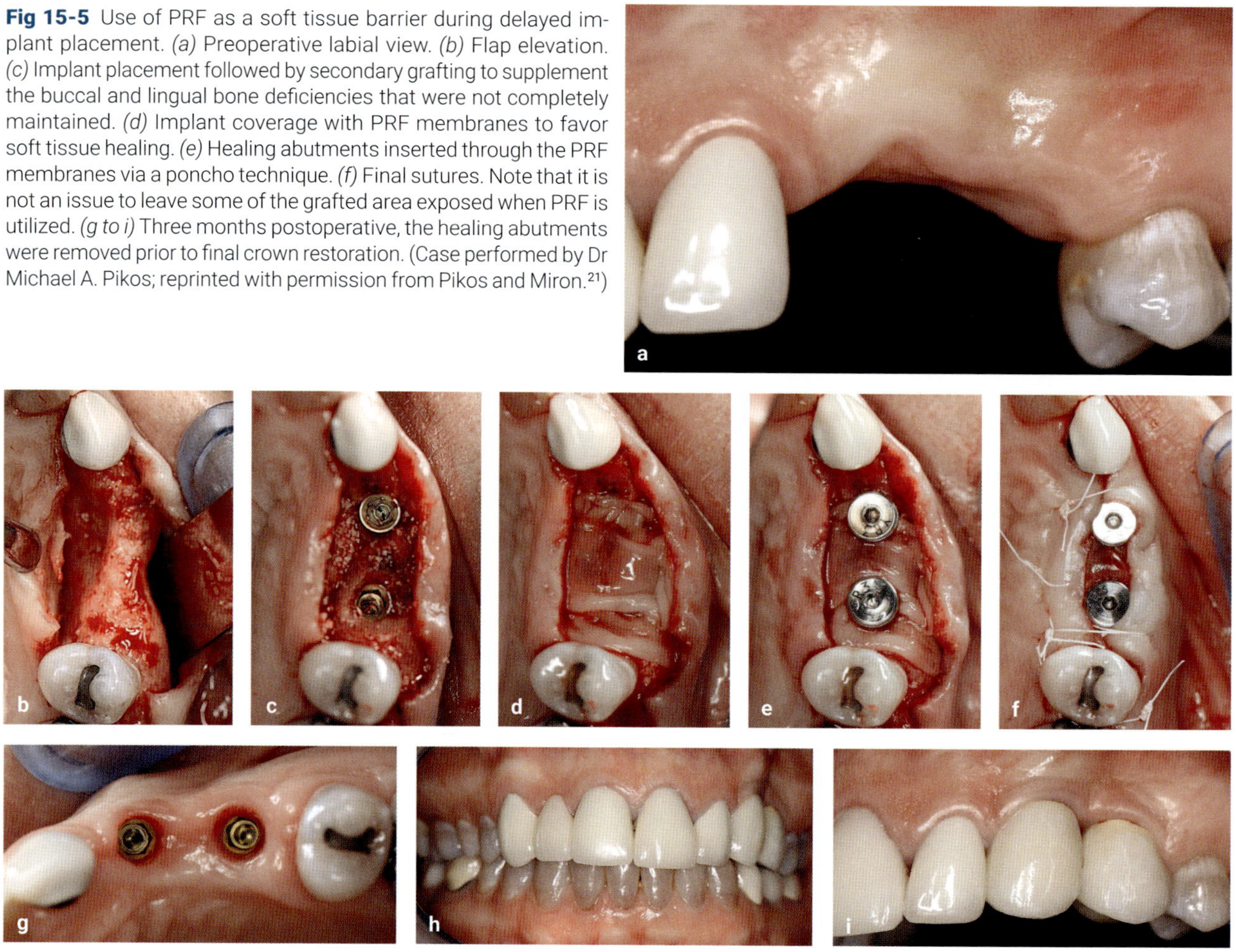

Fig 15-5 Use of PRF as a soft tissue barrier during delayed implant placement. *(a)* Preoperative labial view. *(b)* Flap elevation. *(c)* Implant placement followed by secondary grafting to supplement the buccal and lingual bone deficiencies that were not completely maintained. *(d)* Implant coverage with PRF membranes to favor soft tissue healing. *(e)* Healing abutments inserted through the PRF membranes via a poncho technique. *(f)* Final sutures. Note that it is not an issue to leave some of the grafted area exposed when PRF is utilized. *(g to i)* Three months postoperative, the healing abutments were removed prior to final crown restoration. (Case performed by Dr Michael A. Pikos; reprinted with permission from Pikos and Miron.[21])

As reviewed earlier in this chapter, Hehn et al experimented with the insertion of PRF within a split flap at implant placement and reported this to reduce soft tissue thickness.[14] These findings here suggest that splitting the flap may unnecessarily strain the soft tissue healing in addition to a full mucoperiosteal flap raised for implant insertion. Therefore, when PRF is utilized, it must be placed beneath the full-thickness flap without additionally dividing the tissue.

Figure 15-5 demonstrates implant placement into a deficient ridge. Sticky bone was utilized as the grafting material to better contour and manage the augmentation area, due to its sticky and stable consistency, which aids in the handling of the bone. Note that before soft tissue closure, PRF membranes were laid over the defect as an outer barrier to protect the bone-augmented site from potential incoming pathogens and further facilitate soft tissue wound closure and healing.

Similarly, Fig 15-6 represents a narrow ridge where implant placement was conducted following a ridge split technique. The internal cancellous area and the outer cortical buccal layers were grafted using the sticky bone concept. Following coverage with a collagen barrier membrane, the entire site was then covered with PRF membranes to promote soft tissue healing and minimize potential infection to the grafted site.

Figure 15-7 demonstrates a large full-arch case, including tooth extraction followed by simultaneous immediate implant placement using a full-arch guided approach. Such cases, especially in immunocompromised patients, are excellent cases for PRF use. Once again, PRF membranes may be utilized to cover the grafted sites to promote faster soft tissue healing and minimize the potential risk of infection in such complex cases.

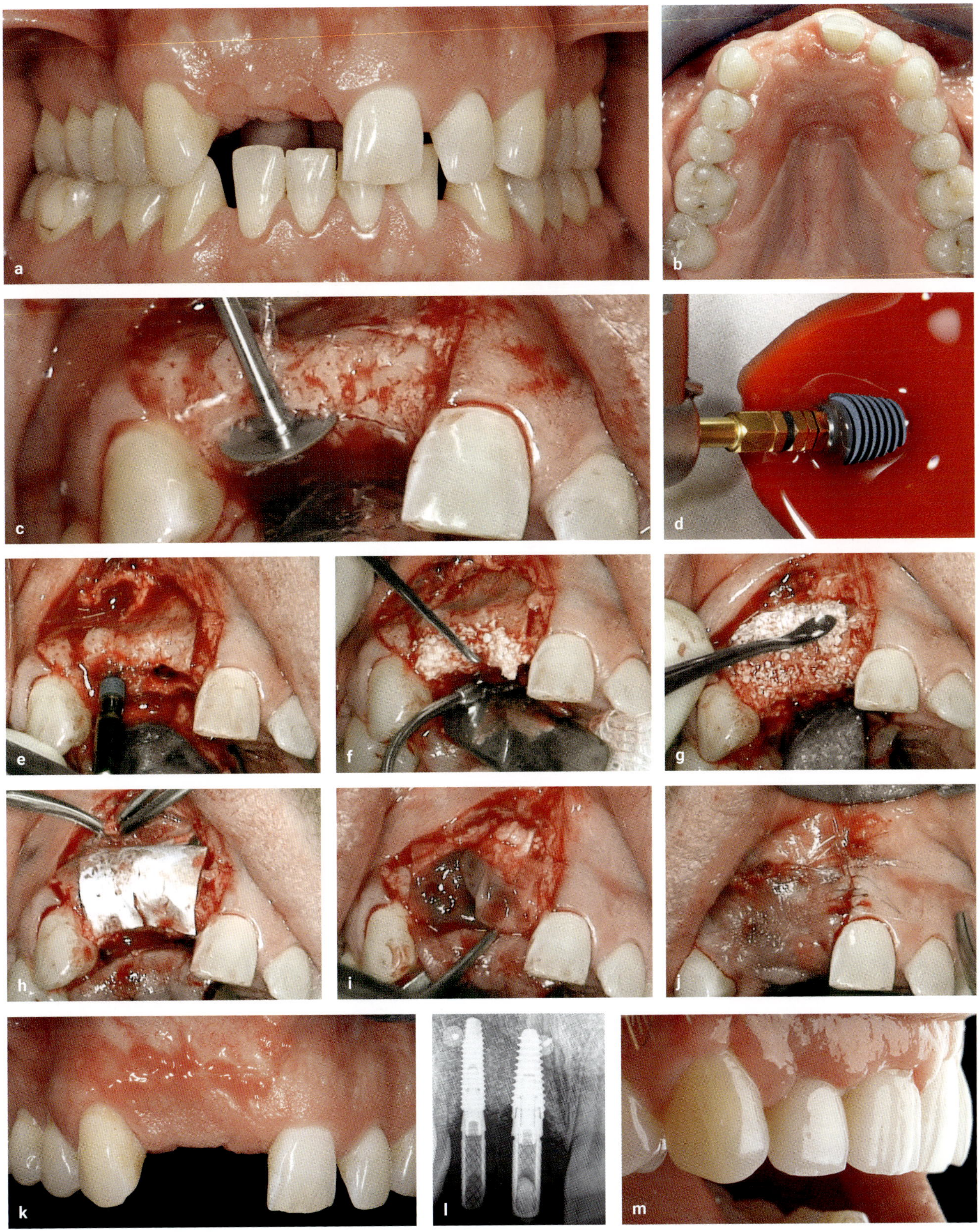

Fig 15-6 Use of PRF during immediate implant placement with a simultaneous ridge split. *(a and b)* Clinical images of a deficient ridge with missing maxillary anterior teeth. *(c)* Following flap elevation, a ridge split technique was performed. *(d and e)* Implants were then coated with liquid-PRF prior to implant placement. *(f and g)* Both the inner and outer regions of the bone defects were grafted with sticky bone (liquid-PRF mixed with bone particles). *(h and i)* Following collagen membrane placement over the defect, the site was coated with PRF membranes. *(j)* Final sutures. *(k)* Postsurgical healing; note the excellent soft tissue healing. *(l)* Periapical radiograph. *(m)* Final restorations. See Video 15-2 for complete case presentation. (Case performed by Dr Howard Gluckman.)

Video 15-2

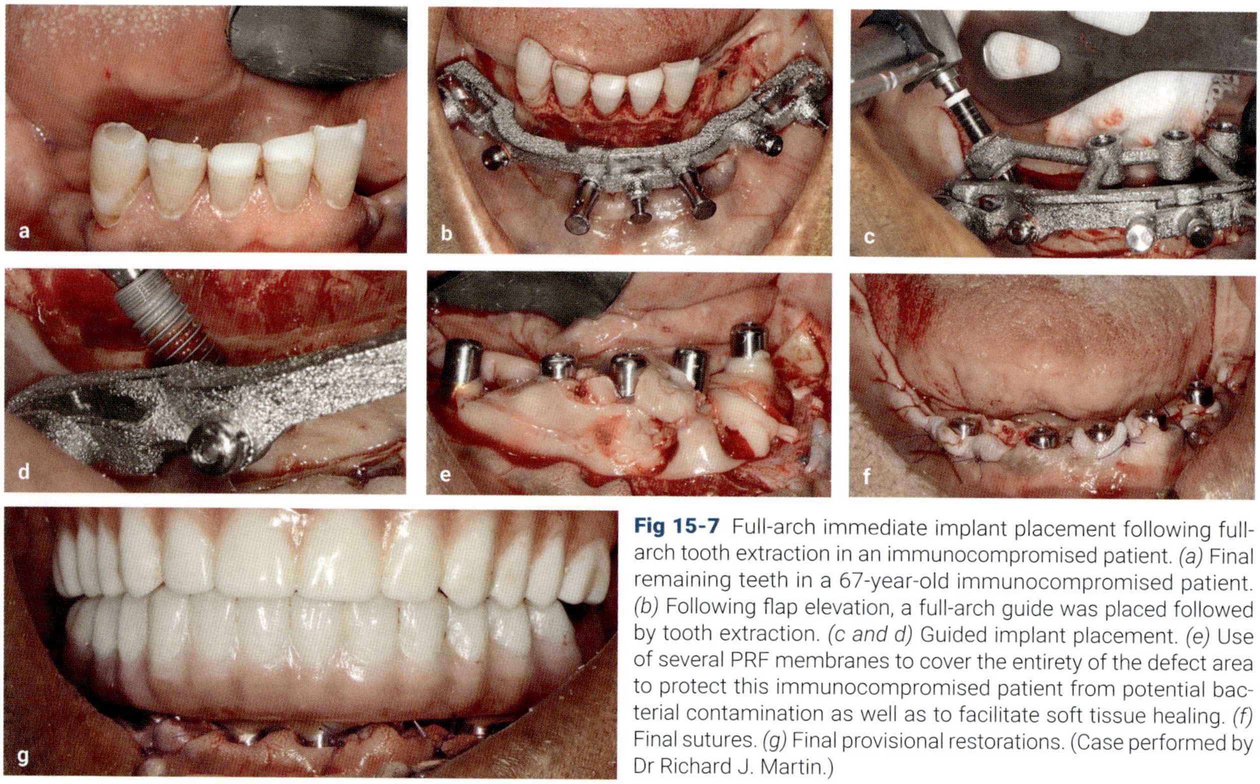

Fig 15-7 Full-arch immediate implant placement following full-arch tooth extraction in an immunocompromised patient. *(a)* Final remaining teeth in a 67-year-old immunocompromised patient. *(b)* Following flap elevation, a full-arch guide was placed followed by tooth extraction. *(c and d)* Guided implant placement. *(e)* Use of several PRF membranes to cover the entirety of the defect area to protect this immunocompromised patient from potential bacterial contamination as well as to facilitate soft tissue healing. *(f)* Final sutures. *(g)* Final provisional restorations. (Case performed by Dr Richard J. Martin.)

Use of PRF at osteotomy preparation

The high predictability of osseointegration has prompted clinicians and researchers to push the boundary to accelerate healing and expedite the completion of treatment. Developments in microroughened and hydrophilic implant surface technology have largely facilitated this, successfully showing increased ISQ values at shorter time intervals.[29] This means that patients can be restored earlier, restoring function and esthetics as soon as possible. However, the downside is that microroughened implant surfaces are potentially more susceptible to bacterial colonization and peri-implantitis.[30]

Numerous studies over the years have investigated implant surfaces enhanced with growth factors with varying results.[31] Some studies utilizing cell adhesion molecules or BMPs showed that they can increase osteoblastic differentiation and functional integration, reporting increases in BIC values.[32] PRF delivers platelets and leukocytes to the wound or osteotomy and releases growth factors locally (namely PDGF, TGF-β, IGF, and VEGF) that accelerate the healing process by attracting undifferentiated endothelial cells and mesenchymal cells to the injured site.[28,33,34] This could theoretically improve the healing response around implants. For these reasons, implants may be coated with liquid-PRF before implant placement (see Fig 15-6d). However, limited data exists supporting its ability to favor implant osseointegration or to increase BIC values. A recent study reported increased ISQ values during the early healing period when PRF was applied inside the osteotomy during insertion. The implant itself coated in plasma extruded from the PRF.[35] Though these studies have shown statistically significant improvement in type 2 bone, limited data exists supporting other types of bone. It is typically types 3 and 4 bone with poorer density that present a clinical challenge. The significance of PRF and implants is thus limited, with further research needed.

Conclusion

A consistent theme in discussions regarding PRF is that experts on the topic continue to recommend more research on the subject for its ability to promote bone regeneration. In implant dentistry, while some studies have investigated PRF's value in accelerating osseointegration, the evidence has provided only marginal improvements, with the majority of clinical studies to date having focused on its ability to facilitate soft tissue wound healing. PRF provides the benefit of faster soft tissue healing and wound closure and presents an array of host immune-fighting leukocytes to the defect area. Therefore, its use has been primarily favored around soft tissue, especially in immediate implant dentistry, where larger gaps are often encountered and readily filled with PRF.

References

1. Scala A, Lang NP, Schweikert MT, Oliveira JA, Rangel-Garcia I, Botticelli D. Sequential healing of open extraction sockets. An experimental study in monkeys. Clin Oral Implants Res 2014; 25:288–295.
2. Esposito M, Grusovin MG, Felice P, Karatzopoulos G, Worthington HV, Coulthard P. Interventions for replacing missing teeth: Horizontal and vertical bone augmentation techniques for dental implant treatment. Cochrane Database Syst Rev 2009;2009(4): CD003607.
3. Esposito M, Maghaireh H, Gabriella Grusovin M, Ziounas I, Worthington HV. Soft tissue management for dental implants: What are the most effective techniques? A Cochrane systematic review. Eur J Oral Implantol 2012;5:221–238.
4. Puisys A, Linkevicius T. The influence of mucosal tissue thickening on crestal bone stability around bone-level implants. A prospective controlled clinical trial. Clin Oral Implants Res 2015;26:123–129.
5. Levine RA, Huynh-Ba G, Cochran DL. Soft tissue augmentation procedures for mucogingival defects in esthetic sites. Int J Oral Maxillofac Implants 2014;29(suppl):155–185.
6. Linkevicius T, Puisys A, Steigmann M, Vindasiute E, Linkeviciene L. Influence of vertical soft tissue thickness on crestal bone changes around implants with platform switching: A comparative clinical study. Clin Implant Dent Relat Res 2015;17:1228–1236.
7. Bassetti RG, Stähli A, Bassetti MA, Sculean A. Soft tissue augmentation around osseointegrated and uncovered dental implants: A systematic review. Clin Oral Investig 2011;21:53–70.
8. Fu JH, Lee A, Wang HL. Influence of tissue biotype on implant esthetics. Int J Oral Maxillofac Implants 2011;26:499–508.
9. Lee A, Fu JH, Wang HL. Soft tissue biotype affects implant success. Implant Dent 2011;20:e38–e47.
10. Boora P, Rathee M, Bhoria M. Effect of platelet rich fibrin (PRF) on peri-implant soft tissue and crestal bone in one-stage implant placement: A randomized controlled trial. J Clin Diagn Res 2015;9: Zc18–Zc21.
11. Öncü E, Alaaddinoğlu EE. The effect of platelet-rich fibrin on implant stability. Int J Oral Maxillofac Implants 2015;30: 578–582.
12. Tabrizi R, Arabion H, Karagah T. Does platelet-rich fibrin increase the stability of implants in the posterior of the maxilla? A split-mouth randomized clinical trial. Int J Oral Maxillofac Surg 2018;47:672–675.
13. Diana C, Mohanty S, Chaudhary Z, Kumari S, Dabas J, Bodh R. Does platelet-rich fibrin have a role in osseointegration of immediate implants? A randomized, single-blind, controlled clinical trial. Int J Oral Maxillofac Surg 2018;47:1178–1188.
14. Hehn J, Schwenk T, Striegel M, Schlee M. The effect of PRF (platelet-rich fibrin) inserted with a split-flap technique on soft tissue thickening and initial marginal bone loss around implants: Results of a randomized, controlled clinical trial. Int J Implant Dent 2016;2:13.
15. Hamzacebi B, Oduncuoglu B, Alaaddinoglu EE. Treatment of peri-implant bone defects with platelet-rich fibrin. Int J Periodontics Restorative Dent 2015;35:415–422.
16. Degidi M, Daprile G, Nardi D, Piattelli A. Buccal bone plate in immediately placed and restored implant with Bio-Oss collagen graft: A 1-year follow-up study. Clin Oral Implants Res 2013;24:1201–1205.
17. Maia LP, Reino DM, Muglia VA, et al. Influence of periodontal tissue thickness on buccal plate remodelling on immediate implants with xenograft. J Clin Periodontol 2015;42:590–598.
18. Lee JW, Kim SG, Kim JY, et al. Restoration of a peri-implant defect by platelet-rich fibrin. Oral Surg Oral Med Oral Pathol Oral Radiol 2012;113:459–463.
19. Hao PJ, Wang ZG, Xu QC, et al. Effect of umbilical cord mesenchymal stem cell in peri-implant bone defect after immediate implant: An experiment study in beagle dogs. Int J Clin Exp Pathology 2014;7:8271.
20. Şimşek S, Özeç İ, Kürkçü M, Benlidayi E. Histomorphometric evaluation of bone formation in peri-implant defects treated with different regeneration techniques: An experimental study in a rabbit model. J Oral Maxillofac Surg 2016;74:1757–1764.
21. Pikos MA, Miron RJ. Bone Augmentation in Implant Dentistry: A Step-by-Step Guide to Predictable Alveolar Ridge and Sinus Grafting. Chicago: Quintessence, 2019.
22. Esposito M, Grusovin MG, Worthington HV. Treatment of peri-implantitis: What interventions are effective? A Cochrane systematic review. Eur J Oral Implantol 2012;5(suppl):S21–S41.
23. Albrektsson T, Cannullo L, Cochran D, De Bruyn H. "Peri-implantitis": A complication of a foreign body or a man-made "disease." Facts and fiction. Clin Implant Dent Relat Res 2016;18: 840–849.
24. Guler B, Uraz A, Yalım M, Bozkaya S. The comparison of porous titanium granule and xenograft in the surgical treatment of peri-implantitis: A prospective clinical study. Clin Implant Dent Relat Res 2017;19:316–327.
25. Schwarz F, Herten M, Sager M, Bieling K, Sculean A, Becker J. Comparison of naturally occurring and ligature-induced peri-implantitis bone defects in humans and dogs. Clin Oral Implants Res 2007;18:161–170.
26. Monje A, Aranda L, Diaz K, et al. Impact of maintenance therapy for the prevention of peri-implant diseases: A systematic review and meta-analysis. J Dent Res 2016;95:372.
27. Panda S, Sankari M, Satpathy A, et al. Adjunctive effect of autologus platelet-rich fibrin to barrier membrane in the treatment of periodontal intrabony defects. J Craniofac Surg 2016;27:691–696.
28. Miron RJ, Fujioka-Kobayashi M, Bishara M, Zhang Y, Hernandez M, Choukroun J. Platelet-rich fibrin and soft tissue wound healing: A systematic review. Tissue Eng Part B Rev 2017;23:83–99.
29. Smeets R, Stadlinger B, Schwarz F, et al. Impact of dental implant surface modifications on osseointegration. Biomed Res Int 2016;2016:6285620.

30. Esposito M, Coulthard P, Thomsen P, Worthington H. The role of implant surface modifications, shape and material on the success of osseointegrated dental implants. A Cochrane systematic review. Eur J Prosthodont Restor Dent 2005;13:15–31.
31. Scheller EL, Krebsbach PH. Using soluble signals to harness the power of the bone marrow microenvironment for implant therapeutics. Int J Oral Maxillofac Implants 2011;26:70.
32. Liu Y, Enggist L, Kuffer AF, Buser D, Hunziker EB. The influence of BMP-2 and its mode of delivery on the osteoconductivity of implant surfaces during the early phase of osseointegration. Biomaterials 2007;28:2677–2686.
33. Terheyden H, Lang NP, Bierbaum S, Stadlinger B. Osseointegration—Communication of cells. Clin Oral Implants Res 2012;23:1127–1135.
34. Terheyden H, Stadlinger B, Sanz M, Garbe AI, Meyle J. Inflammatory reaction—Communication of cells. Clin Oral Implants Res 2014;25:399–407.
35. Öncü E, Bayram B, Kantarcı A, Gülsever S, Alaaddinoğlu EE. Positive effect of platelet rich fibrin on osseointegration. Med Oral Patol Oral Cirugía Bucal 2016;21:e601.

16

Use of PRF in Guided Bone Regeneration

Contributors

Richard J. Miron
Michael A. Pikos
Howard Gluckman
Omid Moghaddas
Hom-Lay Wang

Chapter Highlights

- *History of GBR including commonly utilized bone grafting materials and barrier membranes*
- *Presentation of RCTs utilizing PRF for various applications in GBR*
- *Clinical case presentations utilizing PRF during GBR*
- *Future perspectives including the potential use of heated e-PRF membranes for GBR applications*

Guided bone regeneration (GBR) has been utilized in dentistry since the mid 1990s with the aim of preventing the faster-growing soft tissues from infiltrating into the slower-growing bony defects. Over the years, the concept of GBR has expanded tremendously, with entire textbooks now dedicated to this topic alone, and much improvement has been made with respect to the biocompatibility and strength of barrier membranes. As such, while original expanded polytetrafluoroethylene (PTFE) nonresorbable membranes were first utilized in regenerative dentistry, a variety of biodegradable collagen membranes, biodegradable polymers, titanium-reinforced nonexpanded PTFE, as well as titanium meshes (Ti-mesh) have since been introduced and utilized. More recently, PRF has also been utilized during GBR procedures either mixed with bone grafting material particles or as an outer membrane. This chapter begins by reviewing standard GBR principles and then discusses the advantages and limitations of utilizing PRF as a barrier membrane during routine GBR applications as well as in combination with bone grafting particles to favor their handling. Clinical guidelines and recommendations are thereafter provided following a detailed review of the topic.

This chapter begins by reviewing the literature related to GBR and drawing conclusions based on comparative studies conducted in RCTs on the topic. PRF typically degrades within a few weeks. While much focus to date has been placed on the biologic properties of PRF as a natural fibrin membrane, we discuss the clinical relevance of utilizing PRF as a sole barrier membrane and examine biologically as well as clinically the pros and cons of combining PRF with a standard collagen barrier membrane for routine GBR procedures in everyday clinical practice.

GTR/GBR procedures and techniques have been utilized by oral surgeons, periodontists, and more frequently general dentists for the regeneration of either periodontal or bony tissues.[1] The concept was developed in the mid 1980s by Gottlow et al, where a series of studies were undertaken based on the hypothesis that in order to optimize the slower-growing bone or periodontal cells, you had to limit or slow the faster-growing overlaying soft tissues.[2] Based on all positive reported findings, the GBR technique is now widely accepted in the bone regeneration field and has since exploded in popularity, with various barrier membranes and bone grafting materials being developed along the way to favor regenerative outcomes.[3–6]

PRF has been utilized over the past 20 years due to its ability to favor angiogenesis and tissue healing utilizing 100% naturally derived autologous growth factors obtained at relatively low cost. While initial studies on PRP were developed to favor angiogenesis and regeneration using platelets, PRF has since been more frequently utilized due to its fibrin mesh formation with entrapment of cells and growth factors favoring their slow and gradual release (shown to increase cellular bioactivity; see chapter 2). PRF has therefore been utilized much like PRP, though fibrin clots are typically cut into smaller PRF fragments and mixed with various bone grafting materials or subsequently flattened and utilized as a barrier membrane in GBR procedures. PRF offers some advantages when compared to traditional collagen barrier membranes in that it contains autologous growth factors as well as living host defense-fighting leukocytes. These cells act to protect host tissues against incoming pathogens and, as reviewed in chapter 14, may reduce potential infection by as much as 10-fold (third molar sites).[7] Nevertheless, while the effects of PRF on angiogenesis and soft tissue wound healing have been well documented, studies on its influence on bone regeneration procedures have been scarce. Furthermore, PRF has limited potential to exclude soft tissues over the course of bone regeneration because it typically degrades within a 2- to 3-week period.

PRF has limited potential to exclude soft tissues over the course of bone or periodontal regeneration because it typically degrades within a 2- to 3-week period.

Overview of GBR

The concept of GBR and the use of barrier membranes were first introduced to the field of periodontology and implant dentistry nearly 30 years ago.[8–11] These concepts were derived from the fact that different rates of cell growth and migration properties were observed between alveolar soft tissues when compared to underlying hard tissues.[8] Thus, the concept originally attempted to hamper fast-growing nonfunctional epithelial cells from infiltrating into bone defects to allow for the unimpaired healing of slower-growing bone cells[12] (Fig 16-1). Over the years, a variety of additional parameters including space maintenance, ability to form a blood clot, mechanical stabilization, cellular infiltration potential, biocompatibility, and resorption properties have been investigated to achieve more favorable bone tissue healing.[13]

Although originally synthetic expanded polytetrafluoroethylene (ePTFE) membranes were shown to provide successful outcomes,[8,9] it was later argued that a second surgical intervention needed to remove the barrier membrane provided additional patient morbidity, which could be prevented by utilizing biodegradable materials.[14,15] For these reasons, absorbable collagen and synthetic membranes have become available in the last decade, practically replacing original ePTFE membranes[1] (see Table 16-1).

Overall, GBR has seen extremely high predictable results with well-established and well-documented protocols leading to high implant survival rates over 90% even following 10 years postsurgery.[16] One key concept described often in the literature is that sufficient primary wound closure is mandatory to prevent soft tissue ingrowth, bacterial contamination, early membrane degradation, soft tissue dehiscence, and potential graft exposure. This concept is discussed later with respect to PRF.

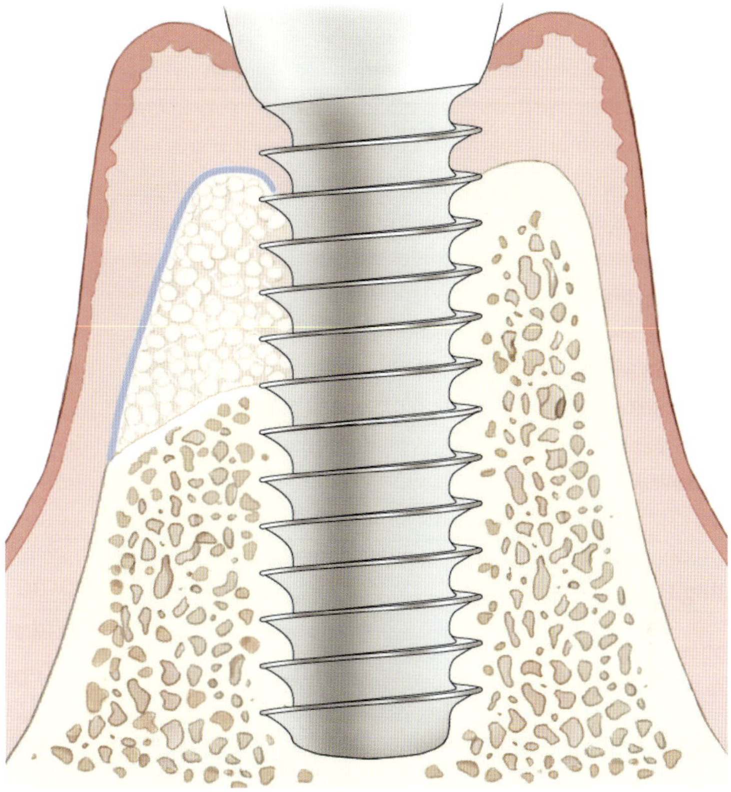

Fig 16-1 Representative diagram illustrating the principles of GBR procedures. The faster-growing soft tissues are excluded via a barrier membrane from the slower-growing underlying bony tissue. This allows more time for bone to properly regenerate. Furthermore, in larger defects, the mechanical strength of the barrier membrane is advantageous because space maintenance and tension-free closure on the grafted area are also important parameters for bone regeneration.

Hundreds of commercially available barrier membranes now exist on the market for GBR procedures, with new products constantly being brought to market on a yearly basis. Furthermore, bone grafting materials have become a multibillion-dollar industry, with an exponential growth in their use.[1] Following we briefly summarize available options on the market for both barrier membranes and bone grafting materials.

Available Options for GBR

The use of barrier membranes has been discussed in the literature with several parameters being relevant to their application.[1,17] Fundamental requirements for biomaterials utilized in GBR are biocompatibility of the material to prevent an adverse host reaction and certain degradability properties to allow adequate bone regeneration and eventual replacement with native host bone. Ideally, GBR materials must provide proper mechanical strength to guarantee space creation and maintenance for migrating cells from the surrounding bone to facilitate bone regeneration. Therefore, various types of barrier membranes are utilized depending on the defect type and size. The greater the defect, in particular in the vertical dimension, the greater the barrier strength requirements (titanium-reinforced). Barrier membranes must also prevent the infiltration of fibrous tissue (the so-called cell-occlusiveness properties) to avoid impaired bone healing, whereas the bone grafting material must facilitate the migration of osteogenic cells such as osteoblasts to the material surface.[17] A wide range of barrier membranes have been made commercially available for various clinical approaches, with distinct advantages and disadvantages of each presented in Table 16-1.[17–19] For example, the currently available Ti-mesh does provide the property of space maintenance and acts like a barrier membrane, but due to the holes that exist within the Ti-mesh, it cannot completely prevent cell migration and has often been termed *protected bone augmentation*.

Table 16-1 Commercially available membranes for GBR procedures

Membranes		Brand name	Manufacturer	Material	Properties	Comments
Non-resorbable membranes	ePTFE	Gore-Tex	W. L. Gore	ePTFE	Good space maintainer; easy to handle	Longest clinical experience
		Gore-Tex-TI	W. L. Gore	Ti-ePTFE	Most stable space maintainer; filler material unnecessary	Titanium should not be exposed; commonly used in ridge augmentation
	dPTFE	High-Density Gore-Tex	W. L. Gore	dPTFE	0.2-µm pores	Avoids a secondary surgery
		Cytoplast	Osteogenics		< 0.3-µm pores	Primary closure unnecessary
		TefGen FD	Lifecore		0.2- to 0.3-µm pores	Easy to detach
		Nonresorbable	ACE Surgical Supply		< 0.2-µm pores; 0.2 mm thick	Limited cell proliferation
	Titanium mesh (protected bone regeneration)	Ti-Micromesh	ACE Surgical Supply	Ti	1,700-nm pores; 0.1 mm thick	Ideal long-term survival rate
		Tocksystem Mesh	Tocksystem		0.1- to 6.5-µm pores; 0.1 mm thick	Minimal resorption and inflammation
		Frios BoneShields	Dentsply		0.03-mm pores; 0.1 mm thick	Sufficient bone to regenerate
		M-TAM			1,700-nm pores; 0.1 to 0.3 mm thick	Excellent tissue compatibility
Synthetic absorbable membranes		OsseoQuest	W. L. Gore	Hydrolyzable polyester	Resorption: 16–24 weeks	Good tissue integration
		Biofix	Bioscience Oy	Polyglycolic acid	Resorption: 24–48 weeks	Isolates the space from cells from soft tissue and bacteria
		Vicryl	Johnson & Johnson	Polyglactin 910 Glycolide/L-lactide 9:1	Well adaptable; resorption: 4–12 weeks	Woven membrane; four prefa ricated shapes
		Atrisorb	Tolmar	Poly-DL-lactide and solvent	Resorption: 36–48 weeks; interesting resorptive characteristics	Custom-fabricated membran "barrier kit"
		EpiGuide	Kensey Nash	Poly-DL-lactic acid	Three-layer membrane; resorption: 6–12 weeks	Self-supporting; support-developed blood clot
		Resolut	W. L. Gore	Poly-DL-lactid/Co-glycolid	Resorption: 10 weeks; good space maintainer	Good tissue integration; separate suture material
		Vivosorb	Polyganics	DL-lactide-ε-caprolactone (PLCL)	Anti-adhesive barrier; up to 8 weeks' mechanical properties	Acts as a nerve guide
Natural biodegradable material		Plasma rich in growth factors (PRGF-Endoret)	BTI Biotechnology Institute	Patient's own blood	Abundant growth factors and proteins mediate cell behaviors; different formulations for various usages; total resorption	Enhances osseointegration a initial implant stability; promc new bone formation; encourages soft tissue recovery
		Bio-Gide	Geistlich	Porcine I and III	Resorption: 6–8 weeks; mechanical strength: 7.5 MPa	Usually used in combination with filler materials
		BioMend BioMend Extend	Zimmer	Bovine I	Resorption: 6–8 weeks but can be extended to 16–20 weeks; mechanical strength: 3.5–22.5 MPa	Fibrous network; modulates c activities
		BioSorb	3M ESPE	Bovine I	Resorption: 26–38 weeks	Tissue integration
		Neomem	Citagenix	Bovine I	Double-layer product; resorption: 26–38 weeks	Used in severe cases
		OsseoGuard	Biomet 3i	Bovine I	Resorption: 24–32 weeks	Improves the esthetics of the final prosthetics
		Ossix Plus	OraPharma	Porcine I	Resorption: 16–24 weeks	Increases the woven bone

Modified with permission from Zhang et al.[17]

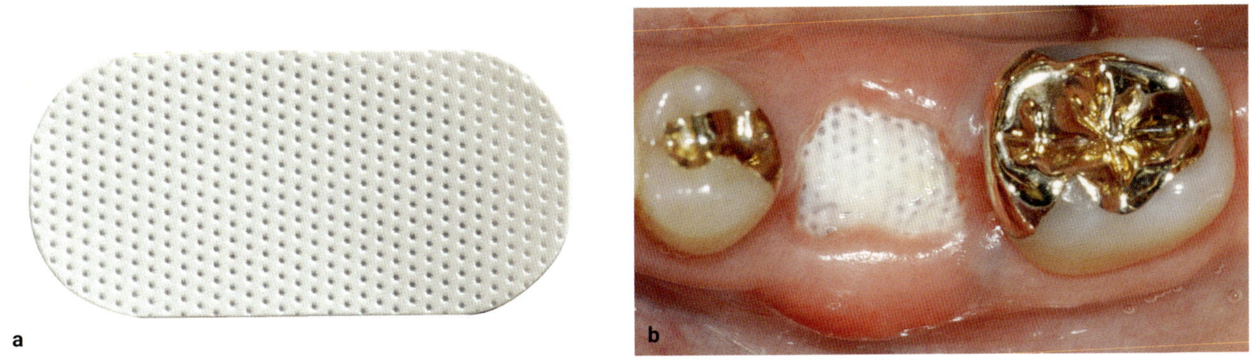

Fig 16-2 *(a)* Clinical image of dPTFE membrane. *(b)* Use of dPTFE membrane for tooth extraction. The primary advantage of these membranes is their better mechanical strength when compared to conventional collagen membranes. (Reprinted with permission from Pikos and Miron.[21])

Nonresorbable PTFE membranes

Nonresorbable membranes include expanded, high-density, and titanium-reinforced PTFE (ePTFE, dPTFE, and PTFE-TR) membranes.[19] In general, the main disadvantage of synthetic nonresorbable membranes is the requirement for a second surgical intervention to remove the membrane because they are not biodegradable[20,21] (Fig 16-2). Although the use of ePTFE membranes has demonstrated higher levels of new bone formation and clinical attachment levels around teeth,[22] the requirement of a second surgical intervention to remove the barrier after implantation is a significant drawback, with the potential to reinjure tissues.[17] Furthermore, the second surgical procedure leads to extra surgical time and therefore incurs additional costs and patient discomfort.

Collagen-based absorbable membranes

The most utilized barrier material in regenerative dentistry is the collagen membrane[21,22] (Fig 16-3). Its main advantages are its excellent biocompatibility as well as its absorbable properties, thereby alleviating patient discomfort and additional morbidity/tissue damage potentially caused by a second surgery. Over the years, a variety of collagen membranes have been developed, each with various resorption rates and cross-linking properties derived from human skin, bovine Achilles tendon, or porcine skin (see Table 16-1). While they are the most frequently utilized barrier membranes in dentistry, their main drawbacks include their lack of space-maintaining ability/rigidity, high costs, potential for creating a foreign body reaction, and risk of infection if left exposed to the oral cavity.

Synthetic absorbable membranes

In addition to collagen resorbable membranes, a series of synthetic resorbable membranes fabricated from polyesters (eg, polyglycolic acid, polylactic acid, polycaprolactone) and their copolymers have been introduced[17,24] (see Table 16-1). Their main advantage is that polyglycolide or polylactide can be made in large quantities at low costs with different physical, chemical, and mechanical properties. Their main disadvantage has been the fact they are more prone to foreign body reactions, thereby influencing their wound healing properties. For these reasons, collagen barrier membranes are much more frequently utilized.

Titanium mesh

Due to the high biocompatibility and additional strength of titanium, titanium-reinforced barrier membranes were introduced as an option for bone regeneration (Fig 16-4).[21] This allows for superior mechanical support, which favors a larger space maintenance for bone regrowth without compression on the underlying bone. Based on these advantages, Ti-meshes have been more frequently utilized, especially for large bone regeneration procedures.[21]

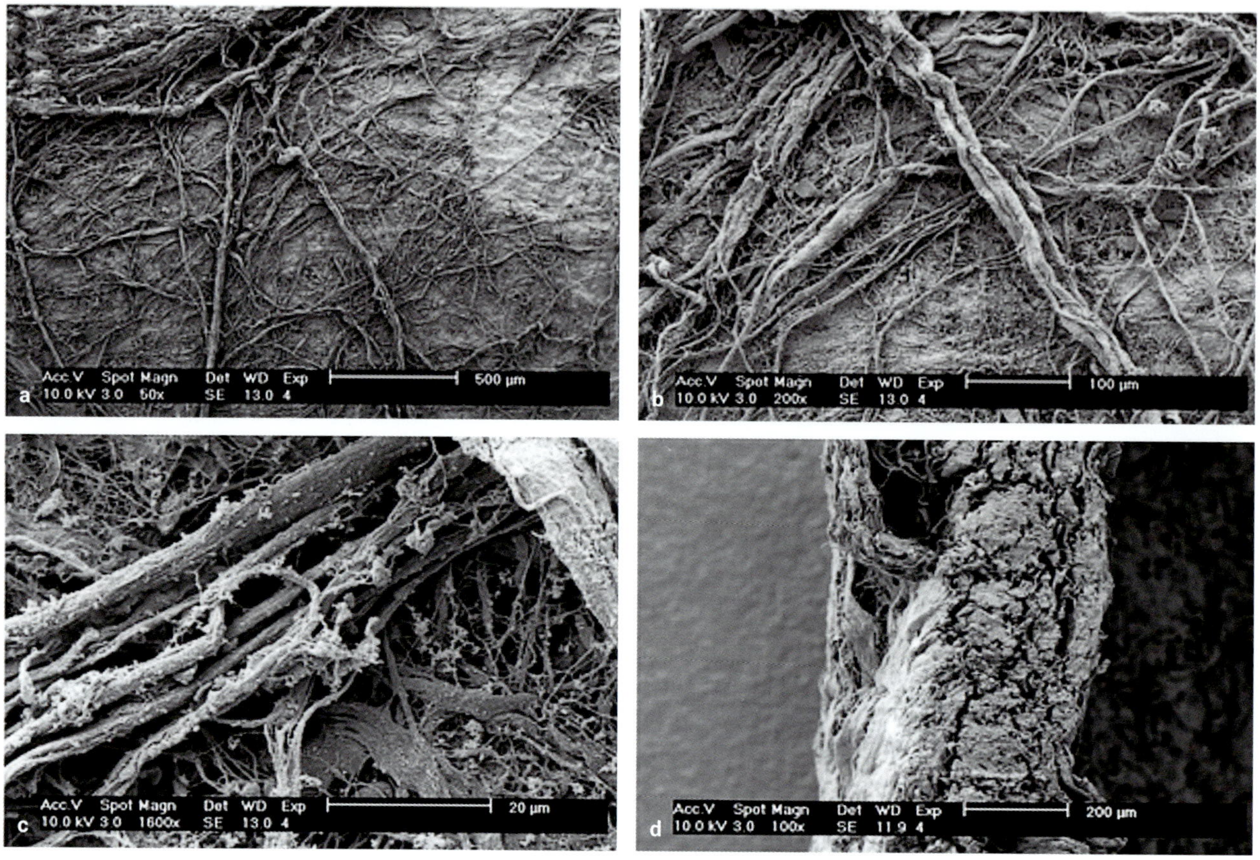

Fig 16-3 Scanning electron microscopy (SEM) analysis of a collagen barrier membrane. *(a and b)* The membrane surface reveals many collagen fibrils that are intertwined with one another with various diameters and directions (magnification: *a*, ×50; *b*, ×200). *(c)* High-magnification SEM demonstrates collagen fibrils ranging in diameter between 1 and 5 µm (magnification ×1,600). *(d)* Cross-sectional view of collagen barrier membrane of approximately 300 µm (magnification ×100). (Reprinted with permission from Miron et al.[23])

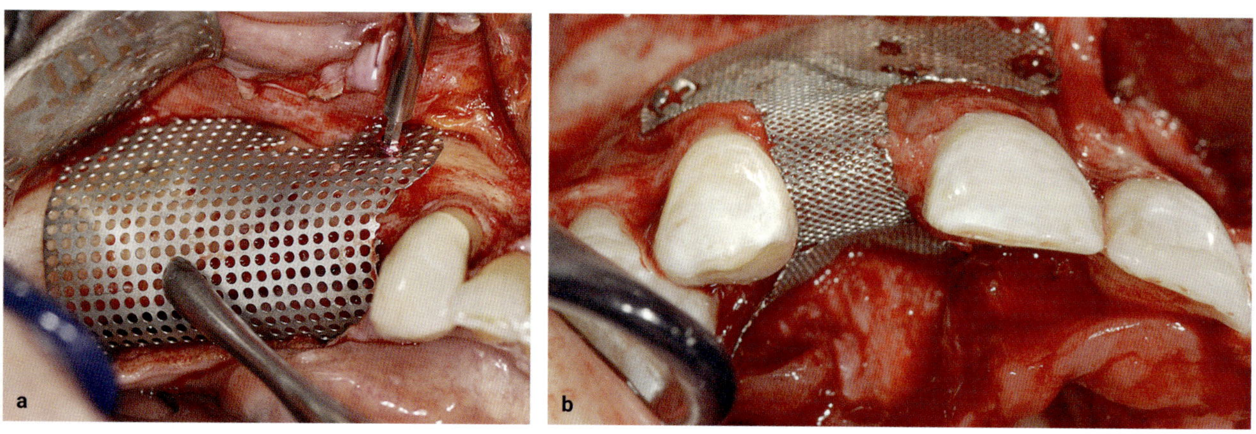

Fig 16-4 *(a and b)* Ti-meshes are adapted according to the defect morphology. Typically, two 3- to 4-mm screws are utilized for fixation. (Reprinted with permission from Pikos and Miron.[21])

PRF as a barrier membrane

One area of research that has gained more popularity over the past two decades has been the use of platelet concentrates as a potential adjunct to bone grafting procedures as well as barrier membranes. Since the 1990s, it has been known that platelets are critical during the wound healing process. PRF brought a modification to the standard PRP protocols whereby a fibrin clot was produced. Since then, it has been compressed into flattened fibrin able to serve as a barrier membrane during GBR procedures. This fibrin scaffold can then be used as a natural barrier membrane alone or subsequently cut and combined with bone grafts, as discussed later in this chapter.

Choice of bone grafting materials during GBR procedures

Much like barrier membranes, an array of bone grafting materials is also currently available on the market. While the goal of this chapter is not to detail all available options, GBR procedures are routinely augmented with either autografts, allografts, xenografts, or alloplasts (synthetic bone grafts), and each has their regenerative properties.[25] Simply, autogenous bone remains the gold standard due to its osteoconductive, osteoinductive, and osteogenic properties. Allografts are the most widely utilized osteoinductive replacement graft, which is attributed to the release of BMPs from their matrix.[26] Xenografts are typically devoid of growth factors (typically deproteinized) but offer the advantage of being better for space maintenance, while synthetic materials (hydroxyapatite, tricalcium phosphate, biphasic calcium phosphate, and bioactive glass) are less frequently utilized.[25,27,28]

Alternative strategies to induce new bone formation during GBR procedures with growth factors

All bone grafting materials provide the essential osteoconductive properties in that they facilitate 3D bone regrowth.[1] Nevertheless, the main feature sought by many clinicians is the ability for a bone grafting material to be osteoinductive.[29] Two main strategies exist: *(1)* bone grafting materials are combined with osteogenic cells/mesenchymal stem cells/allograft scaffolds with BMPs or *(2)* bone grafting materials are combined with bioactive growth factors.[30–32] The use of mesenchymal stem cells for regenerative procedures has shown positive outcomes in RCTs[33–35]; however, the complexity of such procedures limits their current use as a viable treatment option for everyday dental practice.[36] The more utilized scenario has therefore been the combination of bone grafting materials with the use of biologic agents/growth factors, including BMPs, enamel matrix derivative, or platelet concentrates.[20,37,38] Within this group, rhBMP-2 has been shown to have the most potent ability to induce new bone formation; for years, it has been commercially available with FDA approval.[31,39–43] Interestingly, PRP and PRF have also been investigated as bioactive modifiers as potential sources of growth factors for bone regeneration. The rest of this chapter focuses specifically on the use of PRF during GBR procedures.

Evaluation of Clinical Studies Using PRF for GBR

To date, there remains a very limited amount of data supporting the use of PRF for bone regeneration, with even fewer studies investigating its use in well-conducted RCTs.[44] Table 16-2 summarizes all data to date on the topic from a recent systematic review investigating the impact of PRF on bone formation in GBR procedures.

> *There remains a very limited amount of data supporting the use of PRF for bone regeneration.*

In both RCTs,[45,46] the authors investigated the impact of PRF for GBR procedures as a replacement option to collagen. No single study has yet compared the addition of PRF into a bone grafting material in a comparative study (bone graft+PRF vs bone graft alone), and therefore it remains impossible to determine if PRF influences new bone formation during GBR procedures.

In a study by Moussa et al (2016), lateral augmentation was performed with bone blocks covered with either PRF or nothing.[45] There was a statistically significant increase in the buccopalatal bone width in both groups measured by CBCT as well as the manual caliper with no differences reported between the groups. The PRF group did, however, show statistically significantly lower mean

Table 16-2 Characteristics of RCTs evaluating the use of PRF for GBR procedures

Authors (year)	Study design Follow-up	No. of participants Sex Mean age	Groups	Smokers (Y, N)	Conclusions
Moussa et al[45] (2016)	Controlled study (lateral augmentation in the anterior maxila with bone blocks covered with PRF vs alone) 4 months	12 (14 sites) M: 5 / F: 7 26.2	C: 7, Bone block alone T: 7, Bone block+PRF	NR	There was no statistically significant difference between the two groups either preaugmentation (T: 2.9 ± 1.2 mm; C: 3.7 ± 1.0 mm; *P* = .121) or after 4 months (T: 5.7 ± 1.3 mm; C: 6.1 ± 0.8 mm; *P* =.291). However, autogenous bone block surface resorption was significantly decreased in the PRF group.
Hartlev et al[46] (2019)	RCT (lateral augmentation with bone blocks covered with either CM or PRF) 6 months	27 (27 sites) M: 15 / F: 12 50.0	C: 13, Bone block+CM T: 14, Bone block+PRF	NR	The test and control group demonstrated no overall difference in volumentric bone changes of the augmented bone at the 6-month follow-up.

				Methods for PRF preparation		
Authors (year)	**Bone width gain (mm)**	**Augmented bone volume**	**Bone volumetric loss**	**Centrifugation system**	**Volume of tubes for blood drawn**	**Centrifugation parameters speed (rpm) × time (min)**
Moussa et al[45] (2016)	3.8 ± 1.1 (C) 3.4 ± 0.6 (T)	2.2 ± 0.8 mm (C) 2.7 ± 0.9 mm (T)	1.6 ± 0.9 mm (C) 0.8 ± 0.6 mm (T)	Centrifuge 800	NR	3500 × 12–15
Hartlev et al[46] (2019)	NR	465 ± 232 mm³ (C) 426 ± 144 mm³ (T)	17.8% ± 13.3% (C) 14.7% ± 8.9% (T)	Process for PRF	10-mL glass-coated plastic tubes (80 mL total)	200g × 14 (rpm NR)

NR, not reported; C, control group; T, test group; M, male; F, female; CM, collagen membrane.

graft resorption than the control group (test: 0.8 ± 0.6 mm; control: 1.6 ± 0.9 mm; *P* = .006).[45] In a second study by Hartlev et al (2019), a staged lateral ridge augmentation was performed with an autogenous bone block covered with either PRF or an absorbable collagen membrane.[46] A total of 27 partially edentulous patients (test n = 14, control n = 13) were included. CBCT assessment was performed prior to grafting and 2 weeks and 6 months after grafting. It was determined that the mean bone volumetric loss in the PRF test group was 14.7% ± 8.9% and 17.8% ± 13.3% in the control collagen membrane group. Basically no difference in volumetric bone changes of the augmented bone was observed between the two groups at the 6-month follow-up.[46]

Two case series with no control groups have utilized PRF membranes cut into small fragments and mixed with bone grafting material particles. In a pivotal proof-of-concept study,[47] Cortellini et al mixed fragmented PRF membranes with a bone graft in a 50:50 ratio followed by the addition of liquid fibrinogen (ie, sticky bone or PRF block). Horizontal augmentation was assessed linearly and volumetrically immediately after surgery and 5 to 8 months later by matching consecutive CBCTs. The results from this 10-patient study revealed that 15 sites showed an average linear horizontal bone gain of 4.6 mm (±2.3), 5.3 mm (±1.2), and 4.4 mm (±2.3), measured at 2, 6, and 10 mm from the alveolar crest, respectively. The resorption rate after 5 to 8 months was around 15.6% (±6.7) on average.[47]

In a second case series with seven patients,[48] 3D bone augmentation using particulate bone substitute materials combined with autologous bone was used as the graft materials to correct alveolar ridge deficiencies in an open-healing wound approach by employing a collagen matrix loaded with liquid-PRF and then covered by either a PTFE-based membrane or sterile latex. All cases healed successfully using this approach, though no quantifiable data was presented.[48] After a healing period of 4 to 8 months, all patients received dental implants as virtually planned. Additionally, open healing with PRF resulted in newly formed soft tissue without any signs of scar formation or fibrosis. The open-healing concept of the regeneration of the soft tissue with PRF along with bone tissue to regenerate a harmonic implantation bed is a minimally invasive intervention without periosteum splitting or large flap mobilization.[48]

In summary, the present data demonstrates the successful incorporation of PRF utilized either as a replacement to collagen membranes or mixed with bone grafting particles, though no data has yet to demonstrate any advantage in either new bone formation, faster bone formation/shorter healing times, or final amount of augmented bone (reduced bone resorption).

Evaluation of Animal Studies Using PRF for GBR

Previously, comparative work has investigated the use of PRF for bone formation in various animal models. As previously reported, much variability exists on the topic.[44] While some studies seemed to demonstrate better new bone formation when PRF was combined with a bone grafting material in various animal models,[49–51] others failed to demonstrate an advantage.[52–55] Two studies did, however, demonstrate that PRF improved the vascularization of bone grafts by demonstrating higher expression of VEGF and neoangiogenesis.[56,57] In summary, much like data presented in chapter 14, there seems to be greater variability on whether PRF improves bone formation, with half the studies demonstrating a positive outcome and the other half demonstrating no advantage.

Clinical bottom line

1. Limited data in general exists from comparative RCTs evaluating PRF for GBR procedures. Therefore, at present it is not possible to substantiate the claims that PRF improves new bone formation due to a lack of well-conducted studies.
2. Animal studies, much like with extraction socket management, demonstrate variability when evaluating new bone formation when PRF is combined with a bone grafting material.
3. PRF was shown to increase the early vascularization of bone tissues, important for new bone formation.
4. When utilizing PRF as a barrier membrane, it should be applied *over* collagen barrier membranes in contact with soft tissues (as reviewed later in this chapter).

Clinical Procedures

Several important discussion points must first be considered prior to reviewing the surgical application of PRF during GBR procedures. First, PRF's main advantage when combined with a bone grafting material is its improved handling properties, as well as its ability to increase vascularization of bony tissues. Limited data to date exists on the topic demonstrating its ability to convincingly promote new bone formation.

It is important to note that all cases presented in this chapter utilize some form of a barrier membrane (whether it be collagen or titanium) in order to exclude the faster-growing soft tissue from infiltrating the planned bone augmentation area or because of its superior mechanical strength (titanium). PRF itself does not minimize soft tissue infiltration because it generally will be fully resorbed within 2 to 3 weeks and has extremely limited mechanical strength. Therefore, larger augmentation procedures are typically achieved with either a Ti-mesh or Ti-reinforced PTFE membrane (PTFE-TR).

Figure 16-5 demonstrates a horizontal ridge augmentation procedure grafted with a combination of autogenous bone and a xenograft (Bio-Oss, Geistlich) mixed in a 1:1 ratio. Note in Fig 16-5c that a collagen barrier membrane was utilized with overlapping layers of PRF membranes

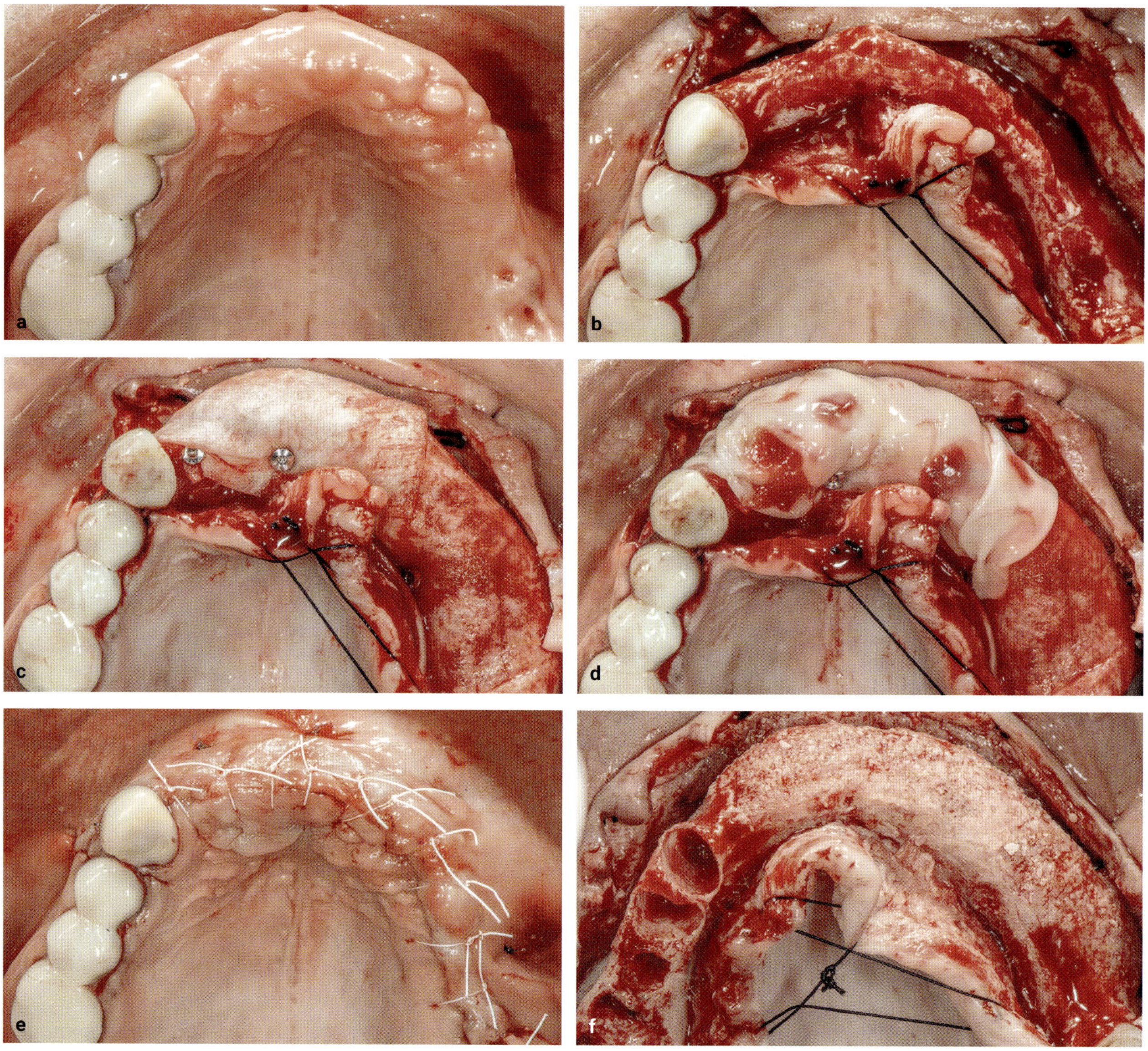

Fig 16-5 Case regenerated using autogenous bone with DBBM and PRF via the "sausage technique." *(a)* Occlusal view of a thin ridge extending about two-thirds of the maxillary arch in a systemically and periodontally healthy, nonsmoking woman. *(b)* Occlusal view of the combined horizontal and sinus graft area. Autogenous bone chips were collected using bone scrapers and bone collector drills and mixed with DBBM in a 40%:60% mixture. A sticky composite bone was packed under the membrane. *(c)* The collagen membrane was fixed and stretched in place using multiple titanium pins. *(d)* Due to the extensive grafted area, only the anterior region of the graft was covered with three layers of PRF membranes. *(e)* Tension-free flap closure was achieved with blunt instruments and apical mattress sutures to minimize trauma to the flap's blood supply. *(f)* Occlusal view of the grafted area 5 months postoperatively. Note the increased ridge width.

(see Fig 16-5d). Following 5 months of healing, note the excellent bone formation.

Figure 16-6 represents a vertical augmentation procedure in the posterior maxilla performed by the same clinician. A PTFE-TR membrane was utilized to favor mechanical stability of the grafted area. The grafting procedure was once again performed using a combination of autogenous bone and a xenograft mixed in a 1:1 ratio with PRF (see Fig 16-6d). Following graft placement, the PTFE-TR membrane was utilized to stabilize the grafted area and tacked in place. Note the excellent bone formation taking place 9 months postoperatively.

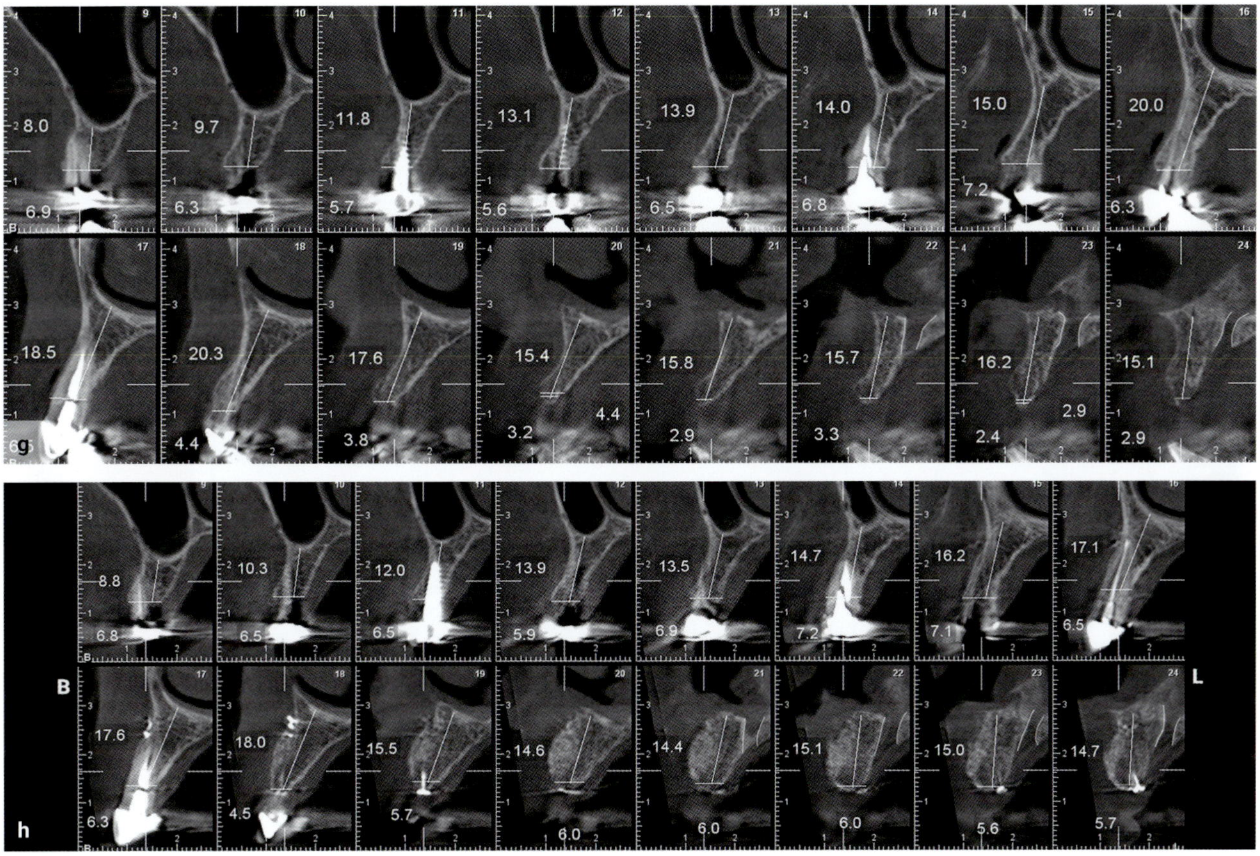

Fig 16-5 *(cont)* *(g and h)* Sections of CBCT view showing alveolar ridge before (very thin) and after bone augmentation surgery at 5 months postsurgery. (Case performed by Dr Omid Moghaddas.)

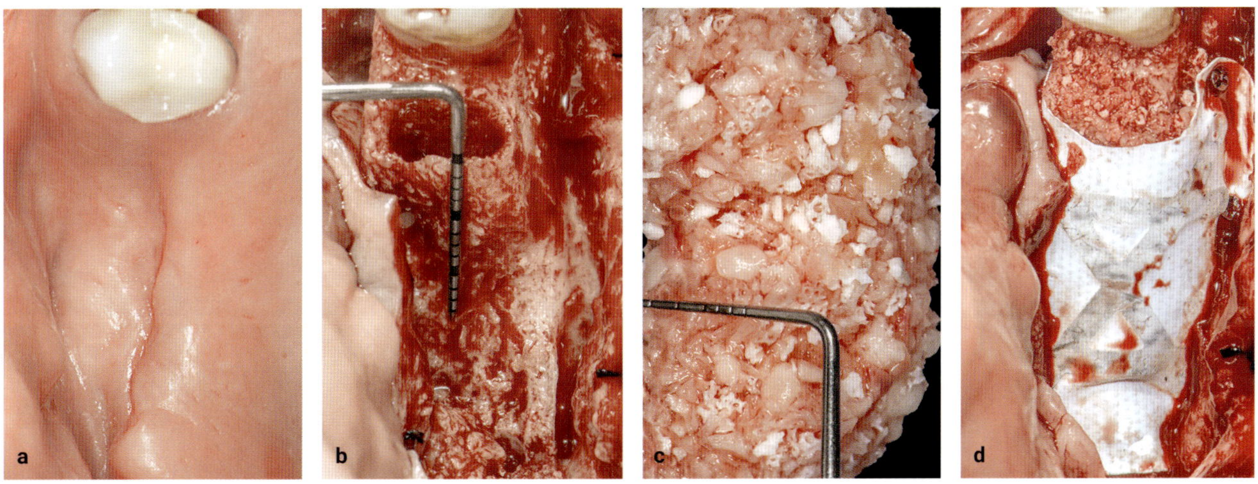

Fig 16-6 Case showing vertical ridge augmentation in right posterior maxillary region after severe bone loss in a male patient with chronic periodontitis. *(a)* Occlusal view of the atrophied posterior maxilla 2 months after extraction of periodontally hopeless teeth. *(b)* Following flap elevation, extensive bone loss was observed at the second premolar site (extracted site) as well as the remaining edentulous area. *(c)* Sticky bone was created with PRF using a composite graft of autogenous bone and DBBM. Particulated autogenous bone chips were harvested from the chin area using bone collector drills and mixed in a 70%:30% ratio with DBBM. *(d)* The alveolar bone was perforated, and a PTFE-TR membrane (Cytoplast) was secured to the palate area by fixation screws, and then sticky bone was packed in the defect area before fixing the membrane to the buccal side using titanium pins. The second premolar socket was grafted and covered by a collagen membrane. The collagen membrane was also extended over the PTFE-TR membrane to reduce the chance of soft tissue growth beneath the nonresorbable membrane.

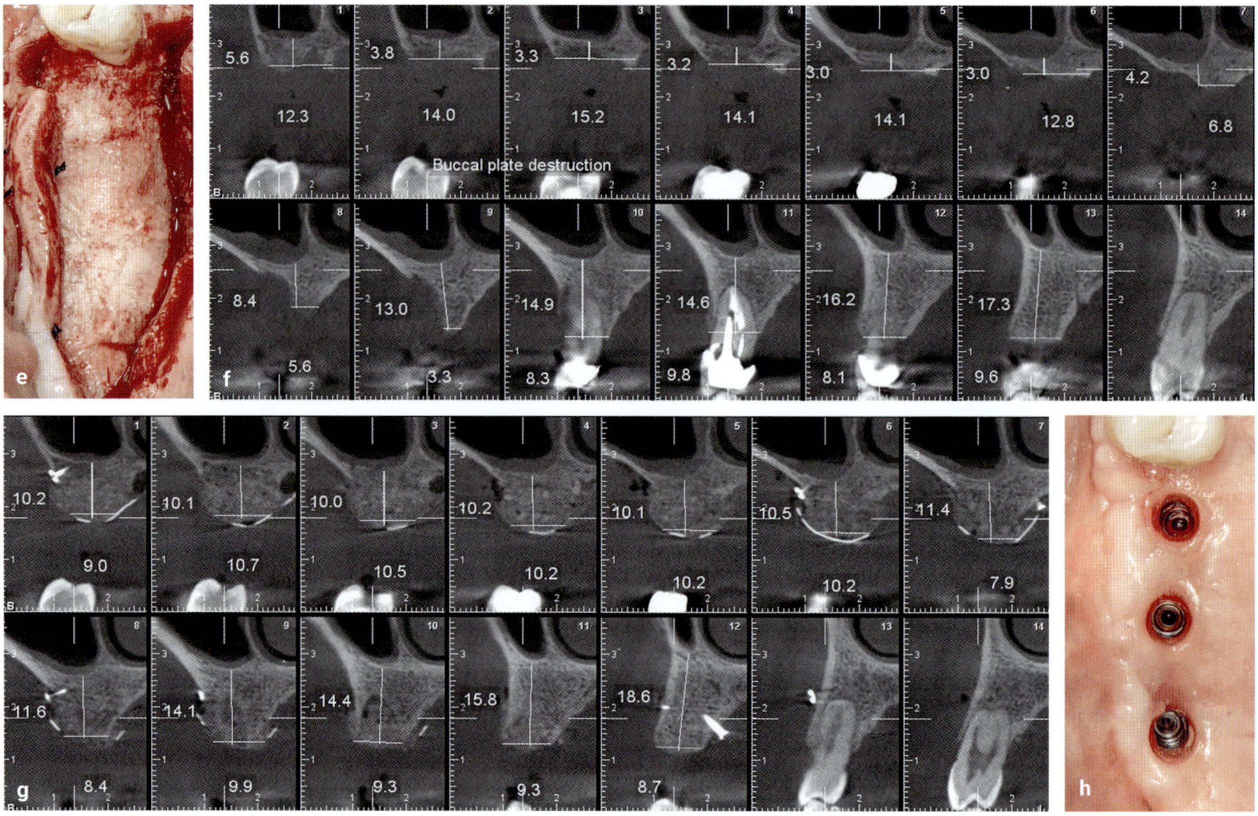

Fig 16-6 *(cont)* *(e)* Occlusal view following 9 months of uneventful healing. *(f)* CBCT demonstrating severe bone loss close to the maxillary sinus floor. *(g)* CBCT view of the site after 9 months of healing. Note the increase in bone volume. *(h)* Final case with implants integrated and good soft tissue healing/maintenance. (Case performed by Dr Omid Moghaddas.)

Figure 16-7 represents a severely deficient narrow ridge with recession and mobility of the mandibular central incisors. A grafting procedure was performed with rhBMP-2 and PRF in combination with a mineralized allograft to maximize bone regeneration without having to utilize autogenous bone. For such large augmentation procedures, PRF was utilized overtop of the Ti-mesh to favor better soft tissue wound healing and minimize potential Ti-mesh exposure (reported up to 50%).[21] Video 16-1 demonstrates a grafting technique using rhBMP-2 and PRF with a bone grafting material complex composed of allograft/xenograft for a large lingual and buccal bone deficiency in the anterior maxilla (Fig 16-8).

Video 16-1

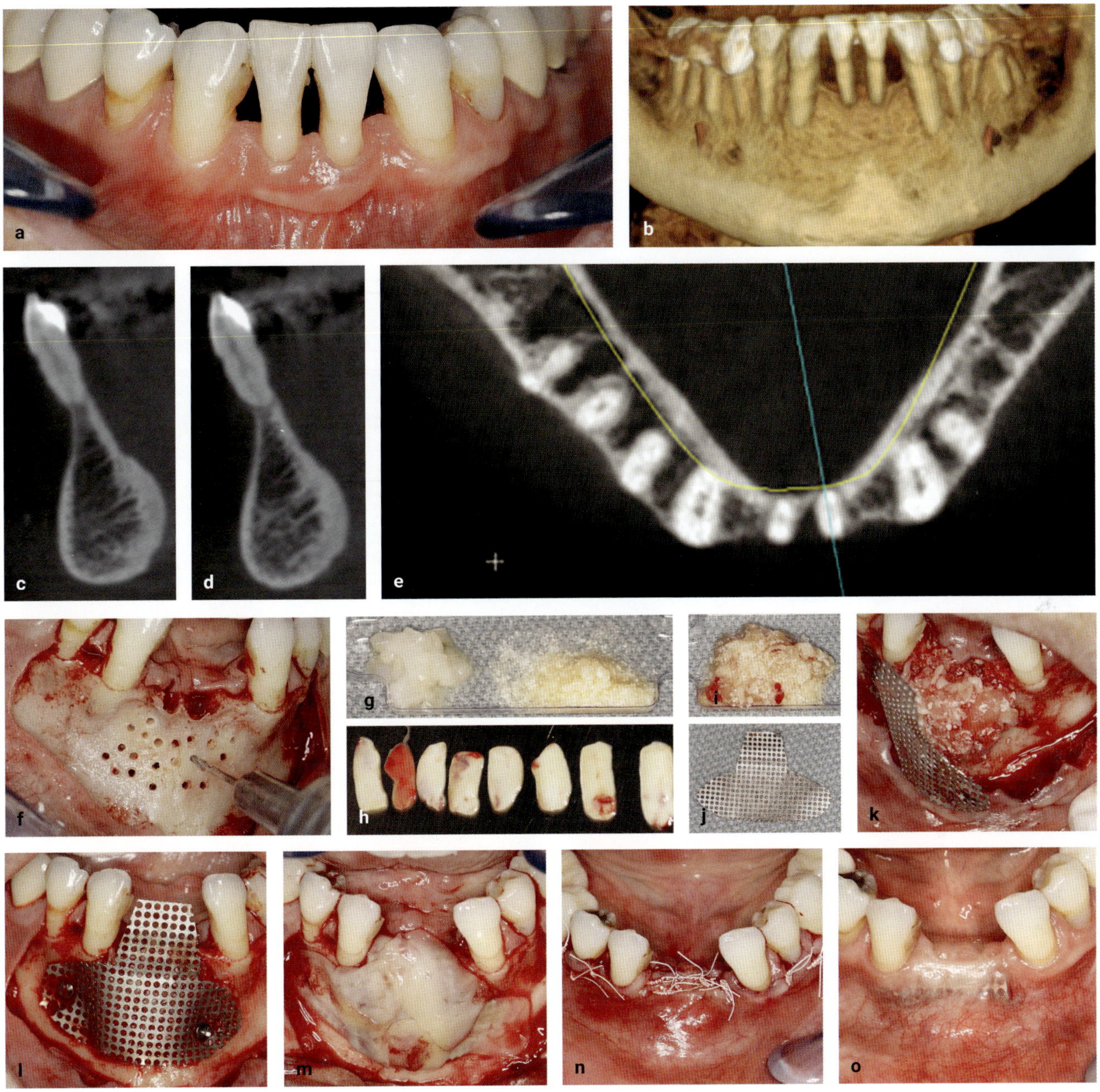

Fig 16-7 GBR in the anterior mandible. *(a)* Clinical photograph demonstrating a narrow ridge with recession and mobility of the mandibular central incisors. *(b to e)* CBCT imaging demonstrating extensive bone loss and narrow buccal and lingual plates with horizontal bone recession. *(f)* Following tooth extraction, intramarrow penetration was performed. *(g to i)* rhBMP-2 was mixed with a graft complex composed of mineralized allograft (MinerOss, BioHorizons) and PRF exudate. *(j)* Prefabrication of a Ti-mesh with ideal contour. *(k)* Graft placement. *(l)* Ti-mesh fixation. *(m)* L-PRF utilized to cover the Ti-mesh. *(n)* Final sutures. *(o)* Clinical photograph 7 months postoperative.

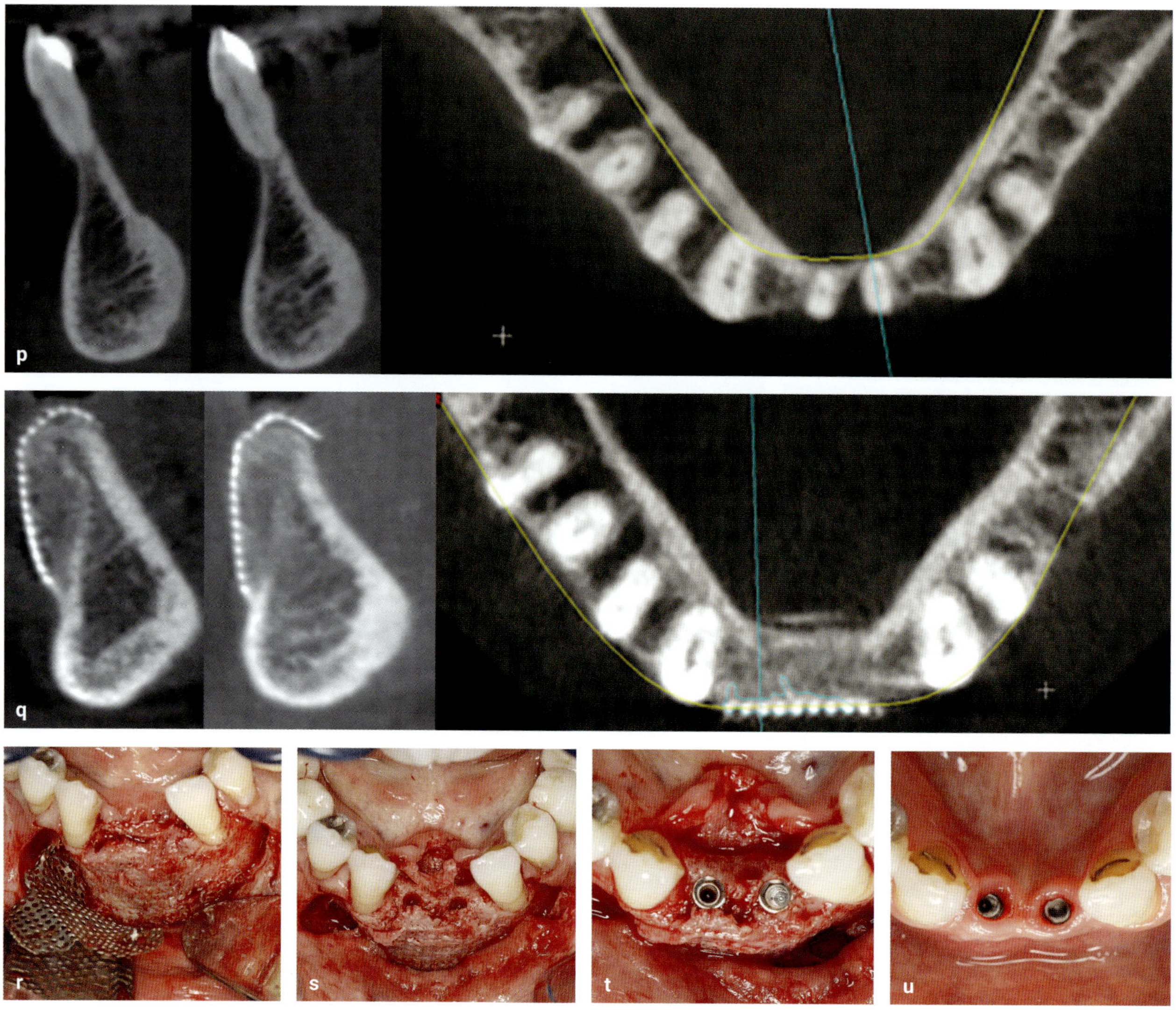

Fig 16-7 *(cont)* *(p and q)* CBCT imaging at 0 days and 7 months postoperative. Notice the gain in bone volume. *(r and s)* Ti-mesh removal. *(t)* Implant placement. *(u)* Soft tissue healing at 4 months. (Case performed by Dr Michael A. Pikos; reprinted with permission from Pikos and Miron.[21])

In a final case (Fig 16-9 and Video 16-2), a Khoury plate technique was utilized with simultaneous implant placement. Note that the grafting procedure was performed using a bone graft mixed with PRF. The advantage of utilizing PRF with such an approach is mainly to favor its improved grafting material handling as well as its ability to potentially increase angiogenesis. More importantly, however, following collagen membrane placement, several layers of PRF membranes were applied to facilitate soft tissue wound healing and protect the augmented site from bacterial invasion.

Video 16-2

Fig 16-9 ➤ Ridge augmentation with simultaneous implant placement grafted with sticky bone. *(a)* Following Khoury plate fixation with simultaneous implant placement, the site was grafted with synthetic bone grafting particles mixed with PRF fragments (sticky bone) within the bony defects. *(b)* Thereafter, the entire grafted site was covered with a collagen barrier membrane and sutured. *(c)* Prior to wound closure, the entire soft tissue area was covered with PRF membranes to facilitate soft tissue healing. *(d)* Soft tissue closure. *(e)* Note the excellent healing 7 days postoperatively. See Video 16-2 for complete case presentation. (Case performed by Dr Howard Gluckman.)

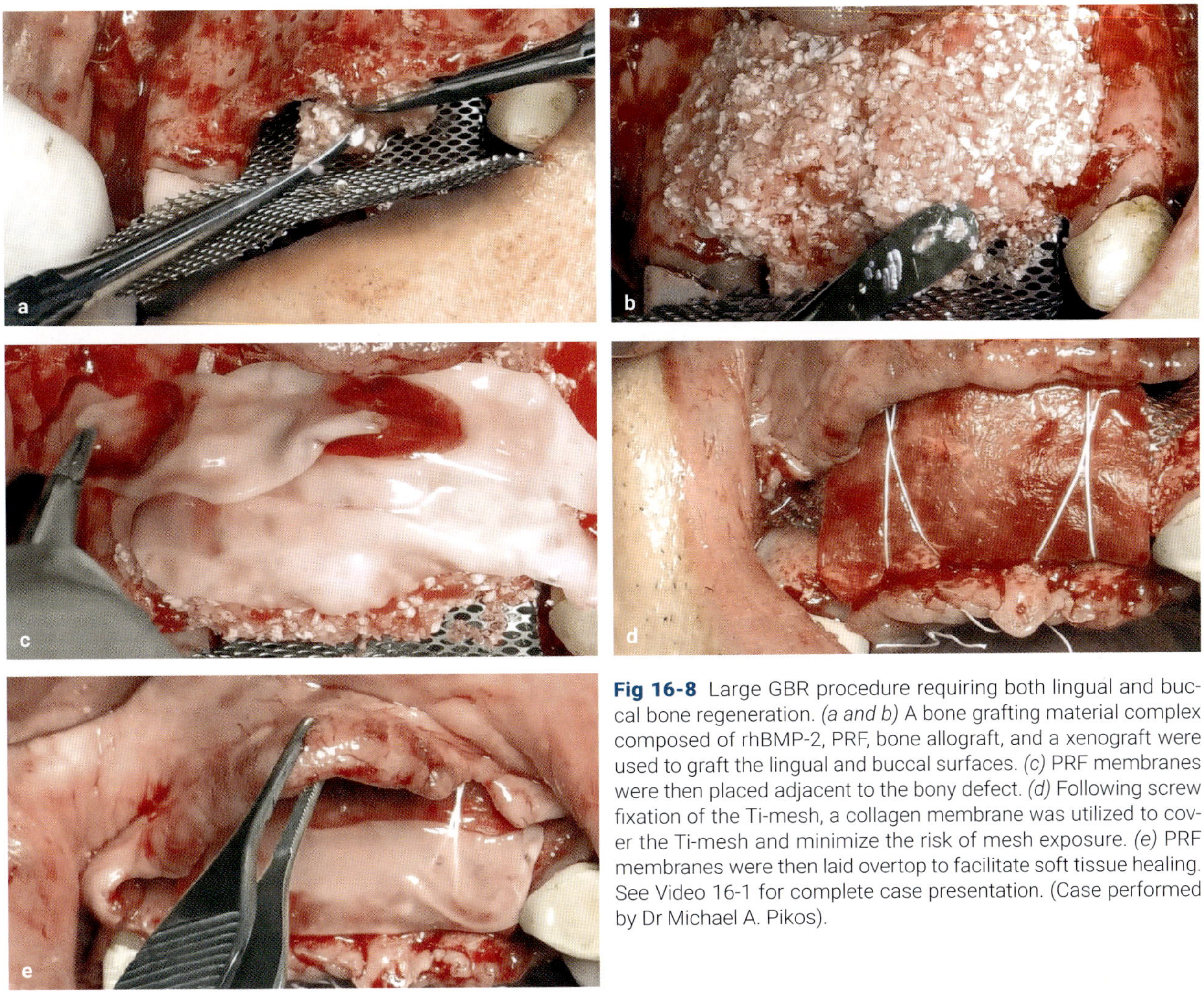

Fig 16-8 Large GBR procedure requiring both lingual and buccal bone regeneration. *(a and b)* A bone grafting material complex composed of rhBMP-2, PRF, bone allograft, and a xenograft were used to graft the lingual and buccal surfaces. *(c)* PRF membranes were then placed adjacent to the bony defect. *(d)* Following screw fixation of the Ti-mesh, a collagen membrane was utilized to cover the Ti-mesh and minimize the risk of mesh exposure. *(e)* PRF membranes were then laid overtop to facilitate soft tissue healing. See Video 16-1 for complete case presentation. (Case performed by Dr Michael A. Pikos).

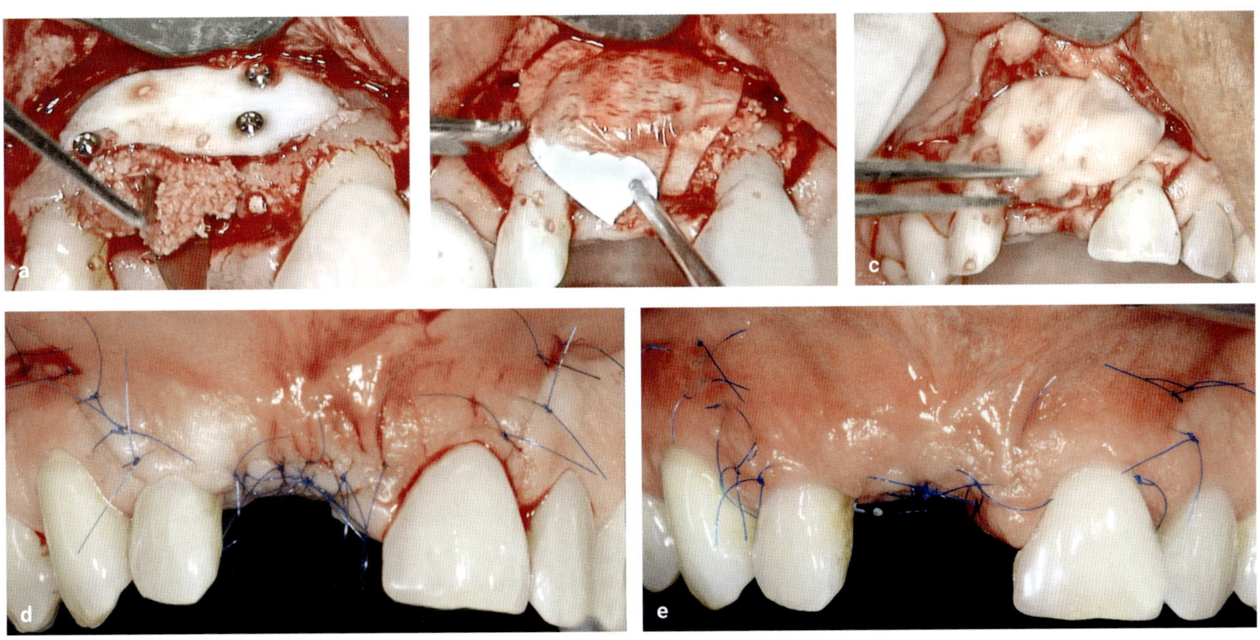

Discussion

Despite the increasing popularity of PRF and its use in regenerative dentistry as highlighted throughout this textbook, there remains a lack of clinical evidence supporting its benefit in GBR procedures. Studies to date have demonstrated how PRF can improve soft tissue wound healing, enhance gingival recession coverage, and promote intrabony defect regeneration, but limited data is available investigating its impact on bone formation during GBR procedures. The use of PRF as a barrier membrane is hypothesized to contribute to soft tissue healing and/or regeneration in GBR procedures, yet no study to date has characterized its regenerative potential on bone formation in a well-conducted RCT. These missing studies are therefore crucial.

While initially PRF was utilized for a variety of biologic procedures in dentistry, its main effects on soft tissues led to the hypothesis that PRF could potentially serve as a barrier membrane in GBR procedures. It is now well understood that PRF is able to accomplish soft tissue regeneration of overlaying tissue, protect the underlying bone tissues from incoming pathogens (due to its accumulation of host defense immune cells such as leukocytes), and facilitate angiogenesis to the underlying bone structure. For these reasons, PRF is typically utilized as an outer barrier membrane overtop either collagen or titanium membranes and further combined with bone grafting material particles to improve their handling, as reviewed in all presented cases. In this way, collagen barrier membranes are additionally protected from membrane exposure, leading to a potential reduction in the risk of infection. This is especially important in medically compromised patients (eg, patients with diabetes, smokers, immunocompromised patients, or patients with advanced age, etc).

Use of PRF as a sole barrier membrane during GBR procedures

Barrier membranes were originally developed to prevent fast-growing soft tissues from entering the slowly growing compartment containing bone.[8] While theoretically their role has been debated over the past decade, common questions include when and if PRF can be utilized alone and whether it should be routinely combined with a barrier membrane. To date, however, while some clinicians have demonstrated excellent results grafting with PRF alone as an exclusive barrier membrane, as reviewed within this chapter, most international colleagues favor its use with a collagen membrane, PTFE-TR membrane, or Ti-mesh, especially during larger GBR procedures, with an expressed desire to favor more predictable approaches. Thus, the use of PRF alone as an outer barrier membrane may be limited to single sites requiring limited bone volume gain and most importantly in sites not requiring mechanical stability. More frequently, PRF membranes are combined with either a collagen barrier membrane or titanium-reinforced membrane when used in the GBR approach.

Use of PRF membranes with collagen barrier membranes

Another frequently arising question is whether PRF membranes should be utilized over or under barrier membranes. While there certainly is no issue in utilizing PRF both over and under barrier membranes (should they be available), to address this question, it is best that the treating clinician understand the biologic advantages and disadvantages to each scenario. One of the main advantages of PRF scaffolds is their incorporation of defense-fighting leukocytes. During large GBR procedures, one of the concerns with utilizing nonvital barrier membranes (such as collagen) is their risk of exposure and subsequent infection. From this point of view, if a PRF membrane is placed underneath a collagen barrier membrane, the avascular and nonvital collagen membrane runs the risk of being left exposed to the oral cavity with a potential risk of infection. In the event that a suture is accidentally dislodged or tension within the flap causes slight exposure to the underlying membrane, the collagen barrier poses a risk of infection because bacteria can readily attach and proliferate on collagen.

In the reverse order, if a PRF membrane is placed over a collagen barrier membrane, even if PRF is left exposed to the oral cavity, due to its supraphysiologic high quantity of pathogen-fighting leukocytes, the chance of infection is reduced dramatically. This is pivotal especially for the ever-increasing number of medically compromised patients such as patients with diabetes, smokers, as well as the large population of patients taking medications that may alter wound healing. A second noted advantage is that PRF is known to rapidly promote greater soft tissue wound healing/regeneration when compared to hard tissues. Therefore, from a biologic point of view, it is strategically more rewarding to utilize PRF in contact with soft tissues on the outer surface of barrier membranes. As a rule of thumb, it is always advantageous to utilize PRF on the outer surface of GBR procedures overtop barrier membranes.

As a rule of thumb, PRF should be utilized on the outer surface of barrier membranes in contact with soft tissues.

Use of PRF membranes overtop titanium membranes

It has been well documented in the literature that one of the frequent complications associated with utilizing Ti-meshes is their high risk of exposure, documented in the 20% to 50% range.[21] Therefore, the additional use of PRF overtop Ti-meshes has been proposed to limit the rate of exposure/complication by favoring faster soft tissue healing (thereby decreasing the chance of exposure). Because PRF is adequately available as a low-cost regenerative agent, it is strongly advised to add PRF to GBR procedures performed with Ti-meshes in order to limit and minimize potential complications associated with their use (such as mesh exposure). Several layers may even be applied. While future comparative studies investigating the rates of Ti-mesh exposure when grafting procedures are performed with or without PRF remain needed, biologically this combination seems both practical and logical.

Future use of PRF for grafting procedures

The data thus far investigating the impact of PRF on bone formation remains scarce. While some data has shown that PRF does promote new bone formation postextraction healing when utilized alone, other studies show either extremely limited or no advantage at all (see chapter 14). Therefore, more research is needed to determine why under one clinical scenario PRF may lead to enhanced clinical outcomes with respect to bone formation, whereas in others it has no impact.

With respect to GBR procedures, ongoing attempts have been made to minimize the raising of flaps and potentially performing various grafting procedures via tunneling, with trends of minimally invasive surgery being at the forefront of research attempts over the past decade (not only in dentistry but in medicine as a whole). Figure 16-10 demonstrates a case whereby a bone grafting procedure with a bone plate was performed via a tunneling technique. Note how PRF was utilized both in combination with the bone grafting material particles and as membranes to favor soft tissue closure/soft tissue healing (Video 16-3). This case highlights currently utilized novel grafting procedures, with ongoing research attempts favoring more minimally invasive surgeries. The future field will benefit from such attempts with the hopes of finding more reproducible and predictable ways to regenerate missing tissues in as minimally invasive procedures as possible.

Another avenue specific to the GBR field as reviewed in the final chapter of this textbook is the growing use of the e-PRF membranes (see chapter 7). By utilizing one additional heating step of plasma, PRF membranes (which typically last anywhere from 2 to 3 weeks) can be extended in their resorption properties upward of 4 to 6 months.[58] Therefore, and based on preliminary animal data, it would then be theoretically possible to utilize a heated e-PRF membrane as a barrier capable of excluding soft tissues. Future research in that field is ongoing, with its use overtop Ti-meshes currently being investigated.

Lastly, it is well known that the graft handling of PRF can also be improved via heating in an incubator. As reviewed in chapter 22, custom bone grafts that are created via 3D printing and personalized to each individual are now being utilized. For such application, the use of heating is an absolute requirement to fabricate denser bone "blocks" or custom PRF blocks that can be utilized rapidly for surgical applications according to individualized preplanned 3D augmentation preferences. That field remains new and in its infancy, but ongoing future research in this area remains exciting with many new possibilities.

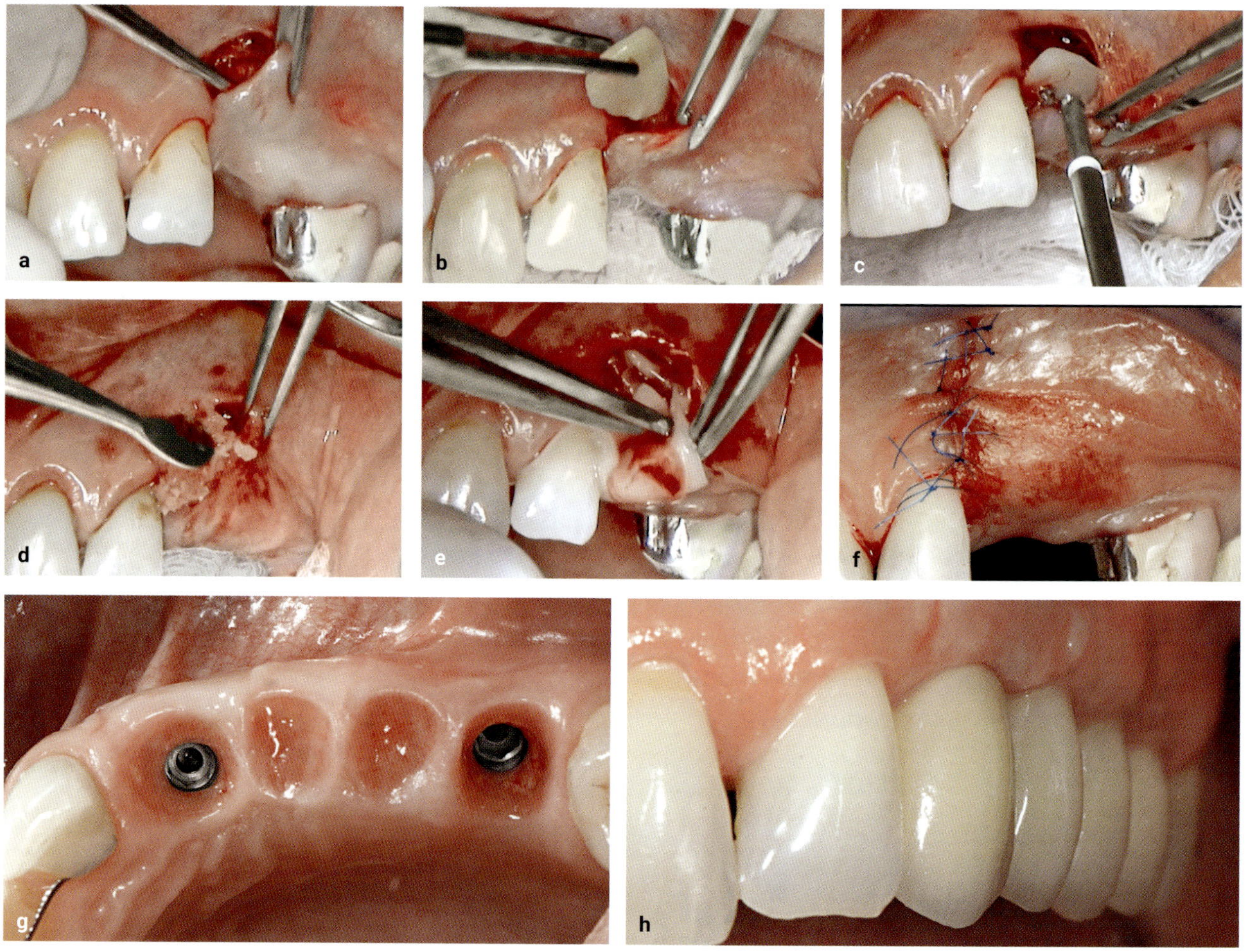

Fig 16-10 A horizontal ridge augmentation performed via a vestibular tunneling approach. *(a)* Creation of a vestibular incision and tunnel for bone grafting material placement. *(b and c)* Autogenous bone block inserted into the area and fixated with screws. *(d)* Sticky bone entered into the grafted area to further graft the area. *(e)* Use of PRF membranes overtop the grafted area to promote soft tissue healing. *(f)* Final sutures. *(g)* Occlusal view of implant placement. Note the excellent soft tissue healing and ridge contour. *(h)* Final view after restoration. See Video 16-3 for complete case presentation. (Case performed by Dr Howard Gluckman.)

Video 16-3

Conclusion

The use of PRF has gained tremendous momentum as a low-cost biologic scaffold capable of improving tissue healing. To date, only limited data currently supports the ability for PRF to promote new bone formation, and therefore future study remains needed. Overall, studies do demonstrate that the use of PRF favors rapid angiogenesis of tissues, an area highly relevant to bone regeneration. Additionally, the handling properties of bone grafts may be improved when PRF is combined with bone particles (sticky bone concept). Two main concluding factors essential for the clinician may be derived. First, PRF alone as a barrier membrane should only be utilized for small GBR procedures when space maintenance is not an issue. In large GBR procedures, it is recommended to combine PRF with another barrier membrane (collagen, PTFE, or Ti-mesh). Second, in such combination cases, it is always advised to place PRF membranes *over* the other barrier membranes. In this way, PRF membranes can be utilized to speed soft tissue defect closure, improve soft tissue wound healing/regeneration, and further protect against infection with its defense-fighting leukocytes.

References

1. Miron RJ, Zhang Y (eds). Next-Generation Biomaterials for Bone & Periodontal Regeneration: Chicago: Quintessence, 2019.
2. Gottlow J, Nyman S, Karring T, Lindhe J. New attachment formation as the result of controlled tissue regeneration. J Clin Periodontol 1984;11:494–503.
3. Briguglio F, Falcomatà D, Marconcini S, Fiorillo L, Briguglio R, Farronato D. The use of titanium mesh in guided bone regeneration: A systematic review. Int J Dent 2019;2019:9065423.
4. Elnayef B, Monje A, Gargallo-Albiol J, Galindo-Moreno P, Wang HL, Hernández-Alfaro F. Vertical ridge augmentation in the atrophic mandible: A systematic review and meta-analysis. Int J Oral Maxillofac Implants 2017;32:291–312.
5. Kinaia BM, Kazerani S, Korkis S, Masabni OM, Shah M, Neely AL. Effect of guided bone regeneration on immediately placed implants: Meta-analyses with at least 12 months follow up after functional loading [epub ahead of print 14 January 2019]. J Periodontol doi:10.100/JPER.18-0543.
6. Salvi GE, Monje A, Tomasi C. Long-term biological complications of dental implants placed either in pristine or in augmented sites: A systematic review and meta-analysis. Clin Oral Implants Res 2018;29(suppl 16):294–310.
7. Hoaglin DR, Lines GK. Prevention of localized osteitis in mandibular third-molar sites using platelet-rich fibrin. Int J Dent 2013; 2013:875380.
8. Dahlin C, Linde A, Gottlow J, Nyman S. Healing of bone defects by guided tissue regeneration. Plast Reconstr Surg 1988;81:672–676.
9. Buser D, Dula K, Belser U, Hirt HP, Berthold H. Localized ridge augmentation using guided bone regeneration. 1. Surgical procedure in the maxilla. Int J Periodontics Restorative Dent 1993;13:29–45.
10. Buser D, Dahlin C, Schenk R. Guided Bone Regeneration in Implant Dentistry. Chicago: Quintessence, 1994.
11. Hardwick R, Dahlin C. Healing pattern of bone regeneration in membrane-protected defects: A histologic study in the canine mandible. Int J Oral Maxillofac Implants 1994;9:13–29.
12. Fujihara K, Kotaki M, Ramakrishna S. Guided bone regeneration membrane made of polycaprolactone/calcium carbonate composite nano-fibers. Biomaterials 2005;26:4139–4147.
13. Toffler M. Guided bone regeneration (GBR) using cortical bone pins in combination with leukocyte-and platelet-rich fibrin (L-PRF). Compend Contin Educ Dent 2014;35:192–198.
14. Salata LA, Hatton PV, Devlin AJ, Craig GT, Brook IM. In vitro and in vivo evaluation of e-PTFE and alkali-cellulose membranes for guided bone regeneration. Clin Oral Implants Res 2001;12:62–68.
15. Carbonell J, Martín IS, Santos A, Pujol A, Sanz-Moliner J, Nart J. High-density polytetrafluoroethylene membranes in guided bone and tissue regeneration procedures: A literature review. Int J Oral Maxillofac Surgery 2014;43:75–84.
16. Jung RE, Fenner N, Hämmerle CH, Zitzmann NU. Long-term outcome of implants placed with guided bone regeneration (GBR) using resorbable and non-resorbable membranes after 12–14 years. Clin Oral Implants Res 2013;24:1065–1073.
17. Zhang Y, Zhang X, Shi B, Miron R. Membranes for guided tissue and bone regeneration. Ann Oral Maxillofac Surg 2013;1:10.
18. Gentile P, Chiono V, Tonda-Turo C, Ferreira AM, Ciardelli G. Polymeric membranes for guided bone regeneration. Biotechnol J 2011;6:1187–1197.
19. Rakhmatia YD, Ayukawa Y, Furuhashi A, Koyano K. Current barrier membranes: Titanium mesh and other membranes for guided bone regeneration in dental applications. J Prosthodont Res 2013;57:3–14.
20. Nguyen TT, Mui B, Mehrabzadeh M, et al. Regeneration of tissues of the oral complex: Current clinical trends and researrch advances. J Can Dent Assoc 2013;79:d1.
21. Pikos MA, Miron RJ. Bone Augmentation in Implant Dentistry: A Step-by-Step Guide to Predictable Alveolar Ridge and Sinus Grafting. Chicago: Quintessence, 2019.
22. AlGhamdi AS, Ciancio SG. Guided tissue regeneration membranes for periodontal regeneration—A literature review. J Int Acad Periodontol 2009;11:226–231.
23. Miron RJ, Saulacic N, Buser D, Iizuka T, Sculean A. Osteoblast proliferation and differentiation on a barrier membrane in combination with BMP2 and TGFβ. Clin Oral Investig 2013;17:981–988.
24. Sculean A, Nikolidakis D, Schwarz F. Regeneration of periodontal tissues: Combinations of barrier membranes and grafting materials—Biological foundation and preclinical evidence: A systematic review. J Clin Periodontol 2008;35:106–116.
25. Miron RJ, Sculean A, Shuang Y, et al. Osteoinductive potential of a novel biphasic calcium phosphate bone graft in comparison with autographs, xenografts, and DFDBA. Clin Oral Implants Res 2016;27:668–675.
26. Sanz M, Vignoletti F. Key aspects on the use of bone substitutes for bone regeneration of edentulous ridges. Dental Materials 2015;31:640–647.
27. Fretwurst T, Gad LM, Nelson K, Schmelzeisen R. Dentoalveolar reconstruction: Modern approaches. Curr Opin Otolaryngol Head Neck Surg 2015;23:316–322.
28. Donos N, Kostopoulos L, Tonetti M, Karring T, Lang NP. The effect of enamel matrix proteins and deproteinized bovine bone mineral on heterotopic bone formation. Clin Oral Implants Res 2006;17: 434–438.
29. Miron RJ, Zhang YF. Osteoinduction: A review of old concepts with new standards. J Dent Res 2012;91:736–744.
30. Lioubavina-Hack N, Carmagnola D, Lynch SE, Karring T. Effect of Bio-Oss with or without platelet-derived growth factor on bone formation by "guided tissue regeneration": A pilot study in rats. J Clin Periodontol 2005;32:1254–1260.
31. Zhang Y, Yang S, Zhou W, Fu H, Qian L, Miron RJ. Addition of a synthetically fabricated osteoinductive biphasic calcium phosphate bone graft to BMP2 improves new bone formation. Clin Implant Dent Relat Res 2016;18:1238–1247.
32. Hämmerle CH, Giannobile WV. Biology of soft tissue wound healing and regeneration–Consensus report of Group 1 of the 10th European Workshop on Periodontology. J Clin Periodontol 2014;41(suppl 15):S1–S5.
33. Bajestan MN, Rajan A, Edwards SP, et al. Stem cell therapy for reconstruction of alveolar cleft and trauma defects in adults: A randomized controlled, clinical trial. Clin Implant Dent Relat Res 2017;19:793–801.
34. Kaigler D, Avila-Ortiz G, Travan S, et al. Bone engineering of maxillary sinus bone deficiencies using enriched CD90+ stem cell therapy: A randomized clinical trial. J Bone Miner Res 2015;30: 1206–1216.
35. Polymeri A, Giannobile WV, Kaigler D. Bone marrow stromal stem cells in tissue engineering and regenerative medicine. Horm Metab Res 2016;48:700–713.
36. Kaigler D, Pagni G, Park CH, et al. Stem cell therapy for craniofacial bone regeneration: A randomized, controlled feasibility trial. Cell Transplant 2013;22:767–777.
37. Turri A, Elgali I, Vazirisani F, et al. Guided bone regeneration is promoted by the molecular events in the membrane compartment. Biomaterials 2016;84:167–183.
38. Shuang Y, Yizhen L, Zhang Y, Fujioka-Kobayashi M, Sculean A, Miron RJ. In vitro characterization of an osteoinductive biphasic calcium phosphate in combination with recombinant BMP2. BMC Oral Health 2016;17:35.
39. Ryoo HM, Lee MH, Kim YJ. Critical molecular switches involved in BMP-2-induced osteogenic differentiation of mesenchymal cells. Gene 2006;366:51–57.

40. Wang D, Tabassum A, Wu G, Deng L, Wismeijer D, Liu Y. Bone regeneration in critical-sized bone defect enhanced by introducing osteoinductivity to biphasic calcium phosphate granules. Clin Oral Implants Res 2017;28:251–260.
41. Torrecillas-Martinez L, Monje A, Pikos MA, et al. Effect of rhBMP-2 upon maxillary sinus augmentation: A comprehensive review. Implant Dent 2013;22:232–237.
42. Misch CM, Jensen OT, Pikos MA, Malmquist JP. Vertical bone augmentation using recombinant bone morphogenetic protein, mineralized bone allograft, and titanium mesh: A retrospective cone beam computed tomography study. Int J Oral Maxillofac Implants 2015;30:202–207.
43. Misch C, Wang H-L. Clinical applications of recombinant human bone morphogenetic protein-2 for bone augmentation before dental implant placement. Clin Adv Periodontics 2011;1:118–131.
44. Miron RJ, Choukroun J. Platelet Rich Fibrin in Regenerative Dentistry. Hoboken, NJ: Wiley, 2017.
45. Moussa M, El-Dahab OA, El Nahass H. Anterior maxilla augmentation using palatal bone block with platelet-rich fibrin: A controlled trial. Int J Oral Maxillofac Implants 2016;31:708–715.
46. Hartlev J, Spin-Neto R, Schou S, Isidor F, Norholt SE. Cone beam computed tomography evaluation of staged lateral ridge augmentation using platelet-rich fibrin or resorbable collagen membranes in a randomized controlled clinical trial. Clin Oral Implants Res 2019;30:277–284.
47. Cortellini S, Castro AB, Temmerman A, et al. Leucocyte- and platelet-rich fibrin block for bone augmentation procedure: A proof-of-concept study. J Clin Periodontol 2018;45:624–634.
48. Ghanaati S, Al-Maawi S, Conrad T, Lorenz J, Rossler R, Sader R. Biomaterial-based bone regeneration and soft tissue management of the individualized 3D-titanium mesh: An alternative concept to autologous transplantation and flap mobilization. J Craniomaxillofac Surg 2019;47:1633–1644.
49. Fahimipour F, Bastami F, Khoshzaban A, et al. Critical-sized bone defects regeneration using a bone-inspired 3D bilayer collagen membrane in combination with leukocyte and platelet-rich fibrin membrane (L-PRF): An in vivo study. Tissue Cell 2020;63:101326.
50. do Lago ES, Ferreira S, Garcia IR, Jr., Okamoto R, Mariano RC. Improvement of bone repair with I-PRF and bovine bone in calvaria of rats. Histometric and immunohistochemical study. Clin Oral Investig 2020;24:1637–1650.
51. Engler-Pinto A, Siéssere S, Calefi A, et al. Effects of leukocyte- and platelet-rich fibrin associated or not with bovine bone graft on the healing of bone defects in rats with osteoporosis induced by ovariectomy. Clin Oral Implants Res 2019;30:962–976.
52. Ozdemir H, Ezirganli S, Isa Kara M, Mihmanli A, Baris E. Effects of platelet rich fibrin alone used with rigid titanium barrier. Arch Oral Biol 2013;58:537–544.
53. Knapen M, Gheldof D, Drion P, Layrolle P, Rompen E, Lambert F. Effect of leukocyte- and platelet-rich fibrin (L-PRF) on bone regeneration: A study in rabbits. Clin Implant Dent Relat Res 2015; 17(suppl 1):e143–e152.
54. Karayürek F, Kadiroğlu ET, Nergiz Y, et al. Combining platelet rich fibrin with different bone graft materials: An experimental study on the histopathological and immunohistochemical aspects of bone healing. J Craniomaxillofac Surg 2019;47:815–825.
55. Faot F, Deprez S, Vandamme K, et al. The effect of L-PRF membranes on bone healing in rabbit tibiae bone defects: Micro-CT and biomarker results. Sci Rep 2017;7:46452.
56. Yoon JS, Lee SH, Yoon HJ. The influence of platelet-rich fibrin on angiogenesis in guided bone regeneration using xenogenic bone substitutes: A study of rabbit cranial defects. J Craniomaxillofac Surg 2014;42:1071–1077.
57. Ratajczak J, Vangansewinkel T, Gervois P, et al. Angiogenic properties of 'leukocyte- and platelet-rich fibrin.' Sci Rep 2018;8:14632.
58. Gheno E, Mourão C, Mello-Machado RC, et al. In vivo evaluation of the biocompatibility and biodegradation of a new denatured plasma membrane combined with liquid PRF (Alb-PRF). Platelets 2020:1–13.

17

Use of PRF for Sinus Grafting

Contributors

Richard J. Miron
Michael A. Pikos
Dwayne Karateew
Samer Shaltoni
Luis Bessa
Hom-Lay Wang

It is well known that the edentulous maxilla demonstrates a decreasing vascularity as bone resorption progresses. Regenerative therapies with platelet concentrates have therefore been utilized over the past two decades to speed the revascularization to damaged/defective tissues. Over the years, PRF has been used as a sole grafting material or in combination with various bone grafting materials for sinus augmentation procedures. This chapter first highlights the theoretical, biologic, and clinical advantages of utilizing PRF with or without a bone grafting material for sinus augmentation procedures. Thereafter, an in-depth literature update is provided encompassing 10 human controlled studies as well as 12 human case series studies conducted to date on the topic. Clinical guidelines from a recent consensus report detailing when, where, and why to use PRF alone versus in combination with a bone graft is provided along with its use for the repair of sinus membrane perforations and as a barrier membrane for lateral window closure. Conservative, predictable, and evidence-based clinical recommendations are thereafter provided with indications supporting its effective use for sinus grafting procedures when utilized in conjunction with either a crestal or lateral sinus grafting procedure.

Chapter Highlights

- *Anatomical considerations of the sinus during PRF grafting*
- *Using PRF as a sole grafting material versus in combination with a bone grafting material*
- *Using PRF alone for the repair of sinus membrane perforations*
- *The potential use of PRF to close the lateral window*

It is well known that the edentulous maxilla demonstrates a decreasing vascularity as bone resorption progresses.[1] As a result, bone augmentation procedures specifically in the sinus have generally been associated with lengthier healing times and higher failure rates.[2–4] Not only is the rate of new bone formation within the area slower, taking nearly twice as long (6–8 months) compared to extraction site grafting (3–4 months), but generally higher infection rates and/or postoperative complications are observed. Reduced blood supply also entails a decrease in defense-fighting immune cells/leukocytes.[5,6]

While sinus elevation procedures are routinely augmented with bone grafting materials, a growing number of clinicians have proposed various platelet concentrates as a means to speed revascularization to the sinus. One systematic review[7] showed that the mean implant survival rate was 97.9% in a total of 864 implants placed simultaneously during a graftless sinus augmentation procedure with a mean residual bone height (RBH) of 5.7 ±1.7 mm. This suggested that the implant can be successfully placed in the sinus cavity with a blood clot instead of a bone graft material.[7]

PRP was the first proposed therapy introduced for regenerative procedures in the field of oral and maxillofacial surgery.[8,9] PRP contains a number of growth factors including VEGF, PDGF, and TGF-β, among others, that are concentrated six to eight times normal physiologic doses.[10,11] While PRP has been utilized extensively in the field of oral and maxillofacial surgery for regenerative procedures (also combined with bone marrow stromal cells), two main reported drawbacks have commonly been reported. First, the protocols to prepare PRP are lengthy, requiring two centrifugation cycles sometimes lasting upward of 30 minutes. Second, in order to prevent coagulation during the lengthy centrifugation cycles, the additional use of anticoagulants is necessary, which alters the natural wound healing/clotting processes. For these reasons, a second-generation platelet concentrate termed *platelet-rich fibrin* (PRF) was proposed with complete anticoagulant removal.[12] Because anticoagulants are not utilized, centrifugation takes place immediately for 8 to 12 minutes (with just one centrifugation cycle).[13] Growth factors released from PRF also demonstrate a slower and more gradual release over time when compared to PRP, which releases the majority of growth factors within the first few minutes/hours.[14]

Due to the biologic advantages of PRF, it has frequently been utilized for sinus augmentation procedures both as a sole grafting material as well as in combination with bone grafts.[15,16] While it is known that PRF alone can be utilized as a grafting material during sinus augmentation procedures with simultaneous implant placement, this chapter discusses the biologic and dimensional properties of the sinus cavity to make clear when and how to use PRF alone versus in combination with a bone grafting material for sinus augmentation procedures.

Evaluation of Clinical Studies Using PRF as a Sole Grafting Material

Table 17-1 demonstrates the results from a recent systematic review investigating the use of PRF for sinus grafting procedures in various case series studies. Of the 12 studies evaluating PRF in a case series,[15,17–27] 10 evaluated PRF as a sole grafting material (Fig 17-1 and Video 17-1), whereas the studies by Meyer et al[18] and Inchingolo et al[20] evaluated the use of PRF in combination with β-TCP and Bio-Oss, respectively (a case series with no controls). Of the former 10 studies, 5 investigated the use of PRF as a sole grafting material utilizing a crestal/osteotome approach, 3 utilized exclusively a lateral approach, and 2 utilized one or the other depending on the clinical situation. In each of these studies, only PRF was utilized as a sole grafting material. Thus, it may be concluded that PRF may be utilized as a sole grafting material with various success rates observed (see Table 17-1).

Key discussed parameters for utilizing PRF as a sole grafting material

- It is an absolute requirement that implants be placed simultaneously. This is because the use of an implant by itself will act as a space creator (ie, tenting) to provide needed space for bone ingrowth. Thus far, studies have supported the notion that implants can be placed alone without use of biomaterials or PRF as long as a stable blood clot is formed around the implant surface. New bone formation will then follow.
- While several authors have demonstrated that PRF alone may be utilized to promote sinus grafting with vertical bone height gain with bone gains sometimes ranging greater than 7.5 mm from the sinus floor to the surrounding bone at the apex of the implant, such cases were typically limited to bone increases ranging in height (3–4 mm), and results were particularly favored in narrower sinuses, as discussed later in this chapter. In wider sinuses and sinuses requiring greater bone gain, a bone grafting material is subsequently needed.

Table 17-1 Use of PRF for sinus grafting in human case series studies

Authors (year)	Study design Follow-up	No. of participants Sex Mean age	Smokers (Y, N)	Conclusions
Diss et al[17] (2008)	Case series; PRF added as a sole grafting material during osteotome sinus floor elevation with simultaneous implant placement	20 M: 6 / F: 14 54.8	Y	The mean endosinus gain was 3.2 ± 1.5 mm at 1 year postoperative.
Meyer et al[18] (2009)	Case series; PRF combined with β-TCP followed by delayed implant placement at 6 months	20 (33 sinuses) M: 14 / F: 6 54.8	NR	After a mean follow-up of 4.5 years, the mean resorption rate of the grafted site was 20.3%, and the mean RBH gain was 16.9 mm.
Mazor et al[19] (2009)	Case series; PRF added as a sole grafting material during osteotome sinus floor elevation with simultaneous implant placement	20 M: 6 / F: 14 54.1	Y	The use of PRF as the sole filling material during a simultaneous sinus elevation resulted in a high volume of natural regenerated bone in the subsinus cavity up to the tip of the implants (10.1 ± 0.9 mm).
Inchingolo et al[20] (2010)	Case series; PRF combined with Bio-Oss followed by delayed implant placement at 6–9 months	23 (31 sinuses) M: 10 / F: 13 31–59	Y	In all cases included in this study, the authors observed a successful implant-prosthetic rehabilitation using PRF combined with Bio-Oss. No reported RBH was provided.
Toffler et al[21] (2010)	Case series; PRF added as a sole grafting material during osteotome sinus floor elevation with simultaneous implant placement	110 (138 sinuses) M: 40 / F: 70 58.4	Y	The mean increase in the height of implant sites was 3.4 mm (range, 2.5–5 mm).
Simonpieri et al[15] (2011)	Case series; PRF added as a sole grafting material during osteotome sinus floor elevation with simultaneous implant placement	20 (23 sinuses) M: 8 / F: 12 59.8	NR	The maximum follow-up was 6 years, and all patients were followed for a minimum of 2 years. No implant was lost during this 6-year experience, and the vertical bone gain was always substantial, between 8.5 and 12 mm bone gain (10.4 ± 1.2).
Tajima et al[22] (2013)	Case series; PRF added as a sole grafting material during lateral sinus floor elevation with simultaneous implant placement	6 (9 sinuses) M: 0 / F: 6 67.8	NR	The use of PRF for lateral sinus augmentation with simultaneous implant placement led to an average 7.5-mm gain in RBH. All implants were clinically stable at the time of abutment insertion, 6 months after sinus augmentation.
Kanayama et al[23] (2016)	Case series; PRF added as a sole grafting material during osteotome sinus floor elevation with simultaneous implant placement	27 (9 sinuses) M: 12 / F: 15 54.2	NR	The average gain in RBH was 4.19 ± 1.60 mm when two different implant systems were utilized.
Barbu et al[24] (2018)	Case series; PRF added as a sole grafting material during lateral sinus floor elevation with simultaneous implant placement	14 (14 sinuses) M: 10 / F: 4 49.5	NR	The mean vertical bone height gain was 10.12 mm 6 months after surgery, and the mean postoperative follow-up time was 43.79 months.
Aoki et al[25] (2018)	Case series; PRF added as a sole grafting material during crestal and lateral sinus floor elevation with simultaneous implant placement	34 (34 sinuses) M: 17 / F: 17 57.6	Y	The mean RBH was 4.26 mm. This retrospective study showed that sinus floor elevation with PRF alone could be applied in cases of lower RBH. However, it should be carefully performed in cases of RBH < 4 mm before surgery.
Kumar et al[26] (2018)	Case series; PRF added to Bio-Oss during lateral sinus floor elevation with simultaneous implant placement	14 (14 sinuses) M: 10 / F: 4 NR (between 18 and 65)	N	Twelve months postoperatively, the endosinus bone gain noted was 7 mm, which indicated the use of PRF with bovine bone graft as a reliable filling material during simultaneous sinus elevation and implantation.
Molemans et al[27] (2019)	Case series; PRF added as a sole grafting material during crestal and lateral sinus floor elevation with simultaneous implant placement	26 (28 sinuses) M: 14 / F: 12 55	N	The mean vertical bone gain was 3.4 ± 1.2 mm and 5.4 ± 1.5 mm for transalveolar sinus floor elevation and lateral sinus floor elevation, respectively.

M, male; F, female; NR, not reported; RBH, residual bone height.

Table 17-1 *(cont)* Use of PRF for sinus grafting in human case series studies

		Methods for PRF preparation		
Authors (year)	**RBH gain (mm)**	**Centrifugation system**	**Volume of tubes for blood drawn**	**Centrifugation parameters speed (rpm) × time (min)**
Diss et al[17] (2008)	3.5 ± 1.4	NR	10-mL BD tubes (80 mL total)	3000 × 10 min
Meyer et al[18] (2009)	16.9 (10–22)	NR	NR	NR
Mazor et al[19] (2009)	10.1 ± 0.9	Intra-Lock	9-mL glass-coated plastic tubes (72 mL total)	400g × 12 (rpm NR)
Inchingolo et al[20] (2010)	NA	NA	10-mL tubes	3000 × 10
Toffler et al[21] (2010)	3.4 (2.5–5)	Intra-Lock (Process for PRF)	9-mL glass-coated plastic tubes (18–54 mL total)	2700 × 12
Simonpieri et al[15] (2011)	10.4 ± 1.2	Intra-Lock (Process for PRF)	9-mL glass-coated plastic tubes (72 mL total)	400g × 12 (rpm NR)
Tajima et al[22] (2013)	7.5 ± 1.51	Medifuge MF200 (Silfradent)	9-mL sterile glass tubes (20–40 mL total)	30 seconds of acceleration, 2 minutes at 2700 rpm, 4 minutes at 2400 rpm, 4 minutes at 2700 rpm, 3 minutes at 3000 rpm, and 36 seconds to decelerate and stop
Kanayama et al[23] (2016)	4.19 ± 1.60	NR	10-mL glass tubes (20 mL total)	400g × 10 (rpm NR)
Barbu et al[24] (2018)	10.12 ± 1.81	NR	NR	NR
Aoki et al[25] (2018)	4.26 ± 2.11	NR	NR	NR
Kumar et al[26] (2019)	6.87 ± 1.48	NR	NR	NR
Molemans et al[27] (2019)	Crestal: 3.4 ± 1.2 Lateral: 5.4 ± 1.5	IntraSpin	9-mL glass-coated plastic tubes	408g × 12 (rpm NR)

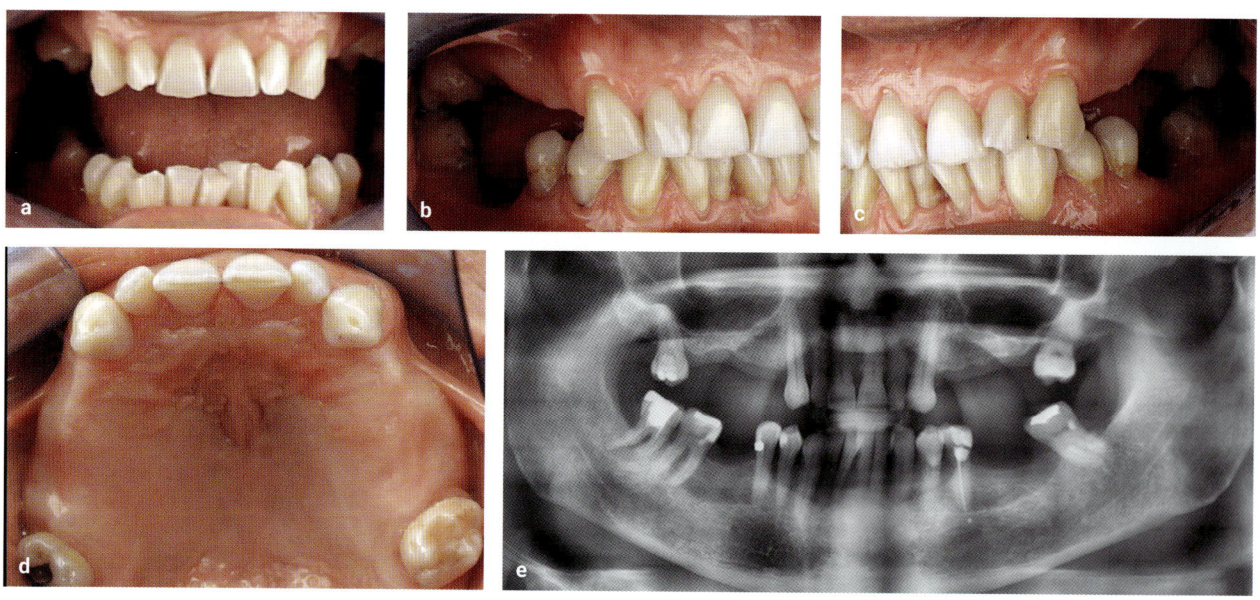

Fig 17-1 Narrow lateral sinus grafting procedure performed with PRF alone. *(a to d)* Clinical images demonstrating a bilateral edentulous posterior maxilla. *(e)* Panoramic radiograph demonstrating bilateral reduction in RBH.

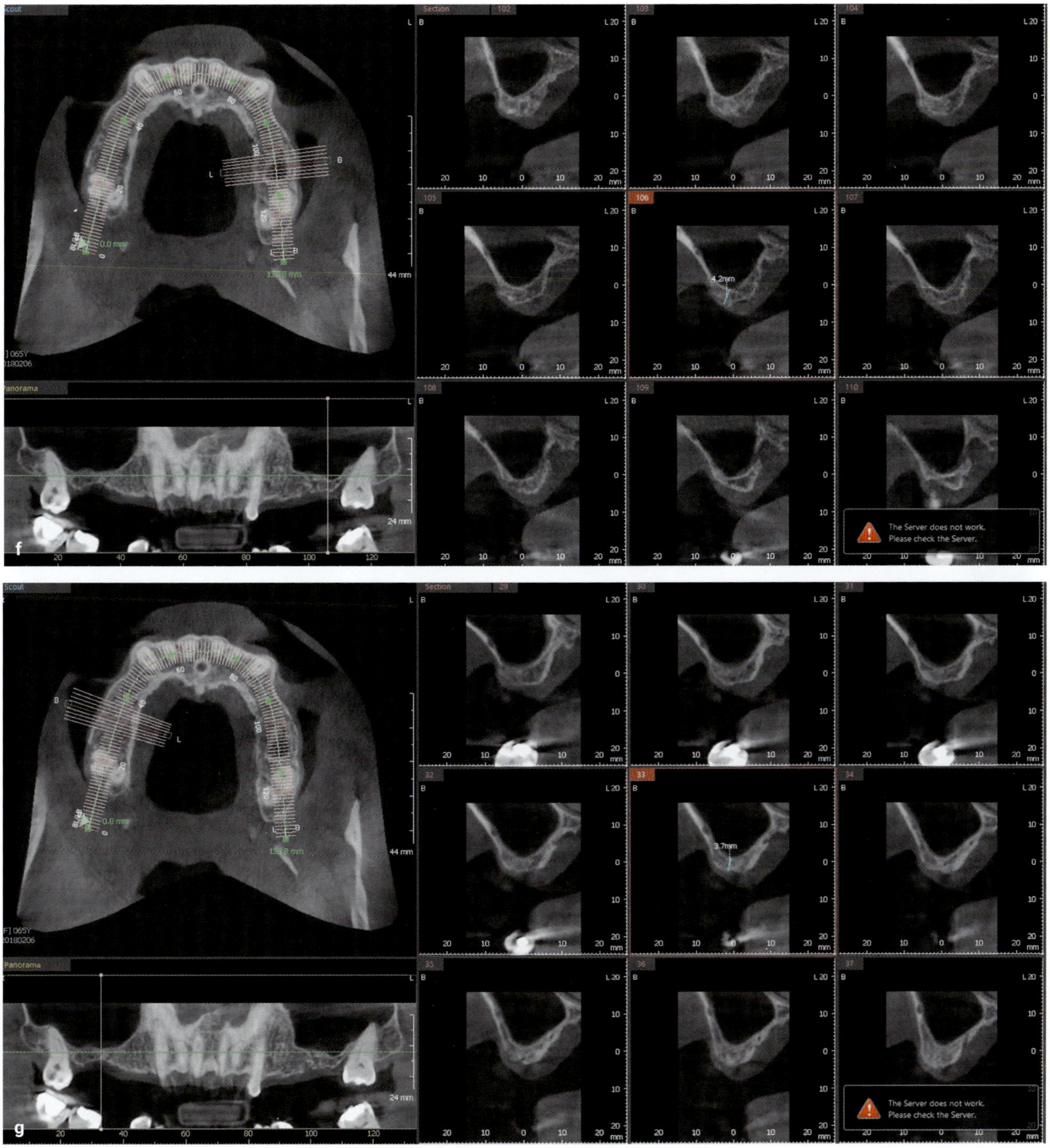

Fig 17-1 *(cont)* *(f and g)* CBCT analysis demonstrating the sinus anatomy including measurements of the remaining RBH.

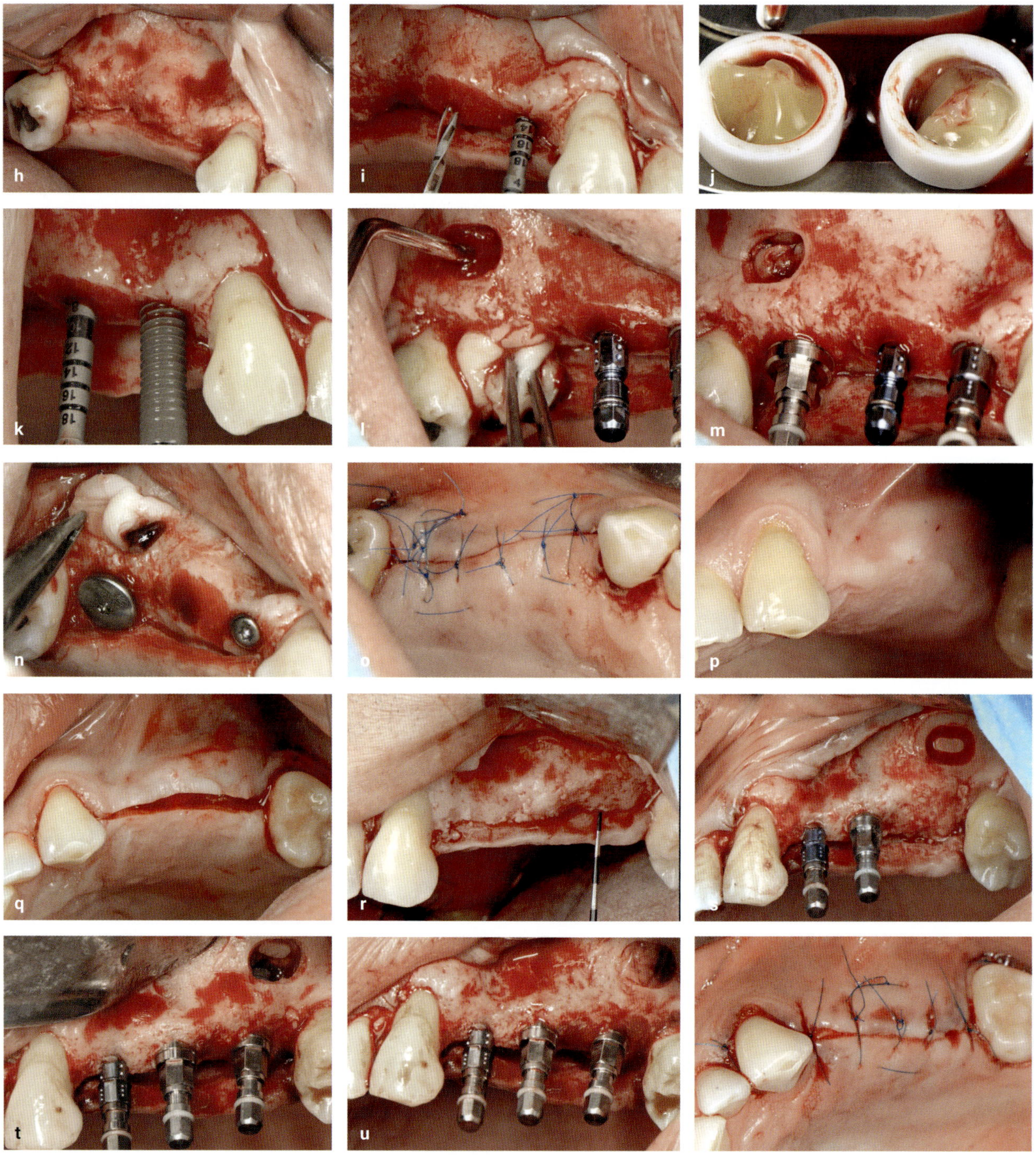

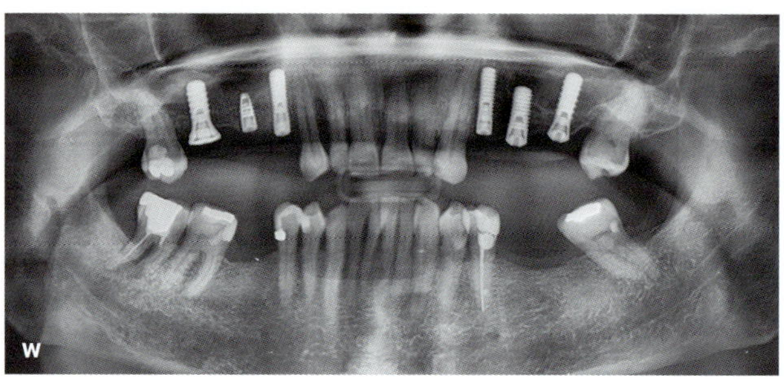

Fig 17-1 ***(cont)*** *(h)* Flap elevation. *(i)* Implant osteotomy. *(j)* PRF plugs being formed. *(k)* Implant placement. *(l)* PRF membranes and plugs being placed within the lateral wall of the sinus cavity as well as through the implant osteotomy. *(m)* Implant placement. *(n)* Use of PRF to close the lateral window. *(o)* Soft tissue closure. *(p)* Clinical image of the maxillary left quadrant. *(q)* Midcrestal incision. *(r)* Flap elevation. Note the deficient ridge. *(s and t)* Implant placement and lateral window contouring. *(u)* Insertion of PRF membranes into the sinus. *(v)* Soft tissue closure. *(w)* Final panoramic radiograph.

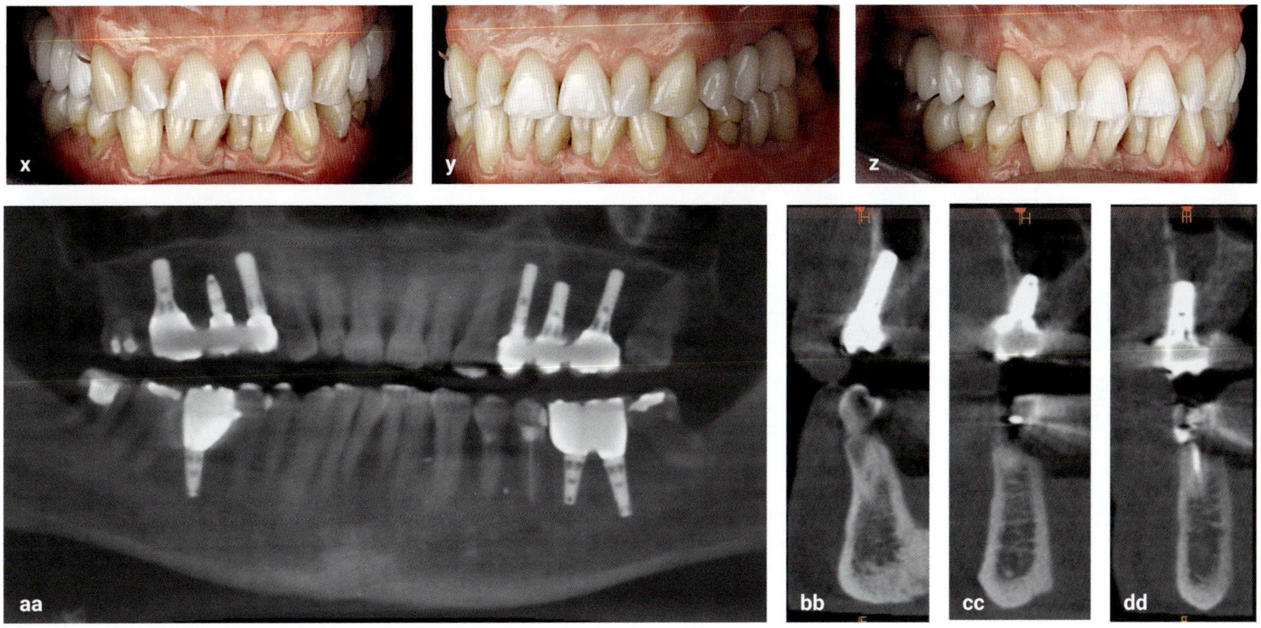

Fig 17-1 *(cont)* *(x to z)* Final restorations 2 years postoperative. *(aa)* Final radiograph 2 years postoperative with final restoration in place. *(bb)* Note that each of the implants placed into the sinus with PRF alone demonstrates bone gain up the apical aspect of the implant 2 years postoperative. See Video 17-1 for complete case presentation. (Case performed by Dr Luis Bessa.)

Video 17-1

Evaluation of Clinical Studies Using PRF in Combination with a Bone Graft

More commonly, PRF has been utilized with a bone grafting material (Fig 17-2) using the sticky bone concept for improved clinical handling as well as obtaining many advantages that PRF possesses. Table 17-2 presents each of the clinical trials from a recent systematic review comparing the use of bone grafting material alone versus in combination with PRF for sinus augmentation procedures.[16,28–36] In total, 10 clinical trials were investigated, with 7 studies investigating the use of a bone graft with or without PRF,[16,28–33] 2 evaluating the use of PRF as a membrane for lateral wall closure,[34,35] and 1 study evaluating the use of PRF on early pain management and swelling postoperative.[36]

Of the seven studies evaluating bone formation, five demonstrated no significant improvement in bone formation as evaluated histologically when PRF was added to a bone grafting material,[28–32] and two demonstrated favorable results with respect to being capable of placing implants earlier in PRF-grafted sites (with bone grafts).[16,33] Thus, in summary, it may be concluded that there is weak evidence demonstrating any long-term beneficial effect of adding PRF to sinus grafting procedures, though implants may potentially be placed at earlier time points when grafted with PRF. Noteworthy is that various authors and

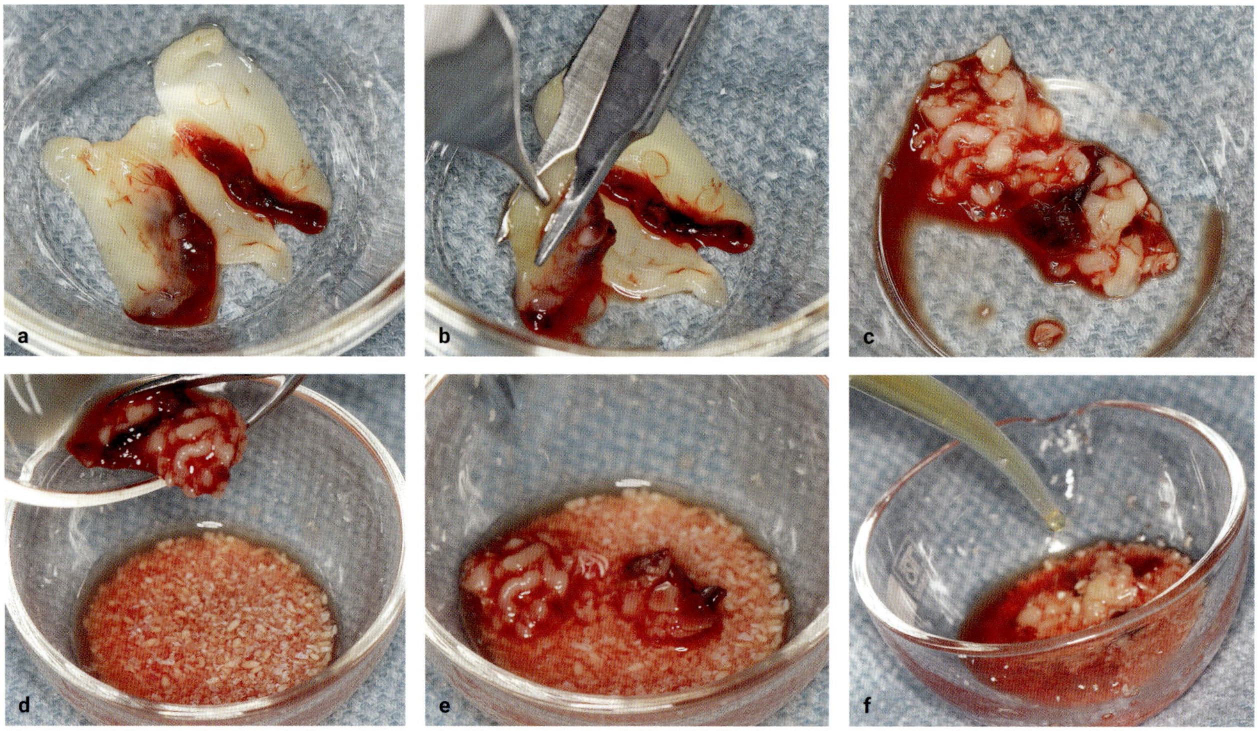

Fig 17-2 *(a and b)* Typical PRF membranes that are obtained after centrifugation being cut into small PRF fragments. *(c)* PRF cut into ~1-mm-sized fragments. *(d and e)* PRF fragments are then added to a bone grafting material and mixed. *(f)* Liquid-PRF is added to the grafting material complex to produce a sticky-consistency bone grafting material (ie, sticky bone).

systematic reviews on the topic have reported criticism regarding the various study designs, protocols, lack of appropriate controls, as well as limited available information regarding patient selection in comparison to other sinus grafting studies.[37,38] Furthermore, most studies to date were not randomized, with data rarely describing the importance or impact of implant loading protocols and the dimensional size of the sinus (most notably the buccopalatal width of the sinus) prior to implant placement.

A recent systematic review and meta-analysis published in 2019 by Lui et al investigated the effectiveness of PRF as an adjunctive material to bone grafts in maxillary sinus augmentation.[37] It was concluded that there were no statistical differences in survival rate, new bone formation, contact between newly formed bone and bone substitute, percentage of residual bone graft, and soft tissue area between the non-PRF and PRF groups. Current evidence supporting the advantages of adding PRF to bone grafts in sinus augmentation is therefore limited.[37]

At present, there remains limited data confirming the advantages of PRF for sinus grafting procedures.

Table 17-2 Human clinical trials investigating the use of PRF for sinus grafting procedures

Authors (year)	Study design Follow-up	No. of participants Sex Mean age	Groups	Smokers (Y, N)	Conclusions
Use of PRF in sinus augmentation procedure					
Choukroun et al[16] (2006)	Lateral approach, delayed implant placement (parallel) 4 –8 months	NR	C: 3, FDBA T: 6, FDBA+PRF	NR	Sinus floor augmentation with FDBA+PRF leads to a reduction of healing time prior to implant placement.
Zhang et al[28] (2012)	Lateral approach, delayed implant placement (parallel) 6 months	10 M: 8 / F: 2 43.5	C: 5, Bio-Oss T: 6, Bio-Oss+PRF	NR	The preliminary result demonstrated neither an advantage nor disadvantage of the application of PRF in combination with DBBM in sinus augmentation after a healing period of 6 months.
Tatullo et al[29] (2012)	Crestal approach, delayed implant placement (parallel) 106–150 days	60 M: 12 / F: 48 43–62	C: 36, Bio-Oss T: 36, Bio-Oss+PRF	N	The histologic findings demonstrated no differences between the control and test groups at all investigated time points.
Bolukbasi et al[30] (2015)	Lateral approach, delayed implant placement (parallel) 6 months	25 M: 10 / F: 15 50.06	C: 15, Bio-Oss T: 17, Bio-Oss+PRF	N	There were no statistical differences in terms of new bone formation (*P* = .61) and biomaterial remnant (*P* = .87).
Cömert Kiliç et al[31] (2017)	Lateral approach, delayed implant placement (parallel) 6 months	26 M: 17 / F: 9 22–51	C: 9, β-TCP T1: 9 β-TCP+PRP T2: 8, β-TCP+PRF	NR	These findings suggested that adding PRP or PRF to β-TCP graft substitute was not beneficial for new bone formation or regeneration, and PRP+β-TCP or PRF+β-TCP is not superior to β-TCP alone.
Nizam et al[32] (2018)	Lateral approach, delayed implant placement (split-mouth) 6 months	13 M: 9 / F: 4 49.92	C: 13, Bio-Oss T: 13, Bio-Oss+PRF	N	The addition of L-PRF to DBBM did not improve the amount of regenerated bone or the amount of the graft integrated into the newly formed bone under histologic and histomorphometric evaluation.
Pichotano et al[33] (2019)	Lateral approach, delayed implant placement (split-mouth) 4 and 8 months	12 M: 6 / F: 6 54.17	C: 12, Bio-Oss T: 12, Bio-Oss+PRF	N	The addition of PRF to the DBBM in the maxillary sinus allowed early implant placement (4 months) with increased new bone formation compared to DBBM alone after 8 months of healing.
Use of PRF to cover the lateral window					
Gassling et al[34] (2013)	Lateral approach, delayed implant placement (split-mouth) 5 months	6 NR 61	C: 6, Bio-Gide T: 6, PRF	NR	Within the limits of the study, the coverage of the lateral sinus window with two different absorbable membranes was shown to result in a similar amount of vital bone formation and residual bone substitute.
Bosshardt et al[35] (2014)	Lateral approach, delayed implant placement (split-mouth) 7–11 months	8 M: 1 / F: 7 41–64	C: 3, Bio-Gide T: 5, PRF	N	There was no additional benefit of the PRF membrane over the non-cross-linked collagen membrane.
Use of PRF on early pain and wound healing					
Gurler et al[36] (2016)	Lateral approach, delayed implant placement (parallel) 6 months	24 M: 14 / F: 10 47.8	C: 27, allograft T: 27, allograft+PRF	Y (< 10 cigarettes per day)	The use of L-PRF and allogeneic bone graft in combination with L-PRF membrane does not significantly improve postoperative complications following direct sinus elevation.

NR, not reported; C, control group; T, test group; M, male; F, female; FDBA, freeze-dried bone allograft; DBBM, deproteinized bovine bone mineral; β-TCP, β-tricalcium phosphate.

Table 17-2 *(cont)* Human clinical trials investigating the use of PRF for sinus grafting procedures

				Methods for PRF preparation		
Authors (year)	**New bone formation (%)**	**Residual bone graft (%)**	**Implant survival (%)**	**Centrifugation system**	**Volume of tubes for blood drawn**	**Centrifugation parameters speed (rpm) × time (min)**
Use of PRF in sinus augmentation procedure						
Choukroun et al[16] (2006)	20.31 (C) 20.95 (T)	10.93 (C) 9.41 (T)	100 (C) 100 (T)	Process for PRF	10 mL	2500 × 10
Zhang et al[28] (2012)	12.95 ± 5.33 (C) 18.35 ± 5.62 (T)	28.54 ± 12.01 (C) 19.16 ± 6.89 (T)	100 (C) 100 (T)	Labofuge 300 (Kendro Laboratory)	10 mL	300g × 10 (rpm NR)
Tatullo et al[29] (2012)	33.46 ± 6.18 (C) 35.08 ± 5.87 (T)	NR	100 (C) 100 (T)	NR	10 mL	3000 × 10
Bolukbasi et al[30] (2013)	32.97 ± 9.71 (C) 35.0 ± 8.60 (T)	33.79 ± 8.57 (C) 33.05 ± 6.29 (T)	100 (C) 100 (T)	Process for PRF	10 mL	2700 × 12
Cömert Kiliç et al[31] (2017)	33.40 ± 10.43 (C) 34.83 ± 10.12 (T1) 32.03 ± 6.34 (T1)	30.39 ± 10.29 (C) 28.98 ± 7.94 (T1) 32.66 ± 7.46 (T1)	NR	NR	5 mL	3000 × 10
Nizan et al[32] (2018)	21.25 ± 5.59 (C) 21.38 ± 8.78 (T)	32.79 ± 5.89 (C) 25.95 ± 9.54 (T)	100 (C) 100 (T)	NF 200 (Nüve)	NR	400g × 12 (rpm NR)
Pichotano et al[33] (2019)	30.02 ± 8.42 (C) 44.58 ± 13.90 (T)	13.75 ± 9.99 (C) 3.59 ± 4.22 (T)	100 (C) 100 (T)	Kasvi K14-0815	10 mL	3000 × 10
Use of PRF to cover the lateral window						
Gassling et al[34] (2013)	17.2 (8.5–24.2) (C) 17.0 (7.8–27.8) (T)	17.3 (0.7–33.5) (C) 15.9 (0.9–33.4) (T)	100 (C) 100 (T)	NR	10 mL	400g × 12 (rpm NR)
Bosshardt et al[35] (2014)	28.74 ± 5.44 (C) 28.59 ± 6.90 (T)	25.50 ± 7.64 (C) 25.67 ± 8.75 (T)	NR	NR	NR	NR
Use of PRF on early pain and wound healing						
Gurler et al[36] (2016)	*Pain (7 days):* 0.5 ± 0.85 (C) 0.1 ± 0.32 (T)	*Swelling (7 days):* 0.2 ± 0.42 (C) 0.2 ± 0.42 (T)	NA	IntraSpin	10 mL	2700 × 12

Use of PRF for Lateral Window Closure

Because PRF has been shown to more favorably improve soft tissue healing, attempts have also been made to utilize PRF as a sole barrier membrane to close lateral windows during sinus augmentation procedures (Fig 17-3; see also Table 17-2).[34,35] Two studies to date from separate groups investigated the use of PRF in comparison to a collagen barrier membrane as a replacement material for lateral window closure. In both studies, PRF was shown to lead to similar results when compared to a collagen barrier membrane.[34,35] These studies also investigated implant stability and new bone formation around implants and/or implant success rates. Nevertheless, it remains difficult to evaluate whether PRF should be utilized as a complete barrier to soft tissue invasion because the resorption period is estimated at 10 to 14 days. For these reasons, lateral window closure is most frequently performed with a collagen barrier membrane. An additional clinical study demonstrated that PRF led to significant reduction in patient-reported pain and swelling when PRF was combined with the sinus grafting procedure.[36]

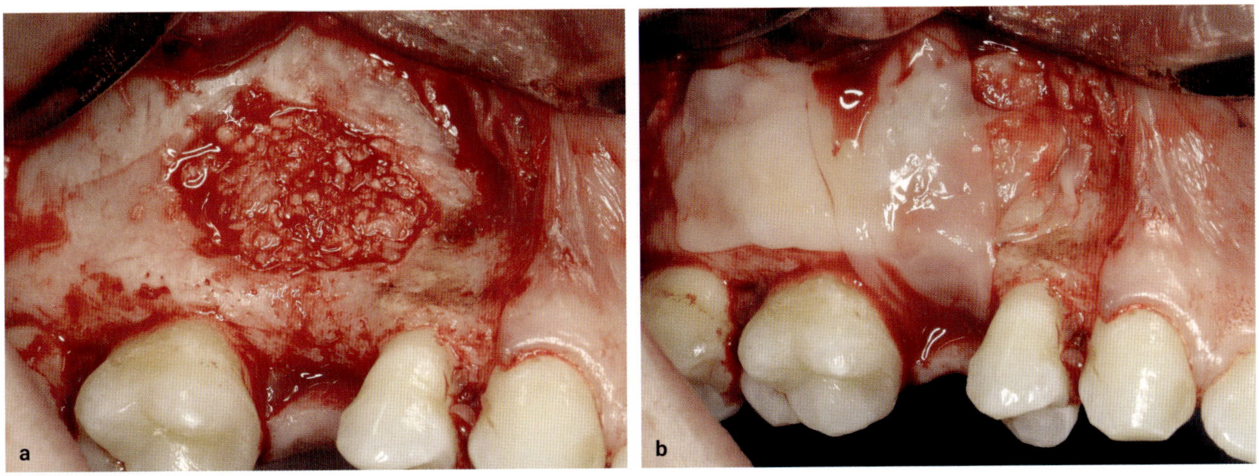

Fig 17-3 *(a)* Lateral window sinus augmentation filled with a combination bone graft with PRF. *(b)* PRF membranes are utilized to cover the lateral window prior to flap closure. (Case performed by Dr Michael A. Pikos.)

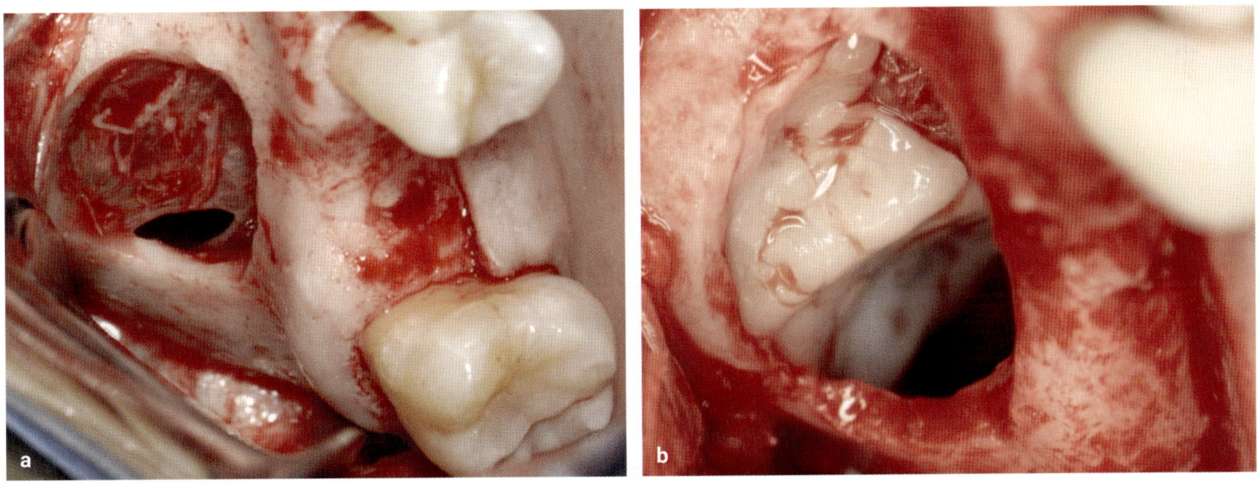

Fig 17-4 *(a)* Small sinus membrane perforation during a standard lateral approach. *(b)* Sinus membrane tear covered with a PRF membrane. (Case performed by Dr Michael A. Pikos.)

PRF as a Sole Material for the Repair of Sinus Membrane Perforations

Another frequent use of PRF has been for the repair of sinus membrane perforations (Fig 17-4). While major advancements have been made with respect to surgical techniques, tools, and instruments to decrease the rate of membrane perforation, typical frequencies of tears have generally been reported in the 20% range.[39,40] While over the years absorbable collagen membranes have most commonly been utilized, more recently, attempts have been made to cover perforations with PRF (see Fig 17-4). Due to the general "sticky" consistency of PRF, this 100% natural fibrin scaffold may be utilized as a low-cost replacement to standard collagen membranes. In light of its natural properties, it is known that PRF, unlike a collagen barrier membrane, will not induce a foreign body reaction.

Figure 17-4 demonstrates a small tear that was covered with PRF alone. A recent consensus report recommended that tears larger than 5 mm be treated with a standard approach utilizing a collagen membrane.[4,41] Nevertheless, the advantage of utilizing PRF for perforations smaller or equal to 5 mm in diameter is that the reparative process may take place with better handling properties due to the sticky consistency of PRF without potentially inducing a foreign body reaction. As a rule of thumb, multiple layers may be utilized for larger tears, at relatively no additional cost. Because resorption periods are typically 10 to 14 days, it is advised to utilize PRF in a double layer to ensure complete coverage. And, again, tears greater than 5 mm should be covered with a collagen membrane.

Clinical bottom line

1. PRF has been utilized as a sole grafting material in nearly a dozen studies. While favorable results have been observed, it is generally recommended to favor this approach for narrow sinuses and/or those requiring small (3–4 mm) sinus elevations.
2. PRF has more commonly been utilized in combination with a bone grafting material. In such cases, no significant advantages in new bone formation have been reported when compared to bone graft alone.
3. PRF has also been utilized to cover the lateral window. In both controlled studies, implant ISQ values were equivalent between sites grafted with either PRF or a collagen barrier membrane. Nevertheless, the ability to extend the working properties of PRF with Bio-Heat (see chapter 7) will logically more favorably exclude potential soft tissue infiltration.
4. PRF has also been utilized to repair sinus membrane perforations. Results from a recent consensus report generally recommended its use in smaller tears (limited to 5 mm or less), with larger tears being more favorably covered with a collagen barrier membrane.
5. PRF has also been demonstrated in one clinical study to favor a significant reduction in patient-reported postoperative pain and swelling. This is likely due to improvements in soft tissue wound healing/closure following grafting procedures when placed in contact with soft tissues.

Clinical Recommendations for Sinus Grafting with PRF

The biologic advantages of utilizing PRF for sinus augmentation procedures have previously been documented.[42] PRF serves as a scaffold capable of protecting/repairing the sinus membrane from potential tears/perforations, may be combined with bone grafting materials to improve their handling and stability, and has been shown to increase angiogenesis to grafted sites. As a result, many clinicians now utilize PRF for sinus augmentation procedures, and the size of the sinus cavity has been one of the deciding factors for when to utilize PRF alone versus combined with a bone grafting material.

Interestingly, a number of studies have now demonstrated that PRF can be utilized as a sole grafting material, with some reports even demonstrating bone augmentations greater than 7.5 mm from the floor of the sinus to the apex of the implant utilizing PRF alone (see Table 17-1).[15,19,22] Despite these reports, criticism has been raised regarding their protocols, lack of appropriate controls, and limited available information regarding patient selection in comparison to other regenerative modalities. When PRF is utilized as a sole grafting material with extensive bone regeneration required, often a very thin bone plate is observed around the apex of the implant (or no bone at all). For these reasons, larger grafting procedures are generally better augmented in combination with a bone grafting material.

Several essential factors must therefore be taken into account for the inexperienced clinician wishing to augment lost or missing bone in the sinus with PRF, as highlighted in Fig 17-5. First, if PRF is to be utilized as a sole grafting material, implant placement *must* be performed simultaneously. When PRF is utilized as a sole grafting material for sinus augmentation utilizing a delayed approach without its combination with a bone grafting material or an implant surface, resorption will quickly follow with limited to no new bone formation occurring. This is due to the limited osteoinductive properties of PRF (its main function is to rapidly stimulate new blood flow and not new bone formation), and therefore the absence of an osteoconductive implant surface or bone grafting material results in no new bone formation. Additionally, the implant geometry also acts as a space creator to provide needed volume for bone tissue ingrowth. These findings reflect the long list

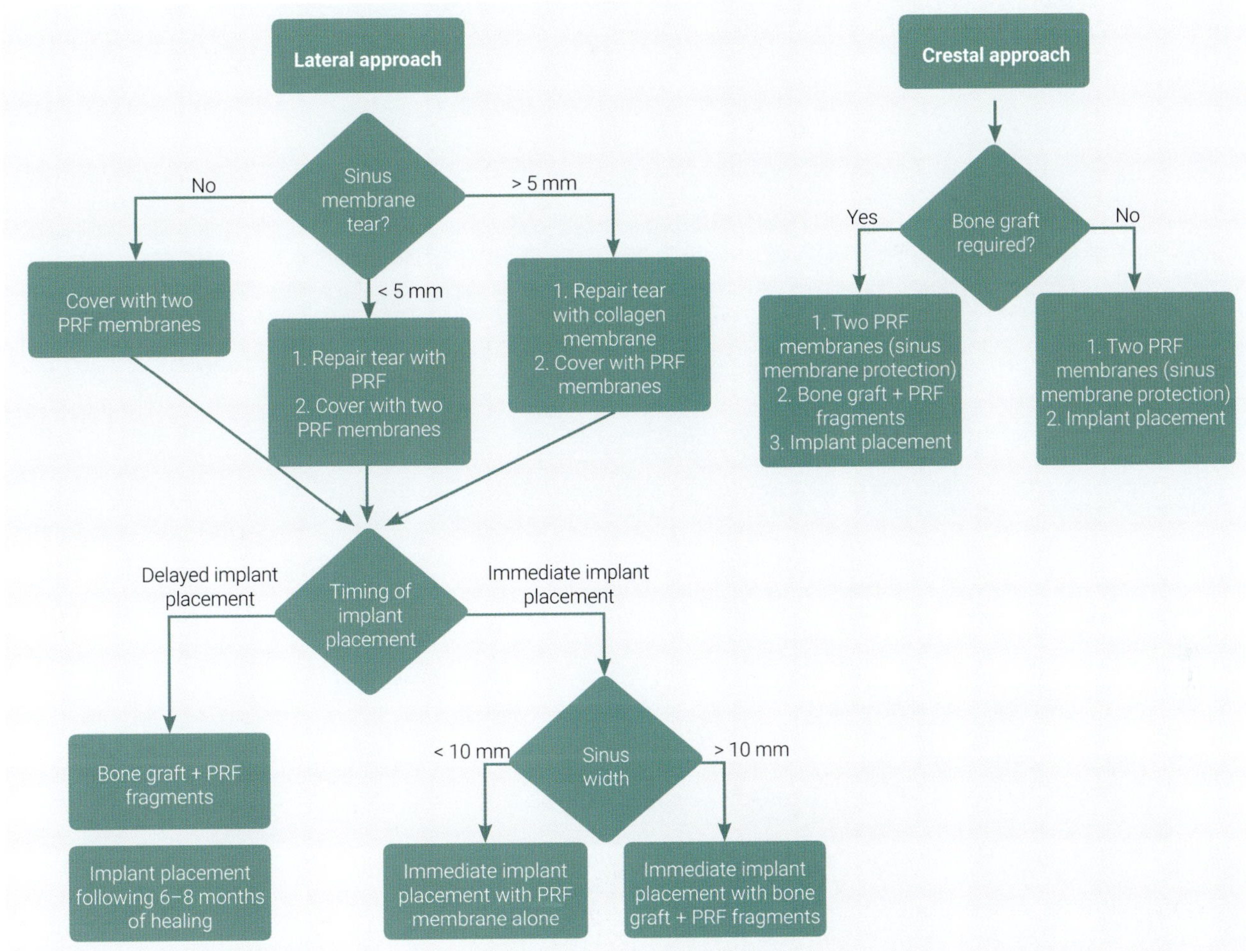

Fig 17-5 Decision-making flowchart describing the use of PRF for either a lateral or crestal sinus augmentation procedure.

of studies to date and recent systematic review that have shown that within the sinus cavity, formation of a blood clot alone around an implant surface may result in new bone if primary stability is achieved.[7,43–46]

When PRF is utilized as a sole grafting material, implant placement must be performed simultaneously.

One of the limitations reported to date with utilizing PRF alone for sinus augmentation procedures has been its higher risk when used in wide sinuses. Previously, Avila et al published an article titled, "The influence of the bucco-palatal distance on sinus augmentation outcomes."[47] It was demonstrated that lateral sinus augmentation procedures performed with an allograft in narrow sinuses (< 10 mm) and medium sinuses (10–15 mm) demonstrated roughly three times more vital bone after a 6-month healing period when compared to wide sinuses (> 15 mm[47]; Fig 17-6). In response to these findings as well as the clinical experiences of numerous oral surgeons utilizing PRF alone for sinus augmentation procedures, it has been recommended that sinuses greater than 10 mm be regenerated in a combination approach with a bone grafting material. Therefore, all sinuses greater than 10 mm (buccopalatally) should be regenerated utilizing a combination approach using PRF cut into small fragments and mixed with a bone grafting

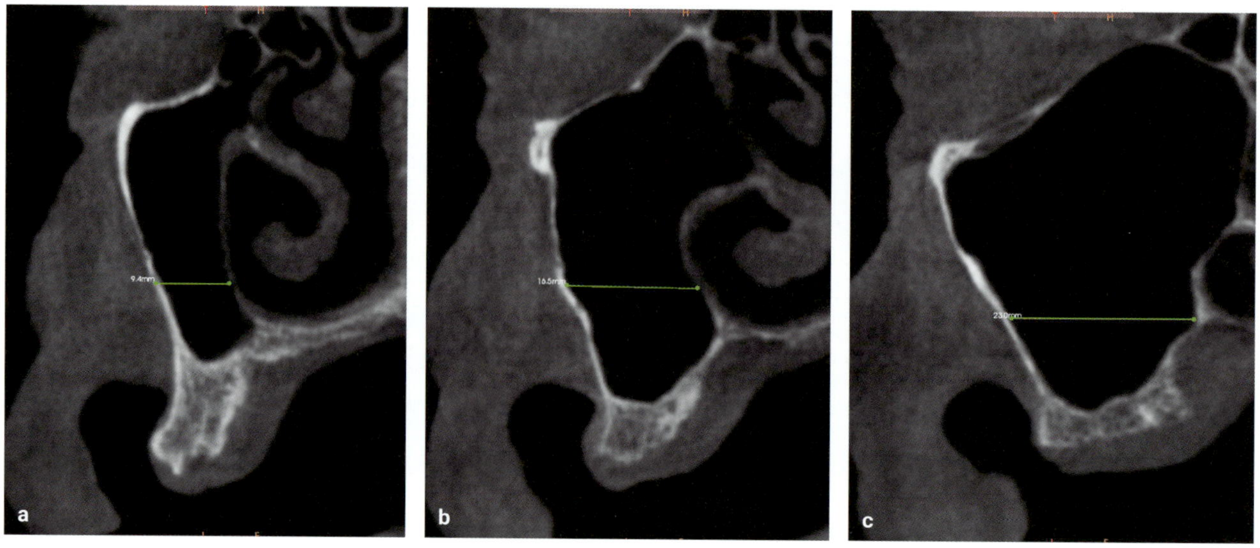

Fig 17-6 CBCT of morphologic size differences between *(a)* narrow (< 10 mm), *(b)* medium (10–15 mm), and *(c)* wide (> 15 mm) sinuses. The buccopalatal distance has been a predictable tool for determining when to use PRF alone versus in combination with a bone grafting material.

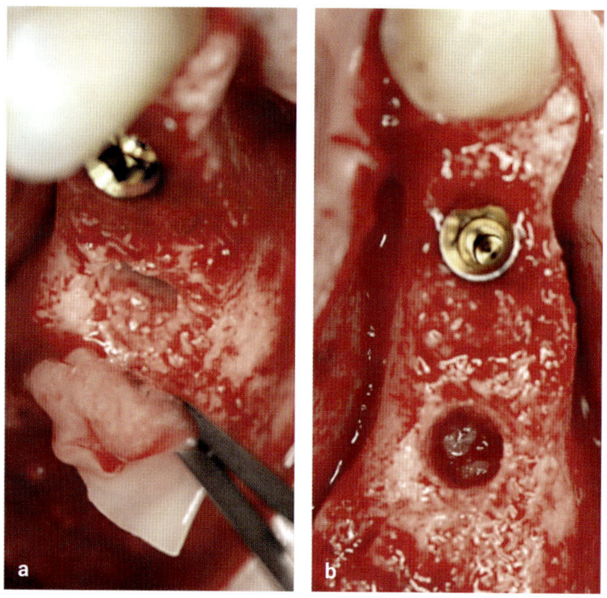

Fig 17-7 *(a and b)* Use of PRF for a crestal sinus augmentation procedure. PRF is passed through the osteotomy site to protect the sinus membrane prior to bone grafting material placement and implant placement. (Case performed by Dr Michael A. Pikos.)

material. For narrow sinuses (< 10 mm), PRF alone has been shown to lead to high success rates with predictable new bone formation taking place in the sinus. Utilizing either a crestal or a lateral approach, PRF may be placed at the apex of the implant prior to insertion (Figs 17-7 and 17-8). This acts as a safety element where the fibrin "cushion" may serve to lower potential direct implant contact with the thin sinus membrane (often characterized as being 0.3 to 0.8 mm thick).

There remains interest to investigate sinus augmentation procedures for sinus width in the 10- to 15-mm range. While reports and clinical case presentations have shown that PRF alone can be utilized, caution must be advised. Additional surgical modalities, for example specially designed surgical burs (osseodensification) or specialized sinus elevation kits, may further be utilized to optimize implant bed preparation and primary stability of implants. Figure 17-9 demonstrates a sinus grafting

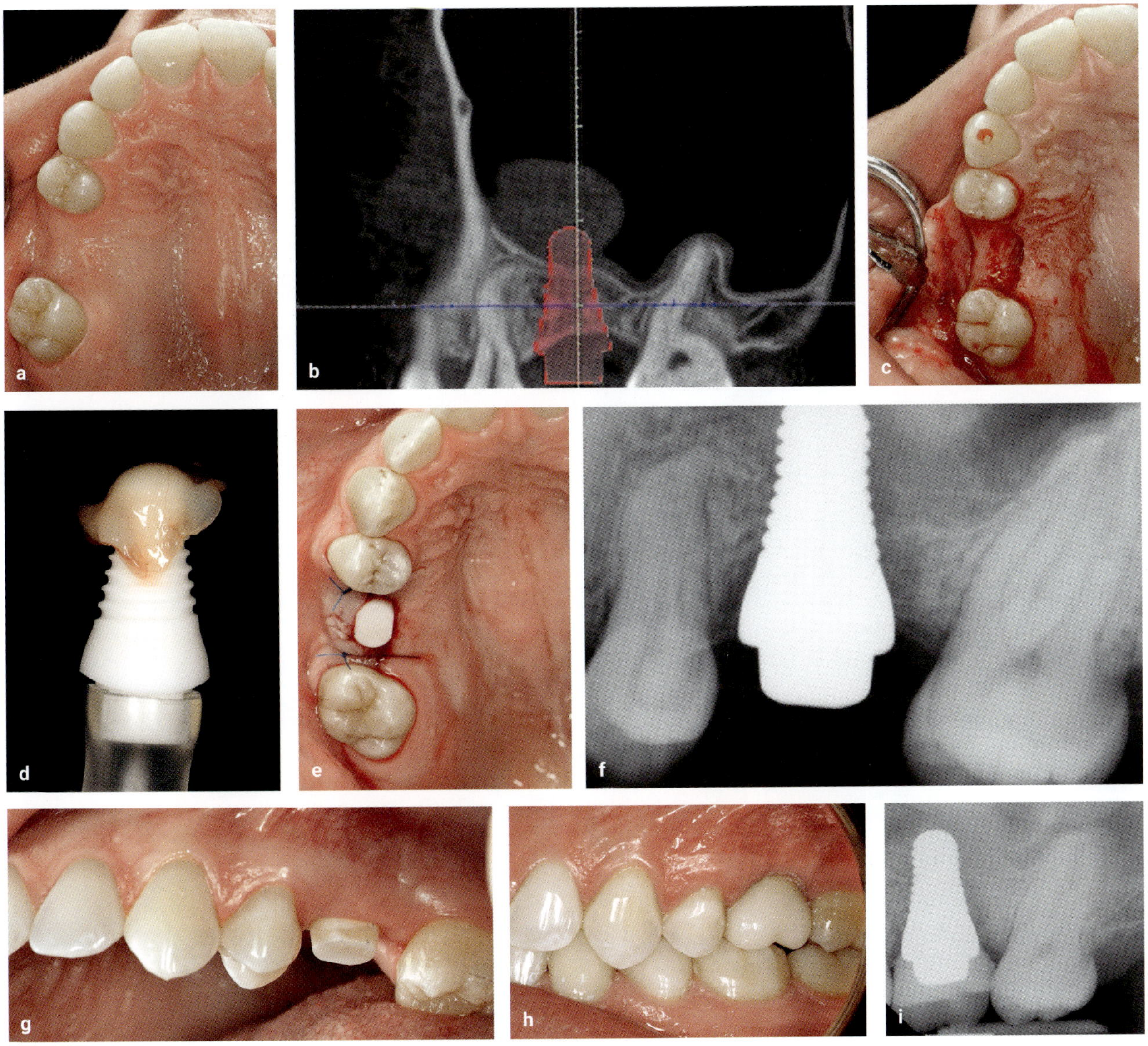

Fig 17-8 Use of a ceramic implant placed utilizing a crestal approach with PRF. *(a)* Clinical image of a missing maxillary second premolar. *(b)* CBCT planning demonstrating the need for a few additional millimeters of bone gain. *(c)* Flap elevation. *(d)* Use of a PRF membrane placed at the apex of the implant. *(e)* Implant placement and final soft tissue closure. *(f)* Postoperative periapical radiograph. *(g)* Healing at 4 months postoperative. *(h)* Final crown. *(i)* Final periapical radiograph during final crown placement. Note the bone formation that took place around the apex of the implant. (Case performed by Dr Yuriy May.)

procedure requiring additional bone volume within the sinus. Following a crestal approach, a bone grafting putty (NovaBone) was placed into the apical portion of the sinus and facilitated using osseodensification burs. Such strategies may be combined with PRF during soft tissue closure to favor better soft tissue healing around the crestal portion of the implant.

When a lateral approach is needed for sinuses greater than 10 mm in the buccopalatal dimension, it is generally recommended to combine PRF with a bone grafting material. In 2015, a systematic review was performed by our group evaluating the use of PRF for sinus elevation procedures.[48] Of the 290 initial studies searched, only 8 met the inclusion criteria, half of which did not utilize appropriate controls to compare their findings.[48] It was further noted that there was a large heterogeneity in the results due to the major reported differences in the surgical techniques utilized (lateral sinus augmentation versus

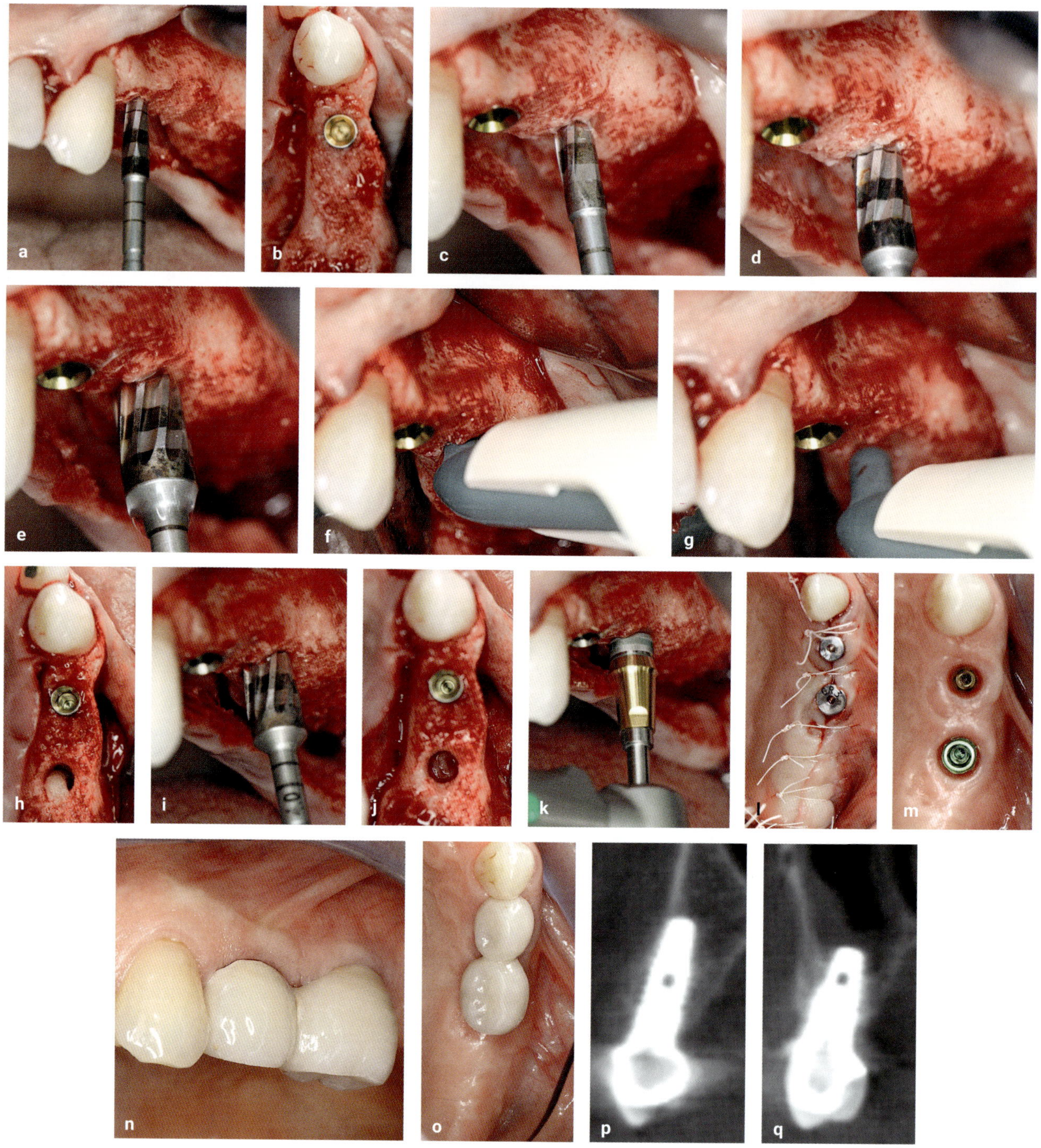

Fig 17-9 Distal extension sinus grafting utilizing a crestal approach with osseodensification (OD) burs. *(a to e)* Use of OD burs for osteotomy preparation and implant placement. *(f to h)* An alloplast putty (NovaBone) is utilized to add bone volume to the apical portion of the osteotomy. *(i and j)* OD burs are utilized to propel the bone putty apically. *(k)* Implant placement. *(l)* Soft tissue closure. *(m)* Clinical view 3 months postoperative. *(n and o)* Clinical views at 3 years. *(p and q)* CBCT imaging demonstrating maintenance of adequate buccal bone volume. (Reprinted with permission from Pikos and Miron.[41])

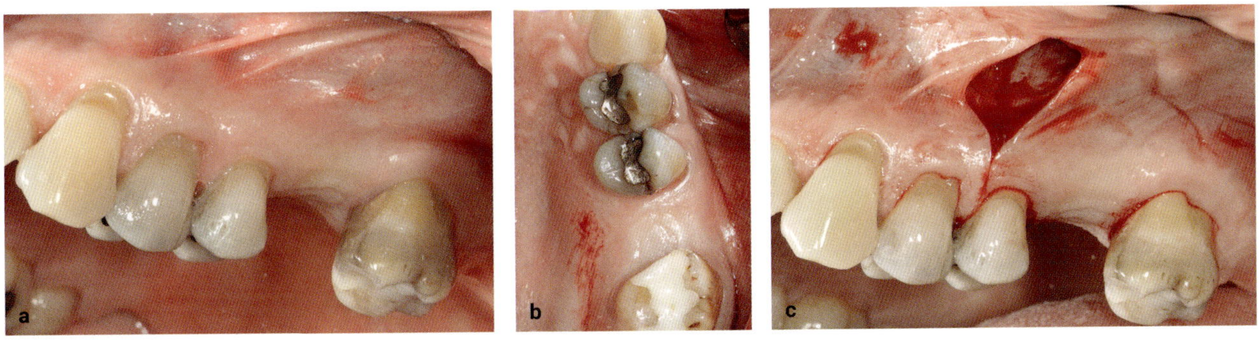

Fig 17-10 Lateral sinus floor elevation with PRF membranes followed by implant placement. *(a)* Lateral view of the missing maxillary right first molar. *(b)* Occlusal view of the area. *(c)* Incision design showing sulcular incisions at the adjacent second molar and second premolar with a palatally based incision in the edentulous area. Note the vertical releasing incision mesial to the buccal line angle of the second premolar.

crestal approach), time of implant placement (simultaneous versus delayed), outcomes measured, biopsy analysis, and whether histologic evaluation was even performed at all.[48] Based on this variability, it remains difficult to assess the "ideal" treatment protocol utilizing PRF in larger sinuses requiring large bone augmentation, and as a result, conservative clinical recommendations are generally provided. A decision-making flowchart provided in Fig 17-5 highlights the fact that sinuses smaller to or equal to 10 mm in buccopalatal width may predictably be regenerated with PRF alone. However, when the sinus is greater than 10 mm, its combination with a bone grafting material is highly advised.

The clinician's experience as well as the surgical protocols utilized may further guide selection for the most appropriate regenerative procedure. Within this chapter, safe, effective, and predictable protocols are provided. Nevertheless, future research is ongoing to determine optimal conditions to better characterize when to utilize PRF alone versus in combination with a bone grafting material. Conservative, predictable protocols are always recommended (see Fig 17-5).

Clinical Procedure

Based on the above-mentioned conclusions, sinus grafting procedures are generally performed combining bone grafting materials with PRF. The following clinical procedure and recommendations are excerpted from a recent textbook detailing sinus grafting procedures.[41]

The role of bone allografts is generally to provide a more osteoinductive bone graft with the ability to facilitate bone formation, while xenografts are generally utilized because of their space-maintaining ability (ie, nonresorbable). For these reasons, bone allografts are generally mixed with xenografts in a 1:1 ratio for sinus grafting procedures in order to take advantage of both grafting properties. PRF membranes cut into fragments are then mixed with this bone graft complex and coated with liquid-PRF to create sticky bone (detailed protocol provided in chapter 10).

When a lateral approach is performed, typically hand instruments and/or a Piezosurgery device is utilized to create the lateral window and minimize trauma to the sinus membrane (Fig 17-10 and Video 17-2). Following flap reflection, two PRF membranes may be introduced into the sinus to minimize potential sinus membrane damage during and/or following grafting. Alternatively, a small collagen membrane may also be utilized. When larger or complete tears/perforations are encountered, a large collagen barrier membrane is recommended with external fixation when needed. For further clinical protocols, the reader is encouraged to view detailed protocols in a recent textbook on the topic.[41]

Video 17-2

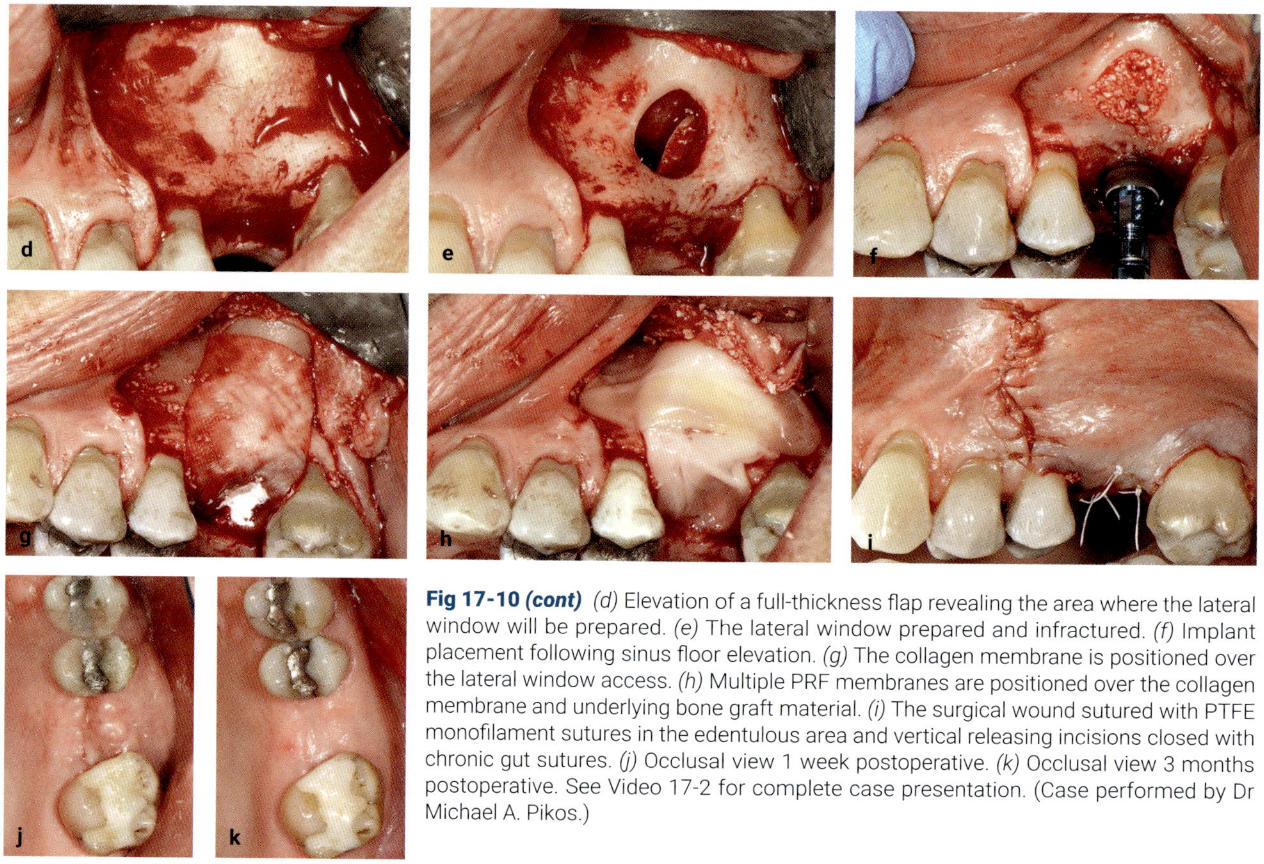

Fig 17-10 *(cont)* *(d)* Elevation of a full-thickness flap revealing the area where the lateral window will be prepared. *(e)* The lateral window prepared and infractured. *(f)* Implant placement following sinus floor elevation. *(g)* The collagen membrane is positioned over the lateral window access. *(h)* Multiple PRF membranes are positioned over the collagen membrane and underlying bone graft material. *(i)* The surgical wound sutured with PTFE monofilament sutures in the edentulous area and vertical releasing incisions closed with chronic gut sutures. *(j)* Occlusal view 1 week postoperative. *(k)* Occlusal view 3 months postoperative. See Video 17-2 for complete case presentation. (Case performed by Dr Michael A. Pikos.)

Once the sinus membrane is covered with PRF and/or collagen membranes, the "sticky bone" may then be entered within the sinus cavity. The general sticky consistency favors graft material handling and placement. At this stage, it is generally recommended to confirm that all bony walls are covered with grafting particles and moderately condensed. Simultaneous implant placement may also be performed at this time, with the osteotomy created prior to placement of any bone grafting material into the sinus. In such cases, protocols typically follow this sequence: *(1)* lateral window creation and sinus membrane elevation, *(2)* implant osteotomy preparation, *(3)* placement of the bone grafting material complex into the sinus, *(4)* implant placement, *(5)* final grafting around the implant, and *(6)* soft tissue closure with PRF membrane use.[41] If the implant is placed prior to bone grafting, it is generally much more difficult to enter the grafting material in and around the implant within the sinus cavity with potentially voided/nongrafted areas.

Advantages of using PRF for sinus grafting

1. It favors better bone grafting material handling.
2. It often requires less bone grafting material because several PRF membranes/fragments are utilized to hold volume (thereby slightly lowering grafting material costs).
3. It may facilitate earlier implant placement when compared to bone grafted sites alone as demonstrated in two controlled studies (though no long-term bone gain has been shown).
4. It may be utilized to further support better soft tissue healing at incision lines as well as around implant healing sites.
5. It reduces pain and swelling postgrafting as demonstrated in one controlled clinical study.

Discussion and Future Research

Regeneration of a pneumatized sinus following tooth loss is a frequently performed surgical procedure that requires lengthier healing times and carries higher potential complication rates when compared to other grafting procedures such as extraction sites. As a result, the use of PRF has been increasingly popular to speed angiogenesis to the defective tissues and also favor improved soft tissue healing. The purpose of this chapter was not to discuss the surgical concepts for sinus augmentation procedures (eg, when to perform a sinus augmentation immediately versus utilizing a delayed approach, or when to utilize a lateral window versus crestal approach for sinus augmentation procedures). Discussion over these surgical topics and regenerative approaches is extensive elsewhere and can be found in other sources on the topic.[41] Instead the focus of this chapter was to highlight when, where, and how to utilize PRF in the above-mentioned clinical scenarios, either alone or in combination with a bone grafting material or barrier membrane, and more importantly to equip clinicians for decision-making in this regard.

Generally speaking, when a delayed approach for implant placement is necessary, PRF must *always* be combined with a bone grafting material due to the fast resorption time of PRF scaffolds. Under no circumstances can PRF be utilized alone without the use of a bone grafting material or an implant. When implant placement is performed simultaneously with sinus augmentation, the factor most predictive regarding when to utilize PRF alone versus in combination with a bone grafting material is the buccopalatal width of the sinus cavity. When sinuses are larger than 10 mm, it is generally recommended to combine PRF with a bone grafting material, while for the narrower sinuses (< 10 mm), PRF may be utilized as a sole grafting material predictably (see Fig 17-5). Similarly, when a perforation of the sinus is encountered, PRF may be utilized alone when the sinus membrane tear is considered small (< 5 mm), whereas a collagen barrier membrane is more often recommended when perforations are larger.

PRF may be utilized following a crestal sinus grafting procedure simply to minimize potential damage to the sinus membrane. Interestingly, and as reviewed in chapter 14, PRF has been shown to reduce postoperative pain and infections of third molars by as much as 10-fold due to its incorporation of defense-fighting leukocytes.[49,50] This may be one of the reasons why patients report less postoperative pain and swelling when PRF is used. It remains relatively unknown to what extent supraphysiologic doses of leukocytes may favor reduced infection rates in sinus grafting procedures, as this topic has never been addressed in a large clinical study.

Furthermore, the recent ability to dramatically extend the resorption properties of PRF from 2 to 3 weeks to 4 to 6 months utilizing a simple heating process may provide additional benefits in particular sinus grafting procedures. Due to its longer resorption period, it may be utilized *(1)* as a barrier for soft tissue closure following a lateral window approach or *(2)* potentially over a large sinus membrane perforation, especially now that custom PRF trays can create custom membranes virtually of any size and shape. Much future research in this space is needed and ongoing.

Conclusion

The use of PRF for sinus augmentation procedures has been widely utilized in recent years due to its ability to support vascularization to the sinus cavity as well as improve soft tissue healing. This chapter summarized the clinical indications for when to utilize PRF alone versus in combination with a bone grafting material or collagen barrier membrane for such procedures, with some discussion on its use when utilized during a crestal versus a lateral sinus grafting procedure. Conservative, predictable, and evidence-based clinical recommendations were provided. Further research in this field is presently ongoing.

References

1. Solar P, Geyerhofer U, Traxler H, Windisch A, Ulm C, Watzek G. Blood supply to the maxillary sinus relevant to sinus floor elevation procedures. Clin Oral Implants Res 1999;10:34–44.
2. Wallace SS, Froum SJ. Effect of maxillary sinus augmentation on the survival of endosseous dental implants. A systematic review. Ann Periodontol 2003;8:328–343.
3. Chambrone L, Preshaw PM, Ferreira JD, Rodrigues JA, Cassoni A, Shibli JA. Effects of tobacco smoking on the survival rate of dental implants placed in areas of maxillary sinus floor augmentation: A systematic review. Clin Oral Implants Res 2014;25:408–416.
4. Miron RJ, Pikos MA. Sinus augmentation using platelet-rich fibrin with or without a bone graft: What is the consensus? Compend Contin Educ Dent 2018;39:355–361.

5. Kozuma A, Sasaki M, Seki K, Toyoshima T, Nakano H, Mori Y. Preoperative chronic sinusitis as significant cause of postoperative infection and implant loss after sinus augmentation from a lateral approach. Oral Maxillofac Surg 2017;51:193–200.
6. Ghasemi S, Fotouhi A, Moslemi N, Chinipardaz Z, Kolahi J, Paknejad M. Intra-and postoperative complications of lateral maxillary sinus augmentation in smokers vs nonsmokers: A systematic review and meta-analysis. Int J Oral Maxillofac Implants 2017;32: 759–767.
7. Duan DH, Fu JH, Qi W, Du Y, Pan J, Wang HL. Graft-free maxillary sinus floor elevation: A systematic review and meta-analysis. J Periodontol 2017;88:550–564.
8. Marx RE. Platelet-rich plasma: Evidence to support its use. J Oral Maxillofac Surg 2004;62:489–496.
9. Marx RE, Carlson ER, Eichstaedt RM, Schimmele SR, Strauss JE, Georgeff KR. Platelet-rich plasma: Growth factor enhancement for bone grafts. Oral Surg Oral Med Oral Pathol Oral Radiol Endod 1998;85:638–646.
10. Leach JK, Kaigler D, Wang Z, Krebsbach PH, Mooney DJ. Coating of VEGF-releasing scaffolds with bioactive glass for angiogenesis and bone regeneration. Biomaterials 2006;27:3249–3255.
11. Kobayashi E, Fluckiger L, Fujioka-Kobayashi M, et al. Comparative release of growth factors from PRP, PRF, and advanced-PRF. Clin Oral Investig 2016;20:2353–2360.
12. Choukroun J, Adda F, Schoeffler C, Vervelle A. Une opportunité en paro-implantologie: Le PRF. Implantodontie 2001;42:e62.
13. Miron RJ, Fujioka-Kobayashi M, Bishara M, Zhang Y, Hernandez M, Choukroun J. Platelet-rich fibrin and soft tissue wound healing: A systematic review. Tissue Eng Part B Rev 2017;23:83–99.
14. Fujioka-Kobayashi M, Miron RJ, Hernandez M, Kandalam U, Zhang Y, Choukroun J. Optimized platelet-rich fibrin with the low-speed concept: Growth factor release, biocompatibility, and cellular response. J Periodontol 2017;88:112–121.
15. Simonpieri A, Choukroun J, Del Corso M, Sammartino G, Ehrenfest DMD. Simultaneous sinus-lift and implantation using microthreaded implants and leukocyte-and platelet-rich fibrin as sole grafting material: A six-year experience. Implant Dent 2011;20: 2–12.
16. Choukroun J, Diss A, Simonpieri A, et al. Platelet-rich fibrin (PRF): A second-generation platelet concentrate. Part V: Histologic evaluations of PRF effects on bone allograft maturation in sinus lift. Oral Surg Oral Med Oral Pathol Oral Radiol Endod 2006;101: 299–303.
17. Diss A, Dohan DM, Mouhyi J, Mahler P. Osteotome sinus floor elevation using Choukroun's platelet-rich fibrin as grafting material: A 1-year prospective pilot study with microthreaded implants. Oral Surg Oral Med Oral Pathol Oral Radiol Endod 2008;105: 572–579.
18. Meyer C, Chatelain B, Benarroch M, Garnier JF, Ricbourg B, Camponovo T. Massive sinus-lift procedures with beta-tricalcium phosphate: Long-term results [in French]. Rev Stomatol Chir Maxillofac 2009;110:69–75.
19. Mazor Z, Horowitz RA, Del Corso M, Prasad HS, Rohrer MD, Dohan Ehrenfest DM. Sinus floor augmentation with simultaneous implant placement using Choukroun's platelet-rich fibrin as the sole grafting material: A radiologic and histologic study at 6 months. J Periodontol 2009;80:2056–2064.
20. Inchingolo F, Tatullo M, Marrelli M, et al. Trial with platelet-rich fibrin and Bio-Oss used as grafting materials in the treatment of the severe maxillar bone atrophy: Clinical and radiological evaluations. Eur Rev Med Pharmacol Sci 2010;14:1075–1084.
21. Toffler M, Toscano N, Holtzclaw D. Osteotome-mediated sinus floor elevation using only platelet-rich fibrin: An early report on 110 patients. Implant Dent 2010;19:447–456.
22. Tajima N, Ohba S, Sawase T, Asahina I. Evaluation of sinus floor augmentation with simultaneous implant placement using platelet-rich fibrin as sole grafting material. Int J Oral Maxillofac Implants 2013;28:77–83.
23. Kanayama T, Horii K, Senga Y, Shibuya Y. Crestal approach to sinus floor elevation for atrophic maxilla using platelet-rich fibrin as the only grafting material: A 1-year prospective study. Implant Dent 2016;25:32–38.
24. Barbu HM, Andreescu CF, Comaneanu MR, Referendaru D, Mijiritsky E. Maxillary sinus floor augmentation to enable one-stage implant placement by using bovine bone substitute and platelet-rich fibrin. Biomed Res Int 2018;2018:6562958.
25. Aoki N, Maeda M, Kurata M, et al. Sinus floor elevation with platelet-rich fibrin alone: A clinical retrospective study of 1–7 years. J Clin Exp Dent 2018;10:e984–e991.
26. Kumar M, Chopra S, Das D, Gupta M, Memoalia J, Verma G. Direct Maxillary sinus floor augmentation for simultaneous dental implant placement. Ann Maxillofac Surg 2018;8:188–192.
27. Molemans B, Cortellini S, Jacobs R, Pinto N, Teughels W, Quirynen M. Simultaneous sinus floor elevation and implant placement using leukocyte- and platelet-rich fibrin as a sole graft material. Int J Oral Maxillofac Implants 2019;34:1195–1201.
28. Zhang Y, Tangl S, Huber CD, Lin Y, Qiu L, Rausch-Fan X. Effects of Choukroun's platelet-rich fibrin on bone regeneration in combination with deproteinized bovine bone mineral in maxillary sinus augmentation: A histological and histomorphometric study. J Craniomaxillofac Surg 2012;40:321–328.
29. Tatullo M, Marrelli M, Cassetta M, et al. Platelet Rich Fibrin (P.R.F.) in reconstructive surgery of atrophied maxillary bones: Clinical and histological evaluations. Int J Med Sci 2012;9:872–880.
30. Bolukbasi N, Ersanlı S, Keklikoglu N, Basegmez C, Ozdemir T. Sinus augmentation with platelet-rich fibrin in combination with bovine bone graft versus bovine bone graft in combination with collagen membrane. J Oral Implantol 2015;41:586–595.
31. Cömert Kılıç S, Güngörmüş M, Parlak SN. Histologic and histomorphometric assessment of sinus-floor augmentation with beta-tricalcium phosphate alone or in combination with pure-platelet-rich plasma or platelet-rich fibrin: A randomized clinical trial. Clin Implant Dent Relat Res 2017;19:959–967.
32. Nizam N, Eren G, Akcalı A, Donos N. Maxillary sinus augmentation with leukocyte and platelet-rich fibrin and deproteinized bovine bone mineral: A split-mouth histological and histomorphometric study. Clin Oral Implants Res 2018;29:67–75.
33. Pichotano EC, de Molon RS, de Souza RV, Austin RS, Marcantonio E, Zandim-Barcelos DL. Evaluation of L-PRF combined with deproteinized bovine bone mineral for early implant placement after maxillary sinus augmentation: A randomized clinical trial. Clin Implant Dent Relat Res 2019;21:253–262.
34. Gassling V, Purcz N, Braesen JH, et al. Comparison of two different absorbable membranes for the coverage of lateral osteotomy sites in maxillary sinus augmentation: A preliminary study. J Craniomaxillofac Surg 2013;41:76–82.
35. Bosshardt DD, Bornstein MM, Carrel JP, Buser D, Bernard JP. Maxillary sinus grafting with a synthetic, nanocrystalline hydroxyapatite-silica gel in humans: Histologic and histomorphometric results. Int J Periodontics Restorative Dent 2014; 34:259–267.
36. Gurler G, Delilbasi C. Effects of leukocyte-platelet rich fibrin on postoperative complications of direct sinus lifting. Minerva Stomatol 2016;65:207–212.
37. Liu R, Yan M, Chen S, Huang W, Wu D, Chen J. Effectiveness of platelet-rich fibrin as an adjunctive material to bone graft in maxillary sinus augmentation: A meta-analysis of randomized controlled trails. Biomed Res Int 2019;2019:7267062.

38. Miron RJ, Zucchelli G, Pikos MA, et al. Use of platelet-rich fibrin in regenerative dentistry: A systematic review. Clin Oral Investig 2017;21:1913–1927.
39. Hernández-Alfaro F, Torradeflot MM, Marti C. Prevalence and management of Schneiderian membrane perforations during sinus-lift procedures. Clin Oral Implants Res 2008;19:91–98.
40. Schwartz-Arad D, Herzberg R, Dolev E. The prevalence of surgical complications of the sinus graft procedure and their impact on implant survival. J Periodontol 2004;75:511–516.
41. Pikos MA, Miron RJ. Bone Augmentation in Implant Dentistry: A Step-by-Step Guide to Predictable Alveolar Ridge and Sinus Grafting. Chicago: Quintessence, 2019.
42. Castro AB, Meschi N, Temmerman A, et al. Regenerative potential of leucocyte-and platelet-rich fibrin. Part B: Sinus floor elevation, alveolar ridge preservation, and implant therapy. A systematic review. J Clin Periodontol 2016;44:225–234.
43. Chen TW, Chang HS, Leung KW, Lai YL, Kao SY. Implant placement immediately after the lateral approach of the trap door window procedure to create a maxillary sinus lift without bone grafting: A 2-year retrospective evaluation of 47 implants in 33 patients. J Oral Maxillofac Surg 2007;65:2324–2328.
44. Lambert F, Léonard A, Drion P, Sourice S, Layrolle P, Rompen E. Influence of space-filling materials in subantral bone augmentation: Blood clot vs. autogenous bone chips vs. bovine hydroxyapatite. Clin Oral Implants Res 2011;22:538–545.
45. Thor A, Sennerby L, Hirsch JM, Rasmusson L. Bone formation at the maxillary sinus floor following simultaneous elevation of the mucosal lining and implant installation without graft material: An evaluation of 20 patients treated with 44 Astra Tech implants. J Oral Maxillofac Surg 2007;65:64–72.
46. Duan DH, Fu JH, Qi W, Du Y, Pan J, Wang HL. Graft-free maxillary sinus floor elevation: A systematic review and meta-analysis. J Periodontol 2017;88:550–564.
47. Avila G, Wang HL, Galindo-Moreno P, et al. The influence of the bucco-palatal distance on sinus augmentation outcomes. J Periodontol 2010;81:1041–1050.
48. Ali S, Bakry SA, Abd-Elhakam H. Platelet-rich fibrin in maxillary sinus augmentation: A systematic review. J Oral Implantol 2015;41:746–753.
49. Everts P, Overdevest EP, Jakimowicz JJ, et al. The use of autologous platelet-leukocyte gels to enhance the healing process in surgery, a review. Surg Endosc 2007;21:2063–2068.
50. Hoaglin DR, Lines GK. Prevention of localized osteitis in mandibular third-molar sites using platelet-rich fibrin. Int J Dent 2013; 2013:875380.

18

Use of PRF in Oral and Maxillofacial Surgery

Today, the use of PRF has been growing in the field of oral and maxillofacial surgery. Many studies have now demonstrated the effectiveness of local application of PRF as an adjunctive therapy to surgical protocols for the management of various common defect types, such as osteonecrosis of the jaw (ONJ). The first aim of this chapter is therefore to present the many research articles reporting the clinical use of PRF for the surgical management of ONJ for the healing of bone and/or soft tissue at the surgical site. These findings report on over 200 patients with a follow-up ranging from 1 to 76 months, where the application of PRF has led to an 87% complete defect healing rate, significantly better than previously reported therapies. Furthermore, this chapter also highlights the use of PRF for other related oral and maxillofacial surgery applications such as temporomandibular joint (TMJ) disorders, bone defects after cystectomy or tumor resection, alveolar cleft/palate bone defects, and oroantral communications. While there remains a shortage of RCTs that fully validate the various treatment strategies using PRF in the field of oral and maxillofacial surgery, there is growing evidence that PRF is a cost-effective regenerative modality for various difficult-to-manage cases due to its ability to speed vascularization and also enhance soft tissue defects.

Contributors

Richard J. Miron
Masako Fujioka-Kobayashi
Stefan Gerber
Arun K. Garg
Michael A. Pikos
Maria Elisa Galarraga-Vinueza
David Lee Hill
Delia Tuttle
Nikola Saulacic
Benoît Schaller

Chapter Highlights

- *Use of PRF for the management of ONJ*
- *Use of liquid-PRF for TMJ injections*
- *Use of PRF for other surgical applications in the maxillofacial region*

In 2003, Marx first observed various maxillary lesions specifically associated with bisphosphonate-treated patients and subsequently termed this new type of defect as *bisphosphonate-related osteonecrosis of the jaw* (BRONJ).[1] These defects were difficult to manage, presenting necrotic exposed bone with obvious soft tissue defects (Fig 18-1).[2] ONJ presents some of the most difficult cases to treat for oral surgeons, and the number of cases presenting this condition has continued to rise among the population.

Bisphosphonates are a group of medications that slow bone loss due to their antiosteoclastic activity.[3,4] In particular, however, they also act as potent inhibitors of angiogenesis and significantly reduce tissue oxygenation, favoring invasion of oral anaerobic bacteria in deeper planes.[5] This constitutes the largest and most prevalent group of bacteria in opportunistic infections, and the management of such infections remains complex. Both in vitro and in vivo studies have shown convincingly that bisphosphonates inhibit angiogenesis, decrease capillary tube formation, and reduce VEGF levels, all leading to poor regenerative potential and an increased rate of complications.[6,7]

Due to the growing number of cases over a decade ago, the American Association of Oral and Maxillofacial Surgeons (AAOMS) first published a position paper in 2007 on the topic.[8] Additional medications (such as denosumab) launched with a completely different mode of action (antibody against RANKL)[9,10] also led to ONJ.[3]

Today, *medication-related osteonecrosis of the jaw* (MRONJ) is defined clinically as necrotic bone exposure, but the following characteristics must also be present: *(1)* actual or previous treatment using antiresorptive or antiangiogenic agents; *(2)* exposed bone or bone that can be probed through an intraoral and/and extraoral fistula that persists for more than 8 weeks; and *(3)* no history of radiotherapy or evident metastatic lesions in the jaw. Clinical manifestations of necrotic bone but without bone exposure, such as deep periodontal pockets, loose teeth, trismus, lower lip hypoesthesia/numbness, and nonodontogenic pain, are classified as "unexposed MRONJ."[11,12]

Currently, there is no specific gold standard treatment protocol for MRONJ. Often, the choice of treatment remains a conservative approach including local debridement, bone sequestrum removal, and a combination of systemic antibiotic treatment and/or antiseptic solutions (eg, chlorhexidine). Nevertheless, several studies have demonstrated the success of early-stage surgical therapy and have further shown that in all stages of necrosis, surgical treatment is more effective when compared to conservative approaches.[13,14] While a number of therapies have been proposed for the management of MRONJ, none provide full or optimal defect resolution.[15] Furthermore, recurrence of defects is high.[14,16,17]

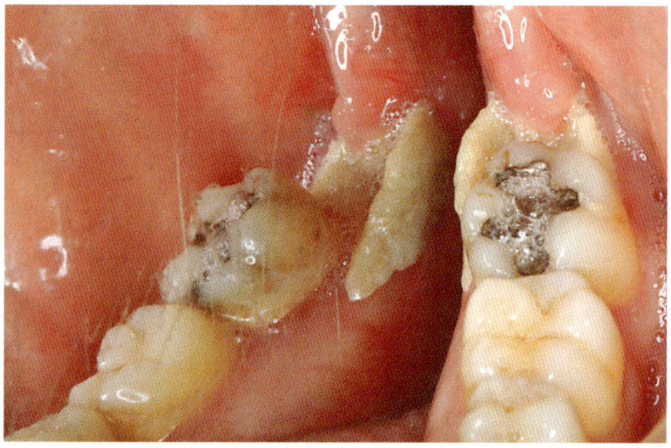

Fig 18-1 Presentation of a typical ONJ case. Necrotic bone is visualized with an open soft tissue defect. (Reprinted with permission from Marx.[2])

In recent years, adjunctive therapeutic protocols have been proposed to enhance the healing of MRONJ and the predictability of surgical treatment, including the use of low-level laser therapy,[18] ozone therapy,[19] hyperbaric oxygen,[20] and autologous platelet concentrates.[11,21,22] Of these groups, platelet concentrates including PRP have been utilized most frequently and have demonstrated improved outcomes in a number of case reports and case series.[5,11,21–24] In particular, PRF has been shown to provide superior biologic potential due to its fibrin clot formation, leading to the entrapment of platelets and leukocytes as well as a longer and greater release of growth factors over time.[11,22–35] In general, the addition of platelet concentrates provides the primary objective of adding supraphysiologic doses of autologous blood-derived growth factors to the surgical site, favoring faster angiogenesis of the defect site caused by ONJ. Furthermore, PRF favors soft tissue wound healing and closure, which are particularly relevant for the management of ONJ.

Evaluation of Clinical Studies Using PRF for ONJ Management

Tables 18-1 and 18-2 present a series of 15 studies investigating the use of PRF for the management and

Table 18-1 RCTs evaluating the effectiveness of PRF in treating ONJ

Authors (year)	Model and number of osteoporosis vs metastasis users	Medication	Healing period	Centrifugation protocol (rpm × min)	Findings/conclusions
Kim et al[28] (2014)	Case series; 34 patients evaluated: 32 for osteoporosis, 2 for bony metastases	Alendronate, risedronate, pamidronate, zoledronate	6 months	3000 × 10	26 patients (77%) showed complete resolution, 6 (18%) had delayed resolution, and 2 (6%) did not resolve at all. This study may contribute to a better understanding based on the biologic effects of L-PRF but cannot judge the effectiveness of L-PRF because there was no control group and only a small sample.
Dinca et al[32] (2014)	Case series; 10 patients all taking bisphosphonates for metastasis	Zoledronic acid, ibandronate	1 month	1300 × 14	PRF led to favorable clinical improvement in recurrent BRONJ.
Norholt et al[22] (2016)	Case series; 15 patients: 7 patients for osteoporosis and 8 for malignant disease	Alendronate, ibandronate, pamidronate, denosumab	7–20 months	1300 × 14	The outcome of the surgical treatment was successful in 14 of the 15 patients (93%).
Park et al[34] (2017)	Clinical trial; 55 patients: 25 patients were treated with L- PRF alone, and 30 patients were treated with L-PRF and rhBMP-2 (48 for osteoporosis and 7 for malignant disease)	Alendronate zoledronic acid	4 months	3000 × 10	The simultaneous application of L-PRF and BMP-2 effectively contributed to the successful treatment outcome of MRONJ. The combined treatment led to the early resolution of MRONJ.
Inchingolo et al[26] (2017)	Case series; 23 patients	Bisphosphonates (not specifically reported)	30 days	3000 × 10	The results of this study must be cautiously interpreted due to the limited sample size. The technique could represent a safe alternative approach aimed at treating this increasing osteoimmunology pathology.
Giudice et al[31] (2018)	Clinical trial; 47 patients: 24 treated with PRF and 23 treated without PRF (12 patients for osteoporosis and 35 for malignancy)	Alendronate, ibandronate, zoledronic acid, denosumab	12 months	1300 × 8	The present study (a small study sample, monocentric study, and heterogenous drug administrations) results are not sufficient to establish an objective advantage in the use of PRF as a treatment option in the surgical management of MRONJ.
Valente et al[25] (2019)	Retrospective study; 15 patients: 8 patients for osteoporosis and 7 for metastasis reasons	Zoledronate, denosumab, alendronate, ibandronate	6–74 months	Not reported	Total resolution in 11 of 15 cases in which it was used, showing potential benefits in the resolution of MRONJ to be investigated in future studies.
Szentpeteri et al[15] (2020)	Retrospective study; 101 patients: 73 control therapy, 28 therapy with PRF (15 patients for osteoporosis and 86 for metastasis reasons)	Bisphosphonates (not specifically reported)	12 months	3000 × 8	The results in the treatment group (with PRF) were significantly better than those in the control group regarding recovery (P = .022), stage improvement (P = .005), and relapse rate (P < .001).
Mourão et al[11] (2020)	Case series; 11 patients, all for osteoporosis	Alendronate	12–36 months	2700 × 12	The PRF membranes were effective to close the soft tissue and to remove the pain in MRONJ, being regarded as a promising alternative for MRONJ treatment.

treatment of ONJ.[11,15,22–34] In a recent systematic review of the literature, nine studies were selected for qualitative analysis (see Table 18-1),[11,15,22,25,26,28,31,32,34] and the others were presented as case reports (see Table 18-2). Of the nine studies included for quantitative evaluation (two clinical trials, two retrospective studies, and five case series), 238 patients were treated with PRF with a follow-up time ranging from 1 to 76 months. Overall, the mandible was most affected (72.9% of cases), and an 82.0% success rate was observed. The two most common prescribed drugs were alendronate (~41.0%) and zoledronate (~24.0%). Females were more affected than males in a 3:1 ratio.

Table 18-2 Case reports evaluating the use of PRF in treating ONJ

Authors (year)	Aim	No. of cases	Disease	Medication	Treatment	Follow-up	Outcomes/ conclusions
Soydan et al[29] (2014)	To describe the treatment of moderate bone exposure with a PRF membrane in MRONJ	1	Oncologic disease	Zoledronate	Removal of necrotic bone and application of PRF membranes	1 and 6 months	PRF membranes are an easy, cheap, and rapid alternative treatment approach for the closure of bone exposure in MRONJ.
Tsai et al[24] (2016)	To describe a protocol used for an MRONJ patient	1	Osteoporosis	Alendronate and zoledronate	Surgical debridement, sequestrectomy, and simultaneous application of PRF membranes and clots	10 days and regularly until 10 months	The application of PRF for MRONJ also revealed total bone closure and new bone regeneration.
Kim et al[30] (2016)	To describe the treatment of MRONJ with rhBMP-2 and PRF	1	Osteoporosis	Alendronate	Surgical debridement, sequestrectomy, and simultaneous application of rhBMP-2 and PRF membranes	11 weeks	Adjunctive simultaneous application of both PRF and rhBMP-2 was effective for the management of MRONJ.
Maluf et al[23] (2016)	To describe the response to L-PRF combined with bone resection as a treatment for MRONJ	2	Oncologic diseases	Denosumab	L-PRF membrane was inserted into the viable bone tissue immediately after debridement	4 and 6 months	Based on the literature and the findings of the reported cases, we can conclude that the use of L-PRF can be considered in the treatment of patients with ONJ related to denosumab.
Maluf et al[27] (2018)	To describe the response to L-PRF combined with bone resection as a treatment for MRONJ	2	Oncologic diseases	Zoledronate	L-PRF membrane was inserted into the viable bone tissue immediately after debridement	Regularly after surgery until 29 and 52 months, respectively	The combination of surgical resection and L-PRF may be advantageous in the treatment of MRONJ, resulting in less morbidity for patients.
Sahin et al[33] (2019)	Surgical management of ONJ observed at the posterior maxilla, using ultrasonic piezoelectric bone surgery combined with L-PRF and pedicled buccal fat pad flap	1	Osteoporosis	Denosumab	Surgical debridement, sequestrectomy, and simultaneous application of L-PRF covered with a pedicled buccal fat pad flap	12 months	The use of ultrasonic bone surgery in combination with L-PRF and pedicled buccal fat pad flap is an alternative treatment method that can be effective in exposed bone coverage and hard and soft tissue healing at the posterior maxillary region following the discontinuation of medication in ONJ patients.

Interestingly, one study reported a combination therapy using PRF with rhBMP-2, a known and potent osteoinductive growth factor.[30,34] In that RCT, a 96.7% success rate was observed with rhBMP-2 when compared to 88.0% with PRF alone. Thus, it was hypothesized that rhBMP-2 may further improve healing in more difficult-to-treat cases due to its greater osteoinductive potential.[34]

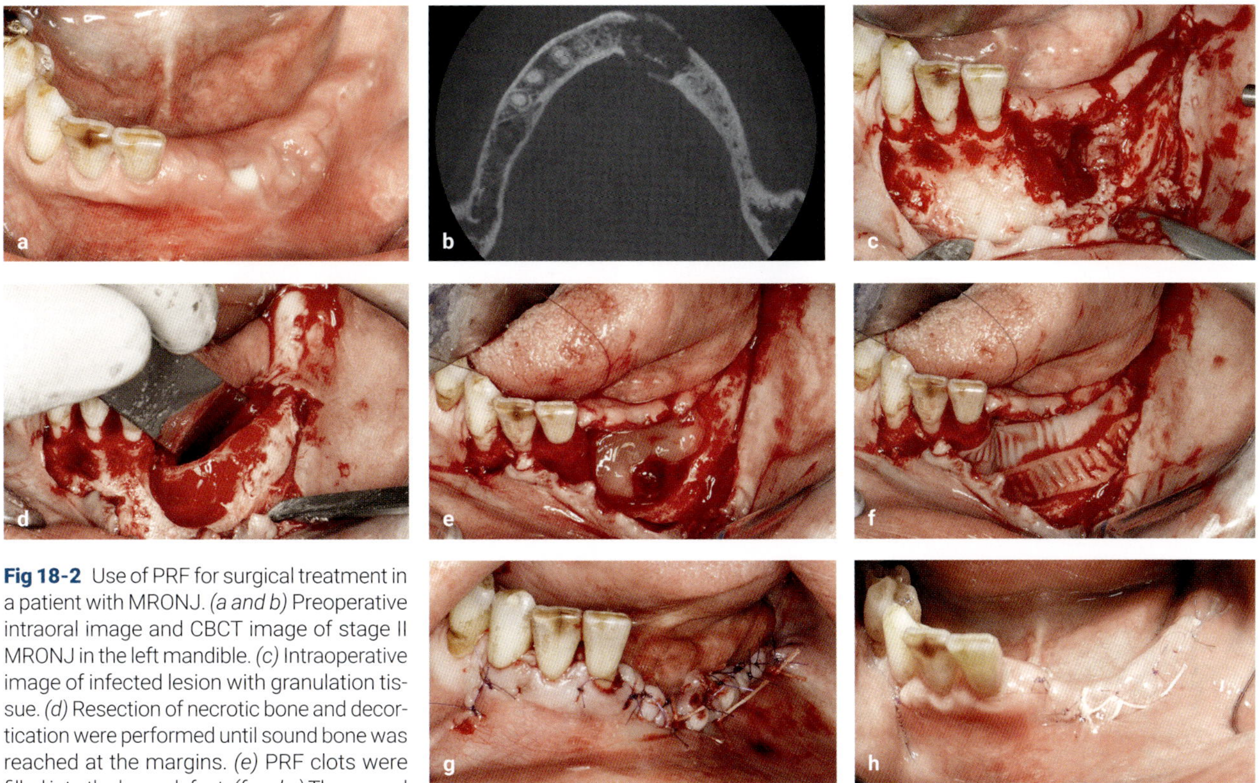

Fig 18-2 Use of PRF for surgical treatment in a patient with MRONJ. *(a and b)* Preoperative intraoral image and CBCT image of stage II MRONJ in the left mandible. *(c)* Intraoperative image of infected lesion with granulation tissue. *(d)* Resection of necrotic bone and decortication were performed until sound bone was reached at the margins. *(e)* PRF clots were filled into the bone defect. *(f and g)* The wound was covered with PRF membranes and sutured. *(h)* Intraoral image at 10 days postsurgery. (Case performed by Professor Benoît Schaller.)

Clinical bottom line

1. Using PRF in conjunction with standard treatment therapies of ONJ leads to improved success rates at roughly 90%.
2. Still, defect resolution is not 100%, and recurrence may occur.
3. Further clinical evaluation is needed to determine precisely why complications occur and what surgical techniques may further enhance outcomes.
4. rhBMP-2 may lead to further improvements, and future research investigating when this additional growth factor may further benefit such cases may serve valuable in future research.

Clinical Procedure

Patients should be treated conservatively before surgery with antibiotics, analgesics, daily irrigation using an antiseptic solution, and professional dental prophylaxis 1 week prior to surgery.

The surgical procedure is performed using local anesthesia with or without intravenous sedation or general anesthesia, depending on the extent of the lesion and the patient's general condition (Fig 18-2). Bone sequestration and granulation tissue should be removed using surgical curettes until fresh bleeding is confirmed from bone. Rotary instruments should be used to smooth all sharp bony margins and remove debris and foreign bodies, followed by copious saline irrigation.

For the preparation of PRF, the peripheral blood must be collected into glass tubes without anticoagulant and immediately centrifuged at 700g for 8 minutes.[36,37] Then, PRF is placed over the defect area following surgical debridement prior to soft tissue closure. It is generally recommended to utilize a larger custom-sized membrane (Bio-Graft), as reviewed in chapter 10. Tension-free primary closure of the mucoperiosteal flap is needed.

For postoperative management, an antibacterial mouthrinse is used in all patients until complete healing is confirmed.

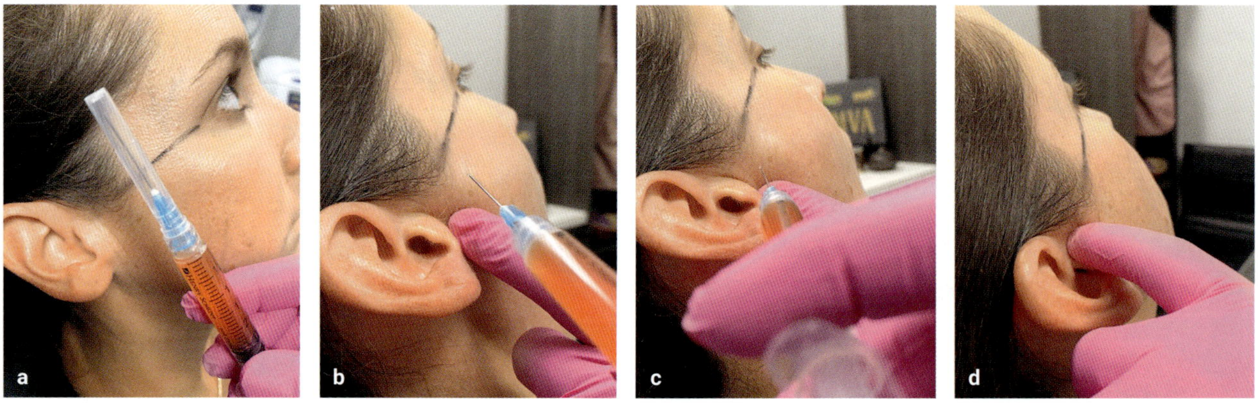

Fig 18-3 TMJ injection with liquid-PRF. *(a)* Line drawn from the corner of the eye to the tragus of the ear. *(b)* Feeling by palpation the condyle of the mandible. *(c)* Injection of liquid-PRF into the TMJ space. *(d)* Compression and massage to the area. See Video 18-1 for complete case presentation. (Case performed by Dr Delia Tuttle.)

Video 18-1

Use of PRF for the Management of TMJ Disorders

The TMJ is a key component of the stomatognathic system and is built as a biarthrodial joint combining a hinge joint and a gliding joint. These two joint components are separated by the disc and its posterior and anterior attachment that form the recesses. These are lined with a metabolically active tissue, known as the *synovium*, that secretes the synovial fluid containing hyaluronic acid and proteoglycans. It provides nutrition and oxygenation of the avascular articular tissues of the TMJ and plays an important role in joint lubrication.[38] It has been shown that biosynthesis of hyaluronic acid in synovial fluid is inhibited by temporary hypoxia due to increased intra-articular pressure that exceeds the capillary perfusion in overloaded joints (eg, parafunctional habits).[38,39] The collapse of the lubrication system causes an increase in joint friction between the joint surface and the disc,[40,41] resulting in degenerative changes including chronic inflammation with subsequent fibrous adhesion and disruption of cartilage (chondromalacia).

Disorders of the TMJ (ie, TMDs) are common types of musculoskeletal conditions found in the orofacial region involving the masticatory muscles, TMJ, and associated structures. The typical features are pain, restriction of mandibular movement, TMJ clicking/popping, and potential facial deformities. Currently, the primary goals of treatment for TMDs are to increase the normal range of mandibular motion and relieve the functional pain in the TMJ itself. Conservative treatments for TMDs include physiotherapy, the use of a stabilization splint, and medications to relieve symptoms; surgical interventions include arthrocentesis, disc repositioning, or discectomy for patients with resistant internal derangement.[42,43] For patients with persisting symptoms or severe degeneration, minimally invasive treatments may be applied, such as lavage, hyaluronic acid/corticosteroid injections, arthrocentesis, and arthroscopy.[43]

The treatment protocol for such TMDs consists of a conservative treatment regime including counseling, physiotherapy, dental splint, nonsteroidal anti-inflammatory gel, and soft diet. If there is no or insufficient improvement after a treatment period of 3 to 6 months, TMJ arthroscopy, injection of hyaluronic acid or PRF, and various combination approaches may be utilized (Fig 18-3 and Video 18-1).

TMJ arthroscopy was first described by Onishi in 1975.[44] Nitzan et al[45] described a more refined technique with two injection needles to be introduced in the upper joint space, which permitted enhanced lavage of the joint as well as aspiration and injection. Lavage of the upper joint space forces the disc apart from the fossa, washes away inflammatory mediators and degraded particles, and decreases the pressure in the joint. This may further release intra-articular adhesions. Apart from lysis and lavage (level I arthroscopy), arthroscopic surgery can also be performed in the TMJ, necessitating additional ports for instruments (level II and III arthroscopy). Intra-articular administration of drugs can also be performed in combination with arthroscopy or alone. Different drugs such as steroids, opioids, hyaluronic acid, and NSAIDs have been widely studied.[46–48] However, none of them seems to be superior over the other.

For these reasons, recent trends in the field have shifted to utilizing regenerative strategies that are more natural by improving angiogenesis with PRP as a first-generation and PRF as a second-generation blood concentrate. The combination of TMJ arthroscopy and infiltration of PRP was previously shown to be most effective for short- and intermediate-term improvement for arthrogenous TMJ disorders compared to hyaluronic acid or arthroscopy alone.[49] Nevertheless, PRP lacks leukocytes from blood concentrates as well as an ability to form a stable fibrin clot after injection, both of which can play important roles in growth factor release and contribute to different phases of wound healing.

Albilia et al investigated the analgesic effects of injectable-PRF (i-PRF) in 48 TMJs in 37 patients with painful internal derangement (ID; Wilkes I–V).[50] Patients were injected with 1.5 to 2 mL of i-PRF within the superior joint space at 2-week intervals. Interestingly, 69% of cases showed significant reduction in pain at 8 weeks. The authors also suggested i-PRF injections were more effective for the patients with ID stages Wilkes IV and V when compared to Wilkes I to III.[50]

Advanced treatment protocol for TMJ injections

The treatment is performed under full anesthesia. On a line between the tragus and the outer canthus (Holmlund-Hellsing line),[51] 10 mm in front of the tragus and 2 mm caudal to this point, the maximum concavity of the glenoid fossa lies,[52] and a 23G needle is inserted in the upper joint space in order to inflate the joint with 2 to 3 mL of ringer lactate (Fig 18-4a). The needle is then removed, the cannula with a pointed trocar is inserted in the upper joint space, and the correct position is confirmed with the arthroscope inserted through the cannula. A 30-degree arthroscope with 1.9 mm diameter is used. A 23G outflow needle is inserted in the upper joint space. After a diagnostic sweep, the joint is washed out with 200 mL of ringer lactate (level I arthroscopy; Fig 18-4b). If necessary, a second port is placed in the anterior recess of the joint in order to perform level II arthroscopy (Fig 18-4c). Through this additional port, a radiofrequency probe is inserted in order to release intra-articular adhesions, to perform anterior myotomy of the medial pterygoid muscle, and for the retrodiscal tissue to be coagulated.

At this point, 20 mL of venous blood is collected from the patient using a butterfly needle set and two liquid PET tubes without additives. The collected blood is then immediately centrifuged. The centrifugation protocol utilized is 300g for 5 minutes or 2000 g for 8 minutes utilizing horizontal centrifugation (Bio-PRF centrifugation machine; Figs 18-4d and 18-4e; see chapter 10). A total of 1 to 2 mL of liquid-PRF is withdrawn into a 2-mL syringe and injected in the upper joint space either through the outflow needle (in level I arthroscopy) or through the arthroscopy port in the case of level II arthroscopy (Figs 18-4f and 18-4g). All instruments are then removed from the joint, and the skin is closed with nonresorbable sutures. The patient is advised to maintain a soft diet for 4 weeks.

In summary, intra-articular injections of PRF in the TMJ as part of level I and II TMJ arthroscopy aim for a shift from catabolic to anabolic state, with liquid-PRF allowing the joint to remodel damaged synovial, cartilage, and bone surfaces. More evidence from RCTs certainly remains needed to prove the effectiveness of liquid-PRF injection into TMJ spaces; nevertheless, this protocol represents a novel minimally invasive option for the relief and improvement of patient-reported TMJ pain.

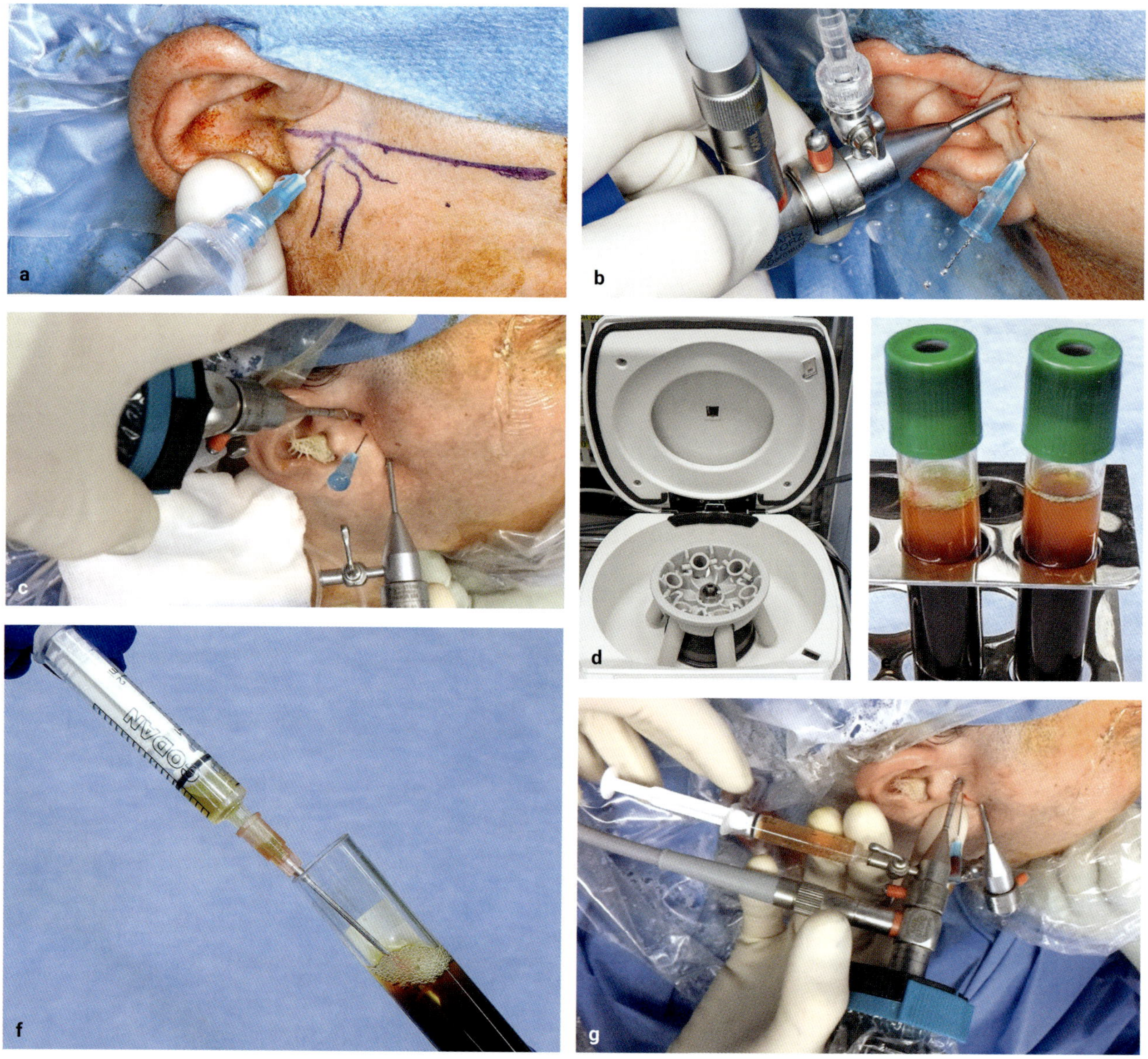

Fig 18-4 *(a)* Landmarks for TMJ puncture are mapped and insufflation of the upper joint space is performed with a 23G needle. *(b)* Level I arthroscopy with the arthroscope and an outflow needle in the upper joint space for irrigation with ringer lactate. *(c)* Level II arthroscopy with a second port used as a working cannula in the anterior recess. A radiofrequency unit is used for tissue coagulation and cutting. *(d)* Bio-PRF centrifuge machine used for horizontal centrifugation. *(e)* The yellowish liquid plasma layer/fraction represents the PRF after centrifugation. *(f)* This layer is withdrawn into a 2-mL syringe for intra-articular injection. *(g)* Injection of liquid-PRF through the arthroscopy cannula in level II arthroscopy. (Case performed by Dr Stefan Gerber.)

Use of PRF in Other Areas of Oral and Maxillofacial Surgery

PRF has also been utilized in other areas of oral and maxillofacial surgery, including for the treatment of bone defects after cystectomy (Fig 18-5) or tumor resections as well as for filling large voids,[53,54] cleft alveolar bone defects, or oroantral communications.

Often, the surgical site may benefit from the angiogenic potential of PRF and/or improved healing potential, for example following large bone voids as well as for relatively straightforward applications. Figure 18-6 demonstrates a case where autogenous bone was harvested using a rotary bone harvester. Note that following harvesting of bone, a bone void is typically created and often left unfilled. One application of PRF is simply to place PRF membranes

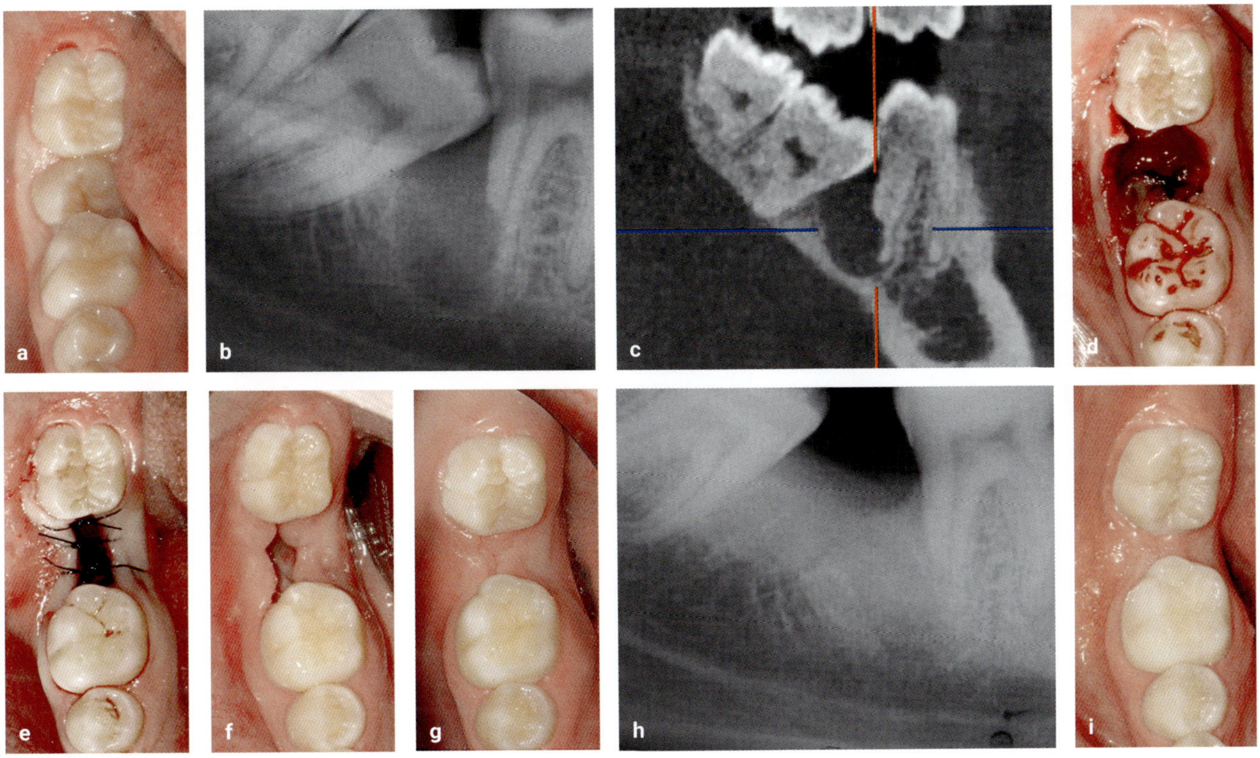

Fig 18-5 Treatment of a mandibular cyst with PRF. *(a)* Clinical view of a mandibular second molar out of occlusion. *(b)* Periapical radiograph demonstrating a large cyst, confirmed by CBCT *(c)*. *(d)* Tooth extraction followed by debridement of the area. *(e)* Placement of a PRF clot followed by suture closure. *(f)* Soft tissue healing at 2 weeks. *(g)* Soft tissue healing at 6 months. *(h)* Periapical radiograph demonstrating complete bone fill. *(i and j)* Clinical and radiographic images at 2 years demonstrating complete case resolution. The adjacent first molar responded positively to sensitivity testing. (Case performed by Dr Nikola Saulacic.)

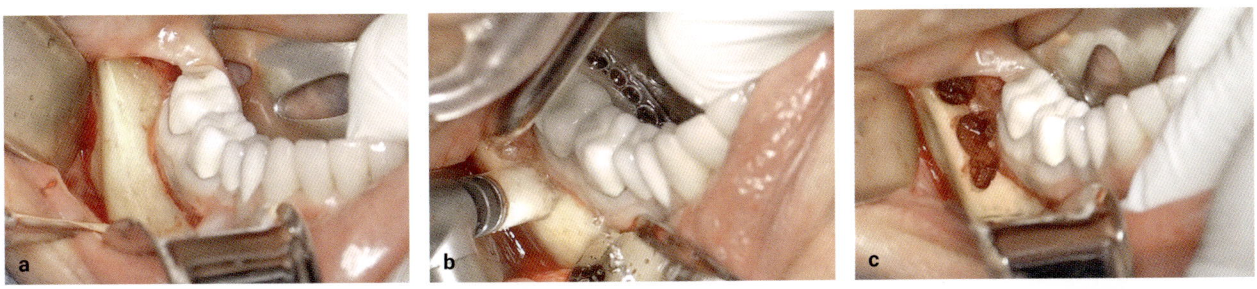

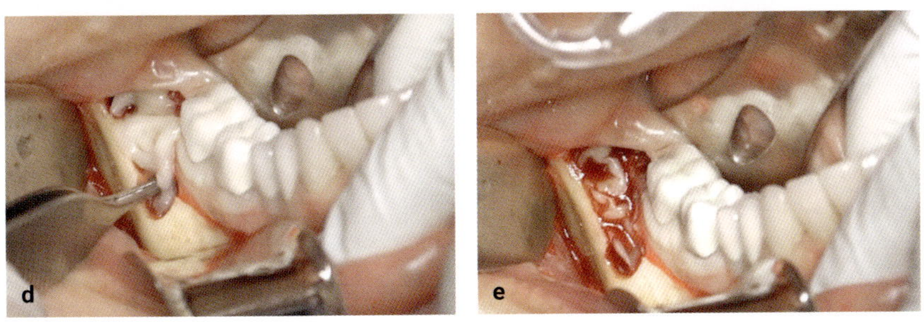

Fig 18-6 Use of PRF for the filling of a bone void following autogenous bone collection harvested with a rotary bone harvester system. *(a)* Clinical image of the ramus. *(b)* Use of a rotary bone harvester to collect autogenous bone. *(c)* Remaining bone void following autogenous bone collection. *(d)* Placement of a PRF membrane into the bone void. *(e)* Final placement prior to suturing. (Case performed by Dr Michael A. Pikos.)

Video 18-2

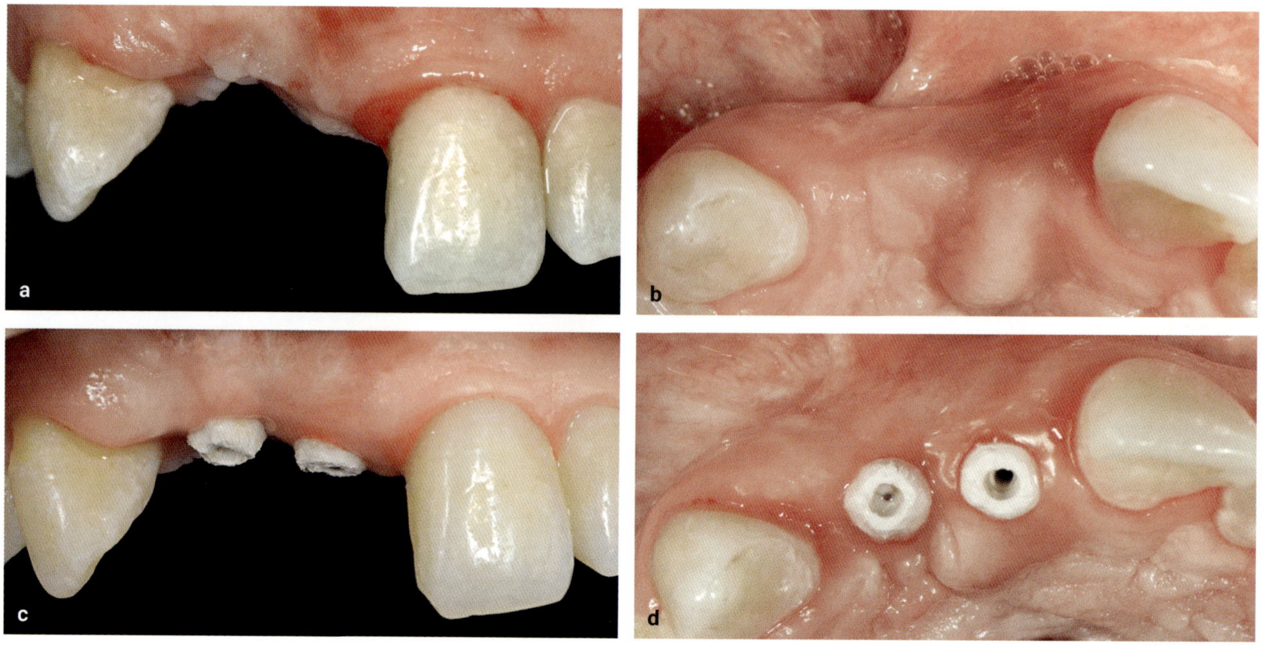

Fig 18-7 *(a and b)* Frontal and occlusal intraoral views of the patient's initial situation showing absence of the maxillary right central and lateral incisors with horizontal and vertical tissue deficiency due to a cleft lip and palate birth defect. *(c and d)* Frontal and occlusal intraoral views 6 months after implant placement surgery exhibiting the polyetheretherketone (PEEK) personalized healing abutments at each implant site and a significant horizontal tissue gain at the buccal aspect of both implants.

within the bone defect to support healing and decrease postoperative pain (see Fig 18-6d). Such applications can be utilized following harvesting of autogenous bone (either as demonstrated in Fig 18-6 or as bone blocks) or for other areas in which bone voids occur to minimize patient morbidity.

For the reconstruction of the alveolar process in patients with cleft lip and/or palate, PRF can also be utilized as an adjunctive option (Fig 18-7). Several RCTs have thus far investigated PRF for such applications where patients underwent a maxillary alveolar cleft reconstruction with or without PRF.[55–59] Of these, four studies investigated the efficacy of PRF with autogenous anterior iliac bone graft when compared to iliac bone alone.[56–59] One study showed that PRF did not result in a significant increase in the thickness, height, and density of the alveolar bone graft,[57] whereas the other three studies suggested that PRF favored higher amounts of newly formed bone.[56,58,59]

Another application of PRF in oral and maxillofacial surgery has been the use of PRF membranes/plugs for flap-free treatment of oroantral perforations after tooth extraction (Fig 18-8).[60–62] While oroantral communications with larger perforations may require a mucoperiosteal buccal sliding flap or buccal fat pad, several case series studies have now demonstrated that PRF can be an alternative option with an ability to be sutured to local tissues due to its 3D architecture.[60–62] In such cases, the use of PRF enables the closure of oroantral communications without primary flap closure, resulting in a less traumatic and easier treatment modality for such limited-size defects.

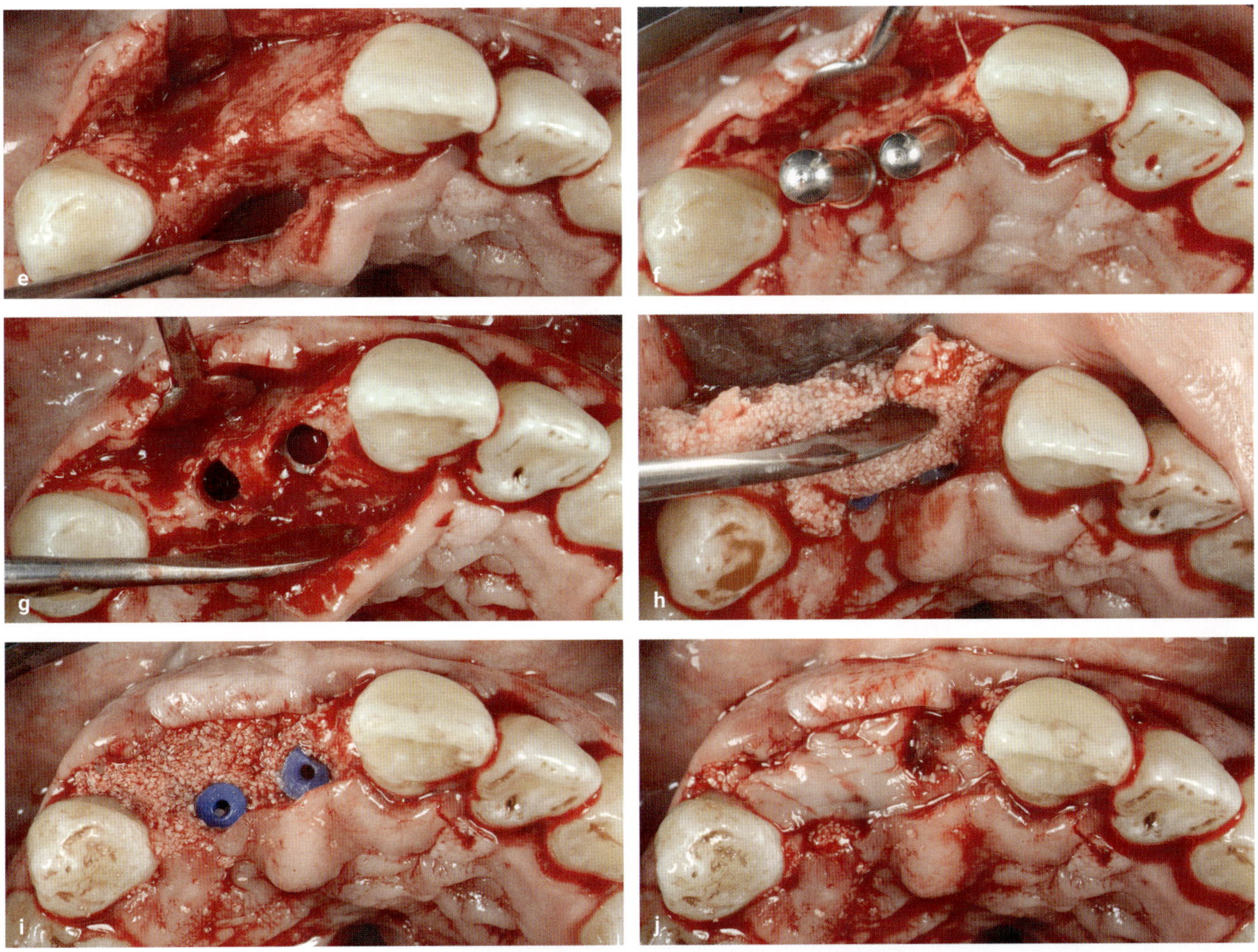

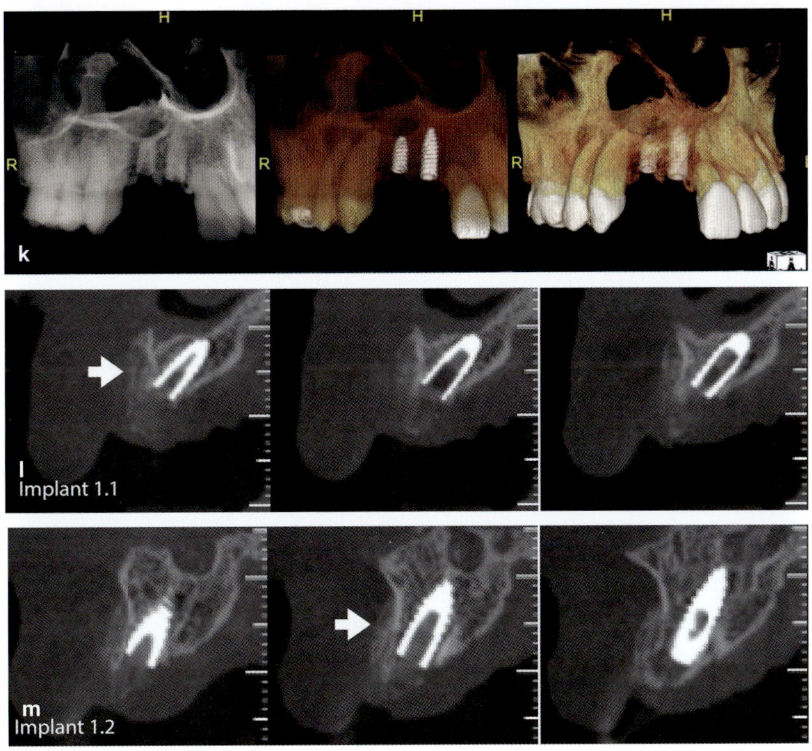

Fig 18-7 *(cont)* *(e to j)* The surgical protocol for implant placement and horizontal ridge augmentation included elevation of buccal and palatal mucoperiostal flaps and bone defect exposure at the right anterior region showing an evident horizontal ridge deficiency, implant placement preserving the buccal wall, and PRF block (sticky bone) application to the buccal and occlusal region, followed by placement of the healing abutments and PRF membranes over the grafted area. *(k)* CBCT scan performed 6 months after surgery showing 3D reconstruction after implant placement. *(l and m)* Implants presenting a satisfactory buccal bone volume (2–3 mm) where the PRF bone graft was placed *(white arrows)*. (Case performed by Dr Maria Elisa Galarraga-Vinueza.)

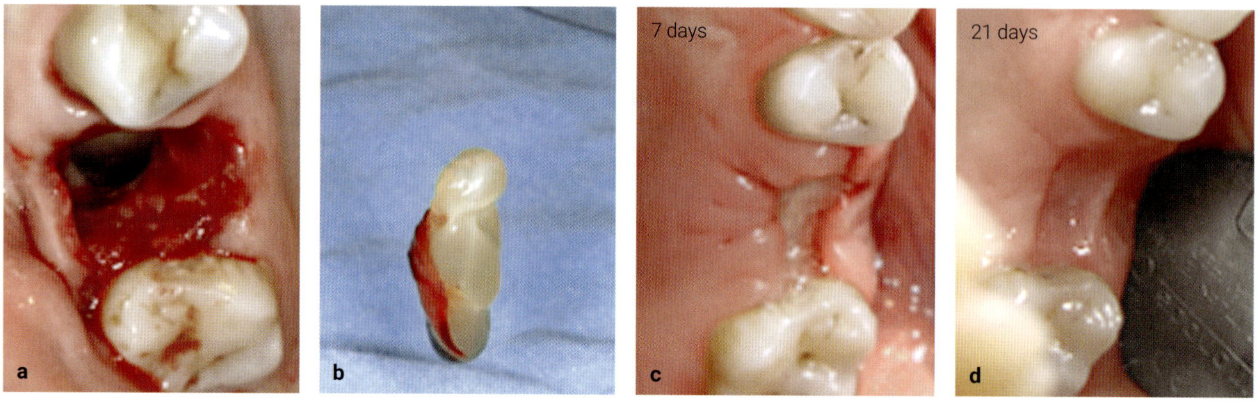

Fig 18-8 *(a to d)* Use of PRF for the healing of an oroantral communication. (Reprinted with permission from Gülşen et al.[60])

Implications for Clinical Practice and Future Directions

In this chapter, we specifically focused on two growing avenues of use for PRF: *(1)* for the management of ONJ and *(2)* for the management of TMDs. To date, no systematic review on either topic has been presented. It is also impossible to perform a meta-analysis due to the lack of quality RCTs on either topic as well. The majority of studies conducted thus far are case series or case reports without the use of appropriate controls. Furthermore, while most studies with respect to ONJ categorized patients on bisphosphonates as either osteoporosis-related or metastasis-related, no summary or report on complications or differences between high-dose and low-dose medication use was typically reported.

Although PRF has been discussed for decades and its use is widespread especially in dentistry, in the field of oral and maxillofacial surgery, the most common use of PRF in studies is related to the treatment of ONJ. Many of the authors from these studies report the need for better-quality RCTs but comment on the difficulty of executing well-structured clinical trials given the obvious preexisting condition, the clinical severity of the condition, and the absolute need to manage the situation as rapidly and effectively as possible. ONJ may be considered life-threatening in certain instances and almost always presents with severe complications, including the additional loss of teeth, patient morbidity, increased fracture rates, and so forth. For these reasons, primarily retrospective analyses have been reported to date following years of clinical use and observed additional benefits. Future research aims to better understand which situation PRF proves to be valuable for in the treatment of ONJ related to the potential medication dose, use of antibiotics, centrifugation protocols, and disease advancement. Additionally, different classes of drugs are now associated with the onset of ONJ, including bisphosphonates, RANKL inhibitors, and antiangiogenics, which would leave a much more heterogenous sample population whereby, at the present time, it is not possible to determine the gold standard treatment in each of these presented cases.[25,26]

It is also important to note that the use of PRF did not lead to a 100% success rate in the management of ONJ. Within one systematic review, 31 cases resulted in nonresolution of ONJ. Of these cases, ~92% occurred in the mandible. Unfortunately, there was no way to clarify or determine the reasons for each of these failures, and much research interest exists to improve patient care in future cases. Different possibilities may have led to failure of the procedure, including adjuvant antibiotic therapy, topical chlorhexidine use, or various other modalities utilized by specific clinician preferences.[11,22–32] For example, the use of chlorhexidine application, while effective for the reduction of incoming pathogens, is also detrimental to host fibroblasts (especially when used over extended periods of time) and may contribute to graft failure.[63,64] Several factors for improving treatment could be put forth, including the density of bone in the treatment region, the vascularization of the remaining viable bone, compromised blood supply due to aging and/or medications, and potentially the PRF centrifugation protocols utilized.

All related cases included a surgical phase whereby debridement of necrotic tissue was performed in an attempt to improve blood supply to the affected tissue. The relevance of PRF in the operated site is of particular interest because the collection of blood-derived proteins such as VEGF has been demonstrated to further improve neovascularization. Additional secondary benefits of PRF reported in the literature include its ability to improve soft tissue wound healing and the presence of leukocytes that favor the release of growth factors and local host defense (antimicrobial action).[65–69] For example, it was recently demonstrated that PRF improves host defense against periodontal pathogens, including *Prevotella intermedia*, *Fusobacterium nucleatum*, *Aggregatibacter actinomycetemcomitans*, and especially *Porphyromonas gingivalis*.[65] Because MRONJ is associated with microorganisms such as *Actinomyces israelii*, PRF may provide additional benefits as a local antimicrobial agent. However, future studies are needed to explore/confirm the antimicrobial action of PRF.

Lastly, it may be possible that the use of rhBMP-2 may further be combined with PRF and serve useful for the more difficult-to-treat cases of ONJ due to its superior osteoinductivity. One randomized study did show a significant improvement in healing when rhBMP-2 was combined with PRF.[34] Future research on this topic is warranted.

With respect to the use of PRF in TMJ cases as well as for other applications in oral surgery such as large bone block harvests and large Le Fort type surgeries, future well-conducted clinical studies are needed to further evaluate the benefit of PRF, as these research topics remain in their infancy.

Conclusion

This chapter overviewed the use of PRF in various applications in the field of oral and maxillofacial surgery. PRF is a known enhancer of vascularization and soft tissue healing/closure and therefore may be utilized as a successful adjunctive regenerative strategy for the management of ONJ cases, TMDs, and various other applications such as cleft lip/palate. While there is a shortage of clinical studies (especially RCTs) on the subject, PRF has been shown in over 20 studies to be a cost-effective therapy leading to better success outcomes/defect resolution for the management of various cases in oral surgery. Future studies are needed to provide better evidence supporting its use for the management of ONJ and other applications in oral surgery to provide clinicians with optimal guidelines and clinical recommendations regarding its use.

References

1. Marx RE. Pamidronate (Aredia) and zoledronate (Zometa) induced avascular necrosis of the jaws: A growing epidemic. J Oral Maxillofac Surg 2003;61:1115–1117.
2. Marx RE. Oral and Intravenous Bisphosphonate-Induced Osteonecrosis of the Jaws: History, Etiology, Prevention, and Treatment, ed 2. Chicago: Quintessence, 2011.
3. Ruggiero SL, Dodson TB, Fantasia J, et al. American Association of Oral and Maxillofacial Surgeons position paper on medication-related osteonecrosis of the jaw—2014 update. J Oral Maxillofac Surg 2014;72:1938–1956.
4. Ruggiero SL, Mehrotra B, Rosenberg TJ, Engroff SL. Osteonecrosis of the jaws associated with the use of bisphosphonates: A review of 63 cases. J Oral Maxillofac Surg 2004;62:527–534.
5. De Ponte FS, Favaloro A, Siniscalchi EN, et al. Sarcoglycans and integrins in bisphosphonate treatment: Immunohistochemical and scanning electron microscopy study. Oncol Rep 2013;30: 2639–2646.
6. Ziebart T, Pabst A, Klein MO, et al. Bisphosphonates: Restrictions for vasculogenesis and angiogenesis: Inhibition of cell function of endothelial progenitor cells and mature endothelial cells in vitro. Clin Oral Investig 2011;15:105–111.
7. Kos M, Junka A, Smutnicka D, Szymczyk P, Gluza K, Bartoszewicz M. Bisphosphonates enhance bacterial adhesion and biofilm formation on bone hydroxyapatite. J Craniomaxillofac Surg 2015;43:863–869.
8. American Association of Oral and Maxillofacial Surgeons position paper on bisphosphonate-related osteonecrosis of the jaws. J Oral Maxillofac Surg 2007;65:369–376.
9. Taguchi Y, Kiga Y, Gohda J, Inoue J. Identification and characterization of anti-osteoclastogenic peptides derived from the cytoplasmic tail of receptor activator of nuclear factor kappa B. J Bone Miner Metab 2012;30:543–553.
10. Bogado CE, Zanchetta MB, Boailchuk JA, Massari FE, Zanchetta JR. Denosumab: What's new? Curr Osteoporos Rep 2011;9:12–19.
11. de Almeida Barros Mourão CF, Calasans-Maia MD, Del Fabbro M, et al. The use of platelet-rich fibrin in the management of medication-related osteonecrosis of the jaw: A case series. J Stomatol Oral Maxillofac Surg 2020;121:84–89.
12. Diniz-Freitas M, Fernandez-Feijoo J, Diz Dios P, Pousa X, Limeres J. Denosumab-related osteonecrosis of the jaw following non-surgical periodontal therapy: A case report. J Clin Periodontol 2018;45:570–577.
13. Rupel K, Ottaviani G, Gobbo M, et al. A systematic review of therapeutical approaches in bisphosphonates-related osteonecrosis of the jaw (BRONJ). Oral Oncol 2014;50:1049–1057.
14. Stockmann P, Burger M, von Wilmowsky C, et al. The outcome after surgical therapy of bisphosphonate-associated osteonecrosis of the jaw—Results of a clinical case series with an average follow-up of 20 months. Clin Oral Investig 2014;18:1299–1304.
15. Szentpeteri S, Schmidt L, Restar L, Csaki G, Szabo G, Vaszilko M. The effect of platelet-rich fibrin membrane in surgical therapy of medication-related osteonecrosis of the jaw. J Oral Maxillofac Surg 2020;78:738–748.

16. Fliefel R, Tröltzsch M, Kühnisch J, Ehrenfeld M, Otto S. Treatment strategies and outcomes of bisphosphonate-related osteonecrosis of the jaw (BRONJ) with characterization of patients: A systematic review. Int J Oral Maxillofac Surgery 2015;44: 568–585.
17. Mücke T, Koschinski J, Deppe H, et al. Outcome of treatment and parameters influencing recurrence in patients with bisphosphonate-related osteonecrosis of the jaws. J Cancer Res Clin Oncol 2011;137:907–913.
18. Poli PP, Souza FA, Ferrario S, Maiorana C. Adjunctive application of antimicrobial photodynamic therapy in the prevention of medication-related osteonecrosis of the jaw following dentoalveolar surgery: A case series. Photodiagnosis Photodyn Ther 2019;27:
117–123.
19. Silva LF, Curra C, Munerato MS, et al. Surgical management of bisphosphonate-related osteonecrosis of the jaws: Literature review. Oral Maxillofac Surg 2016;20:9–17.
20. Beth-Tasdogan NH, Mayer B, Hussein H, Zolk O. Interventions for managing medication-related osteonecrosis of the jaw. Cochrane Database Syst Rev 2017;10:CD012432.
21. Fabbro M, Taschieri S, Goker F. Platelet concentrates as an adjunctive therapy for medication-related osteonecrosis of the jaw: A systematic review and meta-analysis. Int J Growth Factors Stem Cells Dent 2018;1:48–57.
22. Norholt SE, Hartlev J. Surgical treatment of osteonecrosis of the jaw with the use of platelet-rich fibrin: A prospective study of 15 patients. Int J Oral Maxillofac Surg 2016;45:1256–1260.
23. Maluf G, Pinho MC, Cunha SR, Santos PS, Fregnani ER. Surgery Combined with LPRF in denosumab osteonecrosis of the jaw: Case report. Braz Dent J 2016;27:353–358.
24. Tsai LL, Huang YF, Chang YC. Treatment of bisphosphonate-related osteonecrosis of the jaw with platelet-rich fibrin. J Formos Med Assoc 2016;115:585–586.
25. Valente NA, Chatelain S, Alfonsi F, Mortellaro C, Barone A. Medication-related osteonecrosis of the jaw: The use of leukocyte-platelet-rich fibrin as an adjunct in the treatment. J Craniofac Surg 2019;30:1095–1101.
26. Inchingolo F, Cantore S, Dipalma G, et al. Platelet rich fibrin in the management of medication-related osteonecrosis of the jaw: A clinical and histopathological evaluation. J Biol Regul Homeost Agents 2017;31:811–816.
27. Maluf G, Caldas RJ, Silva Santos PS. Use of leukocyte- and platelet-rich fibrin in the treatment of medication-related osteonecrosis of the jaws. J Oral Maxillofac Surg 2018;76:88–96.
28. Kim JW, Kim SJ, Kim MR. Leucocyte-rich and platelet-rich fibrin for the treatment of bisphosphonate-related osteonecrosis of the jaw: A prospective feasibility study. Br J Oral Maxillofac Surg 2014;52:854–859.
29. Soydan SS, Uckan S. Management of bisphosphonate-related osteonecrosis of the jaw with a platelet-rich fibrin membrane: Technical report. J Oral Maxillofac Surg 2014;72:322–326.
30. Kim JW, Kim SJ, Kim MR. Simultaneous application of bone morphogenetic protein-2 and platelet-rich fibrin for the treatment of bisphosphonate-related osteonecrosis of jaw. J Oral Implantol 2016;42:205–208.
31. Giudice A, Barone S, Giudice C, Bennardo F, Fortunato L. Can platelet-rich fibrin improve healing after surgical treatment of medication-related osteonecrosis of the jaw? A pilot study. Oral Surg Oral Med Oral Pathol Oral Radiol 2018;126:390–403.
32. Dinca O, Zurac S, Staniceanu F, et al. Clinical and histopathological studies using fibrin-rich plasma in the treatment of bisphosphonate-related osteonecrosis of the jaw. Rom J Morphol Embryol 2014; 55:961–964.
33. Sahin O, Odabasi O, Ekmekcioglu C. Ultrasonic piezoelectric bone surgery combined with leukocyte and platelet-rich fibrin and pedicled buccal fat pad flap in denosumab-related osteonecrosis of the jaw. J Craniofac Surg 2019;30:e434–e436.
34. Park JH, Kim JW, Kim SJ. Does the addition of bone morphogenetic protein 2 to platelet-rich fibrin improve healing after treatment for medication-related osteonecrosis of the jaw? J Oral Maxillofac Surg 2017;75:1176–1184.
35. Şahin O, Aliyev T, Tatar B. Leucocyte-rich and platelet-rich fibrin (L-PRF) for the Treatment of medication-related osteonecrosis of the jaw (MRONJ) with long term follow-up. Biomed Res 2018;2:3.
36. Miron RJ, Xu H, Chai J, et al. Comparison of platelet-rich fibrin (PRF) produced using 3 commercially available centrifuges at both high (~ 700 g) and low (~ 200 g) relative centrifugation forces. Clin Oral Investig 20120;24:1171–1182.
37. Miron RJ, Chai J, Zheng S, Feng M, Sculean A, Zhang Y. A novel method for evaluating and quantifying cell types in platelet rich fibrin and an introduction to horizontal centrifugation. J Biomed Mater Res A 2019;107:2257–2271.
38. Nitzan DW. The process of lubrication impairment and its involvement in temporomandibular joint disc displacement: A theoretical concept. J Oral Maxillofac Surg 2001;59:36–45.
39. Merry P, Williams R, Cox N, King JB, Blake DR. Comparative study of intra-articular pressure dynamics in joints with acute traumatic and chronic inflammatory effusions: Potential implications for hypoxic-reperfusion injury. Ann Rheum Dis 1991;50:917–920.
40. Tanaka E, Iwabe T, Dalla-Bona DA, et al. The effect of experimental cartilage damage and impairment and restoration of synovial lubrication on friction in the temporomandibular joint. J Orofac Pain 2005;19:331–336.
41. Forster H, Fisher J. The influence of loading time and lubricant on the friction of articular cartilage. Proc Inst Mech Eng H 1996;210:109–119.
42. Hancı M, Karamese M, Tosun Z, Aktan TM, Duman S, Savaci N. Intra-articular platelet-rich plasma injection for the treatment of temporomandibular disorders and a comparison with arthrocentesis. J Craniomaxillofac Surg 2015;43:162–166.
43. Bousnaki M, Bakopoulou A, Koidis P. Platelet-rich plasma for the therapeutic management of temporomandibular joint disorders: A systematic review. Int J Oral Maxillofac Surg 2018;47:188–198.
44. Onishi M. Arthroscopy of the temporomandibular joint [in Japanese]. Kokubyo Gakkai Zasshi 1975;42:207–213.
45. Nitzan DW, Dolwick MF, Martinez GA. Temporomandibular joint arthrocentesis: A simplified treatment for severe, limited mouth opening. J Oral Maxillofac Surg 1991;49:1163–1167.
46. Moldez MA, Camones VR, Ramos GE, Padilla M, Enciso R. Effectiveness of intra-articular injections of sodium hyaluronate or corticosteroids for intracapsular temporomandibular disorders: A systematic review and meta-analysis. J Oral Facial Pain Headache;32:53–66.
47. Aktas I, Yalcin S, Sencer S. Intra-articular injection of tenoxicam following temporomandibular joint arthrocentesis: A pilot study. Int J Oral Maxillofac Surg 2010;39:440–445.
48. Brennan PA, Ilankovan V. Arthrocentesis for temporomandibular joint pain dysfunction syndrome. J Oral Maxillofac Surg 2006;64: 949–951.
49. Al-Moraissi EA, Wolford LM, Ellis E 3rd, Neff A. The hierarchy of different treatments for arthrogenous temporomandibular disorders: A network meta-analysis of randomized clinical trials. J Craniomaxillofac Surg 2020;48:9–23.
50. Albilia JB, Vizcaíno CH-, Weisleder H, Choukroun J, Ghanaati S. Liquid platelet-rich fibrin injections as a treatment adjunct for painful temporomandibular joints: Preliminary results. Cranio 2018:1–13.

51. Holmlund A, Hellsing G. Arthroscopy of the temporomandibular joint. An autopsy study. Int J Oral Surg 1985;14:169–175.
52. McCain JP, de la Rua H, LeBlanc WG. Puncture technique and portals of entry for diagnostic and operative arthroscopy of the temporomandibular joint. Arthroscopy 1991;7:221–232.
53. Ghanaati S, Al-Maawi S, Conrad T, Lorenz J, Rossler R, Sader R. Biomaterial-based bone regeneration and soft tissue management of the individualized 3D-titanium mesh: An alternative concept to autologous transplantation and flap mobilization. J Craniomaxillofac Surg 2019;47:1633–1644.
54. Lorenz J, Al-Maawi S, Sader R, Ghanaati S. Individualized titanium mesh combined with platelet-rich fibrin and deproteinized bovine bone: A new approach for challenging augmentation. J Oral Implantol 2018;44:345–351.
55. Francisco I, Fernandes MH, Vale F. Platelet-rich fibrin in bone regenerative strategies in orthodontics: A systematic review. Materials 2020;13:1866.
56. Saruhan N, Ertas U. Evaluating of platelet-rich fibrin in the treatment of alveolar cleft with iliac bone graft by means of volumetric analysis. J Craniofac Surgery 2018;29:322–326.
57. Omidkhoda M, Jahnabin A, Khoshandam F, et al. Efficacy of platelet-rich fibrin combined with autogenous bone graft in the quality and quantity of maxillary alveolar cleft reconstruction. Iran J Otorhinolaryngol 2018;30:329.
58. Desai AK, Kumar N, Dikhit P, Koikude SB, Bhaduri S. Efficacy of platelet-rich fibrin in secondary cleft alveolar bone grafting. Craniomaxillofac Trauma Reconstr Open 2019;3:e43–e50.
59. Shawky H, Seifeldin SA. Does platelet-rich fibrin enhance bone quality and quantity of alveolar cleft reconstruction? Cleft Palate Craniofac J 2016;53:597–606.
60. Gülşen U, Şentürk MF, Mehdiyev İ. Flap-free treatment of an oroantral communication with platelet-rich fibrin. Br J Oral Maxillofac Surg 2016;54:702–703.
61. Agarwal B, Pandey S, Roychoudhury A. New technique for closure of an oroantral fistula using platelet-rich fibrin. Br J Oral Maxillofac Surg 2016;54:e31–e32.
62. Bilginaylar K. The use of platelet-rich fibrin for immediate closure of acute oroantral communications: An alternative approach. J Oral Maxillofac Surg 2018;76:278–286.
63. Sawada K, Fujioka-Kobayashi M, Kobayashi E, Schaller B, Miron RJ. Effects of antiseptic solutions commonly used in dentistry on bone viability, bone morphology, and release of growth factors. J Oral Maxillofac Surg 2016;74:247–254.
64. Sawada K, Nakahara K, Haga-Tsujimura M, Fujioka-Kobayashi M, Iizuka T, Miron RJ. Effect of irrigation time of antiseptic solutions on bone cell viability and growth factor release. J Craniofac Surg 2018;29:376–381.
65. Castro AB, Herrero ER, Slomka V, Pinto N, Teughels W, Quirynen M. Antimicrobial capacity of leucocyte- and platelet rich fibrin against periodontal pathogens. Sci Rep 2019;9:8188.
66. Bayer A, Lammel J, Tohidnezhad M, et al. The antimicrobial peptide human beta-defensin-3 is induced by platelet-released growth factors in primary keratinocytes. Mediators Inflamm 2017;2017:6157491.
67. Bayer A, Lammel J, Rademacher F, et al. Platelet-released growth factors induce the antimicrobial peptide human beta-defensin-2 in primary keratinocytes. Exp Dermatol 2016;25:460–465.
68. Kour P, Pudakalkatti PS, Vas AM, Das S, Padmanabhan S. Comparative evaluation of antimicrobial efficacy of platelet-rich plasma, platelet-rich fibrin, and injectable platelet-rich fibrin on the standard strains of *Porphyromonas gingivalis* and *Aggregatibacter actinomycetemcomitans*. Contemp Clin Dent 2018;9(suppl 2): S325–S330.
69. Hansen T, Kunkel M, Springer E, et al. Actinomycosis of the jaws—Histopathological study of 45 patients shows significant involvement in bisphosphonate-associated osteonecrosis and infected osteoradionecrosis. Virchows Arch 2007;451:1009–1017.

19

Use of PRF in Regenerative Endodontics

Contributors

Valerie Kanter

Richard J. Miron

Chapter Highlights

- *Use of PRF for direct pulp capping*
- *Use of PRF for pulpotomy*
- *Use of PRF for apexification and revascularization*
- *Use of PRF for apical surgery*
- *Use of PRF for the treatment of avulsed teeth*

While many systematic reviews with meta-analysis have demonstrated the successful incorporation of PRF in periodontal procedures, for gingival recession coverage, and for various bone augmentation procedures, the use of PRF in endodontics only began relatively recently. As such, both preclinical and human studies investigating its use on pulpal cells and tissues began many years after PRF was initially discovered. Nevertheless, PRF demonstrates the advantages of being regenerative, improving blood supply, limiting inflammation, and being antimicrobial toward incoming pathogens. Thus, it was hypothesized that should early detection of pulpal inflammation be observed, PRF could potentially *(1)* help stimulate pulpal cells and tissues and *(2)* reverse the inflammatory condition. This chapter highlights the research and clinical concepts to date on this topic, presenting numerous human clinical studies. These include use of PRF for direct pulp capping, pulpotomy, apexification, revascularization, apical surgery, and treatment of avulsed teeth. The chapter concludes with research topics presently being investigated and an overall future direction of the field.

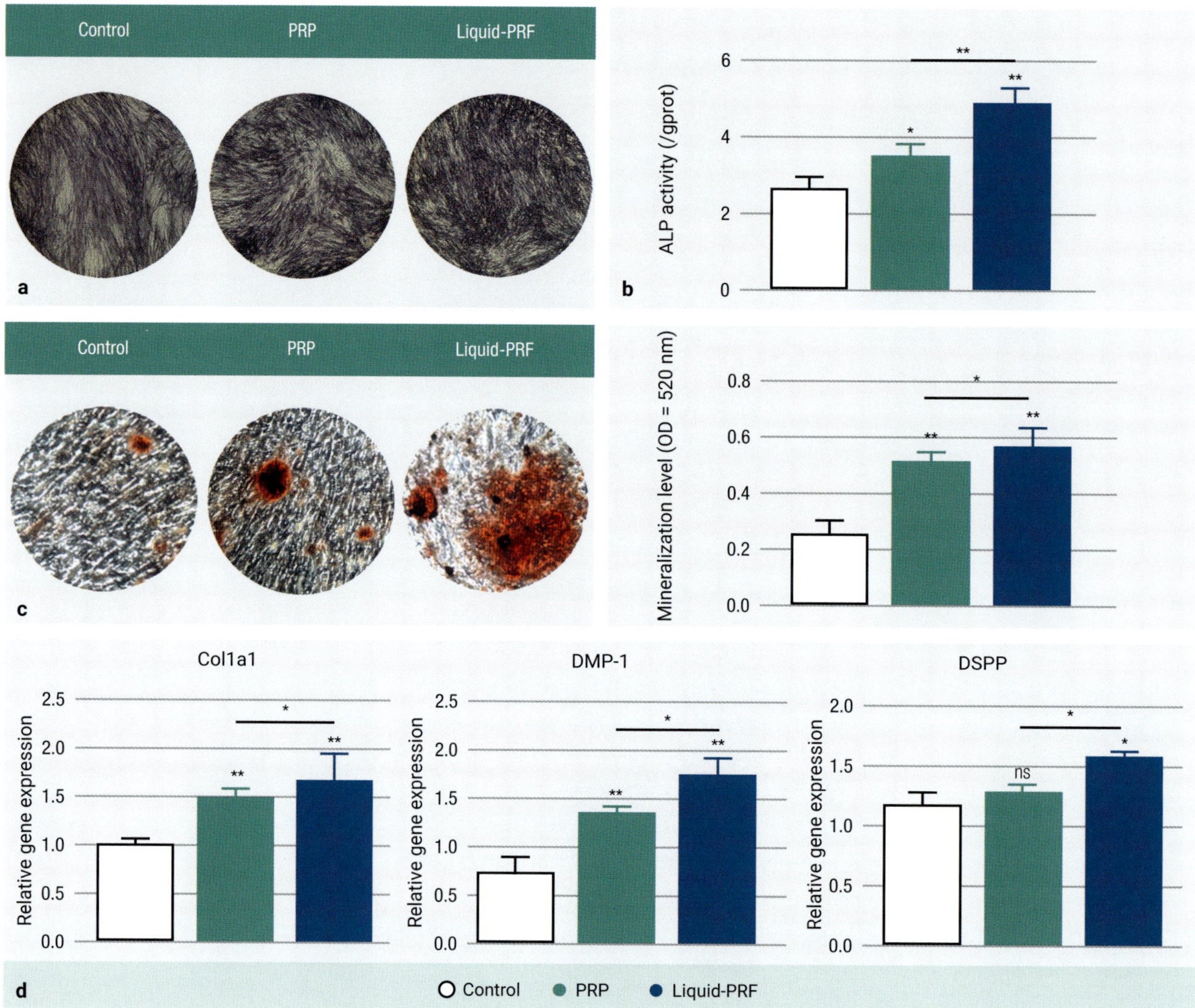

Fig 19-1 The effect of PRP and liquid-PRF on odontoblast differentiation of human dental pulp cells. The effects of PRP and liquid-PRF on ALP activity were detected by ALP staining *(a)* and ALP activity test *(b)*. *(c)* Alizarin Red S staining showed the mineralized nodules in each group after induction for 14 days, right panel indicating the semiquantification of mineralization level. *(d)* Relative gene expression levels of *Col1a1*, *DMP1* (dentin matrix protein1), and *DSPP* (dentin sialophosphoprotein) after culture with PRP or liquid-PRF for 14 days. Error bars correspond to the mean ± SD; significant differences are indicated: $*P < .05$; $**P < .01$; ns, not statistically significant vs control group. (Reprinted with permission from Chai et al.[4])

PRF is a bioactive autologous material that may also be utilized in clinical endodontics to enhance soft and hard tissue healing. The unique and biocompatible properties of PRF make it applicable in a wide range of endodontic procedures, spanning from regenerative to surgical endodontics. The slow polymerization of fibrinogen into fibrin during centrifugation creates a flexible, elastic, and resilient biomaterial that slowly releases growth factors including PDGFs, TGFs, VEGF, FGF, and more.[1] These essential elements have a synergistic effect on the wound healing process and facilitate a variety of activities, such as cell migration, cell attachment, cell proliferation, and cell differentiation.[1]

Regenerative endodontics is defined as biologically based procedures designed to replace damaged structures, including dentin and root structures, as well as cells of the pulp-dentin complex.[2] The release of growth factors needed for regeneration of dentin can be triggered endogenously by carious demineralization or acid etching.[2] An enhanced understanding of the biologic processes mediating tissue repair has created an opportunity to investigate bioactive materials that may mimic the natural ability

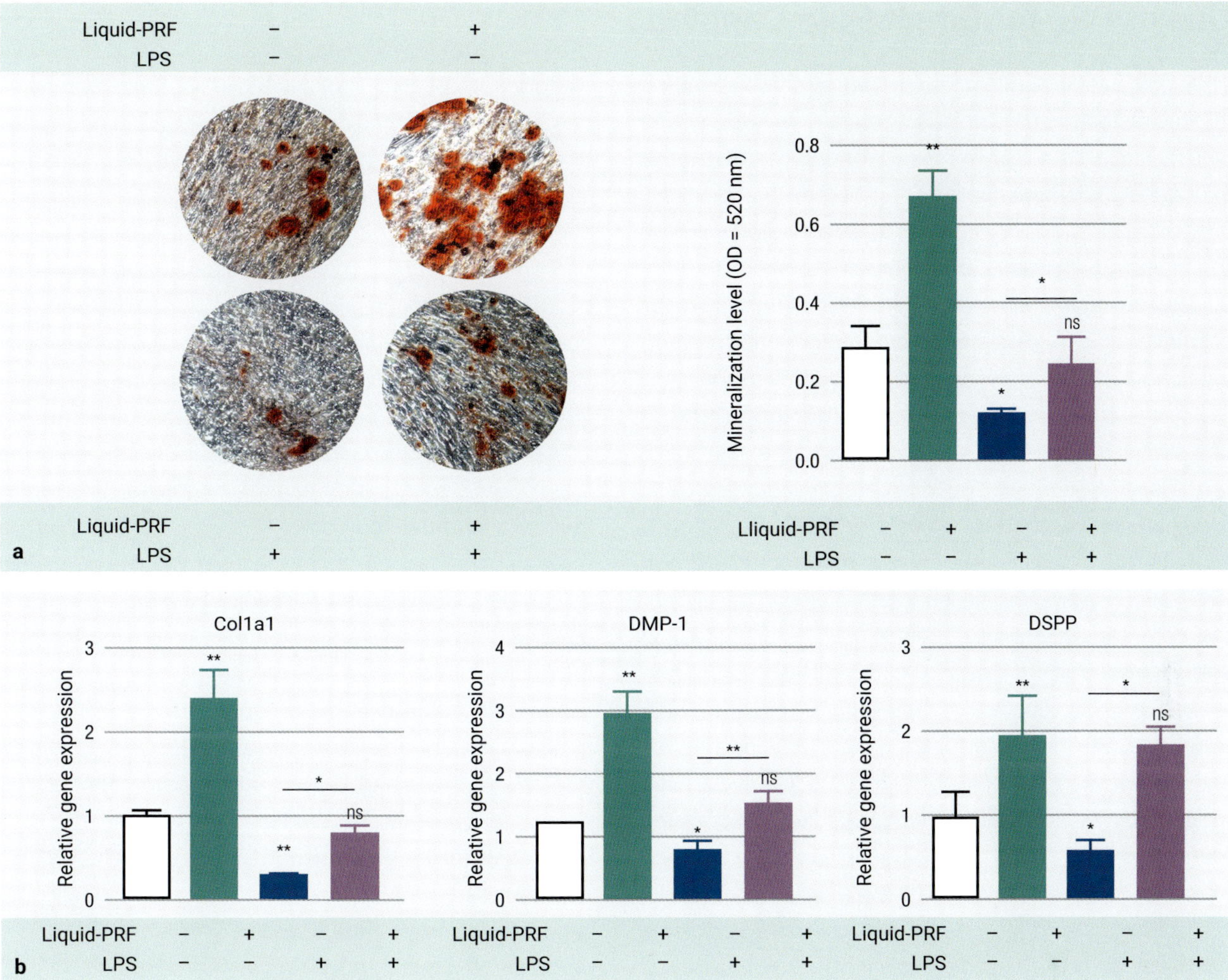

Fig 19-2 Odontoblast differentiation of human dental pulp cells cultured in LPS inflammatory conditions with liquid-PRF. *(a)* Alizarin Red S staining of human dental pulp cells cultured in the presence of LPS and/or liquid-PRF. *(b)* Gene expression levels of *Col1a1*, *DMP1*, and *DSPP* in inflammatory conditions. Error bars correspond to the mean ± SD; significant differences are indicated: $*P < .05$; $**P < .01$; ns, not statistically significant vs control group. (Reprinted with permission from Chai et al.[4])

of dentin to release growth factors and stimulate tissue responses. Both in vitro and in vivo growth factors can induce regenerative events in the dentin-pulp complex in a similar way to dentin matrix extracts that naturally contain these growth factors.[3]

PRF offers several advantages when added to the pulp because it is autologous and nontoxic to cells. Because the pulp is an enclosed space, it is important to consider the potential inflammatory response generated when foreign body materials enter this space. PRF stimulates the migration and proliferation of various cell types that secrete collagenous proteins, the main macromolecules in dental pulp extracellular matrix. It also releases various growth factors that may assist in odontogenesis and has demonstrated dentin formation after vital pulp therapy procedures in vivo. As observed in the preclinical studies by our research team, liquid-PRF can be applied as a pulp capping agent directly to pulp cells, demonstrating significantly greater ability to induce alkaline phosphatase (ALP) activity, collagen matrix synthesis, and mineralization when compared to PRP (Fig 19-1).[4] In addition, when dental pulp cells were cultured with lipopolysaccharide (LPS) to stimulate an inflammatory environment, a marked decrease in dentin-related repair was observed; however, when liquid-PRF was cultured within this inflammatory environment, a marked improvement in regeneration was observed with a decrease in inflammatory markers including TNF-α and IL-1β (Fig 19-2).[4]

Use of PRF for Direct Pulp Capping

The addition of PRF prior to pulp capping or incorporation into the restorative protocol is an exciting new topic in vital pulp therapy. Vital pulp therapy is one subset of regenerative endodontics and includes direct pulp capping and pulpotomy procedures. The objectives of these procedures include maintaining pulp vitality and sensibility to optimize function and oral-motor behaviors, improving long-term survival, preventing adverse clinical signs or symptoms, and preventing breakdown of the periradicular supporting tissue.[5]

Direct pulp capping has variable outcomes that depend on operator technique, material properties, and the host pulpal responses.[6] A successful pulp capping procedure can also prevent more invasive and expensive dental treatment, such as root canal therapy.[6] Therefore, it is important to optimize direct pulp capping techniques, improve biocompatibility of the materials, and enhance biologic responses of the pulp tissues to maximize regeneration of reparative dentin. Historically, direct pulp capping has been reported to have a high failure rate, with the 5- and 10-year groups resulting in 44.5% and 79.7% failure, respectively.[7] A more recent systematic review and meta-analysis concluded that the success rates are affected by the type of capping material used.[7] Mineral trioxide aggregate (MTA) cements have a higher success rate, with a lower inflammatory response and a more predictable hard dentin barrier formation than calcium hydroxide cements. Other factors play a role such as degree of microbial invasion and sterile technique while excavating decay.

Research in the field of molecular diagnostics has demonstrated that the pulp is a reactive tissue that extends its biologic products into the outside environment, and a sample from nearby gingival crevicular fluid may give us a snapshot of the biologic mechanisms that drive this immunocompetent tissue toward healing or necrosis.[8] For these reasons, the addition of an autologous biologic agent such as liquid-PRF is being investigated in the field of regenerative endodontics because it may present the ability to shift the balance toward healing. PRF has the ability to *(1)* enhance the immune defense in inflammatory tissues by a slow release of growth factors, *(2)* mitigate LPS-stimulated expression of inflammatory cytokines, and *(3)* enhance dentinogenesis through dentin sialophosphoprotein (DSPP), dentin matrix protein1 (DMP1), ALP activity, and mineralization.[4]

Standard treatment includes placement of a radiopaque capping material directly onto the surface of vital pulp tissue at the site of the pulp exposure followed by a base or liner. The combination of PRF with materials such as MTA, which is known for its ability to induce pulp repair and the formation of dentin bridges with minimal inflammation, has a synergistic effect on the stimulation of odontoblastic differentiation of human dental pulp cells via the modulation of the BMP/Smad signaling pathway.[9] Direct pulp capping is a regenerative treatment that has traditionally been indicated after a mechanical or traumatic exposure of a clinically vital and asymptomatic pulp.[5] Ultimately, the goal is to maintain vitality of the pulp-dentin complex, facilitate the formation of reparative dentin, and prevent breakdown of pulpal and periradicular tissues.[10,11] Controlling contamination of the pulp exposure site and choosing the appropriate biocompatible capping material are important factors in treatment outcome.[11,12] If aseptic conditions are maintained in these types of exposures, the underlying pulp is less likely to become inflamed or infected.[13] The ideal balance between pro- and anti-inflammatory cytokines is thus critically important to promote the healing of pulpal tissues and reduce disease-associated tissue destruction. Proinflammatory cytokines, such as IL-1β, play an essential role in both acute and chronic inflammatory processes. When human dental pulp stem cells were challenged with LPS, PRF significantly reduced levels of IL-1β and TNF-α.[4]

In summary, PRF has been investigated as a direct pulp capping agent in numerous studies[14–16] (Table 19-1) and is demonstrated in the following case presentation.

Clinical case example

A 38-year-old man presented with a deep carious lesion in his mandibular left first molar (Fig 19-3 and Video 19-1). He reported that his filling came out over 10 years ago but he never experienced pain in this tooth. The pulpal and periapical tests led to a diagnosis of normal pulp and normal periapical tissues. The radiographs, however, revealed a carious lesion extending into the pulp chamber and a widened periodontal ligament (PDL) on the mesial apex (Figs 19-3a and 19-3b).

Video 19-1

After thorough presentation of various treatment options, the patient selected a treatment plan that included laser therapy and direct pulp capping should the lesion not

Table 19-1 Use of PRF in pulp capping

Authors (year)	Study design	Follow-up	No. of participants/ sample	Groups/ treatment method	Results	Conclusions
Tabatabayi et al[14] (2017)	Animal study: A cervical cavity was prepared in three maxillary and mandibular incisors. When the pink color of pulp was seen through the thin layer of dentin, a no. 2 explorer was used to expose the pulp.	Weekly for up to 2 months	One male, mixed-breed dog	1. No treatment was performed and the cavities were filled only with zonalin cement. 2. PRF was placed in the exposure site with zonalin cement over it.	A dentin bridge was observed only in the PRF group.	Dentin bridge formation occurred in the pulpal tissue of the dog's incisor teeth capped with PRF, which shows promising effects in healing of the wounded pulp.
Bakshi et al[15] (2017)	Case report: Direct pulp capping was performed with PRF in a mature permanent molar with irreversible pulpitis.	1 day and 6 months	One human patient	Direct pulp capping was performed with PRF in a mature permanent molar with irreversible pulpitis.	At 1 day follow-up, the patient's spontaneous symptoms had resolved. The 6-month follow-up demonstrated pulp vitality, clinical function, as well as the absence of pain/tenderness to percussion/palpation/ cold sensitivity tests; the periapical radiograph showed normal periodontium.	The favorable results indicate that PRF shows promising results in cases of pulpal inflammation with irreversible signs and may be a good treatment option in comparison to endodontic treatment.
Dou et al[16] (2020)	In vitro study: Investigation of the proliferation, viability, apoptosis, and mineralization of human dental pulp cells.	24 hours, 3 days, and 7 days	Freshly extracted impacted third molars (human)	$Ca(OH)_2$ MTA iRoot BP PRF CGF	Cell viability and proliferation were improved with PRF.	PRF and CGF may be potential pulp capping materials for vital pulp therapy.

CGF, concentrated growth factor; $Ca(OH)_2$, calcium hydroxide.

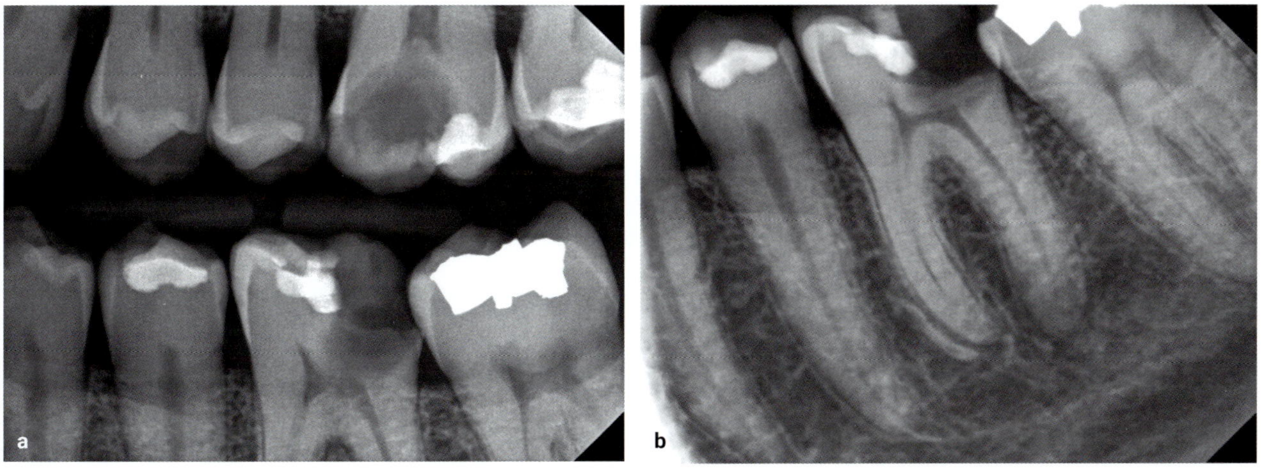

Fig 19-3 Direct pulp capping with PRF using the EndOzLase protocol. *(a and b)* Preoperative bitewing and periapical radiographs revealing a coronal radiolucency at the mandibular left first molar extending toward the pulp chamber. A widened PDL is noted around the mesial apex.

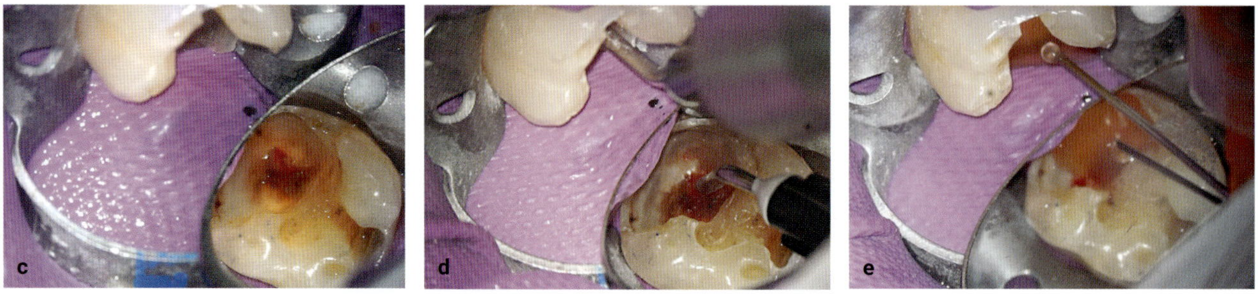

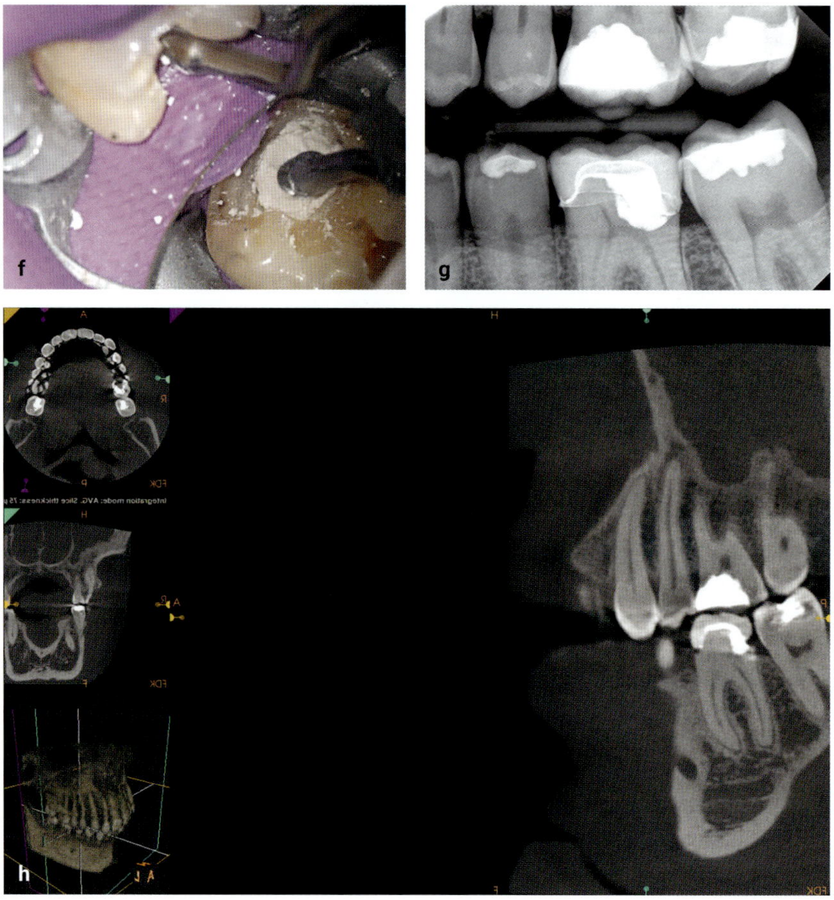

Fig 19-3 *(cont)* *(c)* Pulp exposure during caries removal with an Er:YAG laser. Remaining tooth structure over the pulp chamber is classified as arrested decay because it is completely remineralized. *(d)* Use of ozone therapy applied directly on the pulpal exposure for decontamination. *(e)* Addition of liquid-PRF to cover pulp exposure, stimulate dental pulp stem cells, and neutralize the effect of endotoxins. *(f)* Placement of a fast-setting bioceramic putty was used as the pulp capping agent. *(g)* Six-month bitewing radiograph of the final restoration. *(h)* Six-month recall CBCT showing the PDL to be within normal limits. After clinical testing, the diagnosis was confirmed: vital pulp and normal periapical tissues. (Case performed by Dr Valerie Kanter.)

extend extensively into the pulp. Following isolation with a dental dam, the initial outline preparation was performed with a high-speed drill followed by use of a laser for cavity preparation.[17,18] An Er:YAG laser (LightWalker, Fotona) was used first, because the 2940-nm wavelength is rapidly absorbed by water/hydroxyapatite and has selectivity for the carious tissue, which permits minimally invasive preparations.[17] Biofilms present on dental hard tissues are also destroyed because of their high water content (targeted by Er:YAG).[17,19] In this case, the LightWalker laser was used in QSP mode at 100 mJ, 20 Hz with air and water spray to remove the deepest layers of dentin and disinfect the floor of the preparation.

During the preparation, a pulp exposure was observed due to the depth of the cavity as the laser was removing the carious dentin (Fig 19-3c). Thereafter, an Nd:YAG laser was utilized because it penetrates more readily into dentin and exhibits bactericidal effects of dentin up to a depth of 1000 µm.[18] This was utilized to treat the pulpal floor of the cavity preparation and was followed by ozone therapy for disinfection and hemostasis of the pulp tissue (Fig 19-3d). Due to the production of free radicals, ozone

O_2/O_3 gas has antimicrobial effects, as demonstrated in several research articles.[20] Its main advantage is its high level of biocompatibility to fibroblasts, cementoblasts, and epithelial cells. A few drops of liquid-PRF were placed over the pulp exposure with a syringe to mitigate the effects of LPS and increase potential for dentinogenesis (Fig 19-3e). Thereafter, a fast-setting bioceramic putty (calcium silicate cement with zirconium oxide, Brasseler) was used as the pulp capping agent (Fig 19-3f). A well-sealed restoration was placed, and the patient was referred for a full-coverage restoration.

At the 6-month recall appointment, a limited CBCT was taken. A sagittal slice of the mandibular left first molar shows resolution of the inflammation that was originally present in the PDL and lamina dura (Figs 19-3g and 19-3h). This regenerative strategy combining laser therapy, ozone, and liquid-PRF follows a set protocol (EndOzLase) that is geared toward their synergistic use to favor the likelihood of a successful outcome.

Use of PRF for Pulpotomy

A pulpotomy is the surgical removal of the coronal portion of vital pulp tissue, followed by the placement of a biologically acceptable material in the pulp chamber and a final restoration. According to the American Association of Endodontists' *Guide to Clinical Endodontics*, a pulpotomy is only indicated in primary teeth or as an emergency or interim procedure in permanent teeth.[5] However, a recent systematic review suggested that full coronal pulpotomy had a favorable success rate in treating carious vital pulp exposure of permanent mature teeth with closed root apices, showing a 92% success rate after 2 years.[21] Another study evaluating long-term success of 273 cases in cariously exposed permanent teeth showed success rates of 75% after 5 years and 63% after 10 years,[22] concluding that pulpotomy may be a successful treatment option at any age and not only for young permanent teeth. The benefit of retaining the radicular pulp includes continued supply of essential factors via the apical foramen, which plays a key role in tooth maintenance.[23] Blood vessels that supply these nutrients also remove waste products and lymphatics and house the neural network that indicates the presence of harmful stimuli through pain.

Criteria for a successful pulpotomy include the following: lack of periapical radiolucency or widening of PDL apically; no pain following vertical or horizontal percussion; radiographic evidence of dentin bridge; lack of clinical or radiographic signs and symptoms of root resorption or apical periodontitis; and positive response to sensitivity testing, whenever possible.[22]

In a carious exposure, it is important to consider disinfection of radicular pulp stumps. The ability to control bleeding is usually considered an indicator to assess the extent of pulpal inflammation, but there may still be presence of inflammation in residual radicular pulp. Pulp tissue in teeth clinically diagnosed with irreversible pulpitis still has putative cells with stem cell properties.[24] Pulp regeneration can be achieved by harnessing the regenerative potential of these cells if provided with a suitable environment.[24] A study comparing the use of calcium hydroxide, MTA, and a combination of PRF and MTA for pulpotomy procedures in cases of irreversible pulpitis showed poor overall success rates (combined clinical and radiographic) of all three groups at 1 year of 37.5%, 44.4%, and 35.7%, respectively.[24] As mentioned previously, the status of pulp before vital pulp therapy is a key factor in determining the success rate of this technique.[25]

While the ability to control bleeding is generally used as an indicator to assess the extent of pulpal inflammation,[24] this does not necessarily correlate with the extent of inflammation in all cases. Studies have reported the use of chemicals such as sodium hypochlorite, pyrogen peroxide, and ozone to disinfect the remaining tissue and achieve homeostasis; however, these have varying levels of cytotoxicity. One study reported the use of a corticosteroid-antibiotic paste, placed in contact with the surface of the remaining radicular pulp for 5 minutes.[22] PRF may offer a more natural immune defense, as it releases cytokines such as IL-4 by activating a subpopulation of T-cells that aid in healing by moderating inflammation and inhibiting IL-1β–mediated stimulation of MMP-1 and MMP-3 and synthesis of prostaglandin E2[26] (Table 19-2).[24,27–30] No matter the approach chosen, the final restoration is placed over the base, and the status of the pulp and periradicular tissues should be assessed through periodic recall examinations.

Table 19-2 Use of PRF in pulpotomy

Authors (year)	Study design	Follow-up	No. of participants/ sex/age	Groups/ treatment method	Results	Conclusions
Hiremath et al[27] (2012)	Case report: Pulpotomy with PRF in a mature permanent molar tooth	1 day and 6, 12, 18, and 22 months	19-year-old female	PRF	At the first recall (1 day), no postoperative pain was reported. At 6, 12, 18, and 22 months, the tooth responded positively to pulp sensibility tests, and radiographic examination revealed a normal PDL space.	The use of PRF for the treatment of pulpitis in human permanent molar teeth was successful in this case report.
Keswani et al[28] (2014)	RCT: Clinical and radiographic comparison of the effects of PRF and MTA as pulpotomy agents in permanent teeth with incomplete root development.	6, 12, and 24 months	62 children	1. MTA 2. PRF	There was no significant difference between the two groups in terms of clinical and radiologic success. Radiographically, all available cases (53 teeth) showed evidence of root growth and canal narrowing. Complete apical closure was observed in 88.8% of teeth in the PRF group (experimental group) and 80.07% of roots in the MTA group (control group) at 24 months.	PRF could be used as a suitable biologic and economic alternative to MTA in pulpotomy procedures of permanent teeth with incomplete root development.
Pathak et al[29] (2014)	Case report: Pulpotomy with PRF in a human immature permanent molar tooth.	24 hours and 6 months	14-year-old male	PRF	No postoperative pain was reported. At the 6-month recall, the tooth responded positively to pulp sensibility tests, and radiographic examination revealed a normal PDL space.	There is a reasonable biologic argument to carry out pulpotomy with PRF as a possible alternative treatment in mature permanent teeth with pulpitis.
Kumar et al[24] (2016)	RCT: Investigation of various pulpotomy agents in mature permanent molars with irreversible pulpitis.	24 hours, 7 days, 6 months, and 1 year	54 permanent mandibular molars	1. $Ca(OH)_2$ 2. MTA 3. PRF	Clinical success rate was 94.4% at 7 days, which dropped to 85.4% at 12 months. All three agents were equally effective in providing pain relief at all the intervals tested, with no significant difference between them.	Pulpotomy exhibited a high clinical success rate in mature molars with irreversible pulpitis, and the selection of biomaterial did not affect its outcome.
Patil et al[30] (2017)	Case report: Evaluation of the use of autologous substances such as platelet concentrates and calcium silicate–based materials in promoting the healing and regeneration of the inflamed pulp.	24 hours, 1 week, and 1, 3, and 6 months	25-year-old male	PRF	PRF was effective as a vital pulp therapy agent for treating chronic irreversible pulpitis in this case.	The present case report showed that PRF is an effective vital pulp therapy agent for treating human permanent teeth with irreversible pulpitis, relieving associated pulpal symptoms, and preserving pulp vitality.

RCT, randomized controlled trial.

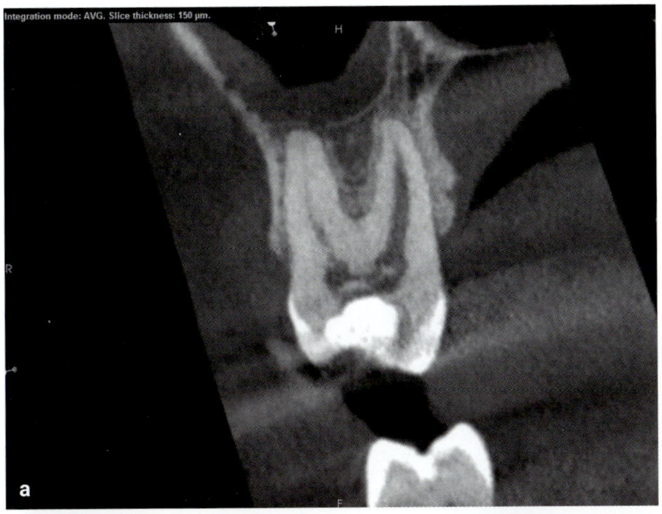

Fig 19-4 Use of PRF for pulpotomy treatment. *(a)* Preoperative CBCT demonstrating carious lesions under the current restoration in close proximity to pulpal tissue of the maxillary right second molar. *(b and c)* Inflamed pulp. *(d)* Following the GentleWave procedure using 3% sodium hypochlorite for 90 seconds and distilled water for 15 seconds, the coronal one-third to one-half of each canal was removed, leaving healthy-looking pulp tissue in the remainder of each canal.

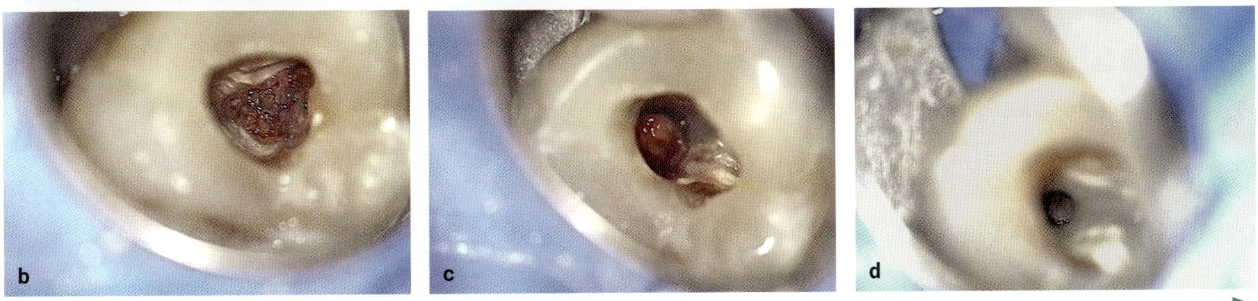

Clinical case example

A 32-year-old man presented with nonlingering hypersensitivity to cold on his maxillary right second molar. A CBCT demonstrated an existing deep filling with close proximity to the pulp (Fig 19-4a). On a scale of 1 to 10, the pain level to cold was described as 6 to 7. Treatment options were discussed, and the patient decided to pursue vital pulp therapy using the GentleWave protocol with PRF.

Local anesthesia was achieved with 3% carbocaine plain and 4% articaine 1:100,000 epinephrine. With rubber dam isolation, the pulp chamber was accessed through the existing composite restoration. The pulp was vital, hyperemic, and upon entry demonstrated an inflammatory condition (Figs 19-4b and 19-4c). Without instrumentation, the GentleWave (Sonendo) was used with 3% sodium hypochlorite for 90 seconds and then distilled water for 15 seconds (Fig 19-4d). This atraumatically removed the inflamed pulp tissue in the coronal one-third to one-half of each canal, leaving healthy-looking pulp tissue in the remainder of each canal.

PRF was then placed into the coronal portion of each canal, as well as the floor of the pulp chamber (Fig 19-4e). A 2-mm-thick layer of bioceramic putty (Brasseler) was placed over the PRF, and glass ionomer was flowed over the bioceramic putty and allowed to set for 3 minutes (Fig 19-4f). A long-term provisional composite restoration was placed over the glass ionomer (Fig 19-4g).

The patient was recalled the next day and reported no pain or cold sensitivity, and at 3 months he reported that the tooth was completely asymptomatic. Cold tests were negative. Percussion and bite tests were within normal limits. The radiographic appearance was also within normal limits (Fig 19-4h). At the 15-month recall, the patient reported that the tooth had been asymptomatic since the day it was treated. Bite, percussion, and cold tests were still negative. The CBCT showed what appeared to be formation of a dentinal bridge in the coronal third of the palatal root (Fig 19-4i).

This case highlights the potential use of PRF for pulpotomy cases. Future research remains needed to further investigate success rates and potential improvements to current protocols.

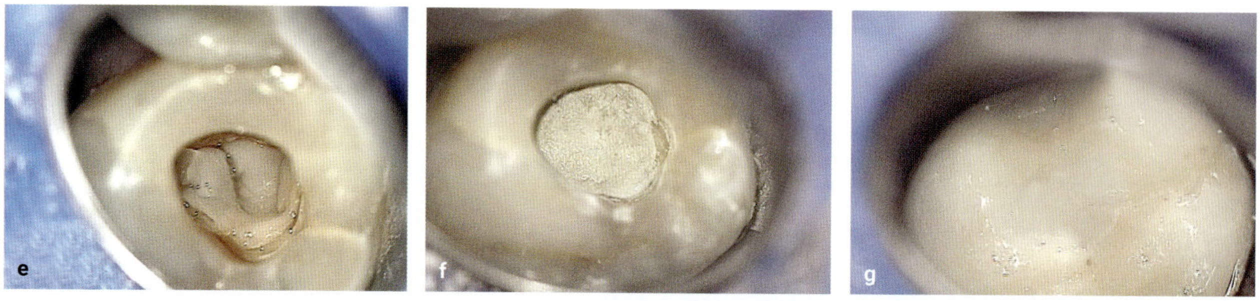

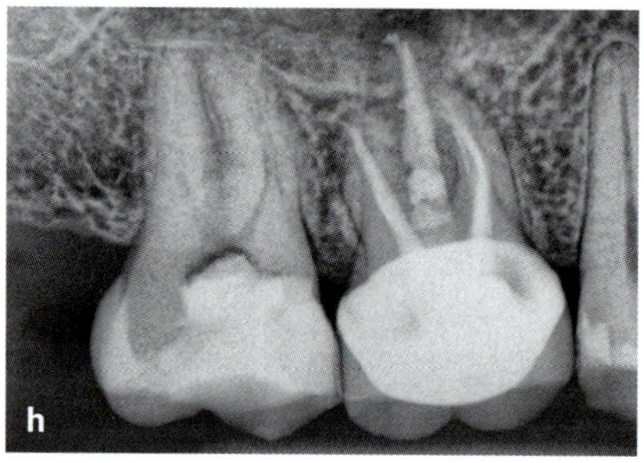

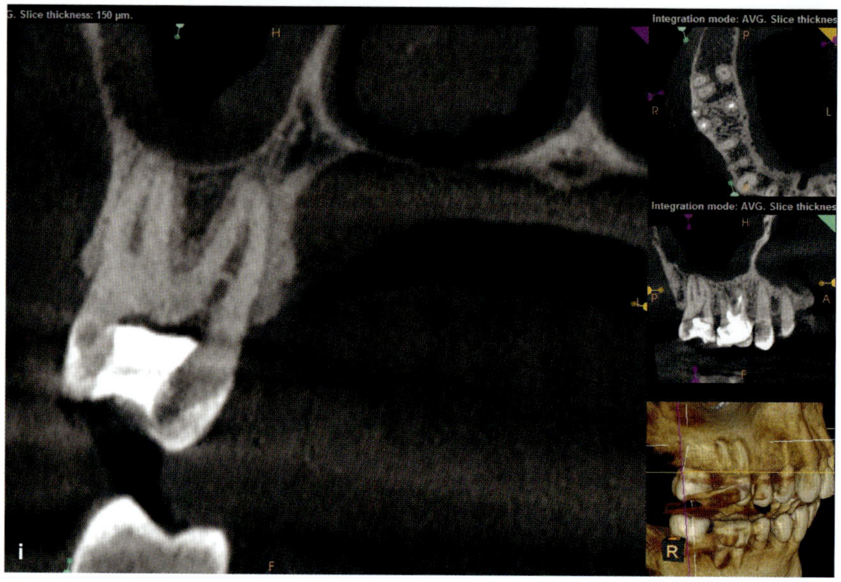

Fig 19-4 *(cont)* *(e)* Placement of PRF. *(f)* Placement of a 2-mm-thick layer of bioceramic putty followed by glass ionomer. *(g)* Final composite restoration. *(h)* Periapical radiograph at the 3-month recall visit. *(i)* Fifteen-month postoperative CBCT. Note the dentin bridge that is formed in the coronal third of the buccal and palatal canals. (Case performed by Dr Randy Garland.)

Use of PRF for Apexification

Immature permanent teeth with necrotic pulps have very thin walls, incomplete root formation, and a wide apical foramen, which makes them very difficult to treat.[31] Apexification is a treatment designed to induce a calcified barrier that facilitates root canal obturation. Historically, this treatment involves long-term calcium hydroxide therapy over multiple appointments[31] (Fig 19-5). One drawback of this technique is the effect that calcium hydroxide can have on the mechanical properties of dentin, making these teeth even more susceptible to root fracture.

More recently, MTA has been recommended as an apical barrier material that also results in root closure due to its regenerative potential and biocompatibility. The time required for the formation of the barrier is significantly reduced compared to calcium hydroxide.[31] MTA has the capacity to induce bone, dentin, and cementum formation as well as regeneration of periapical tissues with minimal cytotoxicity.[32] In a large cohort study of 252 MTA apexification cases, a 90% success rate was reported.[32]

Proper length control can be challenging when packing MTA at the apex of an immature tooth,[31] and some clinical studies have identified the extrusion of root filling material beyond the radiographic apex as a negative factor for treatment outcome.[33,34] The extrusion of MTA can be prevented by placement of an internal matrix as an apical stop.[35] Various materials have been suggested for this purpose, such as tricalcium phosphate, collagen, hydroxyapatite-based materials, and more recently, PRF. While there were no statistical differences between the groups in prevention of MTA extrusion with different apical matrices, PRF has plausible advantages over all other materials. PRF serves as a vehicle in carrying cells involved in tissue regeneration and provides a sustained release of growth factors.[36] Leukocytes present in PRF act as anti-inflammatory, anti-infective, and immune-stimulating agents.[31,35] They also provide VEGF to promote angiogenesis.[31,35] Mature teeth with open apices from resorption overinstrumentation may also benefit from PRF apical plug placement in order to avoid extrusion of obturation materials[37] (see Fig 19-6).

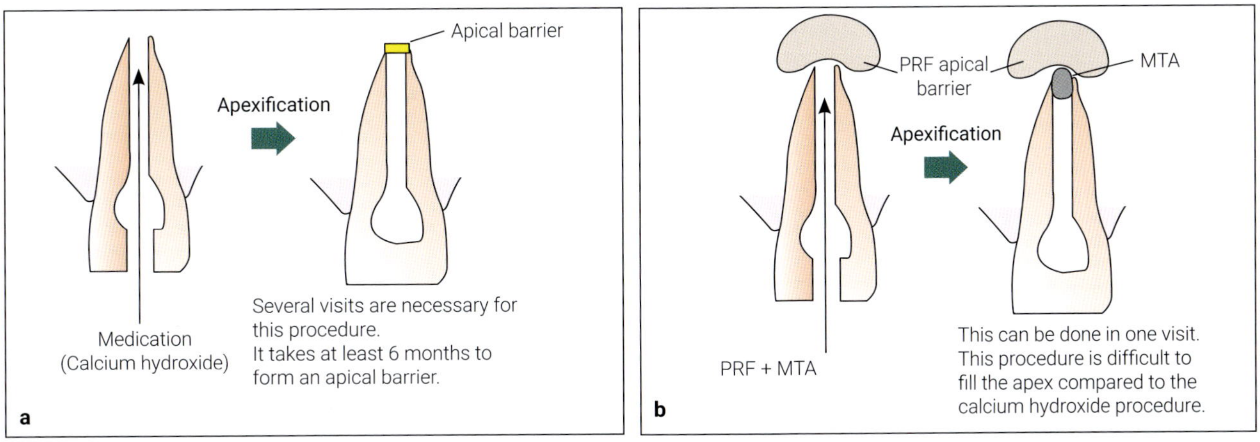

Fig 19-5 Diagram of standard apexification protocol. *(a)* Immature tooth with an open apex. Traditional protocols include the entry of calcium hydroxide for long periods of time to promote mineralization. *(b)* Following traditional protocols, the goal is to form an apical barrier in order to obturate the remainder of the canal.

Table 19-3 Use of PRF for apexification

Authors (year)	Study design	Follow-up	No. of participants/ sex/age	Groups/ treatment method	Results	Conclusions
Kathuria et al[42] (2011)	Case report: Management of single-rooted immature mandibular second molar with single canal using MTA and PRF as a membrane barrier.	1 week	15-year-old male	PRF+MTA	Faster periradicular healing and complete resolution of symptoms occurred with combined use of MTA and PRF membrane as an apical barrier.	The combination of PRF membrane as a matrix and MTA can prove to be an effective alternative for creating artificial root-end barriers and to induce faster periapical healing with large periapical lesions.
Rudagi and Rudagi[43] (2012)	Case report: Healing and apexification with combined use of MTA as an apical barrier and autologous PRF membrane as an internal matrix.	1 week, 2 months, and 1 year	20-year-old female	PRF	Follow-up radiographs at 2 months showed reduction in size of radiolucency and at 1 year showed further healing with calcified barrier at the apex.	The combination of PRF as a matrix and MTA as an apical barrier could be considered a good option for creating artificial root-end barriers.
Kumar et al[44] (2014)	Case report: One-step apexification using MTA as an apical barrier and autologous PRF as an internal matrix.	1 and 3 months and 1 year	18-year-old female	PRF+MTA	1 month: Asymptomatic 3 months: Reduction in radiolucency 1 year: Complete healing	The combination of PRF as a matrix and MTA as an apical barrier can be considered a good option for one-step apexification.

The success of the PRF protocol depends on the handling, mainly related to blood collection time and its transference to the centrifuge.[38,39] Previous studies have reported underfilling of the obturation material observed in patients treated with PRF as apical barriers. This may be due to improper compaction force because of an inability to feel resistance from the softness of PRF membranes.[35] With proper techniques, studies have more recently demonstrated that this new protocol with PRF and MTA is a predictable and reproducible clinical procedure for one-visit apexification.[38,40] Therefore, the combination of PRF with MTA is an effective method for management of difficult cases of open apex[38,41–49] (Table 19-3).

Table 19-3 *(cont)* Use of PRF for apexification

Authors (year)	Study design	Follow-up	No. of participants/ sex/age	Groups/ treatment method	Results	Conclusions
Yadav et al[45] (2015)	Case report: Management of a failed revascularization/ revitalization case, which could be due to inadequate removal of biofilm and bacteria in dentinal tubules.	6 months and 2 years	One patient	PRF	One-visit apexification techniques provide an alternative treatment for failed revascularization cases. Follow-up confirmed complete healing periradicularly.	Apexification in one step using an apical barrier of PRF and a plug of MTA can be considered a predictable treatment and may be an alternative to long-term revascularization failures.
Sharma et al[41] (2016)	Case series: Immature permanent teeth with open apices using internal matrix of autologous PRF membrane and one-step apical barrier placement of MTA.	2 weeks and 6 months	26-year-old male and 16-year-old male	PRF+MTA	Asymptomatic teeth were isolated and reaccessed.	The combination of PRF as a matrix and MTA as an apical barrier is a good option for creating artificial root-end barriers.
Khanduri and Kurup[46] (2018)	Case report: Discolored and fractured maxillary anterior tooth with open apex.	12 months	19-year-old male	Biodentine+PRF	Biodentine, a new calcium silicate–based material, with PRF was effective as an apical barrier.	Use of PRF in combination with Biodentine was effective for apexification.
Prasad et al[47] (2018)	Case report: The maxillary right incisors were treated with PRF, while the maxillary left incisors were treated with PRP.	Every 6 months for 2 years	13-year-old male	1. PRP 2. PRF	Dentinal wall thickening, root lengthening, and apical closure had occurred in all the incisors treated with PRP and PRF. However, apical bridge formation was observed only in those incisors treated with PRF.	PRF and PRP are efficient in inducing pulp regeneration and apexogenesis in teeth that have undergone endodontic treatment failure. However, apical bridge formation was observed only in teeth that were treated with PRF.
Santhakumar et al[48] (2018)	Triple-blinded clinical trial: Undertaken to check whether PRF in its membrane form is as reliable as it is in its gel form.	12 months	40 patients	1. PRF gel 2. PRF membrane	Regenerative endodontics with PRF membrane is easier and less time-consuming. Both groups showed similar rates of success.	It was preferable to utilize PRF in its membrane form.
Rathod et al[49] (2019)	Case report: Broken maxillary anterior tooth. The treatment plan included apexogenesis using PRF followed by MTA.	3, 6, and 9 months and 1 year	9-year-old male	PRF+MTA	Gradual closure of apex.	PRF can be a compatible material for apexogenesis and root maturation.

Clinical case example

A 45-year-old woman presented with continued pain in the maxillary right first premolar 3 years after completion of endodontic therapy (Fig 19-6a). Once all the gutta-percha was removed during the retreatment procedure, a large open apex with signs of apical resorption was identified through the microscope. After thorough irrigation and disinfection, a PRF membrane was placed into the chamber (Figs 19-6b and 19-6c) and then pushed down the canal and out of the apex (Fig 19-6d), just into the periapical tissues. This created a barrier to predictably proceed with the obturation of the root canal (Figs 19-6e and 19-6f).

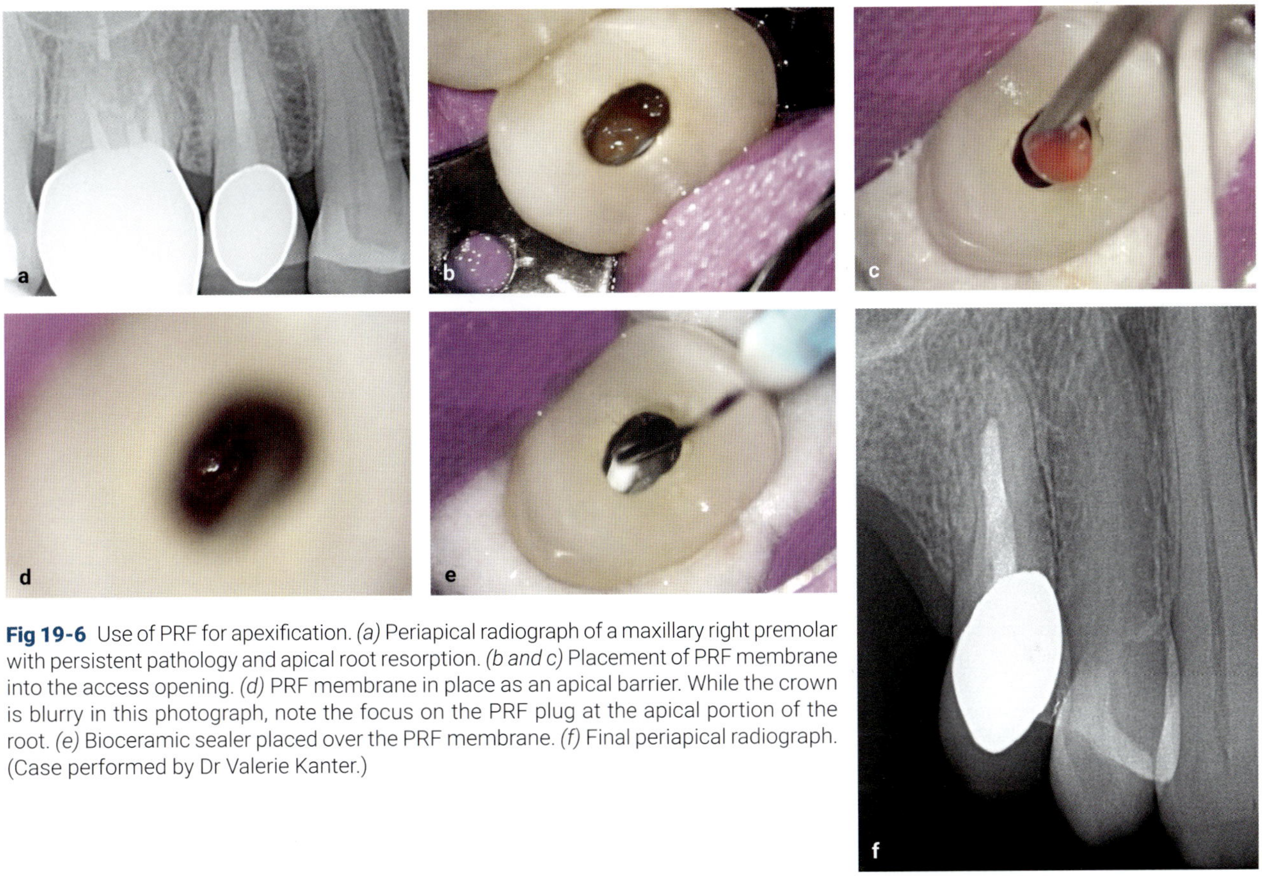

Fig 19-6 Use of PRF for apexification. *(a)* Periapical radiograph of a maxillary right premolar with persistent pathology and apical root resorption. *(b and c)* Placement of PRF membrane into the access opening. *(d)* PRF membrane in place as an apical barrier. While the crown is blurry in this photograph, note the focus on the PRF plug at the apical portion of the root. *(e)* Bioceramic sealer placed over the PRF membrane. *(f)* Final periapical radiograph. (Case performed by Dr Valerie Kanter.)

Use of PRF for Endodontic Revascularization

In 2004, Banchs and Trope proposed a new regenerative treatment protocol for immature permanent teeth with apical periodontitis called *revascularization*.[50] Root canal revascularization was proposed to overcome the obstacles related to the clinical management of immature necrotic teeth with open apices.[50] The goal is to attain pulplike tissue within the root canal after the induction of a blood clot inside the root canal system, which will serve as a scaffold for stem cells from the apical region (Fig 19-7).

Revascularization has many advantages over apexification, including continued root development and reinforcement, which may prevent root fractures by increasing root length and thickening the dentin wall.[51] In animal studies, the type of tissue growth seen in these cases is an altered form of dentin that also shows characteristics of PDL, cementum, and bone. It is not truly a process of pulp revascularization. In human studies, we are limited to radiographic assessment, pulp sensibility tests, as well as laser Doppler flowmetry.[52]

The first published case of revascularization described the process of disinfection followed by a scaffold with blood clot formation induced from the apex, as a source of stem cells from the apical papilla.[52] Over time, this treatment protocol has been modified and improved upon with the advent of autologous platelet aggregates.[50,52]

In 2017, a systematic review and meta-analysis on outcomes of regenerative endodontic therapy quantified success rates through periapical healing, apical closure, increase in root length, and wall thickness.[53] Success rates were high across all comparison groups, although autologous platelet aggregates such as PRP proved to have a slight advantage compared with an induced blood clot, as far as the healing of periapical tissues (94% vs 91%), apical closure (82% vs 76%), and root length and thickness (94% vs 80%).[53] Because PRF releases more growth factors and has more effective leukocyte concentrations, ongoing studies should also lead to similar outcomes as PRP[51,54–72] (Table 19-4).

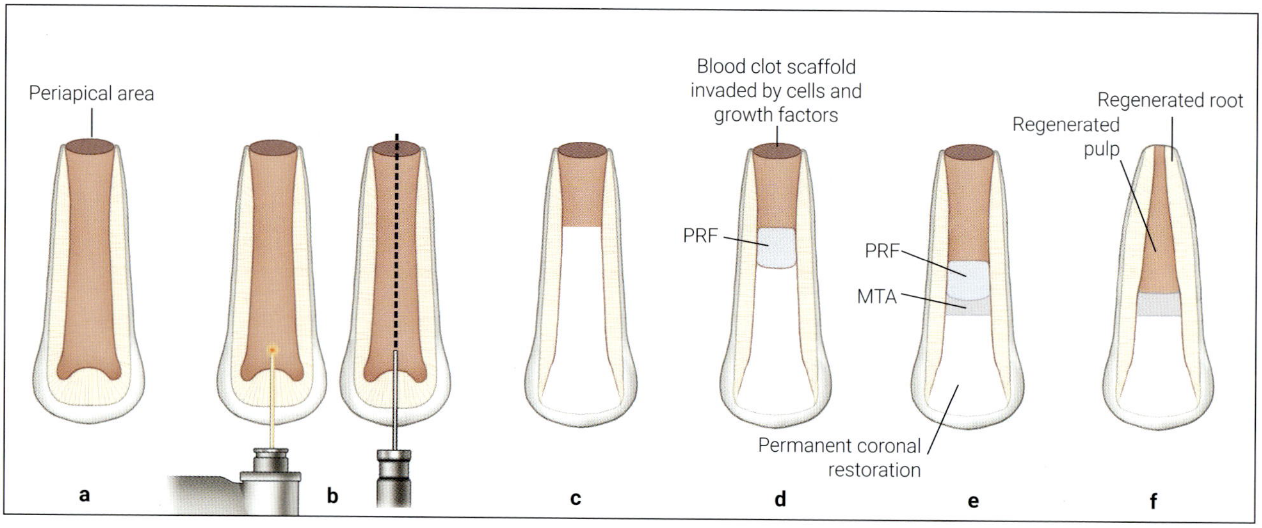

Fig 19-7 Use of PRF for revascularization. *(a)* Immature tooth with an open apex. *(b)* Stimulating bleeding by overinstrumentation or laser activation. *(c)* Blood forms in the canal to form a clot. *(d)* Placement of liquid-PRF and solid-PRF to induce additional blood clot formation and introduce a supraconcentration of growth factors and cells. *(e)* Sealing the tooth using a biocompatible material such as a bioceramic. *(f)* Root formation.

Table 19-4 Use of PRF in endodontic revascularization

Authors (year)	Study design	Follow-up	No. of participants/ sex/age	Groups/ treatment method	Results	Conclusions
Shivashankar et al[54] (2012)	Case report: Use of PRF for the revitalization of a tooth with necrotic pulp and open apex.	3, 6, 9, and 12 months	9-year-old male	PRF+MTA	Asymptomatic, periapical healing, apical closure.	Autologous platelet concentrates contribute to the success.
Keswani and Pandey[55] (2013)	Case report: Revascularization of an immature tooth with a necrotic pulp using PRF.	7, 12, and 15 months	7-year-old male	PRF+MTA	Radiographic examination demonstrated continued thickening of root canal walls, root lengthening, and apical closure. The tooth was asymptomatic with no sensitivity to percussion, palpation, or cold.	PRF might serve as a potentially ideal scaffold in revascularization of immature permanent teeth with necrotic pulps as it is rich in growth factors, enhances cellular proliferation and differentiation, and acts as a matrix for tissue ingrowth.
Mishra et al[56] (2013)	Case report: Revitalization of an immature necrotic tooth with PRF.	6 and 12 months	11-year-old male	PRF+MTA	Asymptomatic, periapical healing.	Autologous platelet concentrates contribute to the success.
Johns et al[57] (2014)	Case report: Photoactivated disinfection of necrotic pulp followed by application of PRF for regenerative endodontics.	6 and 10 months	9-year-old male	PRF+MTA+ Coltosol+ Biodentine	Clinical examination revealed no sensitivity to percussion or palpation tests. Radiography revealed continued thickening of the dentinal walls, root lengthening, regression of the periapical lesion, and apical closure. Both the roots showed complete apical closure at the 10-month follow-up.	This report of pulp revascularization shows that disinfection with photodynamic therapy combined with PRF leads to satisfactory root development in necrotic immature teeth.

Table 19-4 *(cont)* Use of PRF in endodontic revascularization

Authors (year)	Study design	Follow-up	No. of participants/ sex/age	Groups/ treatment method	Results	Conclusions
Faizuddin et al[58] (2015)	Case report: Revitalization of a traumatized immature tooth with PRF.	3, 6, 9, 12, and 14 months	14-year-old male	PRF+MTA+-glass ionomer+ Biodentine	Asymptomatic, root lengthening.	Autologous platelet concentrates contribute to the success.
Jadhav et al[59] (2015)	Case report: Revascularization potential of PRF in an immature, nonvital permanent tooth with apical periodontitis.	12 and 18 months	16-year-old male	PRF+MTA+ resin-modified glass	Asymptomatic, periapical healing, apical closure.	Autologous platelet concentrates contribute to the success.
Narang et al[51] (2015)	Controlled trial: Evaluation and comparison of the regenerative potential of PRF, PRP, and blood clots in immature necrotic permanent teeth with or without associated apical periodontitis.	6 and 18 months	20 patients	1. MTA 2. Blood clot 3. PRF 4. PRP+collagen	The PRF group showed the best outcomes of all treatment groups.	PRF has huge potential to accelerate the growth characteristics in immature necrotic permanent teeth as compared to PRP and blood clots.
Sharma and Mittal[60] (2016)	Clinical trial: Evaluation and comparison of the regenerative potential of natural autologous scaffolds (blood clot and PRF) with artificial scaffolds (commercially available collagen and PLGA polymer) in inducing apexogenesis in necrotic immature permanent teeth.	6 and 12 months	Patients 10–25 years old	1. Blood clot 2. PRF 3. Collagen 4. PLGA	PRF and collagen gave better results than blood clots and PLGA in terms of periapical healing, apical closure, and dentinal wall thickening.	The revascularization procedure is more effective and conservative over apexification in the management of necrotic immature permanent teeth. This study has shown that PRF and collagen are better scaffolds than blood clots and PLGA for inducing apexogenesis in immature necrotic permanent teeth.
Subash et al[61] (2016)	Case report: Revitalization of an immature permanent mandibular molar with a necrotic pulp.	3, 6, 9, and 12 months	13-year-old male	PRF+ Biodentine	Asymptomatic, periapical healing, apical closure.	Revitalization was successfully achieved using PRF as a scaffold in an immature mandibular molar tooth.
Bakhtiar et al[62] (2017)	Case series: Clinical and radiographic results of a pulp regenerative procedure using PRF in immature teeth with necrotic pulps.	1, 3, 6, 12, and 18 months	Four immature teeth	PRF	Clinical examinations revealed that all cases were asymptomatic.	On the basis of the short-term results up to 12 months, PRF clots acted as successful scaffolds for the regeneration of pulpal contents in immature teeth with necrotic pulps.
Shivashankar et al[63] (2017)	RCT: Radiographic evaluation of root development, increase in the dentin wall thickness and narrowing of canal space, apical closure, and resolution of the periapical lesion, and clinical evaluation of the response to pulp sensibility testing as well as percussion and palpation tests.	3, 6, 9, and 12 months	60 patients (6–28 years old)	1. Conventional bleeding 2. PRP 3. PRF	PRP was better than PRF and induced bleeding technique with respect to periapical wound healing when used in the regenerative endodontic procedures. The groups were comparable on the grounds of root lengthening and lateral wall thickening.	On analyzing the pros and cons of the techniques performed, it is wise to establish induced bleeding technique as the standard endodontic procedure for revascularization of a nonvital immature permanent tooth.

PLGA, polylactic coglycolic acid.

Table 19-4 *(cont)* Use of PRF in endodontic revascularization

Authors (year)	Study design	Follow-up	No. of participants/ sex/age	Groups/ treatment method	Results	Conclusions
Zhou et al[64] (2017)	Animal study: Investigation of pulp revascularization in a dog model.	3 months	3 beagle dogs	1. Control 2. Blood clot 3. Blood clot+PRF	There was a significant difference among the periapical radiolucency decreasing rate of the control (43.75%) and the blood clot (100%) and PRF+blood clot (100%) groups ($P < .05$). The increase in root length and thickness in both the blood clot and PRF+blood clot groups was significantly greater than that in the control group ($P < .05$).	Either a combination of PRF with blood clot or blood clot alone could improve periapical healing, induce root development, and reinforce tooth structure. No additional benefit of PRF to blood clots in endodontic revascularization was found.
Adhikari and Gupta[65] (2018)	Case report: PRF-mediated revascularization of an immature tooth with histopathologic evaluation.	12 months	29-year-old female	PRF	It was revealed that a collagenous matrix for forming hard tissue similar to acellular cementum with a diffuse area of calcification could be seen in the pulp space, being attached with root canal dentin walls.	PRF may yield favorable outcomes in regenerative endodontic procedures.
Nageh et al[66] (2018)	Case series: Evaluation of the possibility of regaining pulp sensibility of immature necrotic teeth using modified regenerative endodontic procedures by inducing bleeding in root canals and using PRF.	3–12 months	15 patients	PRF	Readings of tooth sensibility revealed a highly significant difference ($P < .0001$) between baseline and the 12-month follow-up period.	The presence of sensibility is indicative of the formation of vital pulplike tissue. Reestablishing real pulp tissue after regenerative endodontic treatment is debatable and still needs a higher level of evidence.
Abd El Nasser[67] (2019)	Animal study: Histologic assessment of regenerative procedures in immature dog teeth.	1 month	Three mixed-breed dogs; 60 roots total	1. Blood clot 2. PRF 3. Negative control	There was an increase in the mean area fraction of blood vessels and in the mean number of odontoblasts in the study group compared to the other two groups with a statistical significant difference ($P < .001$). There was an increase in the number of inflammatory cells in the negative control group compared to the study and control group.	Blood clot and PRP show comparative results in periapical healing in terms of vascularity and number of odontoblasts.
Metlerska et al[68] (2019)	Systematic review of human studies.	The meta-analysis literature search included all publications without a year limit. The last search was performed on January 31, 2018.	Electronic search	Platelet concentrates	Only three cases from randomized trials were unsuccessful. Ten of the case reports found positive results for pulp vitality. In randomized trials, the teeth treated with platelet concentrates showed better results for pulp vitality. In almost all cases, they also described thickening and lengthening of the root wall and closure of the apical foramen, which are important for successful treatment of permanent teeth.	Procedures using autologous platelet concentrates were successful in treating permanent teeth with root development.

Table 19-4 *(cont)*

Authors (year)	Study design	Follow-up	No. of participants/ sex/age	Groups/ treatment method	Results	Conclusions
Meza et al[69] (2019)	Case report: PRF was introduced into the instrumented and disinfected root canal, and expanded dental pulp cells were inoculated into the clot. The cervical part of the root canal was sealed with Biodentine and a composite resin.	6 months and 3 years	50-year-old male	PRF	Normal responses to percussion and palpation tests. The tooth had a delayed response to cold, and the electric pulp test was responsive. The PAI and CBCT imaging revealed that the periapical area remained normal with a PAI score of 1 and a CBCT PAI score of 0. The vitality test performed indicated low blood perfusion units.	This case study reveals the potential use of a patient's own dental pulp stem cells and L-PRF as an alternative procedure for the treatment of pulpitis in mature permanent teeth. It also paves the way for the design of personalized cell-based clinical trials in regenerative endodontics.
Mittal et al[70] (2019)	Controlled trial: Comparison of the regenerative potential of a natural scaffold (PRF) and artificial scaffolds (commercially available collagen, placentrex, and chitosan) in necrotic immature permanent teeth.	3, 6, and 12 months	16 necrotic immature permanent maxillary incisors	1. PRF 2. Collagen 3. Placentrex 4. Chitosan	Clinically, patients were completely asymptomatic throughout the study period. Radiographically, all cases showed an improvement in terms of periapical healing, apical closure, root lengthening, and dentinal wall thickening. PRF and collagen gave better results than placentrex and chitosan in terms of periapical healing, apical closure, and dentinal wall thickening.	The revascularization procedure is more effective and conservative over apexification in the management of necrotic immature permanent teeth. This study showed that PRF and collagen are better scaffolds than placentrex and chitosan for inducing apexogenesis in immature necrotic permanent teeth.
Ulusoy et al[71] (2019)	Prospective RCT: Evaluation of blood clot, PRP, PRF, and platelet pellet as potential scaffolds for regenerative endodontics.	Average follow-up time of 28.2 ± 1.2 months	67 healthy children (aged 8–11 years) with 88 immature necrotic incisors	1. PRP 2. PRF 3. Platelet pellet 4. Blood clot	Except for two teeth in the PRF and blood clot groups, all teeth showed similar and high success scores (periapical healing, radiographic root development, and positive response to sensitivity tests) after an average follow-up time of 28.2 ± 1.2 months. Of all teeth, 73.9% showed complete apical closure, with similar closure rates among groups.	PRP, PRF, and platelet pellet can yield similar clinical and radiographic outcomes to blood clots without the need for prior apical bleeding and with significantly less tendency for root canal obliteration.
Nagaveni et al[72] (2020)	Case report: Two methods of revascularization done in traumatized immature, nonvital anterior teeth using PRF and natural blood clot as a scaffold material.	1, 3, 6, 9, and 12 months	11-year-old male	1. Blood clot + MTA 2. PRF+MTA	After 12 months, on clinical examination both teeth showed negative response to percussion and palpation tests but positive response to cold and electric pulp tests. On radiographic examination, the tooth treated with PRF exhibited comparatively faster root lengthening, complete closure of the root apex, more thickening of the root dentinal walls, and narrowing of root canal space compared to the conventionally revascularized tooth.	PRF proves to be more effective than traditional strategies.

RCT, randomized controlled trial.

Table 19-5 Use of PRF in apical surgery

Authors (year)	Study design	Follow-up	No. of participants/ sex/age	Groups/ treatment method	Results	Conclusions
Singh et al[77] (2013)	Case series: Conventional endodontic therapy failed to resolve the problem, and periapical root-end surgery was required; application of PRF in surgical management of periapical lesions.	6 months	15 patients aged 20–50 years	PRF	At the end of 6 months, all patients showed complete bone regeneration.	PRF proves effective for use in apical surgery cases.
Angerame et al[79] (2015)	RCT: Patients with chronic apical periodontitis were randomly assigned to either PRF (n = 7) or control (n = 4).	2 and 3 months	11 patients	1. Control 2. PRF	In the PRF group, the patients experienced less pain in the 2–6 hours postoperatively as well as edema, which never exceeded moderate intraoral swelling. Radiographic healing was detectable earlier in the PRF group, with the majority of cases scored as complete healing after 2–3 months.	The adjunctive use of PRF might promote the acceleration of radiographic healing and may reduce postoperative discomfort.

Use of PRF in Endodontic Microsurgery

Since the 1990s, endodontic surgery has transformed into a more advanced microsurgical procedure with specialized instruments and more biologically acceptable materials. Typically, it is often a last resort to surgically maintain a tooth with a periapical lesion that cannot be managed with conventional endodontic retreatment.[73] The main goal of apical surgery is to prevent bacterial leakage from the root canal system into the periradicular tissues by placing a tight root-end filling from the apical portion of the canal following root-end resection.[73] A recent systematic review discussing the healing of periapical lesions after surgical endodontics concluded that surgical intervention reduced the healing time needed for large periapical lesions, with success rates as high as 92%.[74]

The procedure involves surgical elevation of a mucoperiosteal flap, removal of targeted tissue and foreign material, root-end resection, preparation, and retrofill with a biocompatible material.[5] The use of regenerative techniques in apical surgery has been reviewed in the literature and can have significant benefits in periapical lesions larger than 10 mm, through-and-through lesions, and apicomarginal lesions.[75] The goals of tissue regeneration in apical surgery include *(1)* regrowth of alveolar and periradicular bone, *(2)* reestablishment of a PDL at the resection plane and at the surgically exposed root surface, and *(3)* formation of new cementum at the cut root face.[75]

A variety of grafting materials have been used for GBR, such as autogenous bone, allografts, xenografts, and synthetic materials.[75] These materials act as a space holder until the osteogenic cells and cytokines responsible for bone regeneration migrate into the site and these undifferentiated mesenchymal cells differentiate into PDL cells and cementoblasts.[75] Each of these grafting materials has different resorption rates and properties, and any commercial graft product has the potential to form a foreign body reaction that could affect treatment outcomes.

PRF was since proposed as a regenerative agent in apical endodontic microsurgery.[76,77] A randomized controlled trial evaluating the benefit of PRF in the treatment of apicomarginal defects showed only a slight increase in success with the use of PRF (86.7%) compared to controls (80%).[76] However, periodontal probing depths showed a statistically significant reduction in the PRF group.[76] A systematic review on the use of PRF in apical surgery concluded that PRF was beneficial for patients' quality of life during the (early) postoperative stage of healing and displayed accelerated bone healing.[78] Table 19-5 shows the results of studies on the use of PRF in apical surgery.[76,77,79–82]

Table 19-5 *(cont)*

Authors (year)	Study design	Follow-up	No. of participants/ sex/age	Groups/ treatment method	Results	Conclusions
Dhiman et al[76] (2015)	RCT: Patients with suppurative chronic apical periodontitis and apicomarginal communication were randomly assigned to either the PRF or the control group.	Every 3 months for a period of 12 months	30 patients	1. Control 2. PRF	The overall success rate was 83.33%, with a success rate of 86.66% (13 of 15 teeth) for the PRF group and 80% (12 of 15 teeth) for the control group. Both groups exhibited a significant reduction in probing depth, clinical attachment level, gingival marginal position, and size of periapical lesion at 12-month follow-up.	The adjunctive use of regenerative techniques may not promote healing of apicomarginal defects of endodontic origin.
Vidhale et al[80] (2015)	Case report: Management of a radicular cyst using PRF and iliac bone graft.	3 months	22-year-old male	PRF	Panoramic radiography showed increased radiopacity as compared to earlier radiography and a good amount of regeneration; repair at the site of injury was also appreciated, which could be connected to the use of PRF and iliac bone grafting.	With the present case report, it can be confirmed that PRF along with bone grafting presented regenerative benefits.
Anantula et al[81] (2016)	Case reports: Management of periapical inflammatory lesions using PRF.	3 and 12 months	23-year-old male and 22-year-old male	PRF	Follow-up radiographs at 3 and 12 months showed satisfactory bone regeneration in the periapical defect.	PRF features all the necessary parameters permitting optimal healing, but numerous perspectives of PRF have still to be clinically tested.
Pinto et al[82] (2017)	Case report: Asymptomatic apical periodontitis associated with dens invaginatus type II.	6 months and 1 year	20-year-old female	PRF	The radiographs and CBCT images after 1 year showed a repair of the periapical lesion and reestablishment of the vestibular cortical bone.	PRF can be used as a complement in apical surgery.

RCT, randomized controlled trial.

Clinical case example

A 53-year-old woman presented with pain and swelling associated with a necrotic maxillary left lateral incisor (Fig 19-8 and Video 19-2). An impacted canine was present and masked any radiographic signs of pathology (Fig 19-8a). A CBCT was acquired, and a large lesion was identified at the apex of the maxillary left lateral incisor that had ruptured the buccal cortical plate (Fig 19-8b). Nonsurgical therapy was completed, followed by a surgical approach to remove all infected tissues. A submarginal periapical flap was elevated, and the inflammatory tissue was removed with curettes (Fig 19-8c) and laser-activated irrigation. A retrofill was placed with bioceramic fast-setting putty (Fig 19-8d). Multiple PRF membranes were placed into the surgical site (Fig 19-8e), and the flap was reapproximated and sutured with primary closure (Fig 19-8f). A follow-up CBCT was taken at the 9-month recall and showed complete healing of the site.

Video 19-2

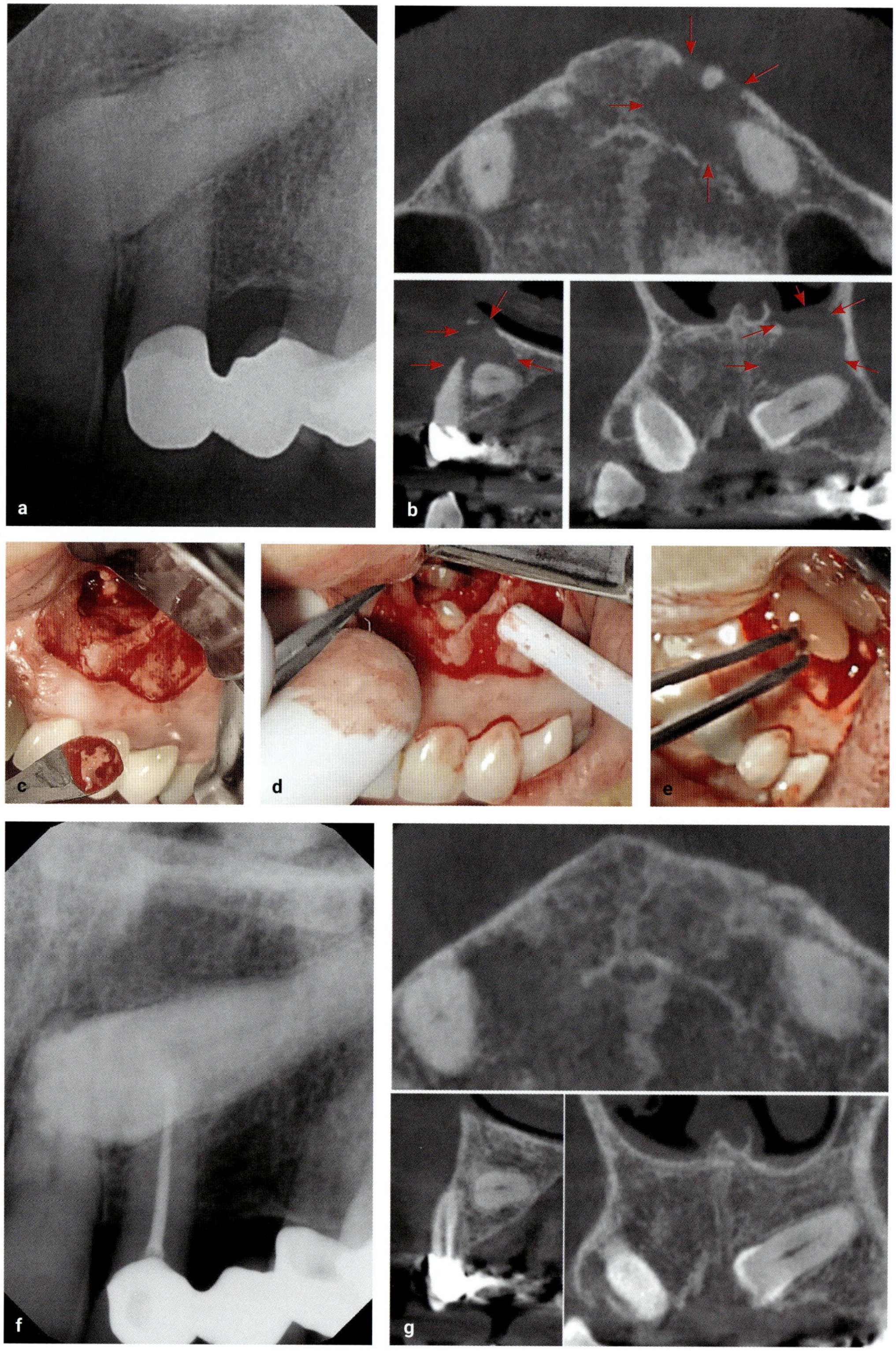

Fig 19-8 Use of PRF for apical endodontic microsurgery. *(a)* Periapical radiograph demonstrating an impacted canine. *(b)* A follow-up CBCT was taken the same day. Note the large periapical lesion present and hardly observable on the periapical radiograph. *(c)* Removal of inflammatory granulation tissue. *(d)* Following apicoectomy, bioceramic retrofill of the canal. *(e)* Placement of PRF membranes into the defect. *(f)* Periapical radiograph immediately postoperative. *(g)* Nine-month CBCT recall showing complete healing. (Case performed by Dr Valerie Kanter.)

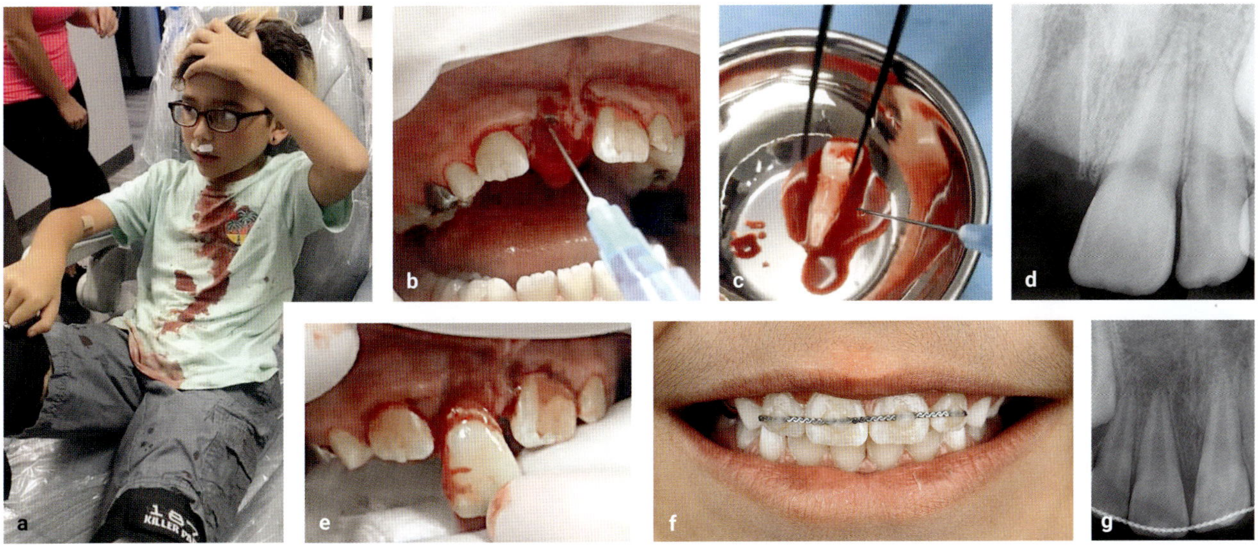

Fig 19-9 Use of PRF for the treatment of traumatic avulsion of a maxillary right central incisor. *(a)* Traumatic tooth avulsion following an accident. *(b)* Intraoral photograph demonstrating the avulsed tooth. *(c)* The tooth was placed in PRF as soon as possible. Once the patient entered with an avulsed tooth and clinical diagnosis was confirmed, blood was immediately drawn and centrifuged to get the tooth into PRF as rapidly as possible. *(d)* Periapical radiograph demonstrating the avulsed tooth. *(e)* Tooth replantation. *(f)* Clinical photograph of the splinted tooth in place. *(g)* Periapical radiograph showing the splint.

Use of PRF in the Treatment of Avulsed Teeth

A traumatic injury that results in avulsion—the complete removal of the tooth from its alveolus—severely damages the tooth's neurovascular support and PDL. The treatment of choice is immediate reimplantation; however, postoperative complications include loss of pulp vitality, ankylosis, replacement or inflammatory resorption, and tooth loss. The extent of postoperative complications depends on several factors such as elapsed time between trauma and repositioning, the maturity of the root, and the preservation medium.[83,84]

The use of PRF as a root surface treatment prior to reimplantation has been proposed due to its proven success in tissue repair.[84,85] In vitro studies have shown that the high levels of PDGF, TGF-β, and VEGF contained in platelet concentrates modulate the cell-specific proliferation of cultures of gingival fibroblasts, oral epithelial cells, PDL cells, and osteoblasts.[85,86]

Until the use of PRF was investigated, Hank's balanced salt solution was deemed the best solution available because it contains essential nutrients for the maintenance of cells in an avulsed tooth and was able to preserve cells and tissues for a 24-hour period. Traditionally, if a tooth was dry for more than 1 hour, the prognosis was drastically reduced because the PDL cells were nonviable. Various studies have shown that platelet concentrates increase the proliferation and viability of various cell types including PDL cells (see chapter 2). This makes PRF a valuable option both as a medium for preserving the avulsed tooth and for placing in the alveolus prior to replantation. PRF may reduce postplantation complications such as replacement and inflammatory resorption,[87] and basic research studies have shown that it inhibits bone-resorbing osteoclasts.[88]

A recent study on PRF showed its effectiveness in maintaining and recovering PDL cells from extracted teeth, even when kept dry up to 2 hours.[87] In this study, the teeth that were dry for 2 hours and then stored in PRF for 30 minutes to 1 hour showed no significant difference in total number of cells and viability compared to the controls. These results support the use of PRF as an effective technique with low cost and easy application for clinical practice.[87] Figure 19-9 demonstrates a traumatic avulsion of a maxillary right central incisor treated with PRF and endodontic therapy (Video 19-3), and Table 19-6 shows the results of studies on the topic.[83,85,89,90]

Video 19-3

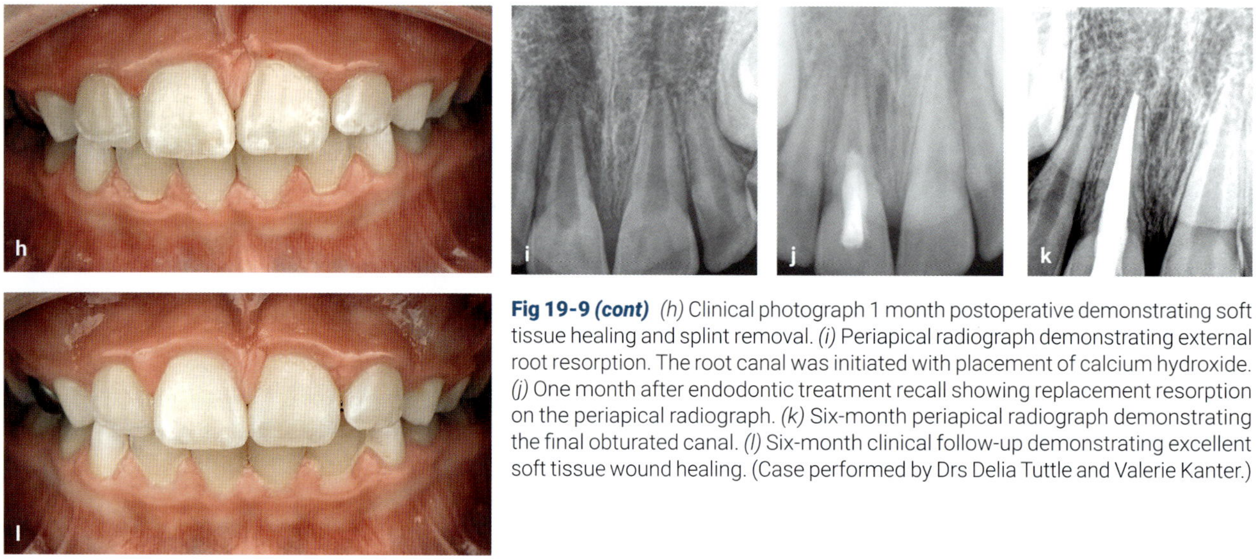

Fig 19-9 *(cont)* *(h)* Clinical photograph 1 month postoperative demonstrating soft tissue healing and splint removal. *(i)* Periapical radiograph demonstrating external root resorption. The root canal was initiated with placement of calcium hydroxide. *(j)* One month after endodontic treatment recall showing replacement resorption on the periapical radiograph. *(k)* Six-month periapical radiograph demonstrating the final obturated canal. *(l)* Six-month clinical follow-up demonstrating excellent soft tissue wound healing. (Case performed by Drs Delia Tuttle and Valerie Kanter.)

Table 19-6 Use of PRF in traumatized/avulsed teeth

Authors (year)	Study design	Follow-up	No. of participants/ sex/age	Groups/ treatment method	Results	Conclusions
Johns et al[89] (2013)	Case report: Emergency and rehabilitation treatment of an avulsed maxillary anterior tooth using PRF.	6 months	15-year-old male	PRF	During follow-up, no clinical signs and symptoms were present. After the initial 6 months, no further bone loss and attachment loss were observed. The tooth remained functional and was esthetically acceptable.	If the PDL that is left attached to the root surface does not dry out using PRF, the consequences of tooth avulsion are usually minimal.
Zhao et al[85] (2013)	Animal study: Assessment of the potential application of the PDLSCs/PRF construct in tooth reimplantation; 36 incisors were extracted from 6 dogs.	56 days	6 dogs	1. Control 2. PRF 3. PDLSCs 4. PRF/ PDLSCs	Eight weeks post-reimplantation, the PDLSCs/PRF group achieved a more effective periodontal healing, characterized by the regeneration of PDL-like tissues and a reduction of ankylosis and inflammation, compared with the other testing groups.	These overall results suggest that the PDLSCs/PRF construct may be a useful tool for alveolar surgery that has the potential to improve the clinical outcomes in future avulsed tooth reimplantations.
Hiremath et al[90] (2014)	In vitro study: 30 teeth with an intact crown and healthy PDL (advised for orthodontic extraction).	0 minutes to 1 hour	30 extracted teeth	1. PPP 2. PRF	Viability of PDL cells was significantly higher in the PRF group.	Within the parameters of this study, a combination of PRF and PPP demonstrated a higher number of viable PDL cells and hence could be a good biologic rejuvenating medium for avulsed teeth.
Bucchi et al[83] (2016)	Systematic review	–	Electronic search of animal and human studies	All forms of platelet concentrates	Platelet concentrates increase the proliferation and viability of PDL cells in avulsed teeth.	PRF is superior in terms of neoformation of tissue similar to PDL. Evidence exists to support the use of these concentrates; however, no RCTs were found that compared the effectiveness of the various concentrates with one another.

PDLSC, periodontal ligament stem cell; RCT, randomized controlled trial.

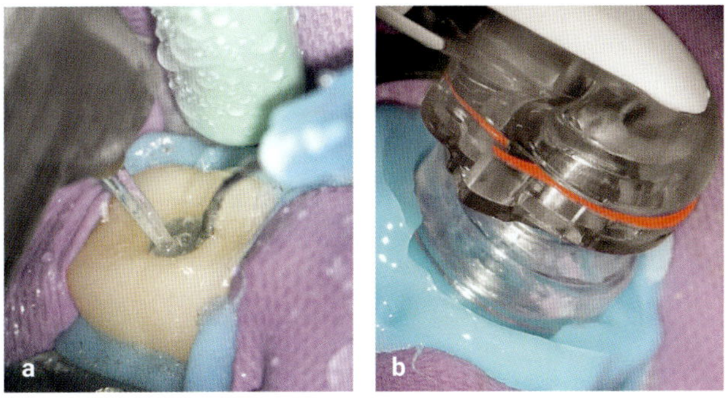

Fig 19-10 *(a and b)* Use of laser-activated irrigation and multisonic irrigation for endodontic procedures.

Discussion

This chapter demonstrated the uses of PRF in pulp tissue engineering applications as a scaffold of autogenous sources. Typically, when endodontic treatments are underway, the pulp is under an inflammatory condition. The introduction of foreign body materials and other biomaterials derived from animal/human sources may potentially further increase inflammation. Therefore, PRF was hypothesized some years ago to act as a regenerative agent to stimulate pulpal tissues because it has been shown in preclinical studies to counteract LPS, decrease inflammatory cytokines, and improve overall healing.

Some major advancements in recent years have focused on the ability to utilize new modalities that facilitate various aspects of endodontic procedures such as root canal irrigation and disinfection. These include the incorporation of laser-activated irrigation as well as multisonic irrigation (Fotona GentleWave; Fig 19-10). These procedures have facilitated the preparation, irrigation, and disinfection of canals when compared to classical/traditional endodontic techniques.

One additional therapy investigated more recently has been that of ozone therapy. When ozone is produced (O_3), it becomes a highly active/unstable molecule that disrupts bacterial cell walls and membranes, thereby destroying them. The combination of such modalities has more recently been investigated in novel protocols for endodontic procedures such as the EndOzLase. Future research continuing to combine laser therapy with ozone and new modalities that surely will be introduced in upcoming years will certainly increase the success rates of various regenerative endodontic procedures.

One interesting avenue of research has been on the additional incorporation of various small biomolecules into liquid-PRF.[91] As presented in previous chapters, antibiotics have been added to liquid-PRF for treatment of intrabony defects, and liquid-PRF has similarly been dosed with rhBMP-2 to facilitate bone formation. In the regenerative endodontic space, many small molecules have been proposed as potential contributors to pulp regeneration. Therefore, future research aimed at incorporation of small biomolecules such as vitamins, growth factors, melatonin, exosomes, and peptides could additionally lead to more predictable and regenerative outcomes.[91–93] Therefore, the future field is filled with many ongoing research endeavors.

Conclusion

As highlighted throughout the chapter, many case reports, case series, and randomized controlled trials have now been performed evaluating PRF in regenerative endodontic procedures. PRF proves to be a low-cost, easily procurable regenerative strategy with much upside due to its ability to revascularize tissues and its incorporation of defense-fighting immune cells. This chapter highlighted the research and application of PRF for pulp capping, pulpotomy, apexification, revascularization, apical surgery, and treatment of avulsed teeth. Future research in this field is certainly ongoing, with many randomized studies planned for the upcoming years.

References

1. Hotwani K, Sharma K. Platelet rich fibrin-a novel acumen into regenerative endodontic therapy. Rest Dent Endod 2014;39:1–6.
2. Murray PE, Garcia-Godoy F, Hargreaves KM. Regenerative endodontics: A review of current status and a call for action. J Endod 2007;33:377–390.
3. Mao X, Liu Y, Chen C, Shi S. Mesenchymal stem cells and their role in dental medicine. Dent Clin North Am 2017;61:161–172.
4. Chai J, Jin R, Yuan G, Kanter V, Miron RJ, Zhang Y. Effect of liquid platelet-rich fibrin and platelet-rich plasma on the regenerative potential of dental pulp cells cultured under inflammatory conditions: A comparative analysis. J Endod 2019;45:1000–1008.
5. American Association of Endodontists. Guide to Clinical Endodontics: Chicago: AAE, 2013.
6. Song M, Yu B, Kim S, et al. Clinical and molecular perspectives of reparative dentin formation: Lessons learned from pulp-capping materials and the emerging roles of calcium. Dent Clin North Am 2017;61:93–110.

7. Paula AB, Laranjo M, Marto CM, et al. Direct pulp capping: What is the most effective therapy?—Systematic review and meta-analysis. J Evid Based Dent Pract 2018;18:298–314.
8. Rechenberg D-K, Galicia JC, Peters OA. Biological markers for pulpal inflammation: A systematic review. PloS one 2016;11.
9. Woo SM, Kim WJ, Lim HS, et al. Combination of mineral trioxide aggregate and platelet-rich fibrin promotes the odontoblastic differentiation and mineralization of human dental pulp cells via BMP/Smad signaling pathway. J Endod 2016;42:82–88.
10. Schröder U. Effects of calcium hydroxide-containing pulp-capping agents on pulp cell migration, proliferation, and differentiation. J Dent Res 1985;64:541–548.
11. Bal C, Alacam A, Tuzuner T, Tirali RE, Baris E. Effects of antiseptics on pulpal healing under calcium hydroxide pulp capping: A pilot study. Eur J Dent 2011;5:265–272.
12. Andelin WE, Shabahang S, Wright K, Torabinejad M. Identification of hard tissue after experimental pulp capping using dentin sialoprotein (DSP) as a marker. J Endod 2003;29:646–650.
13. Komabayashi T, Zhu Q. Innovative endodontic therapy for anti-inflammatory direct pulp capping of permanent teeth with a mature apex. Oral Surg Oral Med Oral Pathol Oral Radiol Endod 2010;109: e75–e81.
14. Tabatabayi MH, Tavakoli A, Ameghani BA. Regenerative property of PRF used as capping material in pulpotomoy in dogs. Biomed Res 2017;28:4634–4639.
15. Bakshi C, Bogra P, Singh SV, Gupta S, Makhija C. PRF in vital pulp therapy: Case report. Baba Farid Univ Dent J 2017;7:61–66.
16. Dou L, Yan Q, Yang D. Effect of five dental pulp capping agents on cell proliferation, viability, apoptosis and mineralization of human dental pulp cells. Exp Ther Med 2020;19:2377–2383.
17. Moritz A, Schoop U, Goharkhay K, et al. The bactericidal effect of Nd:YAG, Ho:YAG, and Er:YAG laser irradiation in the root canal: An in vitro comparison. J Clin Las Med Surg 1999;17:161–164.
18. Olivi G, Genovese M, Maturo P, Docimo R. Pulp capping: Advantages of using laser technology. Eur J Paediatr Dent 2007;8: 89–95.
19. Ozses Ozkaya B, Gulsahi K, Ungor M, Gocmen JS. A comparison of Er:YAG laser with photon-initiated photoacoustic streaming, Nd:YAG laser, and conventional irrigation on the eradication of root dentinal tubule infection by *Enterococcus faecalis* biofilms: A scanning electron microscopy study. Scanning 2017;2017:6215482.
20. Re L, Martínez-Sánchez G, Perez-Davison G, Sirito M. Role of ozone/oxygen in fibroblast growth factor activation: Discovering the facts. Int J Ozone Ther 2010;9:8.
21. Alqaderi H, Lee CT, Borzangy S, Pagonis TC. Coronal pulpotomy for cariously exposed permanent posterior teeth with closed apices: A systematic review and meta-analysis. J Dent 2016;44: 1–7.
22. Kunert GG, Kunert IR, da Costa Filho LC, de Figueiredo JAP. Permanent teeth pulpotomy survival analysis: Retrospective follow-up. J Dent 2015;43:1125–1131.
23. Morotomi T, Washio A, Kitamura C. Current and future options for dental pulp therapy. Jpn Dent Sci Rev 2019;55:5–11.
24. Kumar V, Juneja R, Duhan J, Sangwan P, Tewari S. Comparative evaluation of platelet-rich fibrin, mineral trioxide aggregate, and calcium hydroxide as pulpotomy agents in permanent molars with irreversible pulpitis: A randomized controlled trial. Contemp Clin Dent 2016;7:512–518.
25. Ward J. Vital pulp therapy in cariously exposed permanent teeth and its limitations. Aust Endod J 2002;28:29–37.
26. Dohan DM, Choukroun J, Diss A, et al. Platelet-rich fibrin (PRF): A second-generation platelet concentrate. Part III: Leucocyte activation: A new feature for platelet concentrates? Oral Surg Oral Med Oral Pathol Oral Radiol Endodontol 2006;101:e51–e55.
27. Hiremath H, Saikalyan S, Kulkarni S, Hiremath V. Second-generation platelet concentrate (PRF) as a pulpotomy medicament in a permanent molar with pulpitis: A case report. Int Endod J 2012;45:105–112.
28. Keswani D, Pandey RK, Ansari A, Gupta S. Comparative evaluation of platelet-rich fibrin and mineral trioxide aggregate as pulpotomy agents in permanent teeth with incomplete root development: A randomized controlled trial. J Endod 2014;40:599–605.
29. Pathak S, Bansode P, Ahire C. PRF as a pulpotomy medicament in a permanent molar with pulpitis: A case report. IOSR J Dent Med Sci 2014;13:5–9.
30. Patil PAB, Kamatagi L, Patil J, Saraf P. Pulpotomy using platelet rich fibrin and biodentin in a permanent molar tooth: A case report. IOSR J Dent Med Sci 2017;16:38–41.
31. Guerrero F, Mendoza A, Ribas D, Aspiazu K. Apexification: A systematic review. J Conserv Dent 2018;21:462–465.
32. Mente J, Leo M, Panagidis D, et al. Treatment outcome of mineral trioxide aggregate in open apex teeth. J Endod 2013;39:20–26.
33. Sjögren U, Hägglund B, Sundqvist G, Wing K. Factors affecting the long-term results of endodontic treatment. J Endod 1990;16: 498–504.
34. Ørstavik D, Qvist V, Stoltze K. A multivariate analysis of the outcome of endodontic treatment. Eur J Oral Sci 2004;112:224–230.
35. Praveena N, Subbarayan Y, Deepthi R. Comparative evaluation of prevention of apical extrusion of MTA apical plug with different internal matrices: A clinical study. Int J Oral Health Med Res 2017;4:1–5.
36. Kobayashi E, Fluckiger L, Fujioka-Kobayashi M, et al. Comparative release of growth factors from PRP, PRF, and advanced-PRF. Clin Oral Investig 2016;20:2353–2360.
37. Pawar AM, Pawar SM, Pawar MG, Kokate SR. Retreatment of endodontically failed tooth with wide-open apex using platelet rich fibrin membrane as matrix and an apical plug of Biodentine. Eur J Gen Dent 2015;4:150.
38. Khetarpal A, Chaudhry S, Talwar S, Verma M. Endodontic management of open apex using MTA and platelet-rich fibrin membrane barrier: A newer matrix concept. J Clin Exp Dent 2013;5:e291.
39. Borie E, Oliví DG, Orsi IA, et al. Platelet-rich fibrin application in dentistry: A literature review. Int J Clin Exp Med 2015;8:7922.
40. Hiremath H, Gada N, Kini Y, Kulkarni S, Yakub SS, Metgud S. Single-step apical barrier placement in immature teeth using mineral trioxide aggregate and management of periapical inflammatory lesion using platelet-rich plasma and hydroxyapatite. J Endod 2008;34:1020–1024.
41. Sharma V, Sharma S, Dudeja P, Grover S. Endodontic management of nonvital permanent teeth having immature roots with one step apexification, using mineral trioxide aggregate apical plug and autogenous platelet-rich fibrin membrane as an internal matrix: Case series. Contemp Clin Dent 2016;7:67–70.
42. Kathuria A, Chaudhry S, Talwar S, Verma M. Endodontic management of single rooted immature mandibular second molars with single canal using MTA and platelet-rich fibrin membrane barrier: A case report. J Clin Exp Dent 2011;3:e487–e490.
43. Rudagi KB, Rudagi BM. One-step apexification in immature tooth using grey mineral trioxide aggregate as an apical barrier and autologous platelet rich fibrin membrane as an internal matrix. J Conserv Dent 2012;15:196–199.
44. Kumar A, Yadav A, Shetty N. One-step apexification using platelet rich fibrin matrix and mineral trioxide aggregate apical barrier. Indian J Dent Res 2014;25:809–812.
45. Yadav P, Pruthi PJ, Naval RR, Talwar S, Verma M. Novel use of platelet-rich fibrin matrix and MTA as an apical barrier in the management of a failed revascularisation case. Dent Traumatol 2015;31:328–331.

46. Khanduri N, Kurup D. Single-visit apexification with Biodentine and platelet-rich fibrin. Endodontology 2018;30:181–183.
47. Prasad J, de Noronha de Ataide I, Chalakkal P, Likhyani LK. Comparison between the outcomes of two platelet-rich concentrates on apexogenesis in young permanent incisors requiring endodontic retreatment. Contemp Clin Dent 2018;9(suppl 1):S156–S159.
48. Santhakumar M, Yayathi S, Retnakumari N. A clinicoradiographic comparison of the effects of platelet-rich fibrin gel and platelet-rich fibrin membrane as scaffolds in the apexification treatment of young permanent teeth. J Indian Soc Pedod Prev Dent 2018;36:65–70.
49. Rathod S, Patil AT, Sandhyarani B, Bahirat S, Singh DM. Platelet rich fibrin: A boon to apexogenesis. Int J Curr Res 2019;11:3161–3164.
50. Banchs F, Trope M. Revascularization of immature permanent teeth with apical periodontitis: New treatment protocol? J Endodontics 2004;30:196–200.
51. Narang I, Mittal N, Mishra N. A comparative evaluation of the blood clot, platelet-rich plasma, and platelet-rich fibrin in regeneration of necrotic immature permanent teeth: A clinical study. Contemp Clin Dent 2015;6:63–68.
52. Conde M, Chisini L, Sarkis-Onofre R, Schuch H, Nör J, Demarco F. A scoping review of root canal revascularization: Relevant aspects for clinical success and tissue formation. Int Endod J 2017;50:860–874.
53. Tong HJ, Rajan S, Bhujel N, Kang J, Duggal M, Nazzal H. Regenerative endodontic therapy in the management of nonvital immature permanent teeth: A systematic review—Outcome evaluation and meta-analysis. J Endod 2017;43:1453–1464.
54. Shivashankar VY, Johns DA, Vidyanath S, Kumar MR. Platelet rich fibrin in the revitalization of tooth with necrotic pulp and open apex. J Conserv Dent 2012;15:395–398.
55. Keswani D, Pandey R. Revascularization of an immature tooth with a necrotic pulp using platelet-rich fibrin: A case report. Int Endod J 2013;46:1096–1104.
56. Mishra N, Narang I, Mittal N. Platelet-rich fibrin-mediated revitalization of immature necrotic tooth. Contemp Clin Dent 2013;4:412–415.
57. Johns DA, Shivashankar VY, Krishnamma S, Johns M. Use of photoactivated disinfection and platelet-rich fibrin in regenerative endodontics. J Conserv Dent 2014;17:487–490.
58. Faizuddin U, Solomon RV, Mattapathi J, Guniganti SS. Revitalization of traumatized immature tooth with platelet-rich fibrin. Contemp Clin Dent 2015;6:574–576.
59. Jadhav GR, Shah D, Raghvendra SS. Autologus platelet rich fibrin aided revascularization of an immature, non-vital permanent tooth with apical periodontitis: A case report. J Nat Sci Biol Med 2015;6:224.
60. Sharma S, Mittal N. A comparative evaluation of natural and artificial scaffolds in regenerative endodontics: A clinical study. Saudi Endod J 2016;6:9.
61. Subash D, Shoba K, Aman S, Bharkavi SK. Revitalization of an immature permanent mandibular molar with a necrotic pulp using platelet-rich fibrin: A case report. J Clin Diagn Res 2016;10:Zd21–Zd23.
62. Bakhtiar H, Esmaeili S, Tabatabayi SF, Ellini MR, Nekoofar MH, Dummer PM. Second-generation platelet concentrate (platelet-rich fibrin) as a scaffold in regenerative endodontics: A case series. J Endod 2017;43:401–408.
63. Shivashankar VY, Johns DA, Maroli RK, et al. Comparison of the effect of PRP, PRF and induced bleeding in the revascularization of teeth with necrotic pulp and open apex: A triple blind randomized clinical trial. J Clin Diagn Res 2017;11:ZC34.
64. Zhou R, Wang Y, Chen Y, et al. Radiographic, histologic, and biomechanical evaluation of combined application of platelet-rich fibrin with blood clot in regenerative endodontics. J Endod 2017;43:2034–2040.
65. Adhikari HD, Gupta A. Report of a case of platelet-rich fibrin-mediated revascularization of immature 12 with histopathological evaluation. J Conserv Dent 2018;21:691–695.
66. Nageh M, Ahmed GM, El-Baz AA. Assessment of regaining pulp sensibility in mature necrotic teeth using a modified revascularization technique with platelet-rich fibrin: A clinical study. J Endod 2018;44:1526–1533.
67. Abd El Nasser AK. Evaluation of the success rate of revascularization technique using leukocyte-platelet-rich fibrin (L-PRF) concentrate compared to blood clot as a scaffold. EC Dent Sci 2019;18:623–630.
68. Metlerska J, Fagogeni I, Nowicka A. Efficacy of autologous platelet concentrates in regenerative endodontic treatment: A systematic review of human studies. J Endod 2019;45:20–30.e21.
69. Meza G, Urrejola D, Saint Jean N, et al. Personalized cell therapy for pulpitis using autologous dental pulp stem cells and leukocyte platelet-rich fibrin: A case report. J Endod 2019;45:144–149.
70. Mittal N, Parashar V, Paul S, et al. Regenerative Evaluation of immature roots using PRF and artificial scaffolds in necrotic permanent teeth: A clinical study. J Contemp Dent Pract 2019;20:720–726.
71. Ulusoy AT, Turedi I, Cimen M, Cehreli ZC. Evaluation of blood clot, platelet-rich plasma, platelet-rich fibrin, and platelet pellet as scaffolds in regenerative endodontic treatment: A prospective randomized trial. J Endod 2019;45:560–566.
72. Nagaveni NB, Poornima P, Mathew MG, Soni AJ, Khan MM. A comparative evaluation of revascularization done in traumatized immature, necrotic anterior teeth with and without platelet-rich fibrin: A case report. Int J Clin Pediatr Dent 2020;13:98–102.
73. Torabinejad M, Corr R, Handysides R, Shabahang S. Outcomes of nonsurgical retreatment and endodontic surgery: A systematic review. J Endod 2009;35:930–937.
74. Alghamdi F, Alhaddad AJ, Abuzinadah S. Healing of periapical lesions after surgical endodontic retreatment: A systematic review. Cureus 2020;12:e6916.
75. von Arx T, AlSaeed M. The use of regenerative techniques in apical surgery: A literature review. Saudi Dent J 2011;23:113–127.
76. Dhiman M, Kumar S, Duhan J, Sangwan P, Tewari S. Effect of platelet-rich fibrin on healing of apicomarginal defects: A randomized controlled trial. J Endod 2015;41:985–991.
77. Singh S, Singh A, Singh S, Singh R. Application of PRF in surgical management of periapical lesions. Natl J Maxillofac Surg 2013;4:94–99.
78. Meschi N, Castro AB, Vandamme K, Quirynen M, Lambrechts P. The impact of autologous platelet concentrates on endodontic healing: A systematic review. Platelets 2016;27:613–633.
79. Angerame D, De Biasi M, Kastrioti I, Franco V, Castaldo A, Maglione M. Application of platelet-rich fibrin in endodontic surgery: A pilot study [in Italian]. Giornale Italiano di Endodonzia 2015;29:51–57.
80. Vidhale G, Jain D, Jain S, Godhane AV, Pawar GR. Management of radicular cyst using platelet-rich fibrin & iliac bone graft: A case report. J Clin Diagn Res 2015;9:ZD34.
81. Anantula K, Annareddy A. Platelet-rich fibrin (PRF) as an autologous biomaterial after an endodontic surgery. J NTR Univ Health Sci 2016;5:49.
82. Pinto N, Harnish A, Cabrera C, Andrade C, Druttman T, Brizuela C. An innovative regenerative endodontic procedure using leukocyte and platelet-rich fibrin associated with apical surgery: A case report. J Endod 2017;43:1828–1834.

83. Bucchi C, Arias A, Fuentes R. Platelet concentrates in the replantation of avulsed teeth: A systematic review. Int J Clin Exp Med 2016;9:22985–22992.
84. Reichert da Silva Assunção L, Colenci R, Ferreira do-Amaral CC, et al. Periodontal tissue engineering after tooth replantation. J Periodontol 2011;82:758–766.
85. Zhao YH, Zhang M, Liu NX, et al. The combined use of cell sheet fragments of periodontal ligament stem cells and platelet-rich fibrin granules for avulsed tooth reimplantation. Biomaterials 2013;34:5506–5520.
86. Arikan F, Becerik S, Sonmez S, Gurhan I. Effect of platelet-rich plasma on gingival and periodontal ligament fibroblasts: New in-vitro growth assay. Braz J Oral Sci 2007;6:1432–1437.
87. Nene KS, Bendgude V. Prognosis of replanted avulsed permanent incisors: A systematic review. Int J Pedo Rehabil 2018;3:87.
88. Kargarpour Z, Nasirzade J, Strauss FJ, et al. Platelet-rich fibrin suppresses in vitro osteoclastogenesis. J Periodontol 2020;91:413–421.
89. Johns DA, Shivashankar VY, Maroli RK, Vidyanath S. Novel management of avulsed tooth by pulpal and periodontal regeneration. J Endod 2013;39:1658–1662.
90. Hiremath H, Kulkarni S, Sharma R, Hiremath V, Motiwala T. Use of platelet-rich fibrin as an autologous biologic rejuvenating media for avulsed teeth—An in vitro study. Dent Traumatol 2014;30:442–446.
91. Miron RJ, Zhang Y. Autologous liquid platelet rich fibrin: A novel drug delivery system. Acta Biomater 2018;75:35–51.
92. Li JG, Lin JJ, Wang ZL, et al. Melatonin attenuates inflammation of acute pulpitis subjected to dental pulp injury. Am J Transl Res 2015;7:66–78.
93. Huang CC, Narayanan R, Alapati S, Ravindran S. Exosomes as biomimetic tools for stem cell differentiation: Applications in dental pulp tissue regeneration. Biomaterials 2016;111:103–115.

20

Use of PRF in Facial Esthetics

Contributors

Catherine Davies
Richard J. Miron
Ana Paz
Miguel Stanley

Chapter Highlights

- *Use of PRF with microneedling*
- *Injection techniques using PRF*
- *Hair regeneration with PRF*

Facial esthetics has become one of the fastest-growing industries in the world. While commonly utilized products including Botox and Juvederm have dominated the market, platelet concentrates offer a safe, easy-to-obtain, and completely immune-biocompatible method for the regeneration of aging skin. As trends continue to support minimally invasive esthetic procedures, it is clear that both beginners as well as advanced practitioners seek convenient, safe, and effective therapies. This chapter describes the use of PRF to augment/regenerate various facial tissues and to treat hair loss. The chapter provides an overview of the global and regional treatment strategies to rejuvenate skin using a combination of PRF, microneedling, and lasers. Recommendations regarding needle gauge, length, and type are also presented. Finally, a brief overview of PRF use for hair regeneration is provided with a video demonstrating its use in clinical practice.

In our modern lifestyle, a youthful yet natural appearance has been the desired goal for many individuals. Yet aging skin is an inevitable process that occurs as we gradually get older.[1,2] Several genetic and environmental factors have been associated with this process.[3] Exposure to sun, pollution, and various chemicals has been known to cause skin and/or DNA damage, speeding the aging process.[3] A number of changes to the skin may occur as a result, including skin atrophy, telangiectasia, fine and deep wrinkles, yellowing (solar elastosis), and dyspigmentation.[3] Furthermore, other physical and environment factors including poor diet, lack of exercise, caffeine intake, smoking, and drug use are known to speed the aging process.[4]

A variety of treatment options have been proposed over the years for the aging skin to improve esthetic appearance. One of the first methods proposed for facial rejuvenation incorporated acupuncture.[5] This concept was derived based on accumulating evidence that trauma to the skin in the form of a needle and/or syringe, dermal roller, or more recently microneedling could induce slight tissue damage leading to new angiogenesis, growth factor release, and subsequent new tissue regeneration, thereby offering a more youthful appearance. Because of the popularity of facial esthetics and rapidly evolving trends in the field, more invasive techniques were also proposed. These include facelifts, aggressive laser treatment modalities, and various grafting procedures.[6–8] One of the advantages of platelet therapies is their ability to be combined with microneedling, lasers, other facial esthetic treatment modalities, plastic surgery, and hair restoration simply to improve healing outcomes. PRF therapy in facial esthetics has become so prominent that the authors of this chapter have written an entire textbook on the topic.[9]

Autologous PRF is an excellent source of growth factors and fibrin due to its biologic properties and endogenous origin. Apart from the wide uses of PRF in medicine, there is substantial clinical evidence regarding its use in the esthetic field for the stimulation of the superficial dermis as well as for the deep layers of the dermis. For superficial stimulation, the injection must be performed in the superficial dermis. This can be achieved using microneedling in order to enhance the skin texture, glow, and hydration. When used as a filler, PRF must be injected into the deep dermis or into the subdermal tissues in a manner similar to the techniques commonly used for fillers. This is done to volumize and reshape the skin. This type of PRF application augments the skin and increases its volume.

Dentists are well adept at injection techniques as well as PRF therapy, making them well suited for such facial procedures. In the majority of countries including the United States, minimally invasive esthetic procedures are also within the scope of dental practice. Due to the growing demand for such procedures combined with the large profit margins on these therapies, dentists are encouraged to consider implementing such procedures within their dental office. Currently, about 10% of dental offices (~25,000) in the United States perform facial esthetic procedures, and this number is only expected to increase.[9]

PRF in skin regeneration

- Global skin improvement (regenerate skin): Microneedling with PRF
- Regional skin improvement (volumize and reshape): Injection into relevant skin layer/area

Types of PRF Used for Injection

There are two types of PRF that can be used for injection: liquid-PRF and e-PRF (heated plasma). For liquid-PRF, the horizontal centrifuge should be used at 300g for 5 minutes. This creates a cell-rich layer with growth factors. For e-PRF, the Bio-Heat (Bio-PRF) protocol is used: 2000g for 8 minutes with white tubes. Remove the upper layer and heat at 75°C for 10 minutes, then reconstitute with the cell-rich layer. This material is used when a higher volume and "filler" type of effect is needed (ie, for volumizing injections).

Tips for successful PRF injections

PRF injections are time dependent!

- All treatment planning and discussions must take place before blood is drawn.
- All equipment must be ready and in place before blood is drawn.
- Ensure that the centrifuge is in close proximity to the patient and switched on.
- After blood withdrawal, insert the tubes into the centrifuge within 90 seconds.

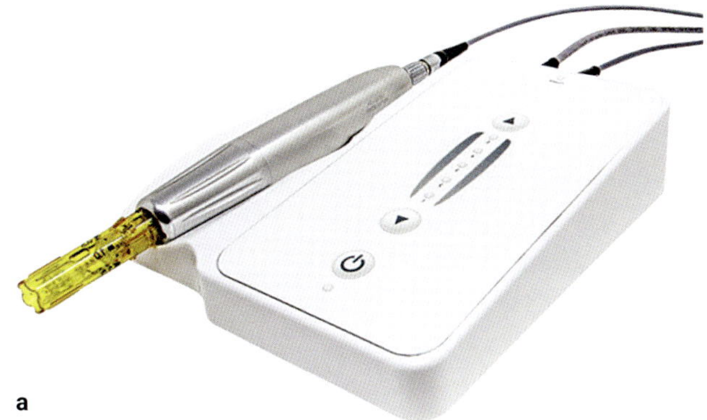

Fig 20-1 *(a)* Dermapen microneedling device. *(b)* Dermapen microneedling tip. Note the 12 small microneedles within the tip, which repeatedly penetrate 0.25 to 2.5 mm deep into facial tissues at roughly 1,000 to 5,000 rotations per minute. (Reprinted with permission from Davies and Miron.[9])

- The only activity that should be left to complete after withdrawal of blood is removal of topical anesthesia from the patient.
- After spin completion, allow the PRF to rest in the centrifuge for a few seconds before removing the tubes.
- Once removed from the centrifuge, place the tubes in a stand to prevent shaking.
- Extract PRF without removing the cap on the tube to minimize oxygenation, which will speed clotting.
- Be sure to get the bottommost layer of PRF.
- Inject PRF into the patient as soon as possible to avoid clotting.

Global Skin Regeneration: PRF and Microneedling

Microneedling therapy, also known as *collagen induction therapy* or *percutaneous collagen induction*, is an increasingly popular treatment modality for minimally invasive skin rejuvenation with minimal risk and downtime. As the term implies, a number of microneedles (typically 12) are utilized to create minimally invasive, nonsurgical, and nonablative therapy of facial tissues (see Fig 20-1b). Microneedling relies on the principle of neovascularization that occurs as a result of minimal trauma causing rapid neocollagenesis and tissue repair. This is performed in an automated fashion with a microneedling device—eg, the Dermapen (Fig 20-1). The Dermapen is an electrically powered medical device that delivers a vibrating stamplike motion to the skin, resulting in a series of microchannels. These channels are then filled with PRF. The device may also be utilized to "push" PRF into the skin at specific depths to facilitate facial rejuvenation via autologous growth factor release.[9]

Microneedling therapy is a safe skin resurfacing therapy and results in minimal damage to the skin. The downtime is usually approximately 24 to 48 hours, which is significantly less than other comparable methods. It also has a lower risk of side effects such as hyperpigmentation and scarring (when compared to lasers, for instance), making it an ideal treatment choice for all individuals, especially those with thin, sensitive, or specific skin types.[10] It is also effective for smokers and other individuals with significant exposure to external pollutants.[11]

Indications and contraindications

The original indication for microneedling was simple wrinkle reduction, but its use has since expanded into many additional indications. These include various types of scars, stretch marks (striae rubra), androgenetic alopecia and alopecia areata (with and without PRP/PRF), pigmentations, and acne.[9]

The contraindications of microneedling have been reviewed previously and include the following[9]:

- Dermatosis like vitiligo, lichen planus, and psoriasis, because trauma leading to koebnerization can aggravate the dermatosis; nevertheless, some authors have used microneedling with topical latanoprost to treat vitiligo
- Blood clotting disorders and anticoagulant therapy like warfarin and heparin, as this can cause uncontrolled bleeding
- Rosacea
- Skin malignancy, moles, warts, and solar keratosis, because the needles may disseminate abnormal cells by implantation
- Other chronic skin diseases like eczema
- History of having taken isotretinoin within 6 months
- Skin infection such as impetigo or herpes labialis
- Extreme keloidal tendency
- Chemotherapy or radiotherapy

Preparation and technique

Topical anesthesia with lidocaine and prilocaine are utilized and applied to the area to be treated. While various companies market and sell over-the-counter topical creams, the authors recommend pharmaceutical preparations of topicals. This is achieved typically utilizing a lidocaine 23% and tetracaine 7% formulation (prepared at a compounding pharmacy via prescription). This local anesthetic cream is applied to the face and neck of the patient 30 to 60 minutes before the procedure. Alternatively, the patient could apply the cream at home 1 hour before the procedure.

Prior to microneedling, the skin must be cleansed of any creams or makeup. Any areas to be treated should be cleaned again within the office using a disinfectant (ideally alcohol-free). A headband or cap should be applied to keep the patient's hair out of the treatment area. Once the face is prepared and clean, a layer of liquid-PRF is applied to the skin surface. With one hand slightly stretching the skin, the microneedling device is then passed over various areas in a directional motion following the guidelines shown in Fig 20-2. Typically, three passes are performed on each area of the face at various recommended depths, as highlighted in Fig 20-2. Pinpoint bleeding will typically result. Following the procedure, any remaining liquid-PRF is layered on all surfaces of the skin that have been microneedled.

Postoperative care

The treated area often demonstrates pinpoint bleeding. Some superficial bruising may also occur.

Following treatment, a face-cooling mask is applied. This utilizes a natural hydrogel without any additional drugs/agents. The patient is then presented with postoperative instructions. These entail avoiding sun exposure and harsh chemicals or any cosmetic procedure to the face for at least 72 hours posttreatment.

Clinical significance and advantages of microneedling

Several reported advantages have been discussed in the literature for microneedling. These include the following:

- Short healing times when compared to other modalities (typically 24 hours)
- Easy-to-master technique
- Can be utilized on all skin types, unlike lasers and deep peels
- Convenient office procedure with minimal overhead cost
- Well tolerated by patients
- Minimal risk of postinflammatory hyperpigmentation or bruising due to shallow needle penetration (max 2.5 mm)

Regional Injection Techniques

The techniques performed to volumize and reshape the skin, thereby augmenting the skin and/or increasing its volume, may be done by area following a regional approach.

PRF is a safe and reliable treatment option for soft tissue rejuvenation and dermal augmentation, with the ability to restore lost volume over time. Although it will not replace dermal fillers in esthetic practice, it is an effective alternative for patients who might feel they are too young for dermal fillers or for those who would prefer a more natural treatment.

Where there are fine lines and regeneration is required, liquid-PRF is used, and where more volume is required, the Bio-Heat protocol may be used (see Video 20-1 and chapter 10).

Understanding facial anatomy for injections

Before planning any regional injections of the face, it is critical to understand the relevant anatomy of the underlying

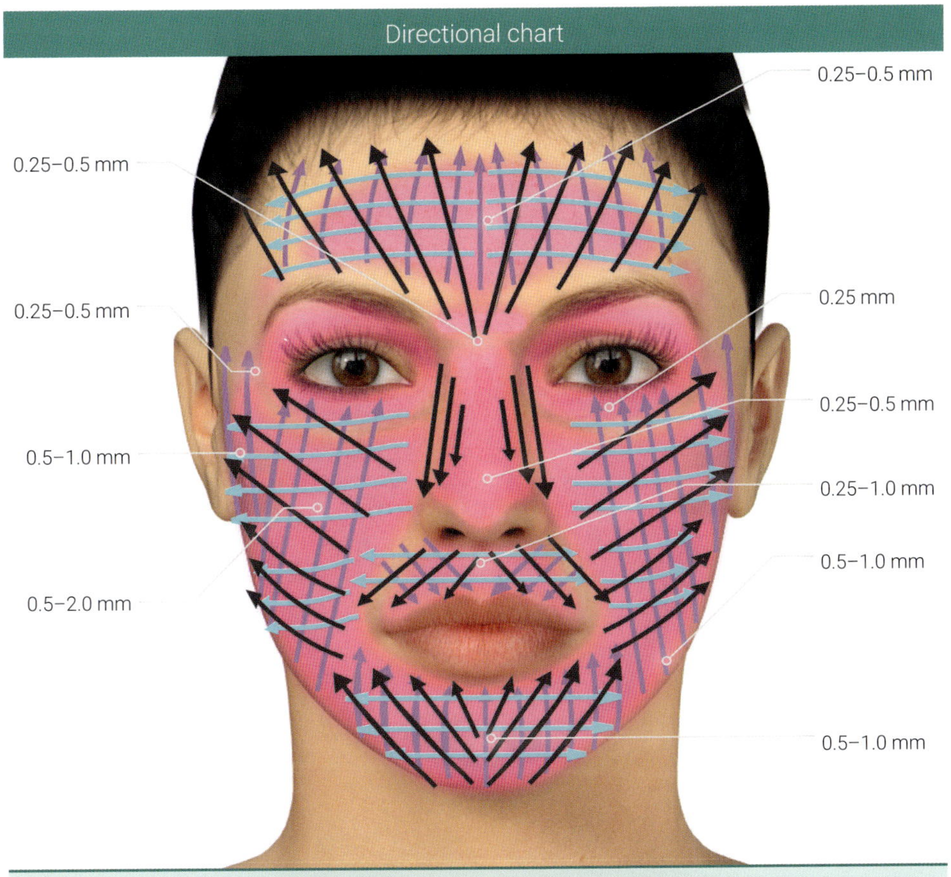

First pass	*(Purple)* Treat using the "striping technique" and follow the direction bottom to top. Start at the bottom of the movement and treat with upward strokes.
Second pass	*(Blue)* Making constant contact, start medially and work laterally from inside toward the outer face. Pick up the tip from the face at the end of the outer face and then repeat starting from the inside to the outer face. This creates a "striping" movement that follows protocol.
Third pass	***(Black)*** Outward and upward (when treating nose and upper lip on third pass, use downward strokes as indicated in diagram above).

Fig 20-2 Microneedling depth and direction chart. (Reprinted with permission from Davies and Miron.[9])

Video 20-1

structures (Fig 20-3). Certain areas of the face are at higher risk of unforeseen complications due to underlying structures of the skin, blood vessels, and nerves. Also, one needs to understand the subtle 3D manifestations of aging, which reflect the combined effects of gravity, progressive bone loss, decreased tissue elasticity, and redistribution of subcutaneous fullness.

Before planning any regional injections, the clinician must understand (1) the underlying anatomy of aging and (2) the regional anatomy of the face to avoid risk of vascular injury.

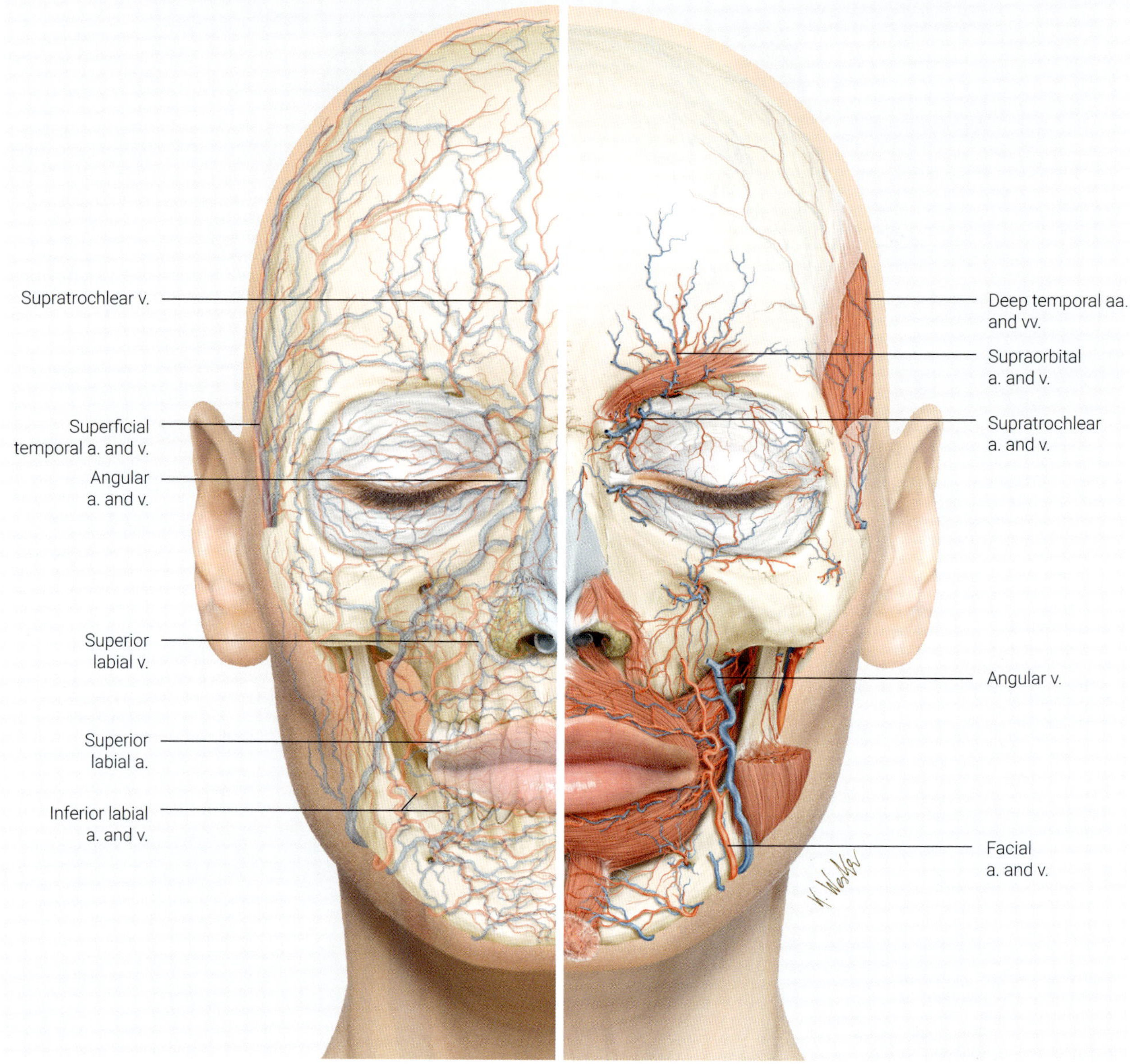

Fig 20-3 Blood vessels of the face projected onto the facial skeleton *(left)* and the position of the deep arteries and veins of the face relative to the deep muscles *(right)*. (Reprinted with permission from Sattler and Gout's *Illustrated Guide to Injectable Fillers* [Quintessence, 2016].)

Understanding the vascular "danger zones"

When injecting the face, certain zones are identified as being higher risk for complications due to the vessels and nerves that lie beneath the skin (Fig 20-4). The most severe potential complication associated with the use of dermal fillers and volume enhancers is arterial/venous occlusion, which leads to ischemia, with subsequent necrosis of the skin and/or vision loss. This is less of a concern when injecting pure PRF; however, care must be taken not to cause vascular or nerve injury by introduction of the needle or by external compression of the blood supply by surrounding PRF or swelling. It is very important to note that if PRF is combined with any form of "filler," there is a risk of arterial/venous occlusion, and all filler safety protocols should be followed.

The 6 facial danger zones

1. Glabella (area of highest risk because the vessels are small and do not have a good source of collateral circulation[6])
2. Temple
3. Nose angular artery (extension of the facial artery), which provides blood to medial cheek, nasal ala and side wall, and dorsum of the nose [Care should be exercised when injecting near the alar groove because excessive compression with large volumes of material or direct injection into the vessel can lead to nasal ala, nasal tip, nasolabial fold, and upper lip necrosis.[7]]
4. Perioral area
5. Infraorbital region
6. Nasolabial fold

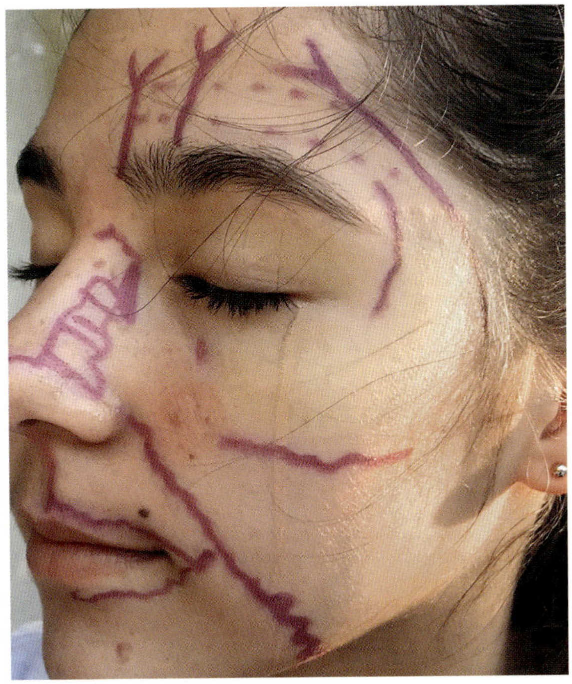

Fig 20-4 The facial arterial/venous system and "danger zones" marked on a live model. (Reprinted with permission from Davies and Miron.[9])

Practical tips when injecting near the "danger zones"

Sound knowledge of facial anatomy, especially the key arterial and venous structures that are vulnerable to damage, and knowing how aging affects the location of these anatomical structures is essential when injecting any area of the face. However, nerves and vessels may often be in atypical anatomical positions.

Each injection entails some degree of risk, and all of the potential complications should be in the informed consent signed by the patient. In addition, the most frequent and significant adverse events should be discussed before the procedure.

The clinician can take the following actions to decrease the potential of complications:

- Be aware of the pertinent anatomy outlined in the danger zones.
- Aspirate before injecting.
- Inject in a retrograde fashion where possible.
- Use small syringes and inject small aliquots of PRF at a time.
- Avoid using anesthesia near a vascular bundle that may induce vascular spasm, because anesthetics containing epinephrine may mask the vascular spasm.
- Use the smallest-gauge needle possible to slow the flow of product.
- Use a cannula where appropriate.
- Pinch/tent the skin to provide more space between superficial branches of the main arteries and to move away from underlying vasculature.[7]
- Assess the level of pain during injection.
- Use a vein light to assist in identifying major vessels.
- Use extreme caution when injecting areas of previous trauma or scarring.

Regions

The regional approach usually uses a combination of techniques, including superficial augmentation technique, horizontal cannula, and vertical augmentation. This is customized per patient and per area. The most commonly requested areas for injection with PRF are horizontal forehead lines, crow's feet, nasolabial folds, midface volume (cheek area), lips, and perioral/marionette lines (Fig 20-5).

It is beyond the scope of this book to include injection techniques for each of these regional areas. Included here is the technique for the cheek area to explain the general protocol and showcase how simple these treatments can

Fig 20-5 Commonly requested regions for facial injections. (Reprinted with permission from Sattler and Gout's *Illustrated Guide to Injectable Fillers* [Quintessence, 2016].)

be. Detailed protocols for each region are available in the recent book *PRF in Facial Esthetics* by Catherine Davies and Richard J. Miron (Quintessence, 2020).

Cheek injection

The aim of treatment to the cheek area is to beautify the cheekbone region, restore youthful volume to the anterior cheek, provide lift within the subzygomatic area, and cause a decrease to the nasolabial fold.

First, draw two intersecting lines: one from the alar groove of the nose to the top of the tragus of the ear and one from the lateral canthus to corner of the mouth (see Fig 20-6). The upper outer quadrant is the appropriate site for deep supraperiosteal injections.

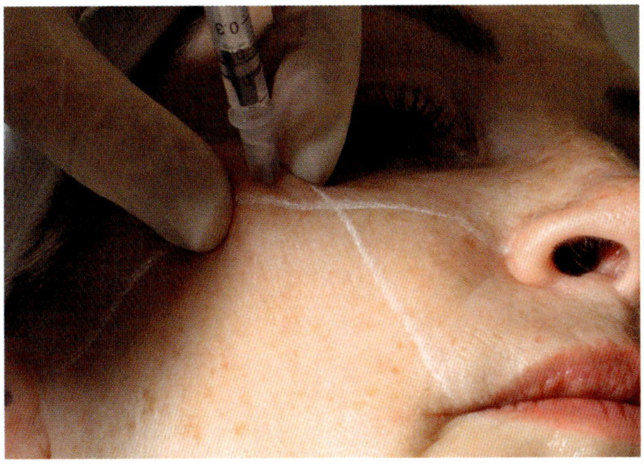

Fig 20-6 The noninjecting hand tents the skin during insertion of the needle onto periosteum. (Reprinted with permission from Davies and Miron.[9])

To enhance cheek volume

Technique: Vertical supraperiosteal depot technique. The clinician gathers up the skin with the noninjecting hand for maximum penetration and pierces the skin at the thinnest entry point using a 90-degree angle at the level of the bone (Fig 20-6). The needle is repositioned and advanced slowly below the soft tissue before material deposition.

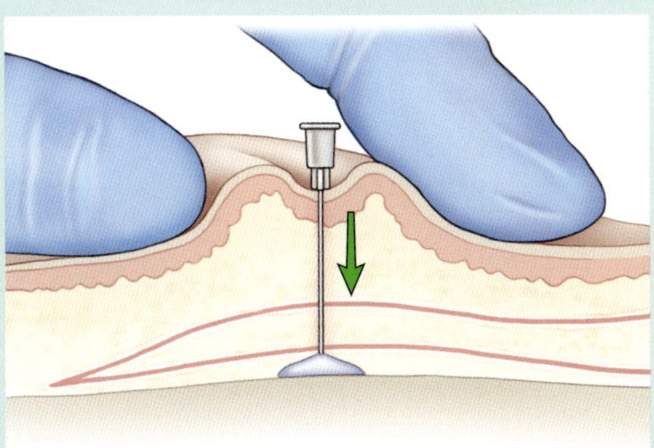

Substance: e-PRF
Entry point: Upper outer quadrant after marking intersecting lines from the ala of the nose to the tragus and from the lateral canthus to the corner of the mouth.
Volume: 0.2 mL per injection point; up to three points per side.
Tool: 27G needle

Treatment steps

1. Apply local anesthetic cream prior to treatment (optional).
2. Remove makeup and/or anesthetic thoroughly.
3. Disinfect the area.
4. Inject PRF.
5. Massage the injected fluid into the area for even distribution.
6. Inform the patient about aftercare instructions.
7. Arrange a follow-up appointment.

Microneedling and/or laser resurfacing can be performed in conjunction with PRF injections to maximize the therapeutic results. Note that volume may be achieved over time with multiple sessions. Where injections overly the bone, the injection should be deep. Elsewhere in the cheek, where there is no bony support, the injections must be superficial.

PRF Injection for Hair Regeneration

Hair loss (alopecia) is a common problem and is often a major source of distress for patients. One minimally invasive treatment for this condition is PRF injection. PRF has been observed in clinical studies to promote hair growth, reverse hair loss, and enhance hair thickness. The procedure is generally safe, tolerable for the patient, and can result in only mild irritation afterward, and no serious allergic reactions have been reported concerning the administration of PRF for hair rejuvenation. Nevertheless, patients should sign a consent form before undergoing any treatment.

Indications

It is important to note that PRF is only indicated for nonscarring alopecia. It is therefore crucial that the first consultation involves a systematic approach in order to correctly diagnose the type of alopecia. The differential diagnosis of alopecia includes both scarring and nonscarring alopecias.

PRF therapy is only indicated for nonscarring alopecia.

PRF will be most successful if used for the following indications:

- Patterned hair loss (female pattern hair loss, male pattern hair loss)
- Diffuse alopecias
- Traction alopecia (if follicles are still present)
- Alopecia areata (may work, but discontinue if no results are observed after 3 months)

Consultation

The hair loss consultation includes a thorough history, examination of the patient, examination of the scalp and hair with trichoscopy, as well as laboratory investigations when needed. The type of hair loss is then identified and, if indicated, treated with PRF injections. It is recommended to note the stage of hair loss on each chart. Typically, the Norwood scale is used for men, and the Ludwig scale is used for women (Fig 20-7).

Examination

The clinical examination first involves looking at the scalp for any skin conditions or lesions and to identify the pattern and distribution of hair loss. Next, the hair pull test and hair tug tests are performed to determine the hair loss activity and fragility of the hair. Cross-sectional trichometry then follows to quantify the hair within a fixed area; tools are available to help the clinician with this process (eg, HairCheck). Trichosopic examination can be useful to view the scalp and follicles and provide high resolution of problematic areas. And of course, photographic documentation is used to establish a baseline for treatment. While not always necessary, a punch biopsy of the scalp can be useful to help with diagnosis.

Treatment

Prior to treatment

Ask the patient to shampoo and detangle their hair prior to treatment. When they arrive to the office, the hair should be clean, without any hair products (such as wax or hair spray) present.

Tools needed

- Antiseptic solution spray
- Anesthetic
- Disposable comb
- 1-mL Luer-Lock syringe
- 30G, 4-mm needle for injection

Step-by-step method

1. Disinfect and anesthetize the scalp. (For detailed information on proper anesthesia, see *PRF in Facial Esthetics*.[9])
2. Identify and mark the affected area.
3. Comb hair into sections vertically, 1 cm apart.
4. Inject approximately 0.1 mL per injection site along the combed part, approximately 2 mm deep (Fig 20-8).
5. After injection, massage the head to disperse the fluid.
6. Microneedling may also be utilized (Video 20-2).

Video 20-2

The patient must not wash the hair or use any product on the hair until the following day.

Protocol

- 1 session per month for 3 months (Fig 20-9)
- Treat any underlying abnormality (such as nutritional deficiencies)
- Annual maintenance may be needed.

Follow-up

Treatment response should be assessed at months 3, 6, and 12, but the patient should only expect to see maximal results at month 12.

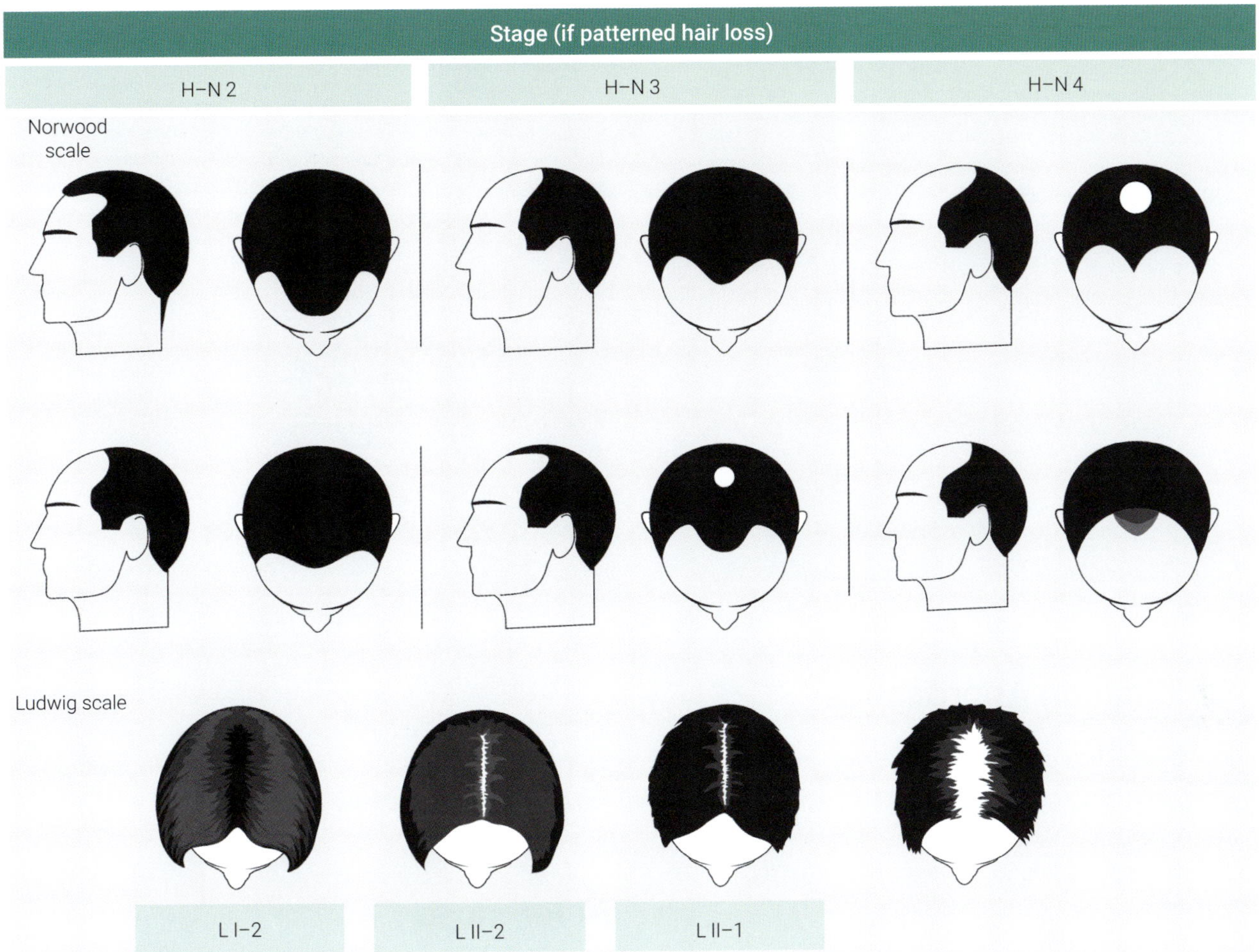

Fig 20-7 Classifications of hair loss. The Norwood scale is used for men, and the Ludwig scale is used for women. (Reprinted with permission from Davies and Miron.[9])

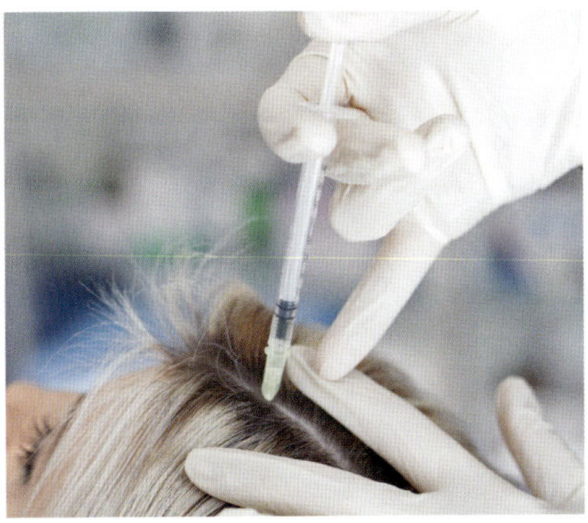

Fig 20-8 Point-by-point injection technique along the combed part. (Reprinted with permission from Davies and Miron.[9])

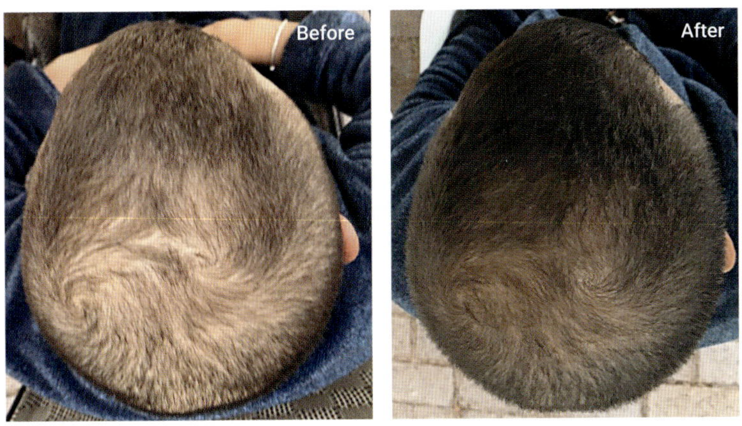

Fig 20-9 Results before and after three sessions of PRF injections. (Reprinted with permission from Davies and Miron.[9])

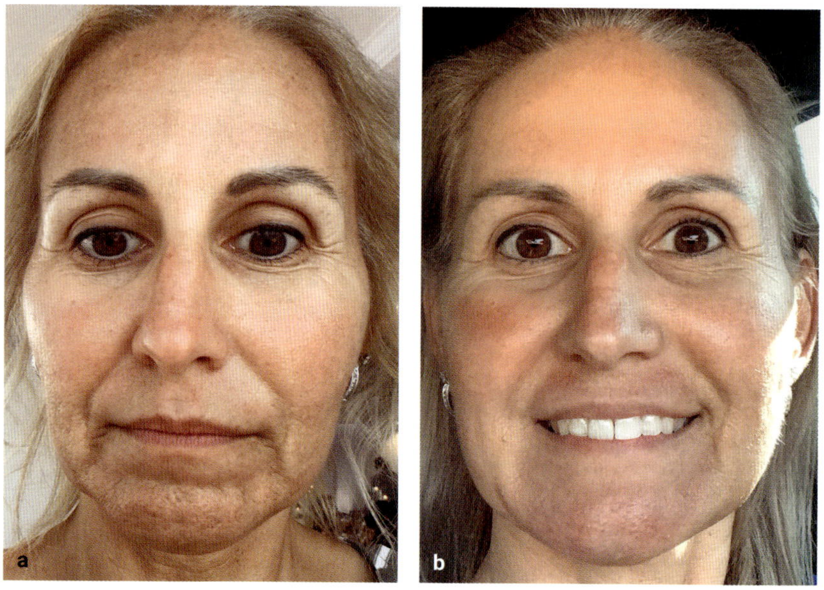

Fig 20-10 *(a)* Woman in her mid 40s with pronounced marionette lines, deep nasolabial folds, and an overall aged facial appearance. *(b)* Final outcome following three treatments with 100% natural approaches, including laser therapy (SmoothLase), microneedling with PRF, and e-PRF injections. (Case performed by Dr Scott Delboccio; reprinted with permission from Davies and Miron.[9])

Video 20-3

Combination Approaches

Many new technologies have been developed that enhance facial rejuvenation and repair. Many can also be utilized in combination approaches. Figure 20-10a presents a woman in her 40s with pronounced marionette lines, deep nasolabial folds, and an overall aged facial appearance. She was treated with three laser therapy treatments (SmoothLase, Fotona) in combination with microneedling with PRF. During the third appointment (evenly spaced 1 month apart), the novel e-PRF (see chapter 7) with extended resorption properties was used for her marionette lines, nasolabial folds, as well as within her midface region to augment volume. Note the final outcome 1 month after therapy (Fig 20-10b and Video 20-3). There was a clear and definite reduction in wrinkles and a dramatic overall improvement in facial appearance. Such combination therapies like the one used in this case are 100% natural. Future research in this field aims to determine which combination approaches are most effective to favor better facial rejuvenation while minimizing the use of chemical/synthetic additives.

Conclusion

This chapter highlighted the growing awareness of facial esthetics and the role of PRF therapy in that field. While it only briefly covered various PRF injection techniques and its combination with microneedling, it is clear that PRF offers patients and practitioners a very safe method for facial and hair rejuvenation.

References

1. Branchet M, Boisnic S, Frances C, Robert A. Skin thickness changes in normal aging skin. Gerontology 1990;36:28–35.
2. Helfrich YR, Sachs DL, Voorhees JJ. Overview of skin aging and photoaging. Dermatol Nurs 2008;20:177.
3. Herbig U, Ferreira M, Condel L, Carey D, Sedivy JM. Cellular senescence in aging primates. Science 2006;311:1257–1257.
4. Puizina-Ivi N. Skin aging. Acta Dermat APA 2008;17:47.
5. Barrett JB. Acupuncture and facial rejuvenation. Aesthet Surg J 2005;25:419–424.
6. Ramirez OM, Maillard GF, Musolas A. The extended subperiosteal face lift: A definitive soft-tissue remodeling for facial rejuvenation. Plast Reconstr Surg 1991;88:227–236.
7. Rohrich RJ, Ghavami A, Lemmon JA, Brown SA. The individualized component face lift: Developing a systematic approach to facial rejuvenation. Plast Reconstr Surg 2009;123:1050–1063.
8. El-Domyati M, Medhat W. Minimally invasive facial rejuvenation: Current concepts and future expectations. Exp Rev Dermatol 2013;8:565–580.
9. Davies C, Miron RJ. PRF in Facial Esthetics. Chicago: Quintessence, 2020.
10. Fernandes D. Minimally invasive percutaneous collagen induction. Oral Maxillofac Surg Clin 2005;17:51–63.
11. Amer M, Farag F, Amer A, ElKot R, Mahmoud R. Dermapen in the treatment of wrinkles in cigarette smokers and skin aging effectively. J Cosmet Dermatol 2018;17:1200–1204.

21

Medical Uses of PRF

While this textbook primarily focuses on the use of PRF in dental medicine, recent systematic reviews have demonstrated that the clinical use of PRF for soft tissue wound healing has been applied far more often in general medicine for a variety of indications, highlighted throughout this chapter. While PRF was first utilized for the treatment of hard-to-heal ulcers, it is now being utilized in a variety of other medical specialties due to its universal mechanism of action. This chapter provides an overview of the many applications of PRF in the medical field.

Contributors

Thomas Lau Hansen
Johan Lenz
Jacob Coakley Meyer
Marco Antonio Castro Pinto
Richard J. Miron
Robert Talac
Scott Delboccio
Tommy Hardon
Thomas Boas

Chapter Highlights

- *PRF in the treatment of hard-to-heal ulcers*
- *Management of inflammation and pain with liquid-PRF*
- *PRF in sports medicine and regeneration of injuries*
- *Use of PRF in equine sports medicine*
- *Dental applications of PRF in veterinary medicine*

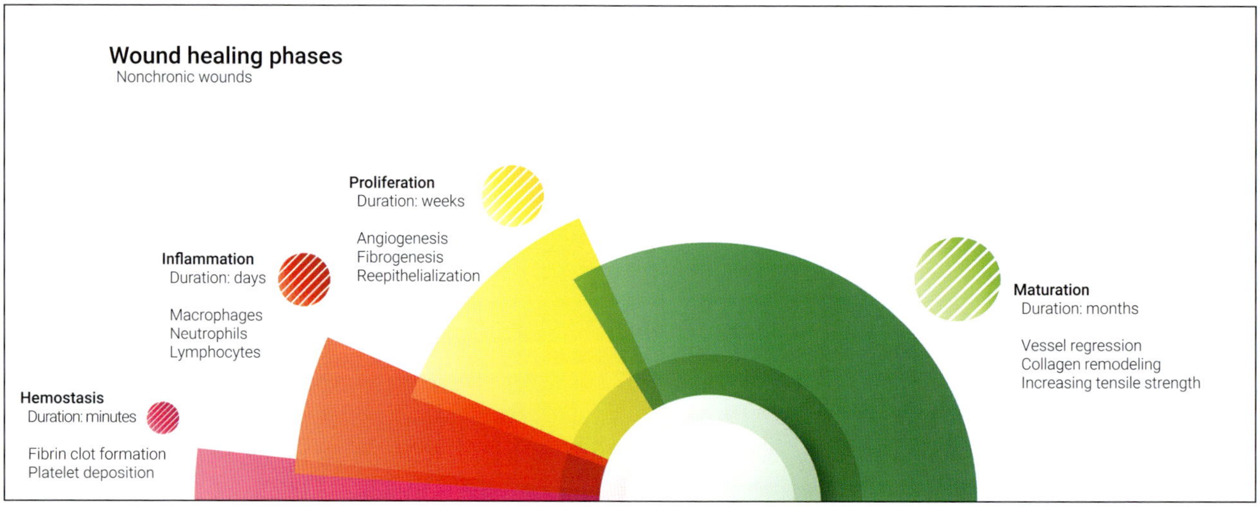

Fig 21-1 Graphic demonstrating the overlapping phases of wound healing.

PRF as we know it today is the result of many years of research in the content and properties of blood concentrates, on both a molecular and a clinical level. Blood concentrate products have now been used in medicine for nearly four decades due to their ability to rapidly secrete growth factors. They have gained tremendous popularity as a regenerative compound, derived from autologous sources, capable of stimulating tissue regeneration in a number of medical fields.

PRF was first introduced as a regenerative solution for hard-to-heal wounds.[1] Typically, regeneration occurs in a well-defined cascade of complex events in four overlapping phases: hemostasis, inflammation, proliferation, and remodeling[2–4] (Fig 21-1; see also chapter 2). These occur via the secretion of soluble mediators and signals capable of influencing the homing of circulating progenitor cells to damaged tissues.[5] Perhaps the main prerequisite for regeneration of tissue is blood supply, which in wound healing is supported by the ingrowth of a vascular network.[6] PRF was introduced with the ability to markedly improve angiogenesis through its accumulation of growth factors derived from the fibrin-bound autologous cells.[1] It has since been utilized for a variety of procedures in medicine and dentistry due to this potential.[6]

In this chapter, we describe the use of PRF in various areas of medicine to provide further insight into the regenerative capacity of PRF and its analogs. Specifically, this chapter focuses on its use and function in five overlapping topics: *(1)* wound healing, *(2)* soft tissue regeneration, *(3)* sports medicine and rheumatology, *(4)* veterinary sports medicine, and *(5)* veterinary dentistry.

PRF in Wound Healing and Regeneration

Chronic nonhealing wounds are a major and often underappreciated burden to the individual, the healthcare system, and society as a whole.[7–10] It is estimated that 1% to 2% of the population in developed countries will experience a chronic wound during their lifetime,[11] and in the United States alone, it is a problem that affects up to 6.5 million patients.[12–14] In 2018, the United States was believed to have spent between 28 and 97 billion dollars on treatments,[15] an expense that in Scandinavian countries is associated with 2% to 4% of the total health care budget.[16] Often disguised as a comorbid condition, the management of chronic wounds poses a significant medical challenge, resulting in prolonged courses of treatment or amputation,[11] and medical staff is therefore very much in need of regenerative treatment options to improve and accelerate the clinical outcomes (Figs 21-2 and 21-3).

Being a modulator of inflammation and an activator of angiogenesis and regeneration, PRF has been used in a

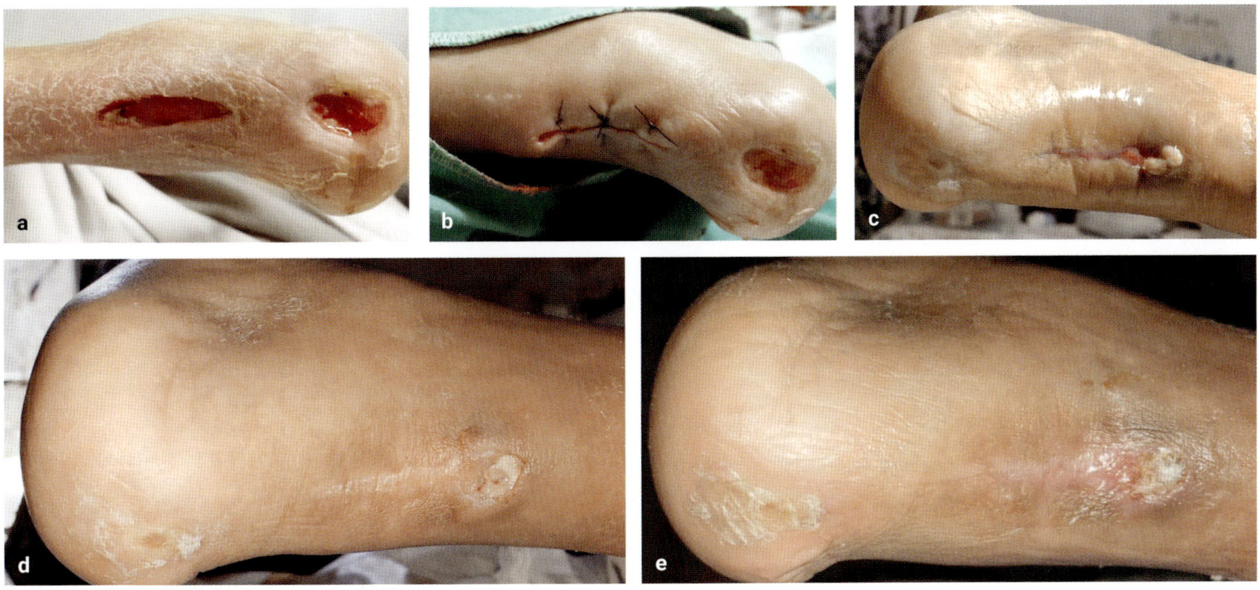

Fig 21-2 A 76-year-old woman with type 2 diabetes presented with chronic renal failure leading to end-stage renal disease. She underwent dialysis every 2 days. The patient was also being treated for hypertension and was prescribed irbesartan (300 mg) and metformin (500 mg). She was scheduled for leg amputation after Doppler analysis determined that she only had 10% of blood flow to her leg. *(a)* The patient had had chronic ulcers (2-year evolution) on her foot and leg, with multiple previous infections of the ulcers. Prior to amputation, the patient elected to attempt therapy with PRF. *(b)* PRF membranes were placed within the upper wound, which was sutured closed and also injected with C-PRF at the wound openings as well as the perimeters of the wound. It was not possible to suture the heel closed, and therefore an e-PRF membrane was utilized with C-PRF injections around the periphery. The woman revisited the clinic every 10 days for new applications of PRF. Note the progression of the wound closure at 1 month *(c)*, 2 months *(d)*, and 2.5 months *(e)*. Following seven treatments, the patient remained without ulcer thereafter. This case demonstrates the ability for PRF alone to favor wound healing in complex chronic wounds otherwise scheduled for amputation. (Case performed by Dr Marco Antonio Castro Pinto and Dr Romeo Leiva.)

Video 21-1

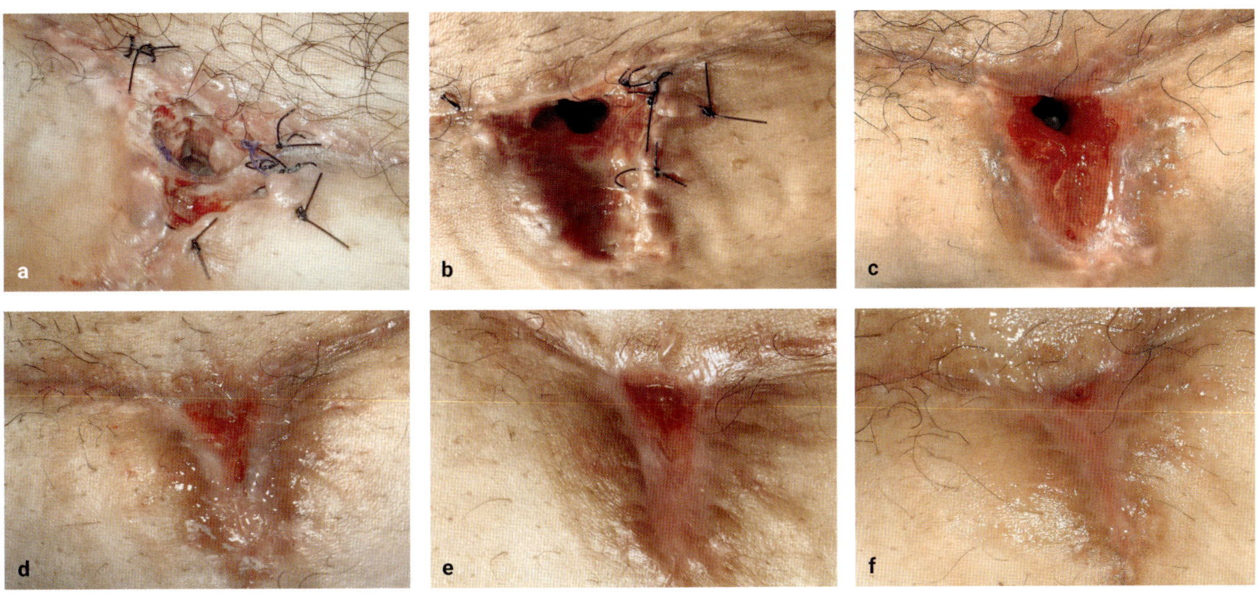

Fig 21-3 A 36-year-old woman had an ulcer developing on her abdomen that did not heal properly after surgery. The patient was obese with a BMI of 36. Upon surgery to resolve the issue, a 25-pound cyst was removed from her abdomen. Due to its extensive size, the surgeon had to modify the incision design and closure during the surgery. Passive tension-free closure was not possible, and as a result, a wound dehiscence was noted. Two separate attempts over a 1-month period were made to resuture the wound. *(a)* The patient was referred thereafter for PRF therapy due to the large opening, which measured 3.5 cm in diameter by 5 cm in depth. *(b to f)* Case after 2, 3, 4, 5, and 6 weeks, respectively. (Case performed by Dr Marco Antonio Castro Pinto and Dr Romeo Leiva.)

variety of wound healing treatments, including hard-to-heal refractory and chronic ulcers,[17–23] otherwise graft-demanding injuries,[24] and surgical wounds.[25–28] The standard procedure in the treatment of chronic wounds has for decades been a surgical debridement to restart the wound healing process, sharp debridement of the wound bed, and induced bleeding; in acute wounds, healing is not initiated until the formation of a blood clot (hemostasis).[29] The idea here has therefore been to stimulate a shift in the wound healing status from a chronic inflammatory state to an acute state, allowing for the restart and progression of wound healing.[18] As platelets, producing a variety of growth factors, are believed to be the initiators of the healing process,[30–35] the use of PRF to initiate or support this shift has been trialed in both human and animal studies[2826,36–41] and is also discussed in recent publications.[42]

Studies have shown that PRF treatment can prevent or improve postsurgical edema or ecchymosis,[26] lead to improvements in wound healing,[36] and promote an increase in type 1 collagen formation in full- and split-thickness flaps during skin grafting procedures in a porcine model.[40] These studies point to the fact that autologous platelet concentrates have wound-modulating properties that accelerate healing and improve the surgical outcome through their angiogenic capacity.[43,44]

Venous leg ulcers

As reviewed by Tuttle in 2015, wounds such as venous leg ulcers (VLUs) are infamously susceptible to microbial invasion, and serious complications can arise without the appropriate control of infection.[45] The use of PRF to increase angiogenesis thus further allows for allocation of circulating immune cells from the host, on top of the concentrated pool of immune cells present in the PRF scaffold. A 2008 prospective study by O'Connell et al also indicated a marked effect from using PRF on VLUs and other chronic lower-extremity ulcers recalcitrant to a variety of standard treatments, with courses of treatment ranging from 4 months to 53 years.[17] Complete closure was achieved in 66.7% of the VLU patients (64.7% of treated ulcers) within 7.1 weeks (median, 6 weeks), with an average of two applications of PRF membranes per patient. In non-VLU patients (31% of treated ulcers), 44% complete closure was achieved.[17] In fact, it was later established in a 2018 autocontrolled prospective cohort study that VLUs as well as diabetic foot ulcers (DFUs), pressure ulcers (PUs), and complex wounds could benefit from topical application of PRF.[22] In this study, all wounds showed significant improvements, and all VLUs ≤ 10 cm^2, all DFUs, and all complex wounds presented full closure within 3 months.[22]

Diabetic foot ulcers

Though the underlying mechanism and treatment needs differ between VLUs and DFUs, studies of these hard-to-heal foot ulcers common to diabetes mellitus patients show that these wounds can greatly benefit from the induction of angiogenesis that PRF can offer.[21,46] DFUs are associated with an increased risk of secondary infection and amputation and affect up to 25% of the 463 million diabetes mellitus patients worldwide[47] (2019 data), with up to 20% of them needing various degrees of amputations.[48,49] In 2013, Game et al initiated a prospective multicenter clinical RCT on the application of a proprietary form of PRF to accelerate the healing of DFUs reluctant to heal.[46] The investigators concluded the trial in 2018 with observations that the application of PRF was associated with a significant enhancement of healing. Of the 269 randomly allocated participants, 45 (34%) of 132 ulcers healed within 20 weeks following the application of PRF, versus 29 (22%) of 134 ulcers in the standard care group (odds ratio 1.58, 96% CI 1.04–2.40; P = .0235). Healing time was also shorter in the PRF group (P = .0246).[23] This increased healing is further supported by xenograft studies in mice, where the use of human PRF was associated with significantly improved wound healing in diabetic mice.[50] Owing to its angiogenic properties, the application of PRF revealed a significant increase in the number of capillaries and CD31+ cells observed, suggesting that PRF was promoting blood vessel formation in the wounds.[50]

The authors have previously conducted several clinical case studies on the use of PRF in treatment of hard-to-heal wounds such as DFUs. One example is presented in Fig 21-4, which illustrates the accelerated healing of an otherwise recalcitrant DFU tormenting the patient for more than 6 months. The topical application of PRF allowed for the accelerated healing and closure of the wound in just 4.5 weeks.

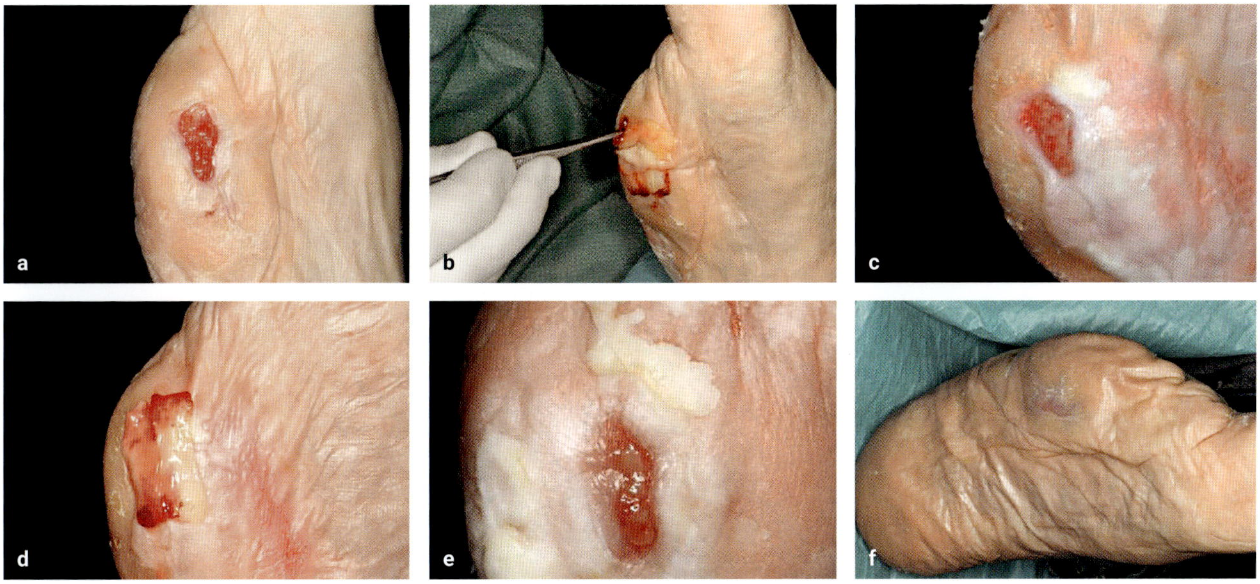

Fig 21-4 Treatment of nonhealing DFU in amputee patient with a long history of hard-to-heal foot ulcers. Topical application of PRF resulted in accelerated wound healing of the DFU that had been refractory to other treatments for more than 6 months. *(a)* DFU after debridement. *(b)* Onlay of PRF membranes, which were protected from being dried out by the application of thin plastic film before supportive wound dressing was applied. *(c)* Reduced wound diameter at the 6-day follow-up. *(d)* Reapplication of fresh PRF membranes, film, and dressing. *(e)* Macerated skin around ulcer at the 14-day follow-up. PRF was reapplied, and the film was cut to smaller size before application of the dressing. *(f)* Six-month follow-up of DFU. The wound took 41 days from first application of PRF to fully heal. (Case performed by Dr Steffen Lund.)

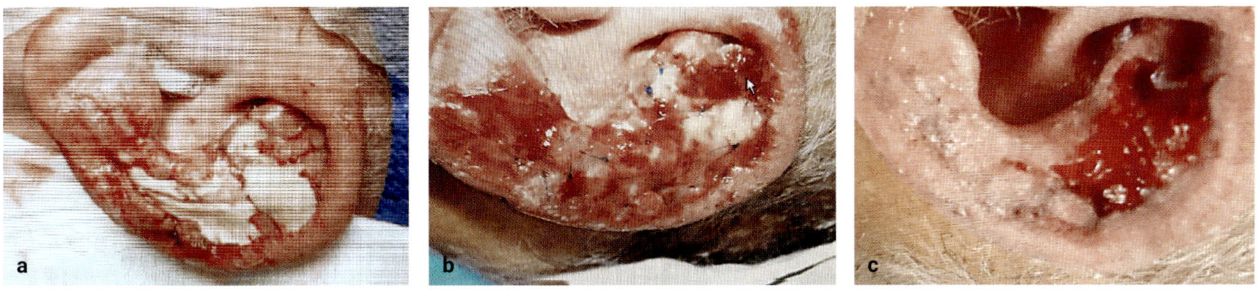

Fig 21-5 Treatment of ear trauma with exposed cartilage refractory to skin grafting.Topical application of PRF resulted in complete reepithelialization of the wound otherwise refractory to skin grafting attempts. *(a)* Wound after trauma but before grafting. *(b)* Grafting with PRF sutured (monofilament suture) in place, covering the exposed cartilage. *(c)* Healing result after 5 days. (Case performed at the Department of Plastic Surgery, Skejby Hospital, Denmark.)

Plastic surgery

Another area of wound healing benefitting from the application of PRF is plastic surgery. An example clinical case from this field involves the reepithelialization of ear trauma with cartilage exposed. The case is presented in Fig 21-5 and shows how PRF is able to ensure complete reepithelialization of exposed cartilage after unsuccessful attempts at skin grafting.

The use of platelet- and leukocyte-rich blood concentrate products for the treatment of chronic and recalcitrant wounds is a promising therapy that could alleviate not only the pain of the numerous patients suffering from VLUs and DFUs but also the enormous burden of managing patients in the health care system with a continuous need for wound care, debridement, and application of fresh dressings, not to forget the cost to society. Despite the many positive results, however, clinical strategies at the hospital level have not yet switched over to this model.

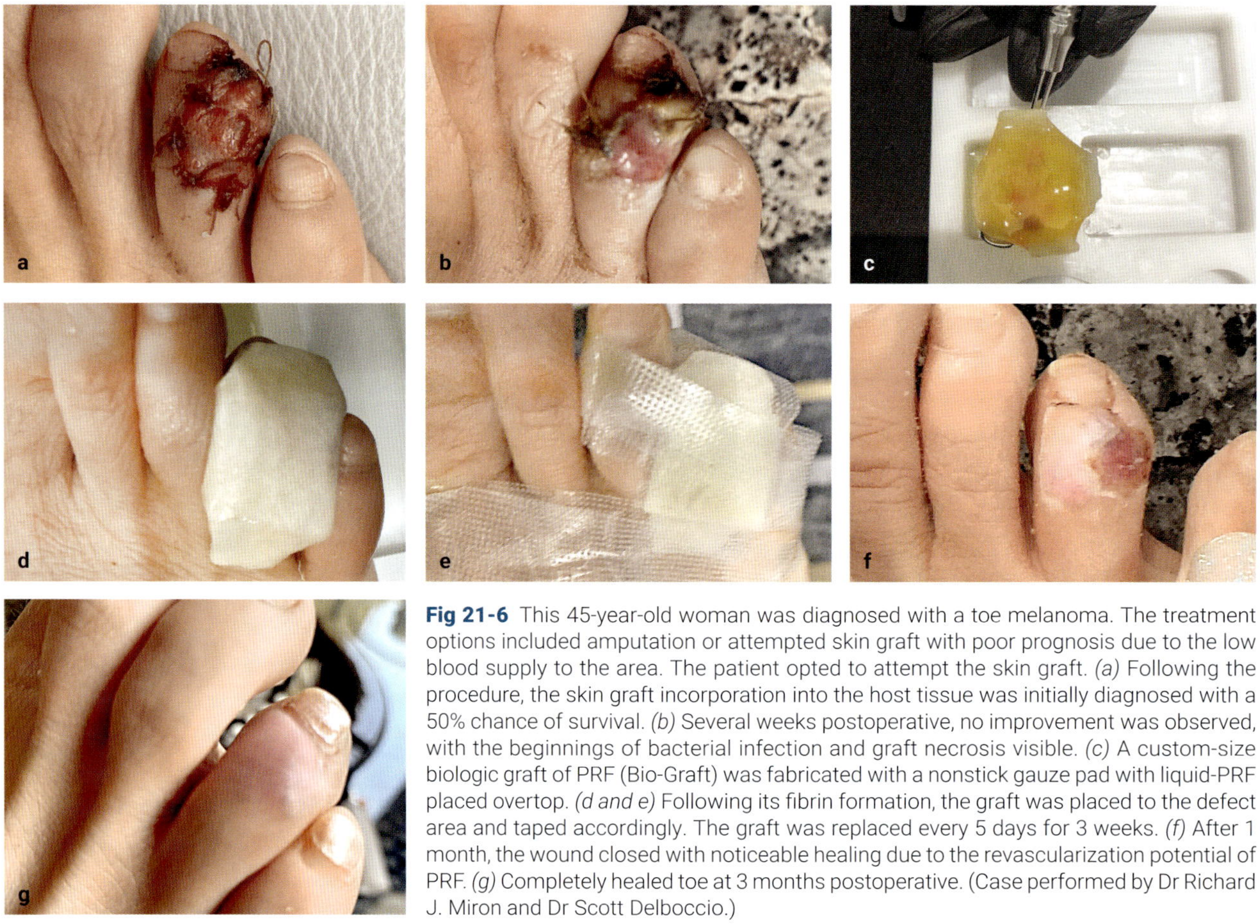

Fig 21-6 This 45-year-old woman was diagnosed with a toe melanoma. The treatment options included amputation or attempted skin graft with poor prognosis due to the low blood supply to the area. The patient opted to attempt the skin graft. *(a)* Following the procedure, the skin graft incorporation into the host tissue was initially diagnosed with a 50% chance of survival. *(b)* Several weeks postoperative, no improvement was observed, with the beginnings of bacterial infection and graft necrosis visible. *(c)* A custom-size biologic graft of PRF (Bio-Graft) was fabricated with a nonstick gauze pad with liquid-PRF placed overtop. *(d and e)* Following its fibrin formation, the graft was placed to the defect area and taped accordingly. The graft was replaced every 5 days for 3 weeks. *(f)* After 1 month, the wound closed with noticeable healing due to the revascularization potential of PRF. *(g)* Completely healed toe at 3 months postoperative. (Case performed by Dr Richard J. Miron and Dr Scott Delboccio.)

PRF in Soft Tissue Regeneration

The concept of wound healing is applicable to many fields of medicine, both surgical and nonsurgical.[3,51] Both in humans and various animal models, PRF has been shown to induce soft tissue regeneration. From the treatment of facial defects,[52] superficial rhytids, acne scars,[28] and induction of dermal collagenesis[53] to the suppressed subcutaneous collagen synthesis after laparoscopic cholecystectomy,[27] vaginal prolapse repair,[54] and urethracutaneous fistula repair,[55–57] PRF has the capacity to modulate inflammation, accelerate healing and regeneration, and at the same time function as a structural graft for scaffolding (Fig 21-6).

Ear, nose, and throat applications

Within the field of ENT, PRF has been used to accelerate the closure of palatal wounds in donor sites after harvesting subepithelial connective tissue grafts[58–60] and excision of superficial but potentially malignant lesions.[61] PRF has furthermore been applied in the healing of acute traumatic eardrum perforations,[62] where patients treated with PRF (in the form of a membrane patch) had a significant greater outcome than the control group (64.3% complete healing at 1 month compared to 22.2% in the control group) with much greater healing areas (9.57 ± 3.93 mm^2 compared to only 5.61 ± 3.74 mm^2).[62] In a 2014 retrospective study on 108 patients of middle ear microsurgery, Garin et al reported similar positive results from the use of PRF as packing material of the external auditory canal or middle ear compared to standard animal products.[63] In a fairly recent RCT by Nair et al, PRF was furthermore used to increase graft take after total myringoplasty due to chronic suppurative otitis media.[64] The study, which included 86 patients divided equally into the two groups, reported an increased graft take of 97.7% of the study group compared to 81% of the control group (P = 0.012).[64]

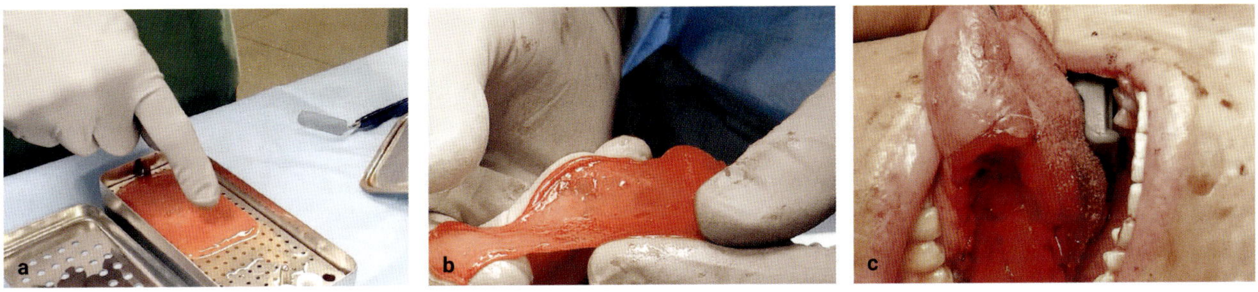

Fig 21-7 *(a to c)* Large PRF membrane used to cover the wound created by tumor excision in the tongue. See Video 21-2 for complete case presentation.

Several clinical case studies have furthermore been conducted within the field of ENT. Large PRF membranes were especially useful in this context to cover surgical wounds after tumor excisions from the tongue and floor of the mouth floor as well as donor sites after harvesting of palatal connective tissue grafts. The case shown in Fig 21-7 (see Video 21-2) was carried out in the tumor section of the ENT Department at Rigshospitalet in Denmark.

Video 21-2

Urology, andrology, and gynecology applications

PRF has also been applied for the advancement of urologic, andrologic, and gynecologic challenges. In 2018, Matz et al did a safety and feasibility study on the use of PRF for various urologic pathologies, such as treatment of erectile dysfunction (ED), Peyronie disease (PD), or stress urinary incontinence (SUI), as the concern with PRP had been that the cells and growth factors would be washed out immediately after injection.[65] The results were suggestive of efficacy and showed that PRF was feasible and safe to use.[65] These data furthermore coincide with results from a previous study on PD therapy using a combination of PRP and hyaluronic acid (HA). Here, repeated intralesional injections of PRP-HA significantly improved penile deformation and thickening of the tunica albuginea.[66]

One of the major applications for PRF within urology and gynecology is fistula repair. PRF has on several occasions been studied in the prevention and repair of urethracutaneous fistulas arising postoperatively from the treatment of hypospadias in young boys.[55–57] In this context, PRF has been used as an intermediate layer or patch to help reduce the incidence of postoperative complications.

In anal fistula repair, fibrin sealant has previously been injected into the fistula tract of patients suffering from trans-sphincteric or suprasphincteric fistulas in an attempt to seal one side of the tract. In the 2004 study from Jurczak et al, fibrin sealant was in this way used to treat 31 patients, with a success rate of 83.9% (75% after only a single application).[67] The use of autologous PRF in the treatment of these types of fistulas as well as those arising from perianal fistulizing Crohn disease[68] remains to be investigated but presents some interesting options to the need for immunomodulating features, a structural scaffold, and accelerated healing.

PRF in Sports Medicine and Rheumatology

Pain in joints, tendons, ligaments, and muscles often arise from inflammation affecting the somatosensory nervous system.[69] Adding to this, injuries to or the entrapment of neurons, either directly or as a bystander effect from swollen and inflamed tissue, can cause a chronic inflammatory state in which recruited macrophages stabilize a continuous pain inflicting stimulation of neurons.[70] The capacity to modulate inflammation, influence macrophage polarization, and stimulate an acute response leading to accelerated healing and regeneration puts PRF in a unique position as an autologous biologic agent for the treatment of inflammation, pain, and damage of joints and tendons arising from injuries, arthrosis, or arthritis.

Within sports medicine and rheumatology, the use of PRF can be applied in both conservative treatments as well as arthroscopic and open surgery. PRP has for many years been utilized for knee injections in a wide range of patients, including professional athletes.[71–76] The more recent introduction of an additive-free liquid-PRF that does not utilize anticoagulants and thus will form a fibrin clot upon injection has been a proposed method to further regenerate cartilage, particularly in the knee.[77]

Cartilage and meniscus are two of the most avascular and low–cell density tissues found in the human body, and thus they have very limited potential for repair and regeneration. Left untreated, defects in the cartilage often do not heal at all. A popular research topic has therefore been the use of platelet derivatives for the regeneration of cartilage. From a wound healing perspective, it makes sense to introduce the damaged cartilage to the regenerative cells present in blood and PRF. The metabolism of mature articular cartilage is regulated by a number of growth factors originating from the cartilage as well as the synovial fluid and surrounding tissues. The specific growth factors IGF and PDGF are especially key to the regulation of cartilage repair. These growth factors are used to encourage cell migration into sites of injury. In fact, in the absence of a supplemental source of chondrocytes, fibroblasts, or their precursors, cartilage repair is completely dependent on this recruitment, retention, and proliferation of cells,[78] further reviewed by Schmidt et al in 2006.[79] Multiple studies in rabbit models have shown liquid-PRF to significantly promote this proliferation of cells and contribute to meniscal repair.[77,79,80] Video 21-3 shows a case collection of the injection of PRF into several types of joints and tissue (Fig 21-8).

Video 21-3

Similar to the joint injections, platelet concentrates have also been heavily utilized for the regeneration of ligaments and tendons such as rotator cuffs[81] and various forms of tendinopathies.[82–85] Because they have a marked impact on the proliferation of VEGF and the proliferation of fibroblasts, platelet concentrates present a powerful adjunct to conventional therapies to improve blood flow and regeneration.[86,87] It is hypothesized that further development of liquid-PRF could allow it to even replace conventional therapies. The fact that PRF allows for the fibrin matrix to be formed, either ex vivo or after injection, makes PRF a versatile solution to tissue healing and remodeling in clinical situations characterized by a poor healing outcome. In a 2006 study by Anitua et al, it was the specific use of fibrin matrices that allowed for an increased local density of cells, which led to the neovascularization of Achilles tendon in a sheep model. The presence of platelets within the fibrin matrices also significantly increased the proliferation of tendon cells. In an additional in vitro assay, cultured tendon cells seeded in fibrin matrices also had a significantly higher level of collagen1 synthesis and VEGF expression compared to cultures not seeded in fibrin.[88] These results closely follow previous results on human tendon cells.[89,90]

In a 2007 study investigating open suture Achilles tendon repair with versus without PRF in 12 athletes, the authors found that athletes receiving the novel treatment including the fibrin matrix recovered their range of motion earlier (7 ± 2 weeks vs 11 ± 3 weeks, $P = .025$), showed no wound complications, and took less time to begin gentle running (11 ± 1 weeks vs 18 ± 3 weeks, $P = .042$) and to resume training activities (14 ± 0.8 weeks vs 21 ± 3 weeks, $P = .004$).[91] Similar to this, a clinical pilot study on PRF in the replacement of ACLs with autologous tendon was carried out on 20 patients with full ACL rupture (Fig 21-9). The application of PRF membranes in the repair resulted in a significantly accelerated course of treatment and recovery for the patients. The early recovery was characterized by ultrasound to be correlated to an accelerated attachment to the bone (Jørgensen U, unpublished study, 2020).

Although progress has been made in understanding the underlying biology and possibilities of PRF in sports medicine and rheumatology, further research in this field is still very much warranted. PRF could provide patients suffering from severe chronic disabilities such as osteoarthritis or transient injuries such as golfer's elbow with the needed relief to restore function. The literature presented in this chapter showcases the anti-inflammatory capacity and associated regenerative capabilities of PRF, from which many areas of medicine, including sports medicine and rheumatology, could benefit.

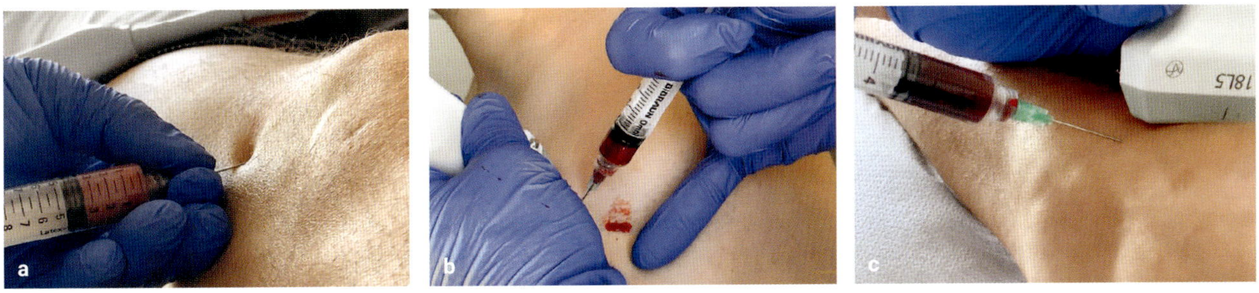

Fig 21-8 *(a to c)* Injection of liquid-PRF into the knee for cartilage and meniscus repair. See Video 21-3 for a case collection of PRF injected into various joints and tissue types.

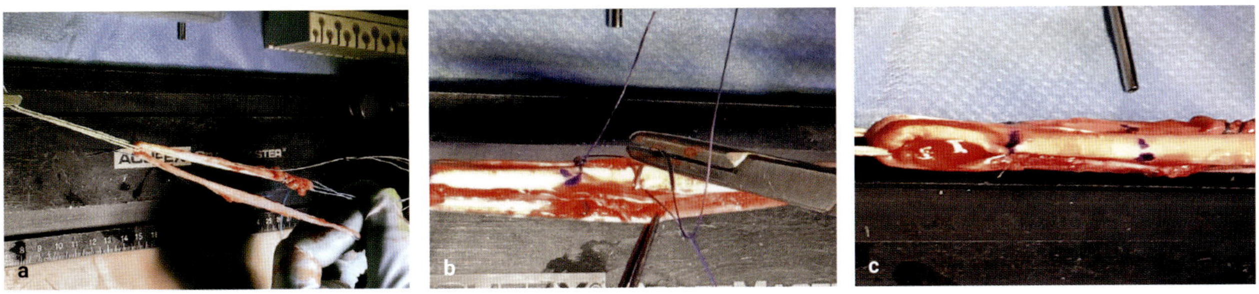

Fig 21-9 Acceleration of ACL regeneration and attachment by PRF-assisted replacement. After the harvest of two autologous tendons and preparation by removal of muscle tissue, the tendons are rigged with several PRF membranes between them. *(a)* Rigging of tendons with PRF membranes. *(b)* The replacement tendons are marked to be divided into three segments. One PRF membrane is fixed in each of the two tendon-to-bone contact segments, using 4-0 suture. *(c)* The finished replacement tendon ready for insertion. (Case performed by Professor Uffe Jørgensen in the Orthopedic Department at Odense Universitetshospital in Denmark.)

PRF in Equine Sports Medicine

In addition to professional sports medicine in humans, the equine industry is a noteworthy market, with an estimated global value of 300 billion US dollars.[92] Orthopedic injuries are one of the main problems when managing sport horses. The most frequent injuries are tendon or ligament injuries and joint arthritis or arthrosis. In the worst-case scenario, these problems can be career ending or even fatal to the horse.[93–96]

Traditional therapies have typically been mid- or short-acting steroids, HA, or polysulfated glycosaminoglycans in combination or as single drugs. These drugs are merely managing the symptoms by reducing inflammation and do not aim to cure or reverse the injuries caused by the ongoing degenerative disease. As a result, regenerative therapies have been used in equine sports medicine for many years. This includes more primitive methods such as harvesting and injecting bone marrow into lesions to more sophisticated methods such as laboratory-cultured, primed tissue-specific stem cells.[97–99]

Among regenerative therapies, autologous conditioned serum (ACS) has been widely used. However, in the context of equine medicine, this procedure is inconvenient and fairly expensive.[100,101] Parallel to ACS, PRP has come into use. From a practical view, an advantage of PRP over ACS is that PRP can be processed at the time of the examination. For both treatments, however, there is a lack of convincing studies of the true effects.[102] Mesenchymal stem cells (MSCs) have also been used. Primary data is promising in the treatment of osteoarthritis, but these products are legally considered a drug in many countries, and the procedure is expensive and logistically difficult due to it containing live donor cells.[99]

The latest regenerative therapy to be introduced in equine sports medicine is PRF, particularly liquid-PRF. Big advantages of liquid-PRF are the easy and quick processing of the final product and the fact that it is entirely safe. The content of leukocytes furthermore adds to the product's safety, as iatrogenic joint infections can be life-threatening in horses.[103] As a fairly new treatment in equine sports medicine, clinical studies are still not in place regarding PRF, but preliminary

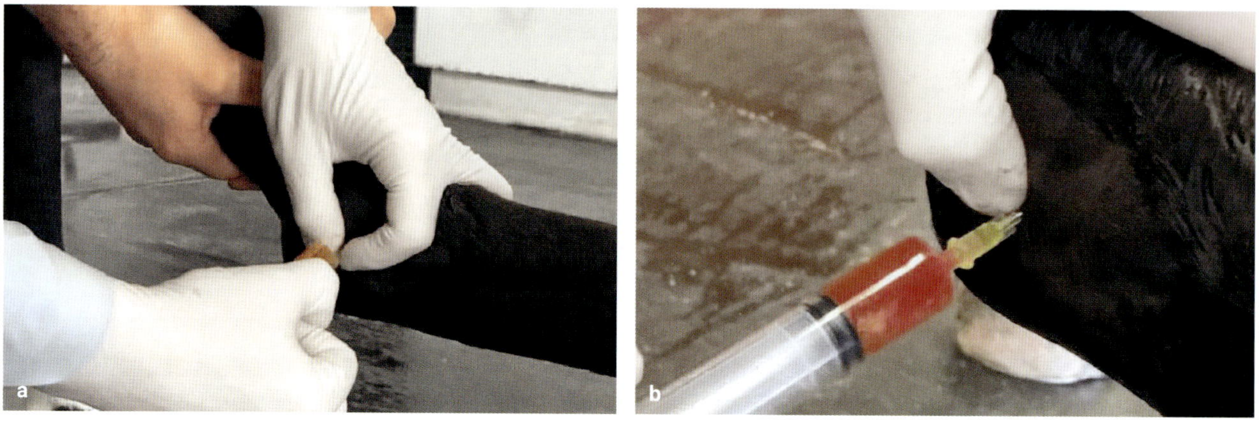

Fig 21-10 *(a and b)* Intra-articular treatment with liquid-PRF of a fetlock on a warmblood trotter with arthritis. (Case performed by DVM Johan Lenz.)

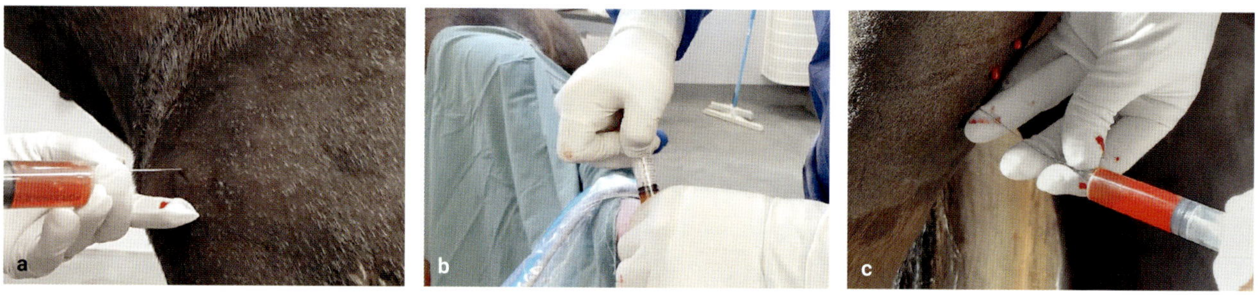

Fig 21-11 *(a to c)* Procedure for blood collection and PRF injection into racehorses. See Video 21-4 for demonstration of the full procedure.

data from other species and other fields is quite promising. At the equine sports medicine clinic Veterinär Jonas Tornell AB in Sweden, liquid-PRF has been injected into over 1,000 joints for the treatment of arthritis and osteoarthrosis in the 1 year it's been in use so far (Fig 21-10).

The procedure, as presented in Fig 21-11 and Video 21-4, involves scrubbing the jugular vein with chlorhexidine and alcohol, retracting blood in the liquid-PRF vacutainer tubes, and centrifuging right before injection. Sterile gloves are used during blood collection. The joint is also scrubbed with chlorhexidine, alcohol, and iodine before injection, and fresh sterile gloves are used for injection. Approximately 5 mL of liquid-PRF is used for fetlocks (metacarpophalangeal joint), coffin joints (distal interphalangeal joint), and carpal joints. To stifle hock joints (tarsal joint), 6 to 8 mL is used. Low-motion joints, such as proximal interphalangeal joints or distal tarsal joints, are injected with lower volumes due to the lack of space in these joints. A handful of tendons and suspensory ligament injuries have also been injected with liquid-PRF. For all of these injections, no adverse reactions have been registered, and no iatrogenic infections have occurred. All clinical experience points to PRF being safe to use in treatments of equine joint disease.

Video 21-4

Liquid-PRF also seems to have just as good an effect in reducing lameness and reducing soreness at flexion tests as the standard treatment of betamethasone/dexamethasone and HA. There is, however, perhaps slightly less reduction in joint effusion than with steroids, at least in the short term (8 weeks). For cases of osteoarthrosis, while fewer cases have been evaluated clinically after

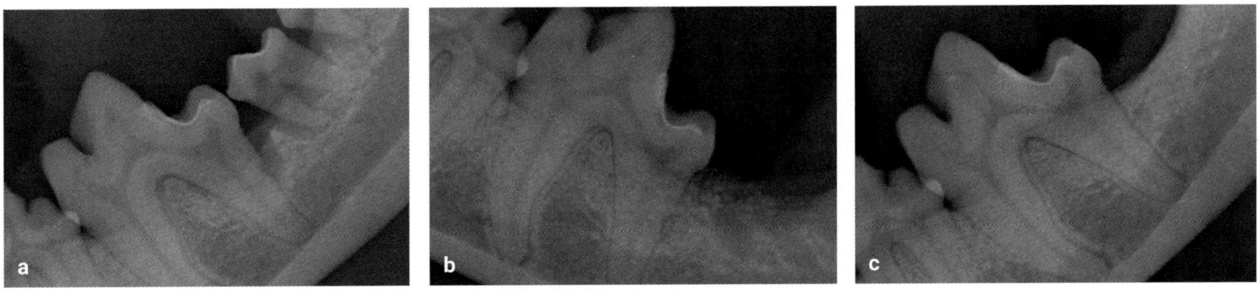

Fig 21-12 Radiograph showing PD4 teeth in a cat treated with liquid-PRF mixed with xenograft. *(a)* Before treatment: The left mandibular first molar has 60% vertical attachment loss at the distal surface of the distal root. *(b)* During treatment, a xenograft was mixed with liquid-PRF and placed in the defect. *(c)* Control radiograph 1 year after the treatment shows complete bone and periodontal ligament healing. (Case performed by DVM Jacob Coakley Meyer.)

liquid-PRF treatment compared to synovitis/arthritis cases, the primary experience has been that when treated with PRF, these joints keep range of motion and are less painful to the horse, which indicates a disease-modifying/regenerative effect. For moderate to severe osteoarthrosis, a protocol with a series of three to four PRF injections with 3- to 4-week intervals has been used.

There are many injuries in equine sports medicine where the known and expected effects of liquid-PRF could be useful. Tendon and ligament injuries with low vascularization and a slow healing process are just one example. Degenerative joint disease with cartilage defects as well as treatment of developmental conditions such as osteochondrosis could perhaps also benefit from PRF injections. Injuries to equine flexor tendons and suspensory ligaments often creates core lesions, which are injectable. Liquid-PRF would have a much better possibility to stay in the lesion due to its fibrin content and coagulation compared to, for example, injections with PRP or stem cells. In concomitance, injections of injuries such as lesions at ligament-bone interfaces are highly interesting. These injuries are difficult to manage when poorly vascularized and are slow to heal.[104,105] Randomized blinded prospective research in equines is often limited by costs. Large groups of the same types of horses in a set environment are so much more expensive and laboring than, for example, laboratory mice. We could hope for smaller randomized prospective studies, but the main equine research must be expected in clinical case control studies.

PRF in Veterinary Dentistry

Over the past 15 to 20 years, the role of dentistry in veterinary practices has increased dramatically and is only expected to steadily keep increasing.[106,107] It has in recent years become more apparent to veterinarians that oral health has a big impact on the welfare of small companion animals.[107] Oral pathology and pain, however, are often difficult to assess for the owner as well as the veterinarian.[108,109]

One of the most frequent diseases in dogs is periodontal disease, with more than 80% prevalence in dogs over 2 years.[107] While previously treatment was primarily extraction of teeth with severe periodontal disease (PD4),[110,111] today many general veterinary practitioners also diagnose and treat mild periodontal disease.[112] Furthermore, client expectations of veterinary treatment of periodontal disease has increased. Clients more often decide to have reconstructive treatments of teeth with moderate to severe periodontal disease when it is an alternative to extraction. As such, the use of PRF in regenerative dentistry in veterinary specialized dental clinics around the world suggest a great potential.[113–115]

One specific area in veterinary dentistry benefiting from the use of PRF is bone augmentation. Experiences indicate that PRF in many cases can supplement or even completely substitute the use of more expensive collagen membranes. A case by DVM Jacob C. Meyer is presented in Fig 21-12. In this cat, a PD4 vertical defect at the distal root of the left mandibular first molar was filled with a xenograft mixed with liquid-PRF. An A-PRF membrane was placed between the gingiva and the graft before suturing the gingiva. Figure 21-12c shows the same tooth 1 year later, and no signs of periodontal disease are apparent.

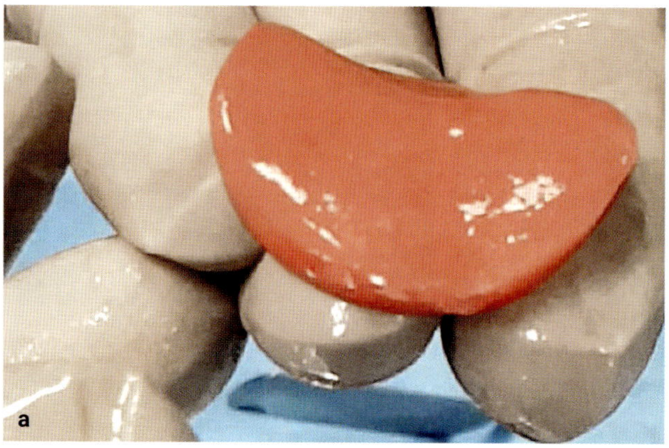

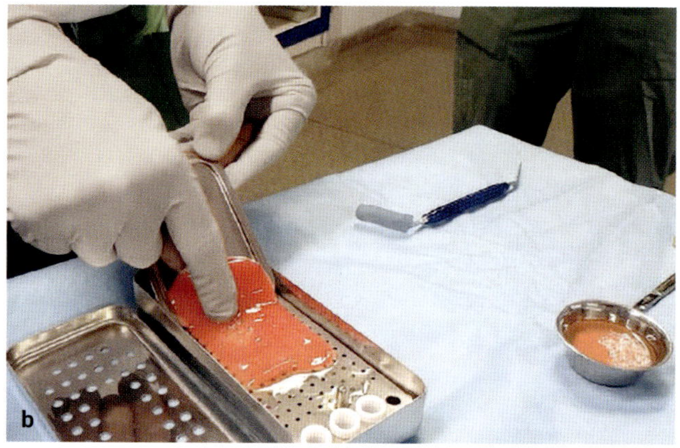

Fig 21-13 Large membranes and blocks created from PRF. *(a)* Block from trapezoidal volume mold. *(b)* Large membrane from flat square mold before dehydration in the PRF box. (Case performed at the ENT Department at Rigshospitalet in Denmark.)

Discussion and Future Perspectives

Many new and longstanding areas of research and medicine currently experiment with PRF to identify novel clinical indications. For instance, on several occasions PRP, PRF, and other PRF analogs have been used in orthopedic surgery for the repair of fractures and large bone defects,[116] as well as related research on osteoblast behavior[117] and the differentiation of bone marrow stromal cells.[118] While this field is very early in its attempts to characterize the potential benefits of utilizing PRF, some possible advances, bridging the gap between maxillofacial and orthopedic surgery, are being uncovered.[119–122]

Another ongoing topic of research regarding PRF and its analogs is bioprinting and tissue engineering, which is used within various fields of medicine and dentistry.[123–128] In 2018, Lei et al described the application of a 3D-printed mold, based on a patient's CBCT data, to form a PRF and xenograft replicate of the patient's bone defect. A 15-month clinical and radiographic follow-up revealed greatly reduced pocket depths and significant alveolar bone fill at the treatment site.[129] In another proof-of-concept paper, Faramarzi et al positively demonstrated the development of a PRP-alginate bioink for 3D bioprinting that had a gradual release of various proteins and growth factors over several days. In vitro experiments demonstrated that the bioink could positively affect the function of MSCs and human umbilical vein endothelial cells.[130]

The specific exploitation of fibrin as a structural scaffold and reservoir of self-contained regenerative cells and soluble growth factors makes PRF an interesting medium for the filling of cavities and migration of cells from adjacent tissue. In a case study from the ENT Department at Rigshospitalet in Denmark, we utilized the possibility to produce large membranes and blocks to fill defects and cover wounds after tumor excision. An example of these molded structures can be seen in Fig 21-13.

The versatility of PRF in tissue engineering is perhaps best illustrated by the distinct types of tissue it has been used to heal and regenerate. In veterinary ophthalmic surgery, the standard technique for repairing corneal ulcers is using discs prepared of either amnion or porcine urinary bladder submucosa to serve as a scaffold for healing.[131,132] In theory, PRF membranes could be an excellent autologous alternative for scaffolding and use as a reservoir of different growth factors, cytokines, and other molecules such as nutrients necessary for proper healing. In a clinical example, an 8-year-old pug presented with a painful deep stromal paracentral ulcer with melting of stromal tissue and diffuse corneal edema (Fig 21-14).

Lastly, more recent research has utilized Alb-PRF (or e-PRF) for spinal disc repair. In a study titled "Repair of an annulus fibrosus tear of the lumbar intervertebral disc using Alb-PRF: A novel autologous technique that may prevent patients more extensive/invasive surgical procedures," Dr Robert Talac (orthopedic surgeon) reports how he has been utilizing the e-PRF formulation and injecting directly into the spine of patients with various pathologies (Fig 21-15). His ongoing research has demonstrated the body's own healing capability in these complex defects, and patients are often without pain within 15 minutes postinjection, as highlighted in Video 21-5.

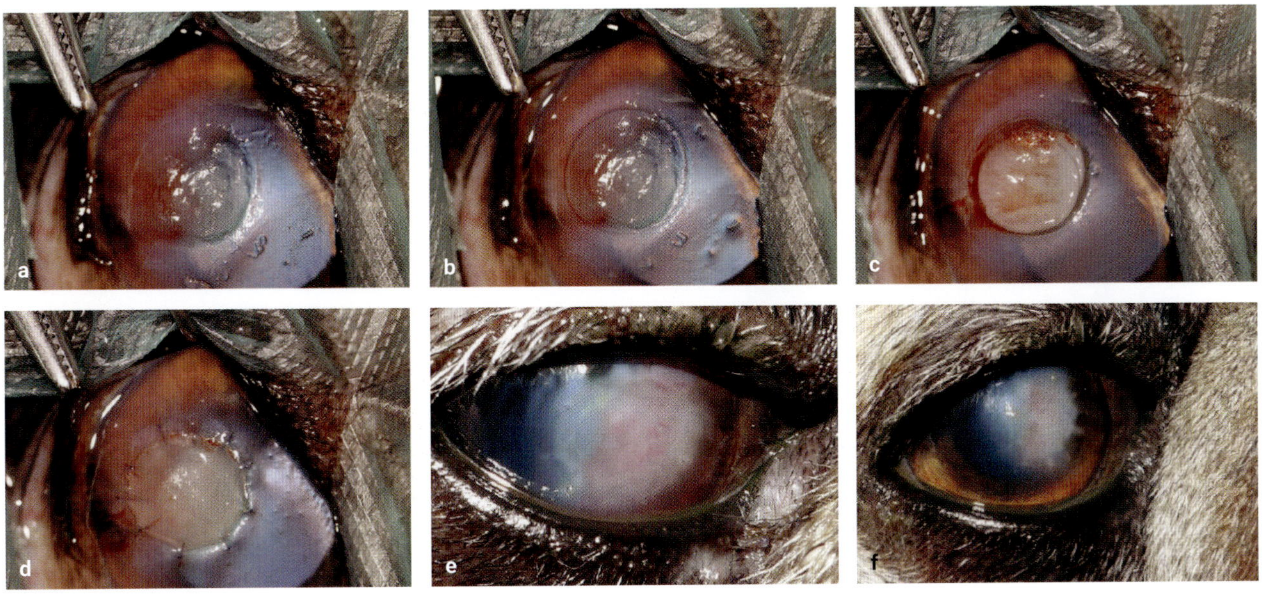

Fig 21-14 Treatment of deep stromal paracentral ulcer with solid-PRF membrane. *(a)* Stromal melting tissue was debrided. *(b)* The ulcer bed was outlined with a 6-mm trephine, and stromal tissue was removed by keratectomy. *(c)* A 6-mm PRF membrane disc was cut to size and placed in the ulcer bed. *(d)* The disc was sutured in place with a monofilament Vicryl 10-0. *(e)* Day 7 postoperative, the disc is well integrated and only small spots of fluorescein-positive staining around the sutures are visible. *(f)* Day 14 postoperative, the cornea is clearing up and only two pin-size areas around the sutures remain fluorescein positive. (Case performed by DVM Tommy Hardon.)

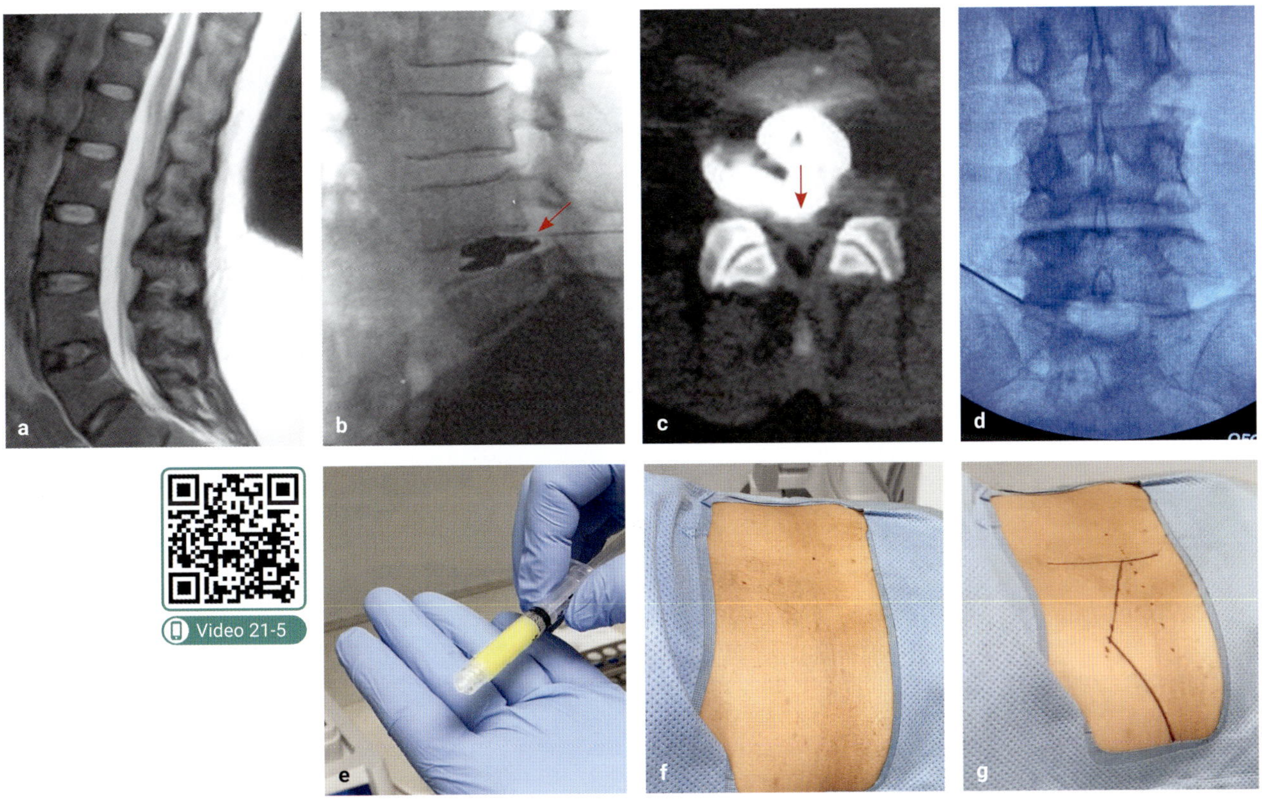

Video 21-5

Fig 21-15 *(a)* Sagittal T2 MRI showing a slight signal change at L4/5 of the intervertebral disc. *(b)* Radiopaque contrast injected into the L4/5 disc (*red arrow* shows annular defect). *(c)* Axial CT scan across L4/5 disc depicting radiopaque contrast spreading through the annular defect. *(d)* Anteroposterior lumbar radiograph showing the tip of the spinal needle on the lateral edge of the L5/S1 facet joint. *(e)* Final e-PRF formulation. *(f)* Patient in prone position before procedure. *(g)* Transforaminal access and marking the entry point.

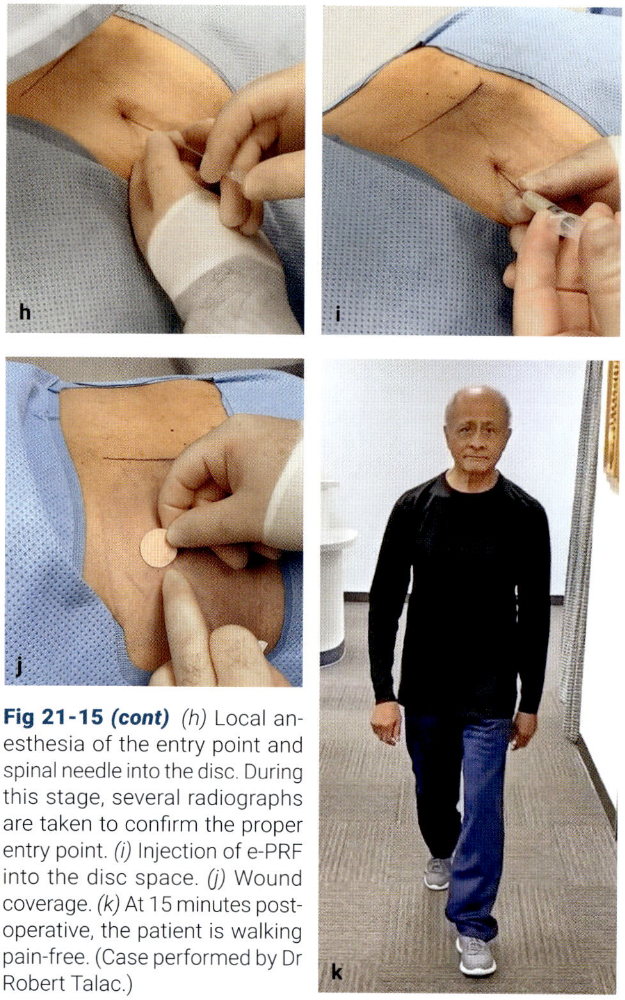

Fig 21-15 *(cont)* *(h)* Local anesthesia of the entry point and spinal needle into the disc. During this stage, several radiographs are taken to confirm the proper entry point. *(i)* Injection of e-PRF into the disc space. *(j)* Wound coverage. *(k)* At 15 minutes postoperative, the patient is walking pain-free. (Case performed by Dr Robert Talac.)

Conclusion

The future holds great potential for the use of PRF in various fields of medicine and research. While the literature in this chapter highlights numerous clinical reports having investigated the medical use of PRF in its core applications as well as in some more fringe topics, future RCTs are still very much needed for most of the prospective indications.

References

1. Choukroun J, Adda F, Schoeffler C, Vervelle A. An opportunity in perio-implantology: The PRF [in French]. Implantodontie 2001;42:55–62.
2. Gosain A, DiPietro LA. Aging and wound healing. World J Surg 2004;28:321–326.
3. Eming SA, Brachvogel B, Odorisio T, Koch M. Regulation of angiogenesis: Wound healing as a model. Prog Histochem Cytochem 2007;42:115–170.
4. Eming SA, Kaufmann J, Löhrer R, Krieg T. Chronische wunde: Neue wege in forschung und therapie. Hautarzt 2007;58:939–944.
5. Guo S, Dipietro LA. Factors affecting wound healing. J Dent Res 2010;89:219–229.
6. Upputuri PK, Sivasubramanian K, Mark CSK, Pramanik M. Recent Developments in vascular imaging techniques in tissue engineering and regenerative medicine. Biomed Res Int 2015;2015:783983.
7. Järbrink K, Ni G, Sönnergren H, et al. Prevalence and incidence of chronic wounds and related complications: A protocol for a systematic review. Syst Rev 2016;5:152.
8. Sen CK, Gordillo GM, Roy S, et al. Human skin wounds: A major and snowballing threat to public health and the economy. Wound Repair Regen 2009;17:763–771.
9. Olsson M, Järbrink K, Divakar U, et al. The humanistic and economic burden of chronic wounds: A systematic review. Wound Repair Regen 2019;27:114–125.
10. Sen CK. Human wounds and its burden: An updated compendium of estimates. Adv Wound Care 2019;8:39–48.
11. Gottrup F. A specialized wound-healing center concept: Importance of a multidisciplinary department structure and surgical treatment facilities in the treatment of chronic wounds. Am J Surg 2004;187:S38–S43.
12. Singer AJ, Clark RAF. Cutaneous wound healing. N Engl J Med 1999;341:738–746.
13. Crovetti G, Martinelli G, Issi M, et al. Platelet gel for healing cutaneous chronic wounds. Transfus Apher Sci 2004;30:145–151.
14. Järbrink K, Ni G, Sönnergren H, et al. The humanistic and economic burden of chronic wounds: A protocol for a systematic review. Syst Rev 2017;6:15.
15. Nussbaum SR, Carter M, Fife CE, et al. An economic evaluation of the impact, cost, and Medicare policy implications of chronic nonhealing wounds. Value Heal 2018;21:27–32.
16. Gottrup F, Holstein P, Jørgensen B, Lohmann M, Karlsmark T. A new concept of a multidisciplinary wound healing center and a national expert function of wound healing Arch Surg 2001;136:765–772.
17. O'Connell SM, Impeduglia T, Hessler K, Wang XJ, Carroll R, Dardik H. Autologous platelet-rich fibrin matrix as cell therapy in the healing of chronic lower-extremity ulcers. Wound Repair Regen 2008;16:749–756.
18. Steenvoorde P, van Doorn LP, Naves C, Oskam J. Use of autologous platelet-rich fibrin on hard-to-heal wounds. J Wound Care 2008;17:60–63.
19. Anitua E, Aguirre JJ, Algorta J, et al. Effectiveness of autologous preparation rich in growth factors for the treatment of chronic cutaneous ulcers. J Biomed Mater Res B Appl Biomater 2008;84:415–421.
20. Jørgensen B, Karlsmark T, Vogensen H, Haase L, Lundquist R. A pilot study to evaluate the safety and clinical performance of Leucopatch, an autologous, additive-free, platelet-rich fibrin for the treatment of recalcitrant chronic wounds. Int J Low Extrem Wounds 2011;10:218–223.
21. Löndahl M, Tarnow L, Karlsmark T, et al. Use of an autologous leucocyte and platelet-rich fibrin patch on hard-to-heal DFUs: A pilot study. J Wound Care 2015;24:172–178.
22. Pinto NR, Ubilla M, Zamora Y, Del Rio V, Dohan Ehrenfest DM, Quirynen M. Leucocyte- and platelet-rich fibrin (L-PRF) as a regenerative medicine strategy for the treatment of refractory leg ulcers: A prospective cohort study. Platelets 2018;29:468–475.
23. Game F, Jeffcoate W, Tarnow L, et al. LeucoPatch system for the management of hard-to-heal diabetic foot ulcers in the UK, Denmark, and Sweden: An observer-masked, randomised controlled trial. Lancet Diabetes Endocrinol 2018;6:870–878.
24. Reksodiputro MH, Harba'i HM, Koento T, Harahap AR. Platelet-rich fibrin enhances wound epithelialization in the skin graft donor site. J Phys Conf Ser 2018;1073:032046.

25. Chignon-Sicard B, Georgiou CA, Fontas E, et al. efficacy of leukocyte- and platelet-rich fibrin in wound healing. Plast Reconstr Surg 2012;130:819e–829e.
26. Powell DM, Chang E, Farrior EH. Recovery from deep-plane rhytidectomy following unilateral wound treatment with autologous platelet gel: A pilot study. Arch Facial Plast Surg 2001;3:245–250.
27. Danielsen PL, Ågren MS, Jorgensen LN. Platelet-rich fibrin versus albumin in surgical wound repair. Ann Surg 2010;251:825–831.
28. Sclafani AP. Safety, efficacy, and utility of platelet-rich fibrin matrix in facial plastic surgery. Arch Facial Plast Surg 2011;13:247–251.
29. Brem H, Stojadinovic O, Diegelmann RF, et al. Molecular markers in patients with chronic wounds to guide surgical debridement. Mol Med 2007;13:30–39.
30. Werner S, Grose R. Regulation of wound healing by growth factors and cytokines. Physiol Rev 2003;83:835–870.
31. Everts PAM, Knape JTA, Weibrich G, et al. Platelet-rich plasma and platelet gel: A review. J Extra Corpor Technol 2006;38:174–187.
32. Rozman P, Bolta Z. Use of platelet growth factors in treating wounds and soft-tissue injuries. Acta Dermatovenerologica Alp Pannonica Adriat 2007;16:156–165.
33. Nurden AT, Nurden P, Sánchez M, Andia I, Anitua E. Platelets and wound healing. Front Biosci 2008;13:3532–3548.
34. Nurden A. Platelets, inflammation and tissue regeneration. Thromb Haemost 2011;105(suppl 1):S13–S33.
35. Khiste SV, Naik Tari R. Platelet-rich fibrin as a biofuel for tissue regeneration. ISRN Biomater 2013;2013:1–6.
36. Roy S, Driggs J, Elgharably H, et al. Platelet-rich fibrin matrix improves wound angiogenesis via inducing endothelial cell proliferation. Wound Repair Regen 2011;19:753–766.
37. Suzuki S, Morimoto N, Ikada Y. Gelatin gel as a carrier of platelet-derived growth factors. J Biomater Appl 2013;28:595–606.
38. Li Q, Pan S, Dangaria SJ, et al. Platelet-rich fibrin promotes periodontal regeneration and enhances alveolar bone augmentation. Biomed Res Int 2013;2013:638043.
39. Horii K. Kanayama T, Miyamoto H, et al. Platelet-rich fibrin has a healing effect on chemotherapy-induced mucositis in hamsters. Oral Surg Oral Med Oral Pathol Oral Radiol 2014;117:445–453.
40. Reksodiputro M, Widodo D, Bashiruddin J, Siregar N, Malik S. PRFM enhance wound healing process in skin graft. Facial Plast Surg 2014;30:670–675.
41. Long DW, Johnson NR, Jeffries EM, Hara H, Wang Y. Controlled delivery of platelet-derived proteins enhances porcine wound healing. J Control Release 2017;253:73–81.
42. Caruana A, Savina D, Macedo JP, Soares SC. From platelet-rich plasma to advanced platelet-rich fibrin: Biological achievements and clinical advances in modern surgery. Eur J Dent 2019;13:280–286.
43. Lundquist R, Holmstrøm K, Clausen C, Jørgensen B, Karlsmark T. Characteristics of an autologous leukocyte and platelet-rich fibrin patch intended for the treatment of recalcitrant wounds. Wound Repair Regen 2013;21:66–76.
44. D'asta F, Halstead F, Harrison P, et al. The contribution of leucocytes to the antimicrobial activity of platelet-rich plasma preparations: A systematic review. Platelets 2018;29:9–20.
45. Tuttle MS. Association between microbial bioburden and healing outcomes in venous leg ulcers: A review of the evidence. Adv Wound Care 2015;4:1–11.
46. Game F, Jeffcoate W, Tarnow L, Day F, Fitzsimmons D, Jacobsen J. The LeucoPatch system in the management of hard-to-heal diabetic foot ulcers: Study protocol for a randomised controlled trial. Trials 2017;18:1–8.
47. Federation ID. Worldwide toll of diabetes. IDF Diabetes Atlas, ed 9. Available at https://diabetesatlas.org/en/sections/worldwide-toll-of-diabetes.html. Accessed 27 April 2020.
48. Singh N, Armstrong DG, Lipsky BA. Preventing foot ulcers in patients with diabetes. J Am Med Assoc 2005;293:217–228.
49. Choukroun J, Diss A, Simonpieri A, et al. Platelet-rich fibrin (PRF): A second-generation platelet concentrate. Part IV: Clinical effects on tissue healing. Oral Surg Oral Med Oral Pathol Oral Radiol Endod 2006;101:e56–e60.
50. Ding Y, Cui L, Zhao Q, Zhang W, Sun H, Zheng L. Platelet-rich fibrin accelerates skin wound healing in diabetic mice. Ann Plast Surg 2017;79:e15–e19.
51. Nanditha S, Chandrasekaran B, Muthusamy S, Muthu K. Apprising the diverse facets of platelet rich fibrin in surgery through a systematic review. Int J Surg 2017;46:186–194.
52. Desai CB, Kini YK, Mahindra UR, Bakshi MK. Use of platelet-rich fibrin over skin wounds: Modified secondary intention healing. J Cutan Aesthet Surg 2013;6:35.
53. Sclafani AP, McCormick SA. Induction of dermal collagenesis, angiogenesis, and adipogenesis in human skin by injection of platelet-rich fibrin matrix. Arch Facial Plast Surg 2012;14:132–136.
54. Gorlero F, Glorio M, Lorenzi P, Bruno-Franco M, Mazzei C. New approach in vaginal prolapse repair: Mini-invasive surgery associated with application of platelet-rich fibrin. Int Urogynecol J 2012;23:715–722.
55. Soyer T, Çakmak M, Aslan MK, Şenyücel MF, Kisa Ü. Use of autologous platelet rich fibrin in urethracutaneous fistula repair: Preliminary report. Int Wound J 2103;10:345–347.
56. Guinot A, Arnaud A, Azzis O, Habonimana E, Jasienski S, Frémond B. Preliminary experience with the use of an autologous platelet-rich fibrin membrane for urethroplasty coverage in distal hypospadias surgery. J Pediatr Urol 2014;10:300–305.
57. El-Sayed IM, Moustafa WA, El-Zamarany EA, Sadaka MS. The use of autologous platelet-rich fibrin membrane in hypospadias surgery: A preliminary study. Tanta Med J 2017;45:161.
58. Mehta D, Jain V, Triveni M, Tarun Kumar A. Role of platelet-rich-fibrin in enhancing palatal wound healing after free graft. Contemp Clin Dent 2012;3(suppl 2):S240–S243.
59. Kulkarni MR, Thomas BS, Varghese JM, Bhat GS. Platelet-rich fibrin as an adjunct to palatal wound healing after harvesting a free gingival graft: A case series. J Indian Soc Periodontol 2014;18:399–402.
60. Femminella B, Iaconi MC, Di Tullio M, et al. Clinical comparison of platelet-rich fibrin and a gelatin sponge in the management of palatal wounds after epithelialized free gingival graft harvest: A randomized clinical trial. J Periodontol 2016;87:103–113.
61. Pathak H, Mohanty S, Urs AB, Dabas J. Treatment of oral mucosal lesions by scalpel excision and platelet-rich fibrin membrane grafting: A review of 26 sites. J Oral Maxillofac Surg. 2015;73:1865–1874.
62. Habesoglu M, Oysu C, Sahin S, et al. Platelet-rich fibrin plays a role on healing of acute-traumatic ear drum perforation. J Craniofac Surg 2014;25:2056–2058.
63. Garin P, Mullier F, Ghledof D, Dogne JM, Putz L, Van Damme JP. Platelet-rich fibrin (PRF): An autologous packing material for middle ear microsurgery. B-ENT 2014;10:27–34.
64. Nair N, Alexander A, Abhishekh B, Hedge JS, Ganesan S, Saxena SK. Safety and efficacy of autologous platelet-rich fibrin on graft uptake in myringoplasty: A randomized controlled trial. Int Arch Otorhinolaryngol 2019;23:77–82.
65. Matz EL, Pearlman AM, Terlecki RP. Safety and feasibility of platelet rich fibrin matrix injections for treatment of common urologic conditions. Investig Clin Urol 2018;59:61–65.
66. Virag R, Sussman H, Lambion S, de Fourmestraux V. Evaluation of the benefit of using a combination of autologous platelet rich-plasma and hyaluronic acid for the treatment of Peyronie's disease. Sex Heal Issues 2017;1:1–8.
67. Jurczak F, Laridon JY, Raffaitin P, Pousset JP. Biological fibrin and anal fistulas: 31 patients in a clinical trial. Ann Chir 2004;129:286–289.

68. Spinelli A, Kotze PG. Anal fistula plug for perianal fistulising Crohn's disease: An important trial for inflammatory bowel disease surgeons. J Crohn's Colitis 2016;10:125–126.
69. Wang T, He C. Pro-inflammatory cytokines: The link between obesity and osteoarthritis. Cytokine Growth Factor Rev 2018;44: 38–50.
70. Kiguchi N, Kobayashi D, Saika F, Matsuzaki S, Kishioka S. Pharmacological regulation of neuropathic pain driven by inflammatory macrophages. Int J Mol Sci 2017;18:2296.
71. Sánchez M, Azofra J, Anitua E, et al. Plasma rich in growth factors to treat an articular cartilage avulsion: A case report. Med Sci Sports Exerc 2003;35:1648–1652.
72. Campbell KA, Saltzman B, Mascarenhas R, et al. Does intra-articular platelet-rich plasma injection provide clinically superior outcomes compared with other therapies in the treatment of knee osteoarthritis? A systematic review of overlapping meta-analyses. Arthroscopy 2015;31:2213–2221.
73. Kanchanatawan W, Arirachakaran A, Chaijenkij K, et al. Short-term outcomes of platelet-rich plasma injection for treatment of osteoarthritis of the knee. Knee Surg Sports Traumatol Arthrosc 2016;24:1665–1677.
74. Meheux CJ, McCulloch PC, Lintner DM, Varner KE, Harris JD. Efficacy of intra-articular platelet-rich plasma injections in knee osteoarthritis: A systematic review. Arthroscopy 2016;32:495–505.
75. Nguyen C, Lefèvre-Colau MM, Poiraudeau S, Rannou F. Evidence and recommendations for use of intra-articular injections for knee osteoarthritis. Ann Phys Rehabil Med 2016;59:184–189.
76. Sánchez M, Delgado D, Sánchez P, et al. Combination of intra-articular and intraosseous injections of platelet rich plasma for severe knee osteoarthritis: A pilot study. Biomed Res Int 2016;2016: 4868613.
77. Abd El Raouf M, Wang X, Miusi S, et al. Injectable-platelet rich fibrin using the low speed centrifugation concept improves cartilage regeneration when compared to platelet-rich plasma. Platelets 2017;30:213–221.
78. Nixon AJ, Fortier LA, Williams J, Mohammed H. Enhanced repair of extensive articular defects by insulin-like growth factor-I-laden fibrin composites. J Orthop Res 1999;17:475–487.
79. Schmidt MB, Chen EH, Lynch SE. A review of the effects of insulin-like growth factor and platelet derived growth factor on in vivo cartilage healing and repair. Osteoarthritis Cartilage 2006;14: 403–412.
80. Wong CC, Kuo TF, Yang TL, et al. Platelet-rich fibrin facilitates rabbit meniscal repair by promoting meniscocytes proliferation, migration, and extracellular matrix synthesis. Int J Mol Sci 2017;18:1722.
81. Zumstein MA, Rumian A, Lesbats V, Schaer M, Boileau P. Increased vascularization during early healing after biologic augmentation in repair of chronic rotator cuff tears using autologous leukocyte- and platelet-rich fibrin (L-PRF): A prospective randomized controlled pilot trial. J Shoulder Elb Surg 2014;23:3–12.
82. Cai Y, Zhang C, Lin X. Efficacy of platelet-rich plasma in arthroscopic repair of full-thickness rotator cuff tears: A meta-analysis. J Shoulder Elb Surg 2015;24:1852–1859.
83. Nourissat G, Ornetti P, Berenbaum F, Sellam J, Richette P, Chevalier X. Does platelet-rich plasma deserve a role in the treatment of tendinopathy? Joint Bone Spine 2015;82:230–234.
84. Saltzman BM, Jain A, Campbell KA, et al. Does the Use of platelet-rich plasma at the time of surgery improve clinical outcomes in arthroscopic rotator cuff repair when compared with control cohorts? A systematic review of meta-analyses. Arthroscopy 2016;32:906–918.
85. Warth RJ, Dornan GJ, James EW, Horan MP, Millett PJ. Clinical and structural outcomes after arthroscopic repair of full-thickness rotator cuff tears with and without platelet-rich product supplementation: A meta-analysis and meta-regression. Arthroscopy 2015;31:306–320.
86. Anitua E, Sánchez M, Zalduendo MM, et al. Fibroblastic response to treatment with different preparations rich in growth factors. Cell Prolif 2009;42:162–170.
87. Anitua E, Sánchez M, Nurden AT, et al. Platelet-released growth factors enhance the secretion of hyaluronic acid and induce hepatocyte growth factor production by synovial fibroblasts from arthritic patients. Rheumatology (Oxford) 2007;46:1769–1772.
88. Anitua E, Sánchez M, Nurden A, et al. Autologous fibrin matrices: A potential source of biological mediators that modulate tendon cell activities. J Biomed Mater Res A 2006;77:285–293.
89. Anitua E, Andía I, Sánchez M, et al. Autologous preparations rich in growth factors promote proliferation and induce VEGF and HGF production by human tendon cells in culture. J Orthop Res 2005;23:281–286.
90. Zhang J, Wang JHC. Platelet-rich plasma releasate promotes differentiation of tendon stem cells into active tenocytes. Am J Sports Med 2010;38:2477–2486.
91. Sánchez M, Anitua E, Azofra J, Andía I, Padilla S, Mujika I. Comparison of surgically repaired Achilles tendon tears using platelet-rich fibrin matrices. Am J Sports Med 2007;35:245–251.
92. Equine Business Association. The Equine Industry: A Global Perspective 2018;5–10. https://www.equinebusinessassociation.com/equine-industry-statistics/. Accessed 6 July 2020.
93. Dyson PK, Jackson BF, Pfeiffer DU, Price JS. Days lost from training by two- and three-year-old thoroughbred horses: A survey of seven UK training yards. Equine Vet J 2008;40:650–657.
94. Egenvall A, Tranquille CA, Lönnell AC, et al. Days-lost to training and competition in relation to workload in 263 elite show-jumping horses in four European countries. Prev Vet Med 2013;112:387–400.
95. Sloet Van Oldruitenborgh-Oosterbaan MM, Genzel W, van Weeren PR. A pilot study on factors influencing the career of Dutch sport horses. Equine Vet J 2010;42:28–32.
96. Murray RC, Dyson SJ, Tranquille C, Adams V. Association of type of sport and performance level with anatomical site of orthopaedic injury diagnosis. Equine Vet J 2006;38:411–416.
97. Fortier LA. Making progress in the what, when and where of regenerative medicine for our equine patients. Equine Vet J 2012;44: 511–512.
98. Mcilwraith CW. Management of joint diseases in horses: Current and future prospects. Equine Vet Educ 2015;27:335–337.
99. Abu-Seida AM. Regenerative therapy for equine osteoarthritis: A concise review. Asian J Anim Vet Adv 2015;10:500–508.
100. Hraha TH, Doremus KM, McIlwraith CW, Frisbie DD. Autologous conditioned serum: The comparative cytokine profiles of two commercial methods (IRAP and IRAP II) using equine blood. Equine Vet J 2011;43:516–521.
101. Wehling P, Moser C, Frisbie D, et al. Autologous conditioned serum in the treatment of orthopedic diseases: The Orthokine therapy. BioDrugs 2007;21:323–332.
102. Hessel LN, Bosch G, van Weeren PR, Ionita JC. Equine autologous platelet concentrates: A comparative study between different available systems. Equine Vet J 2015;47:319–325.
103. Carstanjen B, Boehart S, Cislakova M. Septic arthritis in adult horses. Pol J Vet Sci 2010;13:201–212.
104. Fortier LA, Smith RK. Regenerative medicine for tendinous and ligamentous injuries of sport horses. Vet Clin North Am Equine Pract 2008;24:191–201.
105. Geburek F, Lietzau M, Beineke A, Rohn K, Stadler PM. Effect of a single injection of autologous conditioned serum (ACS) on tendon healing in equine naturally occurring tendinopathies. Stem Cell Res Ther 2015;6:1–14.
106. Lund EM, Armstrong PJ, Kirk CA, Kolar LM, Klausner JS. Health status and population characteristics of dogs and cats examined at private veterinary practices in the United States. J Am Vet Med Assoc 1999;214:1336–1341.

107. Niemiec BA. Dental services: Good medicine for patients & practices. Today's Vet Pract 2011;38–42.
108. Niemiec BA. Fundamentals of endodontics. Vet Clin North Am Small Animal Pract 2005;35:837–868.
109. Snyder LC, Snyder C, Beebe D. Anesthesia and pain management. In: Lobprise H, Dodd JR (eds). Wiggs's Veterinary Dentistry: Principles and Practice, ed 2. Hoboken, NJ: Wiley, 2019:177–192.
110. Holmstrom SE, Bellows J, Colmery B, Conway ML, Knutson K, Vitoux J. 2013 AAHA dental care guidelines for dogs and cats. J Am Anim Hosp Assoc 2013;49:75–82.
111. Niemiec BA, Gawor J, Nemec A, et al. World Small Animal Veterinary Association Global Dental Guidelines 2017. www.wsava.org/wp-content/uploads/2020/01/Dental-Guidleines-for-endorsement_0.pdf. Accessed 6 July 2020.
112. Bellows J, Berg ML, Dennis S, et al. 2019 AAHA Dental Care Guidelines for Dogs and Cats. J Am Anim Hosp Assoc 2019;55:49–69.
113. Hatakeyama I, Marukawa E, Takahashi Y, Omura K. Effects of platelet-poor plasma, platelet-rich plasma, and platelet-rich fibrin on healing of extraction sockets with buccal dehiscence in dogs. Tissue Eng Part A 2014;20:874–882.
114. Wang QL, Yang PP, Ge LH, Liu H. Preliminary evaluation of platelet rich fibrin-mediated tissue repair in immature canine pulpless teeth. Chinese J Dent Res 2016;19:49–54.
115. Feigin K, Shope, B. Use of platelet-rich plasma and platelet-rich fibrin in dentistry and oral surgery: Introduction and review of the literature. J Vet Dent 2019;36:109–123.
116. Anitua E, Sánchez M, Nurden AT, Nurden P, Oriva G, Andía I. New insights into and novel applications for platelet-rich fibrin therapies. Trends Biotechnol 2006;24:227–234.
117. Wang X, Zhang Y, Choukroun J, Ghanaati S, Miron RJ. Effects of an injectable platelet-rich fibrin on osteoblast behavior and bone tissue formation in comparison to platelet-rich plasma. Platelets 2018;29:48–55.
118. Perut F, Filardo G, Mariani E, et al. Preparation method and growth factor content of platelet concentrate influence the osteogenic differentiation of bone marrow stromal cells. Cytotherapy 2013;15:830–839.
119. Park HC, Kim SG, Oh JS, et al. Early bone formation at a femur defect using CGF and PRF grafts in adult dogs: A comparative study. Implant Dent 2016;25:387–393.
120. Kökdere NN, Baykul T, Findik Y. The use of platelet-rich fibrin (PRF) and PRF-mixed particulated autogenous bone graft in the treatment of bone defects: An experimental and histomorphometrical study. Dent Res J (Isfahan) 2015;12:418–424.
121. Sindel A, Dereci Ö, Toru HS, Tozoğlu S. Histomorphometric comparison of bone regeneration in critical-sized bone defects using demineralized bone matrix, platelet-rich fibrin, and hyaluronic acid as bone substitutes. J Craniofac Surg 2017;28:1865–1868.
122. Durmuşlar MC, Balli U, Dede FO, et al. Evaluation of the effects of platelet-rich fibrin on bone regeneration in diabetic rabbits. J Craniomaxillofac Surg 2016;44:126–133.
123. Herrera-Vizcaíno C, Dohle E, Al-Maawi S, et al. Platelet-rich fibrin secretome induces three dimensional angiogenic activation in vitro. Eur Cell Mater 2019;37:250–264.
124. Ozbolat IT, Yu Y. Bioprinting toward organ fabrication: Challenges and future trends. IEEE Trans Biomed Eng 2013;60:691–699.
125. Campbell PG, Weiss LE. Tissue engineering with the aid of inkjet printers. Exp Opin Biol Ther 2007;7:1123–1127.
126. Anil Kumar S, Alonzo M, Allen SC, et al. A visible light-cross-linkable, fibrin-gelatin-based bioprinted construct with human cardiomyocytes and fibroblasts. ACS Biomater Sci Eng 2019;5:4551–4563.
127. Ahmed TAE, Dare EV, Hincke M. Fibrin: A versatile scaffold for tissue engineering applications. Tissue Eng Part B Rev 2008;14:199–215.
128. Ahadian S, Khademhosseini A. A perspective on 3D bioprinting in tissue regeneration. Bio-Design Manuf 2018;1:157–160.
129. Lei L, Sun W, Yu Y, Ke T, Chen L. The application of three-dimensional printing model and platelet-rich fibrin (PRF) technology in guided tissue regeneration surgery for severe bone defects. J Oral Implantol 2018;45:35–43.
130. Faramarzi N, Yazdi I, Nabavinia M, et al. Patient-specific bioinks for 3D bioprinting of tissue engineering scaffolds. Adv Healthc Mater 2018;7:e1701347.
131. Costa D, Leiva M, Sanz F, et al. A multicenter retrospective study on cryopreserved amniotic membrane transplantation for the treatment of complicated corneal ulcers in the dog. Vet Ophthalmol 2019;22:695–702.
132. Balland O, Poinsard AS, Famose F, et al. Use of a porcine urinary bladder acellular matrix for corneal reconstruction in dogs and cats. Vet Ophthalmol 2016;19:454–463.

22

Future Research with PRF

The past 5 years of research on PRF have seen a prolific increase in papers published on the topic, with much advancement being documented in the literature. Over 250 articles each year are now dedicated to PRF, leading the way to new breakthroughs on this topic. Its use in regenerative dentistry has certainly become widespread across many fields in recent years, including periodontology, oral surgery, and implant dentistry, and its use is gradually being incorporated in the fields of endodontics, facial esthetics, and laser dentistry, to name a few. In 2016, 10 areas of research were highlighted as important factors requiring further investigation. Five years later, we revisit each of these areas one by one and highlight the contributions made to each. These include uses of PRF for the management of ONJ, TMJ disorders, and endodontics as well as future optimization of protocols. We also introduce a new series of 10 key areas requiring future research into the 2020s, which include the use of PRF in the fabrication of custom 3D bone grafts, development of safer/better PRF tubes, and new applications of PRF in combination with lasers and zirconia implants. Lastly, novel uses of liquid-PRF as a growth factor/antibiotic–carrying delivery system is discussed as a potent and exciting avenue of future research endeavors.

Contributors

Richard J. Miron
Naheed Mohamed
Delia Tuttle
Fabrice Baudot
Marius Leretter
Victoria Lima
Yuriy May
Yufeng Zhang

Chapter Highlights

- *Necessity for the development of chemical-free PRF tubes*
- *Use of PRF for various novel clinical applications including vestibular extension, fabrication of custom 3D bone blocks, and as a drug-delivery system for growth factors and antibiotics*
- *Combination therapy with lasers*

This chapter takes a look back at some of the previously keen research topics at the forefront of PRF in 2016 and addresses their progress. Almost all of the topics were heavily researched in significant detail, as highlighted next.

PRF and Centrifugation Speeds

In 2016, major progress was made with respect to centrifugation speeds, having observed the launch of the low-speed centrifugation concept and the ability to potentially further improve concentrations of platelets and leukocytes in PRF formulations.[1,2] Since then, the following noteworthy improvements have been made:

1. The low-speed concept has been validated on several centrifugation systems.[3]
2. It has been found that it is possible to centrifuge *too* slowly and for too little time. The originally proposed i-PRF protocols at 60g (700–800 rpm) for 3 minutes needed to be improved.
3. This led to the development of C-PRF formulations (see chapters 2 and 3).
4. More cells and growth factors could be achieved utilizing horizontal centrifugation when compared to fixed-angle.

Over the past 5 years, significant advancements have been made with respect to PRF and centrifugation protocols that will surely favor more highly concentrated liquid-PRF and solid-PRF matrices. The next 5 years will focus more on the development and impact of heat treatment and/or cooling of PRF on final outcomes, as highlighted later in this chapter.

Effect of Hematocrit Count on PRF Scaffolds

Five years ago, it was noted that the formation of PRF was dependent not only on the centrifugal preparation method but also the composition of peripheral blood cells of the patient. At the macroscopic level, drastic differences in cell morphology have been observed between patients when using PRF in routine clinical practice. This variability caused significant confusion among colleagues, hence the launch of the study titled, "The effect of age, gender, and time between blood draw and start of centrifugation on the size outcomes of platelet-rich fibrin (PRF) membranes."[4] This study revealed for the first time the strong correlation between patient hematocrit levels and PRF membrane sizes. As reviewed in chapter 2, it showed that patients with lower hematocrit levels (females and older patients) had significantly larger (up to 30% to 40%) membranes when compared to patients with higher hematocrit levels (younger males). Furthermore, patients living at higher altitudes typically have higher hematocrit counts and therefore require higher centrifugation forces. Protocols would benefit from being modified accordingly, and future research on this topic is ongoing.

Influence of PRF on the Treatment of ONJ

In 2016, it was noted that PRF had been utilized successfully in a few case reports/series for the management of osteonecrosis of the jaw (ONJ). This was certainly logical given that ONJ typically has two major issues: *(1)* a lack of blood supply to the area and necrotizing bone, and *(2)* an open soft tissue defect. In fact, the main actions of PRF are to stimulate angiogenesis and favor soft tissue wound healing. Therefore, the addition of PRF to standard treatment protocols for ONJ cases was hypothesized to lead to superior outcomes.

As reviewed in chapter 18, PRF has now been utilized in over a dozen research articles and in hundreds of patients for the management of ONJ. Today, the additional use of PRF for the treatment of ONJ cases in routine therapy may in fact be considered a gold standard therapy.

i-PRF as a Potential Therapy for TMJ Disorders

At the same time, it was proposed that PRF could also be utilized to manage TMJ disorders. PRF has been proven to increase blood flow, and the ability for liquid-PRF to form a stable fibrin clot following injection was thought to lead to significantly better outcomes when compared to PRP. Unfortunately, only one study has since been published on this topic,[5] as reviewed in chapter 18, with much further

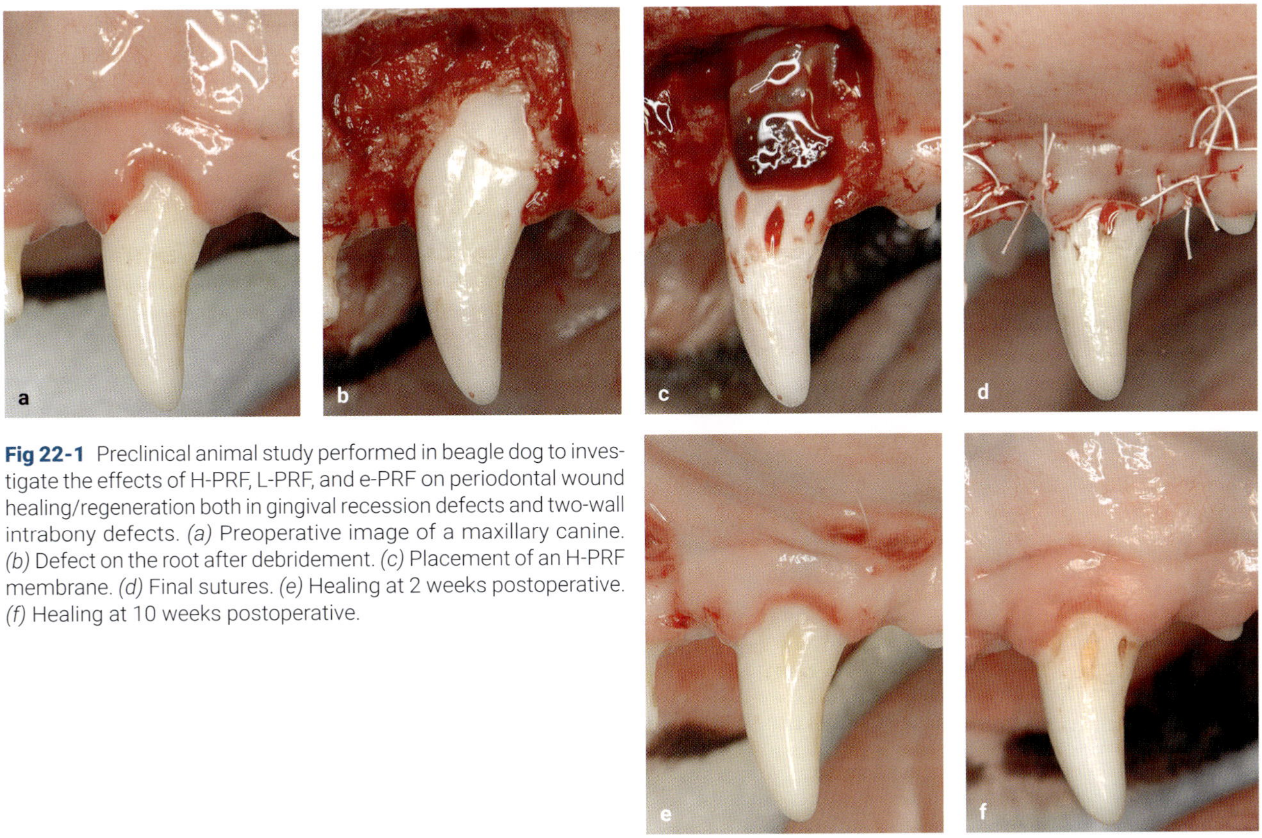

Fig 22-1 Preclinical animal study performed in beagle dog to investigate the effects of H-PRF, L-PRF, and e-PRF on periodontal wound healing/regeneration both in gingival recession defects and two-wall intrabony defects. *(a)* Preoperative image of a maxillary canine. *(b)* Defect on the root after debridement. *(c)* Placement of an H-PRF membrane. *(d)* Final sutures. *(e)* Healing at 2 weeks postoperative. *(f)* Healing at 10 weeks postoperative.

research needed before this can be considered a standard therapy. In that study, it was reported that 69% of patients demonstrated a significant reduction in pain at 2, 3, 6, and 12 months. The best responders to PRF injections were internal derangement stages Wilkes IV (78.5%) and V (100%).[5] Future research on this topic remains needed.

Potential Use of PRF for Pulp Regeneration

This was also a topic highly intriguing to many clinicians 5 years ago. Pulp infections typically result from *(1)* lack of blood flow to the area and *(2)* mild infection/inflammation to the pulpal tissue. Again, PRF was hypothesized to at least in part improve the clinical outcomes of certain cases (specific inclusion criteria reviewed in chapter 19) because of its ability to induce angiogenesis as well as reduce infection/inflammation. Since then, our group has demonstrated convincingly the ability for PRF to reduce both periodontal and pulpal pathogens due to its incorporation of leukocytes.[6] An entire chapter in this book is dedicated to the use of PRF for various endodontic procedures, including vital pulp therapy (direct pulp capping and pulpotomy), endodontic surgery, and revascularization cases (see chapter 19).

PRF and Periodontal Regeneration

In 2016, it was proposed that the regenerative properties of PRF for periodontal regeneration could be investigated at the histologic level. True periodontal regeneration must lead to new periodontal ligament attachment with Sharpey fibers spanning across the periodontium from cementum into alveolar bone. While demonstrated for a variety of other materials such as allografts or enamel matrix derivative, to date no single study has investigated this phenomenon for PRF in humans. While a variety of RCTs have shown the benefit of PRF in reducing probing depths and improving clinical attachment levels (see chapter 13), this remains an open question requiring future data and research. Current animal studies as well as human histologic studies are in progress (Fig 22-1).

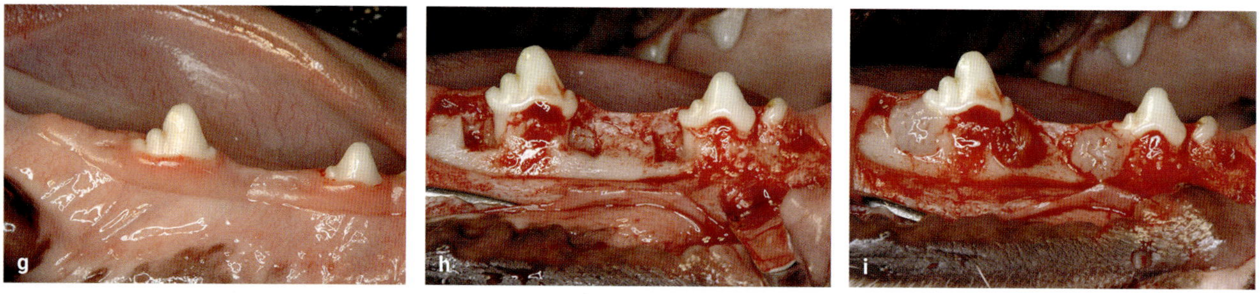

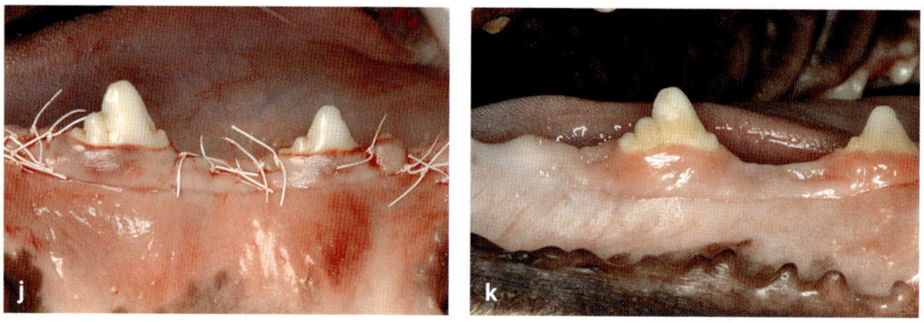

Fig 22-1 ***(cont)*** *(g)* Mandibular clinical aspect before reconstructive surgery. *(h)* Creation of several two-wall intrabony defects. *(i)* Intrabony defects filled with e-PRF (distal aspects of the second and fourth premolars) and L-PRF (mesial aspect of the fourth premolar). *(j)* Final sutures. *(k)* Healing at 8 weeks postoperative. (Study being performed by lead investigator Dr Yoshinori Shirakata.)

Potential Use of Stem Cells Derived from Blood

It was previously hypothesized that stem cells could be potentially harvested from whole blood. While it is known that progenitor cells do exist in blood, their levels remain extremely low, and to date no attempts have been made to further concentrate them with various laboratory methods for future use. While the strategy is plausible, considerable efforts to develop protocols and culture them thereafter in a cellular biology laboratory or incubator would be necessary. Far too often this field has received negative attention because various clinicians and industry representatives have made claims that stem cells are found in PRF and further attempted to market PRF as "stem cell–based regeneration." It is important to understand that while stem cells are actually present in blood and PRF, they are in extremely low levels and should therefore not be marketed to patients as stem cell–based therapy.

PRF: Regeneration by Growth Factors, Leukocytes, or Fibrin?

In 2016, it was not clear if the regeneration potential of PRF was derived simply from growth factors or from leukocytes and fibrin. Since then, much research on this topic has brought new insight into this question. By producing PRF via horizontal centrifugation, up to four-times higher concentrations of leukocytes have been observed.[7] More cellular regenerative potential and activity have also been reported in previous studies as well as an ability to possess antimicrobial properties.[8–10] Furthermore, it was shown in a recent article that osteal macrophages are in fact key players and regulators around bone biomaterials.[11] Therefore, the role and impact of leukocytes and their importance during tissue regeneration has continued to be established.

With respect to the impact of fibrin on the wound healing capability of PRF, it has been demonstrated that the majority of growth factors bind to fibrin, from which they are released slowly and gradually.[2,12,13] Animal and/or recombinant fibrin alone has been more commonly utilized in clinical practice and has demonstrated to be effective when utilized alone for various regenerative procedures.[14–18] Therefore, evidence has accumulated that PRF is not simply an array of autologous growth factors but also possesses additional regenerative potential linked to the fibrin matrix as well as leukocyte content.

PRF and Its Degradation Properties

In 2016, clinical studies were already showing that heating PRF scaffolds, for example with an iron, could significantly extend their degradation rate. This concept has been the topic of intensive research endeavors to bring to market the Bio-Heat technology, whereby the resorption properties of PRF can be dramatically extended from 2–3 weeks to 4–6 months simply by a heat-treatment process of the PRF (see chapter 7). Future research will certainly continue to investigate the use of heat and cold stimulation of PRF, favoring either its resorption properties or its ability to remain liquid for longer.

PRF and Osteoinduction

In 2016, a misconception existed among clinicians that PRF is osteoinductive. This is certainly not the case, and animal implantation experiments have revealed the favored ability for PRF to form soft tissue. In combination with the low evidence that PRF induces bone formation during GBR, PRF remains more conducive toward its benefits for improvements in bone graft handling and soft tissue closure of GBR procedures.

FUTURE RESEARCH: 2021 AND BEYOND

Tube Development of PRF

As highlighted throughout this textbook, the impact of tubes on the formation of PRF is critical. Like dental implant surfaces, the more chemically charged and hydrophilic the surface, the more platelets can aggregate and the better the clot formation. This concept has been shown to induce PRF membranes of bigger size and better strength.

As noted throughout chapter 6, today the majority of tubes utilized for the production of PRF were not designed for PRF but instead have been around for 20+ years for routine laboratory testing. Vacuette tubes, for example, have been utilized for routine testing for years far before PRF was commercialized, but they have since been adopted into PRF therapy. The main issue, however, is the fact that these tubes were not designed for human use and instead contain additives within their tube walls that end up in the PRF membrane and therefore in the patient's body (see chapter 6).

Recently, various tubes utilized commercially for the production of PRF were sent to the RMS Foundation in Switzerland, an independent organization and nonprofit establishment to better serve the community by investigating biomaterials. Following standard biocompatibility testing in the toxicology department according to ISO 10993-18:2020 (biocompatibility standard regarding chemical characterization), it was found that various tubes were in fact filled with various chemical additives, as highlighted below (Fig 22-2):

- The particle counts of plain glass vacuum tubes are insignificant compared to the other tube types.
- The A-PRF tubes show the highest particle counts. They peak around 1.5 μm.
- The Vacuette tubes show the second-highest particle counts. They peak around 15 μm.

The data observed within this study (yet to be published) correlates precisely with the studies published by Professor Kawase's group demonstrating the marked and pronounced "leaching" of various chemical additives, including silica and silicone, found within PRF clots following centrifugation in various tubes.[19,20]

Therefore, collectively researchers, clinicians, and industrial partners need to place efforts into the further optimization of PRF tubes for the betterment of PRF therapy. Reports have now clearly shown that some PRF tubes are releasing various unnecessary chemical additives responsible for causing apoptosis of human cells.[21] PRF can easily be produced with natural chemical-free plain tubes, and efforts should be directed to collectively advance this notion.

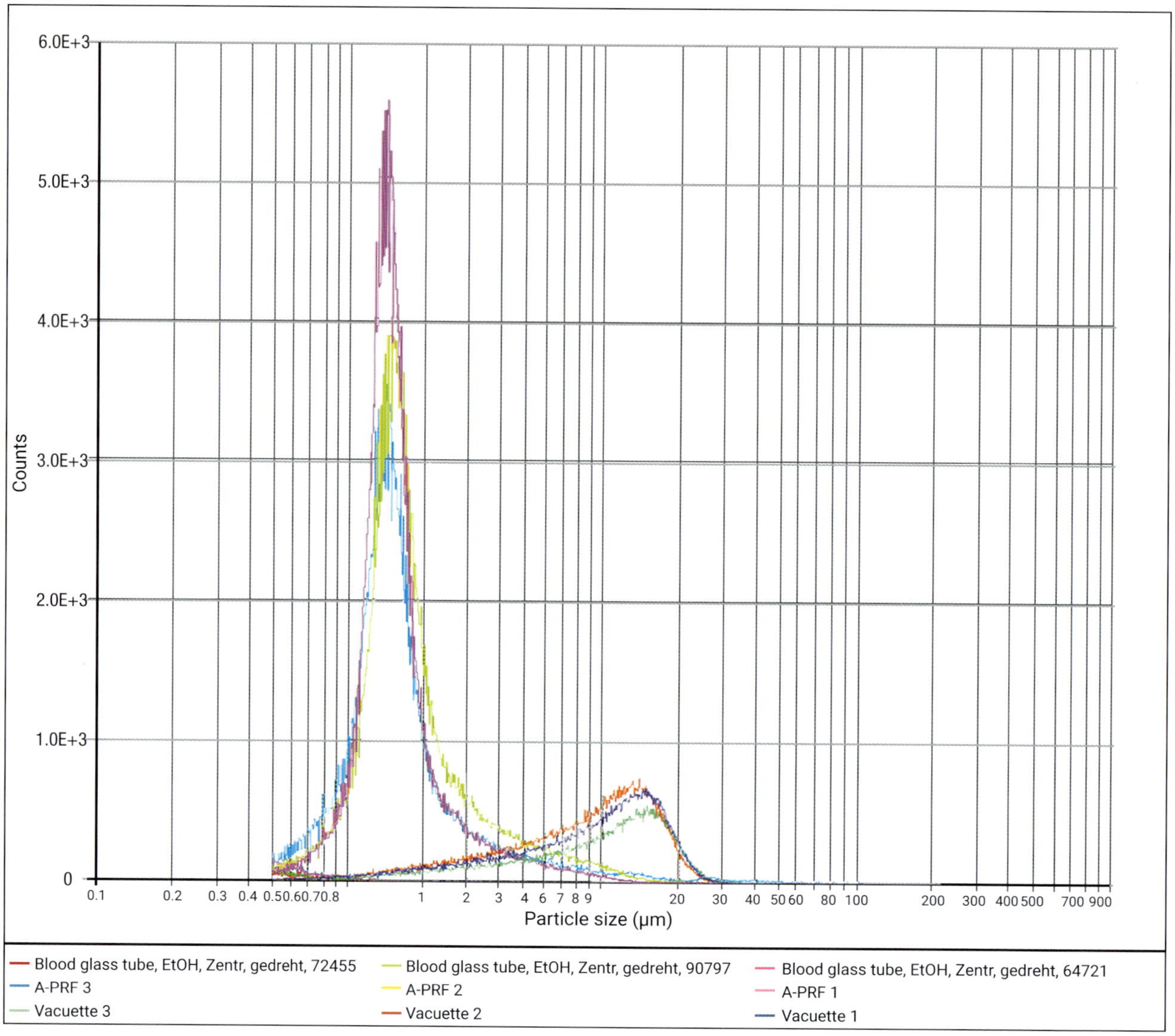

Fig 22-2 Use of an AccuSizer to measure particle sizes of silicone/silica released from PRF tube walls. The A-PRF tubes demonstrated the highest particle counts. They peak around 1.5 μm. The Vacuette tubes show the second-highest particle counts, peaking around 15 μm. (Unpublished data from the RMS Foundation.)

3D Bone Allograft Block Fabrication with Heat-Treated PRF

Digital technology has become a means to favor more accurate treatment planning, minimize patient chairside time, and lead to greater surgical accuracy in modern implant dentistry. Often, following adequate intraoral and CBCT scans, additional bone grafting is required to contour a bone defect prior to implant placement. The field has since evolved with the ability to custom mill accurate 3D bone allograft/xenograft blocks to simplify the surgical grafting procedure. Thus, while expensive, DICOM files may be sent to 3D bone milling facilities with the ability to custom fabricate a near perfect-fitting bone graft.

Recent work by our research team has described a simplified low-cost technique for fabricating a custom 3D bone block made of particulate bone allograft in combination with heat-treated PRF (Fig 22-3). Briefly, following a 3D scan, a digitally planned 3D custom bone mold is created and incorporated with bone allograft and liquid-PRF. Because recent research has demonstrated the ability for heat treatment of PRF to increase its mechanical stability,

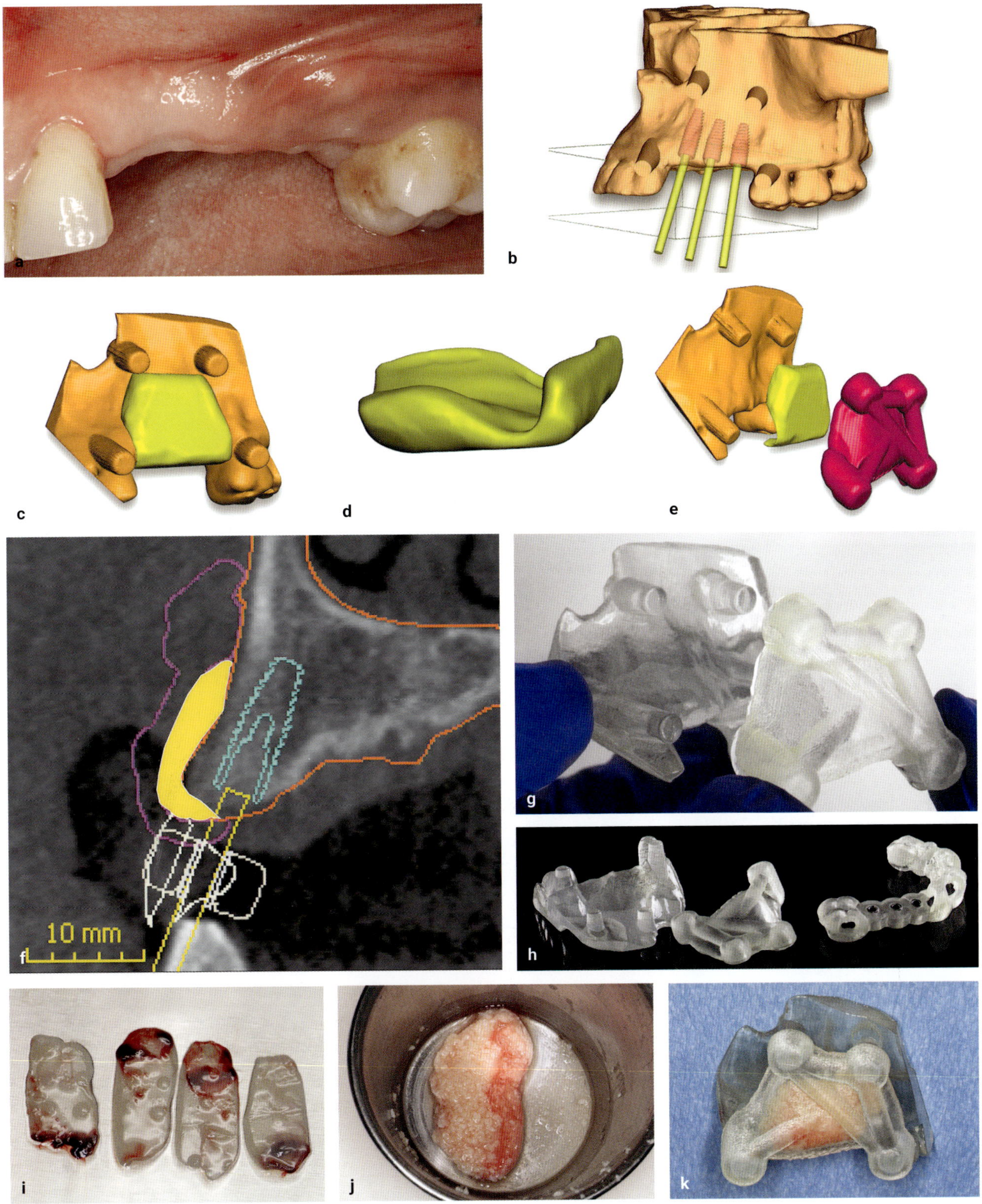

Fig 22-3 *(a)* Preoperative image of the edentulous ridge. *(b)* 3D model of the preoperative ridge and planned implant sites using CBCT software. *(c)* Digitally modeled bone graft over the ridge. *(d)* Custom 3D rendered digital bone graft. *(e)* Digital orientation of the digital bone mold with the bone graft. *(f)* CBCT DICOM cross-sectional view showing the digital planning of the guides, molds, implant, and bone graft shown in *yellow*. *(g)* Bone mold orientation for seating is shown with the 3D printed models. *(h)* 3D printed bone mold in two pieces and implant surgical pilot osteotomy guide. *(i)* Four PRF membranes were prepared along with two liquid-PRF tubes. *(j)* Sticky bone using cut-up PRF fragments and liquid-PRF. *(k)* Sticky bone inserted into the bone mold.

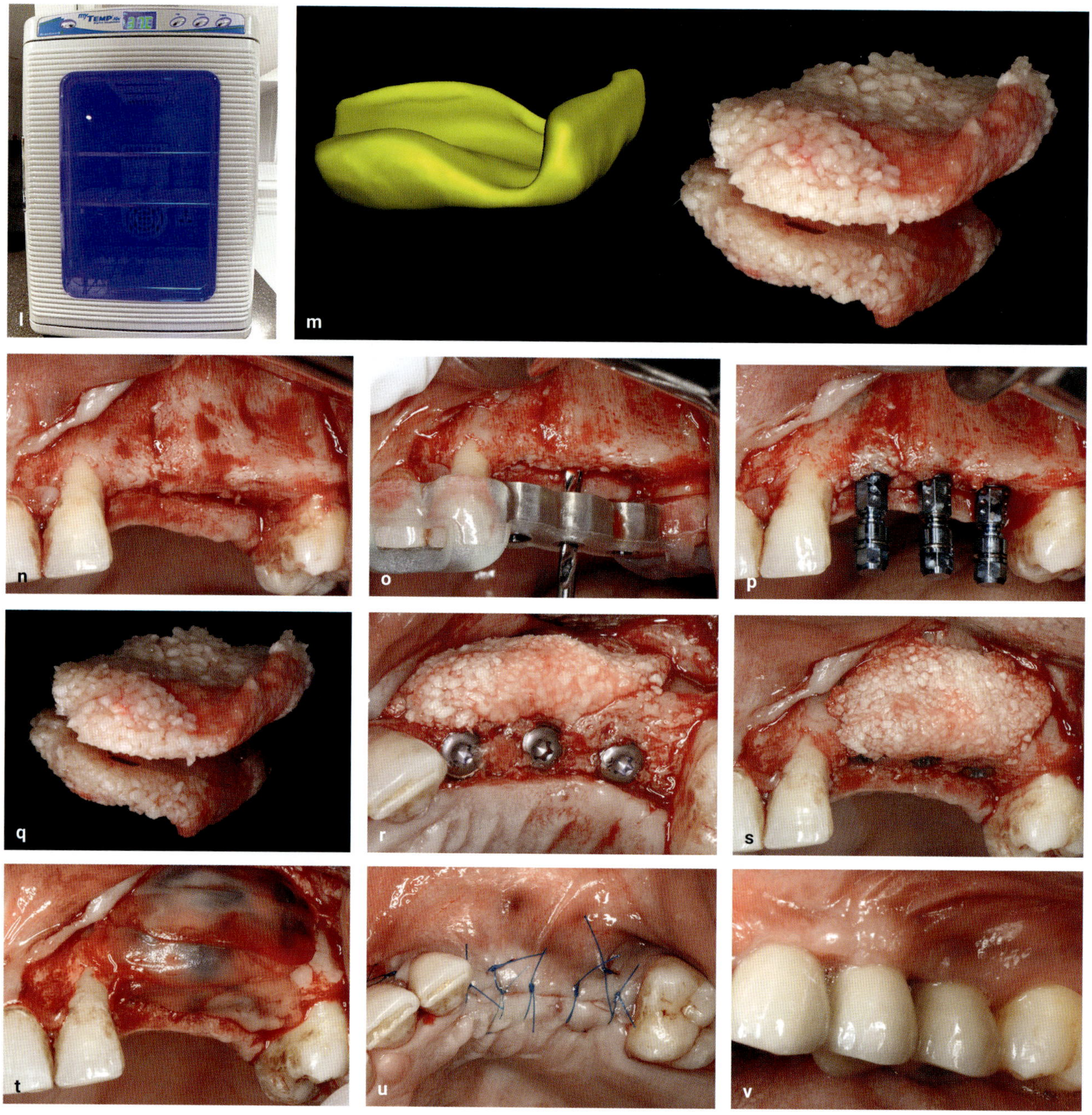

Fig 22-3 *(cont)* *(l)* Bone mold with incorporated sticky bone placed into an incubator at 37°C. *(m)* After 15 minutes, note the custom-made 3D bone graft. *(n)* Elevated flap exposing the ridge. *(o)* Pilot osteotomies being drilled with the surgical guide in place. *(p)* Osteotomies and placement of Straumann Bone Level Tapered Implants placed with Loxim attachments. *(q)* Custom patient-specific bone graft shown here removed from the mold. *(r)* Occlusal view of the custom bone graft in place as planned digitally. *(s)* Buccal view of custom bone graft in planned position. *(t)* PRF membranes placed over the bone graft. *(u)* Flap closure with Prolene sutures. *(v)* Final restorations at 8 months postoperative. (Case performed by Dr Naheed Mohamed.)

the bone mold complex is sent to an incubator at 37°C for 15 minutes to improve the mechanical stability of the graft. Thereafter, the bone mold is removed, and a custom bone graft with precise 3D architecture and improved mechanical strength and comprising supraphysiologic concentrations of platelets and growth factors is created. The use of an incubator at 37°C allows for the same body temperature whereby cells and growth factors can maintain survival, yet it is possible to increase the density of the fibrin mesh.

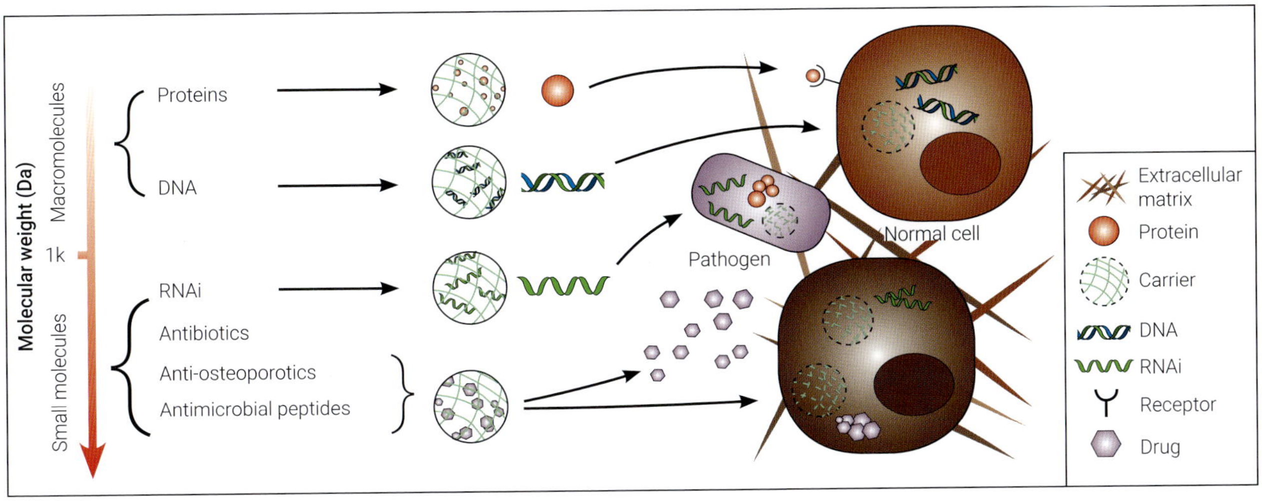

Fig 22-4 Classification and controlled release of biomolecules involved in therapeutic applications in bone regeneration. Macromolecules, primarily including proteins and DNA, can be loaded by carrier devices and released extracellularly or intracellularly to guide cell behaviors and regulate the bone healing process, whereas small molecules, herein defined as biomolecules with a molecular weight lower than 5 kDa, primarily target bone regeneration under compromised conditions such as infection, skeletal malignancies, and metastases in bone and osteoporosis. (Modified from Ji et al.[23])

This ongoing research endeavor has created a step-by-step protocol to create a low-cost custom bone allograft complex including living cells and growth factors with excellent mechanical properties for bone augmentation procedures prior to implant placement (see Fig 22-3). Furthermore, it highlights the growing trend of facilitating various surgical procedures with the help of digital dentistry. It is only expected that this field will continue to expand in the coming years.

PRF as a Drug-Delivery System for Growth Factors and Genes

In 2018, PRF was proposed as a way to better deliver biomolecules in a simplified, efficient, and low-cost biocompatible modality[22,23] (Fig 22-4). There is currently widespread interest within the biomaterial field to locally deliver biomolecules for bone and cartilage regeneration. In that article, a liquid version of PRF composed of liquid fibrinogen/thrombin was utilized as a drug-delivery system.[22] Therein, we introduced the use of liquid-PRF as an advanced local delivery system for small and large biomolecules including growth factors, cytokines, and morphogenetic/angiogenic factors as well as antibiotics, peptides, gene therapy, and antiosteoporotic molecules as potential candidates for enhanced bone/cartilage tissue regeneration.

Today, rhBMP-2 has been successfully incorporated into liquid-PRF prior to clotting and mixed with a bone graft to form sticky bone (reviewed in chapter 10), with rhBMP-2 clotted within the fibrin scaffold. This allows for the slow and gradual release of the growth factor (Fig 22-5). This research field is highly intriguing; unpublished data has demonstrated the more favorable binding of rhBMP-2 and rhPDGF to fibrin when compared to collagen.

Similarly, the use of various antibiotics and chlorhexidine delivery systems such as Arestin and PerioChip focus on their delivery into periodontal pockets with a slow and gradual release of antibiotics by various biomaterials utilized. Some delivery systems utilize polymers, which are quite inflammatory to immune cells found within periodontal pockets yet possess the advantage of delivering a gradual dose of antibiotic. Very recently, our research group has been interested in delivering antibiotics with liquid-PRF/e-PRF to favor a more natural delivery system. The ability to further extend the working properties of PRF also allows for the lengthier delivery of antibiotics/chlorhexidine over time points. Thus, the field is quite exciting with many new research endeavors focused on utilizing liquid-PRF as a delivery system for biomolecules.

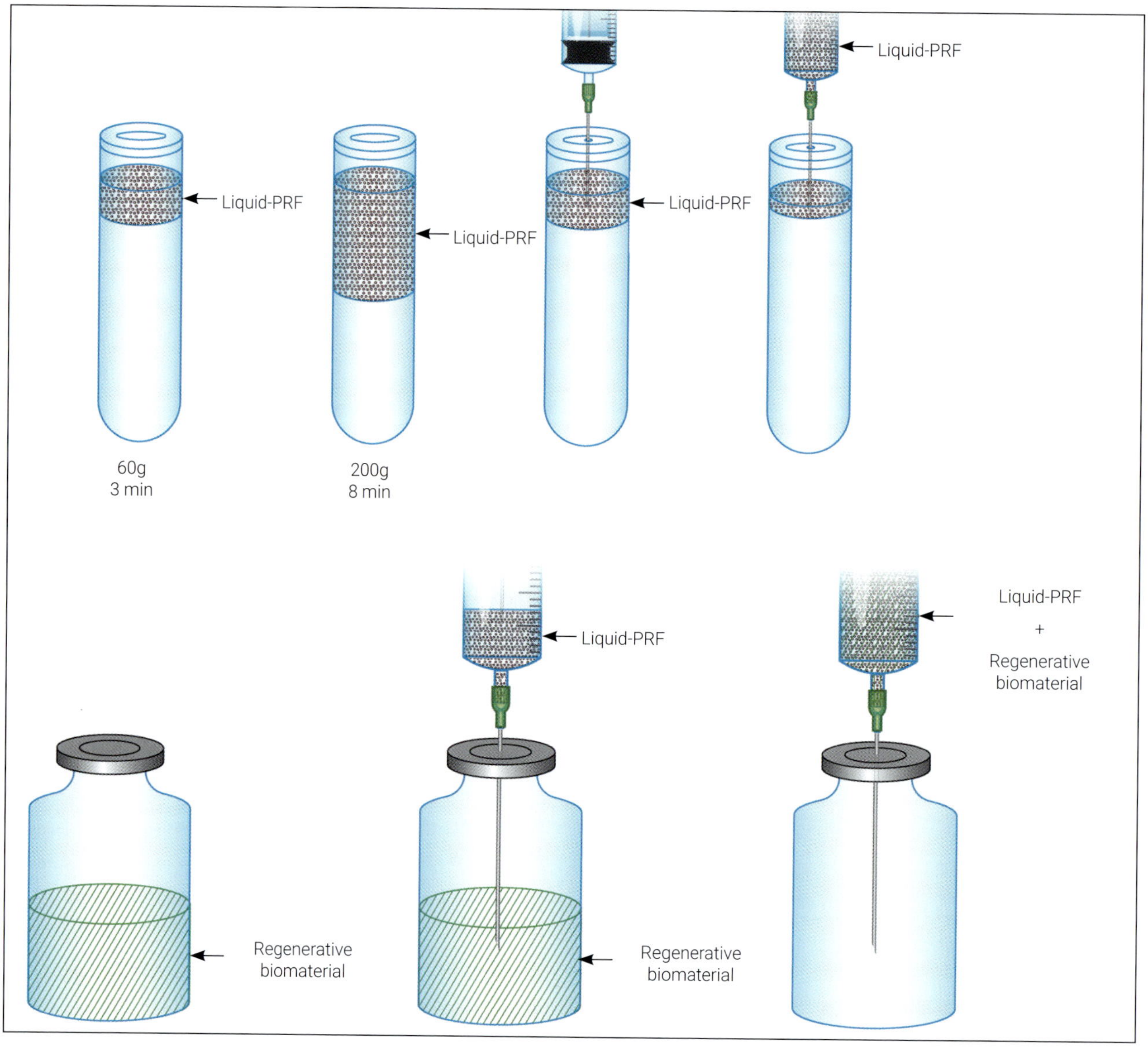

Fig 22-5 Modifications to centrifugation speed and time can alter the volume of liquid-PRF. Shorter and lower centrifugation speeds produce more highly concentrated PRF with higher concentrations of growth factors and cells, whereas longer centrifugation times produce more volume. Thereafter, liquid-PRF can be mixed with various regenerative biomolecules or biomaterials and injected within the human body to regenerate various tissues.

PRF in Combination with Lasers

Adjunctive laser therapy has proven effective for a variety of dental procedures. Specifically in periodontology, the use of appropriate laser protocols has been shown to lead to an additional 1.5 mm of clinical attachment level gain when compared to scaling and root planing alone.[24] Typically, an example of an industry standard protocol such as with the Fotona LightWalker system is the use of Nd:YAG and Er:YAG lasers in a three-step protocol. The third step of that protocol is the use of an Nd:YAG to generate a clot, because clot formation is an important step to stabilize the area and induce periodontal regeneration (Fig 22-6).

With the increasing popularity of utilizing lasers for periodontal regeneration and biofilm disruption, it was hypothesized several years ago that instead of inducing a clot

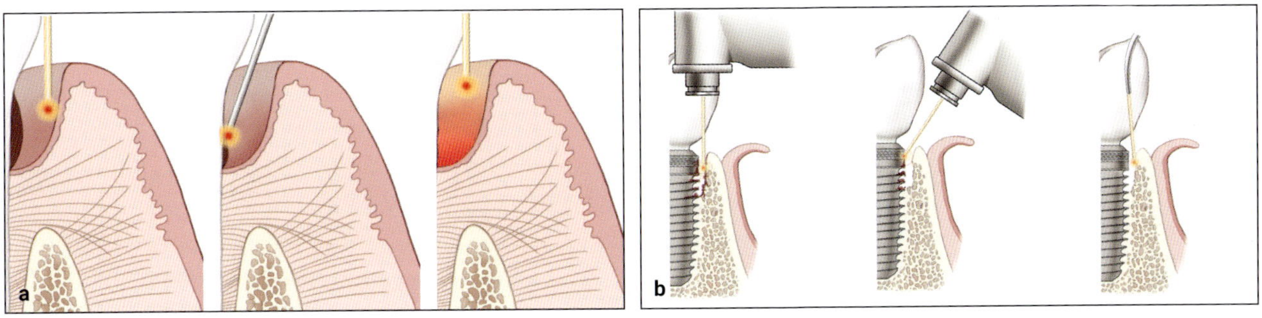

Fig 22-6 *(a and b)* Industry standard protocol of the Fotona LightWalker system for treatment of periodontal pockets with the Nd:YAG and Er:YAG lasers in a three-step protocol. *(1)* Removal of the soft granulation tissue and abiation of the infected bone with Er:YAG. *(2)* Removal of the bacterial biofilm on the implant surfaces with Er:YAG. *(3)* Removal, reduction, and biostimulation of the bone tissue with Nd:YAG. The third step of this protocol is the use of an Nd:YAG laser to generate a clot. This has more recently been modified with the addition of PRF.

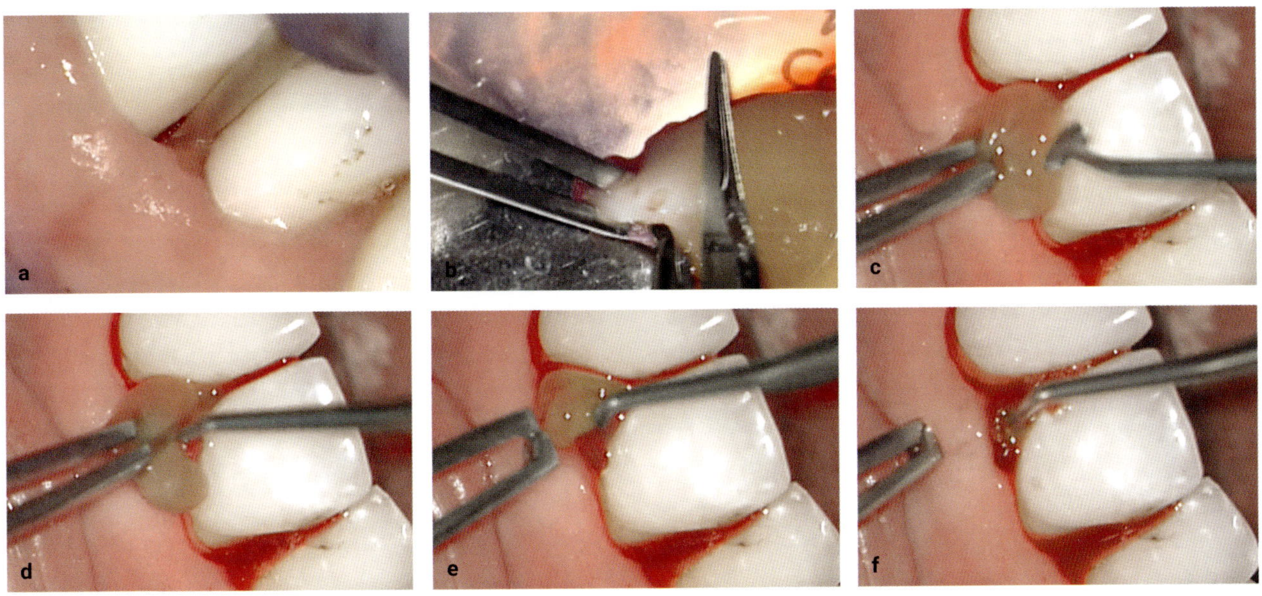

Fig 22-7 Nonsurgical entry of PRF into the periodontal pocket. *(a)* Use of an Er:YAG laser to decontaminate the root surface. *(b)* Cutting of a PRF clot. *(c to f)* Insertion of the PRF clot into the pocket nonsurgically. (Case performed by Dr Fabrice Baudot.)

Video 22-1

with the Nd:YAG laser, PRF could be introduced into the pocket following detoxification with an Er:YAG (Fig 22-7; see chapter 13). By doing so, a supraphysiologic dose of growth factors and defense-fighting leukocytes could be introduced into the pocket to further fight potential pathogens and improve healing/inflammation.

Bio-Filler in Facial Esthetics

The popularity of minimally invasive facial esthetic procedures is ever-growing among dentists and patients alike. PRF offers a natural and safer approach to regenerate facial tissues as opposed to introducing chemical synthetic fillers or toxins. Interestingly, one of the main limitations of PRF several years ago was the fact that it had quicker-than-desired resorption properties, which meant that facial voids filled with PRF would rapidly reappear. The development of Bio-Heat technology has allowed a massive increase in the longevity of PRF, extending its working properties from 2–3 weeks to 4–6 months. In facial esthetics, this Bio-Filler can be used as an alternative to synthetic fillers for a more natural-looking outcome. It can also be used in combination with laser therapy to allow for a significant improvement in youthful appearance utilizing the Fotona LightWalker's protocols (see chapter 20 and Fig 20-9). With these combination protocols, no single biomaterial is needed beyond the 100% naturally derived PRF.

e-PRF and Bio-Bone

Bio-Heat technology has also been adopted in the field of dentistry to create PRF barrier membranes with extended working properties (ie, e-PRF). Initial therapy includes its use over GBR procedures as a membrane capable of greatly improved resorption properties (Video 22-2). It will be interesting to observe the long-term data regarding its effectiveness. One noted difference has been its slightly inferior strength following heating. Various techniques including heat compression with a miniature iron have been proposed as a means to favor better graft stability. Future research will certainly improve the membrane strength and its handling and potentially even further extend its resorption properties.

Video 22-2

PRF in Medicine: FDA Approval

More than 5 years have passed since breakthrough cases were observed when using PRF for wound management (such as diabetic ulcers). Complex wounds were being treated in what has since been termed *natural guided regeneration*. Over the past 5 years, many key opinion leaders and researchers alike have presented this breakthrough application of PRF in many conferences across the world. However, this technology has not yet been adopted in the majority of hospitals in developed countries because it has not yet received FDA or CE clearance in medicine (PRF is only FDA-cleared for dental applications). Thus, to date, no single company has FDA clearance for use of PRF for the management and treatment of such wounds.

Unfortunately, the majority of PRF companies are dental, with no interest in commercializing the technology for medical applications. The next 5 years should gather more colleagues in various avenues of medicine to make contributions in that space focused on FDA and CE clearance. It is a sincere pity that no such efforts have thus far been made, and the future holds such promise if these breakthrough procedures can be adopted in routine medical practice.

PRF in Combination with Zirconia Implants

Culturally, many patients are seeking more biologic materials with fewer chemical additives and metal incorporation. One alternative to titanium dental implants has been the use of zirconia, which has been described as a metal-free ceramic alternative. Over the years, leading research groups have demonstrated the ability for zirconia to equally lead to favorable osseointegration.[25–27] Furthermore, improvements in soft tissue response to zirconia when compared to titanium have also been reported.

Because PRF is a biologic material and autogenous, it will be interesting to see how it advances in the field, especially among holistic colleagues. PRF is not a miracle product, and its role in bone formation has not been well established, yet culturally many patients prefer their own tissues to be utilized for regenerative purposes. Therefore, it will be interesting to create new protocols and biomaterial combinations established specifically for the holistic candidate wishing to avoid the use of human cadaver or animal sources for either personal or religious reasons (Fig 22-8).

PRF and Soft Tissue Vestibular Extension

Following a large vertical bone augmentation procedure, one of the often-reported issues postsurgery is the shortened vestibule requiring a second-stage surgery. For years, this has often been performed utilizing an AlloDerm matrix

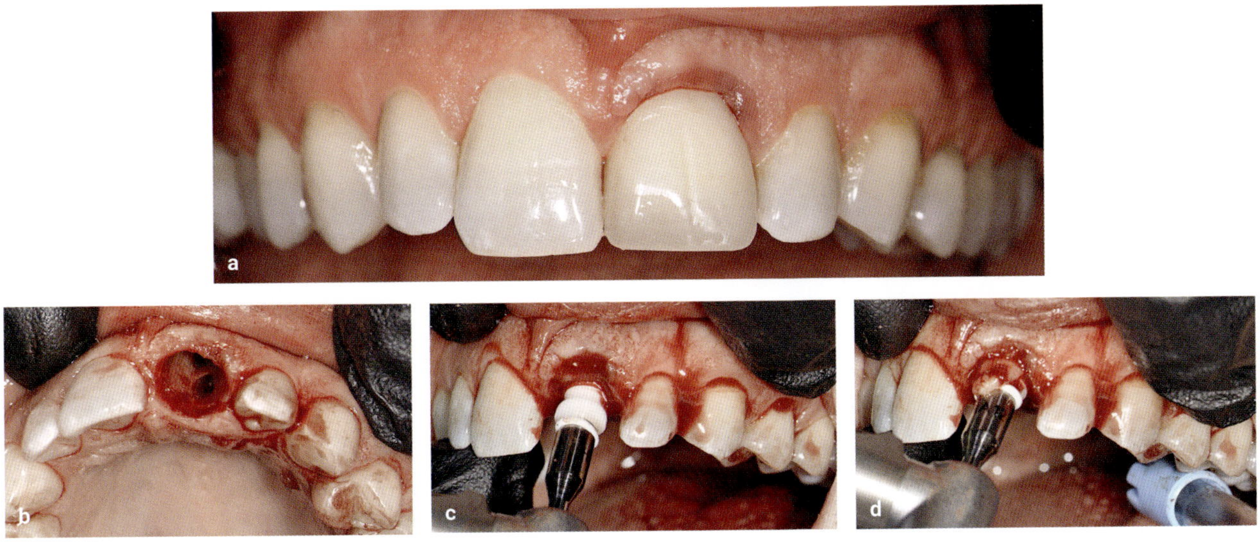

Fig 22-8 Use of PRF with a zirconia implant. *(a)* Hopeless maxillary left central incisor. *(b)* Extraction site following tooth extraction. *(c)* Immediate implant placement of a zirconia implant. *(d)* Placement of a PRF membrane in the coronal third of the implant to facilitate soft tissue healing. (Case performed by Dr Yuriy May.)

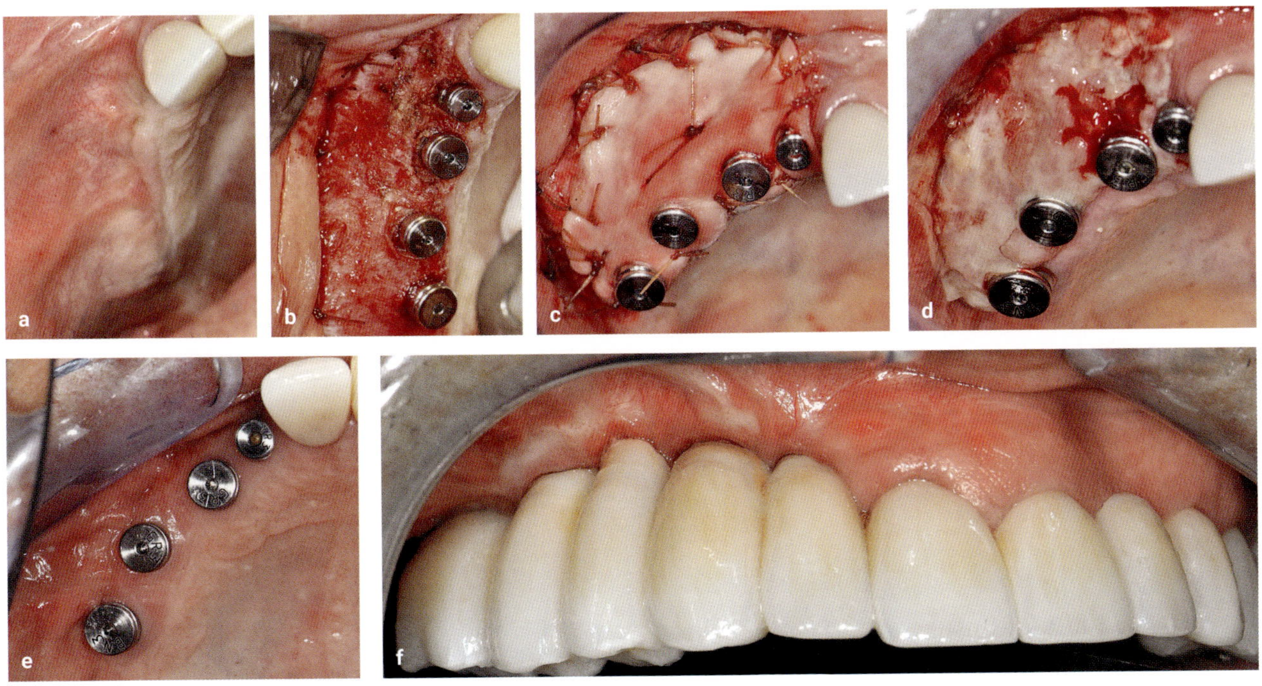

Fig 22-9 *(a)* A shortened vestibule following a vertical bone augmentation procedure requiring vestibuloplasty. *(b and c)* A maxillary vestibuloplasty performed with acellular dermal matrix (AlloDerm). *(d and e)* Clinical photographs 2 weeks and 2 months postoperative, respectively. *(f)* Clinical photograph 13 years postoperative. (Reprinted with permission from Pikos and Miron.[28])

(BioHorizons), as presented in Fig 22-9.[28] Because PRF is an excellent biomaterial specifically for soft tissue wound healing and can be left exposed to the oral cavity without issue, it was hypothesized that PRF could be utilized in a similar fashion to increase the vestibular depth. Primary results are encouraging (Fig 22-10), with future case series needed to further investigate success rates and potential complications. Biologically this concept fits perfectly with the current understanding of PRF and is likely to be one adopted by many clinicians in the coming years.

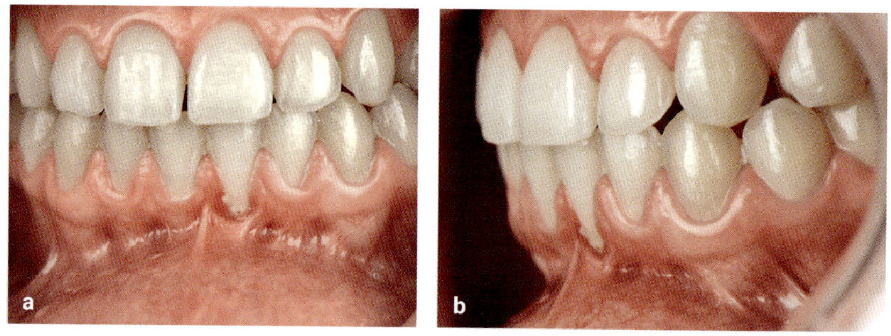

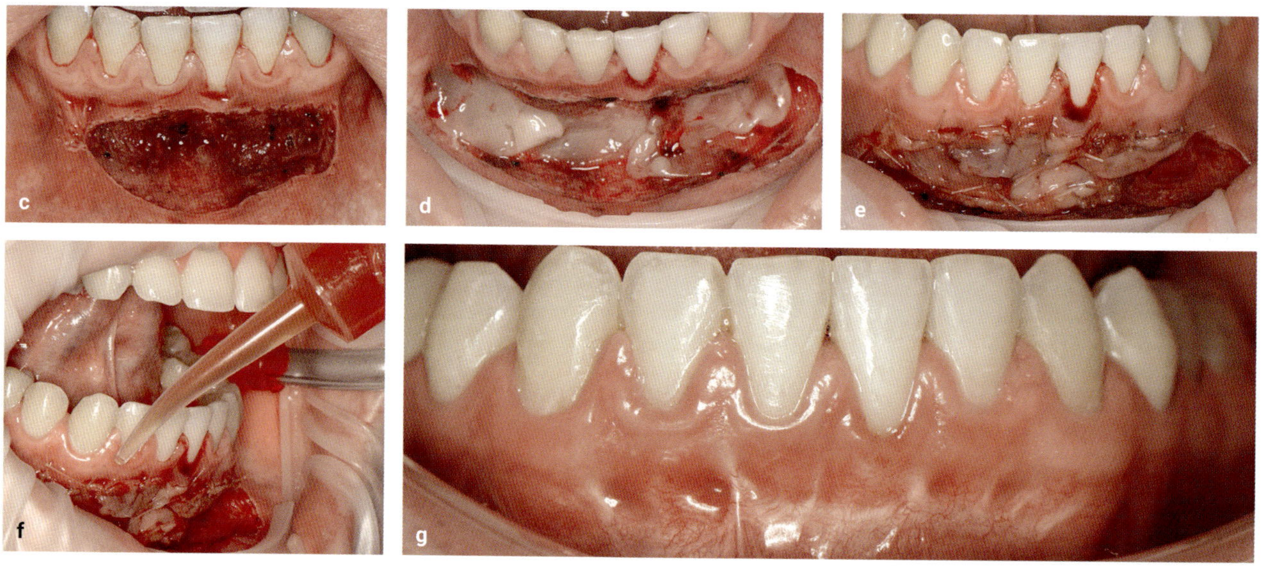

Fig 22-10 *(a)* A shortened vestibule with a strong frenum pull. *(b)* Note the resulting gingival recession. *(c)* A mandibular vestibuloplasty was performed. *(d)* Application of PRF membranes. *(e)* Sutures were performed with natural gut. *(f)* Use of liquid-PRF to improve growth factor delivery in the local area. *(g)* Final case 6 months postoperative. Note the extended vestibule and the improvement in the gingival recession. (Case performed by Dr Delia Tuttle.)

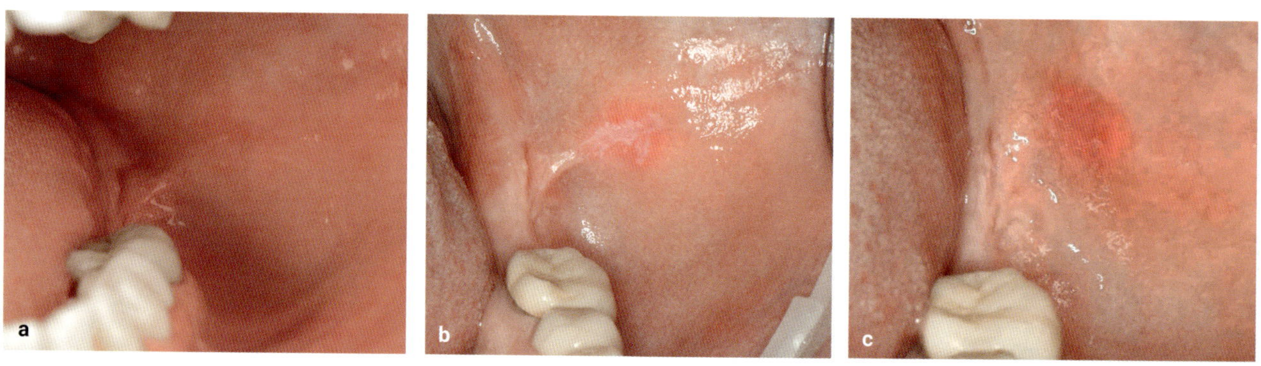

Fig 22-11 *(a to c)* A 60-year-old patient presented with a history of lichen planus. The patient was taking benazepril (20 mg), estradiol (1 mg), and citalopram (20 mg). Three injections with i-PRF were performed over a 6-month period. The lichen planus disappeared and has not returned. The case demonstrates that long-term lichen planus may potentially benefit from PRF infiltrations, though much further research is needed to determine its true effect. (Case performed by Dr Delia Tuttle.)

PRF and Immunocompromised Patients

One interesting avenue of research is related to use of PRF for the healing of immunocompromised patients. Because healing is delayed, in theory the advantages of PRF should additionally promote healing in such patients. Figure 22-11 demonstrates the healing of mucosal tissue in an immunocompromised patient with lichen planus treated with PRF. While the use of PRF led to the rapid and improved healing on the mucosal tissue, application/injection in other areas has not resulted in a similar resolution. Thus, the interplay between immunocompromised cells (ie, immune cells currently attacking the body) and the application of additional immune cells harvested from blood in PRF has yet to be investigated in much detail. Much research can be performed on this topic, with few research teams currently investigating it. Interestingly, PRP has been shown in the literature to lead to improved healing in cases of vitiligo and lichen planus.[29–31] Thus, it remains an open question whether additional applications and uses of PRF within that community of patients may benefit them.

Conclusion

This concluding chapter sets the stage for future research in the field. As highlighted through this book, much advancement has been made over the past 5 years with respect to PRF, and it is only expected that this trend will continue. **The future of PRF is bound to be exciting!**

References

1. Ghanaati S, Booms P, Orlowska A, et al. Advanced platelet-rich fibrin: A new concept for cell-based tissue engineering by means of inflammatory cells. J Oral Implantol 2014;40:679–689.
2. Fujioka-Kobayashi M, Miron RJ, Hernandez M, Kandalam U, Zhang Y, Choukroun J. Optimized platelet rich fibrin with the low-speed concept: Growth factor release, biocompatibility and cellular response. J Periodontol 2017;88:112–121.
3. Miron RJ, Xu H, Chai J, et al. Comparison of platelet-rich fibrin (PRF) produced using 3 commercially available centrifuges at both high (~ 700 g) and low (~ 200 g) relative centrifugation forces. Clin Oral Investig 2020;24:1171–1182.
4. Miron RJ, Dham A, Dham U, Zhang Y, Pikos MA, Sculean A. The effect of age, gender, and time between blood draw and start of centrifugation on the size outcomes of platelet-rich fibrin (PRF) membranes. Clin Oral Investig 2019;23:2179–2185.
5. Albilia JB, Herrera-Vízcaino C, Weisleder H, Choukroun J, Ghanaati S. Liquid platelet-rich fibrin injections as a treatment adjunct for painful temporomandibular joints: Preliminary results. Cranio 2020;38:292–304.
6. Chai J, Jin R, Yuan G, Kanter V, Miron RJ, Zhang Y. Effect of liquid platelet-rich fibrin and platelet-rich plasma on the regenerative potential of dental pulp cells cultured under inflammatory conditions: A comparative analysis. J Endod 2019;45:1000–1008.
7. Miron RJ, Chai J, Zheng S, Feng M, Sculean A, Zhang Y. A novel method for evaluating and quantifying cell types in platelet rich fibrin and an introduction to horizontal centrifugation. J Biomed Mater Res A 2019;107:2257–2271.
8. Fujioka-Kobayashi M, Kono M, Katagiri H, et al. Histological comparison of platelet rich fibrin clots prepared by fixed-angle versus horizontal centrifugation. Platelets 2020:1–7.
9. Miron RJ, Chai J, Zhang P, et al. A novel method for harvesting concentrated platelet-rich fibrin (C-PRF) with a 10-fold increase in platelet and leukocyte yields. Clin Oral Investig 2020;24: 2819–2828.
10. Zhang J, Yin C, Zhao Q, et al. Anti-inflammation effects of injectable platelet-rich fibrin via macrophages and dendritic cells. J Biomed Mater Res A 2020;108:61–68.
11. Miron RJ, Bosshardt DD. OsteoMacs: Key players around bone biomaterials. Biomaterials 2016;82:1–19.
12. Kobayashi E, Fluckiger L, Fujioka-Kobayashi M, et al. Comparative release of growth factors from PRP, PRF, and advanced-PRF. Clin Oral Investig 2016;20:2353–2360.
13. Miron RJ, Fujioka-Kobayashi M, Hernandez M, et al. Injectable platelet rich fibrin (i-PRF): Opportunities in regenerative dentistry? Clin Oral Investig 2017;21:2619–2627.
14. Brittberg M, Sjögren-Jansson E, Lindahl A, Peterson L. Influence of fibrin sealant (Tisseel) on osteochondral defect repair in the rabbit knee. Biomaterials 1997;18:235–242.
15. Spotnitz WD. Fibrin sealant: Past, present, and future: A brief review. World J Surg 2010;34:632–634.
16. Le Nihouannen D, Le Guehennec L, Rouillon T, et al. Micro-architecture of calcium phosphate granules and fibrin glue composites for bone tissue engineering. Biomaterials 2006;27: 2716–2722.
17. Ahmed TA, Dare EV, Hincke M. Fibrin: A versatile scaffold for tissue engineering applications. Tissue Eng Part B Rev 2008;14:199–215.
18. Fortelny RH, Petter-Puchner AH, Glaser KS, Redl H. Use of fibrin sealant (Tisseel/Tissucol) in hernia repair: A systematic review. Surg Endosc 2012;26:1803–1812.
19. Takahashi A, Tsujino T, Yamaguchi S, et al. Distribution of platelets, transforming growth factor-beta1, platelet-derived growth factor-BB, vascular endothelial growth factor and matrix metalloprotease-9 in advanced platelet-rich fibrin and concentrated growth factor matrices. J Investig Clin Dent 2019;10:e12458.
20. Tsujino T, Masuki H, Nakamura M, et al. Striking differences in platelet distribution between advanced-platelet-rich fibrin and concentrated growth factors: Effects of silica-containing plastic tubes. J Funct Biomater 2019;10:43.
21. Masuki H, Isobe K, Kawabata H, et al. Acute cytotoxic effects of silica microparticles used for coating of plastic blood-collection tubes on human periosteal cells. Odontology 2020;108: 545–552.
22. Miron RJ, Zhang Y. Autologous liquid platelet rich fibrin: A novel drug delivery system. Acta Biomater 2018;75:35–51.
23. Ji W, Wang H, van den Beucken JJ, et al. Local delivery of small and large biomolecules in craniomaxillofacial bone. Adv Drug Deliv Rev 2012;64:1152–1164.
24. Grzech-Lesniak K, Sculean A, Gaspirc B. Laser reduction of specific microorganisms in the periodontal pocket using Er:YAG and Nd:YAG lasers: A randomized controlled clinical study. Lasers Med Sci 2018;33:1461–1470.
25. Bosshardt DD, Chappuis V, Buser D. Osseointegration of titanium, titanium alloy and zirconia dental implants: Current knowledge and open questions. Periodontol 2000 2017;73:22–40.
26. Chappuis V, Cavusoglu Y, Gruber R, Kuchler U, Buser D, Bosshardt DD. Osseointegration of zirconia in the presence of multinucleated giant cells. Clin Implant Dent Relat Res 2016;18:686–698.
27. Janner SFM, Gahlert M, Bosshardt DD, et al. Bone response to functionally loaded, two-piece zirconia implants: A preclinical histometric study. Clin Oral Implants Res 2018;29:277–289.
28. Pikos MA, Miron RJ. Bone Augmentation in Implant Dentistry: A Step-by-Step Guide to Predictable Alveolar Ridge and Sinus Grafting. Chicago: Quintessence, 2019.

29. Khattab FM, Abdelbary E, Fawzi M. Evaluation of combined excimer laser and platelet-rich plasma for the treatment of nonsegmental vitiligo: A prospective comparative study. J Cosmet Dermatol 2020;19:869–877.
30. Parambath N, Sharma VK, Parihar AS, Sahni K, Gupta S. Use of platelet-rich plasma to suspend noncultured epidermal cell suspension improves repigmentation after autologous transplantation in stable vitiligo: A double-blind randomized controlled trial. Int J Dermatol 2019;58:472–476.
31. Sardana K, Verma G. Overview of medical therapies and phototherapy in vitiligo based on their pathogenetic action and the role of platelet-rich plasma. J Cutan Aesthet Surg 2018;11:167–168.

Index

Page references followed by "f" denote figures; those followed by "t" denote tables.